Advances in Experimental Medicine and Biology

Volume 898

More information about this series at http://www.springer.com/series/5584

Juan A. Rosado
Editor

Calcium Entry Pathways in Non-excitable Cells

Springer

Editor
Juan A. Rosado
Departamento de Fisiología
University of Extremadura
Cáceres, Spain

ISSN 0065-2598 ISSN 2214-8019 (electronic)
Advances in Experimental Medicine and Biology
ISBN 978-3-319-26972-6 ISBN 978-3-319-26974-0 (eBook)
DOI 10.1007/978-3-319-26974-0

Library of Congress Control Number: 2016939014

Printed on acid-free paper

This Springer imprint is published by Springer Nature
The registered company is Springer International Publishing AG Switzerland

Dedicated to Marcos and Maria

Preface

Cytosolic calcium homeostasis is a crucial event in cell biology as this ion acts as a second messenger for a large number of cellular functions. The observations of Ringer, published in the *Journal of Physiology* in the early 1880s, revealed the importance of calcium in cell physiology but specially focused the attention of the researchers on the essential role of extracellular calcium entry. This issue has attracted the attention of an increasing number of basic scientists interested in cellular biology as well as biomedical researchers and clinicians interested in the pathophysiological implications of calcium entry. Since this is a highly dynamic topic, recapitulation of the historic and current knowledge is usually helpful to understand the concepts and stimulate new approaches and directions that allow significant advances in various aspects of calcium influx.

After the assumption that Ca^{2+} influx was a major event in cell biology, voltage-dependent calcium currents were identified, and, later on, a mechanism for calcium influx operated by the occupation of membrane receptors was reported both in electrically excitable and non-excitable cells. Soon after the identification of the so-called receptor-operated calcium entry, the experimental observations indicated that this process was not a unique mechanism, at least concerning the signaling pathway underlying calcium channel gating. Despite that the occupation of membrane receptors by physiological agonists is a common feature for all the mechanisms of receptor-operated calcium entry, the subsequent activation of calcium-permeable channels involves three different pathways. The simplest mechanism involves direct opening of a receptor-operated channel where agonist-receptor binding leads to a conformational change that results in calcium influx. The other two signaling routes involve indirect channel gating either via generation of a diffusible second messenger that is able to activate channel opening by itself (second messenger-operated calcium entry) or by the discharge of the intracellular calcium stores (store-operated or capacitative calcium entry). The term capacitative calcium entry was proposed by James Putney Jr. to refer to a process whereby the discharge of the intracellular calcium stores within a cell secondarily activates calcium entry into the cell across the plasma membrane. During the past decades, a number of research groups have explored the mechanisms underlying the activation, maintenance, and termination of the different routes for calcium entry in electrically excitable and non-excitable cells, as well as the identity of the molecular components. Concerning the latter, the different members of the Orai and TRPC families of channels have been found to play a relevant role both for capacitative calcium entry and for other calcium influx pathways. Given the importance of calcium influx in cell biology, one would expect that dysfunction of this process might underlie serious pathological consequences. In fact, a number of human disorders have been reported to involve Orai- as well as TRPC-associated channelopathies.

Finally I would like to express my sincere thanks and appreciation to all contributors for their kindly dedicated collaboration in the elaboration of this book and also to Ginés M. Salido for his support editing this book. I would also like to thank Thijs van Vlijmen, from Springer, for his encouraging support in the production of this book.

I hope that the content of this book will be helpful for both PhD students and advanced researchers interested in this fascinating field as well as inspiring for basic and biomedical scientists working in the pathophysiological implications of the calcium entry pathways.

Cáceres, Spain
October 2015

Juan A. Rosado

Contents

Part I
Store-Operated Calcium Entry

Chapter 1
Historical Overview of Store-Operated Ca^{2+} Entry

Letizia Albarran, Jose J. Lopez, Ginés M. Salido, and Juan A. Rosado

Abstract Calcium influx is an essential mechanism for the activation of cellular functions both in excitable and non-excitable cells. In non-excitable cells, activation of phospholipase C by occupation of G protein-coupled receptors leads to the generation of inositol 1,4,5-trisphosphate (IP_3) and diacylglycerol (DAG), which, in turn, initiate two Ca^{2+} entry pathways: Ca^{2+} release from intracellular Ca^{2+} stores, signaled by IP_3, leads to the activation of store-operated Ca^{2+} entry (SOCE); on the other hand, DAG activates a distinct second messenger-operated pathway. SOCE is regulated by the filling state of the intracellular calcium stores. The search for the molecular components of SOCE has identified the stromal interaction molecule 1 (STIM1) as the Ca^{2+} sensor in the endoplasmic reticulum and Orai1 as a store-operated channel (SOC) subunit. Furthermore, a number of reports have revealed that several members of the TRPC family of channels also take part of the SOC macromolecular complex. This introductory chapter summarizes the early pieces of evidence that led to the concept of SOCE and the components of the store-operated signaling pathway.

Keywords SOCE • STIM1 • Orai1 • TRPC channels

1.1 The Concept of Store-Operated Ca^{2+}Entry

A number of physiological agonists activate cellular functions by inducing changes in cytosolic Ca^{2+} concentration ($[Ca^{2+}]_c$). In general, Ca^{2+} mobilization consist of the release of Ca^{2+} from intracellular stores, as well as Ca^{2+} influx from the extracellular medium through Ca^{2+}-permeable channels. Despite the identification of Ca^{2+}as a

L. Albarran • J.J. Lopez • G.M. Salido
Department of Physiology (Cell Physiology Research Group), University of Extremadura, Av. Universidad s/n, 10003 Spain

J.A. Rosado (✉)
Departamento de Fisiología, University of Extremadura, Cáceres, Spain
e-mail: jarosado@unex.es

J.A. Rosado (ed.), *Calcium Entry Pathways in Non-excitable Cells*, Advances in Experimental Medicine and Biology 898, DOI 10.1007/978-3-319-26974-0_1

second messenger in the excitation-contraction coupling took place in the mid-twentieth century [1], the relevance of extracellular Ca^{2+}in cellular physiology was already highlighted in the publications of Ringer in the Journal of Physiology in the early 1880s [2, 3]. Throughout the 1960s and 1970s, different laboratories determined Ca^{2+} signals induced by physiological agonists, first using luminescent photoproteins and later on through the use of fluorescent probes, until Ca^{2+} was recognized as a *bona fide* second messenger [4].

After Ringer's findings, the studies concerning the functional role of Ca^{2+} entry were mainly focused on muscle contraction, especially in the heart and the sartorius muscle, where $^{45}Ca^{2+}$ uptake (influx) by the cell was analyzed under different experimental conditions [5, 6]. Among the first analysis of Ca^{2+} entry in non-electrically excitable cells were performed in HeLa cells in culture, where Borle identified two exchangeable pools, assumed as the extracellular Ca^{2+} and Ca^{2+} stored in intracellular compartments, and two unexchangeable Ca^{2+} pools, corresponding to extracellular and intracellular bound Ca^{2+} [7]. A year earlier, the same author reported that the extracellular medium was a major source of Ca^{2+} [8]. Borle defined Ca^{2+} influx in these cells as a facilitated diffusion process that can be affected by increasing the membrane permeability to this ion and not by stimulating an active or metabolically dependent process [7].

Voltage-dependent Ca^{2+}currents independent of metabolic events were described in the early 70s and a mechanism for Ca^{2+} influx operated by the occupation of membrane receptors was also soon reported in electrically excitable and non-excitable cells [9–11]. But, it was in 1986, when a mechanism for receptor-operated Ca^{2+} influx regulated by the filling state of the intracellular Ca^{2+} stores was proposed [12]. The term capacitative Ca^{2+}entry was coined by James Putney to refer to a process whereby the discharge of Ca^{2+} stores within a cell secondarily activates Ca^{2+}entry into the cell across the plasma membrane [12], a mechanism analog to the function of a capacitor in an electrical circuit, since this process is characterized by the fact that charged or full intracellular Ca^{2+} stores prevent Ca^{2+} current through the plasma membrane, while discharge of the intracellular stores is followed by rapid entry of Ca^{2+} into the store and, in the continued presence of inositol (1,4,5) trisphosphate (IP_3), into the cytosol [12–14].

Analysis of the kinetics of the rises in $[Ca^{2+}]_c$ using the intracellular fluorescent probe fura-2 in human platelets revealed that stimulation with the physiological agonist thrombin, which induces active Ca^{2+} release from the stores via generation of IP_3, resulted in Mn^{2+} entry that starts a few tenths of second after discharge of the intracellular Ca^{2+}stores [15]. This finding was consistent with a mechanism controlled by the agonist-sensitive intracellular Ca^{2+} stores. Further studies revealed that cell treatment with thapsigargin, a sesquiterpene lactone obtained from *Thapsia garganica* [16] that inactivates the sarco-endoplasmic reticulum Ca^{2+}-ATPase (SERCA) [17] and leads to passive store depletion via Ca^{2+} leak, resulted in Ca^{2+} influx across the plasma membrane independently of IP_3 generation [18]. Soon after this publication, it was found that Ca^{2+} influx evoked either by agonist stimulation or by treatment with thapsigargin was mediated via the same mechanism, thus concluding that the mechanism underlying the activation of capacitative Ca^{2+} entry was the discharge of the intracellular Ca^{2+} stores per se and not the cellular levels of inositol phosphates [19]. The concept of a Ca^{2+} entry mechanism regulated by the stores was further confirmed by electrophysiological studies that reported that Ca^{2+} stores discharge activated a Ca^{2+} current in mast cells called Ca^{2+} release-activated Ca^{2+} current, I_{CRAC}, the first store-operated Ca^{2+} current identified [20]. In fact, in addition to the physiological pathway to activate capacitative Ca^{2+} entry via generation of IP_3, this mechanism for Ca^{2+} influx has been reported to be activated by a number of experimental maneuvers leading to a reduction in the amount of free Ca^{2+} in the intracellular stores. These procedures include treatment with thapsigargin, as well as other SERCA inhibitors, such as cyclopiazonic acid or di-*tert*-butylhydroquinone [21, 22] (agents that prevent Ca^{2+} store refilling), loading of Ca^{2+} stores with the metal Ca^{2+} chelator N,N,N,N-tetrakis(2-pyridylmethyl)ethylene diamine (which reduces free intraluminal Ca^{2+} concentration without altering $[Ca^{2+}]_c$ [23, 24]), or dialyzing the cytoplasm with the Ca^{2+} chelators EGTA or BAPTA, which bind Ca^{2+} leaking from the stores, thus preventing store refilling (see [25]). From that point capacitative Ca^{2+}entry was also known as store-operated Ca^{2+} entry (SOCE) [26] or store-mediated Ca^{2+}entry [27]. Since SOCE is the most extended term we will use this denomination throughout the chapter.

The manuscript by Kwan and coworkers [19] also raised a key conceptual feature of SOCE. The initial observations had suggested that the transport of extracellular Ca^{2+} into the cell involved a direct movement of Ca^{2+} into the intracellular stores, since refilling of the Ca^{2+} stores were found to occur with no substantial elevation in $[Ca^{2+}]_c$ [12, 28]. However, this hypothesis somehow limited the role of Ca^{2+} entry to the refilling of the intracellular Ca^{2+} stores. Using the trivalent cation lanthanum, which induces a concentration-dependent inactivation of Ca^{2+} extrusion through the plasma membrane Ca^{2+}-ATPase, and impairs Ca^{2+} entry [29, 30], Kwan and coworkers observed that, in the absence of Ca^{2+} extrusion and entry, Ca^{2+} stores were refilled by Ca^{2+} released into the cytoplasm upon agonist stimulation, providing evidence that store refilling did not involve a direct route into the intracellular Ca^{2+} pools, but rather is the result of a sequential Ca^{2+} entry into the cytoplasm and subsequent reuptake into the Ca^{2+}stores by active SERCA pumping [19, 31].

SOCE has also been described in a number of excitable cells, including neurons [32], cardiomyocytes [33], smooth muscle cells [34] and both endocrine [35–37] and neuroendocrine cells [38, 39]. In these cells, the functional role of SOCE is not limited to store refilling but it plays an important functional role as revealed by a number of disorders observed in the presence of defective SOCE [40–44].

SOCE activation in electrically excitable cells could trigger Ca^{2+} entry through voltage-operated channels by causing membrane depolarization [45, 46], which might result directly from the activation of store-operated Ca^{2+} (SOC) channels, which can be permeable to Ca^{2+}, as well as to other cations, such as Na^+, and have reversal potentials near 0 mV. In addition, indirect mechanisms for membrane depolarization associated to rises in $[Ca^{2+}]_c$ have been reported, such as a Ca^{2+}-dependent inhibition of voltage-dependent K^+ channels [47] reported in pulmonary arterial smooth muscle cells [48]. In contrast to these observations, current evidence supports that the SOCE element STIM1 interacts with the voltage-gated $Ca_v1.2$ channels and suppresses its activity [49, 50]. Since functional expression of voltage-operated Ca^{2+} channels, including $Ca_v1.2$, has been reported not only in excitable cells but also in non-electrically excitable cells, such as T and B lymphocytes and mast cells [51–54], these findings further confirm the role of STIM1 as an essential element in the modulation of agonist-induced Ca^{2+} signals.

The electrophysiological analysis of store-operated currents has revealed the presence of two types of currents: the Ca^{2+} release-activated Ca^{2+} current (I_{CRAC}) and a heterogeneous set of non-CRAC currents, described in different cell types with diverse biophysical properties grouped under the denomination of store-operated currents (I_{SOC}) [25]. The I_{CRAC} current was the first store-operated Ca^{2+} current identified using combined patch-clamp and fura-2 measurements to monitor membrane currents in mast cells [20]. I_{CRAC} is a non-voltage-operated current, unlike Ca^{2+} currents through the Ca_v family of channels, that exhibits large current amplitude at negative potentials and approaches the zero current level at very positive potentials. In addition, the current-voltage relationship for I_{CRAC} reveals a significant inward rectification at negative voltages [20, 25]. The channels conducting I_{CRAC} (CRAC channels) show a single channel conductance <1pS and are highly selective for Ca^{2+} over monovalent cations. The Ca^{2+}:Na^+ permeability ratio has been estimated around 1,000:1. Removal of extracellular Ca^{2+} in the presence of external Na^+ and Mg^{2+} has been reported to abolish this current [55, 56]. However, CRAC channels lose this selectivity in divalent-free solutions, which allow Na^+ to permeate the channels, leading to whole cell currents that are five to eightfold larger than the Ca^{2+} currents [57, 58].

The I_{SOC} currents are mediated by poorly selective cation channels that exhibit a significantly greater conductance than CRAC channels and show different biophysical properties. I_{SOC} currents have been described in different cell types, including vascular endothelial cells [59, 60], pancreatic acinar cells [61], human A431 carcinoma cells [62, 63], smooth muscle cells [64, 65], submandibular and parotid gland cells [66], liver cells [67], skeletal muscle cells [68], neurons [69] and adrenal chromaffin cells [70]. The Ca^{2+}:Na^+ permeability ratio of the channels conducting this currents is very heterogeneous, ranging from 1:0.07 to 50:1.

SOCE has been reported to play a number of important roles in cell physiology, including, among others, Ca^{2+} store refilling upon agonist stimulation, support for the sustained elevations in $[Ca^{2+}]_c$ required for a number of cellular functions and maintenance of the amplitude of Ca^{2+} oscillations [71, 72]. In addition, a variety of functions have been reported to be regulated by SOCE in a number of cell types, including endothelial cell permeability [73], vascular smooth muscle cell proliferation and contractility [34, 74], platelet function [75, 76], immunological response [77] or exocytosis [70, 78, 79], among many others.

1.2 Activation Mechanisms: Initial Studies

Since the discovery of SOCE, the mechanism underlying the regulation of SOC channels in the plasma membrane by distantly located intracellular Ca^{2+} stores has been intensely investigated and debated. The description that STIM1 is the Ca^{2+}

sensor of the intracellular Ca^{2+} stores in 2005 [80, 81] was a milestone in the identification of the events that regulate SOC channels following a reduction of the intraluminal Ca^{2+} concentration and marked an inflection point between the initial models and the current one. The initial studies can be grouped into those supporting the indirect coupling between the stores and SOC channels, those that propose a direct or conformational coupling between SOC channels and elements in the stores, and those reporting the insertion of preformed channels in the plasma membrane [25, 82].

The indirect coupling hypothesis assumes that Ca^{2+} store depletion results in the generation or activation of diffusible molecules that participate in SOC channel gating. This model includes roles for cGMP [83, 84], a product of cytochrome P450 [85], probably the metabolite 5,6-epoxyeicosatrienoic acid [86], tyrosine kinases [87–89], monomeric GTP-binding proteins [90–92], calmodulin [93] and a still uncharacterized Ca^{2+}-influx factor (CIF) [94, 95].

The conformational coupling hypothesis proposes a physical and constitutive interaction between elements in the membrane of the Ca^{2+} stores and SOC channels in the plasma membrane. This model was originally proposed by Irvine as a mechanism for the activation of SOCE involving IP_3 receptors in the endoplasmic reticulum (ER) and inositol 1,3,4,5-tetrakisphosphate (IP_4) receptors in the plasma membrane [96, 97]. Later on, Birnbaumer and colleagues demonstrated that the association of the IP_3 receptor with transiently expressed TRPC1, TRPC3, and TRPC6 plays a relevant role in the activation of SOCE [98]. This hypothesis received support from studies providing evidence for a role in the activation of SOCE of the protein junctate, an ER Ca^{2+}-binding protein that induces and/or stabilize the interaction between the ER and the plasma membrane [99]. Junctate has been proposed to play an important role in the association of the IP_3 receptors and TRPC3 [99]. Consistent with the conformational coupling are a number of studies reporting that TRPC3 channels present in excised patches can be activated by IP_3 [100]. Furthermore, the N-terminus of IP_3 receptors, containing the IP_3-binding domain, is essential for the activation of plasma membrane TRPC3 channels [101].

In addition to the indirect and conformational coupling hypotheses, a number of studies also supported the activation of SOCE by the translocation and insertion of preformed channels, initially located in intracellular vesicles, into the plasma membrane. This model, that was proposed by Penner and colleagues [90], requires the participation of the synaptosome associated protein SNAP-25 [102]. In hippocampal neurons, HEK 293 cells stably expressing TRPC6 and neuronal and epithelial cells expressing TRPC3 depletion of the intracellular Ca^{2+} stores results in the expression of TRPC channels in the plasma membrane [103–105].

The conformational coupling model assumed that the IP_3 receptor in the stores and SOC channels are permanently associated, so that a decrease in the intraluminal Ca^{2+} concentration is communicated to the SOC channels via a conformational change in the IP_3 receptor. Analyzing the role of the actin cytoskeleton in the activation of SOCE in different cell types, including smooth muscle cell lines, human platelets and other non-excitable cells, such as pancreatic acinar cells and the human hepatocellular carcinoma cell line HepG2, an alternative to the constitutive conformational coupling model called secretion-like coupling or de novo conformational coupling was described. This model assumes a reversible physical coupling between the ER and the plasma membrane where the actin cytoskeleton plays an essential role [106–109]. The de novo coupling between elements in the ER and the plasma membrane was found to be strongly dependent on actin filament reorganization. Thus, the actin network located underneath the plasma membrane acts as a negative clamp that prevents constitutive coupling [106, 109, 110]. According to this, stabilization of the cortical actin network has been reported to impair the activation of SOCE in a variety of cell types, including smooth muscle cell lines, human platelets, corneal endothelial cells, the human prostate adenocarcinoma cell line LNCaP, pancreatic acinar cells, smooth muscle cells or neutrophils [39, 45, 106, 107, 109, 111, 112]; however, in pancreatic β cells, thyroid FRTL-5 cells and *Aplisia* bag cell neurons, SOCE has been reported to be insensitive to stabilization of the cortical actin cytoskeleton by treatment with jasplakinolide [113–115], and stabilization of the peripheral cytoskeleton and disassembly of actin microfilaments have been shown to fail to alter the rate or extent of activation of I_{CRAC} in the RBL-1 rat basophilic cell line [116]. This discrepancies might be attributed to the different idiosyncrasy of the cells investigated. The role of the cortical cytoskeleton in the modulation of SOCE has been further supported by analysis of the actin polymerization and Ca^{2+} mobilization on a subsecond time scale. In human platelets stimulated with the physiological agonist thrombin we detected an initial decrease in the actin filament content within 0.1 s after stimulation that reached a minimum 0.9 s after the addition of thrombin. Actin depolymerization, involving the actin-binding protein cofilin, was observed before the initiation of SOCE, which occurred with a latency of 2.1 s after agonist stimulation [117], which is consistent with a role for the actin cortical cytoskeleton in the modulation of SOCE.

After the identification of STIM1 as the ER Ca^{2+} sensor [80, 118] and Orai1 as the prototypical CRAC/SOC channel [77, 119–122] the initial hypotheses had to be necessarily reconsidered and/or reinterpreted. Therefore, as described in a number of cells for SOCE, is the interaction between STIM1 and Orai1 modulated by the cortical actin cytoskeleton? There is a growing body of evidence supporting a role for the cytoskeleton in the interaction between STIM1 and Orai1. In HEK-293 cells and platelets, we have demonstrated that stabilization of the cortical actin cytoskeleton impairs de novo association between the ER Ca^{2+} sensor STIM1 and the SOC channel Orai1 [123, 124], without altering the coupling between plasma membrane-resident STIM1 and Orai1 [125]. These findings might explain why, in the presence of extracellular Ca^{2+}, treatment with jasplakinolide prevented Ca^{2+} entry but not the influx of other cations, such as Mn^{2+} or Na^{+} [126] that might permeate through non-selective cation channels containing TRP subunits. More recently, it has been reported in HeLa cells that calcineurin impairs cytoskeleton remodeling and STIM1/Orai1 puncta-like formation in a KSR-2-dependent manner [127]. Furthermore, in activated T cells, the Rac1 effector protein WAVE2 has been reported to modulate SOCE and STIM1/Orai1 interaction by regulation of actin filament reorganization [128]. Therefore, the participation of the cortical actin cytoskeleton in the modulation of the de novo interaction between STIM1 and Orai1 and, subsequently SOCE, might still be a potential model.

1.3 STIM1, Orai and TRPCs

As we discussed in the previous section, STIM1, originally named GOK, was presented in 2005 as an essential protein involved in the activation of SOCE [80, 81]. The initial studies on STIM1 were published in the mid-1990s, where STIM1 was described as a type-1 transmembrane glycoprotein of 685 amino acids involved in cell-cell interactions in hematopoietic cells [129] and as a cell growth suppressor that plays a pivotal role in the establishment and progression of rhabdomyosarcomas and rhabdoid tumors [130, 131]. Since then, several studies were focused on the study of the structure and function of STIM1 and the determination of the signaling pathways involving STIM1. In 2000, immunofluorescence and cell surface biotinylation analysis revealed that STIM1 is a 90-kDa integral transmembrane phosphoprotein ubiquitously expressed both in the plasma membrane and intracellular membranes of a variety human primary cells, including neonatal foreskin fibroblasts and MG63 osteoblast-like cells, and established tumor cell lines, such as the human leukemic K562, HL60 and U937 cell lines [132]. Later on, the same group demonstrated that the luminal N-terminal region of STIM1 includes an ER signal peptide, EF-hand Ca^{2+}-binding motif, and a single sterile alpha motif (SAM) involved in protein-protein interaction, whereas the cytosolic region consists of two coiled-coil domains, a proline/serine-rich region, and a lysine-rich region [132–134]. These studies also showed that STIM1 is subjected to different post-translational modifications, including phosphorylation on serine residues [133], and N-linked glycosylation in two sites within in the SAM domain [134].

The role of STIM1 in the regulation of SOCE was initially discovered in 2005 by Roos and coworkers [80]. In this study, RNA interference (RNAi)-based screen of more than 170 proteins with known signaling motifs, was designed to identify genes that alter SOCE and the signal pathways controlling them in *Drosophila* S2 cells. With this approach, the *stim* gene and its product, STIM1, were identified as essential regulators of SOCE and CRAC channel activity since RNAi-mediated knockdown of the *stim* gene significantly blocked thapsigargin-induced Ca^{2+} influx in S2 cells. This new promising role of STIM1 was also confirmed in human cells. Suppression of STIM1 expression also significantly reduced thapsigargin-induced SOCE in Jurkat T cells expressing a short RNA hairpin loop (shRNA) targeting human STIM1. Moreover, Ca^{2+} influx was also diminished when STIM1 expression was down-regulated in HEK-293 and SH-SY5Y human cells transfected with STIM1 siRNA [80].

Since STIM1 overexpression in HEK-293 cells was not associated with an increased Ca^{2+} influx and detectable SOCE current, it was proposed that STIM1 could not function as SOC channel itself, and the intraluminal location of the EF-hand motif suggested that STIM1 might act as an ER Ca^{2+} sensor [80]. The report by Roos and coworkers was followed shortly thereafter by two studies from Liou and coworkers [81] and Zhang and coworkers [118]. Cell transfection with a mutated EF-hand motif of STIM1 or *Drosophila* Stim induced a constitutive Ca^{2+} influx mediated by CRAC channels and independent of the filling state of the Ca^{2+}

stores, and treatment with thapsigargin failed to further promote SOCE in HeLa [81], Jurkat and S2 cells [118]. Using different techniques, from immunofluorescence and electron microscopy to surface biotinylation, both studies also demonstrated that, once Ca^{2+} stores are depleted, STIM1 migrates from the ER to the proximity of the plasma membrane. Liou and coworkers first reported that upon store depletion STIM1 redistributes into *punctae* structures, ER regions enriched in STIM1 oligomers and located in close proximity to the plasma membrane [81]. Later on, it was demonstrated that SAM is an essential STIM1 domain for oligomerization and *punctae* formation, while the coiled-coil domains are involved in further stabilization of the STIM1 oligomers and also in functional protein-protein interactions of STIM1 with Ca^{2+} channels located in the plasma membrane [135].

The identification of STIM1 as the intraluminal Ca^{2+} sensor provided an explanation for the communication between the intracellular Ca^{2+} stores and the CRAC/SOC channels located in the plasma membrane, whose identity still remained elusive. In 2006, Orai1 was identified as the pore subunit of the CRAC channels [77, 121, 136–138]. Initially Feske and coworkers [77] and soon thereafter Vig et al. [136] presented Orai1, also named as CRAC modulator 1 (CRACM1), as an essential component regulator of the CRAC channel. Both studies identified *Drosophila* gene *olf186-F* as an essential gene in the activation of thapsigargin-induced SOCE by using a genome-wide RNAi screen in *Drosophila* S2 cells to identify the genes encoding the CRAC channel or other proteins involved in its regulation [77, 136]. The *Drosophila olf186-F* gene has three human homologues that encode the proteins named, for the first time, as Orai1, Orai2 and Orai3 [77]. The term "orai" is referred to the three keepers of heaven's gate in Greek mythology. The gene encoding human Orai1 was mapped on chromosome region 12q24 and was shown to be mutated in the hereditary severe combined immune deficiency (SCID) syndrome patients. Collected T cells from these patients are characterized by an impaired SOCE and CRAC channel activity. This mutation consists in a C→T transition at position 271 of the coding sequence of the human *orai1* gene, leading to replacement of a highly conserved arginine residue by tryptophan at position 91 of the protein (R91W). Expression of an exogenous wild-type Orai1 in SCID T-cells restored store depletion-evoked Ca^{2+} influx, suggesting that this single point mutation R91W in Orai1 sequence was responsible of the defective I_{CRAC}. However, expression of wild-type Orai1 in SCID T-cells did not promote Ca^{2+} influx per se in resting conditions, suggesting that Orai1 is not a Ca^{2+} channel constitutively activated. Furthermore, electrophysiological and pharmacological properties of the restored Ca^{2+} influx current were fully consistent with those observed in I_{CRAC} in normal T cells [77]. According to these results, the selective small interfering RNA (SiRNA)-mediated knockdown of Orai1 reduced both Ca^{2+} influx and I_{CRAC} in response to thapsigargin in human HEK-293 and Jurkat T cells [136]. In addition to the demonstration that Orai1 plays a pivotal role in CRAC current generation, both studies also predicted the topology Orai1 as a membrane protein with four transmembrane domains and cytosolic N- and C- termini, which shows no structural homology to other known ion-channels [77, 136].

Soon after the identification of Orai1 as a the pore forming subunit of the CRAC channel published studies demonstrated the earliest pieces of evidence for the functional interaction between STIM1 and Orai1 in the activation of CRAC currents were reported. Co-expression of STIM1 and Orai1 resulted in a significantly enhanced CRAC current in HEK-293, Jurkat T cells [119, 120, 122] and *Drosophila* S2 cells [139]. The generated current showed Ca^{2+} selectivity, development and inactivation kinetics, pharmacological profiles and biophysical properties indistinguishable from native CRAC current [120, 139].

According to the idea that conserved acidic residues are essential for ion permeation through known Ca^{2+} channels, three simultaneous studies analyzed highly conserved glutamate residues across species and the most interesting results come from the residues E178 and E262 in *Drosophila* and their corresponding human homologues E106 and E190. Mutation of E106 to alanine or glutamine greatly impaired channel function. However, the conservative mutations of E106 to an aspartate (E106D) resulted in a channel with reduced selectivity for Ca^{2+}, thus indicating that this residue functions as part of the selectivity filter, and confirms that Orai1 is a pore forming subunit of the CRAC channel [121, 137, 138]. While mutation of E190 to aspartate or alanine had no significant effect on channel function, the substitution by glutamine (E190Q) resulted in diminished Ca^{2+} selectivity [121, 138]. These studies also demonstrated, for the first time, the STIM1-Orai1 protein interaction upon Ca^{2+} store depletion to initiate CRAC channel activation, by using co-immunoprecipitation in co-transfected HEK-293 cells [138] and *Drosophila* S2 cells [137]. The STIM1/Orai1 complex was reported to be located in the *punctae* region on the plasma membrane after CRAC current activation by fluorescence microscopy [140, 141].

The identification of Orai1 as the pore forming region of the CRAC channel raised the question concerning the architecture of the channel, with reports proposing tetrameric and even higher-order multimeric structures [142–144]. Hou and coworkers [145] reported in 2012 the crystal structure of *Drosophila* Orai channel, which shares 70 % homology with human Orai1. The crystal structure indicates that the pore of *Drosophila* Orai is formed by six subunits, each containing four transmembrane regions (M1 to M4). The transmembrane helices are arranged in three concentric circles around the pore, where the inner ring is formed of six M1 helices, the middle ring consists of M2 and M3 helices and the outer ring is formed of M4 helices located at the periphery of the channel. The side chains of M1 line the pore of Orai, which is composed of four different sections: a selectivity filter formed by a ring of glutamates at the extracellular end of the channel, followed by a hydrophobic section, a basic region spanning near the intracellular side of the channel, which might contribute to the stabilization of the closed state of the channel, and a wider section that extends into the cytosol. The pore region of Orai shows biochemical differences with the known K^+ channels, which emphasizes the distinct mechanisms of ion selectivity and permeation between Orai and K^+ channels.

In addition to the small families of STIM and Orai, another somewhat larger family of ion channels are also linked to SOCE. These are a subfamily of the larger family of TRP channels, called TRPC for classical or canonical TRPs, because it is comprised of those proteins most highly related to the *Drosophila* TRP channel [146].

The origin of the *Drosophila* TRP goes back to 1969, when Cosens and coworkers identified a spontaneously formed mutant on the basis of a behavioral phenotype [147], but it was not until 1975 that Minke described and named it as "Transient Receptor Potential" (TRP) because of the transient depolarization of the photoreceptors due to Na^+ and Ca^{2+} entry [148]. The identity of the mutated protein was reported in 1989 by Montell and coworkers who cloned, sequenced and presented a molecular characterization of the *Drosophila trp* gen and highlighted its similarity with the Ca^{2+} channels [149]. Later studies characterized TRP and its homologue TRPL as Ca^{2+} permeable channels activated downstream of phospholipase C [150].

After the initial identification of the *Drosophila* TRPs, all the attempts were focused on finding its counterpart in mammals. In 1995, regarding its high homology to the TRP protein sequence, an expressed sequence from a human fetal brain cDNA library was reported by two groups as the first homologue in human (TRPC1) [151, 152]. Since the identification of TRPC1, several mammalian homologues have been described and are classified into six subfamilies: TRPC (canonical), TRPV (vanilloid), TRPM (melastatin), TRPA (ankyrin), TRPP (polycystin), TRPML (mucolipin). In addition, a less related TRPN has been reported in worms and fish and TRPY in fungi [153].

The DNA and corresponding amino acid sequence of the previously called TRPC1, the expression pattern as well as another general features of the novel protein were described by Wes and coworkers and Zhu and coworkers, who characterized it as a polypeptide of 793 amino acids expressed at the highest levels in the fetal brain and in the adult heart, brain, testis and ovaries [151, 152]. The primary structure was supposed to present six transmembrane domains with a pore region between the fifth (S5) and sixth (S6) transmembrane segments and both C and N termini located intracellularly [154]. This structure is shared by all the members of the TRP family. Furthermore, Wes also described that the most highly conserved domain among all TRP family members contains three ankyrin repeats which are thought to play key roles in protein-protein interactions [151, 155]. Later on the whole structure of the protein was reported, with the N-terminus containing three to four ankyrin repeats, a predicted coiled-coil motif and a caveolin binding region. By the way, the citoplasmic C-terminus includes an EWKFAR TRP box, a highly conserved proline rich domain, a coiled-coil motif and a *C*aM/*I*P_3 *r*eceptor *b*inding (CIRB) domain [98, 156]. The presence of coiled-coil domains, commonly involved in subunit oligomerization suggest that they are probably involved in homo-and heteromerization of TRPCs or in linking TRPCs to other proteins also containing coiled-coil domains [157, 158].

Despite the key role of TRPs in receptor-operated Ca^{2+} entry (ROCE) and in SOCE has been reported in different cell types, the latter has been a matter of intense debate, and the role of TRPCs in SOCE has not been demonstrated in all the cellular models investigated. Among the first pieces of evidence for a role of TRPC1 in SOCE comes from a study made in Chinese hamster ovary cells (CHO), where expression of TRPC1A, a splice variant of TRPC1, resulted in a linear nonselective cation current with similar permeabilities for Na^+, Ca^{2+}, and Cs^+, activated by intracellular infusion of either IP_3 or thapsigargin to deplete intracellular Ca^{2+} stores [159]. At the same time, Zhu and coworkers demonstrated that expression in COS cells of full-length cDNA encoding human TRPC1 increased SOCE while expression of antisense sequences suppressed it [160]. Further evidence supporting the role of TRPC1 in SOCE comes from studies based on overexpression of TRPC1 proteins or knockdown of the endogenous TRPC1 channels in several cell types, including human cells. In human submandibular gland cells and vascular endothelial cells, the overexpression of TRPC1 increased SOCE while the silencing of the endogenous protein by using antisense oligonucleotides reduced thapsigargin-evoked influx [161, 162]. In general, the key role of TRPC1 in SOCE has been supported by several research groups by means of different experimental maneuvers in many cell types, such as DT40-B-lymphocytes, endothelial cells, rat cardiac myocytes, the human megakaryoblastic cell line MEG01, the C2C12 mouse myoblastic cell line, rat kidney fibroblasts or human platelets [162–166]. Other members of the TRPC family have also been reported to be activated by store depletion or to be involved in SOCE both in excitable and non-excitable cells. For instance, the transient expression of the full length cDNA of mouse TRPC2, the homologue of the human *trp2* pseudogene, was reported to evoke a Ca^{2+} entry that was shown to be readily activated not only after agonist stimulation but also by store depletion in the absence of an agonist [167]. In addition, in 1996 Zhu and coworkers observed that the heterologous expression of the human TRPC3 in COS cells enhanced SOCE [160] and similar results were obtained by Boulay and coworkers by expressing the murine TRPC6 in this cells [168]. Over the years, some studies have highlighted that there is also a growing body of evidence supporting a role of TRPC6 in the conduction of SOCE in different cells [169, 170].

Nevertheless, several studies have failed to observe store-operated behavior of exogenously expressed TRPC channels in different cell types, thus demonstrating that TRPCs did not account for SOCE in all cellular models investigated [171]. Moreover, the mode of expression of TRPCs might determine its involvement in SOCE or ROCE in the same cell type, as it was reported by Lievremont and coworkers in 2004, who described that TRPC7 activation mode was different in cells transiently or stably expressing this channel [172].

With the identification of STIM1 and Orai1 as the key molecular players of CRAC currents, most studies focused on the possible interaction between these proteins and TRPC channels. In 2006, after Spassova and coworkers confirmed the crucial functional role of STIM1 over the regulation of SOC channels by using electrophysiological analysis [173], two independent studies demonstrated for the first time that STIM1 is associated with TRPC1, in the plasma membrane upon Ca^{2+} store depletion, and that the STIM1-TRPC1 interaction promotes channel activation and SOCE [174, 175]. In HEK-293 cells expressing exogenous STIM1, Ca^{2+} store depletion increased the association between STIM1 and TRPC1, while the expression of the EF-hand mutant STIM1 yield a constitutive STIM1-TRPC1 complex formation and subsequent TRPC1 activation and Ca^{2+} influx [174]. According to this, electrotransjection of human platelets with anti-STIM1 antibody, directed toward the EF-hand Ca^{2+}-binding motif, prevented the migration of STIM1 toward the plasma membrane, the interaction between endogenously expressed STIM1 and TRPC1 and reduced SOCE [175]. Functional interaction between STIM1 and TRPC1 has also been reported, among others cellular types, in intestinal epithelial cells [176], salivary and pancreatic acinar cells [79, 177], pulmonary artery cells [178] or messanglial cells [179].

STIM1 also interacts directly with other members of TRPC family involved in SOCE, such as TRPC4 and TRPC5 in HEK293 cells [180] and TRPC6 in human platelets [24], regulating their activation. STIM1 also indirectly regulates TRPC3 and TRPC6 by mediating the heteromultimerization of TRPC3 with TRPC1 and TRPC4 with TRPC6 [180]. Direct STIM1/TRPC interaction has been reported to require interaction of the STIM1 SOAR region with the TRPC C-terminal coiled-coil domains, as well as electrostatic interactions involving the lysine-rich domain of STIM1 [181, 182]. This association is supported by the actin cytoskeleton [175] and plasma membrane lipid rafts domains [183–185].

In addition to the STIM1-TRPC coupling previously described, there is a body of evidence supporting the formation of ternary complexes among the three main elements involved in SOCE, STIM, Orai and TRPCs. In HEK293 cells, SOCs were reported to be built by TRPC pore-forming subunits and Orai regulatory subunits that transduce the Ca^{2+} store depletion signal from STIM1 to TRPCs [186]. Supporting this hypothesis, a dynamic assembly among STIM1, Orai1 and TRPC1 has been proposed to be essential for SOCE in human salivary gland cells [187] and human platelets [123]. Interestingly, the association of these proteins has been reported to occur in specific plasma membrane microdomains known as lipid rafts which provide the adequate environment for the formation of the store-operated Ca^{2+} influx complex (SOCIC) or receptor-operated Ca^{2+} entry complexes [184, 185]. The following chapters will highlight recent advances in the Ca^{2+} entry mechanisms, involving store-operated and receptor-operated signaling pathways, in non-excitable cells.

Acknowledgments This work was supported by MINECO (Grant BFU2013-45564-C2-1-P) and Gobierno de Extremadura-FEDER (GR15029). JJL and LA are supported by Juan de la Cierva Program (JCI-2012-12934) and MINECO fellowship BES-2011-043356, respectively.

References

1. Ebashi S (1961) Calcium binding activity of vesicular relaxing factor. J Chir (Paris) 50:236–244
2. Ringer S (1883) A further contribution regarding the influence of the different constituents of the blood on the contraction of the heart. J Physiol 4(1):29–42.3
3. Ringer S, Buxton DW (1885) Concerning the action of small quantities of calcium, sodium, and potassium salts upon the vitality and function of contractile tissue and the cuticular cells of fishes. J Physiol 6(4–5):154–161
4. Hofer AM, Lefkimmiatis K (2007) Extracellular calcium and cAMP: second messengers as "third messengers"? Physiology (Bethesda) 22:320–327

5. Bianchi CP, Shanes AM (1959) Calcium influx in skeletal muscle at rest, during activity, and during potassium contracture. J Gen Physiol 42(4):803–815
6. Winegrad S, Shanes AM (1962) Calcium flux and contractility in guinea pig atria. J Gen Physiol 45:371–394
7. Borle AB (1969) Kinetic analyses of calcium movements in HeLa cell cultures. I. Calcium influx. J Gen Physiol 53(1):43–56
8. Borle AB (1968) Calcium metabolism in HeLa cells and the effects of parathyroid hormone. J Cell Biol 36(3):567–582
9. Shigenobu K, Sperelakis N (1972) Calcium current channels induced by catecholamines in chick embryonic hearts whose fast sodium channels are blocked by tetrodotoxin or elevated potassium. Circ Res 31(6):932–952
10. Hogestatt ED (1984) Characterization of two different calcium entry pathways in small mesenteric arteries from rat. Acta Physiol Scand 122(4):483–495
11. Hallam TJ, Rink TJ (1985) Agonists stimulate divalent cation channels in the plasma membrane of human platelets. FEBS Lett 186(2):175–179
12. Putney JW Jr (1986) A model for receptor-regulated calcium entry. Cell Calcium 7(1):1–12
13. Holda JR, Klishin A, Sedova M, Huser J, Blatter LA (1998) Capacitative calcium entry. News Physiol Sci 13:157–163
14. Takemura H, Putney JW Jr (1989) Capacitative calcium entry in parotid acinar cells. Biochem J 258(2):409–412
15. Sage SO, Merritt JE, Hallam TJ, Rink TJ (1989) Receptor-mediated calcium entry in fura-2-loaded human platelets stimulated with ADP and thrombin. Dual-wavelengths studies with Mn^{2+}. Biochem J 258(3):923–926
16. Thastrup O, Foder B, Scharff O (1987) The calcium mobilizing tumor promoting agent, thapsigargin elevates the platelet cytoplasmic free calcium concentration to a higher steady state level. A possible mechanism of action for the tumor promotion. Biochem Biophys Res Commun 142(3):654–660
17. Lytton J, Westlin M, Hanley MR (1991) Thapsigargin inhibits the sarcoplasmic or endoplasmic reticulum Ca-ATPase family of calcium pumps. J Biol Chem 266(26):17067–17071
18. Takemura H, Hughes AR, Thastrup O, Putney JW Jr (1989) Activation of calcium entry by the tumor promoter thapsigargin in parotid acinar cells. Evidence that an intracellular calcium pool and not an inositol phosphate regulates calcium fluxes at the plasma membrane. J Biol Chem 264(21):12266–12271
19. Kwan CY, Takemura H, Obie JF, Thastrup O, Putney JW Jr (1990) Effects of MeCh, thapsigargin, and La3+ on plasmalemmal and intracellular Ca2+ transport in lacrimal acinar cells. Am J Physiol 258(6 Pt 1):C1006–C1015
20. Hoth M, Penner R (1992) Depletion of intracellular calcium stores activates a calcium current in mast cells. Nature 355(6358):353–356
21. Okon EB, Golbabaie A, Breemen C (2008) Paracrine effects of endothelial cells in a diabetic mouse model: capacitative calcium entry stimulated thromboxane release. Horm Metab Res 40(9):645–650
22. Rosado JA, Lopez JJ, Harper AG, Harper MT, Redondo PC, Pariente JA, Sage SO, Salido GM (2004) Two pathways for store-mediated calcium entry differentially dependent on the actin cytoskeleton in human platelets. J Biol Chem 279(28):29231–29235
23. Hofer AM, Fasolato C, Pozzan T (1998) Capacitative Ca2+ entry is closely linked to the filling state of internal Ca2+ stores: a study using simultaneous measurements of ICRAC and intraluminal [Ca2+]. J Cell Biol 140(2):325–334
24. Jardin I, Gomez LJ, Salido GM, Rosado JA (2009) Dynamic interaction of hTRPC6 with the Orai1/STIM1 complex or hTRPC3 mediates its role in capacitative or non-capacitative Ca^{2+} entry pathways. Biochem J 420:267–276
25. Parekh AB, Putney JW Jr (2005) Store-operated calcium channels. Physiol Rev 85(2):757–810
26. Montero M, Garcia-Sancho J, Alvarez J (1993) Transient inhibition by chemotactic peptide of a store-operated Ca2+ entry pathway in human neutrophils. J Biol Chem 268(18):13055–13061

27. Davis W, Sage SO, Allen JM (1994) Cytosolic calcium elevation in response to Fc receptor cross-linking in undifferentiated and differentiated U937 cells. Cell Calcium 16(1):29–36
28. Merritt JE, Rink TJ (1987) Regulation of cytosolic free calcium in fura-2-loaded rat parotid acinar cells. J Biol Chem 262(36):17362–17369
29. van Breemen C, De Weer P (1970) Lanthanum inhibition of 45Ca efflux from the squid giant axon. Nature 226(5247):760–761
30. Lansman JB (1990) Blockade of current through single calcium channels by trivalent lanthanide cations. Effect of ionic radius on the rates of ion entry and exit. J Gen Physiol 95(4):679–696
31. Rosado JA (2006) Discovering the mechanism of capacitative calcium entry. Am J Physiol Cell Physiol 291(6):C1104–C1106
32. Emptage NJ, Reid CA, Fine A (2001) Calcium stores in hippocampal synaptic boutons mediate short-term plasticity, store-operated Ca2+ entry, and spontaneous transmitter release. Neuron 29(1):197–208
33. Hunton DL, Lucchesi PA, Pang Y, Cheng X, Dell'Italia LJ, Marchase RB (2002) Capacitative calcium entry contributes to nuclear factor of activated T-cells nuclear translocation and hypertrophy in cardiomyocytes. J Biol Chem 277(16):14266–14273
34. Rodriguez-Moyano M, Diaz I, Dionisio N, Zhang X, Avila-Medina J, Calderon-Sanchez E, Trebak M, Rosado JA, Ordonez A, Smani T (2013) Urotensin-II promotes vascular smooth muscle cell proliferation through store operated calcium entry and EGFR transactivation. Cardiovasc Res 100:297–306
35. Liu YJ, Gylfe E (1997) Store-operated Ca2+ entry in insulin-releasing pancreatic beta-cells. Cell Calcium 22(4):277–286
36. Tornquist K, Malm AM, Pasternack M, Kronqvist R, Bjorklund S, Tuominen R, Slotte JP (1999) Tumor necrosis factor-alpha, sphingomyelinase, and ceramide inhibit store-operated calcium entry in thyroid FRTL-5 cells. J Biol Chem 274(14):9370–9377
37. Zerbes M, Bunn SJ, Powis DA (1998) Histamine causes Ca2+ entry via both a store-operated and a store-independent pathway in bovine adrenal chromaffin cells. Cell Calcium 23(6):379–386
38. Kim JH, Shin SY, Nam JH, Hong EK, Chung YS, Jeong JY, Kang J, Uhm DY, Kim SJ (2003) Adrenergic regulation of the intracellular [Ca2+] and voltage-operated Ca2+ channel currents in the rat prostate neuroendocrine cells. Prostate 57(2):99–110
39. Vanoverberghe K, Lehen'kyi V, Thebault S, Raphael M, Vanden Abeele F, Slomianny C, Mariot P, Prevarskaya N (2012) Cytoskeleton reorganization as an alternative mechanism of store-operated calcium entry control in neuroendocrine-differentiated cells. PLoS ONE 7(9):e45615
40. Feske S (2009) ORAI1 and STIM1 deficiency in human and mice: roles of store-operated Ca2+ entry in the immune system and beyond. Immunol Rev 231(1):189–209
41. McCarl CA, Picard C, Khalil S, Kawasaki T, Rother J, Papolos A, Kutok J, Hivroz C, Ledeist F, Plogmann K, Ehl S, Notheis G, Albert MH, Belohradsky BH, Kirschner J, Rao A, Fischer A, Feske S (2009) ORAI1 deficiency and lack of store-operated Ca2+ entry cause immunodeficiency, myopathy, and ectodermal dysplasia. J Allergy Clin Immunol 124(6):1311–1318, e1317
42. Picard C, McCarl CA, Papolos A, Khalil S, Luthy K, Hivroz C, LeDeist F, Rieux-Laucat F, Rechavi G, Rao A, Fischer A, Feske S (2009) STIM1 mutation associated with a syndrome of immunodeficiency and autoimmunity. N Engl J Med 360(19):1971–1980
43. Ariano P, Zamburlin P, D'Alessandro R, Meldolesi J, Lovisolo D (2010) Differential repression by the transcription factor REST/NRSF of the various Ca2+ signalling mechanisms in pheochromocytoma PC12 cells. Cell Calcium 47(4):360–368
44. Hao B, Lu Y, Wang Q, Guo W, Cheung KH, Yue J (2014) Role of STIM1 in survival and neural differentiation of mouse embryonic stem cells independent of Orai1-mediated Ca2+ entry. Stem Cell Res 12(2):452–466
45. Morales S, Camello PJ, Rosado JA, Mawe GM, Pozo MJ (2005) Disruption of the filamentous actin cytoskeleton is necessary for the activation of capacitative calcium entry in naive smooth muscle cells. Cell Signal 17(5):635–645

46. Chakraborty S, Berwick ZC, Bartlett PJ, Kumar S, Thomas AP, Sturek M, Tune JD, Obukhov AG (2011) Bromoenol lactone inhibits voltage-gated Ca2+ and transient receptor potential canonical channels. J Pharmacol Exp Ther 339(2):329–340
47. Post JM, Gelband CH, Hume JR (1995) [Ca2+]i inhibition of K+ channels in canine pulmonary artery. Novel mechanism for hypoxia-induced membrane depolarization. Circ Res 77(1):131–139
48. Lu W, Wang J, Peng G, Shimoda LA, Sylvester JT (2009) Knockdown of stromal interaction molecule 1 attenuates store-operated Ca2+ entry and Ca2+ responses to acute hypoxia in pulmonary arterial smooth muscle. Am J Physiol Lung Cell Mol Physiol 297(1):L17–L25
49. Wang Y, Deng X, Mancarella S, Hendron E, Eguchi S, Soboloff J, Tang XD, Gill DL (2010) The calcium store sensor, STIM1, reciprocally controls Orai and CaV1.2 channels. Science 330(6000):105–109
50. Park CY, Hoover PJ, Mullins FM, Bachhawat P, Covington ED, Raunser S, Walz T, Garcia KC, Dolmetsch RE, Lewis RS (2009) STIM1 clusters and activates CRAC channels via direct binding of a cytosolic domain to Orai1. Cell 136(5):876–890
51. Suzuki Y, Yoshimaru T, Inoue T, Ra C (2009) Ca v 1.2 L-type Ca2+ channel protects mast cells against activation-induced cell death by preventing mitochondrial integrity disruption. Mol Immunol 46(11–12):2370–2380
52. Yoshimaru T, Suzuki Y, Inoue T, Ra C (2009) L-type Ca2+ channels in mast cells: activation by membrane depolarization and distinct roles in regulating mediator release from store-operated Ca2+ channels. Mol Immunol 46(7):1267–1277
53. Suzuki Y, Inoue T, Ra C (2010) L-type Ca2+ channels: a new player in the regulation of Ca2+ signaling, cell activation and cell survival in immune cells. Mol Immunol 47(4):640–648
54. Matza D, Flavell RA (2009) Roles of Ca(v) channels and AHNAK1 in T cells: the beauty and the beast. Immunol Rev 231(1):257–264
55. Fierro L, Lund PE, Parekh AB (2000) Comparison of the activation of the Ca2+ release-activated Ca2+ current ICRAC to InsP3 in Jurkat T-lymphocytes, pulmonary artery endothelia and RBL-1 cells. Pflugers Arch 440(4):580–587
56. Zweifach A, Lewis RS (1993) Mitogen-regulated Ca2+ current of T lymphocytes is activated by depletion of intracellular Ca2+ stores. Proc Natl Acad Sci U S A 90(13):6295–6299
57. Bakowski D, Parekh AB (2002) Monovalent cation permeability and Ca(2+) block of the store-operated Ca(2+) current I(CRAC) in rat basophilic leukemia cells. Pflugers Arch 443(5–6):892–902
58. Prakriya M, Lewis RS (2002) Separation and characterization of currents through store-operated CRAC channels and Mg2+ -inhibited cation (MIC) channels. J Gen Physiol 119(5):487–507
59. Vaca L, Kunze DL (1994) Depletion of intracellular Ca2+ stores activates a Ca(2+)-selective channel in vascular endothelium. Am J Physiol 267(4 Pt 1):C920–C925
60. Fasolato C, Nilius B (1998) Store depletion triggers the calcium release-activated calcium current (ICRAC) in macrovascular endothelial cells: a comparison with Jurkat and embryonic kidney cell lines. Pflugers Arch 436(1):69–74
61. Krause E, Pfeiffer F, Schmid A, Schulz I (1996) Depletion of intracellular calcium stores activates a calcium conducting nonselective cation current in mouse pancreatic acinar cells. J Biol Chem 271(51):32523–32528
62. Luckhoff A, Clapham DE (1994) Calcium channels activated by depletion of internal calcium stores in A431 cells. Biophys J 67(1):177–182
63. Kiselyov KI, Semyonova SB, Mamin AG, Mozhayeva GN (1999) Miniature Ca2+ channels in excised plasma-membrane patches: activation by IP3. Pflugers Arch 437(2):305–314
64. Trepakova ES, Gericke M, Hirakawa Y, Weisbrod RM, Cohen RA, Bolotina VM (2001) Properties of a native cation channel activated by Ca2+ store depletion in vascular smooth muscle cells. J Biol Chem 276(11):7782–7790
65. Golovina VA, Platoshyn O, Bailey CL, Wang J, Limsuwan A, Sweeney M, Rubin LJ, Yuan JX (2001) Upregulated TRP and enhanced capacitative Ca(2+) entry in human pulmonary artery myocytes during proliferation. Am J Physiol Heart Circ Physiol 280(2):H746–H755

66. Liu X, Groschner K, Ambudkar IS (2004) Distinct Ca(2+)-permeable cation currents are activated by internal Ca(2+)-store depletion in RBL-2H3 cells and human salivary gland cells, HSG and HSY. J Membr Biol 200(2):93–104
67. To MS, Aromataris EC, Castro J, Roberts ML, Barritt GJ, Rychkov GY (2010) Mitochondrial uncoupler FCCP activates proton conductance but does not block store-operated Ca(2+) current in liver cells. Arch Biochem Biophys 495(2):152–158
68. Hopf FW, Reddy P, Hong J, Steinhardt RA (1996) A capacitative calcium current in cultured skeletal muscle cells is mediated by the calcium-specific leak channel and inhibited by dihydropyridine compounds. J Biol Chem 271(37):22358–22367
69. Putney JW Jr (2003) Capacitative calcium entry in the nervous system. Cell Calcium 34(4–5):339–344
70. Fomina AF, Nowycky MC (1999) A current activated on depletion of intracellular Ca2+ stores can regulate exocytosis in adrenal chromaffin cells. J Neurosci 19(10):3711–3722
71. Bird GS, Putney JW Jr (2005) Capacitative calcium entry supports calcium oscillations in human embryonic kidney cells. J Physiol 562(Pt 3):697–706
72. Wedel B, Boyles RR, Putney JW Jr, Bird GS (2007) Role of the store-operated calcium entry proteins Stim1 and Orai1 in muscarinic cholinergic receptor-stimulated calcium oscillations in human embryonic kidney cells. J Physiol 579(Pt 3):679–689
73. Wu S, Cioffi EA, Alvarez D, Sayner SL, Chen H, Cioffi DL, King J, Creighton JR, Townsley M, Goodman SR, Stevens T (2005) Essential role of a Ca2+ -selective, store-operated current (ISOC) in endothelial cell permeability: determinants of the vascular leak site. Circ Res 96(8):856–863
74. Dominguez-Rodriguez A, Diaz I, Rodriguez-Moyano M, Calderon-Sanchez E, Rosado JA, Ordonez A, Smani T (2012) Urotensin-II signaling mechanism in rat coronary artery: role of STIM1 and Orai1-dependent store operated calcium influx in vasoconstriction. Arterioscler Thromb Vasc Biol 32(5):1325–1332
75. Braun A, Varga-Szabo D, Kleinschnitz C, Pleines I, Bender M, Austinat M, Bosl M, Stoll G, Nieswandt B (2009) Orai1 (CRACM1) is the platelet SOC channel and essential for pathological thrombus formation. Blood 113(9):2056–2063
76. Galan C, Zbidi H, Bartegi A, Salido GM, Rosado JA (2009) STIM1, Orai1 and hTRPC1 are important for thrombin- and ADP-induced aggregation in human platelets. Arch Biochem Biophys 490(2):137–144
77. Feske S, Gwack Y, Prakriya M, Srikanth S, Puppel SH, Tanasa B, Hogan PG, Lewis RS, Daly M, Rao A (2006) A mutation in Orai1 causes immune deficiency by abrogating CRAC channel function. Nature 441(7090):179–185
78. Liu X, Cheng KT, Bandyopadhyay BC, Pani B, Dietrich A, Paria BC, Swaim WD, Beech D, Yildrim E, Singh BB, Birnbaumer L, Ambudkar IS (2007) Attenuation of store-operated Ca2+ current impairs salivary gland fluid secretion in TRPC1(-/-) mice. Proc Natl Acad Sci U S A 104(44):17542–17547
79. Pani B, Liu X, Bollimuntha S, Cheng KT, Niesman IR, Zheng C, Achen VR, Patel HH, Ambudkar IS, Singh BB (2013) Impairment of TRPC1-STIM1 channel assembly and AQP5 translocation compromise agonist-stimulated fluid secretion in mice lacking caveolin1. J Cell Sci 126(Pt 2):667–675
80. Roos J, DiGregorio PJ, Yeromin AV, Ohlsen K, Lioudyno M, Zhang S, Safrina O, Kozak JA, Wagner SL, Cahalan MD, Velicelebi G, Stauderman KA (2005) STIM1, an essential and conserved component of store-operated Ca^{2+} channel function. J Cell Biol 169(3):435–445
81. Liou J, Kim ML, Heo WD, Jones JT, Myers JW, Ferrell JE Jr, Meyer T (2005) STIM is a Ca^{2+} sensor essential for Ca^{2+}-store-depletion-triggered Ca^{2+} influx. Curr Biol 15(13):1235–1241
82. Salido GM, Sage SO, Rosado JA (2009) Biochemical and functional properties of the store-operated Ca^{2+} channels. Cell Signal 21(4):457–461
83. Pandol SJ, Schoeffield-Payne MS (1990) Cyclic GMP mediates the agonist-stimulated increase in plasma membrane calcium entry in the pancreatic acinar cell. J Biol Chem 265(22):12846–12853

84. Rosado JA, Porras T, Conde M, Sage SO (2001) Cyclic nucleotides modulate store-mediated calcium entry through the activation of protein-tyrosine phosphatases and altered actin polymerization in human platelets. J Biol Chem 276(19):15666–15675
85. Alvarez J, Montero M, Garcia-Sancho J (1992) Cytochrome P450 may regulate plasma membrane Ca2+ permeability according to the filling state of the intracellular Ca2+ stores. FASEB J 6(2):786–792
86. Ben Amor N, Redondo PC, Bartegi A, Pariente JA, Salido GM, Rosado JA (2006) A role for 5,6-epoxyeicosatrienoic acid in calcium entry by de novo conformational coupling in human platelets. J Physiol 570(Pt 2):309–323
87. Sargeant P, Farndale RW, Sage SO (1993) The tyrosine kinase inhibitors methyl 2,5-dihydroxycinnamate and genistein reduce thrombin-evoked tyrosine phosphorylation and Ca2+ entry in human platelets. FEBS Lett 315(3):242–246
88. Sargeant P, Farndale RW, Sage SO (1993) ADP- and thapsigargin-evoked Ca^{2+} entry and protein-tyrosine phosphorylation are inhibited by the tyrosine kinase inhibitors genistein and methyl-2,5-dihydroxycinnamate in fura-2-loaded human platelets. J Biol Chem 268(24):18151–18156
89. Rosado JA, Graves D, Sage SO (2000) Tyrosine kinases activate store-mediated Ca2+ entry in human platelets through the reorganization of the actin cytoskeleton. Biochem J 351(Pt 2):429–437
90. Fasolato C, Hoth M, Penner R (1993) A GTP-dependent step in the activation mechanism of capacitative calcium influx. J Biol Chem 268(28):20737–20740
91. Bird GS, Putney JW Jr (1993) Inhibition of thapsigargin-induced calcium entry by microinjected guanine nucleotide analogues. Evidence for the involvement of a small G-protein in capacitative calcium entry. J Biol Chem 268(29):21486–21488
92. Rosado JA, Sage SO (2000) Farnesylcysteine analogues inhibit store-regulated Ca^{2+} entry in human platelets: evidence for involvement of small GTP-binding proteins and actin cytoskeleton. Biochem J 347(Pt 1):183–192
93. Vaca L (1996) Calmodulin inhibits calcium influx current in vascular endothelium. FEBS Lett 390(3):289–293
94. Randriamampita C, Tsien RY (1993) Emptying of intracellular Ca2+ stores releases a novel small messenger that stimulates Ca2+ influx. Nature 364(6440):809–814
95. Csutora P, Zarayskiy V, Peter K, Monje F, Smani T, Zakharov SI, Litvinov D, Bolotina VM (2006) Activation mechanism for CRAC current and store-operated Ca2+ entry: calcium influx factor and Ca2+ -independent phospholipase A2beta-mediated pathway. J Biol Chem 281(46):34926–34935
96. Irvine RF, Moor RM (1987) Inositol(1,3,4,5)tetrakisphosphate-induced activation of sea urchin eggs requires the presence of inositol trisphosphate. Biochem Biophys Res Commun 146(1):284–290
97. Irvine RF (1990) 'Quantal' Ca2+ release and the control of Ca2+ entry by inositol phosphates – a possible mechanism. FEBS Lett 263(1):5–9
98. Boulay G, Brown DM, Qin N, Jiang M, Dietrich A, Zhu MX, Chen Z, Birnbaumer M, Mikoshiba K, Birnbaumer L (1999) Modulation of Ca(2+) entry by polypeptides of the inositol 1,4, 5-trisphosphate receptor (IP3R) that bind transient receptor potential (TRP): evidence for roles of TRP and IP3R in store depletion-activated Ca(2+) entry. Proc Natl Acad Sci U S A 96(26):14955–14960
99. Treves S, Franzini-Armstrong C, Moccagatta L, Arnoult C, Grasso C, Schrum A, Ducreux S, Zhu MX, Mikoshiba K, Girard T, Smida-Rezgui S, Ronjat M, Zorzato F (2004) Junctate is a key element in calcium entry induced by activation of InsP3 receptors and/or calcium store depletion. J Cell Biol 166(4):537–548
100. Kiselyov K, Xu X, Mozhayeva G, Kuo T, Pessah I, Mignery G, Zhu X, Birnbaumer L, Muallem S (1998) Functional interaction between InsP3 receptors and store-operated Htrp3 channels. Nature 396(6710):478–482
101. Kiselyov K, Mignery GA, Zhu MX, Muallem S (1999) The N-terminal domain of the IP3 receptor gates store-operated hTrp3 channels. Mol Cell 4(3):423–429

102. Yao Y, Ferrer-Montiel AV, Montal M, Tsien RY (1999) Activation of store-operated Ca2+ current in Xenopus oocytes requires SNAP-25 but not a diffusible messenger. Cell 98(4):475–485
103. Bezzerides VJ, Ramsey IS, Kotecha S, Greka A, Clapham DE (2004) Rapid vesicular translocation and insertion of TRP channels. Nat Cell Biol 6(8):709–720
104. Cayouette S, Lussier MP, Mathieu EL, Bousquet SM, Boulay G (2004) Exocytotic insertion of TRPC6 channel into the plasma membrane upon Gq protein-coupled receptor activation. J Biol Chem 279(8):7241–7246
105. Singh BB, Lockwich TP, Bandyopadhyay BC, Liu X, Bollimuntha S, Brazer SC, Combs C, Das S, Leenders AG, Sheng ZH, Knepper MA, Ambudkar SV, Ambudkar IS (2004) VAMP2-dependent exocytosis regulates plasma membrane insertion of TRPC3 channels and contributes to agonist-stimulated Ca2+ influx. Mol Cell 15(4):635–646
106. Patterson RL, van Rossum DB, Gill DL (1999) Store-operated Ca2+ entry: evidence for a secretion-like coupling model. Cell 98(4):487–499
107. Rosado JA, Jenner S, Sage SO (2000) A role for the actin cytoskeleton in the initiation and maintenance of store-mediated calcium entry in human platelets. Evidence for conformational coupling. J Biol Chem 275(11):7527–7533
108. Rosado JA, Rosenzweig I, Harding S, Sage SO (2001) Tumor necrosis factor-alpha inhibits store-mediated Ca2+ entry in the human hepatocellular carcinoma cell line HepG2. Am J Physiol Cell Physiol 280(6):C1636–C1644
109. Redondo PC, Lajas AI, Salido GM, Gonzalez A, Rosado JA, Pariente JA (2003) Evidence for secretion-like coupling involving pp60src in the activation and maintenance of store-mediated Ca2+ entry in mouse pancreatic acinar cells. Biochem J 370(Pt 1):255–263
110. Rosado JA, Sage SO (2000) The actin cytoskeleton in store-mediated calcium entry. J Physiol 526(Pt 2):221–229
111. Xie Q, Zhang Y, Zhai C, Bonanno JA (2002) Calcium influx factor from cytochrome P-450 metabolism and secretion-like coupling mechanisms for capacitative calcium entry in corneal endothelial cells. J Biol Chem 277(19):16559–16566
112. Wang JP, Hsu MF, Ko HH, Lin CN (2004) Stimulation of cellular free Ca2+ elevation and inhibition of store-operated Ca2+ entry by kazinol B in neutrophils. Naunyn Schmiedebergs Arch Pharmacol 370(6):500–509
113. Gratschev D, Blom T, Bjorklund S, Tornquist K (2004) Phosphatase inhibition reveals a calcium entry pathway dependent on protein kinase A in thyroid FRTL-5 cells: comparison with store-operated calcium entry. J Biol Chem 279(48):49816–49824
114. Kachoei BA, Knox RJ, Uthuza D, Levy S, Kaczmarek LK, Magoski NS (2006) A store-operated Ca(2+) influx pathway in the bag cell neurons of Aplysia. J Neurophysiol 96(5):2688–2698
115. Henquin JC, Mourad NI, Nenquin M (2012) Disruption and stabilization of beta-cell actin microfilaments differently influence insulin secretion triggered by intracellular Ca2+ mobilization or store-operated Ca2+ entry. FEBS Lett 586(1):89–95
116. Bakowski D, Glitsch MD, Parekh AB (2001) An examination of the secretion-like coupling model for the activation of the Ca2+ release-activated Ca2+ current I(CRAC) in RBL-1 cells. J Physiol 532(Pt 1):55–71
117. Redondo PC, Harper MT, Rosado JA, Sage SO (2006) A role for cofilin in the activation of store-operated calcium entry by de novo conformational coupling in human platelets. Blood 107(3):973–979
118. Zhang SL, Yu Y, Roos J, Kozak JA, Deerinck TJ, Ellisman MH, Stauderman KA, Cahalan MD (2005) STIM1 is a Ca2+ sensor that activates CRAC channels and migrates from the Ca2+ store to the plasma membrane. Nature 437(7060):902–905
119. Mercer JC, Dehaven WI, Smyth JT, Wedel B, Boyles RR, Bird GS, Putney JW Jr (2006) Large store-operated calcium selective currents due to co-expression of Orai1 or Orai2 with the intracellular calcium sensor, Stim1. J Biol Chem 281(34):24979–24990

120. Peinelt C, Vig M, Koomoa DL, Beck A, Nadler MJ, Koblan-Huberson M, Lis A, Fleig A, Penner R, Kinet JP (2006) Amplification of CRAC current by STIM1 and CRACM1 (Orai1). Nat Cell Biol 8(7):771–773
121. Prakriya M, Feske S, Gwack Y, Srikanth S, Rao A, Hogan PG (2006) Orai1 is an essential pore subunit of the CRAC channel. Nature 443(7108):230–233
122. Soboloff J, Spassova MA, Tang XD, Hewavitharana T, Xu W, Gill DL (2006) Orai1 and STIM reconstitute store-operated calcium channel function. J Biol Chem 281(30):20661–20665
123. Jardin I, Lopez JJ, Salido GM, Rosado JA (2008) Orai1 mediates the interaction between STIM1 and hTRPC1 and regulates the mode of activation of hTRPC1-forming Ca^{2+} channels. J Biol Chem 283(37):25296–25304
124. Galan C, Dionisio N, Smani T, Salido GM, Rosado JA (2011) The cytoskeleton plays a modulatory role in the association between STIM1 and the Ca2+ channel subunits Orai1 and TRPC1. Biochem Pharmacol 82(4):400–410
125. Jardin I, Lopez JJ, Redondo PC, Salido GM, Rosado JA (2009) Store-operated Ca^{2+} entry is sensitive to the extracellular Ca^{2+} concentration through plasma membrane STIM1. Biochim Biophys Acta 1793(10):1614–1622
126. Harper AG, Sage SO (2007) A key role for reverse Na^+/Ca^{2+} exchange influenced by the actin cytoskeleton in store-operated Ca^{2+} entry in human platelets: evidence against the de novo conformational coupling hypothesis. Cell Calcium 42(6):606–617
127. Giurisato E, Gamberucci A, Ulivieri C, Marruganti S, Rossi E, Giacomello E, Randazzo D, Sorrentino V (2014) The KSR2-calcineurin complex regulates STIM1-ORAI1 dynamics and store-operated calcium entry (SOCE). Mol Biol Cell 25(11):1769–1781
128. Nolz JC, Gomez TS, Zhu P, Li S, Medeiros RB, Shimizu Y, Burkhardt JK, Freedman BD, Billadeau DD (2006) The WAVE2 complex regulates actin cytoskeletal reorganization and CRAC-mediated calcium entry during T cell activation. Curr Biol 16(1):24–34
129. Oritani K, Kincade PW (1996) Identification of stromal cell products that interact with pre-B cells. J Cell Biol 134(3):771–782
130. Parker NJ, Begley CG, Smith PJ, Fox RM (1996) Molecular cloning of a novel human gene (D11S4896E) at chromosomal region 11p15.5. Genomics 37(2):253–256
131. Sabbioni S, Barbanti-Brodano G, Croce CM, Negrini M (1997) GOK: a gene at 11p15 involved in rhabdomyosarcoma and rhabdoid tumor development. Cancer Res 57(20):4493–4497
132. Manji SS, Parker NJ, Williams RT, van Stekelenburg L, Pearson RB, Dziadek M, Smith PJ (2000) STIM1: a novel phosphoprotein located at the cell surface. Biochim Biophys Acta 1481(1):147–155
133. Williams RT, Manji SS, Parker NJ, Hancock MS, Van Stekelenburg L, Eid JP, Senior PV, Kazenwadel JS, Shandala T, Saint R, Smith PJ, Dziadek MA (2001) Identification and characterization of the STIM (stromal interaction molecule) gene family: coding for a novel class of transmembrane proteins. Biochem J 357(Pt 3):673–685
134. Williams RT, Senior PV, Van Stekelenburg L, Layton JE, Smith PJ, Dziadek MA (2002) Stromal interaction molecule 1 (STIM1), a transmembrane protein with growth suppressor activity, contains an extracellular SAM domain modified by N-linked glycosylation. Biochim Biophys Acta 1596(1):131–137
135. Baba Y, Hayashi K, Fujii Y, Mizushima A, Watarai H, Wakamori M, Numaga T, Mori Y, Iino M, Hikida M, Kurosaki T (2006) Coupling of STIM1 to store-operated Ca2+ entry through its constitutive and inducible movement in the endoplasmic reticulum. Proc Natl Acad Sci U S A 103(45):16704–16709
136. Vig M, Peinelt C, Beck A, Koomoa DL, Rabah D, Koblan-Huberson M, Kraft S, Turner H, Fleig A, Penner R, Kinet JP (2006) CRACM1 is a plasma membrane protein essential for store-operated Ca2+ entry. Science 312(5777):1220–1223

137. Yeromin AV, Zhang SL, Jiang W, Yu Y, Safrina O, Cahalan MD (2006) Molecular identification of the CRAC channel by altered ion selectivity in a mutant of Orai. Nature 443(7108):226–229
138. Vig M, Beck A, Billingsley JM, Lis A, Parvez S, Peinelt C, Koomoa DL, Soboloff J, Gill DL, Fleig A, Kinet JP, Penner R (2006) CRACM1 multimers form the ion-selective pore of the CRAC channel. Curr Biol 16(20):2073–2079
139. Zhang SL, Yeromin AV, Zhang XH, Yu Y, Safrina O, Penna A, Roos J, Stauderman KA, Cahalan MD (2006) Genome-wide RNAi screen of Ca(2+) influx identifies genes that regulate Ca(2+) release-activated Ca(2+) channel activity. Proc Natl Acad Sci U S A 103(24):9357–9362
140. Luik RM, Wu MM, Buchanan J, Lewis RS (2006) The elementary unit of store-operated Ca2+ entry: local activation of CRAC channels by STIM1 at ER-plasma membrane junctions. J Cell Biol 174(6):815–825
141. Muik M, Frischauf I, Derler I, Fahrner M, Bergsmann J, Eder P, Schindl R, Hesch C, Polzinger B, Fritsch R, Kahr H, Madl J, Gruber H, Groschner K, Romanin C (2008) Dynamic coupling of the putative coiled-coil domain of ORAI1 with STIM1 mediates ORAI1 channel activation. J Biol Chem 283(12):8014–8022
142. Penna A, Demuro A, Yeromin AV, Zhang SL, Safrina O, Parker I, Cahalan MD (2008) The CRAC channel consists of a tetramer formed by Stim-induced dimerization of Orai dimers. Nature 456:116–120
143. Ji W, Xu P, Li Z, Lu J, Liu L, Zhan Y, Chen Y, Hille B, Xu T, Chen L (2008) Functional stoichiometry of the unitary calcium-release-activated calcium channel. Proc Natl Acad Sci U S A 105(36):13668–13673
144. Mignen O, Thompson JL, Shuttleworth TJ (2008) Orai1 subunit stoichiometry of the mammalian CRAC channel pore. J Physiol 586(2):419–425
145. Hou X, Pedi L, Diver MM, Long SB (2012) Crystal structure of the calcium release-activated calcium channel Orai. Science 338(6112):1308–1313
146. Montell C, Birnbaumer L, Flockerzi V, Bindels RJ, Bruford EA, Caterina MJ, Clapham DE, Harteneck C, Heller S, Julius D, Kojima I, Mori Y, Penner R, Prawitt D, Scharenberg AM, Schultz G, Shimizu N, Zhu MX (2002) A unified nomenclature for the superfamily of TRP cation channels. Mol Cell 9(2):229–231
147. Cosens DJ, Manning A (1969) Abnormal electroretinogram from a Drosophila mutant. Nature 224(5216):285–287
148. Minke B, Wu C, Pak WL (1975) Induction of photoreceptor voltage noise in the dark in Drosophila mutant. Nature 258(5530):84–87
149. Montell C, Rubin GM (1989) Molecular characterization of the Drosophila trp locus: a putative integral membrane protein required for phototransduction. Neuron 2(4):1313–1323
150. Hardie RC, Minke B (1993) Novel Ca2+ channels underlying transduction in Drosophila photoreceptors: implications for phosphoinositide-mediated Ca2+ mobilization. Trends Neurosci 16(9):371–376
151. Wes PD, Chevesich J, Jeromin A, Rosenberg C, Stetten G, Montell C (1995) TRPC1, a human homolog of a Drosophila store-operated channel. Proc Natl Acad Sci U S A 92(21):9652–9656
152. Zhu X, Chu PB, Peyton M, Birnbaumer L (1995) Molecular cloning of a widely expressed human homologue for the Drosophila trp gene. FEBS Lett 373(3):193–198
153. Montell C (2001) Physiology, phylogeny, and functions of the TRP superfamily of cation channels. Sci STKE 2001 (90):re1
154. Vannier B, Zhu X, Brown D, Birnbaumer L (1998) The membrane topology of human transient receptor potential 3 as inferred from glycosylation-scanning mutagenesis and epitope immunocytochemistry. J Biol Chem 273(15):8675–8679
155. Bennett V (1992) Ankyrins. Adaptors between diverse plasma membrane proteins and the cytoplasm. J Biol Chem 267(13):8703–8706

156. Montell C, Birnbaumer L, Flockerzi V (2002) The TRP channels, a remarkably functional family. Cell 108(5):595–598
157. Burkhard P, Stetefeld J, Strelkov SV (2001) Coiled coils: a highly versatile protein folding motif. Trends Cell Biol 11(2):82–88
158. Engelke M, Friedrich O, Budde P, Schafer C, Niemann U, Zitt C, Jungling E, Rocks O, Luckhoff A, Frey J (2002) Structural domains required for channel function of the mouse transient receptor potential protein homologue TRP1beta. FEBS Lett 523(1–3):193–199
159. Zitt C, Zobel A, Obukhov AG, Harteneck C, Kalkbrenner F, Luckhoff A, Schultz G (1996) Cloning and functional expression of a human Ca2+ -permeable cation channel activated by calcium store depletion. Neuron 16(6):1189–1196
160. Zhu X, Jiang M, Peyton M, Boulay G, Hurst R, Stefani E, Birnbaumer L (1996) trp, a novel mammalian gene family essential for agonist-activated capacitative Ca2+ entry. Cell 85(5):661–671
161. Liu X, Wang W, Singh BB, Lockwich T, Jadlowiec J, O'Connell B, Wellner R, Zhu MX, Ambudkar IS (2000) Trp1, a candidate protein for the store-operated Ca(2+) influx mechanism in salivary gland cells. J Biol Chem 275(5):3403–3411
162. Brough GH, Wu S, Cioffi D, Moore TM, Li M, Dean N, Stevens T (2001) Contribution of endogenously expressed Trp1 to a Ca2+ -selective, store-operated Ca2+ entry pathway. FASEB J 15(10):1727–1738
163. Mori Y, Wakamori M, Miyakawa T, Hermosura M, Hara Y, Nishida M, Hirose K, Mizushima A, Kurosaki M, Mori E, Gotoh K, Okada T, Fleig A, Penner R, Iino M, Kurosaki T (2002) Transient receptor potential 1 regulates capacitative Ca(2+) entry and Ca(2+) release from endoplasmic reticulum in B lymphocytes. J Exp Med 195(6):673–681
164. Almirza WH, Peters PH, van Zoelen EJ, Theuvenet AP (2012) Role of Trpc channels, Stim1 and Orai1 in PGF(2alpha)-induced calcium signaling in NRK fibroblasts. Cell Calcium 51(1):12–21
165. Lopez E, Berna-Erro A, Salido GM, Rosado JA, Redondo PC (2013) FKBP52 is involved in the regulation of SOCE channels in the human platelets and MEG 01 cells. Biochim Biophys Acta 1833(3):652–662
166. Rosado JA, Brownlow SL, Sage SO (2002) Endogenously expressed Trp1 is involved in store-mediated Ca2+ entry by conformational coupling in human platelets. J Biol Chem 277(44):42157–42163
167. Vannier B, Peyton M, Boulay G, Brown D, Qin N, Jiang M, Zhu X, Birnbaumer L (1999) Mouse trp2, the homologue of the human trpc2 pseudogene, encodes mTrp2, a store depletion-activated capacitative Ca2+ entry channel. Proc Natl Acad Sci U S A 96(5):2060–2064
168. Boulay G, Zhu X, Peyton M, Jiang M, Hurst R, Stefani E, Birnbaumer L (1997) Cloning and expression of a novel mammalian homolog of Drosophila transient receptor potential (Trp) involved in calcium entry secondary to activation of receptors coupled by the Gq class of G protein. J Biol Chem 272(47):29672–29680
169. Brechard S, Melchior C, Plancon S, Schenten V, Tschirhart EJ (2008) Store-operated Ca2+ channels formed by TRPC1, TRPC6 and Orai1 and non-store-operated channels formed by TRPC3 are involved in the regulation of NADPH oxidase in HL-60 granulocytes. Cell Calcium 44(5):492–506
170. Jardin I, Redondo PC, Salido GM, Rosado JA (2008) Phosphatidylinositol 4,5-bisphosphate enhances store-operated calcium entry through hTRPC6 channel in human platelets. Biochim Biophys Acta 1783(1):84–97
171. DeHaven WI, Jones BF, Petranka JG, Smyth JT, Tomita T, Bird GS, Putney JW Jr (2009) TRPC channels function independently of STIM1 and Orai1. J Physiol 587(Pt 10):2275–2298
172. Lievremont JP, Bird GS, Putney JW Jr (2004) Canonical transient receptor potential TRPC7 can function as both a receptor- and store-operated channel in HEK-293 cells. Am J Physiol Cell Physiol 287(6):C1709–C1716

173. Spassova MA, Soboloff J, He LP, Xu W, Dziadek MA, Gill DL (2006) STIM1 has a plasma membrane role in the activation of store-operated Ca(2+) channels. Proc Natl Acad Sci U S A 103(11):4040–4045
174. Huang GN, Zeng W, Kim JY, Yuan JP, Han L, Muallem S, Worley PF (2006) STIM1 carboxyl-terminus activates native SOC, I(crac) and TRPC1 channels. Nat Cell Biol 8(9):1003–1010
175. Lopez JJ, Salido GM, Pariente JA, Rosado JA (2006) Interaction of STIM1 with endogenously expressed human canonical TRP1 upon depletion of intracellular Ca2+ stores. J Biol Chem 281(38):28254–28264
176. Rao JN, Rathor N, Zou T, Liu L, Xiao L, Yu TX, Cui YH, Wang JY (2010) STIM1 translocation to the plasma membrane enhances intestinal epithelial restitution by inducing TRPC1-mediated Ca2+ signaling after wounding. Am J Physiol Cell Physiol 299(3):C579–C588
177. Hong JH, Li Q, Kim MS, Shin DM, Feske S, Birnbaumer L, Cheng KT, Ambudkar IS, Muallem S (2011) Polarized but differential localization and recruitment of STIM1, Orai1 and TRPC channels in secretory cells. Traffic 12(2):232–245
178. Ng LC, McCormack MD, Airey JA, Singer CA, Keller PS, Shen XM, Hume JR (2009) TRPC1 and STIM1 mediate capacitative Ca2+ entry in mouse pulmonary arterial smooth muscle cells. J Physiol 587(Pt 11):2429–2442
179. Sours-Brothers S, Ding M, Graham S, Ma R (2009) Interaction between TRPC1/TRPC4 assembly and STIM1 contributes to store-operated Ca2+ entry in mesangial cells. Exp Biol Med (Maywood) 234(6):673–682
180. Yuan JP, Zeng W, Huang GN, Worley PF, Muallem S (2007) STIM1 heteromultimerizes TRPC channels to determine their function as store-operated channels. Nat Cell Biol 9(6):636–645
181. Zeng W, Yuan JP, Kim MS, Choi YJ, Huang GN, Worley PF, Muallem S (2008) STIM1 gates TRPC channels, but not Orai1, by electrostatic interaction. Mol Cell 32(3):439–448
182. Lee KP, Choi S, Hong JH, Ahuja M, Graham S, Ma R, So I, Shin DM, Muallem S, Yuan JP (2014) Molecular determinants mediating gating of Transient Receptor Potential Canonical (TRPC) channels by stromal interaction molecule 1 (STIM1). J Biol Chem 289(10):6372–6382
183. Pani B, Ong HL, Liu X, Rauser K, Ambudkar IS, Singh BB (2008) Lipid rafts determine clustering of STIM1 in endoplasmic reticulum-plasma membrane junctions and regulation of store-operated Ca2+ entry (SOCE). J Biol Chem 283(25):17333–17340
184. Jardin I, Salido GM, Rosado JA (2008) Role of lipid rafts in the interaction between hTRPC1, Orai1 and STIM1. Channels (Austin) 2(6):401–403
185. Sampieri A, Zepeda A, Saldaña C, Salgado A, Vaca L (2008) STIM1 converts TRPC1 from a receptor-operated to a store-operated channel: moving TRPC1 in and out of lipid rafts. Cell Calcium 44(5):479–491
186. Liao Y, Erxleben C, Abramowitz J, Flockerzi V, Zhu MX, Armstrong DL, Birnbaumer L (2008) Functional interactions among Orai1, TRPCs, and STIM1 suggest a STIM-regulated heteromeric Orai/TRPC model for SOCE/Icrac channels. Proc Natl Acad Sci U S A 105(8):2895–2900
187. Ong HL, Cheng KT, Liu X, Bandyopadhyay BC, Paria BC, Soboloff J, Pani B, Gwack Y, Srikanth S, Singh BB, Gill DL, Ambudkar IS (2007) Dynamic assembly of TRPC1-STIM1-Orai1 ternary complex is involved in store-operated calcium influx. Evidence for similarities in store-operated and calcium release-activated calcium channel components. J Biol Chem 282(12):9105–9116

Chapter 2
The STIM1: Orai Interaction

Irene Frischauf, Marc Fahrner, Isaac Jardín, and Christoph Romanin

Abstract Ca^{2+} influx via store-operated Ca^{2+} release activated Ca^{2+} (CRAC) channels represents a main signalling pathway for a variety of cell functions, including T-cell activation as well as mast-cell degranulation. Depletion of $[Ca^{2+}]_{ER}$ results in activation of Ca^{2+} channels within the plasmamembrane that mediate sustained Ca^{2+} influx which is required for refilling Ca^{2+} stores and down-stream Ca^{2+} signalling. The CRAC channel is the best characterized store-operated channel (SOC) with well-defined electrophysiological properties. In recent years, the molecular components of the CRAC channel have been defined. The ER – located Ca^{2+}-sensor, STIM1 and the Ca^{2+}-selective ion pore, Orai1 in the membrane are sufficient to fully reconstitute CRAC currents. Stromal interaction molecule (STIM) 1 is localized in the ER, senses $[Ca^{2+}]_{ER}$ and activates the CRAC channel upon store depletion by direct binding to Orai1 in the plasmamembrane. The identification of STIM1 and Orai1 and recently the structural resolution of both proteins by X-ray crystallography and nuclear magnetic resonance substantiated many findings from structure-function studies which has substantially improved the understanding of CRAC channel activation. Within this review, we summarize the functional and structural mechanisms of CRAC channel regulation, present a detailed overview of the STIM1/Orai1 signalling pathway where we focus on the critical domains mediating interactions and on the ion permeation pathway. We portray a mechanistic view of the steps in the dynamics of CRAC channel signalling ranging from STIM1 oligomerization over STIM1-Orai1 coupling to CRAC channel activation and permeation.

Keywords SOCE • STIM1 • Orai1 • CRAC channels

Author contributed equally with all other contributors.

I. Frischauf (✉) • M. Fahrner • C. Romanin (✉)
Institute of Biophysics, University of Linz, Linz, Austria
e-mail: irene.frischauf@jku.at; christoph.romanin@jku.at

I. Jardín
Department of Physiology, University of Extremadura, Cáceres, Spain

J.A. Rosado (ed.), *Calcium Entry Pathways in Non-excitable Cells*, Advances in Experimental Medicine and Biology 898, DOI 10.1007/978-3-319-26974-0_2

Abbreviations

2-APB	2-Aminoethoxydiphenyl borate
Å	Angstrom
Ca^{2+}	Calcium
CAD	CRAC activating domain
CC	Coiled coil
CCb9	Coiled coil containing region b9
CCE	Capacitative calcium entry
CDI	Calcium dependent inactivation
CFP	Cyan fluorescent protein
CMD	CRAC modulatory domain
CRAC	Calcium release activated calcium
Cs^{+}	Caesium
dOrai	Drosophila melanogaster Orai
ER	Endoplasmic reticulum
Et al	Et alii
ETON	Extended transmembrane Orai1 N-terminal region
FRET	Förster resonance energy transfer
IH	Inhibitory helix
IP_3	Inositol(1,4,5)triphosphate
K-rich	Lysine rich
Na^{+}	Sodium
NMR	Nuclear magnetic resonance
OASF	Orai activating small fragment
PLC	Phospholipase C
PM	Plasmamembrane
RNAi	Interference ribonucleic acid
SAM	Sterile alpha motif
SCID	Severe combined immunodeficiency
SERCA	Sarcoplasmic/endoplasmic reticulum calcium ATPase
SHD	STIM1 homomerization domain
SOAP	STIM/Orai association pocket
SOAR	STIM/Orai activating region
SOCE	Store operated calcium entry
STIM	Stromal interaction molecule
TG	Thapsigargin
TM	Transmembrane
TRP	Transient receptor potential
YFP	Yellow fluorescent protein

2.1 Introduction

The discovery of store-operated Ca^{2+} entry has a relatively long history. The concept of a calcium entry pathway that is activated upon store depletion of internal Ca^{2+} reservoirs was first proposed by Putney nearly 30 years ago and initially termed capacitative Ca^{2+} entry (CCE) [1]. This Ca^{2+} influx pathway is now referred to as store-operated Ca^{2+} entry (SOCE). SOCE has been observed in a variety of non-excitable and excitable cells including lymphocytes, pancreatic acinar, vascular endothelial and smooth and skeletal muscle cells [2–7]. A few years later, the biophysical properties of the underlying calcium-release-activated-Ca^{2+} (CRAC) – channel conductance were characterized [5, 8–10] and at the molecular level, store-operated Ca^{2+} channels have only been investigated within the past decade.

Physiologically, SOCE is initiated by Ca^{2+} release from the ER lumen into the cytoplasm upon receptor activation. Once activated, the production of the second messenger inositol-1,4,5-triphosphate (IP_3) is initiated which subsequently binds to its receptors in the ER membrane and activates Ca^{2+} release [11, 12]. In vitro, passive depletion of intracellular Ca^{2+} stores is often used to initiate SOCE by preventing Ca^{2+} re-uptake into the ER using sarcoplasmic/endoplasmic reticulum ATPase (SERCA) pump inhibitors such as thapsigargin (TG). The CRAC channel represents one of the most Ca^{2+} selective channels and currents are defined by a set of characteristic features. It is over 1000 times more permeant for Ca^{2+} than Na^+ at physiological conditions [5] and has a conductance that is among the smallest of any ion channel [10]. CRAC currents exhibit strong inward rectification and essentially positive reversal potentials >60 mV [5]. This unique fingerprint has given rise to years of search for candidate genes involved in CRAC channel function. Most prominently, members of the large family of transient receptor potential (TRP) channels have been proposed [13]. Now, two decades after Putney's concept of SOCE, STIM1 and Orai1, the two key-players of the CRAC channel complex, have finally been identified in parallel by RNAi screens in 2005 and 2006, respectively [14–17]. STIM (stromal interaction molecule) proteins have first been described as regulators of SOCE. In this study, Zhang et al. screened 170 candidates genes in *Drosophila* S2 cells and proved the role of STIM1 in SOC and CRAC channel activity whereby STIM1 controls SOCE by sensing $[Ca^{2+}]_{ER}$ and by translocating to the plasmamembrane (PM) where it activates Orai1 [15]. Orai1 was identified in 2006 by Feske et al. [18] through RNAi screens of patients which exhibited a form of hereditary severe combined immunodeficiency (SCID). They have shown that a single missense mutation in Orai1 (R91W) abrogates CRAC channel function and impacts T-cell effector function leading to a severe immunodeficient phenotype. Altogether, RNAi knockdown of either STIM1 or Orai1 completely suppressed a functional CRAC current, pointing to the indispensable requirement of both proteins in store-operated Ca^{2+} entry. Nowadays, the CRAC channel is the best characterized store-operated channel and is endogenously expressed in T-lymphocytes and other cells of the immune system. CRAC channels in the PM are formed by members of the Orai protein family that are activated upon physical interaction with the

ER-resident STIM1 protein. Subsequently, extracellular Ca^{2+} ions can enter the cell, where they act as second messenger to sustain Ca^{2+} oscillations, to maintain Ca^{2+} homeostasis and to provide long-term (in the range of hours) Ca^{2+} signals. These long-term Ca^{2+} signals regulate a plethora of signalling cascades including mast cell degranulation, cytokine secretion, T-cell differentiation into effector subsets, B-cell differentiation into plasma cells, lysis of infected or cancerous target cells by cytolytic T-cells, gene transcription or cell proliferation [16–28].

2.2 CRAC Activation and Key Players

When ER stores are full, the EF hand of STIM1 is bound with Ca^{2+}, which leads to a uniform distribution of STIM1 in the ER. Store-depletion triggers STIM1 oligomerization and translocation to ER-PM junctions, leading to STIM1 puncta [20, 29–31]. Orai1 accumulates at sites opposite to STIM1. This high proximity of the two proteins allows for direct interaction and leads to channel activation [20, 31, 32]. Details of STIM1-Orai1 interaction will be discussed further below.

There are two mammalian homologs of STIM, i.e. STIM1 and STIM2; and three mammalian homologs of Orai, i.e. Orai1-3. Since their discovery, these molecules have been investigated in great detail on the molecular level for their role in CRAC channel function and in vivo as essential components of SOCE in a variety of tissues. They are distributed widely throughout many tissues in various species, consistent with the widespread prevalence of SOCE and CRAC channel currents in many cell types [21, 33].

2.2.1 STIM Proteins

STIM proteins are type I, single-pass ER transmembrane proteins, primarily localized in the ER membrane, which link Ca^{2+} store-depletion and Ca^{2+} influx through the plasmamembrane. They sense $[Ca^{2+}]_{ER}$ via a tandem EF hand like sequence on their luminal side and physically relay this message to Orai channels in the plasmamembrane by translocating into defined ER-PM junctions [14, 15, 29, 34, 35]. By neutralization of acidic residues within the EF hand structure, STIM1 was no longer able to bind Ca^{2+} and yielded constitutive Ca^{2+} entry, mimicking the store-depleted state [14, 29].

2.2.1.1 Ca^{2+} Sensing, STIM1 Activation and Translocation

In case of full ER Ca^{2+} stores (resting cell conditions), STIM1 is homogenously distributed in the cell interior and rapidly moves as an ER membrane anchored protein along microtubules [36]. Decrease of ER luminal Ca^{2+} concentration (termed store

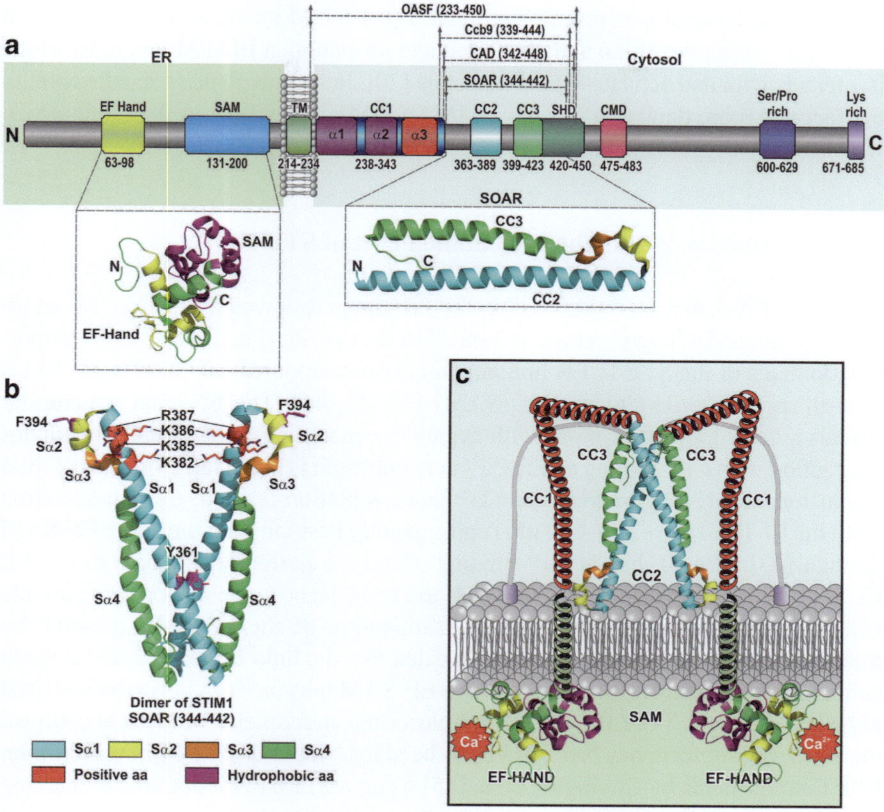

Fig. 2.1 (**a**) Schematic representation of human, full-length STIM1 depicting essential regions for the STIM1/Orai1 signaling cascade. The structure of the EF-SAM domain as well as the STIM1 SOAR (344–442) fragment is shown in detail. (**b**) Cartoon representation of a STIM1 SOAR (344–442) – dimer including coiled-coil domain 2 (CC2) and 3 (CC3) exhibiting a V-shaped structure. Residues mediating dimer interaction and those involved in coupling to Orai1 (positively charged residues) are highlighted. (**c**) Hypothetical model of a STIM1 dimer in the resting state

depletion) leads to STIM1 oligomerization which slows down the movement along microtubules [37]. A former study by Baba et al. has revealed constitutive, dynamic STIM1 movement in full store conditions, while store depletion consequently induced a dramatic redistribution of oligomerized STIM1 forming so called puncta, located at ER – PM junctions [30, 34, 38]. Investigation of STIM1 C-terminal truncation mutants (deletions within the cytosolic strand of STIM1) has shown that STIM1 coiled-coil (CC) domains and the serine-proline rich region (Fig. 2.1a) play a major role for the constitutive movement of STIM1 within the ER membrane under resting cell conditions, whereas STIM1 activation leading to puncta formation involves luminal as well as cytoplasmic STIM1 regions [30]. In line, independent FRET studies revealed that store depletion consequently leads to STIM1 oligomerization and cluster formation [39–41]. Additional experiments with artificially crosslinked

STIM1 at its luminal side revealed luminal STIM-STIM interaction as major signal for STIM1 oligomerization and cluster/puncta formation at ER-PM junctions where it interacts with and activates Orai1 channels [20]. In summary, these results point to the fact that store depletion induces STIM1-STIM1 interaction on the luminal side which finally triggers the STIM1 activation process [14, 15, 34, 38, 42].

2.2.1.2 Domains Within the ER Luminal Part of STIM1

The STIM1 N-terminus resides within the ER lumen followed by a single TM spanning domain and a long C-terminus located in the cytosol (Fig. 2.1a). Highly important domains of the STIM1 ER luminal part are the canonical and hidden EF-hands as well as the sterile-alpha motif (SAM) [35, 43, 44]. The EF-hand structurally reveals a helix-loop-helix motif with negatively charged aspartates and glutamates interacting with Ca^{2+} as long as ER Ca^{2+} concentration is high (full stores). As initial signal for STIM1 oligomerization, Ca^{2+} store depletion leads to Ca^{2+} dissociation from the EF hand (K_d ~200–600 μM) consequently destabilizing the entire EF-SAM domain [43]. The EF-hands Ca^{2+} binding affinity is perfectly adequate to the ER Ca^{2+} concentration range (~400–800 μM) to ensure an accurate response to variable ER Ca^{2+} concentrations. The detailed examination of the EF-SAM complex by Stathopulos et al. [35] (Fig. 2.1a) has revealed that the holo EF-Ca^{2+}- SAM domain contains high α-helicity in contrast to apo EF-SAM (absence of Ca^{2+}) which proved to be less compact. The EF-Ca^{2+}- SAM holoform is monomeric whereas in contrast, the EF-SAM apoform has been proven to be at least a dimer [35, 43]. Furthermore, FRET experiments by Covington et al. [45] using a STIM1 deletion mutant lacking the whole, cytosolic C-terminus result in store depletion mediated di-/oligomerization of the STIM1 luminal parts. The STIM1 homolog STIM2 also contains an ER luminal EF-SAM domain, however, it is shown to have different cellular functions [46]. CRAC currents activated by STIM2 appear upon smaller changes in ER – Ca^{2+} concentrations, suggesting that STIM2 is part of a feedback system keeping ER – Ca^{2+} concentrations within tight limits [46]. The different behaviour of EF-SAM of STIM1 and STIM2 is due to a difference in the N-terminal structural stability of both proteins [47, 48]. Adjacent to the SAM domain, STIM proteins contain a single transmembrane (TM) segment spanning the ER membrane (Fig. 2.1a). Whether the TM domain acts passively or plays an active role in the STIM activation process remains to be resolved.

2.2.1.3 STIM1 C-Terminal, Cytosolic Domains

STIM1 contains a long, cytosolic strand including three coiled-coil (CC1, CC2, CC3) regions, the CRAC modulatory domain CMD, a serine/proline- and a lysine-rich region (Fig. 2.1a). Independent studies performed by Huang et al. [49] and Muik et al. [50] revealed that the STIM1 C-terminus is sufficient to activate CRAC currents through Orai1. In a very close time period, several groups have analyzed short cytosolic STIM1 fragments to identify the Orai channel activating key domains: OASF (233–450), CAD (342–448), SOAR (344–442) and Ccb9 (339–444) [50–53] (Fig. 2.1a). All Orai activating STIM1 fragments have the CC2 (363–389) and CC3 (399–423) regions with additional 19 residues (424–442) in common (Fig. 2.1a) including the STIM1 homomerization domain (SHD), an Orai coupling and activating domain. The SHD corresponds to the segment ~421–450 and plays a major role in the cytosolic STIM1 homomerization process [50]. Its deletion within OASF dramatically decreases FRET signals in homomerization studies and furthermore abrogates Orai1 activation as measured by patch clamp [50].

2.2.1.4 Crystal Structure of the Orai1 Activating Portion of STIM1

The long awaited first crystal structure of a cytosolic portion of STIM1 has finally been published in 2012. It revealed a crystallized hSOAR protein (345–444 $_{L374M, V419A, C437T}$) forming a dimeric assembly with intra- as well as intermolecular interactions [54] (Fig. 2.1b). The "R" shaped structure of the monomeric SOAR protein consists of an antiparallel arrangement of CC2 and CC3 which are linked by two short α-helices. The dimeric assembly is generated by CC interactions as C-terminal residues (R429, W430, I433, L436) from one monomer interact with N-terminal residues (T354, L351, W350, L347) of the other monomer, finally forming an overall V-shaped structure of the SOAR dimer (Fig. 2.1b). Both CC2 cross each other at amino acid Tyr361 forming a stacked interaction at this position. Mutations within the dimeric binding interface in full-length STIM1 as well as SOAR abolished colocalization with and activation of Orai1 channels [54].

2.2.1.5 CC1 Is Highly Involved in Controlling the Activation State of STIM1

SOAR, the Orai1 activating part of STIM1, is under control of the CC1 domain (aa 238–343, Fig. 2.1a, b) as CC1 is proven to play a key role in transducing luminal di-/oligo-merization to the cytosolic part of STIM1 resulting in homomerization and SOAR exposure [54–57]. Covington et al. [45] have examined homomerization using STIM1 C-terminal truncation mutants (STIM1-CC1 and STIM1-CC1-CAD) to study the impact of the respective cytosolic CC domains. Finally, they concluded that the mere presence of CC1 leads to store independent, yet unstable oligomerization, while the larger construct including CC3+SHD (see Fig. 2.1a) enables strong and store dependent oligomerization. Bioinformatic secondary structure predictions of CC1 suggest the presence of three alpha helical parts (α1, α2 and α3) within CC1 [23]. Helix α1 comprises aa 238–271, helix α2 aa 278–304 and helix α3 aa 308–337 (also known as inhibitory helix IH [54]). The structural resolution of STIM1 CC1 proved to be highly challenging [54], however, Cui et al. [58] have presented a crystal structure of STIM1 CC1 revealing a long helix which dimerizes in an antiparallel conformation. The full length CC1 in context with CAD has still not been shown.

Several studies have been performed to elucidate the role of CC1 in the control of STIM1 activation whereas different hypothetical models have been proposed (Fig. 2.1c). Korzeniowski et al. [59] suggested that an autoinhibitory, intramolecular electrostatic interaction between a glutamate-rich segment within α3 of CC1 and a lysine-rich region (aa382–387) within SOAR keeps STIM1-C terminus inactive. However, $\alpha 3_{CC1}$ (the so called inhibitory helix IH) and the K-rich region are not in the close proximity to form the suggested electrostatic interaction, as evident from the CC1+SOAR crystal structure of *C. elegans* (Fig. 2.1b) [54]. Based on this structure, Yang et al. [54] propose that residues within $\alpha 3_{CC1}$ interact with amino acids at the beginning of CC2 as well as the end of CC3 (Fig. 2.1c). Finally, they suggest that STIM1 forms a dimer in the resting state with the SOAR domain responsible for dimerization and $\alpha 3_{CC1}$ stabilizing the inactive state of STIM1. Another hypothesis by Yu et al. [60] presents an intramolecular shielding model keeping STIM1 inactive. They suggest that the acidic region within $\alpha 3_{CC1}$ is not involved in electrostatic interactions. However, they show that multiple mutations within $\alpha 3_{CC1}$ affecting the amphipathic character of the helix have an impact. Therefore, they propose a hypothetical model in which the amphipathic nature of the $\alpha 3_{CC1}$ regulates the STIM1 activation state [60]. To monitor STIM1 C-terminal conformational rearrangements, a double-labeled conformational sensor construct (YFP-OASF-CFP) has been developed [55] to examine STIM1 aa233–474 (OASF), comprising both CC1 and SOAR. FRET results show that OASF folds into a rather compact structure which likely corresponds to the quiescent state of STIM1 when Ca^{2+} stores are full. In line, Tb^{3+}-acceptor energy transfer measurements performed by Zhou et al. [61] revealed a similar compact structure of STIM1 C-terminus (233–685) where the K-rich segment at the C-terminal end is close to residue 233. To elucidate the molecular processes that trigger the cytosolic part of STIM1 from the inactive into the active form, both artificial crosslinking and mutations have been performed. By substituting amino acids in $\alpha 1_{CC1}$ (L251S) or CC3 (L416S L423S) (Fig. 2.1a) within the OASF conformational sensor, FRET measurements have revealed an extended conformation [55]. By mutating full length STIM1 with these CC1 or CC3 point mutations, constitutive CRAC currents were obtained despite resting ER Ca^{2+} store conditions. Hence, the existence of intramolecular CC interactions within the quiescent STIM1 protein has been suggested. Those CC interactions are most likely released by specific $\alpha 1_{CC1}$ or CC3 mutations or physiologically upon store depletion [55]. In line, Zhou et al. [61] have observed an extended conformation of STIM1 C-terminus induced by artificial crosslinking of the CC1 domains to promote CC1 dimerization. Furthermore, STIM1-CT with the "activating" mutation L251S also results in conformational extension [57] which is in accordance with FRET experiments using YFP-OASF_L251S-CFP. Together, this reveals the impact of $\alpha 1_{CC1}$ on the transition of STIM1 from a quiescent into an active state. Furthermore, Zhou et al. have shown the interaction between monomeric CC1 and SOAR suggesting that store depletion induces CC1 dimerization, consequently leading to the extended conformation due to release of the CC1 interaction with SOAR. Finally, Fahrner et al. have identified, by the use of a FRET based technique, the direct CC1α1 – CC3 interaction as a key molecular determinant in setting the quiescent state of

STIM1 [56]. Release of that interaction is a necessary process for STIM1 activation. In summary, CC1 including α1 and α3 fulfils an important mechanistic role in keeping STIM1 in a quiescent, tight structure (see Fig. 2.1c) [54–57]. Additionally, Fahrner et al. have revealed a slight, destabilizing impact of CC1α2 which is apparently drastically enhanced by a R304W CC1α2 mutation that fully activates STIM1 without store depletion and is associated with the Stormorken Syndrome [62–64].

2.2.1.6 STIM1 Activation Model

Based on functional and structural data of human STIM1, a hypothetical STIM1 activation model [56] has been proposed depicting STIM1 conformations as pivotal function of CRAC activation. The model includes interaction of individual C-terminal domains showing the key steps of STIM1 C-terminal CAD exposure as sequential, molecular mechanism integrating helical rearrangement and clustering (Fig.2.1c). The inactive state of STIM1 is represented by a predominant CC1α1-CC3 clamp controlling the tight-structured quiescent STIM1 state. Additionally, CC1α3 supports this clamp, however the presence of CC1α1 is most essential [56]. Combining the recent functional data, the crystal structure by Yang et al. [54] and the NMR structure by Stathopulos et al. [65] suggest a STIM1 activation model where CC1α1 as well as CC1α3 directly interact with CC3/CC2 to stabilize the quiescent, tight conformation. In a first step, Ca^{2+}-depletion resulting in luminal EF-SAM domain destabilization and conformational changes, yields the driving force which induces the structural changes of the cytosolic STIM1 domains [43, 56, 57, 66]. Consequently, the dominant CC1α1 – CC3 clamp is released, triggering the C-terminal STIM1 extended conformation. Release of CAD induces a rearrangement of CC2 and CC3 resulting in formation of SOAP (STIM Orai activation pocket) [65] and exposure of CC3 for higher-order oligomerization and clustering. In this model, CC1α2 does not markedly contribute to the CC1-CC3 clamp, however, a pathophysiological mutation (R304W) has been reported yielding constitutive STIM1 activity associated with the Stormorken syndrome [62–64]. How this mutation affects the overall CC1 structure and how it induces the activated state of STIM1 still remains to be solved. It is tempting to speculate that the R304W mutation promotes the active, extended state of STIM1 thereby leading to constitutive Orai1 activation. STIM1 has also been reported to be a sensor of oxidative stress [67] and temperature variations stress [68] which indicates that STIM1 may generally act as a stress sensor that initiates Ca^{2+} signalling [69].

The second member of this protein family, STIM2, is also widely expressed and differs from STIM1 mainly in its N-terminal region. Differences affect luminal Ca^{2+} sensitivity and self-activation [46, 48, 70, 71]. Coupling of STIM2 to Orai1 is significantly reduced compared to STIM1 and it is assumed that STIM2 functions as competitive inhibitor of STIM1 mediated Ca^{2+} entry [72–74]. Furthermore, it has been shown that distinct Orai1 coupling domains in STIM1 and STIM2 define the Orai activating site [75]. Residue F394 in STIM1, which is substituted by a leucine (F394L) in STIM2 seems to be highly relevant for gating the Orai1 channel. Very recently, Ong et al. [76] have demonstrated that STIM2 promotes recruitment of STIM1 to ER-PM junctions when ER stores are mildly depleted.

2.2.2 Orai

Named after the keepers of heaven's gate, Orai proteins represent the pore-forming subunit of the CRAC channel that opens upon physical interaction with STIM1 to permit Ca^{2+} influx [73, 77, 78]. There are three members of the Orai family expressed in mammals (Orai1-3) within selected tissues exhibiting elevated expression [21, 33]. All three Orai isoforms are tetra-spanning membrane proteins and exhibit the characteristic features of CRAC channels. Nevertheless, they behave distinct in their inactivation profiles and 2-aminoethyldiphenyl-borate (2-APB) sensitivity [79–82]. Orai proteins share no significant sequence homology to other previously identified ion channels and their activation by STIM1 proteins in the ER is highly unique. Members of the Orai family are ~30 kDa cell surface proteins that have four TM-spanning helices, with N- and C-termini located intracellularly. One intracellular and two extracellular loops connect the TM-spanning helices (Fig. 2.2a, c). The CRAC channel itself is composed of an oligomeric assembly of Orai1 subunits, potentially forming homo- as well as heteromeric Orai channels [80, 83–85]. Multimerization of Orai1 subunits is assumed to be established by their transmembrane regions as deletion of the C-terminal strands does not affect aggregation [39, 86]. A cluster of positively charged amino acids within the N-terminal first TM domain is fully conserved among the three Orai isoforms, while only Orai1 contains an N-terminal proline-arginine rich region [87]. Each of the Orai family member contains a coiled-coil domain within their C-terminus, essential for physical interaction with STIM1 [39, 88–90]. Ca^{2+} dependent fast inactivation (CDI) occurs in all three Orai homologs, is supposed to be mediated by cytosolic domains of both STIM1 and Orai1 [91–93], whereas Orai3 exhibits the highest degree of CDI [8, 80, 94]. Orai1 remains the best studied CRAC channel protein and appears to be the predominant isoform mediating SOCE in most cells. In the ensuing years after the identification of Orai1, cysteine-scanning studies first identified the pore-lining residues of the CRAC channel [95, 96] and in agreement with these studies, the crystal structure of the *Drosophila* Orai protein was determined recently, proposing – against all expectations – a hexameric assembly of Orai subunits around a central ion pore [90] (Fig. 2.2c–e). Although preliminary and recent biochemical and fluorescence studies pointed to a tetrameric assembly of Orai subunits [97–103], the evidence of a hexameric dOrai channel complex highly challenged this assumption. Although this is the best structural representation available, it is of note that the closed-state dOrai1 crystal structure, in 3.35 Å resolution, lacks substantial parts of the N- and C-termini, the extracellular TM1-TM2 loop, the intracellular TM2-TM3 loop and is furthermore mutated at positions C224S, P276R, P277R and C283T [90]. Atomic structures of human Orai channels will give further insight into channel architecture.

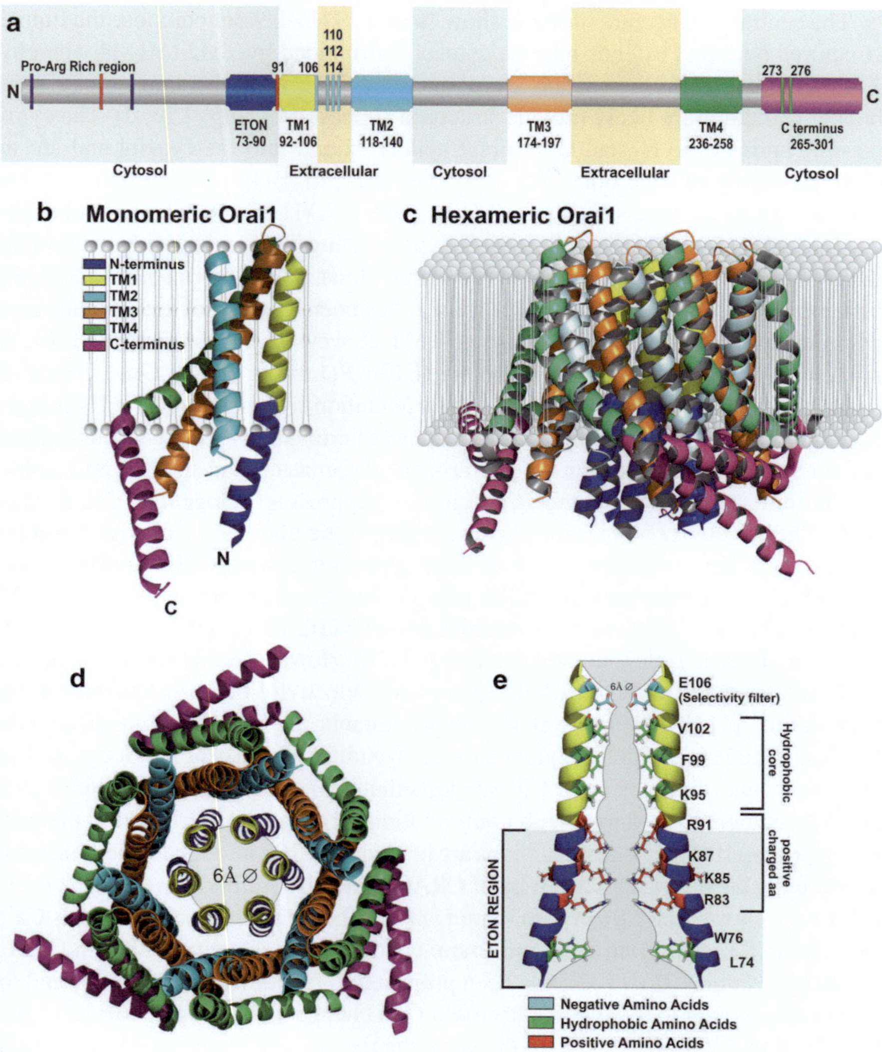

Fig. 2.2 (**a**) Schematic representation of human, full-length Orai1 depicting the overall structure and residues indispensable for Orai1 function. (**b**) Cartoon representing a single Orai1 subunit with the four transmembrane (TM) regions, the N-terminal as well as the C-terminal elongated helices, shown in different colors used throughout (**a–d**). (**c**) Cartoon showing the assembly of six Orai subunits based on the X-ray crystal structure of *Drosophila* Orai. Transmembrane domains of the six subunits are arranged as concentric rings around the ion pore, with TM1 forming the inner ring surrounding the ion pore. (**d**) Assembly of the Orai1 pore seen from the *top*. (**e**) Cartoon representing the human Orai1 pore showing 2 TM1 strands together with the cytosolic, helical extensions including the conserved ETON region. Relevant pore-lining amino acids, especially the selectivity filter, the hydrophobic core as well as residues within the ETON region are highlighted

The central architecture of dOrai shows that six TM1 helices compose the funnel structured pore that extends into the cytosol, surrounded by TM2 to TM4 whereby TM2 and TM3 form the second layer, shielding TM1 from the surrounding lipid bilayer and therefore likely provide structural support [90] (Fig. 2.2c, d). The cytoplasmic ends of the N- and C-terminal helices extend into the cytosol and are in close proximity of each other [90, 104] (Fig. 2.2b, c). Both of these regions have been shown to be necessary for interaction with STIM1 and both are essential for channel activation [39, 59, 105–108]. The pore-lining TM1 helix is identical in all three Orai proteins (Fig. 2.2e), with the extracellular loop between TM3 and TM4 being the most divergent among isoforms. It has been shown that a glutamate residue situated at the extracellular end of TM1 conveys Ca^{2+} selectivity (E106). A mutation to either glutamine or aspartate (E106Q/D in the human Orai1 isoform) drastically alters ion selectivity [109–111]. Mutation to aspartate (E106D) is supposed to widen the pore and therefore increases permeability to larger monovalent cations like Cs^{+}. Substitution of this critical glutamate by an uncharged residue results in non-functional channels and in line suppresses endogenous CRAC currents. The selectivity filter is the narrowest part of the pore with a diameter around 6 Å and it is surrounded by a ring of TM3 glutamate residues (E190) (Fig. 2.2e). The selectivity filter at position E106 may be fine-tuned by the help of negatively charged aspartates in the first extracellular loop (D110, D112, D114) that also alter selectivity upon a triple mutation to alanines [110]. However, single point mutations of these aspartates in the first loop retain Ca^{2+} selectivity but affect inhibition by Gadolinium [111]. Several cysteine loop1-mutants are able to form disulphide bonds and dimerize which suggests a close proximity of adjacent first loops within an Orai channel complex [95]. This suggests that the first loop of Orai channels is a flexible segment that can undergo conformational changes [112]. Overall, the currently proposed model is that Ca^{2+} ions are guided via E106 through the pore thereby maintaining the high Ca^{2+} selectivity of CRAC channels. Similarly to E106 in Orai1, it has been shown that E81 in Orai3 (equivalent to E106 in Orai1) also controls Ca^{2+} selectivity [82, 113]. One helix turn more central to the selectivity filter in Orai1, valine at position 102 (V102), has been proposed to function as a gate [114] and to constitute a barrier to ion flux in the closed Orai channel [115] (Fig. 2.2e). A constitutively active V102C/A channel displays changes in selectivity which leads to a reduced Ca^{2+} selectivity due to an increased pore size. Upon coexpression with STIM1 these mutants regain Ca^{2+} selectivity [114, 116]. Recently, it was shown that STIM1 dependent gating entails a structural change in the vicinity of residue E106 and V102 near the external entrance of the Orai1 pore [114, 115]. The Orai1 pore architecture thus changes upon interaction with STIM1 which underlines the close coupling of gating and ion selectivity in CRAC channels. Glycine at position 98 (G98) has been proposed to function as a gating hinge [116]. While mutations to aspartate or proline (G98D/P) lead to constitutively active channels, substitution to alanine completely abrogated channel function. Residue R91, originally discovered in a human Orai1 channelopathy, as R91W leads to non-functional channels [18], seems to collectively form an electrostatic gate towards the cytosolic portion of TM1. Mutation of position 91 to hydrophilic side chains allowed the channel to function normally [91, 116] while hydrophobic substitution completely abolished channel function. Therefore, arginine at position 91 plays a critical role in channel gating. The constriction site of R91W is bypassed with an additional G98D mutation. The mutation in the gating hinge is able to restore the function of the R91W SCID mutant and seems to extend the channel gate more efficiently than in the non-functional R91W and V102A/C mutant [114, 116]. Residues within TM3 have also been shown to alter gating and permeation of Orai1 channels. Although TM3s do not line the pore they have been shown to modulate both selectivity and gating. E190 in Orai1 contributes to Ca^{2+} selectivity and permeation whereby single point mutations increased the pore diameter to 7 Å [82, 109, 110, 117]. E190 is supposed to affect pore properties by altering intramolecular TM interactions [95, 96]. Residue G183 is also involved in channel gating as a G183A mutant abolished store-operated activation [112, 118]. In summary, the side chains of the following residues protrude into the pore: E106, V102, F99, L95, R91, K87, R83, Y80 and W76 (Fig. 2.2e). This is in accordance with their ability to dimerize upon cysteine substitution in crosslinking experiments except residue F99 [95, 96]. The picture that can be drawn from all the preliminary studies is that the CRAC channel pore can be subdivided into four defined regions (Fig. 2.2e): (1) the selectivity filter that is formed by a ring of glutamates that confer the pore a negative electrostatic potential; (2) a hydrophobic region comprising three α-helical turns with amino acids V102, F99 and L95 that are expected to establish Van der Waals interactions with each other; (3) a basic region comprising three α-helical turns with amino acids R91, K87, R83 that is suggested to coordinate anions. The basic region may be a vital part of the gating mechanism through anion binding that keeps the pore closed in the resting state [90]; and (4) a cytosolic region spanning two α-helical turns. This extension into the cytosol is about 20 Å in length and is called extended TM1 Orai1 N-terminal (ETON) region [119]. Three positively charged residues (R91, K87, R83) directly line the pore at the cytosolic side which creates an unusual environment for a cation channel and they have been suggested to provide both barrier as well as electrostatic stabilization [119]. The ETON region additionally contains two serines (S90, S89) embedded between R91 and K87 which may contribute to the flexibility of the successive N-terminal and TM1 segments [104]. Overall, the inner portion of TM1 (aa91–98) seems to possess more flexibility than the outer region (aa99–104) which forms a more rigid structure [96]. The resolved *Drosophila* crystal structure provides a framework for investigating permeation and selectivity in human CRAC channels. Clearly, it takes further analysis to finally incorporate that knowledge into plausible models for human Orai1 channels. What conformational changes occur during channel opening, closing and inactivation? Channel structures of both closed and open state are needed to provide insight into CRAC channel activation pathways and for drug design of channel blockers and activators.

2.3 STIM1-Orai1 Coupling and CRAC Activation

STIM1 coupling to the Orai1 channel involves the cytosolic strand of STIM1 and both the Orai1 C-terminus as well as the N-terminus (Fig. 2.3). In three independent studies [39, 51, 120] Orai1 C-terminal truncation mutants have for the first time allowed an insight into the STIM1 Orai1 interaction process as it has been shown that Orai1 C-terminus is necessary for the direct coupling to STIM1 C-terminus using co-localization and FRET experiments. As clearly seen in the dOrai crystal structure, the Orai1 TM4 extends into the cytosol (cytosolic part named Orai1 C-terminus) [90] and is therefore accessible to STIM1 C-terminus for coupling. Consequently and based on these data, the focus has been set to elucidate specific residues within Orai C-terminus responsible for coupling to STIM1 C-terminus. Frischauf et al. [89] have suggested a weak coiled-coil probability of Orai1 C-terminus using bioinformatic predictions, while the coiled-coil probability of Orai2 and Orai3 is shown to be 15–17-fold higher suggesting that the affinity for STIM1 increases with the higher probability of Orai C-terminal CC domains. Vice versa, reducing the probability of STIM1 CC2 by an amino acid substitution L373S interferes with Orai1 activation whereas Orai2 and Orai3 still show partial activation [89]. In line, a double mutation in STIM1 CC2 fully disrupts coupling to and activation of all three Orai channels, suggesting that STIM1 – Orai coupling is dependent on these CC domains [89]. The hydrophobic residues L273 and L276 within Orai1 CC domain play an important role in coupling to STIM1. In addition, these positions have been shown to play a role in Orai dimerization within the hexameric Orai structure [90]. Besides coiled coil interactions based on hydrophobic

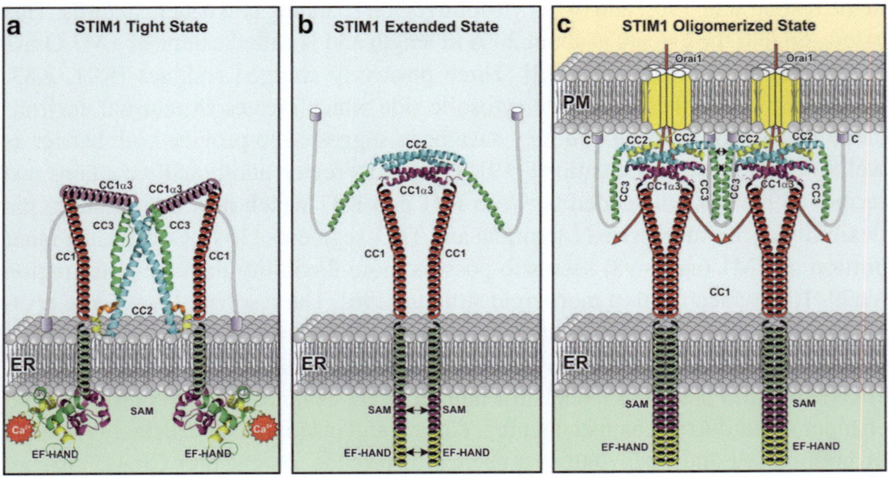

Fig. 2.3 Cartoons representing a model for the oligomerization process of STIM1 (**a**, **b**) and Orai1 coupling (**c**). Upon store depletion, STIM1 proteins adopt an extended conformation (**b**) and oligomerize, processes that are controlled by the CC1 domain. (**c**) Activated Orai1 channel with CC2 involved in the coupling to Orai1 C – terminus

residues, Calloway et al. have proposed electrostatic interactions between STIM1 and Orai1 [105, 121], hypothesizing formation of salt bridges between acidic residues within Orai1 C-terminus and positively charged amino acids in STIM1 (KIKKKR – aa 382–387). Deletion of this basic cluster within STIM1 disrupts store dependent association with Orai1, however, mutating the acidic residues within Orai1 C-terminus still allows for coupling and retained channel function suggesting that the coupling process involves other structural components [65, 105]. Recently, Stathopulos et al. presented two NMR solution structures of a human cytosolic STIM1 fragment (aa 312–387) comprising a part of CC1 and the whole CC2. One structure is presented without Orai1CT and the second structure is shown together with Orai1CT (Orai1 aa272–292) [122]. The STIM1 CC1-CC2 subunits form an antiparallel U-shaped homodimer. Furthermore, the dimeric STIM1 CC2 helix clampes two Orai1 C-terminal fragments in a hydrophobic/basic STIM-Orai association pocket called SOAP. STIM1 mutations interfering with CC1 dimeric interactions inhibit, while substitutions enhancing CC1 stability spontaneously activate Orai1 channels. On the other hand, CC2 mutations have revealed variability in Orai1 activation due to binding Orai1 as well as autoinhibiting STIM1 oligomerization [122]. In summary, STIM1 – Orai1 coupling mediated between Orai1 C-terminus and CC2 of STIM1 predominantly involves hydrophobic interactions as well as ionic bond formation. In another recent paper Wang et al. [75] have studied the differences of SOAR of STIM1 and STIM2 in coupling to and activating Orai1. Herein, they show that SOAR1 and SOAR2 have similar sequences and predicted secondary structures but behave differently in functional coupling to Orai1. By use of SOAR1/SOAR2 chimeras Wang et al. have been able to identify a crucial residue that determines the differences in binding and activation of Orai1. SOAR contains four α-helical parts (see Fig. 2.1b) that are systematically switched and analyzed in these chimeras. Helix Sα4 has been shown to behave similarly in SOAR1 and SOAR2, whereas the residue F394 in Sα2 of SOAR1 seems to be highly crucial for gating Orai1. Introduction of F394 point mutations to L, A or H in full length STIM1 decreases or fully abolishes SOCE, respectively, when coexpressed with Orai1 in HEK293 cells. Patch clamp and FRET analysis finally have revealed that F394 in STIM1 is highly important for both coupling to and gating of Orai1 channels [75]. Park et al. elucidated that additionally to Orai1 C-terminus, the N-terminal strand acts as binding partner for STIM1, however, to a weaker extent [51]. All Orai proteins contain a highly conserved region in the N-terminal part (aa73–90) termed ETON [119] which corresponds to the elongated extension of TM1 into the cytosol. Orai1 deletion mutants lacking ETON or the full N-terminal strand completely lose function. In line, N-terminal truncations of Orai1 (Orai1 Δ1-74 or Δ1-75) still yield store dependent activation whereas the deletion mutant Orai1 Δ1-76 results in loss of function [86, 119]. By introducing point mutations within the ETON region the hydrophobic residues L74 and W76 seem to be part of the Orai1 N-terminal STIM1 binding interface as the double mutant results in highly reduced STIM1 interaction [119]. Furthermore, the positively charged amino acids R83 and K87 have been shown to contribute to the STIM1 binding interface and additionally provide a stabilizing effect to the elongated pore [119]. Another study by McNally et al. [107]

reveals that drastic deletions in the Orai1 N-terminus (Orai1 Δ1-85 and Orai1 Δ73-85) abolished binding to STIM1 as shown with the Orai1-CAD-CAD multiple fusion protein. However, restoration of channel function of Orai1 N-terminal point mutations have been analyzed using the fusion protein but are not visible using the coexpression of Orai1 mutant + STIM1. Additionally, Zheng et al. [108] have analyzed progressive N-terminal truncations yielding the result that Orai1 N-terminus is essential for binding STIM1. However, the presence of both Orai1 N- and C-terminus is crucial for correct STIM1 coupling as they suggest that the STIM1 first binds to Orai1 C-terminus followed by STIM1 – Orai1 N-terminus interaction. In summary, the ETON-region functions as binding partner for STIM1 and provides electrostatic components having an impact on the structure of the elongated pore. ETON is fully conserved between Orai1 and Orai3, however, extensive Orai3 N-terminal truncations still retain STIM1 dependent activation whereas equivalent Orai1 deletion mutants result in loss of function [119, 123]. Therefore, Orai3 may contain additional residues/domains which contribute to STIM1 coupling and Orai3 activation, despite more extensive deletions at the N-terminal strand. Overall, the SOAR domain of STIM1, which is sufficient for Orai1 coupling and activation, most probably involves binding of both Orai1 C- as well as N-terminus. The resulting bridge probably generates a force inducing a conformational change of the ETON and TM1 regions resulting in Ca^{2+} influx probably due to movement of the hydrophobic gate at TM1 V102 as well as removal of the electrostatic hindrance in the ETON region. The mechanistic steps in this process clearly require further analysis.

Acknowledgments This work was supported by the Austrian Science Fund (FWF projects P25172 and P27263 to C.R. and V 286-B21 to I.F.).

References

1. Putney JW Jr (1986) A model for receptor-regulated calcium entry. Cell Calcium 7(1):1–12
2. Abdullaev IF, Bisaillon JM, Potier M, Gonzalez JC, Motiani RK, Trebak M (2008) Stim1 and Orai1 mediate CRAC currents and store-operated calcium entry important for endothelial cell proliferation. Circ Res 103:1289–1299
3. Albert AP, Saleh SN, Peppiatt-Wildman CM, Large WA (2007) Multiple activation mechanisms of store-operated TRPC channels in smooth muscle cells. J Physiol 583(Pt 1):25–36
4. Prakriya M (2009) The molecular physiology of CRAC channels. Immunol Rev 231(1):88–98
5. Hoth M, Penner R (1992) Depletion of intracellular calcium stores activates a calcium current in mast cells. Nature 355(6358):353–356
6. McFadzean I, Gibson A (2002) The developing relationship between receptor-operated and store-operated calcium channels in smooth muscle. Br J Pharmacol 135(1):1–13
7. Mogami H, Nakano K, Tepikin AV, Petersen OH (1997) Ca2+ flow via tunnels in polarized cells: recharging of apical Ca2+ stores by focal Ca2+ entry through basal membrane patch. Cell 88(1):49–55
8. Hoth M, Penner R (1993) Calcium release-activated calcium current in rat mast cells. J Physiol 465:359–386

9. Lewis RS, Cahalan MD (1989) Mitogen-induced oscillations of cytosolic Ca2+ and transmembrane Ca2+ current in human leukemic T cells. Cell Regul 1(1):99–112
10. Zweifach A, Lewis RS (1993) Mitogen-regulated Ca2+ current of T lymphocytes is activated by depletion of intracellular Ca2+ stores. Proc Natl Acad Sci U S A 90(13):6295–6299
11. Berridge MJ (1987) Inositol trisphosphate and diacylglycerol: two interacting second messengers. Annu Rev Biochem 56:159–193
12. Kuno M, Gardner P (1987) Ion channels activated by inositol 1,4,5-trisphosphate in plasma membrane of human T-lymphocytes. Nature 326(6110):301–304
13. Smani T, Shapovalov G, Skryma R, Prevarskaya N, Rosado JA (2015) Functional and physiopathological implications of TRP channels. Biochim Biophys Acta 1853(8):1772–1782
14. Liou J, Kim ML, Heo WD, Jones JT, Myers JW, Ferrell JE Jr, Meyer T (2005) STIM is a Ca2+ sensor essential for Ca2+-store-depletion-triggered Ca2+ influx. Curr Biol 15(13):1235–1241
15. Roos J, DiGregorio PJ, Yeromin AV, Ohlsen K, Lioudyno M, Zhang S, Safrina O, Kozak JA, Wagner SL, Cahalan MD, Velicelebi G, Stauderman KA (2005) STIM1, an essential and conserved component of store-operated Ca2+ channel function. J Cell Biol 169(3):435–445
16. Vig M, Peinelt C, Beck A, Koomoa DL, Rabah D, Koblan-Huberson M, Kraft S, Turner H, Fleig A, Penner R, Kinet JP (2006) CRACM1 is a plasma membrane protein essential for store-operated Ca2+ entry. Science 312(5777):1220–1223
17. Zhang SL, Yeromin AV, Zhang XHF, Yu Y, Safrina O, Penna A, Roos J, Stauderman KA, Cahalan MD (2006) Genome-wide RNAi screen of Ca2+ influx identifies genes that regulate Ca2+ release-activated Ca2+ channel activity. Proc Natl Acad Sci 103(24):9357–9362. doi:10.1073/pnas0603161103
18. Feske S, Gwack Y, Prakriya M, Srikanth S, Puppel SH, Tanasa B, Hogan PG, Lewis RS, Daly M, Rao A (2006) A mutation in Orai1 causes immune deficiency by abrogating CRAC channel function. Nature 441(7090):179–185
19. Hogan PG, Rao A (2007) Dissecting I(CRAC), a store-operated calcium current. Trends Biochem Sci 32:235–245
20. Luik RM, Wang B, Prakriya M, Wu MM, Lewis RS (2008) Oligomerization of STIM1 couples ER calcium depletion to CRAC channel activation. Nature 454(7203):538–542
21. Feske S, Skolnik EY, Prakriya M (2012) Ion channels and transporters in lymphocyte function and immunity. Nat Rev Immunol 12(7):532–547
22. Cahalan MD, Chandy KG (2009) The functional network of ion channels in T lymphocytes. Immunol Rev 231(1):59–87
23. Soboloff J, Rothberg BS, Madesh M, Gill DL (2012) STIM proteins: dynamic calcium signal transducers. Nat Rev Mol Cell Biol 13(9):549–565
24. Kar P, Nelson C, Parekh AB (2012) CRAC channels drive digital activation and provide analog control and synergy to Ca(2+)-dependent gene regulation. Curr Biol 22(3):242–247
25. Di Capite J, Parekh AB (2009) CRAC channels and Ca2+ signaling in mast cells. Immunol Rev 231(1):45–58
26. Pores-Fernando AT, Zweifach A (2009) Calcium influx and signaling in cytotoxic T-lymphocyte lytic granule exocytosis. Immunol Rev 231(1):160–173
27. Lewis RS (2001) Calcium signaling mechanisms in T lymphocytes. Annu Rev Immunol 19:497–521
28. Stathopulos PB, Ikura M (2009) Structurally delineating stromal interaction molecules as the endoplasmic reticulum calcium sensors and regulators of calcium release-activated calcium entry. Immunol Rev 231(1):113–131
29. Zhang SL, Yu Y, Roos J, Kozak JA, Deerinck TJ, Ellisman MH, Stauderman KA, Cahalan MD (2005) STIM1 is a Ca2+ sensor that activates CRAC channels and migrates from the Ca2+ store to the plasma membrane. Nature 437(7060):902–905
30. Baba Y, Hayashit K, Fujii Y, Mizushima A, Watarai H, Wakamori M, Numaga T, Mori Y, Iino M, Hikida M, Kurosaki T (2006) Coupling of STIM1 to store-operated Ca2+ entry through its constitutive and inducible movement in the endoplasmic reticulum. Proc Natl Acad Sci U S A 103(45):16704–16709

31. Navarro-Borelly L, Somasundaram A, Yamashita M, Ren D, Miller RJ, Prakriya M (2008) STIM1-Orai1 interactions and Orai1 conformational changes revealed by live-cell FRET microscopy. J Physiol 586(Pt 22):5383–5401
32. Xu P, Lu J, Li Z, Yu X, Chen L, Xu T (2006) Aggregation of STIM1 underneath the plasma membrane induces clustering of Orai1. Biochem Biophys Res Commun 350(4):969–976
33. Hoth M, Niemeyer BA (2013) The neglected CRAC proteins: Orai2, Orai3, and STIM2. Curr Top Membr 71:237–271
34. Wu MM, Buchanan J, Luik RM, Lewis RS (2006) Ca2+ store depletion causes STIM1 to accumulate in ER regions closely associated with the plasma membrane. J Cell Biol 174(6):803–813. doi:10.1083/jcb200604014
35. Stathopulos PB, Zheng L, Li GY, Plevin MJ, Ikura M (2008) Structural and mechanistic insights into STIM1-mediated initiation of store-operated calcium entry. Cell 135(1):110–122
36. Smyth JT, Dehaven WI, Bird GS, Putney JW Jr (2008) Ca2+-store-dependent and -independent reversal of Stim1 localization and function. J Cell Sci 121(Pt 6):762–772
37. Sochivko D, Pereverzev A, Smyth N, Gissel C, Schneider T, Beck H (2002) The Ca(V)2.3 Ca(2+) channel subunit contributes to R-type Ca(2+) currents in murine hippocampal and neocortical neurones. J Physiol 542(Pt 3):699–710
38. Luik RM, Wu MM, Buchanan J, Lewis RS (2006) The elementary unit of store-operated Ca2+ entry: local activation of CRAC channels by STIM1 at ER-plasma membrane junctions. J Cell Biol 174(6):815–825
39. Muik M, Frischauf I, Derler I, Fahrner M, Bergsmann J, Eder P, Schindl R, Hesch C, Polzinger B, Fritsch R, Kahr H, Madl J, Gruber H, Groschner K, Romanin C (2008) Dynamic coupling of the putative coiled-coil domain of ORAI1 with STIM1 mediates ORAI1 channel activation. J Biol Chem 283(12):8014–8022
40. Malli R, Naghdi S, Romanin C, Graier WF (2008) Cytosolic Ca2+ prevents the subplasmalemmal clustering of STIM1: an intrinsic mechanism to avoid Ca2+ overload. J Cell Sci 121(Pt 19):3133–3139
41. Liou J, Fivaz M, Inoue T, Meyer T (2007) Live-cell imaging reveals sequential oligomerization and local plasma membrane targeting of stromal interaction molecule 1 after Ca2+ store depletion. Proc Natl Acad Sci U S A 104(22):9301–9306
42. Baba Y, Nishida K, Fujii Y, Hirano T, Hikida M, Kurosaki T (2008) Essential function for the calcium sensor STIM1 in mast cell activation and anaphylactic responses. Nat Immunol 9(1):81–88
43. Stathopulos PB, Li G-Y, Plevin MJ, Ames JB, Ikura M (2006) Stored Ca2+ depletion-induced oligomerization of STIM1 via the EF-SAM region: an initiation mechanism for capacitive Ca2+ entry. J Biol Chem 281:35855–35862. doi:10.1074/jbcM608247200:M608247200
44. Stathopulos PB, Ikura M (2010) Partial unfolding and oligomerization of stromal interaction molecules as an initiation mechanism of store operated calcium entry. Biochem Cell Biol 88(2):175–183
45. Covington ED, Wu MM, Lewis RS (2010) Essential role for the CRAC activation domain in store-dependent oligomerization of STIM1. Mol Biol Cell 21(11):1897–1907
46. Brandman O, Liou J, Park WS, Meyer T (2007) STIM2 is a feedback regulator that stabilizes basal cytosolic and endoplasmic reticulum Ca(2+) levels. Cell 131(7):1327–1339
47. Zheng L, Stathopulos PB, Li GY, Ikura M (2008) Biophysical characterization of the EF-hand and SAM domain containing Ca2+ sensory region of STIM1 and STIM2. Biochem Biophys Res Commun 369(1):240–246
48. Zheng L, Stathopulos PB, Schindl R, Li GY, Romanin C, Ikura M (2011) Auto-inhibitory role of the EF-SAM domain of STIM proteins in store-operated calcium entry. Proc Natl Acad Sci U S A 108(4):1337–1342
49. Huang GN, Zeng W, Kim JY, Yuan JP, Han L, Muallem S, Worley PF (2006) STIM1 carboxyl-terminus activates native SOC, I(crac) and TRPC1 channels. Nat Cell Biol 8(9):1003–1010
50. Muik M, Fahrner M, Derler I, Schindl R, Bergsmann J, Frischauf I, Groschner K, Romanin C (2009) A cytosolic homomerization and a modulatory domain within STIM1 C terminus determine coupling to ORAI1 channels. J Biol Chem 284(13):8421–8426

51. Park CY, Hoover PJ, Mullins FM, Bachhawat P, Covington ED, Raunser S, Walz T, Garcia KC, Dolmetsch RE, Lewis RS (2009) STIM1 clusters and activates CRAC channels via direct binding of a cytosolic domain to Orai1. Cell 136(5):876–890
52. Yuan JP, Zeng W, Dorwart MR, Choi YJ, Worley PF, Muallem S (2009) SOAR and the polybasic STIM1 domains gate and regulate Orai channels. Nat Cell Biol 11(3):337–343
53. Kawasaki T, Lange I, Feske S (2009) A minimal regulatory domain in the C terminus of STIM1 binds to and activates ORAI1 CRAC channels. Biochem Biophys Res Commun 385(1):49–54
54. Yang X, Jin H, Cai X, Li S, Shen Y (2012) Structural and mechanistic insights into the activation of Stromal interaction molecule 1 (STIM1). Proc Natl Acad Sci U S A 109(15):5657–5662
55. Muik M, Fahrner M, Schindl R, Stathopulos P, Frischauf I, Derler I, Plenk P, Lackner B, Groschner K, Ikura M, Romanin C (2011) STIM1 couples to ORAI1 via an intramolecular transition into an extended conformation. EMBO J 30(9):1678–1689
56. Fahrner M, Muik M, Schindl R, Butorac C, Stathopulos P, Zheng L, Jardin I, Ikura M, Romanin C (2014) A coiled-coil clamp controls both conformation and clustering of stromal interaction molecule 1 (STIM1). J Biol Chem 289(48):33231–33244
57. Gudlur A, Zhou Y, Hogan PG (2013) STIM-ORAI interactions that control the CRAC channel. Curr Top Membr 71:33–58
58. Cui B, Yang X, Li S, Lin Z, Wang Z, Dong C, Shen Y (2013) The inhibitory helix controls the intramolecular conformational switching of the C-terminus of STIM1. PLoS One 8(9):e74735
59. Korzeniowski MK, Manjarres IM, Varnai P, Balla T (2010) Activation of STIM1-Orai1 involves an intramolecular switching mechanism. Sci Signal 3(148):ra82
60. Yu F, Sun L, Hubrack S, Selvaraj S, Machaca K (2013) Intramolecular shielding maintains the ER Ca(2)(+) sensor STIM1 in an inactive conformation. J Cell Sci 126(Pt 11):2401–2410
61. Zhou Y, Srinivasan P, Razavi S, Seymour S, Meraner P, Gudlur A, Stathopulos PB, Ikura M, Rao A, Hogan PG (2013) Initial activation of STIM1, the regulator of store-operated calcium entry. Nat Struct Mol Biol 20(8):973–981
62. Nesin V, Wiley G, Kousi M, Ong EC, Lehmann T, Nicholl DJ, Suri M, Shahrizaila N, Katsanis N, Gaffney PM, Wierenga KJ, Tsiokas L (2014) Activating mutations in STIM1 and ORAI1 cause overlapping syndromes of tubular myopathy and congenital miosis. Proc Natl Acad Sci U S A 111(11):4197–4202
63. Morin G, Bruechle NO, Singh AR, Knopp C, Jedraszak G, Elbracht M, Bremond-Gignac D, Hartmann K, Sevestre H, Deutz P, Herent D, Nurnberg P, Romeo B, Konrad K, Mathieu-Dramard M, Oldenburg J, Bourges-Petit E, Shen Y, Zerres K, Ouadid-Ahidouch H, Rochette J (2014) Gain-of-function mutation in STIM1 (P.R304W) is associated with Stormorken syndrome. Hum Mutat 35(10):1221–1232
64. Misceo D, Holmgren A, Louch WE, Holme PA, Mizobuchi M, Morales RJ, De Paula AM, Stray-Pedersen A, Lyle R, Dalhus B, Christensen G, Stormorken H, Tjonnfjord GE, Frengen E (2014) A dominant STIM1 mutation causes Stormorken syndrome. Hum Mutat 35(5):556–564
65. Stathopulos PB, Ikura M (2013) Structure and function of endoplasmic reticulum STIM calcium sensors. Curr Top Membr 71:59–93
66. Furukawa Y, Teraguchi S, Ikegami T, Dagliyan O, Jin L, Hall D, Dokholyan NV, Namba K, Akira S, Kurosaki T, Baba Y, Standley DM (2014) Intrinsic disorder mediates cooperative signal transduction in STIM1. J Mol Biol 426(10):2082–2097
67. Hawkins BJ, Irrinki KM, Mallilankaraman K, Lien YC, Wang Y, Bhanumathy CD, Subbiah R, Ritchie MF, Soboloff J, Baba Y, Kurosaki T, Joseph SK, Gill DL, Madesh M (2010) S-glutathionylation activates STIM1 and alters mitochondrial homeostasis. J Cell Biol 190(3):391–405
68. Xiao B, Coste B, Mathur J, Patapoutian A (2011) Temperature-dependent STIM1 activation induces Ca(2+) influx and modulates gene expression. Nat Chem Biol 7:351–358

69. Mancarella S, Wang Y, Gill DL (2011) Signal transduction: STIM1 senses both Ca(2)+ and heat. Nat Chem Biol 7(6):344–345
70. Zhou Y, Mancarella S, Wang Y, Yue C, Ritchie M, Gill DL, Soboloff J (2009) Short N-terminal domains of STIM1 and STIM2 control the activation kinetics of Orai1 channels. J Biol Chem 284:19164–19168
71. Stathopulos PB, Zheng L, Ikura M (2009) Stromal interaction molecule (STIM) 1 and STIM2 calcium sensing regions exhibit distinct unfolding and oligomerization kinetics. J Biol Chem 284(2):728–732
72. Bird GS, Hwang SY, Smyth JT, Fukushima M, Boyles RR, Putney JW Jr (2009) STIM1 is a calcium sensor specialized for digital signaling. Curr Biol 19:1724–1729
73. Soboloff J, Spassova MA, Tang XD, Hewavitharana T, Xu W, Gill DL (2006) Orai1 and STIM reconstitute store-operated calcium channel function. J Biol Chem 281(30):20661–20665
74. Soboloff J, Spassova MA, Hewavitharana T, He LP, Xu W, Johnstone LS, Dziadek MA, Gill DL (2006) STIM2 is an inhibitor of STIM1-mediated store-operated Ca(2+) entry. Curr Biol 16(14):1465–1470
75. Wang X, Wang Y, Zhou Y, Hendron E, Mancarella S, Andrake MD, Rothberg BS, Soboloff J, Gill DL (2014) Distinct Orai-coupling domains in STIM1 and STIM2 define the Orai-activating site. Nat Commun 5:3183
76. Ong HL, de Souza LB, Zheng C, Cheng KT, Liu X, Goldsmith CM, Feske S, Ambudkar IS (2015) STIM2 enhances receptor-stimulated Ca(2)(+) signaling by promoting recruitment of STIM1 to the endoplasmic reticulum-plasma membrane junctions. Sci Signal 8(359):ra3
77. Mercer JC, DeHaven WI, Smyth JT, Wedel B, Boyles RR, Bird GS, Putney JW Jr (2006) Large store-operated calcium selective currents due to co-expression of Orai1 or Orai2 with the intracellular calcium sensor, Stim1. J Biol Chem 281(34):24979–24990.
78. Peinelt C, Vig M, Koomoa DL, Beck A, Nadler MJ, Koblan-Huberson M, Lis A, Fleig A, Penner R, Kinet JP (2006) Amplification of CRAC current by STIM1 and CRACM1 (Orai1). Nat Cell Biol 8(7):771–773
79. DeHaven WI, Smyth JT, Boyles RR, Putney JW Jr (2007) Calcium inhibition and calcium potentiation of Orai1, Orai2, and Orai3 calcium release-activated calcium channels. J Biol Chem 282(24):17548–17556
80. Lis A, Peinelt C, Beck A, Parvez S, Monteilh-Zoller M, Fleig A, Penner R (2007) CRACM1, CRACM2, and CRACM3 are store-operated Ca2+ channels with distinct functional properties. Curr Biol 17(9):794–800
81. Frischauf I, Schindl R, Bergsmann J, Derler I, Fahrner M, Muik M, Fritsch R, Lackner B, Groschner K, Romanin C (2011) Cooperativeness of Orai cytosolic domains tunes subtype-specific gating. J Biol Chem 286(10):8577–8584
82. Schindl R, Bergsmann J, Frischauf I, Derler I, Fahrner M, Muik M, Fritsch R, Groschner K, Romanin C (2008) 2-aminoethoxydiphenyl borate alters selectivity of Orai3 channels by increasing their pore size. J Biol Chem 283(29):20261–20267
83. Schindl R, Frischauf I, Bergsmann J, Muik M, Derler I, Lackner B, Groschner K, Romanin C (2009) Plasticity in Ca2+ selectivity of Orai1/Orai3 heteromeric channel. Proc Natl Acad Sci U S A 106(46):19623–19628
84. Gwack Y, Srikanth S, Feske S, Cruz-Guilloty F, Oh-hora M, Neems DS, Hogan PG, Rao A (2007) Biochemical and functional characterization of Orai proteins. J Biol Chem 282(22):16232–16243
85. Zhang SL, Kozak JA, Jiang W, Yeromin AV, Chen J, Yu Y, Penna A, Shen W, Chi V, Cahalan MD (2008) Store-dependent and -independent modes regulating Ca2+ release-activated Ca2+ channel activity of human Orai1 and Orai3. J Biol Chem 283(25):17662–17671
86. Li Z, Lu J, Xu P, Xie X, Chen L, Xu T (2007) Mapping the interacting domains of STIM1 and Orai1 in Ca2+ release-activated Ca2+ channel activation. J Biol Chem 282(40):29448–29456

87. Takahashi Y, Murakami M, Watanabe H, Hasegawa H, Ohba T, Munehisa Y, Nobori K, Ono K, Iijima T, Ito H (2007) Essential role of the N-terminus of murine Orai1 in store-operated Ca(2+) entry. Biochem Biophys Res Commun 356(1):45–52
88. Cahalan MD, Zhang SL, Yeromin AV, Ohlsen K, Roos J, Stauderman KA (2007) Molecular basis of the CRAC channel. Cell Calcium 42(2):133–144
89. Frischauf I, Muik M, Derler I, Bergsmann J, Fahrner M, Schindl R, Groschner K, Romanin C (2009) Molecular determinants of the coupling between STIM1 and Orai channels: differential activation of Orai1-3 channels by a STIM1 coiled-coil mutant. J Biol Chem 284(32):21696–21706
90. Hou X, Pedi L, Diver MM, Long SB (2012) Crystal structure of the calcium release-activated calcium channel Orai. Science 338(6112):1308–1313
91. Derler I, Fahrner M, Muik M, Lackner B, Schindl R, Groschner K, Romanin C (2009) A Ca2+ release-activated Ca2+ (CRAC) modulatory domain (CMD) within STIM1 mediates fast Ca2+-dependent inactivation of ORAI1 channels. J Biol Chem 284(37):24933–24938
92. Mullins FM, Park CY, Dolmetsch RE, Lewis RS (2009) STIM1 and calmodulin interact with Orai1 to induce Ca2+-dependent inactivation of CRAC channels. Proc Natl Acad Sci U S A 106(36):15495–15500
93. Li J, Sukumar P, Milligan CJ, Kumar B, Ma ZY, Munsch CM, Jiang LH, Porter KE, Beech DJ (2008) Interactions, functions, and independence of plasma membrane STIM1 and TRPC1 in vascular smooth muscle cells. Circ Res 103(8):e97–e104
94. Zweifach A, Lewis RS (1995) Rapid inactivation of depletion-activated calcium current (ICRAC) due to local calcium feedback. J Gen Physiol 105(2):209–226
95. McNally BA, Yamashita M, Engh A, Prakriya M (2009) Structural determinants of ion permeation in CRAC channels. Proc Natl Acad Sci U S A 106(52):22516–22521
96. Zhou Y, Ramachandran S, Oh-Hora M, Rao A, Hogan PG (2010) Pore architecture of the ORAI1 store-operated calcium channel. Proc Natl Acad Sci U S A 107(11):4896–4901
97. Mignen O, Thompson JL, Shuttleworth TJ (2008) Orai1 subunit stoichiometry of the mammalian CRAC channel pore. J Physiol 586(2):419–425
98. Ji W, Xu P, Li Z, Lu J, Liu L, Zhan Y, Chen Y, Hille B, Xu T, Chen L (2008) Functional stoichiometry of the unitary calcium-release-activated calcium channel. Proc Natl Acad Sci U S A 105(36):13668–13673
99. Penna A, Demuro A, Yeromin AV, Zhang SL, Safrina O, Parker I, Cahalan MD (2008) The CRAC channel consists of a tetramer formed by Stim-induced dimerization of Orai dimers. Nature 456:116–120
100. Madl J, Weghuber J, Fritsch R, Derler I, Fahrner M, Frischauf I, Lackner B, Romanin C, Schutz GJ (2010) Resting state Orai1 diffuses as homotetramer in the plasma membrane of live mammalian cells. J Biol Chem 285(52):41135–41142
101. Maruyama Y, Ogura T, Mio K, Kato K, Kaneko T, Kiyonaka S, Mori Y, Sato C (2009) Tetrameric Orai1 is a teardrop-shaped molecule with a long, tapered cytoplasmic domain. J Biol Chem 284:13676–13685
102. Demuro A, Penna A, Safrina O, Yeromin AV, Amcheslavsky A, Cahalan MD, Parker I (2011) Subunit stoichiometry of human Orai1 and Orai3 channels in closed and open states. Proc Natl Acad Sci U S A 108(43):17832–17837
103. Thompson JL, Shuttleworth TJ (2013) How many Orai's does it take to make a CRAC channel? Sci Rep 3:1961
104. Rothberg BS, Wang Y, Gill DL (2013) Orai channel pore properties and gating by STIM: implications from the Orai crystal structure. Sci Signal 6(267):pe9
105. Calloway NT, Holowka DA, Baird BA (2010) A basic sequence in STIM1 promotes Ca2+ influx by interacting with the C-terminal acidic coiled-coil of Orai1. Biochemistry 49:1067–1071
106. Lis A, Zierler S, Peinelt C, Fleig A, Penner R (2010) A single lysine in the N-terminal region of store-operated channels is critical for STIM1-mediated gating. J Gen Physiol 136(6):673–686

107. McNally BA, Somasundaram A, Jairaman A, Yamashita M, Prakriya M (2013) The C- and N-terminal STIM1 binding sites on Orai1 are required for both trapping and gating CRAC channels. J Physiol 591(Pt 11):2833–2850
108. Zheng H, Zhou MH, Hu C, Kuo E, Peng X, Hu J, Kuo L, Zhang SL (2013) Differential roles of the C and N termini of Orai1 protein in interacting with stromal interaction molecule 1 (STIM1) for Ca2+ release-activated Ca2+ (CRAC) channel activation. J Biol Chem 288(16):11263–11272
109. Prakriya M, Feske S, Gwack Y, Srikanth S, Rao A, Hogan PG (2006) Orai1 is an essential pore subunit of the CRAC channel. Nature 443(7108):230–233
110. Vig M, Beck A, Billingsley JM, Lis A, Parvez S, Peinelt C, Koomoa DL, Soboloff J, Gill DL, Fleig A, Kinet JP, Penner R (2006) CRACM1 multimers form the ion-selective pore of the CRAC channel. Curr Biol 16(20):2073–2079
111. Yeromin AV, Zhang SL, Jiang W, Yu Y, Safrina O, Cahalan MD (2006) Molecular identification of the CRAC channel by altered ion selectivity in a mutant of Orai. Nature 443(7108):226–229
112. Srikanth S, Yee MK, Gwack Y, Ribalet B (2011) The third transmembrane segment of orai1 protein modulates Ca2+ release-activated Ca2+ (CRAC) channel gating and permeation properties. J Biol Chem 286(40):35318–35328
113. Yamashita M, Prakriya M (2014) Divergence of Ca(2+) selectivity and equilibrium Ca(2+) blockade in a Ca(2+) release-activated Ca(2+) channel. J Gen Physiol 143(3):325–343
114. McNally BA, Somasundaram A, Yamashita M, Prakriya M (2012) Gated regulation of CRAC channel ion selectivity by STIM1. Nature 482:241–245
115. Gudlur A, Quintana A, Zhou Y, Hirve N, Mahapatra S, Hogan PG (2014) STIM1 triggers a gating rearrangement at the extracellular mouth of the ORAI1 channel. Nat Commun 5:5164
116. Zhang SL, Yeromin AV, Hu J, Amcheslavsky A, Zheng H, Cahalan MD (2011) Mutations in Orai1 transmembrane segment 1 cause STIM1-independent activation of Orai1 channels at glycine 98 and channel closure at arginine 91. Proc Natl Acad Sci U S A 108(43):17838–17843
117. Yamashita M, Navarro-Borelly L, McNally BA, Prakriya M (2007) Orai1 mutations alter ion permeation and Ca2+-dependent fast inactivation of CRAC channels: evidence for coupling of permeation and gating. J Gen Physiol 130(5):525–540
118. Amcheslavsky A, Safrina O, Cahalan MD (2013) Orai3 TM3 point mutation G158C alters kinetics of 2-APB-induced gating by disulfide bridge formation with TM2 C101. J Gen Physiol 142(4):405–412
119. Derler I, Plenk P, Fahrner M, Muik M, Jardin I, Schindl R, Gruber HJ, Groschner K, Romanin C (2013) The extended transmembrane Orai1 N-terminal (ETON) region combines binding interface and gate for Orai1 activation by STIM1. J Biol Chem 288(40):29025–29034
120. Li M, Du J, Jiang J, Ratzan W, Su L-T, Runnels LW, Yue L (2007) Molecular determinants of Mg2+ and Ca2+ permeability and pH sensitivity in TRPM6 and TRPM7. J Biol Chem 282(35):25817–25830.
121. Calloway N, Vig M, Kinet JP, Holowka D, Baird B (2009) Molecular clustering of STIM1 with Orai1/CRACM1 at the plasma membrane depends dynamically on depletion of Ca2+ stores and on electrostatic interactions. Mol Biol Cell 20(1):389–399
122. Stathopulos PB, Schindl R, Fahrner M, Zheng L, Gasmi-Seabrook GM, Muik M, Romanin C, Ikura M (2013) STIM1/Orai1 coiled-coil interplay in the regulation of store-operated calcium entry. Nat Commun 4:2963
123. Bergsmann J, Derler I, Muik M, Frischauf I, Fahrner M, Pollheimer P, Schwarzinger C, Gruber HJ, Groschner K, Romanin C (2011) Molecular determinants within N terminus of Orai3 protein that control channel activation and gating. J Biol Chem 286(36):31565–31575

Chapter 3
The TRPCs, Orais and STIMs in ER/PM Junctions

Dong Min Shin, Aran Son, Seonghee Park, Min Seuk Kim, Malini Ahuja, and Shmuel Muallem

Abstract The Ca^{2+} second messenger is initiated at ER/PM junctions and propagates into the cell interior to convey the receptor information. The signal is maintained by Ca^{2+} influx across the plasma membrane through the Orai and TRPC channels. These Ca^{2+} influx channels form complexes at ER/PM junctions with the ER Ca^{2+} sensor STIM1, which activates the channels. The function of STIM1 is modulated by other STIM isoforms like STIM1L, STIM2 and STIM2.1/STIM2β and by SARAF, which mediates the Ca^{2+}-dependent inhibition of Orai channels. The ER/PM junctions are formed at membrane contact sites by tethering proteins that generate several types of ER/PM junctions, such as $PI(4,5)P_2$-poor and $PI(4,5)P_2$-rich domains. This chapter discusses several properties of the TRPC channels, the Orai channels and the STIMs, their key interacting proteins and how interaction of the STIMs with the channels gates their activity. The chapter closes by highlighting open questions and potential future directions in this field.

Keywords TRPC channels • STIM1 • STIM2 • Orai • ER/PM junction

D.M. Shin (✉)
Department of Oral Biology, BK 21 PLUS Project, Yonsei University College of Dentistry, Seoul 120-752, South Korea
e-mail: dmshin@yuhs.ac

A. Son • M. Ahuja • S. Muallem (✉)
Epithelial Signaling and Transport Section, Molecular Physiology and Therapeutics Branch, National Institute of Dental and Craniofacial Research, National Institute of Health, Bethesda, MD 20892, USA
e-mail: Shmuel.muallem@nih.gov

S. Park
Department of Physiology, School of Medicine, EwhaWomans University, 911-1 Mok-6-dong, Yang Chun-gu, Seoul 158-710, South Korea

M.S. Kim
Department of Oral Physiology, School of Dentistry, Wonkwang University, Iksan City, Jeonbuk, South Korea

J.A. Rosado (ed.), *Calcium Entry Pathways in Non-excitable Cells*, Advances in Experimental Medicine and Biology 898, DOI 10.1007/978-3-319-26974-0_3

3.1 Introduction

Ca^{2+} influx across the plasma membrane (PM) is a central component of the receptor-evoked Ca^{2+} signal. Receptor-stimulated Ca^{2+} influx is activated shortly after and in response to IP_3-mediated Ca^{2+} release from the endoplasmic reticulum (ER) [1] and Ca^{2+} release from acidic compartments [2]. Stored Ca^{2+} is limited and cannot sustain the receptor-evoked Ca^{2+} signal without constant replenishment by Ca^{2+} influx from the vast extracellular pool, both during the physiological Ca^{2+} oscillations and at the end of the stimulation period [3, 4]. As such, Ca^{2+} influx across the PM participates in all Ca^{2+}-mediated cellular functions [4, 5]. At the same time excessive Ca^{2+} influx by the PM channels is highly toxic and causes most of the pathology of Ca^{2+}-related diseases [6, 7].

The receptor-stimulated Ca^{2+} influx is mediated by two families of PM channels, the TRPC and Orai channels [6]. It is now firmly established that the ER Ca^{2+} sensor STIM1 gates these channels, although there are significant differences in the way STIM1 gates the TRPC and Orai1 channels. The properties, regulation and role of the TRPC [8–10] and Orai channels [11, 12] and their interaction with and gating by STIM1 have been extensively discussed and reviewed [13]. In this chapter, we briefly summarize the key features of this topic. Rather, we focus on new information on the regulation of Ca^{2+} influx channels emerging in the last few years and how it may guide future studies of the TRPC channels and their interaction and regulation by STIM1 and STIM1-associated proteins and regulators.

3.2 The SOCs Components

3.2.1 The TRPC Channels

The TRPC channels were discovered as the homologues of the drosophila TRP channel that mediate at least part of the receptor-stimulated Ca^{2+} influx [14, 15]. Their discovery lead to the discovery of all 28 members of the TRP channels superfamily and their role in incredibly diverse physiological roles [16]. Like all other TRP channels, TRPC channels are tetrameric, polymodal channels with each subunit having six transmembrane domains. The TRPCs function as non-selective cation permeable channels with significant Ca^{2+} permeability. No information is available on the structure of the TRPC channels, but recent work used electron cryo-microscopy to report the nearly atomic resolution of the structure of the TRPV1 [17, 18] and TRPA1 channels [19] that highlight several general features applicable to other TRP channels. The channels display tetrameric architecture with the subunits arranged in fourfold symmetry around a central ion permeation path that is formed by transmembrane domains S5 and S6 and the pore helix. The C terminus TRP domain that is found in most TRP channels interacts with both the S4–S5 linker and an N terminus helix located prior to transmembrane domain S1

[17]. In TRPC channels the TRP domain is followed by a coiled-coil domain. Interestingly, many channel modulators and activators interact with or in the region between these domains (see [20] and below). The importance of this region and interaction of the TRP domain with the S4–S5 linker and pre-S1 in TRP channels gating is suggested by the gain of function mutations in TRPV3 and TRPV4 that may disrupt these interactions and lead to Olmsted syndrome [21] and skeletal dysplasia syndromes [22], respectively.

The TRPV1 pore has two, upper and lower gates that restrict conduction. The channel can be in conformations with one or both gates open [18] to allow dynamic regulation of channel conductance with the pore helix coupling the upper and lower gates and perhaps their closing and opening. The dynamic nature of the TRPV1 pore is also suggested by pore dilation in response to agonist stimulation [23]. Similar two gates arrangement may exist in other channels, including TRPML3 [24] and TRPC4/TRPC5 [25], that also undergo pore dilation. Pore opening also involves major conformational changes in the S3–S4 helices, S4–S5 linker and S6 helix [18]. The effect of interaction of TRP domain with S4–S5 linker on these conformational changes remains to be determined.

TRPC Channel Interactors TRPC channels have several domains that interact with other proteins, which keep the channels in defined membrane compartments and regulate channel activities. Comprehensive list of TRPC channels interactors can be found in http://www.trpchannel.org/. Here, we only discuss several interactors pertinent to the topic of this review. All TRPC channels except TRPC7 interact with the scaffolding/adaptor Homer proteins [26]. The Homers have an N terminus EVH domain that mediates protein-protein interaction, and a C-terminus coiled-coil domain that serves to multimerize the Homers into large scaffolds. The EVH domain interacts with proteins having the ligands PPXXF, PPXF and/or LPSSP [27, 28]. The Homers bind to the Homer binding ligands in the N and C terminus of the TRPC channels to ligand the PM TRPC channels with the IP_3 receptors at ER/PM junctions and keep the channels in a close state [26, 29].

An interesting set of TRPC channels interactors are Caveolin, Junctate and Junctophilin. Caveolin interacts with many TRPC channels to recruit and stabilize them in PM lipid rafts [30–32]. This interaction is essential for assembly of TRPC1 into complexes within ER/PM junctions to mediate key function of the channels that are both cell specific and channel specific. This is exemplified well in the action of TRPC1 is salivary glands. Ca^{2+} influx by TRPC1 specifically stimulates salivary glands fluid secretion [33] and gene activation [34]. Other examples include the roles of TRPC1 and TRPC4 in endothelial cells function [35], control of myotube size [36] and axonal guidance [37]. Junctate is a widely expressed integral ER/SR Ca^{2+} binding protein that affects Ca^{2+} signaling in both muscle and non-muscle cells [38]. Notably, junctate expression increases and knockdown decreases the ER/PM junctions [39]. Junctate forms a signaling complex with the IP_3Rs and TRPC3 [39] to regulate both the IP_3-mediated Ca^{2+} release and TRPC3-mediated Ca^{2+} influx. Interestingly, expression of TRPC3 with junctate increases the number and size of the ER/PM junctions [40]. Other TRPC channels shown to interact with Junctate are

TRPC2 and TRPC5, all of which localize in the acrosomal crescent of the sperm head [41]. Junctate also interacts with and regulate Orai1 [42]. This topic is addressed further below.

The four Junctophilins (JPs) are also present at ER/PM junctions, but their expression appears to be more restricted, with JP1 and JP2 found mainly in muscles and JP3 and JP4 in brain, although JP3 and JP4 are likely expressed in other tissues as well [43, 44]. The JPs participate in the formation of ER/PM junctions with the eight conserved C terminus JP-specific MORN motifs tightly interacting with $PI(4,5)P_2$ at the PM [44]. Knockout of JP1 and JP2 in mice disrupts the ER/PM junctions in muscles and the crosstalk between PM Ca^{2+} channels and the ER/SR Ryanodine receptors and E-C coupling. Accordingly, mutations in these JPs are associated with several cardiac and other muscle myopathies in humans [43, 44]. Individual deletion of JP3 and JP4 in mice had no obvious phenotype, but deletion of both JP3 and LP4 inhibited salivary secretion and neuronal control of feeding [45], suggesting a role of these JPs in peripheral tissues.

3.2.2 The Orai Channels

The Orai channels were discovered as the proteins that mediate the highly Ca^{2+} selective Ca^{2+} release activated current (CRAC) [46–48]. Numerous studies used mutations analysis with biochemical, biophysical and electrophysiological studies to determine several key features of the Orai channels and these have been extensively reviewed (for several latest reviews see [11, 12, 49]). Each subunit of the Orai channels has four transmembrane spans with cytoplasmic N and C termini. Understanding the Orai structure-function was dramatically increased with the description of the crystal structure of drosophila Orai channel that is homologous with mammalian Orai1 [50]. The structure revealed that the channel is a hexamer, rather than a tetramer as suggested by biochemical and single molecule photobleaching studies [51–53]. The crystal structure revealed several key features of the Orai channels. It confirmed that the pore is formed by the first transmembrane domain [54] and further showed that the selectivity filter is composed of a ring of glutamates [50]. Proline residue in the middle of the fourth transmembrane domain caused a bend resulting in part of the domain lining perpendicular to the membrane. As a result, the fourth transmembrane domains cytoplasmic extensions of each of two subunits fold into coiled-coil domains. Part of this coiled-coil domain interacts with a STIM1 SOAR dimer to participate in channel opening [55, 56] that is driven by STIM1-mediated conformation transition [57].

Orai Channels Interactors The Orai channels, in particular Orai1, interact with several scaffolding and regulatory proteins. A particularly interesting interaction is heterodimerization of Orai1 with Orai3 that generates STIM1-regulated, but store-independent channel that is activated by receptor-generated arachidonic acid [58–60]. Large number of studies reported that the Orai channels are present in complexes with TRPC channels. However, these interactions are indirect and are largely mediated by STIM1 [9, 13, 30]. A protein that binds both TRPC (see above) and Orai1 channels is caveolin, which interacts with the N terminus of Orai1 to mediate Orai1 endocytosis and recycling [61]. Caveolin is also required for recruitment of the Orai1-STIM1 complex to a $PI(4,5)P_2$-rich domain [62]. The Orai1 cytoplasmic N terminal domain binds calmodulin [63, 64], which was suggested to mediate the fast Ca^{2+}-dependent inactivation (FCDI) of Orai1 [65]. However, the discovery that SARAF mediates the SCDI of Orai1 [66], and that SARAF interacts with the STIM1 SOAR domain [67] open the question of the role of calmodulin in the regulation of Orai1 function. Another protein containing EF-hand that interacts with the N terminus of Orai1 is CRACR2A (for CRAC Regulator 2A) [68]. How CRACR2A regulates channel activity remains to be determined.

3.2.3 The STIMs

A search for the factor that couples Ca^{2+} release from the ER to activation of Ca^{2+} influx channels led to the discovery of ER Ca^{2+} content sensor STIM1 [69, 70]. The original studies noted the presence of STIM2, but its role in Ca^{2+} signaling remains significantly less understood as compared to STIM1. The potential functions of STIM2 are discussed in some details at the end of this review. Extensive deletions and mutations analysis over the last 10 years defined several STIM1 domains, while STIM2 domains are largely inferred based on their homology with STIM1. The STIMs have a single transmembrane domain with N terminus in the ER lumen and C terminus in the cytosol. The STIMs ER lumen has Ca^{2+} binding EF hand and SAM domain. The STIM1 and STIM2 EF-hands are the Ca^{2+} sensors with the STIM1 EF-hand binding Ca^{2+} at higher apparent affinity than the STIM2 EF-hand. The SAM domain participates in STIM1 clustering [71, 72]. Binding of Ca^{2+} to the EF hand keeps STIM1 in a non-clustered form and away from the ER/PM junctions.

The cytosolic domain of STIM1 opens both the TRPC and Orai channels [73]. It starts with the first coiled-coil domain (CC1) that is divided to several sub coiled-coil domains [74] and includes a C terminus short helix that functions as an inhibitory helix [75, 76]. CC1 is followed by the SOAR domain [77], which is also known as CAD [63] or CCb9 [78]. SOAR is the minimal STIM1 domain needed to fully activate Orai1 [63, 77, 78]. The crystal structure of SOAR indicates R-shaped monomers that assemble into a V-shaped dimer [76]. STIM2 splice variants at the tip of the SOAR R curve generates two STIM2 isoforms that can either inhibit or facilitate Orai1 activation [79, 80]. STIM1 SOAR is followed by a C Terminus Inhibitory Domain (CTID) [67] that is followed by a long sequence that may participate in occluding SOAR in a close state [75]. The most C terminal STIM1 and STIM2 sequences are polybasic with multiple lysine residues (K-domains). The STIM1 and likely STIM2 K-domains interact with $PI(4,5)P_2$ at the ER/PM junctions [62, 81] to determine gating of Orai1 by Ca^{2+} [62]. Activation of the Orai and

TRPC channels by STIM1 involves major conformational changes to unocclude the SOAR domain. When STIM1 in not at the ER/PM junctions the SOAR domain is occluded in a sandwich formed by part of CC1 at the N terminus of SOAR and by CTID at the C terminus of SOAR. Ca^{2+} store depletion unwinds STIM1 to an extended conformation to release SOAR and allows its interaction with and activation of the Orai and TRPC channels [67, 75, 82, 83].

STIM Interactors The function of the STIMs is regulated by Ca^{2+} binding and unbinding to the ER hand and by interaction of the open STIMs with plasma membrane $PI(4,5)P_2$. STIM1 also interacts with other proteins that stabilize the open or closed conformations or mediate Ca^{2+}-independent STIM1-dependent activity. An example of the later is interaction of STIM1 with the microtubule-plus-end-tracking protein end binding (EB1) [84]. Interaction of STIM1 with EB1 occurs at the sites of microtubules interaction with the ER and is regulated by phosphorylation of STIM1 in serine/threonine residues [85] and by ERK1/2-mediated phosphorylation at S575, S608 and S621 [86]. The mode of interaction of STIM1 with EB1 that does not affect Ca^{2+} influx appears to participate in ER remodeling [84, 85], while phosphorylation by ERK1/2 leads to dissociation of EB1 to facilitate STIM1 clustering and activation of Ca^{2+} influx channels [86]. The EB1 interaction with STIM1 appears to be regulated by the microtubule binding protein adenomatous polyposis coli (APC) [87]. APC interacts with the STIM1 C terminus (STIM1(650–685)) to dissociate the EB1-STIM1 complex, promotes STIM1 clustering and activation of Orai1 [87]. Another STIM1 function independent of Ca^{2+} influx is the regulation of endothelial barrier function by thrombin stimulated STIM1-mediated activation of RhoA, formation of actin stress fibers and loss of cell-cell adhesion [88].

An interesting regulation of STIM1 is by the ER redox potential. It was reported that oxidative stress leads to STIM1 S-glutathionylation at C56 to promote STIM1 translocation to ER/PM junctions and activation of Orai1 [89]. On the other hand, another study reported that the ER-resident oxidoreductase ERp57 interacts with the ER luminal domain of STIM1 to modify C49 and C56 and inhibits Ca^{2+} influx by reducing STIM1-Orai1 interaction. Mutation of C49 and C56 did not result in spontaneous activation of STIM1 and Ca^{2+} influx [90] as would be predicted by the findings of [89]. Regulation of Ca^{2+} influx by ER redox state has many physiological implications and thus this issue deserves clarifications and further in depth exploration.

A negative regulator of Ca^{2+} influx that interacts with STIM1 is SARAF [66]. SARAF is a single transmembrane span, ER resident protein with the N terminus in the ER lumen and the C terminus in the cytoplasm. A prominent regulation of Ca^{2+} influx by Orai1 is Ca^{2+} itself, which causes two form of inhibition, fast (FCDI) and slow Ca^{2+}-dependent inactivation (SCDI) with time courses of msec and min [4]. Before the discovery of SARAF the only information on the mechanism for Ca^{2+}-dependent inactivation was that a negatively charged STIM1 sequence (^{475}DDVDDMDEE^{483}) mediates the FCDI [65, 91, 92]. However, this may change with the discovery that SARAF mediates SCDI [66] and likely FCDI [62]. Systematic domain analysis showed that SARAF interacts with SOAR and the interaction is regulated by the STIM1 CTID domain [67]. Furthermore, CTID appears to have two lobes, an N terminal lobe (STIM1(447–490)) that restricts access of SARAF to SOAR and a C terminal lobe (STIM1(491–530)) that facilitate interaction of SARAF with SOAR [67]. Significantly, CTID also regulates FCDI of Orai1 since inhibition of SARAF interaction with STIM1 prevents FCDI [67]. Notably, interaction of SARAF with STIM1 take place only when the STIM1-Orai1 complex is in a ER/PM junctions $PI(4,5)P_2$-rich domain [62].

3.3 Gating by STIMs

3.3.1 Gating TRPC Channels by STIM1

Comprehensive reviews of this topic can be found in [9, 10, 13, 30]. A large body of biochemical and functional evidence indicates direct or indirect interaction between TRPC channels and STIM1 in many cell types. However, unlike the Orai channels, the TRPC channels can function both in a STIM1-dependent and STIM1-independent modes and some TRPC channels are gated by STIM1 only when they heteromultemerize with other TRPC channels that directly bind STIM1. This contributed to the confusion and the misunderstanding of how STIM1 gates the TRPC channels. For example, TRPC1, TRPC2, TRPC4 and TRPC5 directly interact with STIM1, while TRPC3, TRPC6 and TRPC7 do not [73]. The STIM1 SOAR domain mediated the interaction with TRPC channels and similar pattern is observed when interaction of TRPC channels with SOAR is examined [93]. However, when TRPC1-TRPC3 [94, 95] and TRPC4-TRPC6 heteromultimerize TRPC3 and TRPC6 gain access to STIM1 and their function changes from STIM1-independent to STIM1-dependent [94]. Functioning of the channels as a heterodimers also occurs in vivo. This is shown in Fig. 3.1 in pancreatic acini taken from $Trpc1^{-/-}$ $Trpc3^{-/-}$ mice [93]. Knockout of TRPC1 or TRPC3 in mice and inhibition of TRPC3 by Pyrazole 10 reduced receptor-stimulated and store depleted Ca^{2+} influx by about 50 % (Fig. 3.1a, b) and inhibition of TRPC3 in $TRPC1^{-/-}$ acini has no further effect. This indicates that the two TRPC channels cooperate to mediate the same Ca^{2+} influx pathway and Ca^{2+} influx by the channels is interdependent. For this the channels and STIM1 have to be present in the same Ca^{2+} signaling complex. This is indeed shown in Fig. 3.1d, which shows that cell stimulation markedly enhances the interaction between STIM1, TRPC1 and TRPC3.

Exploring the molecular mechanism of turning the TRPC channels from STIM1-independent to STIM1-dependent channels revealed that TRPC3 does have a STIM1/SOAR interacting domain within its C terminus coiled-coil domain (CCD). However, this site is hidden by the N terminus CCD that interacts with the C terminus CCD to mask the STIM1/SOAR binding site. Interaction of the TRPC1 CCD with the TRPC3 N terminus CCD exposed the TRPC3 STIM1/SOAR interaction site at the C terminus CCD. In this open conformation the function of TRPC3 is

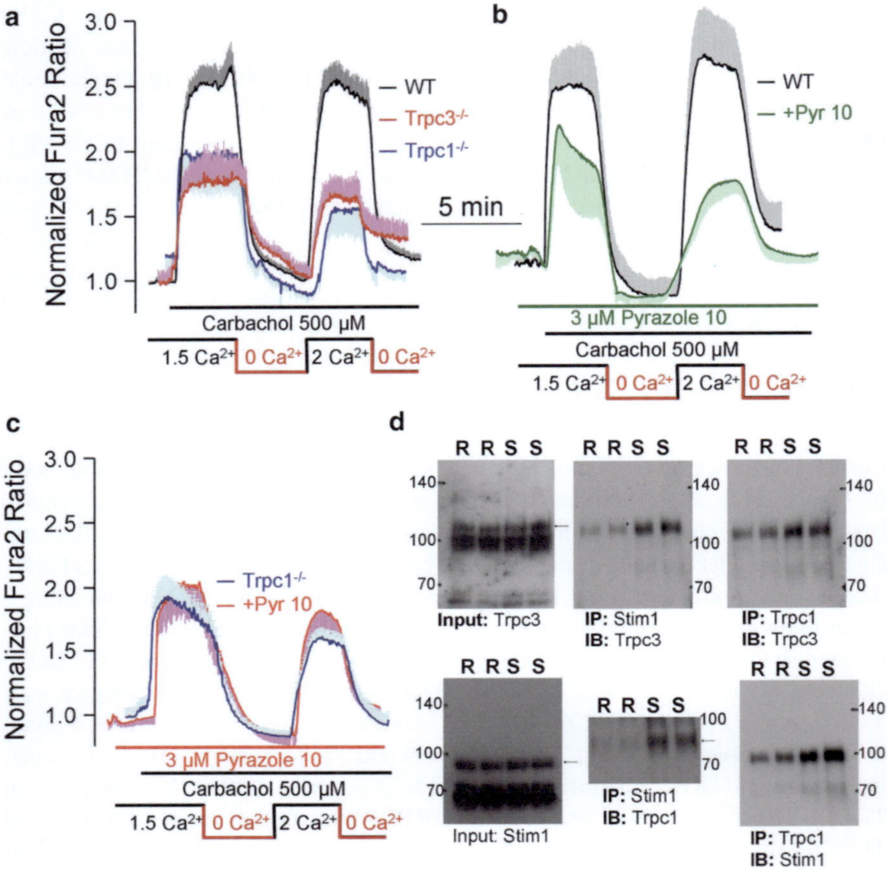

Fig. 3.1 The TRPC1-TRPC3 heterodimer acts to mediate part of the SOCs in vivo. Ca^{2+} was measured in Fura2-loaded salivary gland ducts isolated from wild-type, $Trpc1^{-/-}$ and $Trpc3^{-/-}$ mice. The ducts were perfused with solutions containing 1.5 mM Ca^{2+} or 0.2 mM EGTA (Ca^{2+} free conditions) as indicated by the *bars*. **b**, **c** Ca^{2+} was measured in salivary gland ducts from wild-type **b** and $Trpc1^{-/-}$ mice **c** treated with 3 μM of the Trpc3 inhibitor Pyrazole 10 (see [132]). Note that TRPC1 and TRPC3 mediate the same Ca^{2+} influx pathway and both are needed to keep the pathway active. **d** Mutual co-immunoprecipitation (Co-IP) of Trpc1/Stim1/Trpc3 in resting and stimulated salivary gland cells. Co-IP was tested in resting cells (marked with *R*) and cells stimulated with 100 μM carbachol and 25 μM CPA (marked with *S*) (The results were reproduced from [93])

strictly STIM1-dependent [93]. The two modes of the activity of TRPC1 and TRPC3 channels with respect to STIM1 dependence and independence and how these modes are formed is illustrated in Fig. 3.2.

The STIM1 SOAR domain mediates interaction with the TRPC channels CCDs but the STIM1 polybasic K-domain is essential for the gating [96]. The K-domain is essential for recruiting the STIM1-Orai1 complex to a PI(4,5)P_2-rich ER/PM junctions [62]. It may have the same function in TRPC channels (see below-open questions), but in the case of TRPC channels the K-domain appears to have

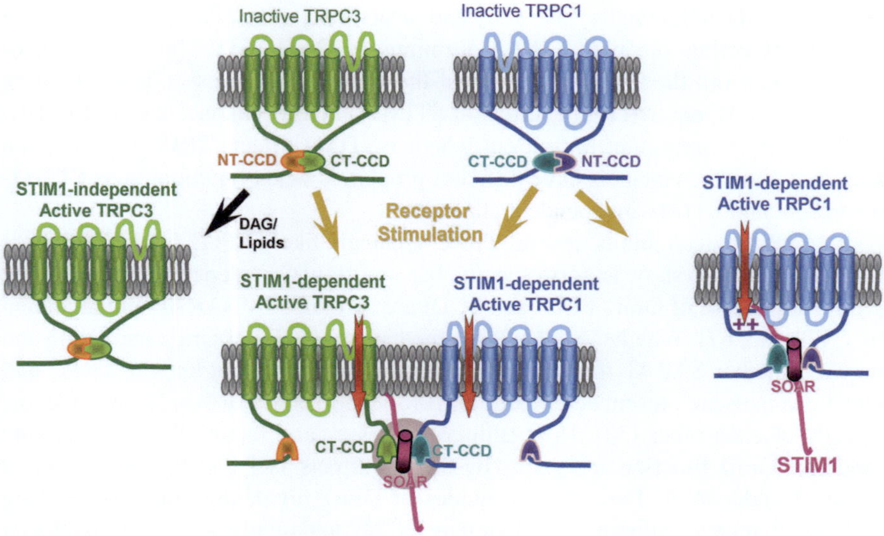

Fig. 3.2 Modes of STIM1-dependent and STIM1-independent functions of TRPC1 and TRPC3. In the inactive state the N and C termini CCDs interact to keep the channel in the inactive state. In cell that express TRPC3 and not TRPC1, TRPC3 functions as a STIM1-independent channel that can be activated by diacylglycerol, other lipids or activators. TRPC1 C terminal CCD has a SOAR binding site and upon cell stimulation and store depletion, this site becomes available for interaction of STIM1 to allow the STIM1 K-domain to facilitate channel activation. In cells expressing TRPC3 and TRPC1, cell stimulation results in formation of TRPC1-STIM1-TRPC3 complexes (see Fig. 3.1). In the complexes, the TRPC1 C terminus CCD interacts with the TRPC3 C terminus CCD to dissociates the interaction between the N and C terminus CCDs of TRPC3, allow binding of SOAR to the TRPC3 C terminus CCD and channel activation. In this mode TRPC3 functions as STIM1-dependent channel

additional function since deletion of the STIM1 K-domain prevents activation of TRPC1 by STIM1 [73] but not activation of Orai1 by STIM1 [62, 96]. The last two lysines (K^{684} and K^{685}) of STIM1 communicate with two negatively charged conserved residues (DD/DE/EE) in the C terminus of TRPC channels [29, 96]. The interaction appears to be electrostatic rather than affecting the channel pore since several combinations of negative-positive charges irrespective of the presence on TRPC channels or STIM1 resulted in active channel [96]. Gating by electrostatic interaction could be observed with TRPC3, TRPC4, TRPC5 and TRPC6 [29].

3.3.2 Gating Orai Channels by STIM1

This topic is reviewed elsewhere in this book and in [11, 12, 97, 98] and is only briefly mentioned here to highlight several points. The first is that physical and functional CCDs interactions of SOAR and the C terminal CCDs of TRPC channels [93] has some similarities with the interaction between STIM1 SOAR and Orai1 C terminus CCD functionally [99, 100] and structurally [49, 50, 74]. SOAR has a second interacting site within the N terminus of Orai1 [63]. The interaction of STIM1 with both the C and N termini of the Orai1 participates in channel gating [56, 101] and is required for the effect of STIM1 on Orai1 channel selectivity [101, 102]. It will be important to find out whether STIM1 affects TRPC channels ion selectivity and the channels have different properties when functioning as STIM1-dependent and STIM1-independent channels.

Ca^{2+} inactivates/inhibits several TRPC channels like TRPC3 [103]. The three Orai channels also show isoform specific fast and slow Ca^{2+}-dependent inactivation [104]. In the case of Orai1 FCDI and SCDI are mediated by SARAF and regulated by CTID [66, 67]. Whether SARAF also mediate the FCDI of Orai2 and Orai3 and how or whether SARAF affects TRPC channels activity is not known. TRPC and Orai1 channels are present at the same ER/PM junctions where they likely affect the activity of each other [30]. How cellular processes and factors that regulate and modulate Orai1 function and gating by STIM1 affects TRPC channels function is not well understood. Two of these modes of Orai1 regulations that are drawing attention in recent studies are regulation by STIM2 and regulation by localization at specific ER/PM junctions. These are discussed below with respect to questions and future directions in research of TRPC channels as seen by the authors.

3.4 Open Questions and Future Directions

3.4.1 STIM1L Function in Orai and TRPC Channels

STIM1 has two isoforms, the one usually refer to as STIM1 and a long isoform STIM1L that has an additional C terminus actin binding domain of 105 residues [105]. The differential/complementary roles of the two STIM1 isoform have been examined to a limited extent with Orai1 and not at all with respect to TRPC channels. Moreover, the results of the regulation of Orai1 by STIM1L are not consistent. The original work reported that STIM1L preform clusters with Orai1 at the ER/PM junctions allowing robust and very rapid activation of Ca^{2+} influx by store depletion and enhances Ca^{2+} influx more than STIM1. Significantly, repeated activation of myotubes by depolarization with KCl resulted in rundown of the signal in myotubes with knockdown of STIM1L but not with knockdown of STIM1 [105]. Enhancement of Ca^{2+} influx by STIM1L and its interaction with Orai1 was subsequently observed in transfected HEK cells [106]. A more recent study confirmed enhanced pre-clustering of and enhanced Ca^{2+} influx by STIM1L, however, these authors concluded these clusters do not reflect co-clusters of STIM1L with Orai1. The puncta formed by STIM1L appear smaller than those formed by STIM1 and their size does not increase upon store depletion [107]. Finally the recent study concluded that STIM1L slows down Orai1 recruitment to the ER/PM junctions and the rate of activation of Ca^{2+} influx [107], the opposite of the findings of using virtually the same assay and approach [105].

There is no obvious explanation for the opposite observations with respect to activation of Orai1 and Ca^{2+} influx by STIM1L indicated above. In addition, both studies used Mn^{2+} quench rather than current measurements, so that the exact effect of STIM1L on Orai1 current properties is not known at this time. This is particularly important since STIM1 regulates Orai1 selectivity [102]. Although it was reported that STIM1L interacts with TRPC3 and TRPC6 [106], STIM1L interaction with other TRPC channels is not known, let alone the effect of STIM1L on TRPC channels activation and activity. The role of STIM1L in Ca^{2+} signaling and activation/regulation of the Orai1 and TRPC channels deserves further and thorough examination.

3.4.2 STIM2 Function in Orai and TRPC Channels

Another STIM isoform is STIM2. Originally, it was suggested that the primary role of STIM2 is to set resting Ca^{2+} levels [108]. However, the role of STIM2 in Ca^{2+} signaling is more complex and recent work started to explore the role of STIM2 in Ca^{2+} signaling obtaining very interesting results. STIM2 on its own is not as potent as STIM1 in activation of Orai1 channels or Ca^{2+} influx. However, it has profound effect on Ca^{2+} influx through regulation of STIM1 function. The most interesting finding is that STIM2 controls the recruitment of STIM1 to ER/PM junctions by responding to minimal Ca^{2+} depletion from the ER [109]. Clustering of STIM1 and activation of Ca^{2+} influx in the absence of STIM2 required massive pathological stimulation and ER Ca^{2+} store depletion. STIM2 shifts to the left the dose response to receptor stimulation and store depletion dependent STIM1-Orai1 co-clustering and activation of Ca^{2+} influx. Hence, the main function of STIM2 is to be a better sensor of ER Ca^{2+} than STIM1 and to escort STIM1 to the ER/PM junctions at physiological stimulus intensity [109].

Another mode of regulation of STIM1 function by STIM2 was discovered recently with the identification of two STIM2 splice variants [79, 80]. STIM2.1/STIM2β has a modified SOAR domain that reduces its interaction with Orai1. However, STIM2.1/STIM2β binds strongly to STIM1 and is targeted to the puncta together with STIM1. When in the puncta, STIM2.1/STIM2β strongly inhibits Orai1 activation by STIM1 and Ca^{2+} influx [79, 80]. No information is available on the potential interaction and regulation of TRPC channels by STIM2 or by STIM2.1/STIM2β. However, such a potential regulation is suggested by a recent study reporting that depending on the ration of STIM1/STIM2 in a given cell STIM1 and STIM2 may activate two different channels, possibly Orai1 and TRPC-dependent channels [110]. It is clear that much more work is needed to understand the role of STIM2 in Ca^{2+} signaling, in particular its role in regulation TRPC channels activity.

3.4.3 Orai and TRPC Channels in the ER/PM Junctions

The interaction of STIM1-Orai1, STIM2-Orai1 and TRPC channels-STIM1 to activate Ca^{2+} influx take place at the ER/PM junctions. This topic is being reviewed extensively with several recent reviews on Ca^{2+} signaling in membrane contact sites [12, 30, 98, 111, 112] and on the formation and properties of ER/PM junctions [113–115]. In specialized cells like muscle, neurons and epithelia, the junctions are formed at distinct plasma membrane domains. For example, in polarized acinar cells the Ca^{2+} signal initiates at the apical pole and propagates to the basal pole in the form of Ca^{2+} waves [116, 117]. All Ca^{2+} signaling proteins are clustered at the apical pole including G proteins-coupled receptors, the IP_3Rs, TRPC channels, Orai1, STIM1, PMCA and homer proteins [98]. How the proteins are targeted to these domains is not known at present. However, this problem may be resolved in the near future as proteins that tether, stabilize and regulate the function of the ER/PM junctions are being discovered. One set of proteins are the three Extended synaptotagmins (E-Syts) [118] and their yeast homologues the tricalbins [119]. All three E-Syts can tether the ER and the PM to form ER/PM junctions [120]. Another protein that is at the ER/PM junctions in yeast is Ist2 [121]. The mammalian homologue of Ist2 is not known at this time, but it can be one of the ANO proteins [122]. Septins 4 and 5 were shown to localize at the junctions and regulate $PI(4,5)P_2$ around Orai1 [123], and likely control the size of the ER/PM junctions [124]. Nir2 is an exchange protein that exchanges phosphatidic acid with PI(4)P [125, 126] to maintains plasma membrane $PI(4,5)P_2$ at rest and during stimulation [126, 127]. Junctophilins participate in tethering the ER to the PM [44, 45]. SARAF mediates the slow Ca^{2+}-dependent inactivation of Orai1 only when it is at the $PI(4,5)P_2$-rich ER/PM junctions [62]. A model of the proteins that tether the ER/PM junctions and localization of STIM1, the Orai and TRPC channels at the $PI(4,5)P_2$-poor and $PI(4,5)P_2$-rich junctions is illustrated in Fig. 3.3.

The function of proteins in Ca^{2+} signaling and its related lipids highlight the important of the junctions in the fine-tuning of Ca^{2+} influx. As indicated above, septin 4 and 5 control the $PI(4,5)P_2$ environment around Orai1 [123]. Further studies reveal that the STIM1-Orai1 complex is formed first in a domain that is independent of E-Syt1, Septins and $PI(4,5)P_2$. Then, within a minute of channel activation either $PI(4,5)P_2$ moves to the domain or the Orai1-STIM1 complex moves to an E-Syt1-tethered $PI(4,5)P_2$-rich domain that now allows the recruitment of SARAF and initiation of slow Ca^{2+}-dependent inactivation of Orai1 [62]. Interestingly, the ER/PM junctions formed by STIM1 alone in the absence of Orai1 are different from those formed by E-Syt1 [128]. This supports the notion of more than one ER/PM junctions in which the STIM1-Orai1 complex can exist. Furthermore, the ER/PM junctions formed by the different E-Syts are not the same and show significant functional specificity. For example, the Orai1-STIM1 complex is formed only in E-Syt1 junctions while E-Syt2 and E-Syt3 have no role in controlling the function of STIM1-Orai1 [62]. The ER/PM junctions formed by E-Syt2 mediate EGR receptors

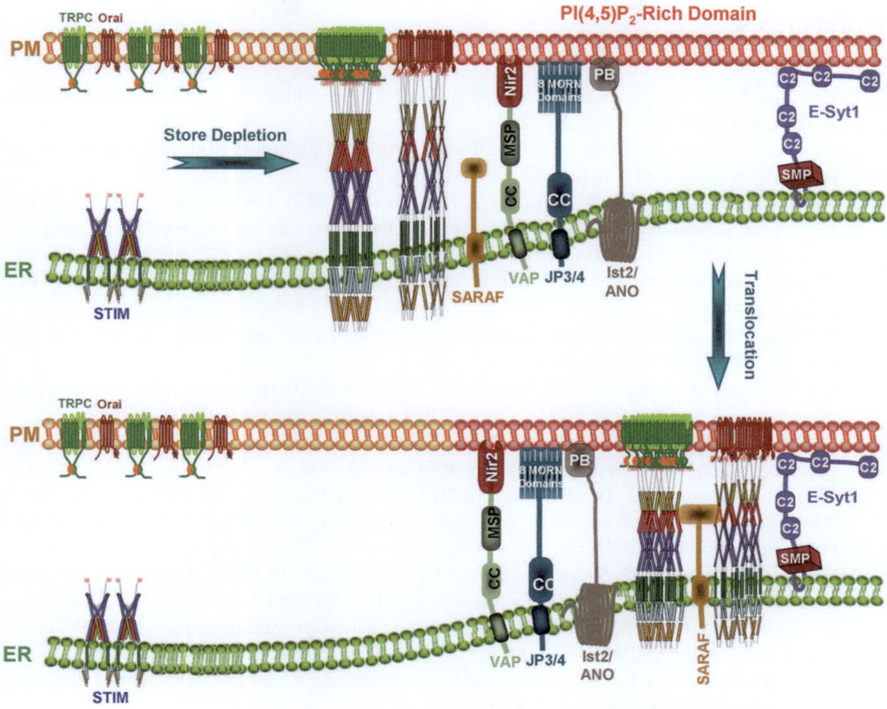

Fig. 3.3 Orai1 and TRPC channels at the ER/PM junctions. The ER/PM junctions are tethered by several proteins that are anchored in the ER and have a phosphatidylinositide (most commonly $PI(4,5)P_2$) binding domain. These include the Extended Synaptotagmins (E-Syt1), the Ist2/ANO, the VAPB-Nir2 complex and likely Junctophilin 3 and or 4 isoforms (JP3/4). At this time only the role of E-Syt1 in regulation of Orai1-STIM1 activity is established. The model shows that in the resting state the distance between the PM and the ER is larger, STIM1 does not access the PM and the TRPC and Orai channels are in a $PI(4,5)P_2$-poor domain and are not clustered (*upper image*). SARAF is in the ER away from STIM1. Upon cell stimulation and store depletion STIM1 unwind and STIM1, Orai1 and the TRPC channels co-cluster at the $PI(4,5)P_2$-poor domain adjacent to the domain formed by the tether proteins and where the channels are fully active but are not inhibited by SARAF. Subsequently the STIM1-Channels complex translocates to the tethers-formed PI(4,5) P_2-rich domain (or $PI(4,5)P_2$ coalesces and the tethers around the channels) (*lower image*), allowing access of SARAF to STIM1 to mediate the Ca^{2+}-dependent inhibition of the channels

endocytosis (and likely other forms of endocytosis) and ERK activation [129]. Similar function is observed with E-Syt3 but not with E-Syt1 [130].

Although it is clear that TRPC channels localize to the ER/PM junctions in native cells [131] and expression systems [30] and this is where they interact with STIM1 and Orai1, we know virtually nothing on the role of the proteins that form and regulate the junctions in modulating TRPC channels activity. Experiments similar to those performed with Orai1-STIM1 are very much needed to address this topic. It is certain that this topic will continue to attract much attention in the coming years to further explore the role of the ER/PM junctions in Ca^{2+} signaling and cell function and in the regulation of both the Orai and TRPC channels.

References

1. Kiselyov K, Shin DM, Muallem S (2003) Signalling specificity in GPCR-dependent Ca2+ signalling. Cell Signal 15(3):243–253
2. Galione A (2015) A primer of NAADP-mediated Ca signalling: from sea urchin eggs to mammalian cells. Cell Calcium 58(1):27–47
3. Pandol SJ, Schoeffield MS, Fimmel CJ, Muallem S (1987) The agonist-sensitive calcium pool in the pancreatic acinar cell. Activation of plasma membrane Ca2+ influx mechanism. J Biol Chem 262(35):16963–16968
4. Parekh AB, Putney JW Jr (2005) Store-operated calcium channels. Physiol Rev 85(2):757–810
5. Berridge MJ (2006) Calcium microdomains: organization and function. Cell Calcium 40(5–6):405–412
6. Lee KP, Yuan JP, Hong JH, So I, Worley PF, Muallem S (2010) An endoplasmic reticulum/plasma membrane junction: STIM1/Orai1/TRPCs. FEBS Lett 584(10):2022–2027
7. Petersen OH, Sutton R, Criddle DN (2006) Failure of calcium microdomain generation and pathological consequences. Cell Calcium 40(5–6):593–600
8. Liao Y, Abramowitz J, Birnbaumer L (2014) The TRPC family of TRP channels: roles inferred (mostly) from knockout mice and relationship to ORAI proteins. Handb Exp Pharmacol 223:1055–1075
9. Smani T, Shapovalov G, Skryma R, Prevarskaya N, Rosado JA (2015) Functional and physiopathological implications of TRP channels. Biochim Biophys Acta 1853(8):1772–1782
10. de Souza LB, Ambudkar IS (2014) Trafficking mechanisms and regulation of TRPC channels. Cell Calcium 56(2):43–50
11. Shim AH, Tirado-Lee L, Prakriya M (2015) Structural and functional mechanisms of CRAC channel regulation. J Mol Biol 427(1):77–93
12. Hogan PG, Rao A (2015) Store-operated calcium entry: mechanisms and modulation. Biochem Biophys Res Commun 460(1):40–49
13. Choi S, Maleth J, Jha A, Lee KP, Kim MS, So I, Ahuja M, Muallem S (2014) The TRPCs-STIM1-Orai interaction. Handb Exp Pharmacol 223:1035–1054
14. Wes PD, Chevesich J, Jeromin A, Rosenberg C, Stetten G, Montell C (1995) TRPC1, a human homolog of a Drosophila store-operated channel. Proc Natl Acad Sci U S A 92(21):9652–9656
15. Zhu X, Chu PB, Peyton M, Birnbaumer L (1995) Molecular cloning of a widely expressed human homologue for the Drosophila trp gene. FEBS Lett 373(3):193–198
16. Gees M, Owsianik G, Nilius B, Voets T (2012) TRP channels. Compr Physiol 2(1):563–608
17. Liao M, Cao E, Julius D, Cheng Y (2013) Structure of the TRPV1 ion channel determined by electron cryo-microscopy. Nature 504(7478):107–112
18. Cao E, Liao M, Cheng Y, Julius D (2013) TRPV1 structures in distinct conformations reveal activation mechanisms. Nature 504(7478):113–118
19. Paulsen CE, Armache JP, Gao Y, Cheng Y, Julius D (2015) Structure of the TRPA1 ion channel suggests regulatory mechanisms. Nature 520(7548):511–517
20. Kiselyov K, Shin DM, Kim JY, Yuan JP, Muallem S (2007) TRPC channels: interacting proteins. Handb Exp Pharmacol 179:559–574
21. Lin Z, Chen Q, Lee M, Cao X, Zhang J, Ma D, Chen L, Hu X, Wang H, Wang X, Zhang P, Liu X, Guan L, Tang Y, Yang H, Tu P, Bu D, Zhu X, Wang K, Li R, Yang Y (2012) Exome sequencing reveals mutations in TRPV3 as a cause of Olmsted syndrome. Am J Hum Genet 90(3):558–564
22. Loukin S, Su Z, Kung C (2011) Increased basal activity is a key determinant in the severity of human skeletal dysplasia caused by TRPV4 mutations. PLoS ONE 6(5):e19533
23. Chung MK, Guler AD, Caterina MJ (2008) TRPV1 shows dynamic ionic selectivity during agonist stimulation. Nat Neurosci 11(5):555–564

24. Kim HJ, Yamaguchi S, Li Q, So I, Muallem S (2010) Properties of the TRPML3 channel pore and its stable expansion by the Varitint-Waddler-causing mutation. J Biol Chem 285(22):16513–16520
25. Jeon JP, Roh SE, Wie J, Kim J, Kim H, Lee KP, Yang D, Jeon JH, Cho NH, Kim IG, Kang DE, Kim HJ, So I (2013) Activation of TRPC4beta by Galphai subunit increases Ca2+ selectivity and controls neurite morphogenesis in cultured hippocampal neuron. Cell Calcium 54(4):307–319
26. Yuan JP, Lee KP, Hong JH, Muallem S (2012) The closing and opening of TRPC channels by Homer1 and STIM1. Acta Physiol 204(2):238–247
27. Beneken J, Tu JC, Xiao B, Nuriya M, Yuan JP, Worley PF, Leahy DJ (2000) Structure of the Homer EVH1 domain-peptide complex reveals a new twist in polyproline recognition. Neuron 26(1):143–154
28. Yuan JP, Kiselyov K, Shin DM, Chen J, Shcheynikov N, Kang SH, Dehoff MH, Schwarz MK, Seeburg PH, Muallem S, Worley PF (2003) Homer binds TRPC family channels and is required for gating of TRPC1 by IP3 receptors. Cell 114(6):777–789
29. Lee KP, Yuan JP, So I, Worley PF, Muallem S (2010) STIM1-dependent and STIM1-independent function of transient receptor potential canonical (TRPC) channels tunes their store-operated mode. J Biol Chem 285(49):38666–38673
30. Ong HL, Ambudkar IS (2015) Molecular determinants of TRPC1 regulation within ER-PM junctions. Cell Calcium 58:376–386
31. Adebiyi A, Narayanan D, Jaggar JH (2011) Caveolin-1 assembles type 1 inositol 1,4,5-trisphosphate receptors and canonical transient receptor potential 3 channels into a functional signaling complex in arterial smooth muscle cells. J Biol Chem 286(6):4341–4348
32. Freichel M, Tsvilovskyy V, Camacho-Londono JE (2014) TRPC4- and TRPC4-containing channels. Handb Exp Pharmacol 222:85–128
33. Pani B, Liu X, Bollimuntha S, Cheng KT, Niesman IR, Zheng C, Achen VR, Patel HH, Ambudkar IS, Singh BB (2013) Impairment of TRPC1-STIM1 channel assembly and AQP5 translocation compromise agonist-stimulated fluid secretion in mice lacking caveolin1. J Cell Sci 126(Pt 2):667–675
34. Ong HL, Jang SI, Ambudkar IS (2012) Distinct contributions of Orai1 and TRPC1 to agonist-induced [Ca(2+)](i) signals determine specificity of Ca(2+)-dependent gene expression. PLoS ONE 7(10):e47146
35. Sundivakkam PC, Freichel M, Singh V, Yuan JP, Vogel SM, Flockerzi V, Malik AB, Tiruppathi C (2012) The Ca(2+) sensor stromal interaction molecule 1 (STIM1) is necessary and sufficient for the store-operated Ca(2+) entry function of transient receptor potential canonical (TRPC) 1 and 4 channels in endothelial cells. Mol Pharmacol 81(4):510–526
36. Antigny F, Koenig S, Bernheim L, Frieden M (2013) During post-natal human myogenesis, normal myotube size requires TRPC1- and TRPC4-mediated Ca(2)(+) entry. J Cell Sci 126(Pt 11):2525–2533
37. Shim S, Zheng JQ, Ming GL (2013) A critical role for STIM1 in filopodial calcium entry and axon guidance. Mol Brain 6:51
38. Treves S, Feriotto G, Moccagatta L, Gambari R, Zorzato F (2000) Molecular cloning, expression, functional characterization, chromosomal localization, and gene structure of junctate, a novel integral calcium binding protein of sarco(endo)plasmic reticulum membrane. J Biol Chem 275(50):39555–39568
39. Treves S, Franzini-Armstrong C, Moccagatta L, Arnoult C, Grasso C, Schrum A, Ducreux S, Zhu MX, Mikoshiba K, Girard T, Smida-Rezgui S, Ronjat M, Zorzato F (2004) Junctate is a key element in calcium entry induced by activation of InsP3 receptors and/or calcium store depletion. J Cell Biol 166(4):537–548
40. Treves S, Vukcevic M, Griesser J, Armstrong CF, Zhu MX, Zorzato F (2010) Agonist-activated Ca2+ influx occurs at stable plasma membrane and endoplasmic reticulum junctions. J Cell Sci 123(Pt 23):4170–4181

41. Stamboulian S, Moutin MJ, Treves S, Pochon N, Grunwald D, Zorzato F, De Waard M, Ronjat M, Arnoult C (2005) Junctate, an inositol 1,4,5-triphosphate receptor associated protein, is present in rodent sperm and binds TRPC2 and TRPC5 but not TRPC1 channels. Dev Biol 286(1):326–337
42. Srikanth S, Jew M, Kim KD, Yee MK, Abramson J, Gwack Y (2012) Junctate is a Ca2 + -sensing structural component of Orai1 and stromal interaction molecule 1 (STIM1). Proc Natl Acad Sci U S A 109(22):8682–8687
43. Takeshima H, Komazaki S, Nishi M, Iino M, Kangawa K (2000) Junctophilins: a novel family of junctional membrane complex proteins. Mol Cell 6(1):11–22
44. Takeshima H, Hoshijima M, Song LS (2015) Ca microdomains organized by junctophilins. Cell Calcium 58:349–356
45. Moriguchi S, Nishi M, Komazaki S, Sakagami H, Miyazaki T, Masumiya H, Saito SY, Watanabe M, Kondo H, Yawo H, Fukunaga K, Takeshima H (2006) Functional uncoupling between Ca2+ release and afterhyperpolarization in mutant hippocampal neurons lacking junctophilins. Proc Natl Acad Sci U S A 103(28):10811–10816
46. Feske S, Gwack Y, Prakriya M, Srikanth S, Puppel SH, Tanasa B, Hogan PG, Lewis RS, Daly M, Rao A (2006) A mutation in Orai1 causes immune deficiency by abrogating CRAC channel function. Nature 441(7090):179–185
47. Vig M, Peinelt C, Beck A, Koomoa DL, Rabah D, Koblan-Huberson M, Kraft S, Turner H, Fleig A, Penner R, Kinet JP (2006) CRACM1 is a plasma membrane protein essential for store-operated Ca2+ entry. Science 312(5777):1220–1223
48. Zhang SL, Yeromin AV, Zhang XH, Yu Y, Safrina O, Penna A, Roos J, Stauderman KA, Cahalan MD (2006) Genome-wide RNAi screen of Ca(2+) influx identifies genes that regulate Ca(2+) release-activated Ca(2+) channel activity. Proc Natl Acad Sci U S A 103(24):9357–9362
49. Stathopulos PB, Ikura M (2013) Structural aspects of calcium-release activated calcium channel function. Channels (Austin) 7(5):344–353
50. Hou X, Pedi L, Diver MM, Long SB (2012) Crystal structure of the calcium release-activated calcium channel Orai. Science 338(6112):1308–1313
51. Demuro A, Penna A, Safrina O, Yeromin AV, Amcheslavsky A, Cahalan MD, Parker I (2011) Subunit stoichiometry of human Orai1 and Orai3 channels in closed and open states. Proc Natl Acad Sci U S A 108(43):17832–17837
52. Penna A, Demuro A, Yeromin AV, Zhang SL, Safrina O, Parker I, Cahalan MD (2008) The CRAC channel consists of a tetramer formed by stim-induced dimerization of Orai dimers. Nature 456(7218):116–120
53. Ji W, Xu P, Li Z, Lu J, Liu L, Zhan Y, Chen Y, Hille B, Xu T, Chen L (2008) Functional stoichiometry of the unitary calcium-release-activated calcium channel. Proc Natl Acad Sci U S A 105(36):13668–13673
54. McNally BA, Yamashita M, Engh A, Prakriya M (2009) Structural determinants of ion permeation in CRAC channels. Proc Natl Acad Sci U S A 106(52):22516–22521
55. Stathopulos PB, Schindl R, Fahrner M, Zheng L, Gasmi-Seabrook GM, Muik M, Romanin C, Ikura M (2013) STIM1/Orai1 coiled-coil interplay in the regulation of store-operated calcium entry. Nat Commun 4:2963
56. Palty R, Stanley C, Isacoff EY (2015) Critical role for Orai1 C-terminal domain and TM4 in CRAC channel gating. Cell Res 25:963–980
57. Tirado-Lee L, Yamashita M, Prakriya M (2015) Conformational changes in the Orai1 C-terminus evoked by STIM1 binding. PLoS ONE 10(6):e0128622
58. Mignen O, Thompson JL, Shuttleworth TJ (2008) Both Orai1 and Orai3 are essential components of the arachidonate-regulated Ca2 + -selective (ARC) channels. J Physiol 586(1):185–195
59. Zhang X, Zhang W, Gonzalez-Cobos JC, Jardin I, Romanin C, Matrougui K, Trebak M (2014) Complex role of STIM1 in the activation of store-independent Orai1/3 channels. J Gen Physiol 143(3):345–359

60. Thompson JL, Mignen O, Shuttleworth TJ (2013) The ARC channel – an endogenous store-independent Orai channel. Curr Top Membr 71:125–148
61. Yu F, Sun L, Machaca K (2010) Constitutive recycling of the store-operated Ca2+ channel Orai1 and its internalization during meiosis. J Cell Biol 191(3):523–535
62. Maleth J, Choi S, Muallem S, Ahuja M (2014) Translocation between PI(4,5)P2-poor and PI(4,5)P2-rich microdomains during store depletion determines STIM1 conformation and Orai1 gating. Nat Commun 5:5843
63. Park CY, Hoover PJ, Mullins FM, Bachhawat P, Covington ED, Raunser S, Walz T, Garcia KC, Dolmetsch RE, Lewis RS (2009) STIM1 clusters and activates CRAC channels via direct binding of a cytosolic domain to Orai1. Cell 136(5):876–890
64. Liu Y, Zheng X, Mueller GA, Sobhany M, DeRose EF, Zhang Y, London RE, Birnbaumer L (2012) Crystal structure of calmodulin binding domain of orai1 in complex with Ca2+ calmodulin displays a unique binding mode. J Biol Chem 287(51):43030–43041
65. Mullins FM, Park CY, Dolmetsch RE, Lewis RS (2009) STIM1 and calmodulin interact with Orai1 to induce Ca2 + -dependent inactivation of CRAC channels. Proc Natl Acad Sci U S A 106(36):15495–15500
66. Palty R, Raveh A, Kaminsky I, Meller R, Reuveny E (2012) SARAF inactivates the store operated calcium entry machinery to prevent excess calcium refilling. Cell 149(2):425–438
67. Jha A, Ahuja M, Maleth J, Moreno CM, Yuan JP, Kim MS, Muallem S (2013) The STIM1 CTID domain determines access of SARAF to SOAR to regulate Orai1 channel function. J Cell Biol 202(1):71–79
68. Srikanth S, Jung HJ, Kim KD, Souda P, Whitelegge J, Gwack Y (2010) A novel EF-hand protein, CRACR2A, is a cytosolic Ca2+ sensor that stabilizes CRAC channels in T cells. Nat Cell Biol 12(5):436–446
69. Roos J, DiGregorio PJ, Yeromin AV, Ohlsen K, Lioudyno M, Zhang S, Safrina O, Kozak JA, Wagner SL, Cahalan MD, Velicelebi G, Stauderman KA (2005) STIM1, an essential and conserved component of store-operated Ca2+ channel function. J Cell Biol 169(3):435–445
70. Liou J, Kim ML, Heo WD, Jones JT, Myers JW, Ferrell JE Jr, Meyer T (2005) STIM is a Ca2+ sensor essential for Ca2 + -store-depletion-triggered Ca2+ influx. Curr Biol 15(13):1235–1241
71. Stathopulos PB, Zheng L, Li GY, Plevin MJ, Ikura M (2008) Structural and mechanistic insights into STIM1-mediated initiation of store-operated calcium entry. Cell 135(1):110–122
72. Zheng L, Stathopulos PB, Li GY, Ikura M (2008) Biophysical characterization of the EF-hand and SAM domain containing Ca2+ sensory region of STIM1 and STIM2. Biochem Biophys Res Commun 369(1):240–246
73. Huang GN, Zeng W, Kim JY, Yuan JP, Han L, Muallem S, Worley PF (2006) STIM1 carboxyl-terminus activates native SOC, I(crac) and TRPC1 channels. Nat Cell Biol 8(9):1003–1010
74. Fahrner M, Muik M, Schindl R, Butorac C, Stathopulos P, Zheng L, Jardin I, Ikura M, Romanin C (2014) A coiled-coil clamp controls both conformation and clustering of stromal interaction molecule 1 (STIM1). J Biol Chem 289(48):33231–33244
75. Yu F, Sun L, Hubrack S, Selvaraj S, Machaca K (2013) Intramolecular shielding maintains STIM1 in an inactive conformation. J Cell Sci 126:2401–2410
76. Yang X, Jin H, Cai X, Li S, Shen Y (2012) Structural and mechanistic insights into the activation of stromal interaction molecule 1 (STIM1). Proc Natl Acad Sci U S A 109(15):5657–5662
77. Yuan JP, Zeng W, Dorwart MR, Choi YJ, Worley PF, Muallem S (2009) SOAR and the polybasic STIM1 domains gate and regulate Orai channels. Nat Cell Biol 11(3):337–343
78. Feske S, Picard C, Fischer A (2010) Immunodeficiency due to mutations in ORAI1 and STIM1. Clin Immunol 135(2):169–182
79. Miederer AM, Alansary D, Schwar G, Lee PH, Jung M, Helms V, Niemeyer BA (2015) A STIM2 splice variant negatively regulates store-operated calcium entry. Nat Commun 6:6899

80. Rana A, Yen M, Sadaghiani AM, Malmersjo S, Park CY, Dolmetsch RE, Lewis RS (2015) Alternative splicing converts STIM2 from an activator to an inhibitor of store-operated calcium channels. J Cell Biol 209(5):653–670
81. Liou J, Fivaz M, Inoue T, Meyer T (2007) Live-cell imaging reveals sequential oligomerization and local plasma membrane targeting of stromal interaction molecule 1 after Ca2+ store depletion. Proc Natl Acad Sci U S A 104(22):9301–9306
82. Korzeniowski MK, Manjarres IM, Varnai P, Balla T (2010) Activation of STIM1-Orai1 involves an intramolecular switching mechanism. Sci Signal 3(148):ra82
83. Muik M, Fahrner M, Schindl R, Stathopulos P, Frischauf I, Derler I, Plenk P, Lackner B, Groschner K, Ikura M, Romanin C (2011) STIM1 couples to ORAI1 via an intramolecular transition into an extended conformation. EMBO J 30(9):1678–1689
84. Grigoriev I, Gouveia SM, van der Vaart B, Demmers J, Smyth JT, Honnappa S, Splinter D, Steinmetz MO, Putney JW Jr, Hoogenraad CC, Akhmanova A (2008) STIM1 is a MT-plus-end-tracking protein involved in remodeling of the ER. Curr Biol 18(3):177–182
85. Smyth JT, Beg AM, Wu S, Putney JW Jr, Rusan NM (2012) Phosphoregulation of STIM1 leads to exclusion of the endoplasmic reticulum from the mitotic spindle. Curr Biol 22(16):1487–1493
86. Pozo-Guisado E, Casas-Rua V, Tomas-Martin P, Lopez-Guerrero AM, Alvarez-Barrientos A, Martin-Romero FJ (2013) Phosphorylation of STIM1 at ERK1/2 target sites regulates interaction with the microtubule plus-end binding protein EB1. J Cell Sci 126(Pt 14):3170–3180
87. Asanov A, Sherry R, Sampieri A, Vaca L (2013) A relay mechanism between EB1 and APC facilitate STIM1 puncta assembly at endoplasmic reticulum-plasma membrane junctions. Cell Calcium 54(3):246–256
88. Shinde AV, Motiani RK, Zhang X, Abdullaev IF, Adam AP, Gonzalez-Cobos JC, Zhang W, Matrougui K, Vincent PA, Trebak M (2013) STIM1 controls endothelial barrier function independently of Orai1 and Ca2+ entry. Sci Signal 6(267):ra18
89. Hawkins BJ, Irrinki KM, Mallilankaraman K, Lien YC, Wang Y, Bhanumathy CD, Subbiah R, Ritchie MF, Soboloff J, Baba Y, Kurosaki T, Joseph SK, Gill DL, Madesh M (2010) S-glutathionylation activates STIM1 and alters mitochondrial homeostasis. J Cell Biol 190(3):391–405
90. Prins D, Groenendyk J, Touret N, Michalak M (2011) Modulation of STIM1 and capacitative Ca2+ entry by the endoplasmic reticulum luminal oxidoreductase ERp57. EMBO Rep 12(11):1182–1188
91. Lee KP, Yuan JP, Zeng W, So I, Worley PF, Muallem S (2009) Molecular determinants of fast Ca2 + -dependent inactivation and gating of the Orai channels. Proc Natl Acad Sci U S A 106(34):14687–14692
92. Derler I, Fahrner M, Muik M, Lackner B, Schindl R, Groschner K, Romanin C (2009) A Ca2(+)release-activated Ca2(+) (CRAC) modulatory domain (CMD) within STIM1 mediates fast Ca2(+)-dependent inactivation of ORAI1 channels. J Biol Chem 284(37):24933–24938
93. Lee KP, Choi S, Hong JH, Ahuja M, Graham S, Ma R, So I, Shin DM, Muallem S, Yuan JP (2014) Molecular determinants mediating gating of transient receptor potential canonical (TRPC) channels by stromal interaction molecule 1 (STIM1). J Biol Chem 289(10):6372–6382
94. Yuan JP, Zeng W, Huang GN, Worley PF, Muallem S (2007) STIM1 heteromultimerizes TRPC channels to determine their function as store-operated channels. Nat Cell Biol 9(6):636–645
95. Liu X, Bandyopadhyay BC, Singh BB, Groschner K, Ambudkar IS (2005) Molecular analysis of a store-operated and 2-acetyl-sn-glycerol-sensitive non-selective cation channel. Heteromeric assembly of TRPC1-TRPC3. J Biol Chem 280(22):21600–21606
96. Zeng W, Yuan JP, Kim MS, Choi YJ, Huang GN, Worley PF, Muallem S (2008) STIM1 gates TRPC channels, but not Orai1, by electrostatic interaction. Mol Cell 32(3):439–448
97. Muik M, Schindl R, Fahrner M, Romanin C (2012) Ca(2+) release-activated Ca(2+) (CRAC) current, structure, and function. Cell Mol Life Sci 69(24):4163–4176

98. Cao X, Choi S, Maleth JJ, Park S, Ahuja M, Muallem S (2015) The ER/PM microdomain, PI(4,5)P and the regulation of STIM1-Orai1 channel function. Cell Calcium 58:342–348
99. Muik M, Frischauf I, Derler I, Fahrner M, Bergsmann J, Eder P, Schindl R, Hesch C, Polzinger B, Fritsch R, Kahr H, Madl J, Gruber H, Groschner K, Romanin C (2008) Dynamic coupling of the putative coiled-coil domain of ORAI1 with STIM1 mediates ORAI1 channel activation. J Biol Chem 283(12):8014–8022
100. Frischauf I, Muik M, Derler I, Bergsmann J, Fahrner M, Schindl R, Groschner K, Romanin C (2009) Molecular determinants of the coupling between STIM1 and Orai channels: differential activation of Orai1-3 channels by a STIM1 coiled-coil mutant. J Biol Chem 284(32):21696–21706
101. McNally BA, Somasundaram A, Jairaman A, Yamashita M, Prakriya M (2013) The C- and N-terminal STIM1 binding sites on Orai1 are required for both trapping and gating CRAC channels. J Physiol 591(Pt 11):2833–2850
102. McNally BA, Somasundaram A, Yamashita M, Prakriya M (2012) Gated regulation of CRAC channel ion selectivity by STIM1. Nature 482(7384):241–245
103. Lichtenegger M, Groschner K (2014) TRPC3: a multifunctional signaling molecule. Handb Exp Pharmacol 222:67–84
104. Lis A, Peinelt C, Beck A, Parvez S, Monteilh-Zoller M, Fleig A, Penner R (2007) CRACM1, CRACM2, and CRACM3 are store-operated Ca(2+) channels with distinct functional properties. Curr Biol 17(9):794–800
105. Darbellay B, Arnaudeau S, Bader CR, Konig S, Bernheim L (2011) STIM1L is a new actin-binding splice variant involved in fast repetitive Ca2+ release. J Cell Biol 194(2):335–346
106. Horinouchi T, Higashi T, Higa T, Terada K, Mai Y, Aoyagi H, Hatate C, Nepal P, Horiguchi M, Harada T, Miwa S (2012) Different binding property of STIM1 and its novel splice variant STIM1L to Orai1, TRPC3, and TRPC6 channels. Biochem Biophys Res Commun 428(2):252–258
107. Sauc S, Bulla M, Nunes P, Orci L, Marchetti A, Antigny F, Bernheim L, Cosson P, Frieden M, Demaurex N (2015) STIM1L traps and gates Orai1 channels without remodeling the cortical ER. J Cell Sci 128(8):1568–1579
108. Brandman O, Liou J, Park WS, Meyer T (2007) STIM2 is a feedback regulator that stabilizes basal cytosolic and endoplasmic reticulum Ca2+ levels. Cell 131(7):1327–1339
109. Ong HL, de Souza LB, Zheng C, Cheng KT, Liu X, Goldsmith CM, Feske S, Ambudkar IS (2015) STIM2 enhances receptor-stimulated Ca(2)(+) signaling by promoting recruitment of STIM1 to the endoplasmic reticulum-plasma membrane junctions. Sci Signal 8(359):ra3
110. Shalygin A, Skopin A, Kalinina V, Zimina O, Glushankova L, Mozhayeva GN, Kaznacheyeva E (2015) STIM1 and STIM2 proteins differently regulate endogenous store-operated channels in HEK293 cells. J Biol Chem 290(8):4717–4727
111. Cooper DM (2015) Store-operated Ca-entry and adenylyl cyclase. Cell Calcium 58:368–375
112. Penny CJ, Kilpatrick BS, Eden ER, Patel S (2015) Coupling acidic organelles with the ER through Ca microdomains at membrane contact sites. Cell Calcium 58:387–396
113. Raiborg C, Wenzel EM, Stenmark H (2015) ER-endosome contact sites: molecular compositions and functions. EMBO J 34:1848–1858
114. Lahiri S, Toulmay A, Prinz WA (2015) Membrane contact sites, gateways for lipid homeostasis. Curr Opin Cell Biol 33:82–87
115. Prinz WA (2014) Bridging the gap: membrane contact sites in signaling, metabolism, and organelle dynamics. J Cell Biol 205(6):759–769
116. Kasai H, Augustine GJ (1990) Cytosolic Ca2+ gradients triggering unidirectional fluid secretion from exocrine pancreas. Nature 348(6303):735–738
117. Thorn P, Lawrie AM, Smith PM, Gallacher DV, Petersen OH (1993) Local and global cytosolic Ca2+ oscillations in exocrine cells evoked by agonists and inositol trisphosphate. Cell 74(4):661–668

118. Min SW, Chang WP, Sudhof TC (2007) E-Syts, a family of membranous Ca2 + -sensor proteins with multiple C2 domains. Proc Natl Acad Sci U S A 104(10):3823–3828
119. Toulmay A, Prinz WA (2012) A conserved membrane-binding domain targets proteins to organelle contact sites. J Cell Sci 125(Pt 1):49–58
120. Giordano F, Saheki Y, Idevall-Hagren O, Colombo SF, Pirruccello M, Milosevic I, Gracheva EO, Bagriantsev SN, Borgese N, De Camilli P (2013) PI(4,5)P(2)-dependent and Ca(2+)-regulated ER-PM interactions mediated by the extended synaptotagmins. Cell 153(7):1494–1509
121. Fischer MA, Temmerman K, Ercan E, Nickel W, Seedorf M (2009) Binding of plasma membrane lipids recruits the yeast integral membrane protein Ist2 to the cortical ER. Traffic 10(8):1084–1097
122. Jang Y, Oh U (2014) Anoctamin 1 in secretory epithelia. Cell Calcium 55(6):355–361
123. Sharma S, Quintana A, Findlay GM, Mettlen M, Baust B, Jain M, Nilsson R, Rao A, Hogan PG (2013) An siRNA screen for NFAT activation identifies septins as coordinators of store-operated Ca2+ entry. Nature 499(7457):238–242
124. Chao JT, Wong AK, Tavassoli S, Young BP, Chruscicki A, Fang NN, Howe LJ, Mayor T, Foster LJ, Loewen CJ (2014) Polarization of the endoplasmic reticulum by ER-septin tethering. Cell 158(3):620–632
125. Kim S, Kedan A, Marom M, Gavert N, Keinan O, Selitrennik M, Laufman O, Lev S (2013) The phosphatidylinositol-transfer protein Nir2 binds phosphatidic acid and positively regulates phosphoinositide signalling. EMBO Rep 14(10):891–899
126. Kim YJ, Guzman-Hernandez ML, Wisniewski E, Balla T (2015) Phosphatidylinositol-phosphatidic acid exchange by Nir2 at ER-PM contact sites maintains phosphoinositide signaling competence. Dev Cell 33(5):549–561
127. Chang CL, Hsieh TS, Yang TT, Rothberg KG, Azizoglu DB, Volk E, Liao JC, Liou J (2013) Feedback regulation of receptor-induced Ca2+ signaling mediated by E-Syt1 and Nir2 at endoplasmic reticulum-plasma membrane junctions. Cell Rep 5(3):813–825
128. Fernandez-Busnadiego R, Saheki Y, De Camilli P (2015) Three-dimensional architecture of extended synaptotagmin-mediated endoplasmic reticulum-plasma membrane contact sites. Proc Natl Acad Sci U S A 112(16):E2004–E2013
129. Jean S, Mikryukov A, Tremblay MG, Baril J, Guillou F, Bellenfant S, Moss T (2010) Extended-synaptotagmin-2 mediates FGF receptor endocytosis and ERK activation in vivo. Dev Cell 19(3):426–439
130. Tremblay MG, Herdman C, Guillou F, Mishra PK, Baril J, Bellenfant S, Moss T (2015) Extended synaptotagmin interaction with the fibroblast growth factor receptor depends on receptor conformation, not catalytic activity. J Biol Chem 290(26):16142–16156
131. Hong JH, Li Q, Kim MS, Shin DM, Feske S, Birnbaumer L, Cheng KT, Ambudkar IS, Muallem S (2011) Polarized but differential localization and recruitment of STIM1, Orai1 and TRPC channels in secretory cells. Traffic 12(2):232–245
132. Schleifer H, Doleschal B, Lichtenegger M, Oppenrieder R, Derler I, Frischauf I, Glasnov TN, Kappe CO, Romanin C, Groschner K (2012) Novel pyrazole compounds for pharmacological discrimination between receptor-operated and store-operated Ca(2+) entry pathways. Br J Pharmacol 167(8):1712–1722

Chapter 4
TRP-Na^+/Ca^{2+} Exchanger Coupling

Alan G.S. Harper and Stewart O. Sage

Abstract Na^+/Ca^{2+} exchangers (NCXs) have traditionally been viewed principally as a means of Ca^{2+} removal from non-excitable cells. However there has recently been increasing interest in the operation of NCXs in reverse mode acting as a means of eliciting Ca^{2+} entry into these cells. Reverse mode exchange requires a significant change in the normal resting transmembrane ion gradients and membrane potential, which has been suggested to occur principally via the coupling of NCXs to localised Na^+ entry through non-selective cation channels such as canonical transient receptor potential (TRPC) channels. Here we review evidence for functional or physical coupling of NCXs to non-selective cation channels, and how this affects NCX activity in non-excitable cells. In particular we focus on the potential role of nanojunctions, where the close apposition of plasma and intracellular membranes may help create the conditions needed for the generation of localised rises in Na^+ concentration that would be required to trigger reverse mode exchange.

Keywords Na^+/Ca^{2+} exchangers • TRPC channels • Sodium entry • Reverse mode NCX activity • Calcium entry

4.1 Introduction

Animal cells maintain a low cytosolic Ca^{2+} concentration ($[Ca^{2+}]_{cyt}$) in the range of 50–100 nM. To maintain this resting $[Ca^{2+}]_{cyt}$ against leakage of Ca^{2+} from intracellular stores or across the plasma membrane, or to restore resting $[Ca^{2+}]_{cyt}$ after the generation of a Ca^{2+} signal, several Ca^{2+} removal mechanisms are used. These include sequestration of Ca^{2+} into organelles including the endoplasmic reticulum

A.G.S. Harper
Institute for Science and Technology in Medicine, Keele University, Guy Hilton Research Centre, Thornburrow Drive, Hartshill, Stoke-on-Trent ST4 7QB, UK
e-mail: a.g.s.harper@keele.ac.uk

S.O. Sage (✉)
Department of Physiology, Development and Neuroscience, University of Cambridge, Cambridge CB2 3EG, UK
e-mail: sos10@cam.ac.uk

J.A. Rosado (ed.), *Calcium Entry Pathways in Non-excitable Cells*, Advances in Experimental Medicine and Biology 898, DOI 10.1007/978-3-319-26974-0_4

or sarcoplasmic reticulum by several isoforms of sarco-endoplasmic reticulum Ca^{2+}-ATPase (SERCA), into mitochondria by the mitochondrial Ca^{2+} uniporter and into acidic organelles by the combined actions of a vacuolar H^+-ATPase (vH^+-ATPase) and a H^+/Ca^{2+} exchanger. Ca^{2+} is also be removed across the plasma membrane either by primary active transport via a plasma membrane Ca^{2+}-ATPase (PMCA), of which four isoforms have been identified [1], or by secondary active transport via Na^+/Ca^{2+} exchange. Two types of Na^+/Ca^{2+} exchangers are found in the plasma membrane: those that exchange Na^+ for Ca^{2+} (NCXs) and those that exchange Na^+ for Ca^{2+} and K^+ (NCKXs) [2]. PMCAs are high affinity Ca^{2+} transporters but have relatively low turnover rates, whilst Na^+/Ca^{2+} exchangers have a lower Ca^{2+} affinity but higher turnover rates than PMCAs [3]. Consequently, PMCAs are important in the maintenance of resting $[Ca^{2+}]_{cyt}$ whilst Na^+/Ca^{2+} exchangers are more important in the restoration of $[Ca^{2+}]_{cyt}$ after it has been elevated during Ca^{2+} signalling. Na^+/Ca^{2+} exchangers influence Ca^{2+} signalling in excitable and non-excitable cells, not just as a means of Ca^{2+} removal, but also as a means of Ca^{2+} entry, since the exchangers can reverse in response to changes in the concentrations of the transported ions and in membrane potential.

4.1.1 Na^+/Ca^{2+} Exchange

Several observations made during the first half of the twentieth century of the effects on cardiac muscle of manipulation of the extracellular Na^+ concentration ($[Na^+]_o$) pointed towards the existence of a Na^+/Ca^{2+} exchange mechanism. During the 1960s several laboratories studying different tissues concluded that a coupled counter transport system for Na^+ and Ca^{2+} existed in the plasma membrane [4–6]. Although earlier studies were generally interpreted in terms of the exchange of two Na^+ ions for one Ca^{2+} ion, evidence suggesting a coupling ratio of $>2Na^+:1Ca^{2+}$ started to emerge in the mid 1970s and over the following decade or so an exchange ratio of $3Na^+:1Ca^{2+}$ was established in several different excitable tissues [e.g. 7, 8]. The NCX first studied in cardiac muscle is now designated NCX1. Two other NCX isoforms, NCX2 and NCX3 are also expressed in some mammalian cells. Given the similar sequence identity in the transport domains of the proteins, NCX2 and NCX3 are also believed to mediate exchange of $3Na^+$ ions for one Ca^{2+} ion [2]. However, there is some uncertainty over the transport ratio, with some electrophysiological experiments suggesting exchange ratios of $4Na^+:1Ca^{2+}$ [9, 10] or even $1Na^+:1Ca^{2+}$ [11].

In the 1980s several groups identified K^+-dependent Na^+/Ca^{2+} exchange in rod outer segments and two groups reported an exchange ratio of $4Na^+:(1Ca^{2+} + 1\ K^+)$ [12, 13]. The K^+-dependent Na^+/Ca^{2+} exchanger first identified in rods is now designated NCKX1. Four related exchangers, NCKX2, NCKX3, NCKX4 and NCKX5 have been identified [2]. The stoichiometry of transport by NCKX2 has been determined to be the same as for NCKX1 [14]. Sequence identity suggests the other NCKXs are likely to also exhibit exchange ratios of $4Na^+:(1Ca^{2+} + 1\ K^+)$.

For a detailed historical account of early work on Na^+/Ca^{2+} exchange see [3].

4.1.2 Na^+/Ca^{2+} Exchanger Families

Na^+/Ca^{2+} exchangers form part of a superfamily of Ca^{2+}/cation antiporters (CaCAs) with almost 150 identified members [15]. Three of the five branches of the CaCA superfamily include Na^+/Ca^{2+} exchangers that are expressed in mammalian cells. As indicated above (Sect. 4.1.1), three isoforms of the NCX family and five of the NCKX family have been identified. A third branch, the Ca^{2+}/cation exchangers (CCXs), includes only one identified mammalian member, rather confusingly known after earlier misidentification as NCKX6 or NCLX ($Na^+/Ca^{2+}/Li^+$ exchanger) as, unlike NCXs or NCKXs, it can transport Na^+ or Li^+ in exchange for Ca^{2+}. The NCLX is now recognised as the Na^+/Ca^{2+} exchanger expressed in the inner mitochondrial membrane [16].

The NCXs form the SLC8 (solute carrier 8) family, with NCX1, NCX2 and NCX3 being coded for by the genes SLC8A1, SLC8A2 and SLC8A3 respectively [2, 17]. NCX1 is widely expressed and found in most tissues whereas NCX2 and NCX3 show a more restricted expression [2, 17]. The transcription products of SLC8A1 and SLCA3 but not SLCA2 are subject to alternative splicing which may be tissue specific [2, 17, 18].

The NCKXs form the SLC24 (solute carrier 24) family, with the five members being coded for by the genes SLC24A1- SLC24A5 and some splice variants exist [19]. All five of the NCKX isoforms appear to be expressed in several tissues with NCKX1-NCKX4 being found in the plasma membrane whilst NCKX5 is only reported to be present in internal membranes [19].

4.1.3 Modes of Exchange

Transport by NCXs and NCKXs is fully reversible, with forward mode exchange removing Ca^{2+} from the cytosol across the plasma membrane and reverse mode exchange transporting Ca^{2+} into the cytosol. The direction of exchange is influenced by the prevailing Ca^{2+} and Na^+ concentrations and, since the transporters are electrogenic, by membrane potential. High $[Ca^{2+}]_{cyt}$ favours forward mode exchange whilst high $[Na^+]$ and more depolarised membrane potentials favour reverse mode exchange. Physiologically $[Ca^{2+}]_{cyt}$ and $[Na^+]_{cyt}$ may change in response to increased activity of ion channels permeable to these ions, and there may be marked local effects in sub-plasma membrane microdomains and in areas of restricted diffusion such as nanojunctions formed by closely apposing membranes. The mode of exchange may be influenced by molecular complexes in which exchangers are functionally coupled to ion channels or other transporters (See Sect. 4.2.1).

Reversal of the direction of exchange by NCX1 in cardiac muscle has long been recognised. Under conditions of resting ionic concentrations and membrane potential, NCX1, with a $3Na^+:1Ca^{2+}$ stoichiometry, operates in forward mode, but early in the cardiac action potential following Na^+ entry through voltage-gated Na^+ channels and with membrane depolarisation, the exchanger switches to reverse mode to mediate Ca^{2+} influx [3, 20]. Similarly, Na^+ entry via non-selective receptor- or store-operated channels can lead to reverse mode exchange in non-excitable cells [21]. Reverse mode exchange by NCKX3 or NCKX4 has been suggested to contribute to stimulated Ca^{2+} entry in arterial smooth muscle.

4.1.4 Ionic Regulation of Exchanger Activity

Transport by NCXs involves alternate occupation of an ion binding site by three Na^+ ions or one Ca^{2+} ion that is sequentially exposed at the two faces of the plasma membrane [2]. The intracellular ion transport site of NCX1 has an affinity for Ca^{2+} in the low micromolar range and an affinity for Na^+ in the order of 10 mM. The externally facing site has Ca^{2+} and Na^+ affinities around 1 mM and 60–80 mM respectively [2]. Activity of the exchanger under resting conditions is thus mainly influenced by occupancy of the binding site for transported Ca^{2+} on the cytosolic face. Transport by NCKXs is believed to occur by an alternate access model as for the NCX. The dissociation constants at the external facing binding sites have been estimated at around 1–3 μM for Ca^{2+}, 2–10 mM for K^+ and 25–45 mM for Na^+. The dissociation constants for Ca^+ and K^+ appear to be similar at the internal facing binding sites [22].

In addition to occupancy of the ion transport sites, NCX activity is allosterically regulated by the binding of Ca^{2+} and Na^+ and by other factors. Ca^{2+} binding at the cytosolic face of NCX1 activates the exchanger [23, 24] whilst Na^+ binding at the cytosolic face is inhibitory [25, 26]. Inactivation of NCKX2 by cytosolic Na^+ has also been reported [27].

4.1.5 Additional Regulation of Exchanger Activity

The activity of Na^+/Ca^{2+} exchangers can be modified through their ability to integrate a wide range of cellular signals other than changes in ion concentrations that alter the rate and directionality of transport. These include factors such as cytosolic pH, nitric oxide, reactive oxygen species, ATP and phosphatidylinositol 4,5-bisphosphate (PIP_2) [28–34]. NCX1 activity is also influenced by the binding of a number of other proteins, including phospholemman, calcineurin and 14-3-3 proteins, the binding of which are all inhibitory [2]. NCX1 function may also be influenced by phosphorylation by protein kinase A and protein kinase C, although this is controversial, particularly for PKA, where a number of conflicting reports exist [2, 35]. NCX activity might also be influenced by phosphorylation of binding partners such as phospholemman. The activity of NCKX2, but not of NCKX3 or NCKX4, is reported to be increased by PKC-dependent phosphorylation [36]. Lastly, the activity of NCX1 and NCX3 may also be influenced by calpain-mediated proteolysis, which acts to favour increased cytosolic calcium concentrations by either downregulating forward mode exchange or upregulating greater reverse mode exchange [37–39].

These regulatory mechanisms allow the NCX to act as a tunable Ca^{2+} transport system which has the capacity to momentarily switch from conducting Ca^{2+} removal through forward mode exchange to generating Ca^{2+} entry through reverse mode exchange. The NCX can potentially act as a mediator of both Ca^{2+} removal and entry within a single cell, and as such can either facilitate or inhibit Ca^{2+}-regulated processes [40, 41]. The ability of the NCX to function in reverse mode and act as a Ca^{2+} entry mechanism has led to investigations of the relative physiological importance of the NCX acting in forward and reverse mode in a range of different cell types.

4.2 Coupling of Localised Na^+ Increases Generated by Activation of Non-selective Cation Channels with Na^+/Ca^{2+} Exchangers Creates Hybrid Ca^{2+} Entry Systems

Reverse mode exchange requires a significant change in the normal resting transmembrane ion gradients and membrane potential, which normally favour exchangers working to remove Ca^{2+} from cells. In non-excitable cells, a shift into reverse mode activity has been suggested to occur principally via the coupling of a localised Na^+ entry through non-selective cation channels that result in local change in the Na^+ gradient and membrane potential. Most reports have suggested members of the Canonical Transient Receptor Potential (TRPC) family as a potential partner for Na^+/Ca^{2+} exchangers. TRPC isoforms are all non-selective cation channels that facilitate significant Na^+ and Ca^{2+} entry into cells [42], and as such could provide a source of Na^+ to allow the reversal potential of the exchanger to be reached resulting in reverse mode exchange and so Ca^{2+} entry. The possibility of spatial coupling of NCX1 and TRPCs has been supported by immunolocalisation studies in which the two proteins have been found to colocalised in specific domains of the plasma membrane of some cells [43–45]. Previous work has suggest that this colocalisation may be the result of direct interaction of the TRPC and NCX as part of a macromolecular complex [43, 46], but interaction may also be indirect through the recruitment of both proteins to lipid rafts or caveolae in some systems [46–51].

Other Na^+ entry routes have also been implicated in the reversal of Na^+/Ca^{2+} exchangers. These include P2X receptors and Na^+/H^+ exchangers in non-excitable cells [52, 53], whilst NMDA and AMPA receptors, as well as glutamate and glycine transporters have additionally been reported to play a role in the reversal of Na^+/Ca^{2+} exchangers in excitable tissues [54–59]. These results suggest that Na^+/Ca^{2+} exchangers can promiscuously couple with a range of different Na^+ transporting proteins to elicit Ca^{2+} entry. Thus when co-expressed, the dissociated Na^+/Ca^{2+} exchangers and Na^+ transporting proteins can work independently to elicit their standard cellular activities, yet when coupled together they can generate a hybrid response and potentiate cytosolic Ca^{2+} signalling. This raises the possibility that subpopulations of Na^+/Ca^{2+} exchangers may exist that can locally modulate the Ca^{2+} signalling system of some cells.

4.2.1 Evidence for Functional Coupling of Na^+ Entry Mechanisms with Reverse Mode Na^+/Ca^{2+} Exchanger Activity

To ensure effective reverse mode exchange, the membrane potential needs to exceed the reversal potential of the exchanger, which is dependent upon both the transmembrane Na^+ and Ca^{2+} gradients [3]. Na^+ and Ca^{2+} entry through non-selective cation channels will alter the reversal potential of the exchanger, as well as favour depolarisation of the membrane potential of a stimulated cell. Thus it would only be possible to accurately predict the effect of Na^+ entry though non-selective cation channels on the rate and mode of Na^+/Ca^{2+} exchanger activity if simultaneous real-time measurements of all three of these parameters in the local vicinity of the protein were performed. In non-excitable cells, where membrane potential changes will be highly-dependent on the proximity of the open channels, this is a near-intractable task and as such investigators will often have to rely on estimates of some of these parameters in order to estimate the change in $[Na^+]_{cyt}$ needed to trigger Ca^{2+} entry by reverse mode exchange. Such an estimate was performed in aortic smooth muscle cells [60], where it was calculated that the NCX would require $[Na^+]_{cyt}$ to be raised to between 16 and 25 mM depending on the extent that the cell is also depolarised by the cation entry. The same group went on to subsequently demonstrate that agonist-evoked opening of a TRPC6-containing receptor operated channel generated a localised subplasmalemmal Na^+ microdomain around the channel, which caused the local $[Na^+]$ to rise to 30 mM [61]. In contrast, the bulk cytosolic $[Na^+]$ rose to only 14 mM, suggesting that close spatial coupling of the TRPC6-containing channel to the NCX would be required to elicit significant Ca^{2+} entry by reverse mode exchange. This possibility was further supported by the observation that pre-treatment with the NCX inhibitor KB-R7943 potentiated the local Na^+ rises but had no significant effect on the global Na^+ signal.

4.2.2 Nanojunctions: A Cellular Architecture to Facilitate Reverse Mode NCX Activity

Whilst the close coupling of NCXs and TRPC channels form primary molecular complexes for eliciting reverse mode exchange, it is becoming clear that the activity of NCX-TRPC complexes may be fine-tuned by their localisation within cellular nanojunctions [62]. Nanojunctions formed by the tight apposition of the membrane of the endoplasmic/sarcoplasmic reticulum with the plasma membrane provide a

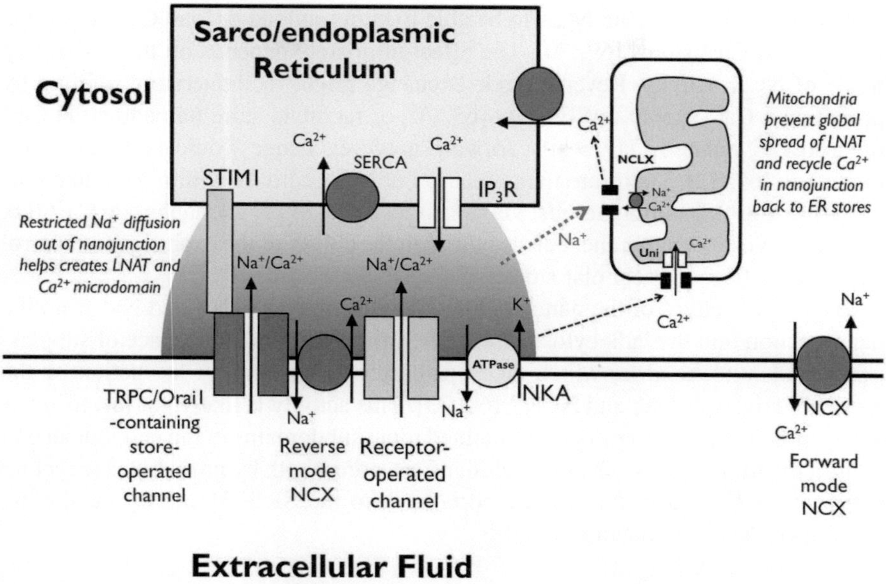

Fig. 4.1 Nanojunctions provide a subcellular environment to favour reverse mode NCX activity. The localisation of non-selective cation channels with the NCX in the restricted environment of the nanojunction creates a highly-concentrated rise in local [Na^+] (Localised Na^+ Transient; LNAT) which can facilitate reversal of the NCX. Through localisation of other cation-transporting proteins such as the Na^+/K^+-ATPase (NKA), IP_3 receptor (IP_3R) and sarco/endoplasmic reticulum Ca^{2+}-ATPase (SERCA), the local levels of Ca^{2+} and Na^+ can be tightly controlled to facilitate both refilling of the ER Ca^{2+} store via reverse mode exchange, or Ca^{2+} unloading through forward mode exchange (not shown). These localised rises in Na^+ and Ca^{2+} can be isolated from the bulk cytosol by the geometry of the nanojunction restricting diffusion of these cations, the actions of the Ca^{2+} and Na^+ transporting proteins and through the actions of nearby mitochondria which can buffer these local cation rises. The isolation of these localised cation transients from the bulk cytosol allows other NCX proteins in the rest of the plasma membrane to continue to act in forward mode as a high-capacity Ca^{2+} removal system

cellular architecture which may facilitate reverse mode Na^+/Ca^{2+} exchange. The ≈20 nm wide cleft creates a tiny isolated volume of cytosol in which Na^+ and Ca^{2+} gradients can be significantly altered in relative isolation from changes in the bulk cytosol [63, 64]. Thus by recruiting the TRPC-NCX complex into nanojunctions, the opening of the non-selective cation channel can markedly raise the local [Na^+] to allow the exchanger to switch into reverse mode exchange (Fig. 4.1).

In addition to the coupled NCX and non-selective cation channels, nanojunctions are also thought to contain a number of other regulators of local Ca^{2+} and Na^+ concentrations. These include the low affinity $\alpha2/\alpha3$ isoforms of the Na^+/K^+-ATPase, SERCAs, inositol 1,4,5-trisphosphate (IP_3) receptors and ryanodine receptors [65–69]. The functional interactions of these transporters within the enclosed cytosolic domain at the nanojunction can bias the ionic conditions surrounding the NCX and therefore can fine tune the rate and directionality of NCX transport depending upon their activity (Fig. 4.1).

Moreover, the close interaction of NCXs with the Ca^{2+} release and sequestration mechanisms in the sarco/endoplasmic reticulum can create a form of linked Ca^{2+}

transport which allows the NCX to be able to either unload or load Ca^{2+} into these intracellular Ca^{2+} stores [69, 70]. The effect ultimately depends on the underlying mode of NCX activity. Reverse mode exchange might facilitate store refilling by providing a Ca^{2+} supply to SERCAs [63, 70] or facilitate store unloading via Ca^{2+} induced Ca^{2+} release [71], whilst forward mode exchange would facilitate store-unloading [69, 72]. Therefore the regulation of NCX activity within nanojunctions can affect Ca^{2+} signalling not just via the direct effect of the exchanger on Ca^{2+} flux across the plasma membrane, but also through the ability of the exchanger to control Ca^{2+} release from intracellular stores.

Beyond the edges of the nanojunction, the movement of Ca^{2+} and Na^{+} from the nanojunction into the bulk cytosol are further restricted by the presence of subplasmalemmal mitochondria, which take up these cations through the action of the mitochondrial uniporter and NCLX [60, 73]. This activity is likely not just to act as a way of helping to create spatially-defined ionic subdomains in the cell, but also to control the filling state of the intracellular Ca^{2+} stores, both by providing a recycling pathway for Ca^{2+} from the mitochondria back to the SR [73] and by facilitating store-operated calcium entry [74–76].

4.2.3 A Role for NCX in Store-Operated Calcium Entry?

In addition to providing an optimal environment for eliciting reverse mode NCX activity, nanojunctions also provide an ideal subcellular location for the activation of store-operated channels through their coupling to the endoplasmic reticulum Ca^{2+} sensor STIM1 [77, 78]. Indeed previous work has demonstrated that store-operated Ca^{2+} entry may refill Ca^{2+} stores by allowing Ca^{2+} entry into a restricted portion of the cytoplasm, as would be provided for by a nanojunction [79, 80]. This begs the question as to whether TRPC-containing store-operated channels could also be coupled to the NCX to elicit reverse mode exchange. In support of this possibility there are some reports that the Ca^{2+} signalling caused by the opening of store-operated channels is further supported by triggering reverse mode NCX activity [81, 82]. This further begs the question as to whether store-operated- and receptor-operated channels could both tune NCX activity within the same cell, either by acting independently in distinct nanojunctions, or by working cooperatively to facilitate reverse mode NCX activity within the same nanojunction.

4.3 Characterising the Physiological Important of Reverse Mode NCX Activity: The Potential Pitfalls of Pharmacological Characterisation

There is a growing appreciation of the ability of reverse mode Na^{+}/Ca^{2+} exchange to mediate Ca^{2+} entry into non-excitable cells. This has been reported under both physiological and pathological conditions in a wide variety of cells including endothelial cells, renal proximal tubule cells, pancreatic duct cells, macrophages, monocytes, neutrophils, platelets and smooth muscle cells of the airways, vasculature, urethra, bladder and gastrointestinal tract [30, 81–90].

4.3.1 *Use of Inhibitors of Na^+/Ca^{2+} Exchange*

Whilst there have been many reports of the contribution of reverse mode NCX activity in controlling Ca^{2+} signalling and cellular activity in a wide range of excitable and non-excitable cells, many of these studies have been pharmacological and have not provided evidence of the co-localisation of proteins suggested to be involved nor have the parameters which alter the mode of exchange been examined. This leaves some of these reports open to question as teasing apart the forward and reverse modes of exchange experimentally is fraught with pitfalls. Examining the effects of inhibition of Na^+/Ca^{2+} exchangers on agonist-evoked Ca^{2+} signals will not unambiguously establish the role of the exchanger because Na^+ entry through a channel could potentiate a rise in $[Ca^{2+}]_{cyt}$ either by triggering Ca^{2+} entry via reverse mode exchange, or by reducing Ca^{2+} removal by forward mode exchange [81, 91, 92]. In addition, inhibitors of Na^+/Ca^{2+} exchangers may have additional effects on Ca^{2+} release from intracellular stores (See Sect. 4.2.2). Most traditional methods for blocking Na^+/Ca^{2+} exchangers are not mode selective and as such it is difficult to tease apart the relative contribution of forward and reverse mode exchange in intact cells. For instance, removing Na^+ from the extracellular medium can provide a physiological blockade of Na^+/Ca^{2+} exchanger function, yet this experimental manipulation is non-selective and will likely impede both forward mode activity (by removing the transmembrane Na^+ gradient) and reverse mode exchange (by preventing Na^+ entry and cellular depolarisation upon activation). Even genetic knockout of Na^+/Ca^{2+} exchangers will not provide a solution to this conundrum as it will remove all the functions of the exchanger.

4.3.2 *Mode-Selective Inhibitors of Na^+/Ca^{2+} Exchange*

The introduction of inhibitors which can more potently block reverse mode exchange has provided a potential mechanism to experimentally determine the role of reverse mode Ca^{2+} exchange in a variety of cell types. However, there is still a need for caution in the interpretation of results obtained with mode selective inhibitors without additional supporting data. For instance, the widely-used inhibitor KB-R7943 is often described as being reverse mode selective based on the initial reports of a significant difference in potency against the two modes of NCX1 operation studied electrophysiologically when ionic conditions were set to favour unidirectional exchange in either forward or reverse mode ($IC_{50} = 0.3$ μM or 17 μM for reverse or forward mode respectively) [93]. Yet when the ionic conditions are set more physiologically and exchange mode is switched by altering the holding potential above or below the reversal potential, the mode selectivity of this inhibitor against NCX1 was not observable [94].

Further examination of the effect of KB-R7943 on net NCX transport at the cellular level have reported that the relative potency of the drug can vary widely depending on the cellular system studied, with IC_{50} values of 1.6 μM and >30 μM being reported for the inhibition of reverse and forward mode exchange in NCX1-transfected CCL39 cells whilst in cardiac sarcolemmal vesicles the IC_{50} values were much less distinct (5 μM and 11 μM respectively) [95]. The cell-dependence of these findings may, in part, reflect the expression of different NCX isoforms and their splice variants as this appears to significantly alter the effectiveness of the different reverse mode inhibitors [96]. For instance, KB-R7943 and YM-244769 are reported to be more a potent inhibitor of NCX3 compared to NCX1 or NCX2 [97, 98], whilst SN-6 and SEA-0400 are better inhibitors of NCX1 [99, 100]. Additionally, due to the NCX sensitivity to metabolic conditions [28, 101–103], differences in the basic cytosolic conditions and activity of different cell types may alter the activity of the NCX and so affect the sensitivity of the exchanger to reverse mode inhibitors. Whilst we have focussed here on the widely-used KB-R7943, it is important to note that reports of minimal mode selectivity have also been demonstrated in studies of other reverse mode selective inhibitors, such as SN-6 [104] and SEA0400 [105]. The potential pitfalls of relying on the reported mode selectivity of drugs have been demonstrated in our work in human platelets, in which we initially utilised a pharmacological methodology to characterise a role for reverse mode exchange in potentiating the Ca^{2+} signal elicited by the opening of store-operated channels (SOCs) [106], yet later reanalysis demonstrated that this potentiation was actually elicited by forward mode NCX activity regulating autocrine potentiation of the initial Ca^{2+} entry through the SOC [92]. These data therefore suggest that when using NCX inhibitors in cellular systems in which the action of these drugs have not been previously characterised, it is important to determine the effects of the doses of inhibitor applied on reverse and forward mode exchange to ensure mode selectivity [e.g. 107].

4.3.3 Selectivity of Na^+/Ca^{2+} Exchange Inhibitors

When assessing inhibitor function in new cell system, consideration needs to be paid to the many reported non-selective effects of these drugs on other components of the Ca^{2+} signalling apparatus and non-specific effects should be ruled out [108–115]. However, when assessing potential non-specific effects of inhibitors of Na^+/Ca^{2+} exchangers it is worth considering that the coupling of the exchanger with a cation conductance creates not only a hybrid protein complex which has altered Ca^{2+} signalling properties, but also a widened pharmacological profile which will encompass inhibitors of either component. This includes the potential of loss of forward mode activity enhancing Ca^{2+}-dependent feedback on the associated channel, as shown with the loss of L-type Ca^{2+} channel activity in cardiac and smooth muscle from NCX1 knockout mice [116, 117].

4.3.4 *Targeting the NCX-TRPC Interaction as a Novel Method for Blocking Reverse Mode Exchange*

Increased reverse mode Na^+/Ca^{2+} exchanger activity has been implicated in a number of pathological conditions including systemic and pulmonary arterial hypertension [45, 117–119], overactive bladder [90], cardiac arrhythmias [120, 121], malignant hyperthermia [122], and cerebral, renal and cardiac reperfusion injury [123–125]. This link to a wide range of pathologies suggests the possibility that a selective reverse-mode inhibitor of Na^+/Ca^{2+} exchange may provide significant therapeutic benefits. However, as described above (Sects. 4.3.2 and 4.3.3), current inhibitors lack the mode- and target-selectivity desired to provide an effective treatment. Recent work has demonstrated promising selectivity of a novel mode-independent NCX inhibitor, ORM-1013, in isolated cell systems [126], yet further work will need to ensure that this drug does not have unintended effects on other important functions for which forward mode activity has been suggested to be essential, including pacemaker activity of the sino-atrial node [127], insulin secretion [91, 128], renal Ca^{2+} and Na^+ retention [129] and platelet activation [72].

As detailed above, the tight interaction of Na^+/Ca^{2+} exchangers with non-selective cation channels appears to be essential for creating the local ionic conditions needed to elicit Ca^{2+} entry by reverse mode exchange in non-excitable cells. By breaking the close apposition of this complex, the Na^+/Ca^{2+} exchanger will likely not be exposed to the local rises in cytosolic [Na^+] required to elicit Ca^{2+} entry by reverse mode exchange, without preventing the exchanger from operating in forward mode. This suggests that selectively blocking reverse mode exchanger function may be possible through disrupting the normal interaction of the exchanger with TRPCs (or other associated Na^+-permeable channels). However, whilst this might provide a promising route for future drug development, there is currently no significant information on the proteins involved in the subcellular targeting of Na^+/Ca^{2+} exchangers, or the proteins scaffolding it with TRPC channels in non-excitable cells. All previous work on scaffolding proteins that interact with Na^+/Ca^{2+} exchangers has come from studies in neuronal or cardiac tissues, with proteins such as caveolin-3 [50], CRMP2 [57], α-fodrin and β-spectrin [68], ankyrin B [130], junctophilin 2 [131] and EHD3 [132] being identified as either binding proteins or critical regulators of the subcellular localisation of Na^+/Ca^{2+} exchangers. Future work defining Na^+/Ca^{2+} exchanger binding proteins and their roles in controlling the interaction with TRPC channels may help identify novel pharmacological targets for blocking reverse mode activity selectively in the cells associated with pathologies of abnormal Ca^{2+} entry via Na^+/Ca^{2+} exchangers.

Acknowledgements The authors work has been supported by the British Heart Foundation.

References

1. Domi T, Di Leva F, Fedrizzi L, Rimessi A, Brini M (2007) Functional specificity of PMCA isoforms? Ann NY Acad Sci 1099:237–246
2. Lytton J (2007) Na^+/Ca^{2+} exchangers: three mammalian gene families control Ca^{2+} transport. Biochem J 406(3):365–382
3. Blaustein MP, Lederer WJ (1999) Sodium/calcium exchange: its physiological implications. Physiol Rev 79(3):763–854
4. Baker PF, Blaustein MP (1968) Sodium-dependent uptake of calcium by crab nerve. Biochim Biophys Acta 150(1):167–170
5. Reuter H, Seitz N (1968) The dependence of calcium efflux from cardiac muscle on temperature and external ion composition. J Physiol 195:451–470
6. Martin DL, de Luca HF (1969) Influence of sodium on calcium transport by the rat small intestine. Am J Physiol 216(6):1351–1359
7. Pitts BJR (1979) Stoichiometry of sodium-calcium exchange in cardiac sarcolemmal vesicles. J Biol Chem 254(14):6232–6235
8. Rasgado-Flores H, Blaustein MP (1987) Na/Ca exchange in barnacle muscle cells has a stoichiometry of 3 Na^+:1 Ca^{2+}. Am J Physiol 252(5 Pt 1):C499–C504
9. Fujioka Y, Komeda M, Matsuoka S (2000) Stoichiometry of Na^+–Ca^{2+} exchange in inside-out patches excised from guinea-pig ventricular myocytes. J Physiol 523(2):339–351
10. Dong H, Dunn J, Lytton J (2002) Stoichiometry of the cardiac Na^+/Ca^{2+} exchanger NCX1.1 measured in transfected HEK cells. Biophys J 82(4):1943–1952
11. Kang TM, Hilgemann DW (2004) Multiple transport modes of the cardiac Na^+/Ca^{2+} exchanger. Nature 427:544–548
12. Schnetkamp PP, Basu DK, Szerencsei RT (1989) Na^+-Ca^{2+} exchange in bovine rod outer segments requires and transports K^+. Am J Physiol 257(1 Pt 1):C153–C157
13. Cervetto L, Lagnado L, Perry RJ, Robinson DW, MaNaughton PA (1989) Extrusion of calcium from rod outer segments is driven by both sodium and potassium gradients. Nature 337:740–743
14. Lytton J, Li XF, Dong H, Kraev A (2002) K^+-dependent Na^+/Ca^{2+} exchangers in the brain. Ann NY Acad Sci 976:382–393
15. Cai X, Lytton J (2004) The cation/Ca^{2+} exchanger superfamily: phylogenetic analysis and structural implications. Mol Biol Evol 21(9):1692–1703
16. Boyman L, Williams GS, Khananshvili D, Sekler I, Lederer WJ (2013) NCLX: the mitochondrial sodium calcium exchanger. J Mol Cell Cardiol 59:205–213
17. Quednau BD, Nicoll DA, Philipson KD (2004) The sodium/calcium exchanger family-SLC8. Pflugers Arch 447(5):543–548
18. Quednau BD, Nicoll DA, Philipson KD (1997) Tissue specificity and alternative splicing of the Na^+/Ca^{2+} exchanger isoforms NCX1, NCX2, and NCX3 in rat. Am J Physiol 272(2 Pt 1):C1250–C1261
19. Schnetkamp PP (2013) The SLC24 gene family of Na^+/Ca^{2+}-K^+ exchangers: from sight and smell to memory consolidation and skin pigmentation. Mol Aspects Med 34(2–3):455–464
20. Bers DM (2002) Cardiac excitation–contraction coupling. Nature 415:198–205
21. Lemos VS, Poburko D, Liao CH, Cole WC, van Breemen C (2007) Na^+ entry via TRPC6 causes Ca^{2+} entry via NCX reversal in ATP-stimulated smooth muscle cells. Biochem Biophys Res Commun 352(1):130–134
22. Schnetkamp PP (2004) The SLC24 Na^+/Ca^{2+}-K^+ exchanger family: vision and beyond. Pflugers Arch 447(5):683–688
23. DiPolo R (1979) Calcium influx in internally dialyzed squid giant axons. J Gen Physiol 73(1):91–113
24. Hilgemann DW (1989) Giant excised cardiac sarcolemmal membrane patches: sodium and sodium–calcium exchange currents. Pflugers Arch 415:247–249

25. Hilgemann DW (1990) Regulation and deregulation of cardiac Na^+–Ca^{2+} exchange in giant excised sarcolemmal membrane patches. Nature 344:242–245
26. Hilgemann DW, Matsuoka S, Nagel GA, Collins A (1992) Steady-state and dynamic properties of cardiac sodium–calcium exchange: sodium-dependent inactivation. J Gen Physiol 100(6):905–932
27. Altimimi HF, Schnetkamp PPE (2007) Na^+-dependent inactivation of the retinal cone/brain Na^+/Ca^{2+}-K^+ exchanger NCKX2. J Biol Chem 282(6):3720–3279
28. Boyman L, Hagen BM, Giladi M, Hiller R, Lederer WJ, Khananshvili D (2011) Proton-sensing Ca^{2+} binding domains regulate the cardiac Na^+/Ca^{2+} exchanger. J Biol Chem 286:28811–28820
29. Zhao J, Majewski H (2009) Endothelial nitric oxide attenuates Na^+/Ca^{2+} exchanger-mediated vasoconstriction in rat aorta. Br J Pharmacol 154(5):982–990
30. Simchowitz L, Foy MA, Cragoe EJ Jr (1990) A role for Na^+/Ca^{2+} exchange in the generation of superoxide radicals by human neutrophils. J Biol Chem 265(23):13449–13456
31. Liu T, O'Rourke B (2013) Regulation of the Na^+/Ca^{2+} exchanger by pyridine nucleotide redox potential in ventricular myocytes. J Biol Chem 288(44):31984–31992
32. Blaustein MP (1977) Effects of internal and external cations and ATP on sodium-calcium exchange and calcium-calcium exchange in squid axons. Biophys J 20(1):79–111
33. Hilgemann DW, Collins A, Matsuoka S (1992) Steady-state and dynamic properties of cardiac sodium–calcium exchange: secondary modulation by cytoplasmic calcium and ATP. J Gen Physiol 100(6):933–961
34. Hilgemann DW, Ball R (1996) Regulation of cardiac Na^+, Ca^{2+} exchange and KATP potassium channels by PIP_2. Science 273:956–959
35. Zhang YH, Hancox JC (2009) Regulation of Na^+-Ca^{2+} exchanger by protein kinase phosphorylation – still a paradox? Cell Calcium 45(1):1–10
36. Lee JY, Visser, Lee JS, Lee KH, Soh JW, Ho WK, Lytton J, Lee SH (2006) Protein kinase C-dependent enhancement of activity of rat brain NCKX2 heterologously expressed in HEK293 cells. J Biol Chem 281:39205–39216
37. Shaikh S, Samanta K, Kar P, Roy S, Chakraborti T, Chakraborti S (2010) m-Calpain-mediated cleavage of Na^+/Ca^{2+} exchanger-1 in caveolae vesicles isolated from pulmonary artery smooth muscle. Mol Cell Biochem 341(1–2):167–180
38. Bano D, Young KW, Guerin CJ, Lefeuvre R, Rothwell NJ, Naldini L, Rizzuto R, Carafoli E, Nicotera P (2005) Cleavage of the plasma membrane Na^+/Ca^{2+} exchanger in excitotoxicity. Cell 120(2):275–285
39. Pannaccione A, Secondo A, Molinaro P, D'Avanzo C, Cantile M, Esposito A, Boscia F, Scorziello A, Sirabella R, Sokolow S, Herchuelz A, Di Renzo G, Annunziato L (2012) A new concept: Aβ1-42 generates a hyperfunctional proteolytic NCX3 fragment that delays caspase-12 activation and neuronal death. J Neurosci 32(31):10609–10617
40. Roberts DE, Matsuda T, Bose R (2012) Molecular and functional characterization of the human platelet Na(+) /Ca(2+) exchangers. Br J Pharmacol 165(4):922–936
41. Berra-Romani R, Raqeeb A, Guzman-Silva A, Torres-Jácome J, Tanzi F, Moccia F (2010) Na^+-Ca^{2+} exchanger contributes to Ca^{2+} extrusion in ATP-stimulated endothelium of intact rat aorta. Biochem Biophys Res Commun 395(1):126–130
42. Nilius B, Owsianik G, Voets T, Peters JA (2007) Transient receptor potential cation channels in disease. Physiol Rev 87(1):165–217
43. Eder P, Probst D, Rosker C, Poteser M, Wolinski H, Kohlwein SD, Romanin C, Groschner K (2007) Phospholipase C-dependent control of cardiac calcium homeostasis involves a TRPC3-NCX1 signaling complex. Cardiovasc Res 73:111–119
44. Goel M, Zuo CD, Sinkins WG, Schilling WP (2007) TRPC3 channels colocalize with Na^+/Ca^{2+} exchanger and Na^+ pump in axial component of transverse-axial tubular system of rat ventricle. Am J Physiol 292(2):H874–H883
45. Zulian A, Baryshnikov SG, Linde CI, Mamlyn JM, Ferrari P, Golovina VA (2010) Upregulation of Na^+/Ca^{2+} exchanger and TRPC6 contributes to abnormal Ca^{2+} homeostasis in

arterial smooth muscle cells from Milan hypertensive rats. Am J Physiol Heart Circ Physiol 299(3):H624–H633
46. Rosker C, Graziani A, Lukas M, Eder P, Zhu MX, Romanin C, Groschner K (2004) Ca(2+) signaling by TRPC3 involves Na(+) entry and local coupling to the Na(+)/Ca(2+) exchanger. J Biol Chem 279(14):13696–13704
47. Teubl M, Groschner K, Kohlwein SD, Mayer B, Schmidt K (1999) Na(+)/Ca(2+) exchange facilitates Ca(2+)-dependent activation of endothelial nitric-oxide synthase. J Biol Chem 274(41):29529–25935
48. Remillard CV, Yuan JX (2006) Transient receptor potential channels and caveolin-1: good friends in tight spaces. Mol Pharmacol 70(4):1151–1154
49. Kuszczak I, Samson SE, Pande J, Shen DQ, Grover AK (2010) Sodium-calcium exchanger and lipid rafts in pig coronary artery smooth muscle. Biochem Biophys Acta 1808(3):589–596
50. Bossuyt J, Taylor BE, James-Kracke M, Hale CC (2002) Evidence for cardiac sodium-calcium exchanger association with caveolin-3. FEBS Lett 511(1–3):113–117
51. Ambudkar IS, Brazer SC, Liu X, Lockwich T, Singh B (2004) Plasma membrane localization of TRPC channels: role of caveolar lipid rafts. Novartis Found Symp 258:63–70, discussion 70–74, 98–102, 263–266
52. Flores-Soto E, Carbajal V, Reyes-García J, García-Hernández LM, Figueroa A, Checa M, Barajas-López C, Montaño LM (2010) In airways ATP refills sarcoplasmic reticulum via P2X smooth muscle receptors and induces contraction through P2Y epithelial receptors. Pflugers Arch 461(2):261–275
53. Chang HB, Gao X, Nepomuceno R, Hu S, Sun D (2015) Na^+/H^+ exchanger in the regulation of platelet activation and paradoxical effects of cariporide. Exp Neurol S0014-4886(15)00008-4 [Epub ahead of print]
54. Cingolani HE, Pérez NG, Pieske B, von Lewinski D, Camilión de Hurtado MC (2003) Stretch-elicited Na+/H+ exchanger activation: the autocrine/paracrine loop and its mechanical counterpart. Cariovasc Res 57(4):953–960
55. de Juan-Sanz J, Núñez E, Zafra F, Berrocal M, Corbacho I, Ibáñez I, Arribas-González E, Marcos D, López-Corcuera B, Mata AM, Aragón C (2014) Presynaptic control of glycine transporter 2 (GlyT2) by physical and functional association with plasma membrane Ca^{2+}-ATPase (PMCA) and Na^+-Ca^{2+} exchanger (NCX). J Biol Chem 289(49):34308–34324
56. Shen JB, Yang R, Pappano A, Liang BT (2014) Cardiac P2X purinergic receptors as a new pathway for increasing Na^+ entry in cardiac myocytes. Am J Physiol Heart Circ Physiol 307(10):H1469–H1477
57. Brustovetsky T, Pellman JJ, Yang XF, Khanna R, Brustovetsky N (2014) Collapsin response mediator protein 2 (CRMP2) interacts with N-methyl-D-aspartate (NMDA) receptor and Na^+/Ca^{2+} exchanger and regulates their functional activity. J Biol Chem 289(11):7470–7482
58. Kim YT, Park YJ, Jung SY, Seo WS, Suh CK (2005) Effects of Na^+-Ca^{2+} exchanger activity on the alpha-amino-3-hydroxy-5-methyl-4-isoxazolone-propionate-induced Ca^{2+} influx in cerebellar Purkinje neurons. Neuroscience 131(3):589–599
59. Rojas H, Colina C, Ramos M, Benaim G, Jaffe EH, Caputo C, DiPolo R (2007) Na^+ entry via glutamate transporter activates the reverse Na^+/Ca^{2+} exchange and triggers Ca_i^{2+}-induced Ca^{2+} release in rat cerebellar Type-1 astrocytes. J Neurochem 100(5):1188–1202
60. Poburko D, Potter K, van Breemen E, Fameli N, Liao C-H, Bassett O, Ruegg UR, van Breemen C (2006) Mitochondria buffer NCX-mediated Ca^{2+}-entry and limit its diffusion into vascular smooth muscle cells. Cell Calcium 40(4):359–371
61. Poburko D, Liao CH, Lemos VS, Lin E, Maruyama Y, William C, Cole WC, Cornelis van Breemen C (2007) Transient receptor potential channel 6–mediated, localized cytosolic [Na^+] transients drive Na^+/Ca^{2+} exchanger–mediated Ca^{2+} entry in purinergically stimulated aorta smooth muscle cells. Circ Res 101:1030–1038

62. van Breemen C, Fameli N, Evans AM (2013) Pan-junctional sarcoplasmic reticulum in vascular smooth muscle: nanospace Ca^{2+} transport for site- and function-specific Ca^{2+} signalling. J Physiol 591(8):2043–2054
63. Fameli N, van Breemen C, Kuo KH (2007) A quantitative model for linking Na+/Ca2+ exchanger to SERCA during refilling of the sarcoplasmic reticulum to sustain [Ca^{2+}] oscillations in vascular smooth muscle. Cell Calcium 42(6):565–575
64. Fameli N, Kuo KH, van Breemen C (2009) A model for the generation of localized transient [Na^+] elevations in vascular smooth muscle. Biochem Biophys Res Commun 389(3):461–465
65. Liang W, Buluc M, van Breemen C, Wang X (2004) Vectorial Ca^{2+} release via ryanodine receptors contributes to Ca^{2+} extrusion from freshly isolated rabbit aortic endothelial cells. Cell Calcium 36(5):431–443
66. Sahin-Erdemli I, Rashed SM, Songu-Mize E (1994) Rat vascular tissues express all three α-isoforms of Na^+-K^+-ATPase. Am J Physiol Heart Circ Physiol 266:H350–H353
67. Juhaszova M, Blaustein MP (1997) Na^+ pump low and high ouabain affinity alpha subunit isoforms are differently distributed in cells. Proc Natl Acad Sci U S A 94:1800–1805
68. Lencesova L, O'Neill A, Resneck WG, Bloch RJ, Blaustein MP (2004) Plasma membrane-cytoskeleton-endoplasmic reticulum complexes in neurons and astrocytes. J Biol Chem 279(4):2885–2893
69. Nazer MA, van Breemen C (1998) Functional linkage of Na^+-Ca^{2+} exchange and sarcoplasmic reticulum Ca^{2+} release mediates Ca^{2+} cycling in vascular smooth muscle. Cell Calcium 24(4):275–283
70. Hirota S, Pertens E, Janssen LJ (2007) The reverse mode of the Na(+)/Ca(2+) exchanger provides a source of Ca(2+) for store refilling following agonist-induced Ca(2+) mobilization. Am J Physiol Lung Cell Mol Physiol 292(2):L438–L447
71. Thomas MJ, Sjaastad I, Andersen K, Helm PJ, Wasserstrom JA, Sejersted OM, Ottersen OP (2003) Localization and function of the Na^+/Ca^{2+}-exchanger in normal and detubulated rat cardiomyocytes. J Mol Cell Cardiol 35(11):1325–1337
72. Sage SO, Pugh N, Farndale RW, Harper AGS (2013) Pericellular Ca^{2+} recycling potentiates thrombin-evoked Ca^{2+} signals in human platelets. Physiol Rep 1:e00085
73. Poburko D, Liao CH, van Breemen C, Demaurex N (2009) Mitochondrial regulation of sarcoplasmic reticulum Ca^{2+} content in vascular smooth muscle cells. Circ Res 104:104–112
74. Gilabert JA, Bakowski D, Parekh AB (2001) Energized mitochondria increase the dynamic range over which inositol 1,4,5-trisphosphate activates store-operated calcium influx. EMBO J 20(11):2672–2679
75. Parekh AB (2008) Mitochondrial regulation of store-operated CRAC channels. Cell Calcium 44(1):6–13
76. Malli R, Frieden M, Osibow K, Zoratti C, Mayer M, Demaurex N, Graier WF (2003) Sustained Ca^{2+} transfer across mitochondria is essential for mitochondrial Ca^{2+} buffering, store-operated Ca^{2+} entry, and Ca^{2+} store refilling. J Biol Chem 278:44769–44779
77. Golovina VA (2005) Visualization of localized store-operated calcium entry in mouse astrocytes. Close proximity to the endoplasmic reticulum. J Physiol 564(3):737–749
78. Luik RM, Wu MM, Buchanan J, Lewis RS (2006) The elementary unit of store-operated Ca^{2+} entry: local activation of CRAC channels by STIM1 at ER–plasma membrane junctions. J Cell Biol 174:815–825
79. Cabello OA, Schilling WP (1993) Vectorial Ca^{2+} flux from the extracellular space to the endoplasmic reticulum via a restricted cytoplasmic compartment regulates inositol 1,4,5-trisphosphate-stimulated Ca^{2+} release from internal stores in vascular endothelial cells. Biochem J 295(2):357–366
80. Alonso MT, Manjarrés IM, García-Sancho J (2012) Privileged coupling between Ca(2+) entry through plasma membrane store-operated Ca(2+) channels and the endoplasmic reticulum Ca(2+) pump. Mol Cell Endocrinol 353(1–2):37–44

81. Seip G, Schultheiss G, Kocks SL, Diener M (2001) Interaction between store-operated non-selective cation channels and the Na(+)-Ca(2+) exchanger during secretion in the rat colon. Exp Physiol 86(4):461–468
82. Baryshnikov SG, Pulina MV, Zulian A, Linde CI, Golovina VA (2009) Orai1, a critical component of store-operated Ca^{2+} entry, is functionally associated with Na^+/Ca^{2+} exchanger and plasma membrane Ca^{2+} pump in proliferating human arterial myocytes. Am J Physiol Cell Physiol 297(5):C1103–C1112
83. Andrikopoulos P, Baba A, Matsuda T, Djamgoz MB, Yaqoob MM, Eccles SA (2011) Ca^{2+} influx through reverse mode Na^+/Ca^{2+} exchange is critical for vascular endothelial growth factor-mediated extracellular signal-regulated kinase (ERK) 1/2 activation and angiogenic functions of human endothelial cells. J Biol Chem 286(44):37919–37931
84. Bradley E, Hollywood MA, Johnston L, Large RJ, Matsuda T, Baba A, McHale NG, Thornbury KD, Sergeant GP (2006) Contribution of reverse Na^+–Ca^{2+} exchange to spontaneous activity in interstitial cells of Cajal in the rabbit urethra. J Physiol 574(3):651–661
85. Zhang S, Yuan JX, Barrett KE, Dong H (2005) Role of Na^+/Ca^{2+} exchange in regulating cytosolic Ca^{2+} in cultured human pulmonary artery smooth muscle cells. Am J Physiol Cell Physiol 288(2):C245–C252
86. Staiano RI, Granata F, Secondo A, Petraroli A, Loffredo S, Frattini A, Annunziato L, Marone G, Triggiani M (2009) Expression and function of Na^+/Ca^{2+} exchangers 1 and 3 in human macrophages and monocytes. Eur J Immunol 39(5):1405–1418
87. Roberts DE, McNicol A, Bose R (2004) Mechanism of collagen activation in human platelets. J Biol Chem 279(19):19421–19430
88. Yang D, Yang D, Jia R, Tan J (2013) Na+/Ca2+ exchange inhibitor, KB-R7943, attenuates contrast-induced acute kidney injury. J Nephrol 26(5):877–885
89. Liu B, Peel SE, Fox J, Hall JP (2010) Reverse mode Na^+/Ca^{2+} exchange mediated by STIM1 contributes to Ca^{2+} influx in airway smooth muscle following agonist stimulation. Respir Res 11(1):168
90. Yamamura H, Cole WC, Kita S, Hotta S, Murata H, Suzuki Y, Ohya S, Iwamoto T, Imaizumi Y (2013) Overactive bladder mediated by accelerated Ca^{2+} influx mode of Na^+/Ca^{2+} exchanger in smooth muscle. Am J Physiol Cell Physiol 305(3):C299–C308
91. Hamming KS, Soliman D, Webster NJ, Searle GJ, Matemisz LC, Liknes DA, Dai XO, Pilinikunnil T, Riedel MJ, Dyck JR, Macdonald PE, Light PE (2010) Inhibition of beta-cell sodium-calcium exchange enhances glucose-dependent elevations in cytoplasmic calcium and insulin secretion. Diabetes 59(7):1686–1693
92. Harper AGS, Mason MJ, Sage SO (2009) A key role for dense granule secretion in potentiation of the Ca^{2+} signal arising from store-operated calcium entry in human platelets. Cell Calcium 45(5):413–420
93. Watano T, Kimura J, Morita T, Nakanishi H (1996) A novel antagonist, No 7943, of the Na^+/Ca^{2+} exchange current in guinea pig cardiac ventricular cells. Br J Pharmacol 119(30):555–563
94. Kimura J, Watano T, Kawahara M, Sakai E, Yatabe J (1999) Direction-independent block of bi-directional Na^+/Ca^{2+} exchange current by KB-R7943 in guinea-pig cardiac myocytes. Br J Pharmacol 128:969–974
95. Iwamoto T, Watano T, Shigekawa M (1996) A novel isothiourea derivative selectively inhibits the reverse mode of Na^+/Ca^{2+} exchange in cells expressing NCX1. J Biol Chem 271(37):22391–22397
96. Algara-Suárez P, Mejia-Elizondo R, Sims SM, Saavedra-Alanis VM, Espinosa-Tanguma R (2010) The 1.3 isoform of Na^+-Ca^{2+} exchanger expressed in guinea pig tracheal smooth muscle is less sensitive to KB-R7943. J Physiol Biochem 66(2):117–125
97. Iwamoto T, Kita S, Uehara A, Inoue Y, Taniguchi Y, Imanaga I, Shigekawa M (2001) Structural domains influencing sensitivity to isothiourea derivative inhibitor KB-R7943 in cardiac Na(+)/Ca(2+) exchanger. Mol Pharmacol 59(3):524–531

98. Iwamoto T, Kita S (2006) YM-244769, a novel Na^+/Ca^{2+} exchange inhibitor that preferentially inhibits NCX3, efficiently protects against hypoxia/reoxygenation-induced SH-SY5Y neuronal cell damage. Mol Pharmacol 70(6):2075–2083
99. Iwamoto T, Inoue Y, Ito K, Sakaue T, Kita S, Katsuragi T (2004) The exchanger inhibitory peptide region-dependent inhibition of Na^+/Ca^{2+} exchange by SN-6 [2-[4-(4-nitrobenzyloxy) benzyl]thiazolidine-4-carboxylic acid ethyl ester], a novel benzyloxyphenyl derivative. Mol Pharmacol 66(1):45–55
100. Iwamoto T, Kita S, Uehara A, Imanaga I, Matsuda T, Baba A, Katsuragi T (2004) Molecular determinants of Na^+/Ca^{2+} exchange (NCX1) inhibition by SEA0400. J Biol Chem 279(9):7544–7553
101. Hamming KS, Riedel MJ, Soliman D, Matemisz LC, Webster NJ, Searle GJ, MacDonald PE (2008) Splice variant-dependent regulation of beta-cell sodium-calcium exchange by acyl-coenzyme As. Mol Endocrinol 22(10):2293–2306
102. Liu T, O'Rourke B (2013) Regulation of the Na^+/Ca^{2+} exchanger by pyridine nucleotide redox in ventricular myocytes. J Biol Chem 288(440):31984–31992
103. Soliman D, Hamming KS, Matemisz LC, Light PE (2009) Reactive oxygen species directly modify sodium-calcium exchanger activity in a splice variant-dependent manner. J Mol Cell Cardiol 47(5):595–602
104. Niu CF, Watanabe Y, Ono K, Iwamoto T, Yamashita K, Satoh H, Urushida T, Hayashi H, Kimura J (2007) Characterization of SN-6, a novel Na^+/Ca^{2+} exchange inhibitor in guinea pig cardiac ventricular myocytes. Eur J Pharmacol 573(1–3):161–169
105. Ozdemir S, Bito V, Holemans P, Vinet L, Mercadier J-J, Varro A, Sipido KR (2008) Pharmacological inhibition of Na/Ca exchange results in increased cellular Ca^{2+} load attributable to the predominance of forward mode block. Circ Res 102:1398–1405
106. Harper AGS, Sage SO (2007) A key role for reverse Na^+/Ca^{2+} exchange influenced by the actin cytoskeleton in store-operated Ca^{2+} entry in human platelets: evidence against the de novo conformational coupling hypothesis. Cell Calcium 42(6):606–617
107. Satoh H, Ginsburg KS, Qing K, Terada H, Hayashi H, Bers DM (2000) KB-R7943 block of Ca(2+) influx via Na(+)/Ca(2+) exchange does not alter twitches or glycoside inotropy but prevents Ca(2+) overload in rat ventricular myocytes. Circulation 101(12):1441–1446
108. Reuter H, Henderson SA, Han T, Matsuda T, Baba A, Ross RS, Goldhaber JI, Philipson KD (2002) Knockout mice for pharmacological screening: testing the specificity of $Na^+–Ca^{2+}$ exchange inhibitors. Circ Res 91(2):90–92
109. Pintado AJ, Herrero CJ, Garcia AG, Montiel C (2000) The novel Na^+/Ca^{2+} exchange inhibitor KB-R7943 also blocks native and expressed neuronal nicotinic receptors. Br J Pharmacol 130:1893–1902
110. Santo-Domingo J, Vay L, Hernandez-Sanmiguel E, Lobaton CD, Moreno A, Montero M, Alvarez J (2007) The plasma membrane Na^+/Ca^{2+} exchange inhibitor KB-R7943 is also a potent inhibitor of the mitochondrial Ca^{2+} uniporter. Br J Pharmacol 151:647–654
111. Birinyi P, Acsai K, Banyasz T, Toth A, Horvath B, Virag L, Szentandrassy N, Magyar J, Varro A, Fulop F, Nanasi PP (2005) Effects of SEA0400 and KB-R7943 on Na^+/Ca^{2+} exchange current and L-type Ca^{2+} current in canine ventricular cardiomyocytes. Naunyn Schmiedebergs Arch Pharmacol 372:63–67
112. Ouardouz M, Zamponi GW, Barr W, Kiedrowski L, Stys PK (2005) Protection of ischemic rat spinal cord white matter: dual action of KB-R7943 on Na^+/Ca^{2+} exchange and L-type Ca^{2+}channels. Neuropharmacology 48:566–575
113. Kraft R (2007) The Na^+/Ca^{2+} exchange inhibitor KB-R7943 potently blocks TRPC channels. Biochem Biophys Res Commun 361:230–236
114. Brustovetsky T, Brittain MK, Sheets PL, Cummins TR, Pinelis V, Brustovetsky N (2011) KB-R7943, an inhibitor of the reverse Na^+/Ca^{2+} exchanger, blocks N-methyl-D-aspartate receptor and inhibits mitochondrial complex I. Br J Pharmacol 162(1):255–270

115. Sobolevsky AI, Khodorov BI (1999) Blockade of NMDA channels in acutely isolated rat hippocampal neurons by the Na^+/Ca^{2+} exchange inhibitor KB-R7943. Neuropharmacology 38(8):1235–1242
116. Pott C, Yip M, Goldhaber JI, Philipson KD (2007) Regulation of cardiac L-type Ca^{2+} current in Na^+-Ca^{2+} exchanger knockout mice: functional coupling of the Ca^{2+} channel and the Na^+-Ca^{2+} exchanger. Biophys J 92(4):1431–1437
117. Zhang J, Ren C, Chen L, Navedo MF, Antos LK, Kinsey SP, Iwamoto T, Philipson KD, Kotlikoff MI, Santana LF, Wier WG, Matteson DR, Blaustein MP (2010) Knockout of Na^+/Ca^{2+} exchanger in smooth muscle attenuates vasoconstriction and L-type Ca^{2+} channel current and lowers blood pressure. Am J Physiol Heart Circ Physiol 298(5):H1472–H1683
118. Blaustein MP, Leenen FH, Chen L, Golovina VA, Hamlyn JM, Pallone TL, Van Huysse JW, Zhang J, Wier WG (2011) How NaCl raises blood pressure: a new paradigm for the pathogenesis of salt-dependent hypertension. Am J Physiol Heart Circ Physiol 302(5):H1031–H1049
119. Zhang S, Dong H, Rubin LJ, Yuan JX (2007) Upregulation of Na^+/ Ca^{2+} exchanger contributes to the enhanced Ca^{2+} entry in pulmonary artery smooth muscle cells from patients with idiopathic pulmonary arterial hypertension. Am J Physiol Cell Physiol 292:C2297–C2305
120. Amran MS, Homma N, Hashimoto K (2003) Effects of SEA0400, a selective Na^+-Ca^{2+} exchange inhibitor, on digitalis and aconitine-induced ventricular arrhythmias in guinea pig. J Pharmacol Sci 91(Suppl 1), 1P267
121. Bourgonje VJ, Vos MA, Ozdemir S, Doisne N, Acsai K, Varro A, Sztojkov-Ivanov A, Zupko I, Rauch E, Kattner L, Bito V, Houtman M, van der Nagel R, Beekman JD, van Veen TA, Sipido KR, Antoons G (2013) Combined Na(+)/Ca(2+) exchanger and L-type calcium channel block as a potential strategy to suppress arrhythmias and maintain ventricular function. Circ Arrhythm Electrophysiol 6(2):371–379
122. Altamirano F, Eltit JM, Robin G, Linares N, Ding X, Pessah IN, Allen PD, López JR (2014) Ca^{2+} influx via the Na^+/Ca^{2+} exchanger is enhanced in malignant hyperthermia skeletal muscle. J Biol Chem 289(27):19180–19190
123. Tomes DJ, Agrawal SK (2002) Role of Na^+-Ca^{2+} exchanger after traumatic or hypoxic/ischemic injury to spinal cord white matter. Spine J 2(1):35–40
124. Yamashita J, Kita S, Iwamoto T, Ogata M, Takaoka M, Tazawa N, Nishikawa M, Wakimoto K, Shigekawa M, Komuro I, Matsumura I (2003) Attenuation of ischemia/reperfusion-induced renal injury in mice deficient in Na^+/Ca^{2+}exchanger. J Pharmacol Exp Ther 304:284–293
125. Hagihara H, Yoshikawa Y, Ohga Y, Takenaka C, Murata KY, Taniguchi S, Takaki M (2004) Na^+/Ca^{2+} exchange inhibition protects the rat heart from ischemia-reperfusion injury by blocking energy-wasting processes. Am J Physiol Heart Circ Physiol 288(4):H1699–H1707
126. Jost N, Nagy N, Corici C, Kohajda Z, Horváth A, Acsai K, Biliczki P, Levijoki J, Pollesello P, Koskelainen T, Otsomaa L, Tóth A, Papp JG, Varró A, Virág L (2013) ORM-10103, a novel specific inhibitor of the Na^+/Ca^{2+} exchanger, decreases early and delayed afterdepolarizations in the canine heart. Br J Pharmacol 170(4):768–778
127. Groenke S, Larson ED, Alber S, Zhang R, Lamp ST, Ren X, Nakano H, Jordan MC, Karagueuzian HS, Roos KP, Nakano A, Proenza C, Philipson KD, Goldhaber JI (2013) Complete atrial-specific knockout of sodium-calcium exchange eliminates sinoatrial node pacemaker activity. PLoS One 8(11):e81633
128. Nguidjoe E, Sokolow S, Bigabwa S, Pachera N, D'Amico E, Allagnat F, Vanderwinden JM, Sener A, Manto M, Depreter M, Mast J, Joanny G, Montanya E, Rahier J, Cardozo AK, Eizirik DL, Schurmans S, Herchuelz A (2011) Heterozygous inactivation of the Na/Ca exchanger increases glucose-induced insulin release, β-cell proliferation, and mass. Diabetes 60(8):2076–2085
129. Gotoh Y, Kita S, Fujii M, Tagashira H, Horie I, Arai Y, Uchida S, Iwamoto T (2015) Genetic knockout and pharmacologic inhibition of NCX2 cause natriuresis and hypercalciuria. Biochem Biophys Res Commun 456(2):670–675

130. Mohler PJ, Davis JQ, Bennett V (2005) Ankyrin-B coordinates the Na/K ATPase, Na/Ca exchanger, and $InsP_3$ receptor in a cardiac T-tubule/SR microdomain. PLoS Biol 3(12):e423
131. Wang W, Landstrom AP, Wang Q, Munro ML, Beavers D, Ackerman MJ, Soeller C, Wehrens XH (2014) Reduced junctional Na^+/Ca^{2+}-exchanger activity contributes to sarcoplasmic reticulum Ca^{2+} leak in junctophilin-2-deficient mice. Am J Physiol Heart Circ Physiol 307(9):H1317–H1326
132. Curran J, Makara MA, Little SC, Musa H, Liu B, Wu X, Polina I, Alecusan JS, Wright P, Li J, Billman GE, Boyden PA, Gyorke S, Band H, Hund TJ, Mohler PJ (2014) EHD3-dependent endosome pathway regulates cardiac membrane excitability and physiology. Circ Res 115(1):68–78

Chapter 5
Role of TRPC Channels in Store-Operated Calcium Entry

Hwei Ling Ong, Lorena Brito de Souza, and Indu S. Ambudkar

Abstract Store-operated calcium entry (SOCE) is a ubiquitous Ca^{2+} entry pathway that is activated in response to depletion of Ca^{2+} stores within the endoplasmic reticulum (ER) and contributes to the control of various physiological functions in a wide variety of cell types. The transient receptor potential canonical (TRPC) channels (TRPCs 1–7), that are activated by stimuli leading to PIP_2 hydrolysis, were first identified as molecular components of SOCE channels. TRPC channels show a miscellany of tissue expression, physiological functions and channel properties. However, none of the TRPC members display currents that resemble I_{CRAC}. Intensive search for the CRAC channel component led to identification of Orai1 and STIM1, now established as being the primary constituents of the CRAC channel. There is now considerable evidence that STIM1 activates both Orai1 and TRPC1 via distinct domains in its C-terminus. Intriguingly, TRPC1 function is not only dependent on STIM1 but also requires Orai1. The critical functional interaction between TRPC1 and Orai1, which determines the activation of TRPC1, has also been identified. In this review, we will discuss current concepts regarding the role of TRPC channels in SOCE, the physiological functions regulated by TRPC-mediated SOCE, and the complex mechanisms underlying the regulation of TRPCs, including the functional interactions with Orai1 and STIM1.

Keywords TRPC channels • SOCE • Trafficking • Function

H.L. Ong (✉) • L.B. de Souza • I.S. Ambudkar (✉)
Secretory Physiology Section, Molecular Physiology and Therapeutics Branch, National Institute of Dental and Craniofacial Research, National Institutes of Health, Building 10, Room 1N-113, NIH, Bethesda, MD 20892, USA
e-mail: ongh@mail.nih.gov; indu.ambudkar@nih.gov

J.A. Rosado (ed.), *Calcium Entry Pathways in Non-excitable Cells*, Advances in Experimental Medicine and Biology 898, DOI 10.1007/978-3-319-26974-0_5

5.1 Introduction

Elevation of cytosolic calcium levels ($[Ca^{2+}]_i$) in response to neurotransmitter stimulation of cells acts as a trigger for activation of many physiological processes, including cell proliferation and differentiation, cell migration, lymphocyte activation, endothelial cell function, as well as protein and fluid secretion from exocrine gland cells. Plasma membrane Ca^{2+} entry channels contribute to $[Ca^{2+}]_i$ elevation and provide critical Ca^{2+} signals that are utilized for regulation of different cell functions. Store-operated calcium entry (SOCE) is a major ubiquitous Ca^{2+} entry pathway that contributes to the control of various physiological functions in a wide variety of cell types. SOCE is a unique mechanism in that it is activated in response to depletion of Ca^{2+} stores within the endoplasmic reticulum (ER). Physiologically, this occurs following stimulation of plasma membrane receptors that lead to phosphatidylinositol 4,5-bisphosphate (PIP_2) hydrolysis and generation of inositol 1,4,5-triphosphate (IP_3). IP_3 binds to its receptor (IP_3R) in the ER membrane and induces Ca^{2+} release from the ER, resulting in a decrease in ER-$[Ca^{2+}]$ and activation of SOCE. This Ca^{2+} entry can also be activated by treating cells with sarco-endoplasmic reticulum (SERCA) pump blockers like thapsigargin or cyclopiazonic acid that inhibit Ca^{2+} uptake into ER, unmasking a passive Ca^{2+} leak pathway that has not yet been clearly elucidated. Importantly, this leads to loss of ER-Ca^{2+} and triggers activation of SOCE in the absence of receptor-stimulated signaling. SOCE is inactivated by refilling the Ca^{2+} stores within the ER, providing further evidence that the activity of channels mediating SOCE are governed by the status of $[Ca^{2+}]$ in the ER [1–3]. The first channel current associated with SOCE, Ca^{2+}-release activated calcium current (I_{CRAC}), was measured in mast cells and T lymphocytes. This highly Ca^{2+}-selective current has a characteristic inward rectification with reversal potential >+40 mV [4–7]. Under the same experimental conditions, currents with varying characteristics and ionic selectivities, ranging from relatively Ca^{2+}-selective to non-selective, have been described in different cell types. These currents have been generally referred to as store-operated calcium current (I_{SOC}) to distinguish them from I_{CRAC} [2, 8].

Search for the molecular components of SOCE, led to the identification of the transient receptor potential canonical (TRPC) channels, part of the superfamily of TRP channels. Mammalian TRPC channels were cloned based on the *Drosophila* TRP channel, which functions as a light-sensitive Ca^{2+}-permeable channel involved in phototransduction. The *Drosophila* phototransduction process, a phospholipase C-mediated pathway [9–11], provided further impetus to the search for mammalian TRP channels. The TRPC subfamily consists of seven members (TRPCs 1–7) that are divided into four subsets based on their amino acid (aa) homology: TRPC1, TRPC2, TRPC3/TRPC6/TRPC7 and TRPC4/TRPC5. All TRPC channels display activation in response to receptor-stimulated PIP_2 hydrolysis and have six transmembrane domains with a pore-forming domain localized between the fifth and sixth domains. The channels contain N-terminal ankyrin repeats, a highly conserved TRP domain in the C-terminus, several calmodulin (CaM)-binding domains and a putative IP_3R binding site [9–11]. TRPC channels show diverse tissue expression, physiological functions and channel properties. Recent reviews have presented a general overview of the molecular components and mechanisms regulating SOCE [1, 12] as well as overviews of the individual TRPC channels: TRPC1 [13], TRPC2 [14], TRPC3 [15], TRPC4 [16], TRPC5 [17], TRPC6 [18], and TRPC7 [19]. We will not be discussing TRPC2, which is a pseudogene in humans [20, 21], in this review. While the discovery of TRPC channels spurred a large number of studies, none of the TRPC family of Ca^{2+}-permeable cation channels generated currents that resembled I_{CRAC}. Thus, identity of the components for this channel, as well as the regulatory proteins in SOCE continued to be a major focus in the field. Intensive search for these finally led to the identification of the CRAC channel component, Orai1, a four-transmembrane domain protein which is assembled as a hexamer to form the pore of the CRAC channel. Two other Orai proteins, Orai2 and Orai3, were also identified and reported to have some similarity with Orai1 and display store-dependent activation. However, since they also contribute to other non-SOCE mechanisms, such as the arachidonic acid-activated channels, further studies are required to fully understand their physiological function. Importantly, the main components involved in sensing ER-[Ca^{2+}] and activating SOCE were also identified. The STIM family of proteins includes two members, STIM1 and STIM2, both of which are Ca^{2+}-sensing proteins that are localized in the ER membrane and sense [Ca^{2+}] within the ER lumen to regulate SOCE. Of the two, STIM1 has been more extensively studied and is now well established as the critical and indispensable regulatory component of SOCE [22]. Furthermore, there is now considerable evidence that STIM1 can activate both Orai1 and TRPC1. The domains of STIM1 involved in gating of these channels are also known. Intriguingly, TRPC1 function is not only dependent on STIM1 but also requires Orai1. The critical functional interaction between TRPC1 and Orai1, which determines the activation of TRPC1, has also been resolved. In the following sections of this review, we will discuss current concepts regarding the role of TRPC channels in SOCE, the physiological functions regulated by TRPC-mediated SOCE, and the complex mechanisms underlying the regulation of TRPCs, including the functional interactions with Orai1 and STIM1.

5.2 Contribution of TRPC Channels to SOCE

All seven TRPC channels have been implicated as components of SOCE. Furthermore, a variety of physiological functions have been associated with TRPC-mediated SOCE. Recent studies also demonstrate that some human diseases are linked to either loss or gain of function of TRPC channels [23–25]. However, not all the TRPC channels consistently display the hallmarks of SOCE, namely (i) activation by store depletion in response to stimulation with an agonist or treatment with SERCA pump blockers, and (ii) inhibition by Gd^{3+} (1 μM) and 2-aminophenyl borate (2-APB; ≤10 μM). A large number of the studies assessing the role of TRPCs have used heterologous expression systems where the channels are relatively overexpressed. This does not always result in generation of a functional channel in cells. Some studies have also demonstrated that the mode of regulation of the channels appears to differ depending on the level of their expression. This led to the suggestion that channel overexpression likely results in an unbalanced stoichiometry between TRPCs and the endogenous accessory proteins that regulate and/or modulate their activities. In contrast, more consistent and conclusive data have been provided by studies which assess the function of endogenous TRPCs in SOCE by modulating their expression and/or function in cell lines, primary cell preparations, as well as animal models. So far, the strongest evidence for the contribution of TRPC channels to SOCE has been provided for TRPC1 and TRPC4, whereas the contribution of TRPC3 to SOCE appears to be dependent on cell type and level of expression. TRPCs 5, 6 and 7 have been generally described to be store-independent, with a few exceptions. Note that unlike with Orai1 or STIM1, TRPC channel contribution to SOCE is not seen in all cell types.

TRPC1 was the first mammalian TRPC channel to be cloned and reported to have a role in SOCE [20, 21, 26, 27]. Among the many studies reported, exogenous expression of TRPC1 did not consistently increase SOCE while knockdown of endogenous TRPC1 significantly decreased SOCE (e.g. in HSG, smooth muscle and endothelial cells, as well as platelets) [26–32]. Further conclusive evidence was provided by studies with mice lacking TRPC1 ($TRPC1^{-/-}$), which despite having normal viability, development, and behavior [33], showed reduced SOCE in cell preparations from several tissues. Among these, salivary gland and pancreatic acinar cells and aortic endothelial cells from $TRPC1^{-/-}$ mice displayed significant reductions in SOCE as well as attenuation in Ca^{2+}-dependent physiological functions [34–36]. SOCE is fundamentally important for fluid secretion in salivary glands and for protein secretion in the exocrine pancreas. $TRPC1^{-/-}$ mice displayed reduction in salivary gland fluid secretion which was associated with a decrease in SOCE and K_{Ca} activity in acinar cells from the mice [36, 37]. Similarly defects in Ca^{2+}-activated Cl^- channel activity and protein secretion, as a consequence of reduced SOCE, were reported in pancreatic acinar cells [34]. Notably while there is no change in Orai1 in salivary gland and pancreatic acinar cells from $TRPC1^{-/-}$ mice, the channel does not appear to compensate for the lack of TRPC1 or support cell function on its own. Hence, decreased secretory function in these exocrine glands is primarily due to the loss of TRPC1-mediated SOCE. In endothelial cells, TRPC1 forms a heteromeric channel with TRP vanilloid 4 (TRPV4) to mediate SOCE. This Ca^{2+} entry was significantly reduced in cells from $TRPC1^{-/-}$ mice which adversely impacted vasorelaxation [35].

The caveolae-residing protein, caveolin-1 (Cav-1), is an important modulator of TRPC1 activity and functions as a plasma membrane scaffold for the channel. In the absence of Cav-1, TRPC1 is mislocalized and is unable to interact with STIM1, which is a requirement for TRPC1 activation [38]. Consistent with this, localization of TRPC1, its interaction with STIM1, as well as SOCE were disrupted in salivary gland acinar cells from Cav-$1^{-/-}$ mice [39]. Together, these findings further establish a role for TRPC1 in mediating SOCE in salivary gland cells. Other physiological functions that are dependent on TRPC1-mediated SOCE are the contractile function of glomerular mesangial cells [40, 41] and osteoclast formation and function [42]. Loss of TRPC1 has also been implicated in aberrant vasorelaxation [43], muscle fatigue and slower regeneration after muscle injury [44, 45], whereas elevated TRPC1 expression has been linked to myopathies such as those observed in patients with Duchenne's Muscular Dystrophy and *mdx* mice lacking dystrophin [46–48]. However, it remains to be established whether these effects are due to changes in TRPC1-SOCE.

TRPC3 is reported to contribute to both the store-operated and receptor-activated calcium entry pathways. Loss of endogenous TRPC3 in cell lines and tissue preparations (pancreatic acinar and submandibular gland cells) from TRPC3$^{-/-}$ mice led to significant reductions in SOCE [49, 50]. In contrast, overexpression of TRPC3 increased SOCE in COS, HEK293 and HEK293T cells, as well as DT40 chicken B-lymphocytes. However, when the channel was expressed to very high levels, the regulatory mode was switched from store-operated to receptor-activated. Cells with relatively lower levels of TRPC3 expression displayed Gd^{3+} (1 μM)-sensitive Ca^{2+} entry, while those with higher levels of channel expression required higher [Gd^{3+}]. Hence, the mechanism by which TRPC3 is regulated appears to be determined by the level of channel expression in the cells [51–54]. TRPC3-mediated Ca^{2+} entry can also contribute to pathology and tissue damage. Pancreatic acini from TRPC3$^{-/-}$ mice showed significant protection from acute pancreatitis induced by hyper-activation of SOCE. Similar effects were seen by blocking channel function in TRPC3$^{+/+}$ mice by treatment with pyrazole 3, a TRPC3 inhibitor [55, 56]. Unlike TRPC3, TRPC6 and TRPC7 channels are largely believed be receptor-activated as both channels are consistently activated by the second messenger, diacylglycerol and its analogs [11, 18, 19, 57].

Both TRPC4 and TRPC5 have been suggested to contribute to SOCE, although there are very few studies reported for either channel. Moreover, TRPC5 can also be directly activated by Ca^{2+}, which makes it difficult to establish conclusively whether TRPC5 is directly regulated by store depletion [58, 59]. Exogenous expression of TRPC4 in HEK293 cells increased SOCE and generated a relatively Ca^{2+}-selective, inwardly rectifying current [60]. Similar results were obtained by overexpression TRPC4 in CHO, RBL cells [61] and *Xenopus laevis* oocytes [62]. Further evidence for TRPC4-mediated SOCE was provided by studies where TRPC4 expression was suppressed in several cell lines or by knockout of the channel in mice (TRPC4$^{-/-}$ mice). Following siRNA treatment, TRPC4-mediated SOCE was diminished in mouse mesangial cells [63], human adrenal cells [64], both mouse and human endothelial cells [65], human gingival keratinocytes [66], human corneal epithelial cells [67] and human pulmonary artery smooth muscle cells [68]. Additionally, TRPC4 forms a heteromeric channel complex with TRPC1 in human mesangial cells [69] as well as human and mouse endothelial cells [65]. Similar to what has been reported for TRPC1$^{-/-}$, knockout of TRPC4 did not adversely impact mortality and fertility of the mice. Nonetheless, TRPC4$^{-/-}$ mice show significantly reduced TRPC4-mediated SOCE in aortic [70] and lung endothelial cells [65, 71], resulting in defective regulation of vascular tone and endothelial permeability, respectively.

TRPC channels interact with numerous proteins which can underlie the diversity of calcium channel activity, their regulation, and specificity of downstream signaling events in the cells (recently reviewed in [1]). Not only do TRPC channels have the ability to undergo homomeric interactions to form functional channels, they also interact with other TRPCs to generate functional heteromeric channels. Most of the available data in this regard comes from studies with exogenously expressed channels. TRPC1 interacts with TRPC4 and TRPC5, whereas TRPC3 interacts with TRPC6 and TRPC7 [72–74]. It is presently not clear whether the resulting heteromeric channels have distinct properties and functions as compared to those of the individual channels. Very few studies have elucidated the status of endogenous store-dependent heteromeric TRPC channels and their physiological function. The contribution of endogenous heteromeric channel complexes to SOCE have been reported for TRPC1/TRPC3 in a human parotid gland ductal cell line [75] and rat H19-7 hippocampal cell lines [76]; TRPC1/TRPC5 in vascular smooth muscle [77]; TRPC1/TRPC4 in endothelial cells [65], and TRPC1/TRPC3/TRPC7 in HEK293 cells [50]. Given the overlapping expression of more than one TRPC channel in different cells and tissues, some physiological functions may involve multiple channels. For example, SOCE mediated by both TRPC1 and TRPC4 has been proposed to control endothelial cell permeability [65] and myogenesis [78, 79]. Multiple TRPC channels have been implicated in cardiac hypertrophy [80, 81], but it is not clear whether aberrant TRPC-mediated SOCE underlies this phenomena. A few studies have utilized double TRPC knockout mouse models to determine the role of heteromeric TRPC channels. Knockout of both TRPC1 and TRPC4 in mice severely impaired neuronal burst firing and caused neurodegeneration [82], whereas loss of both TRPC3 and TRPC6 impaired sensitivity to mechanical pressure and hearing [83]. The underlying basis for creating these double knockout mice models was the preponderance of co-expressed TRPC1 and TRPC4 in the brain [82] and TRPC3 and TRPC6 in sensory neurons and cochlear hair cells [83]. Whether the pathophysiological effects observed from double knockouts of endogenous TRPC heteromeric channel complexes can be conclusively linked to impaired regulation of SOCE also remains to be shown.

TRPC channels have also been found to associate with other TRP channels, including TRPV6, TRPV4, although in most of these cases it is not clear whether the associating channels form a single channel pore and/or contribute to SOCE. Co-expression of TRPC1 and TRPV4 resulted in formation of a heteromeric channel complex that is activated in response to store depletion in HEK293, vascular smooth muscle and endothelial cells. Moreover, the TRPC1/TRPV4 heteromeric channel exhibited distinct current characteristics when compared to currents mediated by either TRPV4 or TRPC1 alone [35, 84, 85]. TRPC1 was reported to interact with TRPV6 and exert negative regulation of TRPV6 function [86]. A critical heteromeric interaction involving TRPC channels is the TRPC-Orai1 interaction. TRPC1, TRPC3 and TRPC6 functionally interact with Orai1. TRPC1-Orai1 interaction has been confirmed by co-immunoprecipitation data as well as TIRFM measurements, where store depletion-dependent clustering of the two channels has been observed in several cell types [87–92]. Further, and more importantly, Orai1 is required for TRPC function, as knockdown of endogenous Orai1 abolished TRPC1 channel activation [88, 92]. A similar requirement for Orai1 has been reported for activation of TRPC3 and TRPC6 in response to ER-Ca^{2+} store depletion [93–95]. The mechanism underlying the Orai1-dependent regulation of TRPC1 has now been resolved (more details will be presented below). Importantly, TRPC1 and Orai1 have been shown to generate two distinct channels that appear to contribute to specific cellular functions [88, 96]. Interestingly, a recent study reported that a splice variant of TRPC1 interacts with and positively regulates Orai1 channel activity in HEK293 cells. This splice variant, TRPC1ε, was first identified in early pre-osteoclasts and together with I-mfa (an inhibitor of MyoD family), has been proposed to function antagonistically to decrease Orai1 channel activity, fine tuning the Ca^{2+} signaling process that regulates osteoclastogenesis [42]. The multiplicity of interactions between various TRPC channels, as well as between TRPC and other channels or regulatory proteins, lead to the generation of a plethora of signaling complexes that can regulate a wide variety of cellular functions. It is possible that the composition of these heteromeric channels as well as the interacting signaling proteins depends on the type of cell and the particular physiological function to be regulated. There are an increasing number of studies that highlight the importance of spatial and temporal aspects as well as the magnitude of Ca^{2+} signals as major determining factors in the regulation of cellular responses to different physiological stimuli. It is important that these should be taken into account when the physiological functions of TRPC channels are being assessed.

5.3 Role of STIM1 and STIM2 in SOCE and TRPC Channel Regulation

STIM1 and STIM2 were discovered in studies using siRNA screening to identify proteins required for SOCE. Both proteins reside within the ER and during resting (unstimulated) conditions, have Ca^{2+} bound to the luminal N-terminal EF hand domains. Following store depletion, Ca^{2+} is released from the EF hand which leads to multimerization of the protein and translocation to the peripheral region of the cells where it concentrates in the form of puncta within distinct ER-plasma membrane (ER-PM) junctions. Within this microdomain, the ER membrane and plasma membrane are in close apposition to each other [22, 97, 98]. More importantly, the proximity between the two membranes allows STIM1 in the ER to interact with and gate both TRPC and Orai1 channels. Different aa regions in the cytosolic C-terminus of STIM1 are involved in activating Orai1 and TRPC1. Orai1 activation is mediated by the STIM1 Orai1 Activating Region (SOAR; aa 344–442) [99], whereas TRPC channel gating occurs via the polybasic domain (aa 672–685) [91, 100]. It has been suggested that TRPC1 and TRPC4 are the main TRPC channels that can interact with and be gated by STIM1. However, if other TRPC channels are assembled in a heteromeric channel complex with either TRPC1 or TRPC4, they appear to become store-dependent due to the activation of TRPC1 and TRPC4 by STIM1. For example, although STIM1 does not interact with TRPC3 and TRPC6, STIM1 can activate TRPC1/TRPC3 or TRPC4/TRPC6 channels [101].

Strong evidence for STIM1 in gating TRPC channels comes from studies showing an effect on channel activity of either knockdown or overexpression of STIM1 or STIM1 with mutations that impair STIM1-TRPC channel interactions. TRPC1 was the first TRPC channel shown to be regulated by STIM1. Knockdown of endogenous STIM1 severely reduced endogenous TRPC1-mediated SOCE and Ca^{2+} currents, whereas co-expression of TRPC1 and STIM1 increased SOCE [87, 88, 91]. Store depletion induced interaction between TRPC1 and STIM1, shown by co-immunoprecipitation experiments as well as FRET and TIRFM measurements. Conversely, store refilling terminated TRPC1 function as well as STIM1-TRPC1 association [91, 100–104]. Thus the TRPC1-STIM1 interaction is dictated by the ER-[Ca^{2+}] status. Conclusive studies by Muallem and co-workers resolved the mechanism by which STIM1 gates TRPC1. Their findings demonstrated that gating of TRPC1 involves electrostatic interactions between the negatively charged aspartate residues in TRPC1 (^{639}DD^{640}) with the positively charged lysines in the STIM1 polybasic domain (^{684}KK^{685}) [100]. Further studies showed that these negatively charged residues are conserved in TRPC3, TRPC4, TRPC5, TRPC6 and TRPC7, suggesting the possibility that STIM1 can also gate these other TRPC channels [100]. Consistent with this study, TRPC4-mediated SOCE in murine and human endothelial cells was suppressed by knockdown of STIM1 or expression of a charge-swap mutant of STIM1 (KK684,685EE). Similarly, disrupting the electrostatic interaction between STIM1 and TRPC4 by mutation of the conserved negatively charged residues on TRPC4 (EE647,648KK) significantly reduced SOCE and STIM1-TRPC4 interactions [65]. Co-immunoprecipitation of STIM1 and TRPC3 was also increased following store depletion in salivary gland duct cells, although this could be due to TRPC1-TRPC3 interaction in these cells [105]. In yet another study, mutations in the conserved negative residues in the C-terminus of TRPC3 (DD697,698KK), TRPC4 (EE648,649KK), TRPC5 (DE651,652KK) and TRPC6 (EE755,756KK) prevented electrostatic interactions with and gating by STIM1. While these mutants did not respond to store depletion induced by cyclopiazonic acid, they could still be activated by muscarinic receptor stimulation in a STIM1-independent manner. However, co-expression of the charge-swap STIM1 mutant (KK684,685EE) restored store responsiveness to these TRPC mutants [106].

Collectively, these data demonstrate that STIM1 has the ability to gate all TRPC channels via similar electrostatic interactions, even though not all channels appear to interact directly with STIM1. Since TRPC channels are widely expressed in tissues and species, it is not yet clear what determines the interaction of any particular TRPC channel with STIM1 and thus their mode of activation. In neuroblastoma cells, STIM1 promoted SOCE mediated via TRPC1 and TRPC6, while inhibiting TRPC6-mediated store-independent Ca^{2+} entry [107]. TRPC5 was also reported to contribute to SOCE in RBL cells. In these cells co-expression of STIM1 with TRPC5 increased, while knockdown of STIM1 abolished, thapsigargin-induced cation entry [108]. Thus, STIM1 might not only be involved in gating some TRPC channels but also determine their recruitment into a store-dependent mode. However, further studies are required to establish this role of STIM1 on the mode of activation of TRPC channels. Involvement of other factors that might contribute to this switch in the mode of regulation also needs further detailed studies. Since several TRP channels show polymodal regulation of their function, possible activation of TRPC channels by mechanisms other than store depletion appears to be quite feasible.

There is considerable information regarding the exact intramolecular rearrangements and molecular domains involved in activation of STIM1. Further, a large number of structure-function studies have now been reported describing the configuration of STIM1 required for binding and activation of Orai1 (recently reviewed in [109]). However, a similar detailed understanding of STIM1-TRPC channel interaction is currently lacking. An ezrin/radixin/moesin (ERM) domain (aa 251–535) was shown to mediate the binding of STIM1 to TRPC channels [91]. However, since the SOAR domain resides within this ERM region of STIM1 it was suggested that the SOAR domain might also be involved in mediating STIM1 binding to TRPC channels [12, 105]. The coiled-coil (CC) motif within the C-terminus of Orai1 is proposed to interact with STIM1 [110]. TRPC channels also have CC domains in both their N- and C-termini. Further, co-immunoprecipitation studies have revealed strong interactions between exogenously expressed SOAR and endogenous TRPC1, TRPC4 and TRPC5, minimal interactions with endogenous TRPC3 and TRPC6, and no interactions with TRPC7 [101, 105]. Mutation of residues in the N-terminal CC domains severely weakened SOAR interactions with TRPC1, TRPC4 and TRPC5 but enhanced association of TRPC3 and TRPC6 with SOAR. Based on this data, it was proposed that interaction of TRPC1 with TRPC3 induces a structural change which exposes a domain in TRPC3 that promotes its binding to STIM1 [105]. A recent study investigating the stoichiometry of TRPC, STIM1, and CaM assembly in a signaling complex reported TRPC channel activation using recombinantly purified SOAR [111]. This study demonstrated that only TRPC channel complexes containing TRPC1, TRPC4 and TRPC5 could be activated by SOAR. Each TRPC tetrameric complex required two SOAR domains for activation and four CaMs for inactivation. SOAR and CaM appeared to reciprocally regulate TRPC channel activity when co-expressed in HEK293 cells. Following application of tenfold higher amounts of CaM, TRPC1 channel activity was reduced, even though SOAR was still bound to the tetramers at the initial stages of inhibition. SOAR eventually detached from TRPC1, which led to further CaM-dependent decline in channel activity [111]. However, to conclusively establish that SOAR directly affects TRPC channel activity, data need to be provided to exclude SOAR-Orai1 effects in the same cell since SOAR domain will also gate Orai1 and Orai1-mediated Ca^{2+} entry will lead to TRPC channel activation. It is worth noting that some studies suggest that Orai1 binding to STIM1 might limit availability of STIM1 for TRPC channels. However, these studies have yet to be confirmed at the level of endogenous proteins. Further studies are required to establish whether STIM1 is indeed a limiting factor for channels contributing to SOCE and whether different physiological conditions favor binding of STIM1 to one type of channel vs the other.

STIM1 shares significant homology with another family member, STIM2. While functional domains such as the EF hand, CC domains, SOAR and polybasic domain are conserved between STIM1 and STIM2 there are several key differences which determine their diverse physiological function and role in SOCE. For example, STIM2 can interact with Orai1 but is a poor activator of the channel compared to STIM1. The difference in the gating efficiency of STIM1 and STIM2 was shown to be due to a single aa difference in their respective SOAR domains; F394 in SOAR1 vs L485 in SOAR2 [112]. The EF hand of STIM2 has a lower affinity for Ca^{2+} than STIM1. Thus, STIM2 can sense and respond to small changes in ER-[Ca^{2+}]. The triggering threshold level of $[Ca^{2+}]_{ER}$ for STIM2 is >400 μM, while STIM1 responds when ER-[Ca^{2+}] is around 200 μM. Based on this, STIM2 has been suggested to aggregate and translocate to ER-PM junction under conditions when there is minimal depletion ER-Ca^{2+} [113] One reported function for STIM2 is the regulation of Orai1 in resting cells to maintain $[Ca^{2+}]_i$ [114]. Another study suggests that STIM2 gates Orai1 in cells stimulated with low agonist where there is less depletion of ER-Ca^{2+} stores while STIM1 is involved in gating Orai1 at high agonist concentration when there is greater depletion [115]. A recent study provides a novel role for STIM2 showing that STIM2 associates with STIM1 and promotes the clustering of STIM1 in ER-PM junctions in cells stimulated with low [agonist]. STIM2 co-clusters with Orai1 and promotes STIM1-Orai1 interactions at low levels of stimulation while STIM1 aggregates efficiently, in a STIM2-independent manner, and interacts with Orai1 in cells stimulated with high [agonist] [116]. Knockdown of STIM2 in HEK293 cells or targeted knockout of STIM2 in mouse salivary glands attenuated STIM1-mediated activation of Orai1 and decreased the agonist sensitivity of SOCE activation. This was especially prominent at lower levels of agonist. On the other hand, knockdown of STIM1 completely eliminated SOCE at low and high levels of stimulus. Hence, STIM2 appears to tune the agonist-sensitivity of the STIM1-Orai1 interactions and associated Ca^{2+} signals [116]. Few studies have investigated the direct contribution of STIM2 to TRPC-mediated SOCE. STIM1 has been proposed to regulate TRPC1 and TRPC3 channel function, whereas STIM2 regulated only TRPC1 function in HEK293 cells [117]. Modulation of the STIM1:STIM2 ratio appears to determine the store responsiveness of TRPC1 channel function in intestinal epithelial cells [118]. Thus, the exact role of STIM2 in TRPC channel function and regulation remains to be determined.

5.4 Orai1-TRPC Channel Interactions in SOCE

The pore-forming component of CRAC channels, Orai1, is indispensable for SOCE. A naturally occurring mutation of the channel, (R91W), which leads to loss of channel function, has been linked to severe combined immune deficiency (SCID) [119–121]. Orai1 has two closely related family members, Orai2 and Orai3, although there is a paucity of data regarding their contribution to SOCE when compared to Orai1 [122]. While endogenous Orai1 function is supported by endogenous

STIM1, exogenously expressed Orai1 does not by itself increase SOCE in cells unless STIM1 is co-expressed with it. The reason for this is not yet clear as cells appear to express STIM1 in excess of Orai1. Subsequent studies identified the pore region of Orai1 by showing that E106Q mutation generates a channel with a non-functional pore, while E106D changes Ca^{2+} selectivity [22, 110, 123, 124]. An additional interesting observation that has been reported is that STIM1 increases the Ca^{2+} selectivity of Orai1 [125]. Again the latter study was carried out with overexpressed protein and needs to be more fully examined using the endogenous channel.

Intriguingly, a number of studies demonstrate that Orai1 is also required for TRPC1 function. Knockdown of endogenous Orai1 abolished SOCE, despite the presence of endogenous or exogenously expressed STIM1 and TRPC1. Further, it was reported that Orai1 and STIM1 form a complex with TRPC1 in response to ER-Ca^{2+} store depletion in HSG cells [103], mouse pulmonary arterial smooth muscle cells [126], human parathyroid cells [127], human liver cells [128] rat kidney fibroblast [129], pancreatic acinar cells and salivary gland acinar cells [34, 39]. Notably assembly of the TRPC1-Orai1 complex requires STIM1 which also gates both channels. Co-localization of the three proteins in ER-PM junctions, as detected by TIRFM, suggests that TRPC1 is also localized in same ER-PM junctions where Orai1-STIM1 complex is assembled. The requirement of Orai1 in TRPC1 function was further revealed by data showing that non-functional Orai1 mutants, either Orai1E106Q or Orai1R91W, abrogated store-dependent activation of TRPC1 [87, 88, 92, 103]. Based on this, it was first proposed that TRPC1 and Orai1 assemble into a heteromeric channel where both proteins contribute to the channel pore. There was also the suggestion that TRPC channel forms the pore while Orai1 serves as a regulator. While this led to an extensive debate regarding the assembly of these putative channels, neither of these proposals was supported by conclusive data. Finally, the mechanism underlying the requirement of Orai1 in TRPC1 function was demonstrated in a study where Ca^{2+} influx mediated by Orai1 triggers plasma membrane insertion of TRPC1 [88]. The insertion presumably occurs within the same ER-PM junctions where the Orai1-STIM1 complex is assembled, to allow for TRPC1 gating by STIM1. Moreover, recruitment of TRPC1 into these junctions brings TRPC1 in close proximity to Orai1, such that Ca^{2+} entry via Orai1 can be sensed locally to trigger plasma membrane recruitment of TRPC1. This requirement of Orai1-mediated Ca^{2+} entry for TRPC1 insertion into the plasma membrane also accounts for the lack of TRPC1 activity when non-functional mutants of Orai1 are expressed. Importantly, Ca^{2+} entry mediated by TRPC1 and Orai1 are utilized by cells to regulate separate functions. Orai1-mediated SOCE is sufficient for activation of NFAT, whereas Ca^{2+} entry via both Orai1 and TRPC1 are required for NFκB expression and function, with TRPC1 contribution being predominant [88, 96]. Thus, Orai1 and TRPC1 form two separate STIM1-regulated channel complexes (Fig. 5.1). TRPC1 and STIM1 form a SOC channel that generates I_{SOC} while Orai1 and STIM1 form the highly Ca^{2+}-selective CRAC channel mediating I_{CRAC}. The smaller I_{CRAC} is masked by the larger I_{SOC} current and unmasked when TRPC1 function is suppressed. It should be noted that true TRPC1 currents

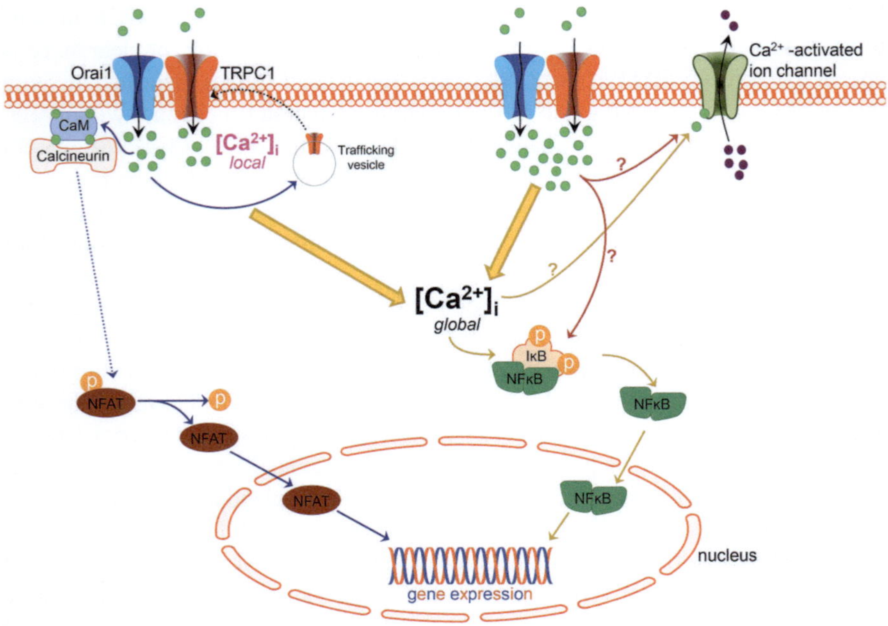

Fig. 5.1 Physiological function of Orai1 and TRPC1 in SOCE. Stimulation with agonists generates $[Ca^{2+}]_i$ changes that occur locally (i.e. close to the channel pore) and globally (i.e. throughout the cell cytosol). Local SOCE mediated by Orai1 has been shown to activate calcineurin, which subsequently induces NFAT translocation into the nucleus to drive gene expression. Local Orai1-SOCE also promotes insertion of TRPC1 into the plasma membrane. Ca^{2+} entry via both Orai1 and TRPC1 contribute to increase in global $[Ca^{2+}]_i$, which has been shown to activate NFκB and NFκB-driven gene expression. While the Ca^{2+}-activated ion channels in the plasma membrane are also activated by global $[Ca^{2+}]_i$, it is not clear whether the activating Ca^{2+} comes from those situated in the vicinity of neighboring Orai1 and TRPC1 channels and/or from the deeper regions of the cell cytosol

have not yet been described as most reported measurements of I_{SOC} include currents generated by both TRPC1+STIM1 and Orai1+STIM1 channels [88]. A requirement for Orai1 in other TRPC channels such as TRPC3 and TRPC6 [93], as well as heteromeric TRPC channels TRPC1/TRPC4 [130], have been reported. Whether TRPC channel trafficking is involved in these cases is not yet known. The exact proteins involved in regulating and mediating exocytosis of TRPC1 have not yet been elucidated.

One interesting suggestion which has been made is that Orai1 can regulate TRPC channels by determining their recruitment into specialized microdomains in the plasma membrane, such as the lipid raft domains (LRDs). This suggestion is consistent with previous studies showing that SOCE requires intact LRDs [39, 88, 90, 104, 131, 132]. Further, STIM1-TRPC1 interaction also takes place within LRDs as disruption of LRD leads to abrogation of STIM1-TRPC1 interaction and loss of SOCE [104]. Similarly, disruption of LRD in human platelets and HEK293 cells reduced interactions between Orai1, TRPC1, TRPC6 and STIM1 [90, 132, 133]. Orai1 interacts with the cytosolic termini of TRPC1 and TRPC6 to modulate their sensitivity to store depletion and STIM1 [93, 134, 135]. Thus, Lutz Birnbaumer and co-workers suggested the hypothesis that recruitment of Orai1, TRPC and STIM1 into LRD confers store-responsiveness to the channels [135]. At this stage, it remains unclear whether Orai1-STIM1 and TRPC-STIM1 complexes are initially formed outside the lipid rafts and subsequently recruited into these microdomains following store depletion or are maintained within this domain by interactions with Orai1 and STIM1. Presence of PIP_2-interacting domains in the C-terminus of STIM1 and STIM2 [132], which are proposed to enable anchoring of the proteins to the plasma membrane within ER-PM junctions, led to several studies examining the role of PIP_2 in SOCE. Effects of PIP_2 depletion on SOCE were inconsistent with some studies showing no effect on SOCE mediated by Orai1 while others demonstrated decreased function and STIM1 clustering [136–139]. Nevertheless, STIM1 or STIM2 lacking the polybasic tail domain do not form puncta within ER-PM junctional domains. However, when Orai1 is expressed with this mutant of STIM1, it rescues STIM1 clustering and CRAC channel activity. It has been suggested that the Orai1-STIM1ΔK complexes might be localized outside the ER-PM junctions, suggesting that the PIP_2 is not required for STIM1-dependent gating of Orai1 [140]. It is unclear whether TRPC-STIM1 interactions can also take place outside the junctional domains. A recent study suggests that dynamic changes in PIP_2 levels within ER-PM junctions mediated by proteins such as septin, impact not only assembly of the Orai1-STIM1 complexes but also regulation of CRAC channel activity [141–143]. It has also been recently reported that the ER-PM junctions might contain different PIP_2 microdomains. This study showed that Orai1-STIM1 complex assembled in ER-PM junctions is transferred from relatively PIP_2-poor to a PIP_2-rich microdomain which dictates the Ca^{2+}-dependent regulation of the channel. Interestingly, this recruitment is determined by Cav-1 and septin [143]. Further studies will be required to fully elucidate the dynamic lipid and protein remodeling that occurs with the ER-PM junctions that critically impact Orai1 and TRPC1 interaction with STIM1 and their function.

Trafficking of TRPC channels has been proposed as a major mode of regulation of their function in the plasma membrane. In addition to the trafficking proteins, scaffolding and regulatory proteins also modulate the magnitude and duration of TRPC-mediated SOCE. The main regulatory pathways that modulate surface expression and function of TRPC channels comprise of constitutive and regulated intracellular trafficking mechanisms. The enhancement of Ca^{2+} influx through TRPC channels can be due to increased exocytosis, retention via interaction with scaffolding proteins, and/or decreased channel endocytosis. As discussed above, TRPC1 function is dependent on LRD [144]. The cholesterol-binding LRD protein Cav-1 is reported to play a pivotal role in plasma membrane localization and activity of TRPC1. TRPC1 interacts with Cav-1 through binding sites located in its N- and C-terminal domains. The N-terminal Cav-1 binding site is involved in scaffolding and localization of TRPC1 in the plasma membrane while the C-terminal domain has been proposed to control channel function and/or inactivation. Knockdown of Cav-1, and mutations in Cav-1 or Cav-1 binding sites in TRPC1 resulted in mislocalization of TRPC1 and impairment of channel activity. Hence, Cav-1 is suggested as an LRD scaffolding protein for TRPC1 that determines plasma membrane localization [38, 39, 131, 145–149]. The current model proposes that in resting cells, constitutive trafficking mechanisms target TRPC1 to cellular regions close to the plasma membrane where the inactive channel interacts with Cav-1 and is retained at that location intracellularly. Following store depletion, STIM1 translocates to ER-PM junctions and activates Orai1. Ca^{2+} entry mediated by Orai1 is a pivotal step in TRPC1 insertion into the plasma membrane and channel activation [88], as it triggers the insertion of TRPC1 into the plasma membrane. Under these conditions, TRPC1 dissociates from Cav-1, interacts with and is gated by STIM1. Following ER-Ca^{2+} store refilling, SOCE is inactivated and TRPC1 disassembles from STIM1. LRDs are essential for STIM1 translocation to the ER-PM junctions as deleting the C-terminal lysine-rich region of STIM1, which contains a PIP_2-binding sequence, impairs puncta formation in these junctions and also alters partitioning of STIM1 into detergent insoluble fractions from cells. In addition, disruption of lipid rafts by cholesterol depletion also affects the ability of STIM1 to interact with TRPC1 [104]. These data demonstrate the importance of structural integrity for caveolar lipid rafts to act as scaffolding platforms for TRPC1-mediated SOCE. Based on the recent study that showed Cav-1 is required for recruitment of Orai1-STIM1 channel to a PIP_2-rich domain [143], it is possible that this event can bring TRPC1 in close proximity to Orai1 such that it can sense local $[Ca^{2+}]_i$ elevation due to Orai1-mediated Ca^{2+} entry. Indeed, Cav-1 might be of utmost importance in Orai1-dependent activation of TRPCs as almost all TRPC channels have once or more Cav-1 binding domains, some of which are fairly well conserved. Nonetheless, Ca^{2+} sensor proteins, as well as the identity of vesicles and intracellular compartments related to TRPC1 trafficking, remain to be determined.

In addition to Cav-1, another scaffolding protein that regulates TRPC1 function is Homer1. The C-terminus of TRPC1 (aa 644–650) forms a complex with Homer1 and IP_3R in resting cells. However, following store depletion, this complex dissociates to enable subsequent TRPC1 interaction with and gating by STIM1 [150, 151]. Further evidence for the contribution of Homer1 came from a study with knockout mice (Homer1$^{-/-}$) that reported impaired SOCE in skeletal muscle cells [47]. Soluble N-ethylmaleimide-sensitive factor attachment receptor (SNARE) proteins, such as synaptosome-associated protein (SNAP-25), are involved in membrane fusion within intracellular compartments or between vesicles and plasma membrane. Interaction of SNAP-25 with TRPC1 is vital for channel function as botulinum toxin treatment, which cleaves and inactivates SNAP-25, decreased SOCE in platelets [152]. Several cytoskeletal and microtubule proteins have also been shown to modulate the TRPC1 channel trafficking and activity. The monomeric GTPase protein, RhoA, regulates TRPC1 translocation to the plasma membrane in endothelial cells [29]. Interaction of β-tubulin with TRPC1 determines surface expression of TRPC1 retinal epithelial cells [153]. Disrupting TRPC1 interaction with either RhoA or β-tubulin significantly decreased SOCE. In aggregate, the data show that proper localization of TRPC1 in the plasma membrane, as well as trafficking to the specific domains where SOCE is regulated, are vital for its interaction with Orai1-STIM1 and its activation.

5.5 Conclusions

The mechanism(s) underlying SOCE involves multiple interactions that allow cells to display dynamic regulatory modes for each physiological stimulus. The multiplicity of channel-protein and protein-protein interactions underscores the variety of signaling complexes that can be generated within a subregion of the cell. Indeed, TRPC channels interact with a wide range of channels and proteins involved in Ca^{2+} signaling, as well as scaffolding and trafficking processes. Such complexity underlies the physiological functions that have been ascribed to TRPC channels. Many studies have investigated the contributions of STIM1 and Orai1 to TRPC channel function. The functional relevance of STIM2, as well as Orai2 and Orai3, in SOCE remains to be resolved. It is worth noting that many cells and tissues express both STIM proteins and more than one Orai protein. Therefore, depending on the type and intensity of the cell stimulus, TRPC channels may also form dynamic signaling complexes with these STIMs and Orais to generate SOCE. Nonetheless, much remains to be elucidated to expand our current understanding of the exact sequence of molecular events involved in the regulation and function of TRPC channels in response to ER-Ca^{2+} depletion. As TRPC channels have been implicated in a number of human diseases, understanding the mechanism(s) involved in regulating and modulating channel function will provide potentially important information and lead to novel targets for the development of effective therapeutic interventions.

Acknowledgements Work in ISA's laboratory is supported by the Intramural Research Program of the NIH, NIDCR.

References

1. Ong HL, de Souza LB, Cheng KT, Ambudkar IS (2014) Physiological functions and regulation of TRPC channels. Handb Exp Pharmacol 223:1005–1034
2. Parekh AB, Putney JW Jr (2005) Store-operated calcium channels. Physiol Rev 85(2):757–810
3. Putney JW Jr (1990) Capacitative calcium entry revisited. Cell Calcium 11(5–6):611–624
4. Parekh AB, Fleig A, Penner R (1997) The store-operated calcium current I(CRAC): nonlinear activation by InsP3 and dissociation from calcium release. Cell 89(6):973–980
5. Penner R, Matthews G, Neher E (1988) Regulation of calcium influx by second messengers in rat mast cells. Nature 334(6182):499–504
6. Matthews G, Neher E, Penner R (1989) Second messenger-activated calcium influx in rat peritoneal mast cells. J Physiol 418:105–130
7. Lewis RS, Cahalan MD (1989) Mitogen-induced oscillations of cytosolic Ca^{2+} and transmembrane Ca^{2+} current in human leukemic T cells. Cell Regul 1(1):99–112

8. Liu X, Groschner K, Ambudkar IS (2004) Distinct Ca^{2+}-permeable cation currents are activated by internal Ca^{2+}-store depletion in RBL-2H3 cells and human salivary gland cells, HSG and HSY. J Membr Biol 200(2):93–104
9. Minke H, Cook B (2002) TRP channel proteins and signal transduction. Physiol Rev 82:429–472
10. Montell C (2005) The TRP superfamily of cation channels. Sci STKE 272:re3
11. Venkatachalam K, Montell C (2007) TRP channels. Annu Rev Biochem 76:387–417
12. Choi S, Maleth J, Jha A, Lee KP, Kim MS, So I, Ahuja M, Muallem S (2014) The TRPCs-STIM1-Orai interaction. Handb Exp Pharmacol 223:1035–1054
13. Nesin V, Tsiokas L (2014) Trpc1. Handb Exp Pharmacol 222:15–51
14. Miller BA (2014) Trpc2. Handb Exp Pharmacol 222:53–65
15. Lichtenegger M, Groschner K (2014) TRPC3: a multifunctional signaling molecule. Handb Exp Pharmacol 222:67–84
16. Freichel M, Tsvilovskyy V, Camacho-Londono JE (2014) TRPC4- and TRPC4-containing channels. Handb Exp Pharmacol 222:85–128
17. Zholos AV (2014) Trpc5. Handb Exp Pharmacol 222:129–156
18. Dietrich A, Gudermann T (2014) TRPC6: physiological function and pathophysiological relevance. Handb Exp Pharmacol 222:157–188
19. Zhang X, Trebak M (2014) Transient receptor potential canonical 7: a diacylglycerol-activated non-selective cation channel. Handb Exp Pharmacol 222:189–204
20. Wes PD, Chevesich J, Jeromin A, Rosenberg C, Stetten S, Montell C (1995) TRPC1, a human homolog of a Drosophila store-operated channel. Proc Natl Acad Sci U S A 92:9652–9656
21. Zhu X, Chu PB, Peyton M, Birnbaumer L (1995) Molecular cloning of a widely expressed human homologue for the Drosophila trp gene. FEBS Lett 373:193–198
22. Hogan PG, Lewis RS, Rao A (2010) Molecular basis of calcium signaling in lymphocytes: STIM and ORAI. Annu Rev Immunol 28:491–533
23. Nilius B, Owsianik G (2010) Transient receptor potential channelopathies. Pflugers Arch 460(2):437–450
24. Greka A, Mundel P (2011) Balancing calcium signals through TRPC5 and TRPC6 in podocytes. J Am Soc Nephrol 22(11):1969–1980
25. Beech DJ (2013) Characteristics of transient receptor potential canonical calcium-permeable channels and their relevance to vascular physiology and disease. Circ J 77(3):570–579
26. Liu X, Singh BB, Ambudkar IS (2003) TRPC1 is required for functional store-operated Ca^{2+} channels. Role of acidic amino acid residues in the S5-S6 region. J Biol Chem 278(13):11337–11343
27. Liu X, Wang W, Singh BB, Lockwich T, Jadlowiec J, O'Connell B, Wellner R, Zhu MX, Ambudkar IS (2000) Trp1, a candidate protein for the store-operated Ca^{2+} influx mechanism in salivary gland cells. J Biol Chem 275(5):3403–3411
28. Dietrich A, Chubanov V, Kalwa H, Rost BR, Gudermann T (2006) Cation channels of the transient receptor potential superfamily: their role in physiological and pathophysiological processes of smooth muscle cells. Pharmacol Ther 112(3):744–760
29. Mehta D, Ahmmed GU, Paria BC, Holinstat M, Voyno-Yasenetskaya T, Tiruppathi C, Minshall RD, Malik AB (2003) RhoA interaction with inositol 1,4,5-triphosphate receptor and transient receptor potential channel-1 regulates Ca^{2+} entry. J Biol Chem 278(35):33492–33500
30. Tiruppathi C, Ahmmed GU, Vogel SM, Malik AB (2006) Ca^{2+} signaling, TRP channels, and endothelial permeability. Microcirculation 13(8):693–708
31. Brownlow SL, Harper AG, Harper MT, Sage SO (2004) A role for hTRPC1 and lipid raft domains in store-mediated calcium entry in human platelets. Cell Calcium 35(2):107–113
32. Rosado JA, Brownlow SL, Sage SO (2002) Endogenously expressed Trp1 is involved in store-mediated Ca^{2+} entry by conformational coupling in human platelets. J Biol Chem 277(44):42157–42163

33. Dietrich A, Fahlbusch M, Gudermann T (2014) Classical transient receptor potential 1 (TRPC1): channel or channel regulator? Cells 3(4):939–962
34. Hong JH, Li Q, Kim MS, Shin DM, Feske S, Birnbaumer L, Cheng KT, Ambudkar IS, Muallem S (2011) Polarized but differential localization and recruitment of STIM1, Orai1 and TRPC channels in secretory cells. Traffic 12(2):232–245
35. Ma X, Cheng KT, Wong CO, O'Neil RG, Birnbaumer L, Ambudkar IS, Yao X (2011) Heteromeric TRPV4-C1 channels contribute to store-operated Ca^{2+} entry in vascular endothelial cells. Cell Calcium 50(6):502–509
36. Liu X, Cheng KT, Bandyopadhyay BC, Pani B, Dietrich A, Paria BC, Swaim WD, Beech D, Yildrim E, Singh BB, Birnbaumer L, Ambudkar IS (2007) Attenuation of store-operated Ca^{2+} current impairs salivary gland fluid secretion in $TRPC1^{-/-}$ mice. Proc Natl Acad Sci U S A 104(44):17542–17547
37. Sun Y, Birnbaumer L, Singh BB (2015) TRPC1 regulates calcium-activated chloride channels in salivary gland cells. J Cell Physiol. doi:10.1002/jcp.25017
38. Pani B, Ong HL, Brazer SC, Liu X, Rauser K, Singh BB, Ambudkar IS (2009) Activation of TRPC1 by STIM1 in ER-PM microdomains involves release of the channel from its scaffold caveolin-1. Proc Natl Acad Sci U S A 106(47):20087–20092
39. Pani B, Liu X, Bollimuntha S, Cheng KT, Niesman IR, Zheng C, Achen VR, Patel HH, Ambudkar IS, Singh BB (2013) Impairment of TRPC1-STIM1 channel assembly and AQP5 translocation compromise agonist-stimulated fluid secretion in mice lacking caveolin1. J Cell Sci 126(Pt 2):667–675
40. Du J, Sours-Brothers S, Coleman R, Ding M, Graham S, Kong DH, Ma R (2007) Canonical transient receptor potential 1 channel is involved in contractile function of glomerular mesangial cells. J Am Soc Nephrol 18(5):1437–1445
41. Sours S, Du J, Chu S, Ding M, Zhou XJ, Ma R (2006) Expression of canonical transient receptor potential (TRPC) proteins in human glomerular mesangial cells. Am J Physiol Renal Physiol 290(6):F1507–F1515
42. Ong EC, Nesin V, Long CL, Bai CX, Guz JL, Ivanov IP, Abramowitz J, Birnbaumer L, Humphrey MB, Tsiokas L (2013) A TRPC1-dependent pathway regulates osteoclast formation and function. J Biol Chem 288(31):22219–22231
43. Kochukov MY, Balasubramanian A, Noel RC, Marrelli SP (2013) Role of TRPC1 and TRPC3 channels in contraction and relaxation of mouse thoracic aorta. J Vasc Res 50(1):11–20
44. Zanou N, Schakman O, Louis P, Ruegg UT, Dietrich A, Birnbaumer L, Gailly P (2012) Trpc1 ion channel modulates phosphatidylinositol 3-kinase/Akt pathway during myoblast differentiation and muscle regeneration. J Biol Chem 287(18):14524–14534
45. Zanou N, Shapovalov G, Louis M, Tajeddine N, Gallo C, Van Schoor M, Anguish I, Cao ML, Schakman O, Dietrich A, Lebacq J, Ruegg U, Roulet E, Birnbaumer L, Gailly P (2010) Role of TRPC1 channel in skeletal muscle function. Am J Physiol Cell Physiol 298(1):C149–C162
46. Sabourin J, Lamiche C, Vandebrouck A, Magaud C, Rivet J, Cognard C, Bourmeyster N, Constantin B (2009) Regulation of TRPC1 and TRPC4 cation channels requires an alpha1-syntrophin-dependent complex in skeletal mouse myotubes. J Biol Chem 284(52):36248–36261
47. Stiber JA, Zhang ZS, Burch J, Eu JP, Zhang S, Truskey GA, Seth M, Yamaguchi N, Meissner G, Shah R, Worley PF, Williams RS, Rosenberg PB (2008) Mice lacking Homer 1 exhibit a skeletal myopathy characterized by abnormal transient receptor potential channel activity. Mol Cell Biol 28(8):2637–2647
48. Vandebrouck A, Sabourin J, Rivet J, Balghi H, Sebille S, Kitzis A, Raymond G, Cognard C, Bourmeyster N, Constantin B (2007) Regulation of capacitative calcium entries by {alpha}1-syntrophin: association of TRPC1 with dystrophin complex and the PDZ domain of {alpha}1-syntrophin. FASEB J 20:136–138

49. Kim EY, Alvarez-Baron CP, Dryer SE (2009) Canonical transient receptor potential channel (TRPC)3 and TRPC6 associate with large-conductance Ca^{2+}-activated K^+ (BKCa) channels: role in BKCa trafficking to the surface of cultured podocytes. Mol Pharmacol 75(3):466–477
50. Zagranichnaya TK, Wu X, Villereal ML (2005) Endogenous TRPC1, TRPC3, and TRPC7 proteins combine to form native store-operated channels in HEK-293 cells. J Biol Chem 280(33):29559–29569
51. Boulay G, Brown DM, Qin N, Jiang M, Dietrich A, Zhu MX, Chen Z, Birnbaumer M, Mikoshiba K, Birnbaumer L (1999) Modulation of Ca2+ entry by polypeptides of the inositol 1,4,5-trisphosphate receptor (IP3R) that bind transient receptor potential (TRP): evidence for roles of TRP and IP3R in store depletion-activated Ca2+ entry. Proc Natl Acad Sci U S A 96(26):14955–14960
52. Kiselyov K, Xu X, Mozhayeva G, Kuo T, Pessah I, Mignery G, Zhu X, Birnbaumer L, Muallem S (1998) Functional interaction between InsP3 receptors and store-operated Htrp3 channels. Nature 396(6710):478–482
53. Vazquez G, Wedel BJ, Trebak M, St John Bird G, Putney JW Jr (2003) Expression level of the canonical transient receptor potential 3 (TRPC3) channel determines its mechanism of activation. J Biol Chem 278(24):21649–21654
54. Zhu X, Jiang M, Peyton M, Boulay G, Hurst R, Stefani E, Birnbaumer L (1996) trp, a novel mammalian gene family essential for agonist-activated capacitative Ca^{2+} entry. Cell 85(5):661–671
55. Kim MS, Hong JH, Li Q, Shin DM, Abramowitz J, Birnbaumer L, Muallem S (2009) Deletion of TRPC3 in mice reduces store-operated Ca^{2+} influx and the severity of acute pancreatitis. Gastroenterology 137(4):1509–1517
56. Kim MS, Lee KP, Yang D, Shin DM, Abramowitz J, Kiyonaka S, Birnbaumer L, Mori Y, Muallem S (2011) Genetic and pharmacologic inhibition of the Ca2+ influx channel TRPC3 protects secretory epithelia from Ca2+-dependent toxicity. Gastroenterology 140(7):2107–2115, 2115 e2101-2104
57. Ambudkar IS, Ong HL (2007) Organization and function of TRPC channelosomes. Pflugers Arch 455(2):187–200
58. Okada T, Shimizu S, Wakamori M, Maeda A, Kurosaki T, Takada N, Imoto K, Mori Y (1998) Molecular cloning and functional characterisation of a novel receptor-activated TRP Ca^{2+} channel from mouse brain. J Biol Chem 273(17):10279–10287
59. Zeng F, Xu SZ, Jackson PK, McHugh D, Kumar B, Fountain SJ, Beech DJ (2004) Human TRPC5 channel activated by a multiplicity of signals in a single cell. J Physiol 559(Pt 3):739–750
60. Philipp S, Cavalie A, Freichel M, Wissenbach U, Zimmer S, Trost C, Marquart A, Murakami M, Flockerzi V (1996) A mammalian capacitative calcium entry channel homologous to Drosophila TRP and TRPL. EMBO J 15(22):6166–6171
61. Warnat J, Philipp S, Zimmer S, Flockerzi V, Cavalie A (1999) P.henotype of a recombinant store-operated channel : highly selective permeation of Ca^{2+}. J Physiol 518(3):631–638
62. Kinoshita M, Akaike A, Satoh M, Kaneko S (2000) Positive regulation of capacitative Ca^{2+} entry by intracellular Ca^{2+} in *Xenopus* oocytes expressing rat TRP4. Cell Calcium 28(3):151–159
63. Wang J, Shimoda LA, Sylvester JT (2004) Capacitative calcium entry and TRPC channel proteins are expressed in rat distal pulmonary arterial smooth muscle. Am J Physiol Lung Cell Mol Physiol 286(4):L848–L858
64. Philipp S, Trost C, Warnat J, Rautmann J, Himmerkus N, Schroth G, Kretz O, Nastainczyk W, Cavalie A, Hoth M, Flockerzi V (2000) TRP4 (CCE1) protein is part of native calcium release-activated Ca^{2+}-like channels in adrenal cells. J Biol Chem 275(21):23985–23972
65. Sundivakkam PC, Freichel M, Singh V, Yuan JP, Vogel SM, Flockerzi V, Malik AB, Tiruppathi C (2012) The Ca^{2+} sensor stromal interaction molecule 1 (STIM1) is necessary and sufficient

for the store-operated Ca^{2+} entry function of transient receptor potential canonical (TRPC) 1 and 4 channels in endothelial cells. Mol Pharmacol 81(4):510–526
66. Fatherazi S, Presland RB, Belton CM, Goodwin P, Al-Qutub M, Trbic Z, Macdonald G, Schubert MM, Izutsu KT (2007) Evidence that TRPC4 supports the calcium selective I_{CRAC}-like current in human gingival keratinocytes. Pflugers Arch 453(6):879–889
67. Yang H, Mergler S, Sun X, Wang Z, Lu L, Bonanno JA, Pleyer U, Reinach PS (2005) TRPC4 knockdown suppresses epidermal growth factor-induced store-operated channel activation and growth in human corneal epithelial cells. J Biol Chem 280(37):32230–32237
68. Zhang S, Remillard CV, Fantozzi I, Yuan JX (2004) ATP-induced mitogenesis is mediated by cyclic AMP response element-binding protein-enhanced TRPC4 expression and activity in human pulmonary artery smooth muscle cells. Am J Physiol Cell Physiol 287(5):C1192–C1201
69. Sours-Brothers S, Ding M, Graham S, Ma R (2009) Interaction between TRPC1/TRPC4 assembly and STIM1 contributes to store-operated Ca^{2+} entry in mesangial cells. Exp Biol Med (Maywood) 234(6):673–682
70. Freichel M, Suh SH, Pfeifer A, Schweig U, Trost C, Weibgerber P, Biel M, Philipp S, Freise D, Droogmans G, Hofmann F, Flockerzi V, Nilius B (2001) Lack of an endothelial store-operated Ca^{2+} current impairs agonist-dependent vasorelaxation in $TRP4^{-/-}$ mice. Nat Cell Biol 3:121–127
71. Tiruppathi C, Freichel M, Vogel SM, Paria BC, Mehta D, Flockerzi V, Malik AB (2002) Impairment of store-operated Ca^{2+} entry in $TRPC4^{-/-}$ mice interferes with increase in lung microvascular permeability. Circ Res 91(1):70–76
72. Goel M, Sinkins WG, Schilling WP (2002) Selective association of TRPC channel subunits in rat brain synaptosomes. J Biol Chem 277(50):48303–48310
73. Hofmann T, Schaefer M, Schultz G, Gudermann T (2002) Subunit composition of mammalian transient receptor potential channels in living cells. Proc Natl Acad Sci U S A 99(11):7461–7466
74. Strubing C, Krapivinsky G, Krapivinsky L, Clapham DE (2003) Formation of novel TRPC channels by complex subunit interactions in embryonic brain. J Biol Chem 278(40):39014–39019
75. Liu X, Bandyopadhyay BC, Singh BB, Groschner K, Ambudkar IS (2005) Molecular analysis of a store-operated and 2-acetyl-sn-glycerol-sensitive non-selective cation channel. Heteromeric assembly of TRPC1-TRPC3. J Biol Chem 280(22):21600–21606
76. Wu X, Zagranichnaya TK, Gurda GT, Eves EM, Villereal ML (2004) A TRPC1/TRPC3-mediated increase in store-operated calcium entry is required for differentiation of H19-7 hippocampal neuronal cells. J Biol Chem 279(42):43392–43402
77. Beech DJ (2005) Emerging functions of 10 types of TRP cationic channel in vascular smooth muscle. Clin Exp Pharmacol Physiol 32(8):597–603
78. Antigny F, Koenig S, Bernheim L, Frieden M (2013) During post-natal human myogenesis, normal myotube size requires TRPC1- and TRPC4-mediated Ca^{2+} entry. J Cell Sci 126(Pt 11):2525–2533
79. Louis M, Zanou N, Van Schoor M, Gailly P (2008) TRPC1 regulates skeletal myoblast migration and differentiation. J Cell Sci 121(Pt 23):3951–3959
80. Seth M, Zhang ZS, Mao L, Graham V, Burch J, Stiber J, Tsiokas L, Winn M, Abramowitz J, Rockman HA, Birnbaumer L, Rosenberg P (2009) TRPC1 channels are critical for hypertrophic signaling in the heart. Circ Res 105(10):1023–1030
81. Wu X, Eder P, Chang B, Molkentin JD (2010) TRPC channels are necessary mediators of pathologic cardiac hypertrophy. Proc Natl Acad Sci U S A 107(15):7000–7005
82. Phelan KD, Mock MM, Kretz O, Shwe UT, Kozhemyakin M, Greenfield LJ, Dietrich A, Birnbaumer L, Freichel M, Flockerzi V, Zheng F (2012) Heteromeric canonical transient receptor potential 1 and 4 channels play a critical role in epileptiform burst firing and seizure-induced neurodegeneration. Mol Pharmacol 81(3):384–392

83. Quick K, Zhao J, Eijkelkamp N, Linley JE, Rugiero F, Cox JJ, Raouf R, Gringhuis M, Sexton JE, Abramowitz J, Taylor R, Forge A, Ashmore J, Kirkwood N, Kros CJ, Richardson GP, Freichel M, Flockerzi V, Birnbaumer L, Wood JN (2012) TRPC3 and TRPC6 are essential for normal mechanotransduction in subsets of sensory neurons and cochlear hair cells. Open Biol 2(5):120068
84. Ma X, Cao J, Luo J, Nilius B, Huang Y, Ambudkar IS, Yao X (2010) Depletion of intracellular Ca^{2+} stores stimulates the translocation of vanilloid transient receptor potential 4-c1 heteromeric channels to the plasma membrane. Arterioscler Thromb Vasc Biol 30(11):2249–2255
85. Ma X, Nilius B, Wong JW, Huang Y, Yao X (2011) Electrophysiological properties of heteromeric TRPV4-C1 channels. Biochim Biophys Acta 1808(12):2789–2797
86. Schindl R, Fritsch R, Jardin I, Frischauf I, Kahr H, Muik M, Riedl MC, Groschner K, Romanin C (2012) Canonical transient receptor potential (TRPC) 1 acts as a negative regulator for vanilloid TRPV6-mediated Ca^{2+} influx. J Biol Chem 287(42):35612–35620
87. Cheng KT, Liu X, Ong HL, Ambudkar IS (2008) Functional requirement for Orai1 in store-operated TRPC1-STIM1 channels. J Biol Chem 283(19):12935–12940
88. Cheng KT, Liu X, Ong HL, Swaim W, Ambudkar IS (2011) Local Ca^{2+} entry via Orai1 regulates plasma membrane recruitment of TRPC1 and controls cytosolic Ca^{2+} signals required for specific cell functions. PLoS Biol 9(3):e1001025
89. Galan C, Zbidi H, Bartegi A, Salido GM, Rosado JA (2009) STIM1, Orai1 and hTRPC1 are important for thrombin- and ADP-induced aggregation in human platelets. Arch Biochem Biophys 490(2):137–144
90. Jardin I, Salido GM, Rosado JA (2008) Role of lipid rafts in the interaction between hTRPC1, Orai1 and STIM1. Channels 2(6):401–403
91. Huang GN, Zeng W, Kim JY, Yuan JP, Han L, Muallem S, Worley PF (2006) STIM1 carboxyl-terminus activates native SOC, I(crac) and TRPC1 channels. Nat Cell Biol 8(9):1003–1010
92. Kim MS, Zeng W, Yuan JP, Shin DM, Worley PF, Muallem S (2009) Native store-operated Ca^{2+} influx requires the channel function of Orai1 and TRPC1. J Biol Chem 284(15):9733–9741
93. Liao Y, Erxleben C, Yildirim E, Abramowitz J, Armstrong DL, Birnbaumer L (2007) Orai proteins interact with TRPC channels and confer responsiveness to store depletion. Proc Natl Acad Sci U S A 104(11):4682–4687
94. Jardin I, Gomez LJ, Salido GM, Rosado JA (2009) Dynamic interaction of hTRPC6 with the Orai1-STIM1 complex or hTRPC3 mediates its role in capacitative or non-capacitative Ca^{2+} entry pathways. Biochem J 420(2):267–276
95. Woodard GE, Lopez JJ, Jardin I, Salido GM, Rosado JA (2010) TRPC3 regulates agonist-stimulated Ca^{2+} mobilization by mediating the interaction between type I inositol 1,4,5-trisphosphate receptor, RACK1, and Orai1. J Biol Chem 285(11):8045–8053
96. Ong HL, Jang SI, Ambudkar IS (2012) Distinct contributions of Orai1 and TRPC1 to agonist-induced $[Ca^{2+}]_i$ signals determine specificity of Ca^{2+}-dependent gene expression. PLoS ONE 7(10):e47146
97. Liou J, Kim ML, Heo WD, Jones JT, Myers JW, Ferrell JE Jr, Meyer T (2005) STIM is a Ca^{2+} sensor essential for Ca^{2+}-store-depletion-triggered Ca^{2+} influx. Curr Biol 15(13):1235–1241
98. Roos J, DiGregorio PJ, Yeromin AV, Ohlsen K, Lioudyno M, Zhang S, Safrina O, Kozak JA, Wagner SL, Cahalan MD, Velicelebi G, Stauderman KA (2005) STIM1, an essential and conserved component of store-operated Ca^{2+} channel function. J Cell Biol 169(3):435–445
99. Yuan JP, Zeng W, Dorwart MR, Choi YJ, Worley PF, Muallem S (2009) SOAR and the polybasic STIM1 domains gate and regulate Orai channels. Nat Cell Biol 11(3):337–343
100. Zeng W, Yuan JP, Kim MS, Choi YJ, Huang GN, Worley PF, Muallem S (2008) STIM1 gates TRPC channels, but not Orai1, by electrostatic interaction. Mol Cell 32(3):439–448
101. Yuan JP, Zeng W, Huang GN, Worley PF, Muallem S (2007) STIM1 heteromultimerizes TRPC channels to determine their function as store-operated channels. Nat Cell Biol 9(6):636–645

102. Lopez JJ, Salido GM, Pariente JA, Rosado JA (2006) Interaction of STIM1 with endogenously expressed human canonical TRP1 upon depletion of intracellular Ca^{2+} stores. J Biol Chem 281(38):28254–28264
103. Ong HL, Cheng KT, Liu X, Bandyopadhyay BC, Paria BC, Soboloff J, Pani B, Gwack Y, Srikanth S, Singh BB, Gill DL, Ambudkar IS (2007) Dynamic assembly of TRPC1-STIM1-Orai1 ternary complex is involved in store-operated calcium influx. Evidence for similarities in store-operated and calcium release-activated calcium channel components. J Biol Chem 282(12):9105–9116
104. Pani B, Ong HL, Liu X, Rauser K, Ambudkar IS, Singh BB (2008) Lipid rafts determine clustering of STIM1 in endoplasmic reticulum-plasma membrane junctions and regulation of Store-operated Ca^{2+} Entry (SOCE). J Biol Chem 283(25):17333–17340
105. Lee KP, Choi S, Hong JH, Ahuja M, Graham S, Ma R, So I, Shin DM, Muallem S, Yuan JP (2014) Molecular determinants mediating gating of transient receptor potential canonical (TRPC) channels by stromal interaction molecule 1 (STIM1). J Biol Chem 289(10):6372–6382
106. Lee KP, Yuan JP, So I, Worley PF, Muallem S (2010) STIM1-dependent and STIM1-independent function of transient receptor potential canonical (TRPC) channels tunes their store-operated mode. J Biol Chem 285(49):38666–38673
107. Albarran L, Dionisio N, Lopez E, Salido GM, Redondo PC, Rosado JA (2014) STIM1 regulates TRPC6 heteromultimerization and subcellular location. Biochem J 463(3):373–381
108. Ma HT, Peng Z, Hiragun T, Iwaki S, Gilfillan AM, Beaven MA (2008) Canonical transient receptor potential 5 channel in conjunction with Orai1 and STIM1 allows Sr^{2+} entry, optimal influx of Ca^{2+}, and degranulation in a rat mast cell line. J Immunol 180(4):2233–2239
109. Shim AH, Tirado-Lee L, Prakriya M (2015) Structural and functional mechanisms of CRAC channel regulation. J Mol Biol 427(1):77–93
110. Muik M, Frischauf I, Derler I, Fahrner M, Bergsmann J, Eder P, Schindl R, Hesch C, Polzinger B, Fritsch R, Kahr H, Madl J, Gruber H, Groschner K, Romanin C (2008) Dynamic coupling of the putative coiled-coil domain of ORAI1 with STIM1 mediates ORAI1 channel activation. J Biol Chem 283(12):8014–8022
111. Asanov A, Sampieri A, Moreno C, Pacheco J, Salgado A, Sherry R, Vaca L (2015) Combined single channel and single molecule detection identifies subunit composition of STIM1-activated transient receptor potential canonical (TRPC) channels. Cell Calcium 57(1):1–13
112. Wang X, Wang Y, Zhou Y, Hendron E, Mancarella S, Andrake MD, Rothberg BS, Soboloff J, Gill DL (2014) Distinct Orai-coupling domains in STIM1 and STIM2 define the Orai-activating site. Nat Commun 5:3183–3193
113. Collins SR, Meyer T (2011) Evolutionary origins of STIM1 and STIM2 within ancient Ca^{2+} signaling systems. Trends Cell Biol 21(4):202–211
114. Brandman O, Liou J, Park WS, Meyer T (2007) STIM2 is a feedback regulator that stabilizes basal cytosolic and endoplasmic reticulum Ca^{2+} levels. Cell 131(7):1327–1339
115. Kar P, Bakowski D, Di Capite J, Nelson C, Parekh AB (2012) Different agonists recruit different stromal interaction molecule proteins to support cytoplasmic Ca^{2+} oscillations and gene expression. Proc Natl Acad Sci U S A 109(18):6969–6974
116. Ong HL, de Souza LB, Zheng C, Cheng KT, Liu X, Goldsmith CM, Feske S, Ambudkar IS (2015) STIM2 enhances receptor-stimulated Ca2+ signaling by promoting recruitment of STIM1 to the endoplasmic reticulum-plasma membrane junctions. Sci Signal 8(359):ra3
117. Shalygin A, Skopin A, Kalinina V, Zimina O, Glushankova L, Mozhayeva GN, Kaznacheyeva E (2014) STIM1 and STIM2 proteins differently regulate endogenous store-operated channels in HEK293 cells. J Biol Chem 290(8): 4717–4727
118. Rao JN, Rathor N, Zhuang R, Zou T, Liu L, Xiao L, Turner DJ, Wang JY (2012) Polyamines regulate intestinal epithelial restitution through TRPC1-mediated Ca^{2+} signaling by differentially modulating STIM1 and STIM2. Am J Physiol Cell Physiol 303(3):C308–C317
119. Vig M, Beck A, Billingsley JM, Lis A, Parvez S, Peinelt C, Koomoa DL, Soboloff J, Gill DL, Fleig A, Kinet JP, Penner R (2006) CRACM1 multimers form the ion-selective pore of the CRAC channel. Curr Biol 16(20):2073–2079

120. Zhang SL, Yeromin AV, Zhang XH, Yu Y, Safrina O, Penna A, Roos J, Stauderman KA, Cahalan MD (2006) Genome-wide RNAi screen of Ca^{2+} influx identifies genes that regulate Ca^{2+} release-activated Ca^{2+} channel activity. Proc Natl Acad Sci U S A 103(24):9357–9362
121. Feske S, Gwack Y, Prakriya M, Srikanth S, Puppel SH, Tanasa B, Hogan PG, Lewis RS, Daly M, Rao A (2006) A mutation in Orai1 causes immune deficiency by abrogating CRAC channel function. Nature 441(7090):179–185
122. Hoth M, Niemeyer BA (2013) The neglected CRAC proteins: Orai2, Orai3, and STIM2. Curr Top Membr 71:237–271
123. Gwack Y, Srikanth S, Feske S, Cruz-Guilloty F, Oh-hora M, Neems DS, Hogan PG, Rao A (2007) Biochemical and functional characterization of Orai proteins. J Biol Chem 282(22):16232–16243
124. Prakriya M, Feske S, Gwack Y, Srikanth S, Rao A, Hogan PG (2006) Orai1 is an essential pore subunit of the CRAC channel. Nature 443(7108):230–233
125. McNally BA, Somasundaram A, Yamashita M, Prakriya M (2012) Gated regulation of CRAC channel ion selectivity by STIM1. Nature 482(7384):241–245
126. Ng LC, McCormack MD, Airey JA, Singer CA, Keller PS, Shen XM, Hume JR (2009) TRPC1 and STIM1 mediate capacitative Ca^{2+} entry in mouse pulmonary arterial smooth muscle cells. J Physiol 587(Pt 11):2429–2442
127. Lu M, Branstrom R, Berglund E, Hoog A, Bjorklund P, Westin G, Larsson C, Farnebo LO, Forsberg L (2010) Expression and association of TRPC subtypes with Orai1 and STIM1 in human parathyroid. J Mol Endocrinol 44(5):285–294
128. Zhang ZY, Pan LJ, Zhang ZM (2010) Functional interactions among STIM1, Orai1 and TRPC1 on the activation of SOCs in HL-7702 cells. Amino Acids 39(1):195–204
129. Almirza WH, Peters PH, van Zoelen EJ, Theuvenet AP (2012) Role of Trpc channels, Stim1 and Orai1 in $PGF_{2\alpha}$-induced calcium signaling in NRK fibroblasts. Cell Calcium 51(1):12–21
130. Cioffi DL, Wu S, Chen H, Alexeyev M, St Croix CM, Pitt BR, Uhlig S, Stevens T (2012) Orai1 determines calcium selectivity of an endogenous TRPC heterotetramer channel. Circ Res 110(11):1435–1444
131. Lockwich TP, Liu X, Singh BB, Jadlowiec J, Weiland S, Ambudkar IS (2000) Assembly of Trp1 in a signaling complex associated with caveolin-scaffolding lipid raft domains. J Biol Chem 275(16):11934–11942
132. Dionisio N, Galan C, Jardin I, Salido GM, Rosado JA (2011) Lipid rafts are essential for the regulation of SOCE by plasma membrane resident STIM1 in human platelets. Biochim Biophys Acta 1813(3):431–437
133. Galan C, Woodard GE, Dionisio N, Salido GM, Rosado JA (2010) Lipid rafts modulate the activation but not the maintenance of store-operated Ca^{2+} entry. Biochim Biophys Acta 1803(9):1083–1093
134. Liao Y, Erxleben C, Abramowitz J, Flockerzi V, Zhu MX, Armstrong DL, Birnbaumer L (2008) Functional interactions among Orai1, TRPCs, and STIM1 suggest a STIM-regulated heteromeric Orai/TRPC model for SOCE/Icrac channels. Proc Natl Acad Sci U S A 105(8):2895–2900
135. Liao Y, Plummer NW, George MD, Abramowitz J, Zhu MX, Birnbaumer L (2009) A role for Orai in TRPC-mediated Ca^{2+} entry suggests that a TRPC:Orai complex may mediate store and receptor operated Ca^{2+} entry. Proc Natl Acad Sci U S A 106(9):3202–3206
136. Chvanov M, Walsh CM, Haynes LP, Voronina SG, Lur G, Gerasimenko OV, Barraclough R, Rudland PS, Petersen OH, Burgoyne RD, Tepikin AV (2008) ATP depletion induces translocation of STIM1 to puncta and formation of STIM1-ORAI1 clusters: translocation and re-translocation of STIM1 does not require ATP. Pflugers Arch: Eur J Physiol 457(2):505–517
137. Korzeniowski MK, Popovic MA, Szentpetery Z, Varnai P, Stojilkovic SS, Balla T (2009) Dependence of STIM1/Orai1-mediated calcium entry on plasma membrane phosphoinositides. J Biol Chem 284(31):21027–21035

138. Walsh CM, Chvanov M, Haynes LP, Petersen OH, Tepikin AV, Burgoyne RD (2010) Role of phosphoinositides in STIM1 dynamics and store-operated calcium entry. Biochem J 425(1):159–168
139. Calloway N, Owens T, Corwith K, Rodgers W, Holowka D, Baird B (2011) Stimulated association of STIM1 and Orai1 is regulated by the balance of PtdIns(4,5)P_2 between distinct membrane pools. J Cell Sci 124(Pt 15):2602–2610
140. Park CY, Hoover PJ, Mullins FM, Bachhawat P, Covington ED, Raunser S, Walz T, Garcia KC, Dolmetsch RE, Lewis RS (2009) STIM1 clusters and activates CRAC channels via direct binding of a cytosolic domain to Orai1. Cell 136(5):876–890
141. Sharma S, Quintana A, Findlay GM, Mettlen M, Baust B, Jain M, Nilsson R, Rao A, Hogan PG (2013) An siRNA screen for NFAT activation identifies septins as coordinators of store-operated Ca^{2+} entry. Nature 499(7457):238–242
142. Wollman R, Meyer T (2012) Coordinated oscillations in cortical actin and Ca^{2+} correlate with cycles of vesicle secretion. Nat Cell Biol 14(12):1261–1269
143. Maleth J, Choi S, Muallem S, Ahuja M (2014) Translocation between PI(4,5)P_2-poor and PI(4,5)P_2-rich microdomains during store depletion determines STIM1 conformation and Orai1 gating. Nat Commun 5:5843
144. Ong HL, Ambudkar IS (2011) The dynamic complexity of the TRPC1 channelosome. Channels 5(5):424–431
145. Brazer SC, Singh BB, Liu X, Swaim W, Ambudkar IS (2003) Caveolin-1 contributes to assembly of store-operated Ca^{2+} influx channels by regulating plasma membrane localization of TRPC1. J Biol Chem 278(29):27208–27215
146. Kwiatek AM, Minshall RD, Cool DR, Skidgel RA, Malik AB, Tiruppathi C (2006) Caveolin-1 regulates store-operated Ca^{2+} influx by binding of its scaffolding domain to transient receptor potential channel-1 in endothelial cells. Mol Pharmacol 70(4):1174–1183
147. Weihuang Y, Chang SJ, Harn HI, Huang HT, Lin HH, Shen MR, Tang MJ, Chiu WT (2015) Mechanosensitive store-operated calcium entry regulates the formation of cell polarity. J Cell Physiol. doi:10.1002/jcp.24936
148. Rathor N, Chung HK, Wang SR, Wang JY, Turner DJ, Rao JN (2014) Caveolin-1 enhances rapid mucosal restitution by activating TRPC1-mediated Ca^{2+} signaling. Physiol Rep 2(11):e12193
149. Sundivakkam PC, Kwiatek AM, Sharma TT, Minshall RD, Malik AB, Tiruppathi C (2009) Caveolin-1 scaffold domain interacts with TRPC1 and IP_3R3 to regulate Ca^{2+} store release-induced Ca^{2+} entry in endothelial cells. Am J Physiol Cell Physiol 296(3):C403–C413
150. Yuan JP, Kiselyov K, Shin DM, Chen J, Shcheynikov N, Kang SH, Dehoff MH, Schwarz MK, Seeburg PH, Muallem S, Worley PF (2003) Homer binds TRPC family channels and is required for gating of TRPC1 by IP3 receptors. Cell 114(6):777–789
151. Yuan JP, Lee KP, Hong JH, Muallem S (2012) The closing and opening of TRPC channels by Homer1 and STIM1. Acta Physiol 204:237–247
152. Redondo PC, Harper AG, Salido GM, Pariente JA, Sage SO, Rosado JA (2004) A role for SNAP-25 but not VAMPs in store-mediated Ca^{2+} entry in human platelets. J Physiol 558(Pt 1):99–109
153. Bollimuntha S, Cornatzer E, Singh BB (2005) Plasma membrane localization and function of TRPC1 is dependent on its interaction with beta-tubulin in retinal epithelium cells. Vis Neurosci 22(2):163–170

Chapter 6
Phospholipase A_2 as a Molecular Determinant of Store-Operated Calcium Entry

Tarik Smani, Alejandro Domínguez-Rodriguez, Paula Callejo-García, Juan A. Rosado, and Javier Avila-Medina

Abstract Activation of phospholipases A_2 (PLA_2) leads to the generation of biologically active lipid products that can affect numerous cellular events. Ca^{2+}-independent PLA_2 ($iPLA_2$), also called group VI phospholipase A_2, is one of the main types forming the superfamily of PLA_2. Beside of its role in phospholipid remodeling, $iPLA_2$ has been involved in intracellular Ca^{2+} homeostasis regulation. Several studies proposed $iPLA_2$ as an essential molecular player of store operated Ca^{2+} entry (SOCE) in a large number of excitable and non-excitable cells. $iPLA_2$ activation releases lysophosphatidyl products, which were suggested as agonists of store operated calcium channels (SOCC) and other TRP channels. Herein, we will review the important role of $iPLA_2$ on the intracellular Ca^{2+} handling focusing on its role in SOCE regulation and its implication in physiological and/or pathological processes.

Keywords Phospholipases A2 • SOCE • TRP channels • Lysophospholipids • STIM1 • Orai1

Abbreviations

AA	Arachidonic acid
$AdPLA_2$	Adipose-specific PLA_2
ARC	Arachidonic acid-regulated calcium channels
BEL	Bromoenol lacotone
CaM	Calmodulin
$cPLA_2$	Cytosolic PLA_2

T. Smani (✉) • A. Domínguez-Rodriguez • P. Callejo-García • J. Avila-Medina
Department of Medical Physiology and Biophysic, Institute of Biomedicine of Seville (IBiS), University Hospital of Virgen del Rocío/CSIC/University of Seville, Sevilla 41013, Spain
e-mail: tasmani@us.es

J.A. Rosado
Departamento de Fisiología, University of Extremadura, Cáceres, Spain

J.A. Rosado (ed.), *Calcium Entry Pathways in Non-excitable Cells*, Advances in Experimental Medicine and Biology 898, DOI 10.1007/978-3-319-26974-0_6

DAG	Diacylglycerol
ER	Endoplasmic reticulum
$iPLA_2$	Calcium-independent PLA_2
LA	Lysophasphatidyl acid
$LyPLA_2$	Lysosomal PLA_2
OAG	1-oleoyl-2-acetyl-sn-glycerol
PAF-AH	Platelet-activating factor acetylhydrolases
PC	Phosphatidylcholine
PE	Phosphatidylethanolamine
PG	Phosphatidylglycerol
PS	Phosphatidylserine
ROC	Receptor operated channels
SMC	Smooth muscle cell
$sPLA_2$	Secretory PLA_2
SOCC/SOCE	Store operated Ca^{2+} channels/entry

6.1 Classification of Phospholipase A_2

The phospholipase A_2 superfamily enzymes are characterized by their ability to catalyze the hydrolysis of glycerophospholipids at the sn-2 position and generate several classes of bioactive lipids, fatty acids and lysophospholipids [1]. Six main families of phospholipases have defined physiological implications. They comprise secretory PLA_2 ($sPLA_2$), cytosolic PLA_2 ($cPLA_2$), lysosomal PLA_2, adipose-specific PLA_2 ($AdPLA_2$); and two major Ca^{2+}-independent groups, calcium-independent PLA_2 ($iPLA_2$) and platelet-activating factor acetylhydrolases (PAF-AH). This subdivision was based on their structures, catalytic mechanisms, localizations and evolutionary relationships, and they are collectively identified as groups, using roman numerals (i.e. Group I to Group XVI), with capital letters to distinguish individual sub-families [2]. Many of PLA_2 have contrasted role in cell signaling that involve intracellular Ca^{2+} homeostasis regulation.

6.1.1 Secretory PLA_2 ($sPLA_2$)

The secretory PLA_2s (belonging to Groups I, II, III, V, IX, X and XII in mammalians) were the first type of PLA_2 enzymes discovered. They were identified in organisms such as snakes and scorpions; in components of pancreatic juices; arthritic synovial fluid; and in many different mammalian tissues [3]. Most $sPLA_2$ isoforms are calcium-dependent, and require millimolar concentrations of the ion to function optimally [2, 4, 5]. Consequently, $sPLA_2$s typically function at the external side of the cell hydrolyzing a wide variety of phospholipids [2, 6]. $sPLA_2$ hydrolyzes the sn-2 ester bond in the glyceroacyl phospholipids presents in lipoproteins and cell membranes, inducing structural and functional changes and forming arachidonic acid (AA), lysopholipids and non-esterified fatty acids with direct proinflammatory effects [7, 8]. In general, $sPLA_2$ isoforms have solid preference for negatively charged phospholipid head groups, in particular phosphatidylserine (PS), phosphatidylglycerol (PG) and phosphatidylethanolamine (PE) [9]. Recent studies have suggested that some $sPLA_2$ isoforms can modify cell functions by binding to receptors and other proteins [5].

6.1.2 Cytosolic PLA_2 ($cPLA_2$)

The $cPLA_2$ family (also named Group IVA–F) is one of the major PLA_2 that contains six isoforms, ranging in size from 60 to 85 kDa, which are generally localized in the cytosol. They are active in the presence of mM levels of Ca^{2+} and, with the exception of $cPLA_2γ$ (Group IVC), contains in their N-terminals a C2 domain for the binding of two Ca^{2+} ions as well as two conserved phosphorylation sites. $cPLA_2$ family members have a catalytic domain characterized by a three-layer architecture employing a conserved Ser/Asp catalytic dyad, instead of the classical catalytic triad, that is similar in structure to that of $iPLA_2$ [10, 11]. The first group IV $cPLA_2$ (Group IVA) was firstly identified in human platelets in 1986 [12] and was cloned and sequenced 5 years later [13, 14]. $cPLA_2$ is perhaps the far most widely studied cytosolic enzyme and, besides transacylase activity, is also known to have PLA_2 and lysophospholipase activities [15]. $cPLA_2$ is activated by several different mechanisms, and is recruited to the membrane by a Ca^{2+} dependent translocation of the C2 domain. A recent work has localized the lipid binding surface of the enzyme in the presence of Ca^{2+} [16].

From the different PLA_2s, $cPLA_2$ is the only one described to have a preference for AA in the sn-2 position of phospholipids [10, 14]. Upon activation and translocation to intracellular membranes, $cPLA_2$ generates and releases AA from membrane phospholipids leading to an active lipoxygenase and cyclooxygenase metabolism [17]. AA, which acts as precursor for the generation of eicosanoids, is a key player in the prostanoid signaling cascades and therefore its activation is important for regulating various physiological and pathological processes including immune and inflammatory-related processes [2, 18, 19]. Furthermore, AA is also considered as an agonist that induces cytosolic Ca^{2+} entry through cationic channels called arachidonic acid-regulated calcium channels (ARC) [20, 21].

6.1.3 PAF Acetyl Hydrolase/Oxidized Lipid (PAF-AH/$LpPLA_2$)

Platelet activating factor (PAF) acetylhydrolases (AH) (PAF-AH, Group VIIA and B, and VIIIA and B) have low molecular weight (26–45 kDa) and represent a unique group of acyl hydrolases with a catalytic serine that is capable of releasing acetate from the sn-2 position of PAF, a 1-*O*-alkyl-PC [22]. However, they can also catalyze the release of oxidized acyl groups from the sn-2 position of PC and PE, not just PAF [2, 4, 23]. Its active site is composed of a serine, histidine, and aspartic acid hydrolase triad, unlike all other PLA_2s, which have dyads [24]. There are four members of this family that specifically catalyze these reactions; one of them is a secreted protein (GVIIA PLA_2), known as plasma-type PAF-AH or "lipoproteinassociated PLA_2" ($LpPLA_2$), that has generated interest as a therapeutic target for atherosclerosis [22, 26–29]. On the other hand, $LpPLA_2$ is a potent phospholipid activator that is secreted by multiple inflammatory cells including monocytes/macrophages, T lymphocytes and mast cells [30, 31]. This enzyme was cloned from human plasma in 1995 and was shown to have anti-inflammatory activity in vivo [25]. The $LpPLA_2$ role in cytosolic Ca^{2+} regulation is still unknown.

6.1.4 Lysosomal PLA_2 ($LyPLA_2$)

Lysosomal PLA_2 was purified from bovine brain as an enzyme that esterifies an acyl group with the hydroxyl group in the C-1 position of ceramide using phospholipids as the acyl group donor, so the enzyme was first named 1-O-acylceramide synthase (ACS). The protein possesses Ca^{2+} independent PLA_2 and transacylase activities. Hiraoka et al. [32] proposed that the hydrolyzed acyl group is transferred through an enzyme-acyl intermediate to ceramide or water, resulting either in the production of either 1-Oacyl- ceramide (ACS activity) or the release of free fatty acids (PLA_2 activity). In terms of catalytic activity, Ly-PLA_2 specifically prefers PC and PE head groups at pH 4.5 in a Ca^{2+}-independent manner. Ly-PLA_2 is ubiquitously expressed in diverse cell types, but highly expressed in alveolar macrophages. In fact, it plays a role in surfactant metabolism, and specifically in the phospholipid catabolism of pulmonary surfactant [33, 34].

6.1.5 Adipose Specific PLA_2 ($AdPLA_2$)

Duncan et al. [35] discovered recently a novel intracellular PLA_2, highly and differentially expressed only in adipocytes and induced during preadipocyte differentiation, that releases sn-2 fatty acid from phospholipids in a Ca^{2+}-dependent manner. This recently discovered enzyme named adipose-specific PLA_2 ($AdPLA_2$, Group XVI), has a molecular weight of 18 KDa. It is found abundantly in white adipose tissue, 40–150 times higher that found in liver. The enzyme is not an acyltransferase, but it functions entirely as a phospholipase, producing lysophosphatidylcholine and AA from the phospholipids. In addition, Duncan and colleagues studied the properties of AdPLA and found its optimal pH was 8.0, requiring cysteine and histidine residues at the active site, with maximal enzymatic activity in the presence of 1.0 mM Ca^{2+} [35]. $AdPLA_2$ have been also implicated in energy regulation as it modultes the release of fatty acids, from stored triglycerides in white adipose

Table 6.1 Isoforms of calcium-independent (Group VI) PLA2

PLA_2 family	Group	Source	MW (KDa)	Alternate name
$iPLA_2$	VIA-1	Human/murine	84–85	$iPLA_2\alpha$
	VIA-2	Human/murine	88–90	$iPLA_2\beta$
	VIB	Human/murine	88–91	$iPLA_2\gamma$
	VIC	Human/murine	146	$iPLA_2\delta$, neuropathy target esterase (NTE)
	VID	Human	53	$iPLA_2\varepsilon$, adiponutrin
	VIE	Human	57	$iPLA_2\zeta$, TTS-2.2
	VIF	Human	28	$iPLA_2\eta$, GS2

tissue, which will be later used as energy source by other tissues. $AdPLA_2$ has been also proposed to play a major role in the supply of AA for prostaglandin E2 (PGE2) synthesis in white adipose tissue [36]. Thus, AdPLA is considered a major regulator of adipocyte lipolysis and is crucial for the development of obesity, although it seems possible that AdPLA could promote obesity through a mechanism distinct from PGE2 signaling [37].

6.1.6 *Calcium Independent PLA_2 ($iPLA_2$)*

The Ca^{2+} independent PLA_2s are members of the GVI family of PLA_2 enzymes. Currently, six isoforms of $iPLA_2$ (Group VIA–F) have been identified as shown in Table 6.1. While their catalytic sites are similar to that of $cPLA_2$, they do not require Ca^{2+} for catalytic activity and they are generally larger in size, with moleculear weights ranging from 55 to 146 kDa except for Group VIF PLA_2 (~28 kDa). $iPLA_2$s are localized either in the cytosol, the endoplasmic reticulum (ER) or in the mitochondrial membrane [38]. $iPLA_2$ are entirely involved in lipid remodeling, in the Land's Cycle, and also mediate cell growth signaling [2, 4]. Members of this family share a protein domain initially discovered in patatin, the most abundant protein of the potato tuber.

In the next part of this chapter, we will go through $iPLA_2$ classification, regulation, and its role in intracellular Ca^{2+} regulation.

6.2 Sub-classification of $iPLA_2$

6.2.1 *GVIA PLA_2 ($iPLA_2\alpha$ and $iPLA_2\beta$)*

Many new $iPLA_2$ (GVI PLA_2) members have been identified in the last years, but the first member and the best characterized of this family is the GVIA PLA_2, which was purified from macrophages in 1994 [39, 40]. GVIA PLA_2 is expressed in multiple different splice variants [41] and, similar to $cPLA_2$ (GIV PLA_2), it catalyzes the cleavage of the sn-2 ester bond. However, it does not show specificity for AA in the sn-2 position and is fully active in the absence of Ca^{2+}. The GVIA PLA_2 also possesses sn-1 lysophospholipase and transacylase activity [41]. The enzyme has a conserved glycine-rich nucleotide-binding motif (GXGXXG) proximal to the catalytic site and it is activated several-fold by ATP [42]. The N-terminal domain of GIVA PLA_2 is composed of seven to eight ankyrin repeats, which are responsible for protein-protein interaction between monomers [43]. It is thought that ankyrin repeats enable the oligomeration of Group VIA monomers required for catalytic activity [39]. In fact, the active form of Group VIA PLA_2 is a tetramer [39].

Several splice variants of GVIA PLA_2 have been identified [39, 44]. Group VIA-1 or $iPLA_2\alpha$, and Group VIA-2 or $iPLA_2\beta$ [44–48], for example, comprise two catalytically active forms of this enzyme [44–48]. Both isoforms are similar in size, 85 and 88 KDa respectively, and contain eight N-terminal ankyrin repeats and a consensus lipase motif ($GXS^{465}XG$), whereas in GVIA-2 PLA_2 the 8 ankyrin repeats are interrupted by an insertion of 54 amino acids and they exhibit a glutamate residue at position 450, while the corresponding position in Group VIA-1 is glutamine.

Three additional splice variants of GVIA $iPLA_2$ have been identified: Group VIA-3 (also known as $iPLA_2$-2); Group VIA Ankyrin-1 (or Ankyrin-$iPLA_2$-1), and Group VIA Ankyrin-2 (or Ankyrin-$iPLA_2$-2). The GroupVIA-3 splice variant encodes an $iPLA_2$ that is identical to Group VIA-2 PLA_2 ($iPLA_2\beta$) at the N-terminus, that retains the $GTS^{519}TG$ active site and that has a truncated C-terminus. However, it is not known whether Group VIA-3 encodes a functional phospholipase A_2. Group VIA Ankyrin-1 is identical to Group VIA-2 at the N-terminus but it ends prior to the $GTS^{519}TG$ active site with a three amino acid modification at the C-terminus; it does not encode a functional PLA_2 enzyme [46]. Similar to Group VIA Ankyrin-1, Group VIA Ankyrin-2 also lacks the GTS519TG active site and additionally present with a 73 amino-acids shorter N-terminus and a 50-amino-acid variation at the C-terminus. Group VIA ankyrin-1 and Group VIA ankyrin-2 may act as negative regulators of Group VIA-1 and Group VIA-2 by precluding catalytically active tetramer aggregation [39, 46]. Processes in which GVIA PLA_2 has been implicated include phospholipids remodeling, AA release causing eicosanoid formation, protein expression, acetylcholine-mediated endothelium-dependent relaxation of the vasculature, secretion, and apoptosis. $iPLA_2$ plays also an important role in lymphocyte proliferation and in Ca^{2+} signaling regulated by calmodulin (CaM) and by a Ca^{2+} influx factor as detailed below [41, 49–52].

6.2.2 *GVIB PLA_2 or $iPLA_2\gamma$*

The $iPLA_2\gamma$ called also GVIB PLA_2 have been less studied. It has been involved in the release of AA that leads to eicosanoid formation [53, 54]. $iPLA_2\gamma$ contains the consensus lipase motif (GXSXG), a C-terminal peroxisome localization signal (SKL), and a glycine-rich nucleotide binding loop motif (GXGXXG). Interestingly, the nucleotide-binding motif commences 34 amino acids upstream of the putative active Ser, which is closely identical to the location of the nucleotide binding loop motif of Group VIA (35 amino acids upstream) [53]. A recent study demonstrated that iPLA$_2\gamma$ is responsible for the release of AA and prostaglandin E2 (PGE2) and inflammatory mediators in cardiac myocytes infected by Chagas' disease parasite [55]. Previously, iPLA$_2\gamma$ was also suggested as a critical participant in the Ca^{2+}-induced opening of the mitochondrial permeability transition pore (mPTP) in Liver [56].

6.2.3 GVIC, GVID, GVIE, GIVF PLA$_2$s

Different Ca^{2+}-independent lipases have been identified newly, and classified according to the terminology of the Group system GVIC, GVID, GVIE and GIVF PLA$_2$s. The GVIC PLA$_2$ enzyme has some sequence similarity to GVIA PLA$_2$ and might play a role in membrane homeostasis. This enzyme was previously known as NEST, the recombinantly expressed esterase domain of NTE (neuropathy target esterase), a membrane protein expressed in neurons of human and mice with physiological function elusive [57, 58] that possesses PLA$_2$ and lysophospholipase activities [59]. NEST might slowly hydrolyze the fatty acid in the sn-2 position of PC and subsequently, in a fast reaction, release the fatty acid in the sn-1 position.

The genes for the three other enzymes have also been identified before. Although, there was no catalytic activity attributed to corresponding proteins. The enzymes were shown to hydrolyze both LA and AA at the sn-2 position in the absence of free Ca^{2+} [60], thus these three enzymes might play a role in the regulation of triacylglycerol homeostasis which implicates the control of energy metabolism in adipocytes. Besides, PLA$_2$ activity, these enzymes possess high triacylglycerol lipase and acylglycerol transacylase activities and all of them were inhibited by bromoenol lacotone (BEL) at sub-micromolar levels [60].

6.3 Regulation of iPLA$_2$

6.3.1 ATP and PKC

The iPLA$_2$ protein contains a lipase consensus sequence and a putative ATP-binding motif. ATP has been reported to stimulate iPLA$_2$ activity in rat islets [61], murine P388D1 cells [45], but not to affect the iPLA$_2$ activity of Chinese Hamster Ovary cells [44]. In an early study, Ackerman et al. discovered that both Triton X-100 and ATP enhanced the activity of iPLA$_2$ in P388D1 cells [39]. The enzyme activity was 1.2–6 fold higher in mixed micelles when assayed in the presence of ATP and other di- or triphosphate nucleotides [39]. In other study, ATP stimulation of an $iPLA_2$ isoform was demonstrated in human pancreatic islet [42]. Interestingly, this same group demonstrated that ATP does not directly activate but rather protects $iPLA_2$ from a loss of its activity [61]. On the other hand, there is no consensus regarding the role of PKC in $iPLA_2$ activation [62]. An early study showed that the activation of PKCα ultimately provoked AA release via $iPLA_2$. This AA release was markedly inhibited by BEL or $iPLA_2$ antisense oligonucleotide [63]. Interestingly, we demonstrated that both diacylglycerol analog 1-oleoyl-2-acetyl-sn-glycerol (OAG) and store depletion with thapsigargin produced a PKCε-dependent activation of $iPLA_2\beta$ in proliferating but not in confluent aortic SMC [64].

6.3.2 Ca^{2+}/Calmodulin Regulation of $iPLA_2$

The first evidence of $iPLA_2$ modulation by CaM came from the observation that Ca^{2+} addition to the cytosol of cardiac myocytes inhibited $iPLA_2$ activity induced by ischemia. This inhibition was demonstrated to be due to CaM [65]. In fact, molecular and structural studies showed that in the absence of CaM, the active site of $iPLA_2$ interacts with the CaM-binding domain, resulting in a catalytically competent enzyme, whereas reversible disruption of this interaction through the binding of CaM abrogates this interaction, resulting in a loss of $iPLA_2$ activity [65–67]. $iPLA_2$ was shown to form a catalytically inactive ternary complex with CaM-Ca^{2+} that could be reversibly dissociated by chelation of Ca^{2+} ion with EGTA to regain full catalytic activity. Although $iPLA_2$ activity is independent of Ca^{2+}, it is able to inhibit the $iPLA_2$ activity by Ca^{2+}-activated CaM and this inhibition is apparently due to the binding to the IQ motif. In fact, the dissociation of CaM from $iPLA_2$ is the main mechanism that changes the Ca^{2+}-independent enzyme into an enzyme that is sensitive to modification in intracellular Ca^{2+} ion homeostasis. Moreover, conformational changes provoked in CaM using agents that inhibited the interaction of CaM with its target proteins resulted in $iPLA_2$ activation. Wolf et al. in 1997 have shown that W7, CaM antagonist, activated $iPLA_2$ in A-10 smooth muscle cells (SMC) [68]. Smani and colleagues also demonstrated that CaM inhibition with calmidazolium and a membrane-impermeable CaM inhibitory peptide, promoted $iPLA_2$ activation in SMC and RBL cell line [69]. Later on, compelling evidences have shown that store depletion with thapsigargin or cyclopiazonic acid stimulated $iPLA_2$ activation through displacement of inhibitory CaM [68–70].

6.3.3 Chemical Inhibition of $iPLA_2$

The most important inhibitor for $iPLA_2$ is BEL, which has specificity 1,000 times higher for $iPLA_2$ over other PLA_2 isoforms [41]. BEL is a suicidal substrate for $iPLA_2$ that is widely used as an irreversible mechanism-based, time- and temperature-dependent, inhibitor. For cell-based studies, it has been described previously that high concentrations of BEL (25 μM) partially inhibit the magnesium-dependent phosphatidate phosphohydrolase (PAP-1), which converts phosphatidic acid to diacylglycerol (DAG) [71, 72]. To some extent it is possible to identify promiscuous effects of BEL on $iPLA_2$ and PAP-1 by performing experiments with BEL and propranolol in parallel [71, 73]. The latter compound inhibits PAP-1 and not $iPLA_2$. Others and we confirmed that $iPLA_2$ activation induced by Ca^{2+} release from the store is inhibited by BEL [68, 70, 73, 74]. Importantly, Jenkins et al. [75] demonstrated that the commonly used BEL is composed of two enantiomers with different specificity for $iPLA_2$ isoforms. S-BEL has higher specificity to $iPLA_2\beta$, and R-BEL is more specific to $iPLA_2\gamma$, which allowed identifying the type of $iPLA_2$ involved in several different cellular processes. Indeed, we confirmed that S-BEL, but not R-BEL, selectively inhibited $iPLA_2\beta$ activity stimulated by intracellular store depletion in SMC and RBL, indicating that S-BEL is a valuable tool to determine the role of $iPLA_2\beta$ in intracellular signaling processes [64, 76].

6.4 $iPLA_2$ Role in the Ca^{2+} Signaling Network

As described above, for long time $iPLA_2$'s main role was especially related to cellular phospholipids remodeling [41]. However, different reports have demonstrated that the specific beta isoform of $iPLA_2$ ($iPLA_2\beta$) is involved in several agonist-stimulated signaling cascades. $iPLA_2$ has several unique features which confused researchers for many years. One of them relies on its activation independently of the presence or absence of Ca^{2+}. $iPLA_2$ is able to function in the presence of strong Ca^{2+} chelators as BAPTA. At the same time $iPLA_2$ is able to bind the Ca^{2+}–CaM complex. Interestingly, conditions for $iPLA_2$ activation are similar to those described for store operated calcium entry (SOCE). In fact, $iPLA_2$ can be activated by depletion of intracellular Ca^{2+} stores caused by vasopressin or by thapsigargin, an inhibitor of Sarco/Endoplamic reticulum Ca^{2+}-ATPase pump [68, 77, 78].

6.4.1 Overview of the Store Operated Ca^{2+} Channels Signaling Pathway

To increase cytoplasmic Ca^{2+} concentration, Ca^{2+} is either released from intracellular stores or enters into the cell by crossing the plasma membrane through ion channels. Store operated Ca^{2+} channels (SOCC) and receptor operated channels (ROC) are considered the main route for Ca^{2+} entry in non-excitable cells, but they also exist in excitable cells such as skeletal muscle, neurons or smooth muscle [79]. In excitable cells, Ca^{2+} entry is achieved largely through opening of voltage and/or voltage independent channels ROC or SOCC that are responsible of SOCE [80]. The concept of SOCE activation seems to be simple: basically upon depletion of ER stores, a signal is produced that activates specific Ca^{2+}-conducting channels SOCC, in plasma membrane that allows Ca^{2+} entry into the cell. SOCC role was originally linked only to refilling the intracellular store. However, now it's widely agreed that these channels provide a sustained Ca^{2+} influx for a variety of important functions in eukaryotic cells. Among those functions are exocytosis, vascular contraction and relaxation, Ca^{2+} oscillations, gene transcription, regulation of enzymatic activity, cell proliferation and apoptosis [80, 81].

6.4.1.1 Mechanism of SOCE Activation: Emerging Role of STIM1 and Orai1

One of the most intriguing mysteries of the store-operated pathway is the mechanism of its activation. Questions of how do the stores communicate with the plasma membrane channels and which is the signal produced by the stores upon their depletion, have been a matter of intense investigation for long time. Hypotheses presented can be mainly grouped into two main categories: those that propose the generation of a diffusible molecule with ability to induce SOCC opening, and those that assume a physical interaction between channel subunit and an element of the ER (for review see [80, 82]). Soon after the identification of SOCE Robin Irvine proposed a physical or conformational coupling between elements in the ER and SOCC in the plasma membrane [83], as a mechanism that resembles the classical excitation-contraction coupling between ryanodine receptors and dihydropyridine receptors in the skeletal muscle [84]. Consequently, most of the early studies focused on the association between inositol-triphosphate receptors (IP_3R) and the subunit channel suggested to form SOCC. This hypothesis received support from studies demonstrating that, under resting conditions; TRPC1, TRPC3 and TRPC6 can be co-immunoprecipitated with IP_3R [85, 86]. However, the major challenge for this model came from the studies in triple IP_3R knockout DT40 cells, in which SOCE seemed completely normal [87–89]. Importantly, in 2005 and 2006 the Ca^{2+} sensor of the ER was identified as the STIM1 (Stromal Interaction Molecule-1) protein, and Orai1 was identified as the structural subunit of the channel conducting the Ca^{2+} selective CRAC [90–92]. Several reports have showed that upon Ca^{2+} depletion, STIM1 lose Ca^{2+} from its EF hand, oligomerize and accumulate into punctate structures in the ER membrane located in close proximity (10–25 nm) to the plasma membrane. Furthermore, STIM1 and Orai1 have been reported to accumulate and colocalize in punctate structures along the plasma membrane and to associate by a reversible and physical coupling mechanism upon depletion of the intracellular Ca^{2+} stores which support the conformational coupling model (for review see [93]). While direct coupling of ER-resident STIM1 to PM-resident Orai1 is considered as the most straightforward mechanism for signal transduction, there is a growing body of evidence for the presence of additional structural and/or functional linker(s) between STIM1 and Orai1. Indeed, Balla's group suggested the presence of additional molecular components within the STIM1-Orai1 complex [94]; meanwhile Rosado and colleagues nicely showed that both STIM1 and Orai1 also co-immunoprecipitate with other TRPC channels when stores are depleted [95–97]. Recently, we have demonstrated that store depletion stimulated STIM1 and iPLA$_2$β colocalization required for SOCE in coronary artery [98].

6.4.1.2 Calcium Influx Factor and SOCE Activation

The other hypothesis focuses on diffusible messengers generated upon intracellular stores depletion. Different signaling molecules have been reported to play an essential role in the activation of SOCE in different cell types, including cGMP [99], tyrosine kinases [100], and small GTP-binding proteins [101], among others. However, special efforts were dedicated to the still uncharacterized molecule known as Ca^{2+} influx factor (CIF) by Victoria Bolotina's group. Refined CIF extract was obtained from different cell lines, including human platelets, which stimulated an extracellular Ca^{2+} influx and I_{CRAC} (CRAC current) sensitive to the well-known SOCC inhibitors [76]. Interestingly, soon after STIM1 discovery, Bolotina and co-workers presented compelling evidences demonstrating that CIF production is tightly linked with STIM1 expression and requires the functional integrity of glycosylation sites in its intraluminal SAM domain [102]. In this study, authors demonstrated that upon store depletion, CIF is produced before STIM1 accumulation in puncta and activation of SOCE. Authors showed that lack of STIM1 in the rare neuronal cell line (NG115-401L), which features virtually no SOCE responses [103], or STIM1 downregulation in cells transfected with siRNAs, dramatically impaired active CIF production confirming CIF and STIM1 relationship [102]. Unfortunately, the molecular identity of CIF is still unknown, although its presence and its biological activity were detected by numerous groups in a wide variety of cell types ranging from yeast to human (for review see [51, 79]). Previously, we have characterized in our earliest studies that iPLA$_2$β is the physiological target of CIF and the mechanism of CIF-induced activation of SOCE was depicted as illustrated in Fig. 6.1 [69, 76, 104].

6.4.2 Essential Role of iPLA$_2$ in Store Operated Calcium Entry

In the last 1990s, several works established an interesting scenario for iPLA$_2$ activation, showing that it could be activated by depletion of Ca^{2+} stores caused by vasopressin or by thapsigargin in A10 SMC line [68, 77]. A10 cells stimulation with thapsigargin induced release of AA that was directly correlated to thapsigargin-induced depletion of intracellular Ca^{2+} stores [68]. Next, iPLA$_2$β, and not iPLA$_2$γ, was identified as the mediator of vasopressin-induced AA release in SMC [75]. Therefore, the role of iPLA$_2$β in SOCE activation was explored and the first evidence of iPLA$_2$ requirement for SOCE activation was obtained by studying SOCE in primary culture of SMC as a model for excitable cells, and RBL cells as a model for non-excitable cells. The functional inhibition of iPLA$_2$ with BEL prevented the

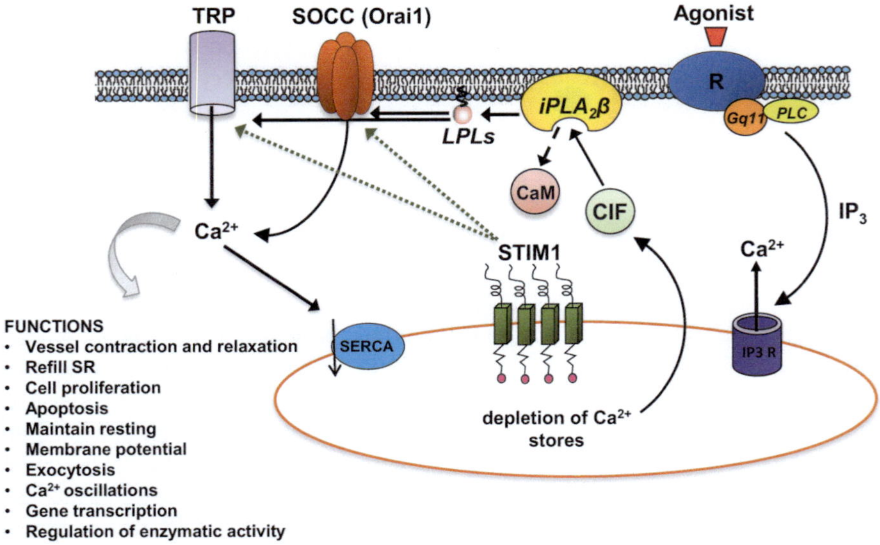

Fig. 6.1 Mechanism suggested for store operated Ca^{2+} entry (*SOCE*), involving Ca^{2+} independent phospholipase $A_2\beta$ (*iPLA₂*) pathway. We propose that agonist binding to R, promotes IP_3-induced Ca^{2+} release, activation/translocation of STIM1, Calcium Influx Factor (*CIF*) release, displacement of calmodulin (*CaM*), activation of $iPLA_2$, lysophospholipids (*LPLs*) production, and activation of store-operated Ca^{2+} channels (*SOCC*) (Orai1) and other transient receptor potential channels (*TRP*). The subsequent Ca^{2+} entry is responsible for a large number of functions as indicated

activation of single SOCC in SMC, and whole cell CRAC currents in RBL induced by TG and/or BAPTA-induced depletion of intracellular stores. In addition, molecular inhibition using antisense against $iPLA_2$, or its functional blocking with BEL impaired dramatically SOCE, while Ca^{2+} release from the stores was not affected, which confirmed the novel role of $iPLA_2$ in SOCE pathway [78]. Furthermore, the use of S-BEL enantiomer confirmed that $iPLA_2\beta$ is the isoform responsible of SOCE in RBL [76] and SMC [64]. Interestingly, cell dialysis with recombinant $iPLA_2\beta$ could substitute the endogenous $iPLA_2\beta$ and rescue activation of I_{CRAC} in the cells in which endogenous $iPLA_2\beta$ was knocked down [76]. One of the most important features of $iPLA_2$ is that it exists in a complex with CaM, which keeps it in a catalytically inactive state; and removal of CaM results in $iPLA_2$ activation [66]. Therefore, the inhibition of CaM was found to mimic the effects of thapsigargin-induced SOCE as it activated $iPLA_2$; it evoked a 2APB and BEL-sensitive Ca^{2+} influx; and finally it stimulated single SOCC in SMC [69]. Similar effect of CaM inhibition was also observed in astrocytes [70] and in rat cerebellar granule [105].

The role of $iPLA_2\beta$ in SOCE was further confirmed by us and by many other investigators in a growing number of cell types, including platelets, Jurkat T lymphocytes [69, 78], RBL-2H3 [104], neuroblastoma/glioma [70], keratinocytes [106], skeletal muscle [107], fibroblasts [108], prostate cancer cells [109] and others. In all these studies molecular knock-down and/or functional inhibition of $iPLA_2\beta$ caused full impairment of SOCE. Strikingly, genetic screening of Drosophila melanogaster performed by Vig et al. indicated that not only STIM1 and Orai1, but also an orthologue of iPLA$_2$β encoded by the CG6718 gene, are gene products with a great impact on SOCE activation [110]. Recently, we have demonstrated that agonist-induced coronary artery contraction involved the activation of SOCE by STIM1, Orai1 and iPLA$_2$ [98]. We have shown that on cells stimulation, STIM1 colocalized with iPLA$_2$β in submembrane compartments suggesting their functional communication and we confirmed that lysophopholipids, product of iPLA$_2$, stimulated an Orai1- but not STIM1- dependent SOCE, suggesting that the functional role of iPLA$_2$β is downstream of STIM1 and upstream of Orai1 in coronary SMC. The complex relationships between the components of the CRAC channel, namely Orai1, STIM1, and iPLA$_2$β in the SOCE pathway have been detailed in a previous review [79].

6.4.3 iPLA$_2$ and Lysophospholipids Activation of Store Operated Calcium Entry

Afterwards, numerous studies focused on the molecular mechanism of iPLA$_2$-dependent signal transduction. Several works from Bolotina's lab in the first decade of this centry, provided compelling evidences demonstrating that SOCC can be activated by CIF produced upon depletion of Ca^{2+} stores in the ER, and it in turn, can displace the inhibitory CaM from iPLA$_2$β. The early studies have shown that CIF activated single SOCC in inside-out membrane patches [111], and the channels remained active even after the membrane patches were excised and CIF was washed away [112], indicating the presence of an additional cascade of plasma membrane-delimited reactions that might be involved in CIF-induced activation of SOCC. In 2004, a major finding has been described by Smani et al. demonstrating that CIF extract can displace inhibitory CaM from iPLA$_2$β leading to lysophopholipids production and the activation of SOCC in membrane-delimited manner in SMC [69]. By contrast, CIF dialysis of RBL cells transfected with antisense to iPLA$_2$β failed to activate I_{CRAC}, confirming the need of functional iPLA$_2$β to stimulate SOCE [76]. Furthermore, the exogenous application of lysophopholipids but not AA, products of iPLA$_2$β activation, were able to stimulate SOCE in intact cells and single SOCC in inside-out membrane patches in SMC [69, 74, 98]. Further studies have confirmed that lysophospholids evoked SOCE in different cell lines such as astrocyte [70], rat cerebellar granule neurons [105], skeletal muscle [113], and keratinocytes [114], among others cells. Thus, several independent works established the need of active iPLA$_2$β, and lysophospholipids to stimulate SOCE in a wide range of cells.

However and independently of its role in SOCE signaling, few reports have shown that iPLA$_2$ might activate some TRP channels. Works from Prevarskaya's lab demonstrated that iPLA$_2$β activated both SOCC and TRPM8 channels [109, 115], and AL-Shawaf and colleagues showed recently that lysophosphatidylcholine and AA generated by iPLA$_2$ are involved in TRPC5 activation by sphingosine-1-phosphate [116].

6.5 Significant Potential as Targets for Novel Therapeutics Strategy

The role of $iPLA_2\beta$, and the consequent activation of SOCE in several physio- and pathological processes have been largely studied. For example, $iPLA_2\beta$-dependent activation of vascular reactivity was demonstrated in aorta, cerebral, mesenteric, carotid and coronary arteries [98, 117, 118]. Furthermore, $iPLA_2\beta$-induced SOCE seems involved in SMC proliferation [119] and in HEK cells migration [120]. Molecular knockdown of Orai1, STIM1 or $iPLA_2\beta$ caused a similar reduction in velocity and distance in migrating HEK cells. Previously, Vanden Abeele et al. demonstrated that $iPLA_2\beta$ activated SOCE in LNCaP prostate cancer proliferative cells [109], and they further showed that $iPLA_2\beta$ is also implicated in the lysophospholipid-dependent gating of TRPM8, a cold sensor [115]. On the other hand, Boittin and Reugg published several interesting studies highlighting the involvement of $iPLA_2$-dependent activation of SOCC in dystrophic muscle fibers [109]. They found that $iPLA_2$ is mainly localized in the vicinity of the sarcolemma, suggesting a close proximity with SOCC, which may be located on the sarcolemma and/or in the T-tubular membranes. These authors have also demonstrated that lysophosphatidylcholine acts downstream of $iPLA_2$ and directly activates SOCC in dystrophic fibers [107, 113]. Interestingly, recent studies have determined that $iPLA_2$ can be targeted by secondary signaling pathway to potentiate or inhibit SOCE such as PKCε [64], and Urocortin through cyclic AMP in SMC [98], in skeletal muscle [121], and in hepatoma carcinoma cell lines [122]. These few examples confirm the important role of $iPLA_2$ and SOCE in several physiological and pathological processes and confirm it as a valuable therapeutic target.

Acknowledgments This work was supported by Spanish Ministry of Economy and Competitiveness [BFU2013-45564-C2-1-P and BFU2013-45564-C2-2-P]; Institute of Carlos III and Cardiovascular Network "RIC" [RD12/0042/0041;PI12/00941]; and from the Andalusia Government [PI-0108-2012; P10-CVI-6095]. A.D.R. is supported by ITRIBIS FP-7-REGPOT.

References

1. Burke JE, Dennis EA (2009) Phospholipase A2 biochemistry. Cardiovasc Drug Ther 23(1):49–59
2. Dennis EA, Cao J, Hsu YH, Magrioti V, Kokotos G (2011) Phospholipase A2 enzymes: physical structure, biological function, disease implication, chemical inhibition, and therapeutic intervention. Chem Rev 111(10):6130–6185
3. Six DA, Dennis EA (2000) The expanding superfamily of phospholipase A 2 enzymes: classification and characterization. Biochim Biophys Acta 1488(1):1–19
4. Murakami M, Taketomi Y, Miki Y, Sato H, Hirabayashi T, Yamamoto K (2011) Recent progress in phospholipase A 2 research: from cells to animals to humans. Prog Lipid Res 50(2):152–192
5. Chakraborti S (2003) Phospholipase A 2 isoforms: a perspective. Cell Signal 15(7):637–665

6. Lambeau G, Gelb MH (2008) Biochemistry and physiology of mammalian secreted phospholipases A2. Annu Rev Biochem 77:495–520
7. Camejo G (2010) Lysophospholipids: effectors mediating the contribution of dyslipidemia to calcification associated with atherosclerosis. Atherosclerosis 211(1):36–37
8. Rodriguéz-Lee M, Bondjers G, Camejo G (2007) Fatty acid-induced atherogenic changes in extracellular matrix proteoglycans. Curr Opin Lipidol 18(5):546–553
9. Singer AG, Ghomashchi F, Le Calvez C, Bollinger J, Bezzine S, Rouault M, Gelb MH (2002) Interfacial kinetic and binding properties of the complete set of human and mouse groups I, II, V, X, and XII secreted phospholipases A2. J Biol Chem 277(50):48535–48549
10. Kramer RM, Sharp JD (1995) Recent insights into the structure, function and biology of cPLA2. Agents Actions Suppl 46:65–76
11. Kramer RM, Sharp JD (1997) Structure, function and regulation of Ca2+-sensitive cytosolic phospholipase A2 (cPLA2). FEBS Lett 410(1):49–53
12. Kramer RM, Checani GC, Deykin A, Pritzker CR, Deykin D (1986) Solubilization and properties of Ca2+-dependent human platelet phospholipase A 2. Biochim Biophys Acta 878(3):394–403
13. Kramer RM, Roberts EF, Manetta J, Putnam JE (1991) The Ca2(+)-sensitive cytosolic phospholipase A2 is a 100-kDa protein in human monoblast U937 cells. J Biol Chem 266(8):5268–5272
14. Clark JD, Lin LL, Kriz RW, Ramesha CS, Sultzman LA, Lin AY, Knopf JL (1991) A novel arachidonic acid-selective cytosolic PLA2 contains a Ca2+-dependent translocation domain with homology to PKC and GAP. Cell 65(6):1043–1051
15. Reynolds LJ, Hughes LL, Louis AI, Kramer RM, Dennis EA (1993) Metal ion and salt effects on the phospholipase A2, lysophospholipase, and transacylase activities of human cytosolic phospholipase A2. Biochim Biophys Acta 1167(3):272–280
16. Burke JE, Hsu YH, Deems RA, Li S, Woods VL, Dennis EA (2008) A phospholipid substrate molecule residing in the membrane surface mediates opening of the lid region in group IVA cytosolic phospholipase A2. J Biol Chem 283(45):31227–31236
17. Leslie CC (1997) Properties and regulation of cytosolic phospholipase A2. J Biol Chem 272(27):16709–16712
18. Kita Y, Ohto T, Uozumi N, Shimizu T (2006) Biochemical properties and pathophysiological roles of cytosolic phospholipase A2s. Biochim Biophys Acta 1761(11):1317–1322
19. Kudo I, Murakami M (2002) Phospholipase A2 enzymes. Prostaglandins Other Lipid Mediat 68–69:3–58
20. Mignen O, Shuttleworth TJ (2000) I(ARC), a novel arachidonate-regulated, noncapacitative Ca(2+) entry channel. J Biol Chem 275(13):9114–9119
21. Mignen O, Thompson JL, Shuttleworth TJ (2009) The molecular architecture of the arachidonate-regulated Ca2+-selective ARC channel is a pentameric assembly of Orai1 and Orai3 subunits. J Physiol 587(17):4181–4197
22. Khakpour H, Frishman WH (2009) Lipoprotein-associated phospholipase A2: an independent predictor of cardiovascular risk and a novel target for immunomodulation therapy. Cardiol Rev 17(5):222–229
23. Stephens JWW, Myers W (1898) The action of cobra poison on the blood: a contribution to the study of passive immunity. J Pathol Bacteriol 5(3):279–301
24. Tjoelker LW, Eberhardt C, Unger J, Le Trong H, Zimmerman GA, McIntyre TM, Gray PW (1995) Plasma platelet-activating factor acetylhydrolase is a secreted phospholipase A2 with a catalytic triad. J Biol Chem 270(43):25481–25487
25. Tjoelker LW, Wilder C, Eberhardt C, Stafforinit DM, Dietsch G, Schimpf B, Gray PW (1995) Anti-inflammatory properties of a platelet-activating factor acetylhydrolase. Nature 374(6522):549–553
26. Chen CH (2004) Platelet-activating factor acetylhydrolase: is it good or bad for you? Curr Opin Lipidol 15(3):337–341
27. Packard CJ (2009) Lipoprotein-associated phospholipase A2 as a biomarker of coronary heart disease and a therapeutic target. Curr Opin Cardiol 24(4):358–363

28. Tsimikas S, Tsironis LD, Tselepis AD (2007) New insights into the role of lipoprotein (a)-associated lipoprotein-associated phospholipase A2 in atherosclerosis and cardiovascular disease. Arterioscler Thromb Vasc Biol 27(10):2094–2099
29. Wilensky RL, Macphee CH (2009) Lipoprotein-associated phospholipase A2 and atherosclerosis. Curr Opin Lipidol 20(5):415–420
30. Stafforini DM, Satoh K, Atkinson DL, Tjoelker LW, Eberhardt C, Yoshida H, Prescott SM (1996) Platelet-activating factor acetylhydrolase deficiency. A missense mutation near the active site of an anti-inflammatory phospholipase. J Clin Invest 97(12):2784–2791
31. Min JH, Wilder C, Aoki J, Arai H, Inoue K, Paul L, Gelb MH (2001) Platelet-activating factor acetylhydrolases: broad substrate specificity and lipoprotein binding does not modulate the catalytic properties of the plasma enzyme. Biochemistry 40(15):4539–4549
32. Hiraoka M, Abe A, Shayman JA (2002) Cloning and characterization of a lysosomal phospholipase A2, 1-O-acylceramide synthase. J Biol Chem 277(12):10090–10099
33. Hiraoka M, Abe A, Lu Y, Yang K, Han X, Gross RW, Shayman JA (2006) Lysosomal phospholipase A2 and phospholipidosis. Mol Cell Biol 26(16):6139–6148
34. Abe A, Poucher HK, Hiraoka M, Shayman JA (2004) Induction of lysosomal phospholipase A2 through the retinoid X receptor in THP-1 cells. J Lipid Res 45(4):667–673
35. Duncan RE, Sarkadi-Nagy E, Jaworski K, Ahmadian M, Sul HS (2008) Identification and functional characterization of adipose-specific phospholipase A2 (AdPLA). J Biol Chem 283(37):25428–25436
36. Jaworski K, Ahmadian M, Duncan RE, Sarkadi-Nagy E, Varady KA, Hellerstein MK, Sul HS (2009) AdPLA ablation increases lipolysis and prevents obesity induced by high-fat feeding or leptin deficiency. Nat Med 15(2):159–168
37. Sanchez-Alavez M, Klein I, Brownell SE, Tabarean IV, Davis CN, Conti B, Bartfai T (2007) Night eating and obesity in the EP3R-deficient mouse. Proc Natl Acad Sci 104(8):3009–3014
38. Cummings BS, McHowat J, Schnellmann RG (2002) Role of an endoplasmic reticulum Ca2+-independent phospholipase A2 in oxidant-induced renal cell death. J Pharmacol Exp Ther 283(3):492–498
39. Ackermann EJ, Kempner ES, Dennis EA (1994) Ca2+-independent cytosolic phospholipase A2 from macrophage-like P388D1 cells Isolation and characterization. J Biol Chem 269(12):9227–9233
40. Schaloske RH, Dennis EA (2006) The phospholipase A2 superfamily and its group numbering system. Biochim Biophys Acta 1761(11):1246–1259
41. Winstead MV, Balsinde J, Dennis EA (2000) Calcium-independent phospholipase A2: structure and function. Biochim Biophys Acta 1488(1):28–39
42. Ma Z, Wang X, Nowatzke W, Ramanadham S, Turk J (1999) Human pancreatic islets express mRNA species encoding two distinct catalytically active isoforms of group VI phospholipase A2 (iPLA2) that arise from an exon-skipping mechanism of alternative splicing of the transcript from the iPLA2 gene on chromosome 22q13 1. J Biol Chem 274(14):9607–9616
43. Sedgwick SG, Smerdon SJ (1999) The ankyrin repeat: a diversity of interactions on a common structural framework. Trends Biochem Sci 24(8):311–316
44. Tang J, Kriz RW, Wolfman N, Shaffer M, Seehra J, Jones SS (1997) A novel cytosolic calcium-independent phospholipase A2 contains eight ankyrin motifs. J Biol Chem 272(13):8567–8575
45. Balboa MA, Balsinde J, Jones SS, Dennis EA (1997) Identity between the Ca2+-independent phospholipase A2 enzymes from P388D1 macrophages and Chinese hamster ovary cells. J Biol Chem 272(13):8576–8580
46. Larsson PK, Claesson HE, Kennedy BP (1998) Multiple splice variants of the human calcium-independent phospholipase A2 and their effect on enzyme activity. J Biol Chem 273(1):207–214
47. Ma Z, Ramanadham S, Wohltmann M, Bohrer A, Hsu FF, Turk J (2001) Studies of insulin secretory responses and of arachidonic acid incorporation into phospholipids of stably trans-

fected insulinoma cells that overexpress group VIA phospholipase A2 (iPLA2beta) indicate a signaling rather than a housekeeping role for iPLA2beta. J Biol Chem 276(16):13198–13208
48. Larsson Forsell PK, Kennedy BP, Claesson HE (1999) The human calcium-independent phospholipase A2 gene multiple enzymes with distinct properties from a single gene. Eur J Biochem 262(2):575–585
49. Akiba S, Sato T (2004) Cellular function of calcium-independent phospholipase A2. Biol Pharm Bull 27(8):1174–1178
50. Balsinde J, Balboa MA (2005) Cellular regulation and proposed biological functions of group VIA calcium-independent phospholipase A2 in activated cells. Cell Signal 17(9):1052–1062
51. Bolotina VM, Csutora P (2005) CIF and other mysteries of the store-operated Ca2+-entry pathway. Trends Biochem Sci 30(7):378–387
52. Turk J, Ramanadham S (2004) The expression and function of a group VIA calcium-independent phospholipase A2 (iPLA2β) in β-cells can. J Physiol Pharmacol 82(10):824–832
53. Mancuso DJ, Jenkins CM, Gross RW (2000) The genomic organization, complete mRNA sequence, cloning, and expression of a novel human intracellular membrane-associated calcium-independent phospholipase A2. J Biol Chem 275(14):9937–9945
54. Murakami M, Masuda S, Ueda-Semmyo K, Yoda E, Kuwata H, Takanezawa Y, Kudo I (2005) Group VIB Ca2+-independent phospholipase A2γpromotes cellular membrane hydrolysis and prostaglandin production in a manner distinct from other intracellular phospholipases A2. J Biol Chem 280(14):14028–14041
55. Sharma J, Eickhoff CS, Hoft DF, Ford DA, Gross RW, McHowat J (2013) The absence of myocardial calcium-independent phospholipase A2γ results in impaired prostaglandin E2 production and decreased survival in mice with acute Trypanosoma cruzi Infection. Infect Immun 81(7):2278–2287
56. Moon SH, Jenkins CM, Kiebish MA, Sims HF, Mancuso DJ, Gross RW (2012) Genetic ablation of calcium-independent phospholipase A(2)γ (iPLA(2)γ) attenuates calcium-induced opening of the mitochondrial permeability transition pore and resultant cytochrome c release. J Biol Chem 287(35):29837–29850
57. Glynn P (1999) Neuropathy target esterase. Biochem J 344:625–631
58. Glynn P (2005) Neuropathy target esterase and phospholipid deacylation. Biochim Biophys Acta 1736(2):87–93
59. Van Tienhoven M, Atkins J, Li Y, Glynn P (2002) Human neuropathy target esterase catalyzes hydrolysis of membrane lipids. J Biol Chem 277(23):20942–20948
60. Jenkins CM, Mancuso DJ, Yan W, Sims HF, Gibson B, Gross RW (2004) Identification, cloning, expression, and purification of three novel human calcium-independent phospholipase A2 family members possessing triacylglycerol lipase and acylglycerol transacylase activities. J Biol Chem 279(47):48968–48975
61. Ma Z, Ramanadham S, Kempe K, Chi XS, Ladenson J, Turk J (1997) Pancreatic islets express a Ca2+-independent phospholipase A2 enzyme that contains a repeated structural motif homologous to the integral membrane protein binding domain of ankyrin. J Biol Chem 272(17):11118–11127
62. Akiba S, Mizunaga S, Kume K, Hayama M, Sato T (1999) Involvement of group VI Ca2+-independent phospholipase A2 in protein kinase C-dependent arachidonic acid liberation in zymosan-stimulated macrophage-like P388D1 cells. J Biol Chem 274:19906–19912
63. Akiba S, Ohno S, Chiba M, Kume K, Hayama M, Sato T (2002) Protein kinase Cα-dependent increase in Ca2+-independent phospholipase A2 in membranes and arachidonic acid liberation in zymosan-stimulated macrophage-like P388D 1 cells. Biochem Pharmacol 63(11):1969–1977
64. Smani T, Patel T, Bolotina VM (2008) Complex regulation of store-operated Ca2+ entry pathway by PKC-ε in vascular SMCs. Am J Physiol Cell Physiol 294(6):C1499–C1508

65. Wolf MJ, Gross RW (1996) The calcium-dependent association and functional coupling of calmodulin with myocardial phospholipase A2. Implications for cardiac cycle-dependent alterations in phospholipolysis. J Biol Chem 271(35):20989–20992
66. Wolf MJ, Gross RW (1996) Expression, purification, and kinetic characterization of a recombinant 80-kDa intracellular calcium-independent phospholipase A2. J Biol Chem 271:30879–30885
67. Jenkins CM, Wolf MJ, Mancuso DJ, Gross RW (2001) Identification of the calmodulin-binding domain of recombinant calcium-independent phospholipase A2β. Implications for structure and function. J Biol Chem 276(10):7129–7135
68. Wolf MJ, Wang J, Turk J, Gross RW (1997) Depletion of intracellular calcium stores activates smooth muscle cell calcium-independent phospholipase A 2. A novel mechanism underlying arachidonic acid mobilization. J Biol Chem 272(3):1522–1526
69. Smani T, Zakharov SI, Csutora P, Leno E, Trepakova E, Bolotina VM (2004) A novel mechanism for the store-operated calcium influx pathway. Nat Cell Biol 6(2):113–120
70. Singaravelu K, Lohr C, Deitmer JW (2006) Regulation of store-operated calcium entry by calcium-independent phospholipase A2 in rat cerebellar astrocytes. J Neurosci 26(37):9579–9592
71. Balsinde J, Dennis EA (1996) Bromoenol lactone inhibits magnesium-dependent phosphatidate phosphohydrolase and blocks triacylglycerol biosynthesis in mouse P388D1 macrophages. J Biol Chem 271(50):31937–31941
72. Balboa MA, Balsinde J, Dennis EA (1998) Involvement of phosphatidate phosphohydrolase in arachidonic acid mobilization in human amnionic WISH cells. J Biol Chem 273(13):7684–7690
73. Johnson CA, Balboa MA, Balsinde J, Dennis EA (1999) Regulation of cyclooxygenase-2 expression by phosphatidate phosphohydrolase in human amnionic WISH cells. J Biol Chem 274(39):27689–27693
74. Smani T, Domínguez-Rodríguez A, Hmadcha A, Calderón-Sánchez E, Horrillo-Ledesma A, Ordóñez A (2007) Role of Ca2+-independent phospholipase A2 and store-operated pathway in urocortin-induced vasodilatation of rat coronary artery. Circ Res 101(11):1194–1203
75. Jenkins CM, Han X, Mancuso DJ, Gross RW (2002) Identification of calcium-independent phospholipase A2 (iPLA2)β, and not iPLA2γ, as the mediator of arginine vasopressin-induced arachidonic acid release in A-10 smooth muscle cells. Enantioselective mechanism-based discrimination of mammalian iPLA2s. J Biol Chem 277(36):32807–32814
76. Csutora P, Zarayskiy V, Peter K, Monje F, Smani T, Zakharov SI, Bolotina VM (2006) Activation mechanism for CRAC current and store-operated Ca2+ entry calcium influx factor and Ca2+-independent phospholipase A2β-mediated pathway. J Biol Chem 281(46):34926–34935
77. Nowatzke W, Ramanadham S, Ma Z, Hsu FF, Bohrer A, Turk J (1998) Mass spectrometric evidence that agents that cause loss of Ca2+ from intracellular compartments induce hydrolysis of arachidonic acid from pancreatic islet membrane phospholipids by a mechanism that does not require a rise in cytosolic Ca2+ concentration. Endocrinology 139(10):4073–4085
78. Smani T, Zakharov SI, Leno E, Csutora P, Trepakova ES, Bolotina VM (2003) Ca2+-independent phospholipase A2 is a novel determinant of store-operated Ca2+ entry. J Biol Chem 278(14):11909–11915
79. Bolotina VM (2008) Orai, STIM1 and iPLA2β: a view from a different perspective. J Physiol 586(13):3035–3042
80. Parekh AB, Putney JW (2005) Store-operated calcium channels. Physiol Rev 85(2):757–810
81. Berridge MJ, Bootman MD, Roderick HL (2003) Calcium signalling: dynamics, homeostasis and remodeling. Nat Rev Mol Cell Biol 4(7):517–529
82. Rosado JA, Redondo PC, Sage SO, Pariente JA, Salido GM (2005) Store-operated Ca2+ entry: vesicle fusion or reversible trafficking and de novo conformational coupling? J Cell Physiol 205(2):262–269

83. Irvine RF (1990) 'Quanta' Ca2+ release and the control of Ca2+ entry by inositol phosphates a possible mechanism. FEBS Lett 263(1):5–9
84. Block BA, Imagawa T, Campbell KP, Franzini-Armstrong C (1988) Structural evidence for direct interaction between the molecular components of the transverse tubule/sarcoplasmic reticulum junction in skeletal muscle. J Cell Biol 107(6):2587–2600
85. Boulay G, Brown DM, Qin N, Jiang M, Dietrich A, Zhu MX, Birnbaumer L (1999) Modulation of Ca2+ entry by polypeptides of the inositol 1,4,5-trisphosphate receptor (IP3R) that bind transient receptor potential (TRP): evidence for roles of TRP and IP3R in store depletion-activated Ca2+ entry. Proc Natl Acad Sci 96(26):14955–14960
86. Rosado J, Sage S (2000) Coupling between inositol 1, 4, 5-trisphosphate receptors and human transient receptor potential channel 1 when intracellular Ca2+ stores are depleted. Biochem J 350(3):631–635
87. Sugawara H, Kurosaki M, Takata M, Kurosaki T (1997) Genetic evidence for involvement of type 1, type 2 and type 3 inositol 1,4,5-trisphosphate receptors in signal transduction through the B-cell antigen receptor. EMBO J 16(11):3078–3088
88. Prakriya M, Lewis RS (2001) Potentiation and inhibition of Ca2+ release-activated Ca2+ channels by 2-aminoethyldiphenyl borate (2-APB) occurs independently of IP3 receptors. J Physiol 536(1):3–19
89. Bakowski D, Glitsch MD, Parekh AB (2001) An examination of the secretion-like coupling model for the activation of the Ca2+ release-activated Ca2+ current ICRAC in RBL1 cells. J Physiol 532(1):55–71
90. Feske S, Gwack Y, Prakriya M, Srikanth S, Puppel SH, Tanasa B, Rao A (2006) A mutation in Orai1 causes immune deficiency by abrogating CRAC channel function. Nature 441(7090):179–185
91. Roos J, DiGregorio PJ, Yeromin AV, Ohlsen K, Lioudyno M, Zhang S, Stauderman KA (2005) STIM1, an essential and conserved component of store-operated Ca2+ channel function. J Cell Biol 169(3):435–445
92. Zhang SL, Yu Y, Roos J, Kozak JA, Deerinck TJ, Ellisman MH, Cahalan MD (2005) STIM1 is a Ca2+ sensor that activates CRAC channels and migrates from the Ca2+ store to the plasma membrane. Nature 437(7060):902–905
93. Berna-Erro A, Redondo PC, Rosado JA (2012) Store-operated Ca2+ entry. Adv Exp Med Biol 740:349–382
94. Várnai P, Tóth B, Tóth DJ, Hunyady L, Balla T (2007) Visualization and manipulation of plasma membrane-endoplasmic reticulum contact sites indicates the presence of additional molecular components within the STIM1-Orai1 complex. J Biol Chem 282(40):29678–29690
95. Albarran L, Lopez JJ, Dionisio N, Smani T, Salido GM, Rosado JA (2013) Transient receptor potential ankyrin-1 (TRPA1) modulates store-operated Ca(2+) entry by regulation of STIM1-Orai1 association. Biochim Biophys Acta 1833(12):3025–3034
96. Jardin I, Lopez JJ, Salido GM, Rosado JA (2008) Orai1 mediates the interaction between STIM1 and hTRPC1 and regulates the mode of activation of hTRPC1-forming Ca2+ channels. J Biol Chem 283(37):25296–25304
97. Jardin I, Gomez L, Salido G, Rosado J (2009) Dynamic interaction of hTRPC6 with the Orai1-STIM1 complex or hTRPC3 mediates its role in capacitative or non-capacitative Ca2+ entry pathways. Biochem J 420:267–276
98. Domínguez-Rodríguez A, Díaz I, Rodríguez-Moyano M, Calderón-Sánchez E, Rosado JA, Ordóñez A, Smani T (2012) Urotensin-II signaling mechanism in rat coronary artery role of STIM1 and Orai1-dependent store operated calcium influx in vasoconstriction. Arterioscler Thromb Vasc Biol 32(5):1325–1332
99. Pandol SJ, Schoeffield-Payne MS (1990) Cyclic GMP mediates the agonist-stimulated increase in plasma membrane calcium entry in the pancreatic acinar cell. J Biol Chem 265(22):12846–12853

100. Rosado J, Graves D, Sage S (2000) Tyrosine kinases activate store-mediated Ca2+ entry in human platelets through the reorganization of the actin cytoskeleton. Biochem J 351(2):429–437
101. Bird GS, Putney JW (1993) Inhibition of thapsigargin-induced calcium entry by microinjected guanine nucleotide analogues. Evidence for the involvement of a small G-protein in capacitative calcium entry. J Biol Chem 268(29):21486–21488
102. Csutora P, Peter K, Kilic H, Park KM, Zarayskiy V, Gwozdz T, Bolotina VM (2008) Novel role for STIM1 as a trigger for calcium influx factor production. J Biol Chem 283(21):14524–14531
103. Bose DD, Rahimian R, Thomas DW (2005) Activation of ryanodine receptors induces calcium influx in a neuroblastoma cell line lacking calcium influx factor activity. Biochem J 386(2):291–296
104. Zarayskiy VV, Monje F, Peter K, Csutora P, Khodorov B, Bolotina VM (2007) Store-operated Orai1 and IP3 receptor-operated TRPC1 channel: separation of the siamese twins. Channels 1(4):246–252
105. Singaravelu K, Lohr C, Deitmer JW (2008) Calcium-independent phospholipase A2 mediates store-operated calcium entry in rat cerebellar granule cells. Cerebellum 7(3):467–481
106. Ross K, Whitaker M, Reynolds NJ (2007) Agonist-induced calcium entry correlates with STIM1 translocation. J Cell Physiol 211(3):569–576
107. Boittin FX, Petermann O, Hirn C, Mittaud P, Dorchies OM, Roulet E, Ruegg UT (2006) Ca2+-independent phospholipase A2 enhances store-operated Ca2+ entry in dystrophic skeletal muscle fibers. J Cell Sci 119(18):3733–3742
108. Martínez J, Moreno JJ (2005) Role of Ca2+-independent phospholipase A2 and cytochrome P-450 in store-operated calcium entry in 3T6 fibroblasts. Biochem Pharmacol 70(5):733–739
109. Vanden Abeele F, Lemonnier L, Thébault S, Lepage G, Parys JB, Shuba Y, Skryma R, Prevarskaya N (2004) Two types of store-operated Ca2+ channels with different activation modes and molecular origin in LNCaP human prostate cancer epithelial cells. J Biol Chem 279(29):30326–30337
110. Vig M, Peinelt C, Beck A, Koomoa DL, Rabah D, Koblan-Huberson M, Kinet JP (2006) CRACM1 is a plasma membrane protein essential for store-operated Ca2+ entry. Science 312(5777):1220–1223
111. Trepakova ES, Csutora P, Hunton DL, Marchase RB, Cohen RA, Bolotina VM (2000) Calcium influx factor directly activates store-operated cation channels in vascular smooth muscle cells. J Biol Chem 275(34):26158–26163
112. Trepakova ES, Gericke M, Hirakawa Y, Weisbrod RM, Cohen RA, Bolotina VM (2001) Properties of a native cation channel activated by Ca2+ store depletion in vascular smooth muscle cells. J Biol Chem 276(11):7782–7790
113. Boittin FX, Shapovalov G, Hirn C, Ruegg UT (2010) Phospholipase A 2-derived lysophosphatidylcholine triggers Ca 2+ entry in dystrophic skeletal muscle fibers. Biochem Biophys Res Commun 391(1):401–406
114. Jans R, Mottram L, Johnson DL, Brown AM, Sikkink S, Ross K, Reynolds NJ (2013) Lysophosphatidic acid promotes cell migration through STIM1-and Orai1-mediated Ca2+(i) mobilization and NFAT2 activation. J Invest Dermatol 133(3):793–802
115. Vanden Abeele F, Zholos A, Bidaux G, Shuba Y, Thebault S, Beck B, Flourakis M, Panchin Y, Skryma R, Prevarskaya N (2006) Ca2+-independent phospholipase A2-dependent gating of TRPM8 by lysophospholipids. J Biol Chem 281(52):40174–40182
116. AL-Shawaf E, Tumova S, Naylor J, Majeed Y, Li J, Beech DJ (2011) GVI phospholipase A2 role in the stimulatory effect of sphingosine-1-phosphate on TRPC5 cationic channels. Cell Calcium 50(4):343–350
117. Guo Z, Su W, Ma Z, Smith GM, Gong MC (2003) Ca2+-independent phospholipase A2 is required for agonist-induced Ca2+ sensitization of contraction in vascular smooth muscle. J Biol Chem 278(3):1856–1863

118. Park KM, Trucillo M, Serban N, Cohen RA, Bolotina VM (2008) Role of iPLA2 and store-operated channels in agonist-induced Ca2+ influx and constriction in cerebral, mesenteric, and carotid arteries. Am J Physiol Heart Circ Physiol 294(3):H1183–H1187
119. Yang B, Gwozdz T, Dutko-Gwozdz J, Bolotina VM (2012) Orai1 and Ca2+-independent phospholipase A2 are required for store-operated Icat-SOC current, Ca2+ entry, and proliferation of primary vascular smooth muscle cells. Am J Physiol Cell Physiol 302(5):C748–C756
120. Schäfer C, Rymarczyk G, Ding L, Kirber MT, Bolotina VM (2012) Role of molecular determinants of store-operated Ca2+ entry (Orai1, phospholipase A2 group 6, and STIM1) in focal adhesion formation and cell migration. J Biol Chem 287(48):40745–40757
121. Reutenauer-Patte J, Boittin FX, Patthey-Vuadens O, Ruegg UT, Dorchies OM (2012) Urocortins improve dystrophic skeletal muscle structure and function through both PKA-and EPAC-dependent pathways. Am J Pathol 180(2):749–762
122. Zhu C, Sun Z, Li C, Guo R, Li L, Jin L, Li S (2014) Urocortin affects migration of hepatic cancer cell lines via differential regulation of cPLA2 and iPLA2. Cell Signal 26(5):1125–1134

Chapter 7
Extracellular Calcium Has Multiple Targets to Control Cell Proliferation

Thierry Capiod

Abstract Calcium channels and the two G-protein coupled receptors sensing extracellular calcium, calcium-sensing receptor (CaSR) and GPRC6a, are the two main means by which extracellular calcium can signal to cells and regulate many cellular processes including cell proliferation, migration and invasion of tumoral cells. Many intracellular signaling pathways are sensitive to cytosolic calcium rises and conversely intracellular signaling pathways can modulate calcium channel expression and activity. Calcium channels are undoubtedly involved in the former while the CaSR and GPRC6a are most likely to interfere with the latter. As for neurotransmitters, calcium ions use plasma membrane channels and GPCR to trigger cytosolic free calcium concentration rises and intracellular signaling and regulatory pathways activation. Calcium sensing GPCR, CaSR and GPRC6a, allow a supplemental degree of control and as for metabotropic receptors, they not only modulate calcium channel expression but they may also control calcium-dependent K+ channels. The multiplicity of intracellular signaling pathways involved, their sensitivity to local and global intracellular calcium increase and to CaSR and GPRC6a stimulation, the presence of membrane signalplex, all this confers the cells the plasticity they need to convert the effects of extracellular calcium into complex physiological responses and therefore determine their fate.

Keywords Calcium channels • Calcium sensing receptors • GPRC6a • Cell proliferation • Orai • TRPC

T. Capiod (✉)
INSERM U1151, INEM (Institut Necker-Enfants Malades), Faculté de Médecine, Université Paris-Descartes, 14 rue Maria Helena Vieira da Silva, Bâtiment Leriche, Porte 14, Paris 75014, France
e-mail: thierry.capiod@inserm.fr

J.A. Rosado (ed.), *Calcium Entry Pathways in Non-excitable Cells*, Advances in Experimental Medicine and Biology 898, DOI 10.1007/978-3-319-26974-0_7

7.1 Introduction

Calcium channels and the two G-protein coupled receptors sensing extracellular calcium, calcium-sensing receptor (CaSR) and GPRC6a, are the two main means by which extracellular calcium can signal to cells and regulate many cellular processes including cell proliferation, migration and invasion of tumoral cells [1–7]. Many intracellular signaling pathways are sensitive to cytosolic calcium rises and conversely intracellular signaling pathways can modulate calcium channel expression and activity. Calcium channels are undoubtedly involved in the former while the CaSR and GPRC6a are most likely to interfere with the latter. Hence it is postulated that extracellular calcium ions may use an ionotropic and metabotropic scheme to modulate calcium fate. Indeed, calcium channels are active at physiological extracellular calcium concentration whereas CaSR and GPRC6a need a rise in extracellular calcium concentration to be activated [8, 9]. Calcium entering the cell may either directly trigger intracellular signaling pathways or modulate the temporal and spatial nature of the cytosolic Ca^{2+} transients enable a cell to tailor its response to a given stimuli [2, 10–12].

Although several calcium entry pathways have been linked to cell proliferation [13, 14], this review is focused on two specific ones, namely the store-operated calcium entry (SOCE) resulting from intracellular calcium store depletion and the store-independent calcium entry (SICE) mainly relying on intracellular second messenger production independently of calcium store depletion [15, 16]. Calcium store depletion is sensed by two proteins, STIM1 and STIM2 (Stromal Interaction Molecule 1 and 2), located on the endoplasmic reticulum membrane [17–19]. The molecular nature of the calcium channels is still a matter of debate but it is clear that SOCE involves a complex of Orai proteins on the plasma membrane and the luminal calcium sensors STIM1 and STIM2 on the endoplasmic reticulum membrane [17–19]. However, transient receptor potential canonical (TRPC) channels are most likely to interfere with this complex [20–22]. SICE was clearly described as a heteropentameric assembly of three Orai1 subunits and two Orai3 subunits [23] coupled to plasma membrane STIM1 [24] and activated by arachidonic acid (AA) [25] but it has also been shown that TRPC channels can be open by another intracellular messenger such as di-acyl-glycerol (DAG) [26]. Although SICE should not be restricted to these two calcium entries only, the few remaining ones were mentioned in a previous review [14].

CaSR and GPRC6a are seven transmembrane domains GPCR sensing increases in extracellular calcium. When activated, they trigger large increase in intracellular signaling pathways and also in cytosolic free calcium concentration [27, 28]. They have been linked to cell proliferation and cancer development [29, 30] and, therefore, they provide another mean for extracellular calcium to modify cell fate.

How extracellular calcium ion rise, calcium channels and CaSR expression and activity are translated by the cell to allow physiological and pathophysiological proliferation will be discussed in this review. The uncoupling of the relation between calcium entry, calcium channel expression and cell proliferation will be the second point addressed here.

7.2 Calcium Influx and Cell Proliferation

Ca^{2+} requirement for cell growth and division has already been addressed in a previous review [14]. It becomes clear that, even if the great majority of cells needs extracellular calcium to grow and proliferate, this requirement may be limited by the degree of cellular transformation and it was suggested that the need for extracellular Ca^{2+} was ruled out during neoplastic transformation. Indeed, we recently confirmed demonstrated that cell proliferation rate was not affected even at very low extracellular calcium concentration (below 10 μM) in human hepatoma cell line Huh-7, in HEK293 and HeLa cells [31]. It was then suggested that intracellular calcium stores were more important than calcium influx to preserve this physiological function. However, one of the strong arguments in favor of a role of calcium influx in cell proliferation still comes from the use of calcium channels blockers. A special effort was made to identify selective I_{CRAC} blockers and several new molecules have now emerged as recently reviewed [14, 32, 33]. Further studies have been performed since to test and characterize the effectiveness of some of these molecules. Two selective small molecules developed by GlaxoSmithKline, GSK-5498A and GSK-7975A, tested for their abilities to inhibit mediator release from mast cells and pro-inflammatory cytokine release from T cells completely inhibited I_{CRAC} currents with an IC_{50} of about 3 μM [34, 35]. Recent studies showed that these two drugs may have potential therapeutic applications in the treatment of asthma and related allergic diseases as well as pancreatitis [35, 36]. Although 2-APB has a wide spectrum of effectors and targets several types of calcium channels [37–43], its structure is still widely used as a platform to design new CRAC blockers. Chemical analogs of 2-APB have proved to be very effective in blocking CRAC currents. The dimeric derivative of 2-aminoethoxydiphenyl borinate (2-APB), DPB162-AE, blocks functional coupling between STIM1 and Orai1 with an IC_{50} (200 nM) 100-fold lower than 2-APB [44]. DPB162-AE does not prevent the SOAR-Orai1 interaction but potently blocks SOAR-mediated Orai1 channel activation, yet its action is not as an Orai1 channel pore blocker [45]. In an effort to discover new inhibitors of TRPV6, 34 novel 2-APB analogs were synthesized. Although several analogs that were more potent inhibitors of TRPV6 than 2-APB, they also inhibited SOCE with comparable IC_{50} [46]. Modifying the two phenyl groups of 2-APB proved to be another way to design more effective CRAC channels blocker. Two molecules, cyclic-APB and dibenzothienyl-APB, inhibited calcium entry with higher sensitivity (IC_{50} of 3 and 0.4 μM, respectively) than 2-APB [47]. This is indeed a notable improvement but there is still a long way to go before any of these drugs based on this molecular structure could be used in clinical trials. Another blocker has been recently identified. SAR7334 is derived from the previous non selective SOCE blocker SKF96365. SAR7334 is rather promising as it blocks TRPC6, TRPC3 and TRPC7 DAG-mediated calcium entry with IC_{50}s of 9.5 nM, 282 nM and 226 nM respectively [48]. SAR7334 has no effect on TRPC4 and TRPC5 mediated calcium entry. Interestingly, this compound was also tested in patch-clamp experiments and application during the response rapidly and totally blocked OAG-induced calcium currents, suggesting a direct effect on the channel itself. Investigating the effect of SAR7334 on hypoxic vasoconstriction in the perfused isolated mouse lung model showed that this drug was able to completely suppress hypoxia-induced increases in pulmonary arterial pressure. Half maximal inhibition was obtained at about 100 nM strongly suggesting that SAR7334 is able to efficiently block native TRPC6 channels in vivo [48]. As TRPC6 is linked to cancer and cell proliferation, this makes SAR7334 a potential drug to block this disease.

A few clinical trials involving calcium influx blockers have already been launched. Carboxyamidotriazole (CAI) was patented almost 20 years ago [49] and is currently in phase III clinical trials in non-small cell lung cancer [50]. However, CAI has potential side effects including inhibition of mitochondrial calcium import [51] and both cAMP-phosphodiesterases and cGMP-phosphodiesterases activity [52]. CalciMedica (La Jolla, CA, USA) has identified multiple structurally-distinct selective CRAC channel inhibitors including CM2489 and CM3457. CM2489 is the first of these compounds to reach the clinic in a Phase 1 clinical study and is being developed for the treatment of moderate-to-severe plaque psoriasis. CM3457, another selective CRAC channel inhibitor that is structurally distinct from CM2489, inhibits CRAC channels, T cell proliferation, and mast cell degranulation and may be used for autoimmune disorders and asthma treatment. Finally, a Phase I clinical trial (Safety and Tolerability Study of SOR-C13 in Subjects With Advanced Cancers Commonly Known to Express the TRPV6 Channel) to test the effect of SOR-C13, an inhibitor of TRPV6 [53], was recently launched. The purpose of this phase I is to determine the safety and tolerability of the drug SOR-C13 when given as an intravenous infusion in patients with ovarian cancer or other cancers known to over express the TRPV6 calcium channel. However, it appears that this drug is not very effective in blocking TRPV6 as calcium entry was only reduced by 25 % at 25 μM SOR-C13 [53].

A new and original approach was based on monoclonal antibodies directly targeting the second extracellular loop of Orai1 channel [54, 55]. They showed in the first study that Anti-Orai1 mAB binds to Orai1 protein leading to its internalization [54]. This mAB was more effective at blocking human peripheral blood mononuclear cells (PBMC) proliferation (80 % at 200 nM) than calcium fluxes in Jurkat cells (60 % at 2.5 μM). However, time scale for both measurements was radically different as calcium entry was monitored 1 h after mAB addition while cell proliferation was assessed after 3 days. The second study [55] refers to another mAB against human Orai1 (2C1.1) with a high efficiency. 2C1.1 (10 nM) was able to reduce calcium entry by 70 % and NFAT activity by 40 % in Jurkat cells.

Although several small molecules have been shown to be potent CRAC channel inhibitors, their selectivity remains questionable. The immunologic approach to CRAC channel inhibition through the development of fully human monoclonal antibodies to Orai1 that antagonize CRAC channel function may represent the best opportunity to overcome this selectivity matter and could be applied to other calcium channels.

7.3 Calcium Influx and Intracellular Signaling Pathways

Calcium influx, once the channels are open, has to be translated into physiological responses and several intracellular signaling pathways are sensitive to cytosolic calcium concentration increases.

Blocking SOCE or silencing TRPC or Orai channels often resulted in a modification of intracellular signaling pathways linked to cell proliferation. Several examples focusing on Orai and TRPC channels are listed below. Knocking down Orai3 or TRPC3 decreased Akt phosphorylation and cell proliferation in non small cell lung adenocarcinoma cells [56] and FSH-stimulated ovarian cancer cells [57] respectively. Down-regulation of this pathway by TRPC silencing was also observed in other physiological processes such as CXCR2-mediated chemotaxis in murine neutrophils (TRPC6) [58], neurite outgrowth in PC-12 cells (TRPC6) [59], myoblast differentiation and muscle regeneration in TRPC1 –/– mice [60]. However, cardiac stress induced by transverse aortic constriction was abolished in TRPC1 –/– mice and associated with an increase in Akt and phospho-mTOR (mammalian target of rapacycin) signaling which are also linked to cell survival [61]. ERK/p-ERK pathway is also controlled by calcium channel expression and calcium influx. SOCE inhibition using YM58483 (BTP-2) [62] reduced CaMKII/Raf-1/ERK signaling pathway and cell proliferation in human metastatic melanoma cell lines [63]. Interestingly, silencing TRPC5 or TRPC6 but not TRPC1 resulted in a decreased ERK/CREB pathway but not Akt in PFGF-mediated neuroprotection [64]. Phospho-CREB was also downregulated in the presence of Sirolimus in human arterial smooth muscle [65]. Sirolimus, a structural derivate of FK506 that reduces cell proliferation rates, inhibits Orai1-mediated SOCE [65] as well as a NFAT dephosphorylation and mTOR phosphorylation [66]. By blocking calcium entry, Sirolimus reduces calmodulin kinase II (CaMKII) activity, hence CREB phosphorylation and gene transcription. STAT5 (Signal Transducer and Activator of Transcription 5) and c-fos expression are also activated by calcium entry in rat mast cells through a tyrosine kinase Syk-dependent pathway [67]. Phosphorylation of AMPK (AMP-activated potein kinase) by CaMKKβ (calmodulin kinase kinase β) is abolished when TRPC1 expression is reduced or in and in TRPC4 –/– mice [68]. Hence, intracellular signaling pathway activation is dependent on various types of calcium channels, suggesting that local increases in cytosolic free calcium concentration or well-defined membrane complexes enable a cell to selectively respond to one stimulus.

7.4 Local and Global Cytosolic Calcium Increase

If calcium entry is important to refill intracellular calcium stores and/or to feed the oscillations observed in stimulated cells, local calcium concentration increases at the mouth of the channel is also important to specifically trigger one signaling pathway. Calcium microdomains arising from the diffusion of calcium through an open channel can easily reach several tens of micromolars, several fold higher than average cytosolic calcium concentration [69]. A first evidence arised from the observation in neurons where L-type calcium channels were able to trigger sustained CREB phosphorylation following KCl-induced membrane depolarization, while other voltage-dependent calcium channels were only able to transiently activate this transcription factor [70]. Since then, several other studies have clearly demonstrated that local calcium signaling allows selective calcium activation of NFAT [67, 71, 72], another calcium sensitive transcription factor. Pro-inflammatory molecule leukotriene C4 (LTC4) production measured following an application of the SERCA inhibitor thapsigargin was larger when external calcium was present despite similar increase in global cytosolic calcium concentration in RBL-1 cells [73]. The same group published a year after another study where they showed that calcium microdomains at the mouth of the channel were activating tyrosine kinase syk which in turn activated STAT5 and c-fos while, again, global increase in cytosolic calcium was unable to do so [67].

This then raised a new question concerning the existence of membrane signaling complexes to account for the selective activation of intracellular signaling pathways and transfection factors by calcium influx. For instances, Orai1 but not Orai3 was inducing NFAT dephosphorylation while an Orai3 chimera with Orai1 N-terminal was able to restore NFAT translocation [72]. The scaffolding protein AKAP79 (A-kinase anchoring protein 79) is a member of a large family of proteins known to foster signal transduction by connecting signaling enzymes with their substrate with well-defined membrane localization. AKAP79 connects with several proteins [74] including Cav1.2 L-type voltage-dependent calcium channels [75]. It was later shown that NFAT signaling was connected to these Cav1.2 channels through AKAP79 [76], and that NFAT nuclear translocation was enabled by AKAP79-Orai1 interaction [72]. Iarc, the major component of SICE, is also regulated by AKAP79 [77] through a PKA mediated phosphorylation of T389 residue of the minor pool of STIM1 located on the plasma membrane whilst phosphorylation of the same residue actually inhibits the ability of STIM1 to activate the CRAC channels [78].

7.5 Plasma Membrane Signalplex

It is indeed trivial to say that the multiplicity of calcium-mediated intracellular outputs and the corresponding diversity of calcium channels make the presence of such dynamic structural membrane signaling complexes. Interactions between Orai, STIM and TRPC proteins and between these proteins and other partners have been thoroughly investigated over the last few years [79, 80]. This first list comprises SPCA2, CRACR2A, calmodulin, caveolin, adenylyl cyclase 8, EBp57, junctate, EB1, Golli MBP, SARAF, polycystin1-P100, stanniocalcin 2, POST, PMCA, Cav1.2, and septin. Additional partners have since been identified. Immunoprecipitation experiments showed that DOCK10, thrombospondin-1, myosin and bestrophin-1 are associated with STIM1 [81, 82], ubiquilin-1 with Orai1 [83] and AKAP79 with Orai1 and Orai3 [72, 78].

In order to get maximal efficiency and specificity, plasma membrane must have a high degree of organization and this could be achieved in specific regions such as lipid raft domains [84]. Early experiments in the 1970s suggested a role of caveolae in regulating Ca^{2+} entry [85]. Later, it was demonstrated that caveolae are preferred sites for SOCE [86, 87]. Interestingly, receptors for two potent mitogens, epidermal growth factors (EGF) and hepatocyte growth factor (HGF), which are known to stimulate TRPC6 expression and cell proliferation in human hepatoma cells [88] are also localized in caveloae [89, 90]. Other growth factors receptors have been found in caveolae [91] and this is known to control and organize Ca^{2+}-dependent signal transduction [92]. Several calcium channels and calcium signaling proteins are found in caveolae [92] and, for example, caveolin has been implicated in TRPC1 targeting the plasma membrane and its association with STIM1 [93–95].

Hence, these complexes may account for the differential activation of intracellular signaling pathways and transcription factor activation following calcium entry. SICE but not SOCE activated by norepinephrine and P2Y-purinergic receptors respectively in primary culture of human prostate cancer epithelial cells, was responsible for nuclear NFAT translocation [96] while it was tightly linked to I_{CRAC} but not to I_{arc} in HEK293 cells [97]. Furthermore, it was recently demonstrated that scaffolding protein caveolin-1 was giving the cell the ability to distinguish between two transcription factors, namely c-fos and NFAT, in response to the same local calcium signal [98]. It was shown in this paper that tyrosine 14 residue of caveolin-1 is the locus for c-fos transcription inhibition. Phosphorylation of Tyr14, which can be triggered by the Src family tyrosine kinases Src, Abl or Fyn [99], unable calcium entry to activate STAT5, hence c-fos expression in RBL cells, while overexpression of the caveolin-1 Y14F mutant oversteps c-fos transcription inhibition [99]. In both conditions, NFAT translocation was not affected. Cells express four members of NFAT family all depending on cytosolic calcium increase to move to the nucleus [100]. Dynamics of nuclear translocation for NFAT1 and NFAT4 are rather different, NFAT4 import and export being almost ten times than for NFAT1 [101]. It was recently demonstrated that Orai1-associated calcium microdomains directly activated NFAT1 while NFAT4 required both local calcium influx and global and nuclear calcium increases allowing a cell to recruit combination of these two isoforms as stimulation strength increases [71]. Altogether, these data emphasize the complex role for local and size in calcium signals in controlling intracellular signaling pathways and, eventually, cell fate (Fig. 7.1).

SOCE and SICE amplitudes increase with membrane potential hyperpolarization and concomitant opening of K+ channels is likely to result in larger rise in calcium microdomains or bulk calcium. It was demonstrated that intracellular signaling pathway activation could depend on how much was entering the cell in membrane microdomains [67]. The presence of K+ channels nearby could therefore increase the driving force for calcium ions. Membrane structural complexes of Orai or TRPC channels and K+ channels precisely exist to modulate calcium entry. There are now evidences that they can be related to cancer development [102].

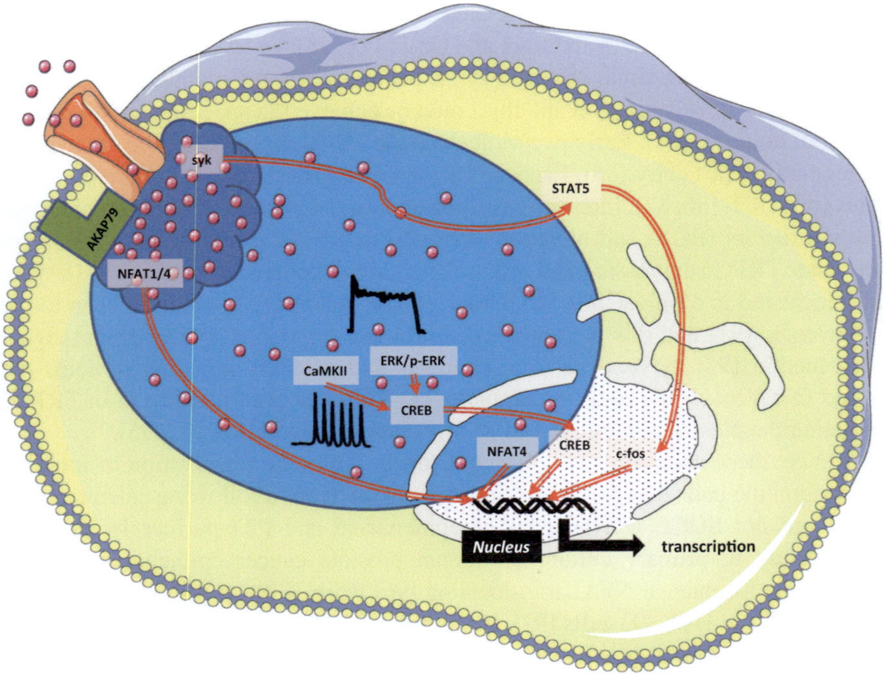

Fig. 7.1 Local and global intracellular calcium increase activates different intracellular signaling pathways to trigger transcription and cell proliferation. NFAT4 activation requires both local and nuclear calcium increase while NFAT1 or syk are sensitive to local rise at the mouth of the channel. Transient or sustained cytosolic increases are likely to act on other signaling pathways

Seventy eight genes over more than 400 for all types of channels, encode for ion channels selectively permeable to potassium [103]. This superfamily of potassium channels can be further divided into four groups based on their mode of gating and the number of structural transmembrane domains. Among them, the complex calcium-dependent K+ channels [104] are the most described as partners for SOCE and SICE channels. SK3 channels are not expressed in normal human prostate and breast epithelial cells as opposed to Orai1 but they appear in tumoral cells. They then form a SK3–Orai1 complex with an essential role in cancer cell migration and bone metastasis development [105]. It was also worth noting that BKCa–Cav3.2 complex also play a major role in the proliferation of LNCaP prostate cancer cells by controlling a constitutive Ca^{2+} entry [106]. BKCa–Cav3.2 complex could maintain membrane potentials within a window of membrane potentials over which a fraction of Cav3.2 are activated. At low voltages, T-type Ca^{2+} channels are known to mediate a phenomenon known as “window current” [107]. The term “window” refers to the voltage overlap between the activation and steady state inactivation at low or resting membrane potentials. As a result, there is a sustained inward calcium current carried by a small portion of channels that are not completely inactivated. Window current allows T-type Ca^{2+} channels to regulate Ca^{2+} homeostasis under non-stimulated or resting membrane conditions [108]. This means that for these membrane potentials which are controlled by BKCa, Cav3.2 channels are open and support constitutive Ca^{2+} entry. Furthermore, this constitutive Ca^{2+} entry would activate BKCa channels that regulate membrane potential working as a positive feedback loop.

In LNCaP prostate cancer cells IKCa channel regulates calcium entry through TRPV6 by controlling the electrochemical gradient for calcium ions [109]. Immunoprecipitation experiments showed a close physical interaction between the IKCa and TRPV6 in these cells. In the same manner, the human ether a′-gogo potassium Channel 1 (hEag1) has been shown to regulate breast cancer cell migration through Orai1-dependent calcium entry [110].

If links between chloride channel expression, cell proliferation and tumorigenesis are now well demonstrated [111, 112], evidence for a role of chloride channels in the control of SOCE or SICE and cell proliferation is sparse. Concavalanin (ConA) induced calcium entry and cell proliferation of human T lymphocytes were reduced in the presence of 4,4′-diisothiocyanostilbene-2,2′-disulfonic acid (DIDS), a known chloride blocker. When combined with calcium channel blocker SKF96365, DIDS enhances the inhibitory effect of SKF96365 on ConA-induced IL-2 T cell proliferation [113].

Hence, the mechanisms relaying calcium influx, intracellular signaling pathways and cell proliferation are dramatically tangled. Differential expression of partners, local versus global increase in cytosolic calcium, direct and rapid regulation by PKA through anchoring proteins, involvement of other intracellular second messengers via PKC [114–118] or PKG [119–123] activation, all this makes the ground for a very tight and selective activation of calcium entry -dependent physiological responses.

7.6 Calcium Channel Expression and Cell Proliferation

The relation between calcium channel expression and cell proliferation was already emphasized in recent reviews [13, 14]. Another interesting aspect is the uncoupling of these two events.

If there are numerous examples of an effect of calcium channel expression and/or calcium entry related to cell proliferation, in some cases this relation is not observed. Our own work unexpectedly demonstrated that HEK293 cells, human hepatoma cell line Huh-7 and Hela cells were not sensitive to external calcium even at concentration as low as a few micromolars [31]. Cell proliferation was however sensitive to Orai1 and Orai3 expression levels and moreover, overexpression of non permeant mutants, E106Q-Orai1 and E81Q-Orai3, resulted in an increase in cell proliferation in HEK293 cells. Expression of these two mutants in the same cell type totally blocked SOCE [124] and SICE currents [125], strongly suggesting that calcium channel expression was more important than calcium entry to control cell proliferation in these cells. In a similar way, it was recently shown that knocking down Orai3 in breast cancer cell line MCF-7 was more effective that withdrawing extracellular calcium on cell proliferation [126]. Although extracellular calcium concentration lower than 200 μM were not tested in their experimental conditions, this was clearly in favor of our hypothesis. It is therefore possible that new functions for TRPC and Orai channels could be uncovered in near future and that a direct link between the channel structure and transcription factors would explain why some cells need the channel more than calcium influx to proliferate.

Knocking down TRPC1 and TRPC4 decreased cell proliferation with no effect on SOCE or I_{CRAC} activity in endothelial cells [127]. This is rather unexpected and could suggest an additional role for these channels. In addition, knocking down STIM1, the ER plasma membrane counterpart for SOCE channels [128, 129], has no effect on cell proliferation in vascular smooth muscle cells [130], human myoblasts [131], HEK293 cells [132] and human umbilical vein endothelial cells (HUVEC) derived cell line EA.hy926 [133]. At the same time I_{CRAC} is greatly reduced, which suggested strongly that, in these cells, this current and cell proliferation is uncoupled. This decoupling between STIM1 expression, calcium entry and cell proliferation was also observed in more recent publications. Silencing STIM1 indeed resulted in a dramatic decrease in SOCE/Icrac while cell proliferation rate and cell cycle distribution were not at all affected in human metastatic renal cellular carcinoma and LNCaP [134, 135]. To conclude here, an event that looked exceptional now seems to run through several cell models. Unfortunately, most experiments using siRNA to silence Orai or STIM proteins are not duplicated in the absence of extracellular calcium to validate this interesting observation.

7.7 Calcium Channels and Cell Cycle

The cell cycle consists of four primary phases: G1, the first gap phase; S phase, in which DNA synthesis occurs; G2, the second gap phase; and M phase, or mitosis, in which the chromosomes and cytoplasmic components are divided between two daughter cells. The transitions between the four cell cycle phases are tightly regulated. Checkpoints during cell cycle allow the cell to determine whether all is within the normal limits before proceeding to the next cell cycle phase [136]. We have discussed the role of calcium channels during cell cycle in a previous review [14]. Briefly, knocking down Orai or TRPC channels resulted in a cell cycle arrest preferably in G1 for TRPC1 and Orai3 and G_2/M phase for Orai1 and TRPC6. Recent literature check confirmed the prevalence of these channels in cell cycle control. Silencing TRPC6 in bone marrow stromal cells induced a significant decrease in G_0/G_1 and increase in S and G_2/M phases [137]. Interestingly, this was associated with 10 mV negative shift in membrane potential and a 50 % increase in CPA-mediated SOCE activation. Silencing TRPC6 in a human renal adenocarcinoma cell line also resulted in an increase in G2/M phase with a concomitant decrease in G0/G1 [138].

A reduction in Orai1, Orai2 and Orai3 expression in LNCaP prostate cancer cells evoked a large increase in G_0/G_1 and a parallel decrease in S and G2/M phase [135].

Silencing Orai2 also induces a small (5 %) but significant decrease in G_0/G_1 phase and cell proliferation in brain capillary endothelial cells [139]. Knocking down Orai3 in lung and breast cancer cells confirmed the previous observations concerning this channel with an increase in G_0/G_1 with a decrease in S and G_2/M phases [56, 140]. Finally, silencing TRPC2, a channel that is not expressed in human, induced an increase in G_0/G_1 phase with a parallel decrease in S and G_2/M, resulting in a reduction in rat thyroid FRTL-5 proliferation [141].

The block in G_2/M observed in Orai1 and TRPC6 knock down experiments is interesting as there is evidence that SOCE inactivates during M phase. The first suggestion that calcium influx is inhibited during cell division was reported more than 20 years ago in a study of HeLa cells [142], probably through uncoupling of store depletion from SOCE [143]. This was confirmed recently in HEK293, Cos-7, RBL-2H3 and HeLa cells where SOCE is dramatically reduced during mitosis [132, 144–146]. STIM1 phosphorylation [145], decrease in Orai1 expression [132] and internalization of Orai1 channel [147, 148] during mitosis were recently suggested to be the reasons for the observed decrease in SOCE. The significance of this has already been discussed [149] and involves microtubule-network remodeling [144] as perturbation of the actin cytoskeleton resulted in a decrease in SOCE amplitude [150].

Cell cycle is regulated by cyclins and cyclin dependent kinases (cdk) successively expressed from the onset of G_1 phase up to mitosis. Briefly, cyclin D is present in G_1 phase coupled to cdk1, cdk4 and cdk6, cyclin E with cdk2 at late G_1 and the transition G_1-S, cyclin A with cdk2 during S and cdk1 at the transition S-G_2, and finally cyclin B with cdk1 during mitosis. Cdk activity is down regulated by selective cyclin dependent kinase inhibitors (cdki) and among them P21 and P27 have been associated with calcium channels. The p21 protein binds to and inhibits the activity of cyclin-cdk1, -cdk2, and – cdk4/6 complexes, and thus functions as a regulator of cell cycle progression at G_1 and S phase while P27 inhibitory activity seems to be restricted to the couple cyclinE/cdk2, hence regulating the decision of cell cycle to enter into S-phase or arrested in G_1-phase by blocking G_1 cyclin dependent kinase activities.

A relation between cyclin D1 and D3, calcium channels and calcium influx has been demonstrated in several studies. Cyclin D1 and D3 expression are linked to Orai1 channels [88, 135, 151, 152], Orai3 channels [56, 126, 135], TRPC1 [153] and TRPC6 channels [88] as their expression decreases when these channels are silenced. Knocking down calcium channels often resulted in an increase in P21, P27 and also in tumor suppressor P53. The trend shows an inverse relationship between calcium channel and cdki and P53 expression. TRPC1 silencing is linked to a decrease in cell proliferation, G_1 arrest and increase in P21 expression in endothelial progenitor cells [154], TRPC2 to an increase in P21, P27 and P53 tumor suppressor in rat thyroid FRTL-5 cells [141]. Finally, overexpression of STIM1/Orai1 has been unexpectedly linked to a decrease in SOCE a G_0/G_1 arrest, cyclin D3 expression decrease and P21 increase in A549 lung cancer cells [152].

To summarize here, these data illustrate the strong link between calcium channels, calcium entry and cell proliferation. It makes these channels perfect targets to stop cancer development.

7.8 Extracellular Calcium Sensing G Protein Coupled Receptors

CaSR was identified as the first seven transmembrane domains G protein coupled receptor (GPCR) able to sense extracellular calcium concentration [8]. GPCR are subdivided in three categories, A, B and C, CaSR being part of group C also including, among others, metabotropic glutamate and GABA receptors and basic amino acid receptors such as GPRC6a [4, 5, 155]. This explains that, although a distinct molecular structure, their pharmacology profiles overlap. CaSR can be stimulated by several other cations and modulated by various physiological stimuli including extracellular pH, L-aromatic amino acids and ionic strength [28]. CaSR maintains calcemia through the control of PTH (parathyroid hormone) secretion by parathyroid glands and through PTH-dependent and PTH-independent regulation of the main target organs, kidney and bone, involved in calcium homeostasis. When it comes to cell proliferation and cancer, CaSR activation causes radically different effects depending on whether one is to the breast and prostate or colon and parathydoid [30, 156–159]. One tempting explanation for this apparent contradiction came from experiments in breast cell lines in response to PTHrP (PTH related protein) stimulation. In normal breast cells, PTHrP secretion is inhibited by extracellular calcium concentration increases while, in breast and prostate cancer cells, PTHrP release is augmented following CaSR activation. A recent study demonstrated that the change between inhibition and stimulation of PTHrP release by extracellular calcium ions occured as a result of a switch in G protein activation by the cancerous cells [160]. In normal breast cells, it was shown that the CaSR couples to $G\alpha i$, leading to inhibition of cAMP formation and, consequently, PTHrP release, whereas in MCF-7 cancerous breast cells, the CaSR was shown to activate $G\alpha s$, thereby promoting PTHrP release.

GPRC6a receptors were cloned and sequenced a little more than 10 years ago [6] and later identified as a calcium-sensing receptor [161]. GPRC6a is activated by calcium, L-arginine, testosterone and osteocalcin [162] and has a strong link with cell proliferation and cancer development. GPRC6a increases proliferation of human dermal fibroblasts stimulated by 6 mM arginine [163] and of osteocalcin stimulated mouse β-cell proliferation [164]. GPRC6a is tightly involved in prostate cancer development and human prostate cancer cell proliferation [162]. Rare variation in protein coding sequence is badly captured by genome-wide association studies arrays and has been hypothesized to contribute to disease heritability. In a large cohort of more than 4000 prostate cancer cases and 7500 control patient, one of the most associated single nucleotide polymorphism (SNP) was in the gene coding for GPRC6a near to known locus for prostate cancer [165]. The functional consequences for Pro91Ser substitution has yet to be investigated.

Regarding cancer, it has been suggested that calcium supplemented diets may increase the progression of prostate cancer. Our own results on two genetically transformed mouse models to develop prostate cancer strongly confirmed this observation (unpublished data). Calcemia is indeed tightly regulated but is has been shown that calcium supplementation in diet results in transient rise of plasma free calcium concentration [166]. The rise lasted for 8 h with a peak amplitude of about 70 μM from a basal level of 1.25 mM 4 h after ingestion of 1 g of calcium citrate or calcium carbonate. Three questions arise from this observation. Is such a small increase in calcemia able to trigger CaSR or GPRC6a activation? Is this transient elevation in calcemia long enough to induce cell proliferation? Do the downstream events remain activated over a long period of calcemia elevation to trigger cell proliferation? Answering these questions may explain how calcium supplementation can be linked to prostate cancer.

All the major intracellular signaling pathways have been at some point associated to CaSR activation [28] while, to date, GPRC6a are connected to ERK and the cAMP-PKA-CREB trio [162, 163, 167]. Interestingly, Orai1, Orai3, TRPC3 and TRPC6 promotor region have CREB binding sites making their expression likely under the control of these two calcium sensing receptors.

There is a clear connection between CaSR stimulation, calcium influx and calcium channel expression. TRPC1 and mainly TRPC3 and TRPC6 are the identified targets identified so far. In breast cancer cell line MCF-7, a 24-h incubation in the presence of 5 mM external calcium induced a clear increase in TRPC1 channel expression which was prevented by the addition of U1026 to block ERK pathway [168]. Other studies have emphasized an effect of CaSR stimulation of calcium channel expression. In human aortic smooth muscle cells, extracellular calcium increased TRPC6 with an EC_{50} of about 0.5 mM [169]. A rise in cytosolic free calcium concentration was also observed when extracellular was increased. The question here is to know whether a pre-incubation in a calcium-free solution containing 500 μM EGTA induced a partial ER calcium store depletion, hence SOCE activation. A measure of ER calcium contents in their conditions or shortly after the switch in calcium-free solution should discriminate between a real CaSR-induced cytosolic calcium rise and SOCE activation. Elevating extracellular calcium concentration from 1 to 5 mM increased TRPC3 and TRPC6 expression but not TRPC1 and TRPC4 in human mesangial cells [170].

To date, there is no evidence for calcium channel enhanced expression following GPRC6a activation but as they increase the same intracellular signaling pathways as CaSR, it is probably just a matter of time before such a link is highlighted.

Finally, CaSR are also known to trigger intracellular calcium increase via the phospholipase C/InsP3 pathway [171] with a threshold concentration for between 1.5 and 2 mM external calcium concentration [27], just above the physiological levels for plasmatic free calcium concentration [172].

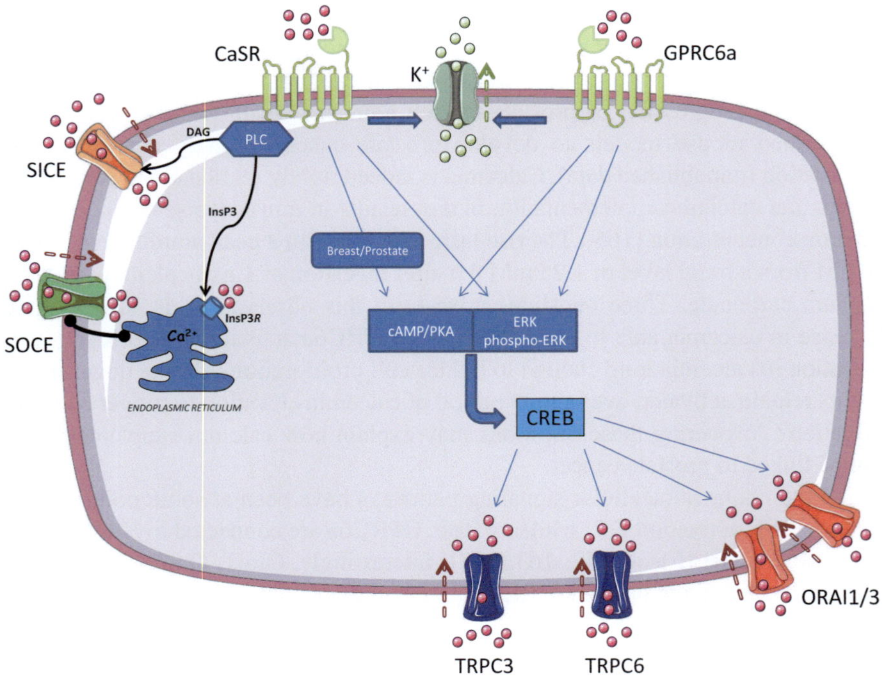

Fig. 7.2 Ionotropic and metabotropic regulation of calcium signaling by CaSR and GPRC6a. CaSR and GPRC6a trigger cAMP/PKA (only in breast and prostate cancer for CaSR) and ERK/phospho-ERK pathways which in turn increase CREB activity. TRPC3, TRPC6, Orai1 and Orai3 have CREB response element resulting in an increase in expression. CaSR also directly increase cytosolic calcium through PLC activation, DAG and InsP3 production, and SOCE and SICE opening. Finally, CaSR and GPRC6a increase K^+ channels activity, hence membrane potential and calcium ions driving force. The *arrows* next to the channel indicate ion flux direction (*red* for calcium and *light green* for potassium ions)

7.9 Conclusion

The data discussed here emphasize the complex relationship between extracellular calcium and cell proliferation. As for neurotransmitters, calcium ions use plasma membrane channels and GPCR to trigger cytosolic free calcium concentration rises and intracellular signaling and regulatory pathways activation (Fig. 7.2). Calcium sensing GPCR, CaSR and GPRC6a, allow a supplemental degree of control and as for metabotropic receptors, they not only modulate calcium channel expression but they may also control calcium-dependent K+ channels [173–175]. The multiplicity of intracellular signaling pathways involved, their sensitivity to local and global intracellular calcium increase and to CaSR and GPRC6a stimulation, the presence of membrane signalplex, all this confers the cells the plasticity they need to convert the effects of extracellular calcium into complex physiological responses and therefore determine their fate.

Conflict of Interest The author declares no potential conflict of interest.

Acknowledgements I thank ARTP (Association pour la Recherche sur les Tumeurs de la Prostate) for support.

References

1. Carafoli E (2002) Calcium signaling: a tale for all seasons. Proc Natl Acad Sci U S A 99(3):1115–1122
2. Berridge MJ, Bootman MD, Roderick HL (2003) Calcium signalling: dynamics, homeostasis and remodelling. Nat Rev Mol Cell Biol 4(7):517–529
3. Berridge MJ, Lipp P, Bootman MD (2000) The versatility and universality of calcium signalling. Nat Rev Mol Cell Biol 1(1):11–21
4. Brown EM (2013) Role of the calcium-sensing receptor in extracellular calcium homeostasis. Best Pract Res Clin Endocrinol Metab 27(3):333–343
5. Riccardi D, Kemp PJ (2012) The calcium-sensing receptor beyond extracellular calcium homeostasis: conception, development, adult physiology, and disease. Annu Rev Physiol 74:271–297
6. Wellendorph P, Brauner-Osborne H (2004) Molecular cloning, expression, and sequence analysis of GPRC6A, a novel family C G-protein-coupled receptor. Gene 335:37–46
7. Alfadda TI, Saleh AM, Houillier P, Geibel JP (2014) Calcium-sensing receptor 20 years later. Am J Physiol Cell Physiol 307(3):C221–C231
8. Brown EM, Gamba G, Riccardi D, Lombardi M, Butters R, Kifor O, Sun A, Hediger MA, Lytton J, Hebert SC (1993) Cloning and characterization of an extracellular Ca(2+)-sensing receptor from bovine parathyroid. Nature 366(6455):575–580
9. Pi M, Quarles LD (2005) Osteoblast calcium-sensing receptor has characteristics of ANF/7TM receptors. J Cell Biochem 95(6):1081–1092
10. Berridge MJ (2009) Inositol trisphosphate and calcium signalling mechanisms. Biochim Biophys Acta 1793(6):933–940
11. Laude AJ, Simpson AW (2009) Compartmentalized signalling: Ca2+ compartments, microdomains and the many facets of Ca2+ signalling. FEBS J 276(7):1800–1816
12. Karlstad J, Sun Y, Singh BB (2012) Ca(2+) signaling: an outlook on the characterization of Ca(2+) channels and their importance in cellular functions. Adv Exp Med Biol 740:143–157
13. Capiod T (2011) Cell proliferation, calcium influx and calcium channels. Biochimie 93(12):2075–2079
14. Capiod T (2013) The need for calcium channels in cell proliferation. Recent Patents Anti-Cancer Drug Discov 8(1):4–17
15. Parekh AB, Putney JW Jr (2005) Store-operated calcium channels. Physiol Rev 85(2):757–810
16. Putney JW Jr (1986) A model for receptor-regulated calcium entry. Cell Calcium 7(1):1–12
17. Stathopulos PB, Li GY, Plevin MJ, Ames JB, Ikura M (2006) Stored Ca2+ depletion-induced oligomerization of stromal interaction molecule 1 (STIM1) via the EF-SAM region: an initiation mechanism for capacitive Ca2+ entry. J Biol Chem 281(47):35855–35862
18. Liou J, Fivaz M, Inoue T, Meyer T (2007) Live-cell imaging reveals sequential oligomerization and local plasma membrane targeting of stromal interaction molecule 1 after Ca2+ store depletion. Proc Natl Acad Sci U S A 104(22):9301–9306
19. Stathopulos PB, Ikura M (2013) Structural aspects of calcium-release activated calcium channel function. Channels (Austin Tex) 7(5):344–353

20. Berna-Erro A, Redondo PC, Rosado JA (2012) Store-operated Ca(2+) entry. Adv Exp Med Biol 740:349–382
21. Choi S, Maleth J, Jha A, Lee KP, Kim MS, So I, Ahuja M, Muallem S (2014) The TRPCs-STIM1-Orai interaction. Handb Exp Pharmacol 223:1035–1054
22. Ong HL, de Souza LB, Cheng KT, Ambudkar IS (2014) Physiological functions and regulation of TRPC channels. Handb Exp Pharmacol 223:1005–1034
23. Mignen O, Thompson JL, Shuttleworth TJ (2009) The molecular architecture of the arachidonate-regulated Ca2+-selective ARC channel is a pentameric assembly of Orai1 and Orai3 subunits. J Physiol 587(Pt 17):4181–4197
24. Thompson JL, Shuttleworth TJ (2013) Molecular basis of activation of the arachidonate-regulated Ca2+ (ARC) channel, a store-independent Orai channel, by plasma membrane STIM1. J Physiol 591(Pt 14):3507–3523
25. Mignen O, Shuttleworth TJ (2000) I(ARC), a novel arachidonate-regulated, noncapacitative Ca(2+) entry channel. J Biol Chem 275(13):9114–9119
26. Gudermann T, Hofmann T, Mederos y Schnitzler M, Dietrich A (2004) Activation, subunit composition and physiological relevance of DAG-sensitive TRPC proteins. Novartis Found Symp 258:103–118, discussion 118-122, 155-109, 263-106
27. Breitwieser GE, Gama L (2001) Calcium-sensing receptor activation induces intracellular calcium oscillations. Am J Physiol Cell Physiol 280(6):C1412–C1421
28. Conigrave AD, Ward DT (2013) Calcium-sensing receptor (CaSR): pharmacological properties and signaling pathways. Best Pract Res Clin Endocrinol Metab 27(3):315–331
29. Pi M, Quarles LD (2012) Multiligand specificity and wide tissue expression of GPRC6A reveals new endocrine networks. Endocrinology 153(5):2062–2069
30. Saidak Z, Mentaverri R, Brown EM (2009) The role of the calcium-sensing receptor in the development and progression of cancer. Endocr Rev 30(2):178–195
31. Borowiec AS, Bidaux G, Tacine R, Dubar P, Pigat N, Delcourt P, Mignen O, Capiod T (2014) Are Orai1 and Orai3 channels more important than calcium influx for cell proliferation? Biochim Biophys Acta 1843(2):464–472
32. Parekh AB (2010) Store-operated CRAC channels: function in health and disease. Nat Rev 9(5):399–410
33. Jairaman A, Prakriya M (2013) Molecular pharmacology of store-operated CRAC channels. Channels (Austin Tex) 7(5):402–414
34. Rice LV, Bax HJ, Russell LJ, Barrett VJ, Walton SE, Deakin AM, Thomson SA, Lucas F, Solari R, House D, Begg M (2013) Characterization of selective calcium-release activated calcium channel blockers in mast cells and T-cells from human, rat, mouse and guinea-pig preparations. Eur J Pharmacol 704(1-3):49–57
35. Ashmole I, Duffy SM, Leyland ML, Morrison VS, Begg M, Bradding P (2012) CRACM/Orai ion channel expression and function in human lung mast cells. J Allergy Clin Immunol 129(6):1628–1635, e1622
36. Gerasimenko JV, Gryshchenko O, Ferdek PE, Stapleton E, Hebert TO, Bychkova S, Peng S, Begg M, Gerasimenko OV, Petersen OH (2013) Ca2+ release-activated Ca2+ channel blockade as a potential tool in antipancreatitis therapy. Proc Natl Acad Sci U S A 110(32):13186–13191
37. Delmas P, Wanaverbecq N, Abogadie FC, Mistry M, Brown DA (2002) Signaling microdomains define the specificity of receptor-mediated InsP(3) pathways in neurons. Neuron 34(2):209–220
38. Trebak M, Bird GS, McKay RR, Putney JW Jr (2002) Comparison of human TRPC3 channels in receptor-activated and store-operated modes. Differential sensitivity to channel blockers suggests fundamental differences in channel composition. J Biol Chem 277(24):21617–21623
39. Xu SZ, Zeng F, Boulay G, Grimm C, Harteneck C, Beech DJ (2005) Block of TRPC5 channels by 2-aminoethoxydiphenyl borate: a differential, extracellular and voltage-dependent effect. Br J Pharmacol 145(4):405–414

40. Hu HZ, Gu Q, Wang C, Colton CK, Tang J, Kinoshita-Kawada M, Lee LY, Wood JD, Zhu MX (2004) 2-aminoethoxydiphenyl borate is a common activator of TRPV1, TRPV2, and TRPV3. J Biol Chem 279(34):35741–35748
41. Li M, Jiang J, Yue L (2006) Functional characterization of homo- and heteromeric channel kinases TRPM6 and TRPM7. J Gen Physiol 127(5):525–537
42. Bootman MD, Collins TJ, Mackenzie L, Roderick HL, Berridge MJ, Peppiatt CM (2002) 2-aminoethoxydiphenyl borate (2-APB) is a reliable blocker of store-operated Ca2+ entry but an inconsistent inhibitor of InsP3-induced Ca2+ release. FASEB J 16(10):1145–1150
43. Chokshi R, Fruasaha P, Kozak JA (2012) 2-aminoethyl diphenyl borinate (2-APB) inhibits TRPM7 channels through an intracellular acidification mechanism. Channels (Austin Tex) 6(5):362–369
44. Goto J, Suzuki AZ, Ozaki S, Matsumoto N, Nakamura T, Ebisui E, Fleig A, Penner R, Mikoshiba K (2010) Two novel 2-aminoethyl diphenylborinate (2-APB) analogues differentially activate and inhibit store-operated Ca(2+) entry via STIM proteins. Cell Calcium 47(1):1–10
45. Hendron E, Wang X, Zhou Y, Cai X, Goto J, Mikoshiba K, Baba Y, Kurosaki T, Wang Y, Gill DL (2014) Potent functional uncoupling between STIM1 and Orai1 by dimeric 2-aminodiphenyl borinate analogs. Cell Calcium 56(6):482–492
46. Hofer A, Kovacs G, Zappatini A, Leuenberger M, Hediger MA, Lochner M (2013) Design, synthesis and pharmacological characterization of analogs of 2-aminoethyl diphenylborinate (2-APB), a known store-operated calcium channel blocker, for inhibition of TRPV6-mediated calcium transport. Bioorg Med Chem 21(11):3202–3213
47. Djillani A, Nusse O, Dellis O (2014) Characterization of novel store-operated calcium entry effectors. Biochim Biophys Acta 1843(10):2341–2347
48. Maier T, Follmann M, Hessler G, Kleemann HW, Hachtel S, Fuchs B, Weissmann N, Linz W, Schmidt T, Löhn M, Schroeter K, Wang L, Rütten H, Strübing C (2015) Discovery and pharmacological characterization of a novel potent inhibitor of diacylglycerol-sensitive TRPC cation channels. Br J Pharmacol 172:3650–3660
49. Kohn CH, Liotta LA, Felder CC (1994) Signal transduction inhibitor compounds. US5359078
50. Johnson EA, Marks RS, Mandrekar SJ, Hillman SL, Hauge MD, Bauman MD, Wos EJ, Moore DF, Kugler JW, Windschitl HE, Graham DL, Bernath AM Jr, Fitch TR, Soori GS, Jett JR, Adjei AA, Perez EA (2008) Phase III randomized, double-blind study of maintenance CAI or placebo in patients with advanced non-small cell lung cancer (NSCLC) after completion of initial therapy (NCCTG 97-24-51). Lung Cancer 60(2):200–207
51. Mignen O, Brink C, Enfissi A, Nadkarni A, Shuttleworth TJ, Giovannucci DR, Capiod T (2005) Carboxyamidotriazole-induced inhibition of mitochondrial calcium import blocks capacitative calcium entry and cell proliferation in HEK-293 cells. J Cell Sci 118(Pt 23):5615–5623
52. Guo L, Luo L, Ju R, Chen C, Zhu L, Li J, Yu X, Ye C, Zhang D (2015) Carboxyamidotriazole: a novel inhibitor of both cAMP-phosphodiesterases and cGMP-phosphodiesterases. Eur J Pharmacol 746:14–21
53. Bowen CV, DeBay D, Ewart HS, Gallant P, Gormley S, Ilenchuk TT, Iqbal U, Lutes T, Martina M, Mealing G, Merkley N, Sperker S, Moreno MJ, Rice C, Syvitski RT, Stewart JM (2013) In vivo detection of human TRPV6-rich tumors with anti-cancer peptides derived from soricidin. PLoS ONE 8(3):e58866
54. Cox JH, Hussell S, Sondergaard H, Roepstorff K, Bui JV, Deer JR, Zhang J, Li ZG, Lamberth K, Kvist PH, Padkjaer S, Haase C, Zahn S, Odegard VH (2013) Antibody-mediated targeting of the Orai1 calcium channel inhibits T cell function. PLoS ONE 8(12):e82944
55. Lin FF, Elliott R, Colombero A, Gaida K, Kelley L, Moksa A, Ho SY, Bykova E, Wong M, Rathanaswami P, Hu S, Sullivan JK, Nguyen HQ, McBride HJ (2013) Generation and characterization of fully human monoclonal antibodies against human Orai1 for autoimmune disease. J Pharmacol Exp Ther 345(2):225–238
56. Ay AS, Benzerdjerb N, Sevestre H, Ahidouch A, Ouadid-Ahidouch H (2013) Orai3 constitutes a native store-operated calcium entry that regulates non small cell lung adenocarcinoma cell proliferation. PLoS ONE 8(9):e72889

57. Tao X, Zhao N, Jin H, Zhang Z, Liu Y, Wu J, Bast RC Jr, Yu Y, Feng Y (2013) FSH enhances the proliferation of ovarian cancer cells by activating transient receptor potential channel C3. Endocr Relat Cancer 20(3):415–429
58. Lindemann O, Umlauf D, Frank S, Schimmelpfennig S, Bertrand J, Pap T, Hanley PJ, Fabian A, Dietrich A, Schwab A (2013) TRPC6 regulates CXCR2-mediated chemotaxis of murine neutrophils. J Immunol 190(11):5496–5505
59. Heiser JH, Schuwald AM, Sillani G, Ye L, Muller WE, Leuner K (2013) TRPC6 channel-mediated neurite outgrowth in PC12 cells and hippocampal neurons involves activation of RAS/MEK/ERK, PI3K, and CAMKIV signaling. J Neurochem 127(3):303–313
60. Zanou N, Schakman O, Louis P, Ruegg UT, Dietrich A, Birnbaumer L, Gailly P (2012) Trpc1 ion channel modulates phosphatidylinositol 3-kinase/Akt pathway during myoblast differentiation and muscle regeneration. J Biol Chem 287(18):14524–14534
61. Seth M, Zhang ZS, Mao L, Graham V, Burch J, Stiber J, Tsiokas L, Winn M, Abramowitz J, Rockman HA, Birnbaumer L, Rosenberg P (2009) TRPC1 channels are critical for hypertrophic signaling in the heart. Circ Res 105(10):1023–1030
62. Ishikawa J, Ohga K, Yoshino T, Takezawa R, Ichikawa A, Kubota H, Yamada T (2003) A pyrazole derivative, YM-58483, potently inhibits store-operated sustained Ca2+ influx and IL-2 production in T lymphocytes. J Immunol 170(9):4441–4449
63. Umemura M, Baljinnyam E, Feske S, De Lorenzo MS, Xie LH, Feng X, Oda K, Makino A, Fujita T, Yokoyama U, Iwatsubo M, Chen S, Goydos JS, Ishikawa Y, Iwatsubo K (2014) Store-operated Ca2+ entry (SOCE) regulates melanoma proliferation and cell migration. PLoS ONE 9(2):e89292
64. Yao H, Peng F, Fan Y, Zhu X, Hu G, Buch SJ (2009) TRPC channel-mediated neuroprotection by PDGF involves Pyk2/ERK/CREB pathway. Cell Death Differ 16(12):1681–1693
65. Konig S, Browne S, Doleschal B, Schernthaner M, Poteser M, Machler H, Wittchow E, Braune M, Muik M, Romanin C, Groschner K (2013) Inhibition of Orai1-mediated Ca(2+) entry is a key mechanism of the antiproliferative action of sirolimus in human arterial smooth muscle. Am J Physiol Heart Circ Physiol 305(11):H1646–H1657
66. Sharma VK, Li B, Khanna A, Sehajpal PK, Suthanthiran M (1994) Which way for drug-mediated immunosuppression? Curr Opin Immunol 6(5):784–790
67. Ng SW, Nelson C, Parekh AB (2009) Coupling of Ca(2+) microdomains to spatially and temporally distinct cellular responses by the tyrosine kinase Syk. J Biol Chem 284(37):24767–24772
68. Bair AM, Thippegowda PB, Freichel M, Cheng N, Ye RD, Vogel SM, Yu Y, Flockerzi V, Malik AB, Tiruppathi C (2009) Ca2+ entry via TRPC channels is necessary for thrombin-induced NF-kappaB activation in endothelial cells through AMP-activated protein kinase and protein kinase Cdelta. J Biol Chem 284(1):563–574
69. Neher E (1998) Vesicle pools and Ca2+ microdomains: new tools for understanding their roles in neurotransmitter release. Neuron 20(3):389–399
70. Dolmetsch RE, Pajvani U, Fife K, Spotts JM, Greenberg ME (2001) Signaling to the nucleus by an L-type calcium channel-calmodulin complex through the MAP kinase pathway. Science 294(5541):333–339
71. Kar P, Parekh AB (2015) Distinct spatial Ca signatures selectively activate different NFAT transcription factor isoforms. Mol Cell 58(2):232–243
72. Kar P, Samanta K, Kramer H, Morris O, Bakowski D, Parekh AB (2014) Dynamic assembly of a membrane signaling complex enables selective activation of NFAT by Orai1. Curr Biol 24(12):1361–1368
73. Chang WC, Di Capite J, Singaravelu K, Nelson C, Halse V, Parekh AB (2008) Local Ca2+ influx through Ca2+ release-activated Ca2+ (CRAC) channels stimulates production of an intracellular messenger and an intercellular pro-inflammatory signal. J Biol Chem 283(8):4622–4631
74. Esseltine JL, Scott JD (2013) AKAP signaling complexes: pointing towards the next generation of therapeutic targets? Trends Pharmacol Sci 34(12):648–655

75. Gao T, Yatani A, Dell'Acqua ML, Sako H, Green SA, Dascal N, Scott JD, Hosey MM (1997) cAMP-dependent regulation of cardiac L-type Ca2+ channels requires membrane targeting of PKA and phosphorylation of channel subunits. Neuron 19(1):185–196
76. Oliveria SF, Dell'Acqua ML, Sather WA (2007) AKAP79/150 anchoring of calcineurin controls neuronal L-type Ca2+ channel activity and nuclear signaling. Neuron 55(2):261–275
77. Mignen O, Thompson JL, Shuttleworth TJ (2005) Arachidonate-regulated Ca2+-selective (ARC) channel activity is modulated by phosphorylation and involves an A-kinase anchoring protein. J Physiol 567(Pt 3):787–798
78. Thompson JL, Shuttleworth TJ (2015) Anchoring protein AKAP79-mediated PKA phosphorylation of STIM1 determines selective activation of the ARC channel, a store-independent Orai channel. J Physiol 593(3):559–572
79. Shim AH, Tirado-Lee L, Prakriya M (2015) Structural and functional mechanisms of CRAC channel regulation. J Mol Biol 427(1):77–93
80. Srikanth S, Ribalet B, Gwack Y (2013) Regulation of CRAC channels by protein interactions and post-translational modification. Channels 7(5):354–363
81. Gomez NM, Tamm ER, Straubeta O (2013) Role of bestrophin-1 in store-operated calcium entry in retinal pigment epithelium. Pflugers Arch 465(4):481–495
82. Ambily A, Kaiser WJ, Pierro C, Chamberlain EV, Li Z, Jones CI, Kassouf N, Gibbins JM, Authi KS (2014) The role of plasma membrane STIM1 and Ca(2+)entry in platelet aggregation. STIM1 binds to novel proteins in human platelets. Cell Signal 26(3):502–511
83. Lee JE, Jeon IS, Han NE, Song HJ, Kim EG, Choi JW, Song KD, Lee HK, Choi JK (2013) Ubiquilin 1 interacts with Orai1 to regulate calcium mobilization. Mol Cells 35(1):41–46
84. Brown DA, London E (1998) Functions of lipid rafts in biological membranes. Annu Rev Cell Dev Biol 14:111–136
85. Popescu LM, Diculescu I, Zelck U, Ionescu N (1974) Ultrastructural distribution of calcium in smooth muscle cells of guinea-pig taenia coli. A correlated electron microscopic and quantitative study. Cell Tissue Res 154(3):357–378
86. Isshiki M, Ying YS, Fujita T, Anderson RG (2002) A molecular sensor detects signal transduction from caveolae in living cells. J Biol Chem 277(45):43389–43398
87. Isshiki M, Anderson RG (2003) Function of caveolae in Ca2+ entry and Ca2+-dependent signal transduction. Traffic (Copenhagen, Den) 4(11):717–723
88. El Boustany C, Bidaux G, Enfissi A, Delcourt P, Prevarskaya N, Capiod T (2008) Capacitative calcium entry and transient receptor potential canonical 6 expression control human hepatoma cell proliferation. Hepatology 47(6):2068–2077
89. Duhon D, Bigelow RL, Coleman DT, Steffan JJ, Yu C, Langston W, Kevil CG, Cardelli JA (2010) The polyphenol epigallocatechin-3-gallate affects lipid rafts to block activation of the c-Met receptor in prostate cancer cells. Mol Carcinog 49(8):739–749
90. Mineo C, James GL, Smart EJ, Anderson RG (1996) Localization of epidermal growth factor-stimulated Ras/Raf-1 interaction to caveolae membrane. J Biol Chem 271(20):11930–11935
91. Pike LJ (2005) Growth factor receptors, lipid rafts and caveolae: an evolving story. Biochim Biophys Acta 1746(3):260–273
92. Pani B, Singh BB (2009) Lipid rafts/caveolae as microdomains of calcium signaling. Cell Calcium 45(6):625–633
93. Pani B, Ong HL, Brazer SC, Liu X, Rauser K, Singh BB, Ambudkar IS (2009) Activation of TRPC1 by STIM1 in ER-PM microdomains involves release of the channel from its scaffold caveolin-1. Proc Natl Acad Sci U S A 106(47):20087–20092
94. Brazer SC, Singh BB, Liu X, Swaim W, Ambudkar IS (2003) Caveolin-1 contributes to assembly of store-operated Ca2+ influx channels by regulating plasma membrane localization of TRPC1. J Biol Chem 278(29):27208–27215
95. Ong HL, Ambudkar IS (2011) The dynamic complexity of the TRPC1 channelosome. Channels (Austin Tex) 5(5):424–431. doi:10.4161/chan.5.5.16471
96. Thebault S, Flourakis M, Vanoverberghe K, Vandermoere F, Roudbaraki M, Lehen'kyi V, Slomianny C, Beck B, Mariot P, Bonnal JL, Mauroy B, Shuba Y, Capiod T, Skryma R,

Prevarskaya N (2006) Differential role of transient receptor potential channels in Ca2+ entry and proliferation of prostate cancer epithelial cells. Cancer Res 66(4):2038–2047
97. Mignen O, Thompson JL, Shuttleworth TJ (2003) Calcineurin directs the reciprocal regulation of calcium entry pathways in nonexcitable cells. J Biol Chem 278(41):40088–40096
98. Yeh YC, Parekh AB (2015) Distinct structural domains of caveolin-1 independently regulate Ca2+ release-activated Ca2+ channels and Ca2+ microdomain-dependent gene expression. Mol Cell Biol 35(8):1341–1349
99. Li S, Seitz R, Lisanti MP (1996) Phosphorylation of caveolin by src tyrosine kinases. The alpha-isoform of caveolin is selectively phosphorylated by v-Src in vivo. J Biol Chem 271(7):3863–3868
100. Hogan PG, Chen L, Nardone J, Rao A (2003) Transcriptional regulation by calcium, calcineurin, and NFAT. Genes Dev 17(18):2205–2232
101. Yissachar N, Sharar Fischler T, Cohen AA, Reich-Zeliger S, Russ D, Shifrut E, Porat Z, Friedman N (2013) Dynamic response diversity of NFAT isoforms in individual living cells. Mol Cell 49(2):322–330
102. Gueguinou M, Chantome A, Fromont G, Bougnoux P, Vandier C, Potier-Cartereau M (2014) KCa and Ca(2+) channels: the complex thought. Biochim Biophys Acta 1843(10):2322–2333
103. Harmar AJ, Hills RA, Rosser EM, Jones M, Buneman OP, Dunbar DR, Greenhill SD, Hale VA, Sharman JL, Bonner TI, Catterall WA, Davenport AP, Delagrange P, Dollery CT, Foord SM, Gutman GA, Laudet V, Neubig RR, Ohlstein EH, Olsen RW, Peters J, Pin JP, Ruffolo RR, Searls DB, Wright MW, Spedding M (2009) IUPHAR-DB: the IUPHAR database of G protein-coupled receptors and ion channels. Nucleic Acids Res 37(Database issue):D680–D685
104. Wei AD, Gutman GA, Aldrich R, Chandy KG, Grissmer S, Wulff H (2005) International Union of Pharmacology. LII. Nomenclature and molecular relationships of calcium-activated potassium channels. Pharmacol Rev 57(4):463–472
105. Chantome A, Potier-Cartereau M, Clarysse L, Fromont G, Marionneau-Lambot S, Gueguinou M, Pages JC, Collin C, Oullier T, Girault A, Arbion F, Haelters JP, Jaffres PA, Pinault M, Besson P, Joulin V, Bougnoux P, Vandier C (2013) Pivotal role of the lipid Raft SK3-Orai1 complex in human cancer cell migration and bone metastases. Cancer Res 73(15):4852–4861
106. Gackiere F, Warnier M, Katsogiannou M, Derouiche S, Delcourt P, Dewailly E, Slomianny C, Humez S, Prevarskaya N, Roudbaraki M, Mariot P (2013) Functional coupling between large-conductance potassium channels and Cav3.2 voltage-dependent calcium channels participates in prostate cancer cell growth. Biol Open 2(9):941–951
107. Crunelli V, Toth TI, Cope DW, Blethyn K, Hughes SW (2005) The 'window' T-type calcium current in brain dynamics of different behavioural states. J Physiol 562(Pt 1):121–129
108. Bean BP, McDonough SI (1998) Two for T. Neuron 20(5):825–828
109. Lallet-Daher H, Roudbaraki M, Bavencoffe A, Mariot P, Gackiere F, Bidaux G, Urbain R, Gosset P, Delcourt P, Fleurisse L, Slomianny C, Dewailly E, Mauroy B, Bonnal JL, Skryma R, Prevarskaya N (2009) Intermediate-conductance Ca2+-activated K+ channels (IKCa1) regulate human prostate cancer cell proliferation through a close control of calcium entry. Oncogene 28(15):1792–1806
110. Hammadi M, Chopin V, Matifat F, Dhennin-Duthille I, Chasseraud M, Sevestre H, Ouadid-Ahidouch H (2012) Human ether a-gogo K(+) channel 1 (hEag1) regulates MDA-MB-231 breast cancer cell migration through Orai1-dependent calcium entry. J Cell Physiol 227(12):3837–3846
111. Peretti M, Angelini M, Savalli N, Florio T, Yuspa SH, Mazzanti M (2015) Chloride channels in cancer: focus on chloride intracellular channel 1 and 4 (CLIC1 AND CLIC4) proteins in tumor development and as novel therapeutic targets. Biochim Biophys Acta 1848(10):2523–2531
112. Qu Z, Yao W, Yao R, Liu X, Yu K, Hartzell C (2014) The Ca(2+) -activated Cl(-) channel, ANO1 (TMEM16A), is a double-edged sword in cell proliferation and tumorigenesis. Cancer Med 3(3):453–461

113. Wang GL, Qian Y, Qiu QY, Lan XJ, He H, Guan YY (2006) Interaction between Cl- channels and CRAC-related Ca2+ signaling during T lymphocyte activation and proliferation. Acta Pharmacol Sin 27(4):437–446
114. Ahmmed GU, Mehta D, Vogel S, Holinstat M, Paria BC, Tiruppathi C, Malik AB (2004) Protein kinase Calpha phosphorylates the TRPC1 channel and regulates store-operated Ca2+ entry in endothelial cells. J Biol Chem 279(20):20941–20949
115. Antigny F, Jousset H, Konig S, Frieden M (2011) Thapsigargin activates Ca(2)+entry both by store-dependent, STIM1/Orai1-mediated, and store-independent, TRPC3/PLC/PKC-mediated pathways in human endothelial cells. Cell Calcium 49(2):115–127
116. Sukumaran P, Lof C, Kemppainen K, Kankaanpaa P, Pulli I, Nasman J, Viitanen T, Tornquist K (2012) Canonical transient receptor potential channel 2 (TRPC2) as a major regulator of calcium homeostasis in rat thyroid FRTL-5 cells: importance of protein kinase C delta (PKCdelta) and stromal interaction molecule 2 (STIM2). J Biol Chem 287(53):44345–44360
117. Venkatachalam K, Zheng F, Gill DL (2003) Regulation of canonical transient receptor potential (TRPC) channel function by diacylglycerol and protein kinase C. J Biol Chem 278(31):29031–29040
118. Venkatachalam K, Zheng F, Gill DL (2004) Control of TRPC and store-operated channels by protein kinase C. Novartis Found Symp 258:172–185, discussion 185-178, 263-176
119. Blackmore PF (2011) Biphasic effects of nitric oxide on calcium influx in human platelets. Thromb Res 127(1):e8–e14
120. Chen J, Crossland RF, Noorani MM, Marrelli SP (2009) Inhibition of TRPC1/TRPC3 by PKG contributes to NO-mediated vasorelaxation. Am J Physiol Heart Circ Physiol 297(1):H417–H424
121. Kwan HY, Huang Y, Yao X (2004) Regulation of canonical transient receptor potential isoform 3 (TRPC3) channel by protein kinase G. Proc Natl Acad Sci U S A 101(8):2625–2630
122. Yao X (2007) TRPC, cGMP-dependent protein kinases and cytosolic Ca2+. Handb Exp Pharmacol 179:527–540
123. Yuasa K, Matsuda T, Tsuji A (2011) Functional regulation of transient receptor potential canonical 7 by cGMP-dependent protein kinase Ialpha. Cell Signal 23(7):1179–1187
124. Vig M, Beck A, Billingsley JM, Lis A, Parvez S, Peinelt C, Koomoa DL, Soboloff J, Gill DL, Fleig A, Kinet JP, Penner R (2006) CRACM1 multimers form the ion-selective pore of the CRAC channel. Curr Biol 16(20):2073–2079
125. Mignen O, Thompson JL, Shuttleworth TJ (2008) Both Orai1 and Orai3 are essential components of the arachidonate-regulated Ca2+-selective (ARC) channels. J Physiol 586(1):185–195
126. Faouzi M, Hague F, Potier M, Ahidouch A, Sevestre H, Ouadid-Ahidouch H (2011) Down-regulation of Orai3 arrests cell-cycle progression and induces apoptosis in breast cancer cells but not in normal breast epithelial cells. J Cell Physiol 226(2):542–551
127. Abdullaev IF, Bisaillon JM, Potier M, Gonzalez JC, Motiani RK, Trebak M (2008) Stim1 and Orai1 mediate CRAC currents and store-operated calcium entry important for endothelial cell proliferation. Circ Res 103(11):1289–1299
128. Cahalan MD (2009) STIMulating store-operated Ca(2+) entry. Nat Cell Biol 11(6):669–677
129. Varnai P, Hunyady L, Balla T (2009) STIM and Orai: the long-awaited constituents of store-operated calcium entry. Trends Pharmacol Sci 30(3):118–128
130. Li J, Sukumar P, Milligan CJ, Kumar B, Ma ZY, Munsch CM, Jiang LH, Porter KE, Beech DJ (2008) Interactions, functions, and independence of plasma membrane STIM1 and TRPC1 in vascular smooth muscle cells. Circ Res 103(8):e97–e104
131. Darbellay B, Arnaudeau S, Konig S, Jousset H, Bader C, Demaurex N, Bernheim L (2009) STIM1- and Orai1-dependent store-operated calcium entry regulates human myoblast differentiation. J Biol Chem 284(8):5370–5380
132. El Boustany C, Katsogiannou M, Delcourt P, Dewailly E, Prevarskaya N, Borowiec AS, Capiod T (2010) Differential roles of STIM1, STIM2 and Orai1 in the control of cell proliferation and SOCE amplitude in HEK293 cells. Cell Calcium 47(4):350–359

133. Antigny F, Girardin N, Frieden M (2012) Transient receptor potential canonical channels are required for in vitro endothelial tube formation. J Biol Chem 287(8):5917–5927
134. Dragoni S, Turin I, Laforenza U, Potenza DM, Bottino C, Glasnov TN, Prestia M, Ferulli F, Saitta A, Mosca A, Guerra G, Rosti V, Luinetti O, Ganini C, Porta C, Pedrazzoli P, Tanzi F, Montagna D, Moccia F (2014) Store-operated Ca2+ entry does not control proliferation in primary cultures of human metastatic renal cellular carcinoma. Biomed Res Int 2014:739494
135. Dubois C, Vanden Abeele F, Lehen'kyi V, Gkika D, Guarmit B, Lepage G, Slomianny C, Borowiec AS, Bidaux G, Benahmed M, Shuba Y, Prevarskaya N (2014) Remodeling of channel-forming ORAI proteins determines an oncogenic switch in prostate cancer. Cancer Cell 26(1):19–32
136. Hanahan D, Weinberg RA (2011) Hallmarks of cancer: the next generation. Cell 144(5):646–674
137. Ichikawa J, Inoue R (2014) TRPC6 regulates cell cycle progression by modulating membrane potential in bone marrow stromal cells. Br J Pharmacol 171(23):5280–5294
138. Song J, Wang Y, Li X, Shen Y, Yin M, Guo Y, Diao L, Liu Y, Yue D (2013) Critical role of TRPC6 channels in the development of human renal cell carcinoma. Mol Biol Rep 40(8):5115–5122
139. Kito H, Yamamura H, Suzuki Y, Yamamura H, Ohya S, Asai K, Imaizumi Y (2015) Regulation of store-operated Ca(2+) entry activity by cell cycle dependent up-regulation of Orai2 in brain capillary endothelial cells. Biochem Biophys Res Commun 459(3):457–462
140. Faouzi M, Kischel P, Hague F, Ahidouch A, Benzerdjeb N, Sevestre H, Penner R, Ouadid-Ahidouch H (2013) ORAI3 silencing alters cell proliferation and cell cycle progression via c-myc pathway in breast cancer cells. Biochim Biophys Acta 1833(3):752–760
141. Sukumaran P, Lof C, Pulli I, Kemppainen K, Viitanen T, Tornquist K (2013) Significance of the transient receptor potential canonical 2 (TRPC2) channel in the regulation of rat thyroid FRTL-5 cell proliferation, migration, adhesion and invasion. Mol Cell Endocrinol 374(1-2):10–21
142. Volpi M, Berlin RD (1988) Intracellular elevations of free calcium induced by activation of histamine H1 receptors in interphase and mitotic HeLa cells: hormone signal transduction is altered during mitosis. J Cell Biol 107(6 Pt 2):2533–2539
143. Preston SF, Sha'afi RI, Berlin RD (1991) Regulation of Ca2+ influx during mitosis: Ca2+ influx and depletion of intracellular Ca2+ stores are coupled in interphase but not mitosis. Cell Regul 2(11):915–925
144. Russa AD, Ishikita N, Masu K, Akutsu H, Saino T, Satoh Y (2008) Microtubule remodeling mediates the inhibition of store-operated calcium entry (SOCE) during mitosis in COS-7 cells. Arch Histol Cytol 71(4):249–263
145. Smyth JT, Petranka JG, Boyles RR, DeHaven WI, Fukushima M, Johnson KL, Williams JG, Putney JW Jr (2009) Phosphorylation of STIM1 underlies suppression of store-operated calcium entry during mitosis. Nat Cell Biol 11(12):1465–1472
146. Tani D, Monteilh-Zoller MK, Fleig A, Penner R (2007) Cell cycle-dependent regulation of store-operated I(CRAC) and Mg2+-nucleotide-regulated MagNuM (TRPM7) currents. Cell Calcium 41(3):249–260
147. Yu F, Sun L, Machaca K (2009) Orai1 internalization and STIM1 clustering inhibition modulate SOCE inactivation during meiosis. Proc Natl Acad Sci U S A 106(41):17401–17406
148. Yu F, Sun L, Machaca K (2010) Constitutive recycling of the store-operated Ca2+ channel Orai1 and its internalization during meiosis. J Cell Biol 191(3):523–535
149. Arredouani A, Yu F, Sun L, Machaca K (2010) Regulation of store-operated Ca2+ entry during the cell cycle. J Cell Sci 123(Pt 13):2155–2162
150. Patterson RL, van Rossum DB, Gill DL (1999) Store-operated Ca2+ entry: evidence for a secretion-like coupling model. Cell 98(4):487–499
151. Bobe R, Hadri L, Lopez JJ, Sassi Y, Atassi F, Karakikes I, Liang L, Limon I, Lompre AM, Hatem SN, Hajjar RJ, Lipskaia L (2011) SERCA2a controls the mode of agonist-induced

intracellular Ca2+ signal, transcription factor NFAT and proliferation in human vascular smooth muscle cells. J Mol Cell Cardiol 50(4):621–633
152. Hou MF, Kuo HC, Li JH, Wang YS, Chang CC, Chen KC, Chen WC, Chiu CC, Yang S, Chang WC (2011) Orai1/CRACM1 overexpression suppresses cell proliferation via attenuation of the store-operated calcium influx-mediated signalling pathway in A549 lung cancer cells. Biochim Biophys Acta 1810(12):1278–1284
153. Tajeddine N, Gailly P (2012) TRPC1 protein channel is major regulator of epidermal growth factor receptor signaling. J Biol Chem 287(20):16146–16157
154. Kuang CY, Yu Y, Wang K, Qian DH, Den MY, Huang L (2012) Knockdown of transient receptor potential canonical-1 reduces the proliferation and migration of endothelial progenitor cells. Stem Cells Dev 21(3):487–496
155. Hofer AM, Brown EM (2003) Extracellular calcium sensing and signalling. Nat Rev Mol Cell Biol 4(7):530–538
156. Brennan SC, Finney BA, Lazarou M, Rosser AE, Scherf C, Adriaensen D, Kemp PJ, Riccardi D (2013) Fetal calcium regulates branching morphogenesis in the developing human and mouse lung: involvement of voltage-gated calcium channels. PLoS ONE 8(11):e80294
157. Rodland KD (2004) The role of the calcium-sensing receptor in cancer. Cell Calcium 35(3):291–295
158. Singh N, Promkan M, Liu G, Varani J, Chakrabarty S (2013) Role of calcium sensing receptor (CaSR) in tumorigenesis. Best Pract Res Clin Endocrinol Metab 27(3):455–463
159. Whitfield JF (2009) Calcium, calcium-sensing receptor and colon cancer. Cancer Lett 275(1):9–16
160. Mamillapalli R, VanHouten J, Zawalich W, Wysolmerski J (2008) Switching of G-protein usage by the calcium-sensing receptor reverses its effect on parathyroid hormone-related protein secretion in normal versus malignant breast cells. J Biol Chem 283(36):24435–24447
161. Pi M, Faber P, Ekema G, Jackson PD, Ting A, Wang N, Fontilla-Poole M, Mays RW, Brunden KR, Harrington JJ, Quarles LD (2005) Identification of a novel extracellular cation-sensing G-protein-coupled receptor. J Biol Chem 280(48):40201–40209
162. Pi M, Quarles LD (2012) GPRC6A regulates prostate cancer progression. Prostate 72(4):399–409
163. Fujiwara T, Kanazawa S, Ichibori R, Tanigawa T, Magome T, Shingaki K, Miyata S, Tohyama M, Hosokawa K (2014) L-arginine stimulates fibroblast proliferation through the GPRC6A-ERK1/2 and PI3K/Akt pathway. PLoS ONE 9(3):e92168
164. Wei J, Hanna T, Suda N, Karsenty G, Ducy P (2014) Osteocalcin promotes beta-cell proliferation during development and adulthood through Gprc6a. Diabetes 63(3):1021–1031
165. Haiman CA, Han Y, Feng Y, Xia L, Hsu C, Sheng X, Pooler LC, Patel Y, Kolonel LN, Carter E, Park K, Le Marchand L, Van Den Berg D, Henderson BE, Stram DO (2013) Genome-wide testing of putative functional exonic variants in relationship with breast and prostate cancer risk in a multiethnic population. PLoS Genet 9(3):e1003419
166. Karp HJ, Ketola ME, Lamberg-Allardt CJ (2009) Acute effects of calcium carbonate, calcium citrate and potassium citrate on markers of calcium and bone metabolism in young women. Br J Nutr 102(9):1341–1347
167. Otani T, Mizokami A, Hayashi Y, Gao J, Mori Y, Nakamura S, Takeuchi H, Hirata M (2015) Signaling pathway for adiponectin expression in adipocytes by osteocalcin. Cell Signal 27(3):532–544
168. El Hiani Y, Lehen'kyi V, Ouadid-Ahidouch H, Ahidouch A (2009) Activation of the calcium-sensing receptor by high calcium induced breast cancer cell proliferation and TRPC1 cation channel over-expression potentially through EGFR pathways. Arch Biochem Biophys 486(1):58–63
169. Chow JY, Estrema C, Orneles T, Dong X, Barrett KE, Dong H (2011) Calcium-sensing receptor modulates extracellular Ca(2+) entry via TRPC-encoded receptor-operated channels in human aortic smooth muscle cells. Am J Physiol Cell Physiol 301(2):C461–C468

170. Meng K, Xu J, Zhang C, Zhang R, Yang H, Liao C, Jiao J (2014) Calcium sensing receptor modulates extracellular calcium entry and proliferation via TRPC3/6 channels in cultured human mesangial cells. PLoS ONE 9(6):e98777.
171. Breitwieser GE, Miedlich SU, Zhang M (2004) Calcium sensing receptors as integrators of multiple metabolic signals. Cell Calcium 35(3):209–216
172. Robertson WG, Marshall RW (1981) Ionized calcium in body fluids. Crit Rev Clin Lab Sci 15(2):85–125
173. Chattopadhyay N, Ye CP, Yamaguchi T, Kifor O, Vassilev PM, Nishimura R, Brown EM (1998) Extracellular calcium-sensing receptor in rat oligodendrocytes: expression and potential role in regulation of cellular proliferation and an outward K+ channel. Glia 24(4):449–458
174. Harno E, Edwards G, Geraghty AR, Ward DT, Dodd RH, Dauban P, Faure H, Ruat M, Weston AH (2008) Evidence for the presence of GPRC6A receptors in rat mesenteric arteries. Cell Calcium 44(2):210–219
175. Vysotskaya ZV, Moss CR 2nd, Gilbert CA, Gabriel SA, Gu Q (2014) Modulation of BK channel activities by calcium-sensing receptor in rat bronchopulmonary sensory neurons. Respir Physiol Neurobiol 203:35–44

Chapter 8
Regulation of Platelet Function by Orai, STIM and TRP

Alejandro Berna-Erro, Isaac Jardín, Tarik Smani, and Juan A. Rosado

Abstract Agonist-induced changes in cytosolic Ca^{2+} concentration ($[Ca^{2+}]_c$) are central events in platelet physiology. A major mechanism supporting agonist-induced Ca^{2+} signals is store-operated Ca^{2+} entry (SOCE), where the Ca^{2+} sensor STIM1 and the channels of the Orai family, as well as TRPC members are the key elements. STIM1-dependent SOCE plays a major role in collagen-stimulated Ca^{2+} signaling, phosphatidylserine exposure and thrombin generation. Furthermore, studies involving Orai1 gain-of-function mutants and platelets from Orai1-deficient mice have revealed the importance of this channel in thrombosis and hemostasis to those found in STIM1-deficient mice indicating that SOCE might play a prominent role in thrombus formation. Moreover, increase in TRPC6 expression might lead to thrombosis in humans. The role of STIM1, Orai1 and TRPCs, and thus SOCE, in thrombus formation, suggests that therapies directed against SOCE and targeting these molecules during cardiovascular and cerebrovascular events could significantly improve traditional anti-thrombotic treatments.

Keywords Orai1 • TRPC1 • STIM1 • Calcium entry • Human platelets • TRPC6

A. Berna-Erro
Laboratory of Molecular Physiology and Channelopathies, Department of Experimental and Health Sciences, Universitat Pompeu Fabra, Barcelona 08003, Spain

I. Jardín
Department of Physiology (Cell Physiology Research Group), University of Extremadura, Cáceres 10003, Spain

T. Smani
Department of Medical Physiology and Biophysic, Institute of Biomedicine of Seville (IBiS), University Hospital of Virgen del Rocío/CSIC/University of Seville, Sevilla 41013, Spain

J.A. Rosado (✉)
Departamento de Fisiología, University of Extremadura, Cáceres, Spain
e-mail: jarosado@unex.es

J.A. Rosado (ed.), *Calcium Entry Pathways in Non-excitable Cells*, Advances in Experimental Medicine and Biology 898, DOI 10.1007/978-3-319-26974-0_8

8.1 Molecular Basis of Platelet Ca^{2+} Influx: Store-Operated Ca^{2+} Entry

8.1.1 Ca^{2+}-Handling Proteins in Platelets

Platelets are anucleated cell fragments derived from bone marrow megakaryocytes that play an important role in hemostasis. Resting platelets are biconvex discoid structures with 0.5–2 μm diameter. In platelets, Ca^{2+}signaling has mostly been investigated in human and mouse, where these cells exhibit a complex machinery for Ca^{2+} homeostasis. In human platelets, Ca^{2+} entry occurs via both store-dependent (store operated or capacitative) and store-independent (or non-capacitative) mechanisms. Among the proteins and channels involved in the mechanisms underlying Ca^{2+} entry, human platelets express the intraluminal Ca^{2+} sensors STIM1 and STIM2 [1, 2], which have been found both in the dense tubular system (the analog of the endoplasmic reticulum (ER) in platelets) and the acidic stores [2] (agonist-releasable Ca^{2+} compartments that maintain a proton gradient across their membranes mostly identified as lysosomal-like organelles [3]). Moreover, human platelets contain significant levels of Orai1 [4–6] and its two identified homologs, Orai2 and Orai3 [7]. In addition, human platelets express different TRPC subfamily members, including TRPC1, TRPC3, TRPC4, TRPC5 and TRPC6 [7, 8], as well as TRPV1 [9]. We have not detected expression of TRPA1 at the protein level in these cells [10] and the expression of other TRP channels is uncertain. Agonist-stimulated Ca^{2+} release from the intracellular stores is mediated by the generation of inositol 1,4,5-trisphosphate (IP_3) or nicotinic acid-adenine dinucleotide phosphate (NAADP) [11, 12] (Fig. 8.1). Human platelets are reported to express three isoforms of the IP_3 receptor (IP_3R): IP_3R types I, II and III [13] and, although there is no direct evidence for the expression of two-pore channels (TPC) in human platelets, RT-PCR analysis has revealed the expression of TPC1 and TPC2 in the human megakaryoblastic cell line MEG01 [14, 15]. By contrast, ryanodine receptors have not been described in these cells. Ca^{2+} clearance from the cytosol has been reported to occur through the activation of a battery of mechanisms, including Ca^{2+} sequestration into the agonist-sensitive Ca^{2+} stores and extrusion to the extracellular medium. Ca^{2+} reuptake into the stores is mediated by different isoforms of the sarco-endoplasmic reticulum Ca^{2+} ATPase (SERCA), including SERCA2b and SERCA3 [16]. Ca^{2+} extrusion might occur either by active pumping by the two isoforms of the plasma membrane Ca^{2+} ATPase (PMCA) identified in human platelets, PMCA1b and PMCA4b [17–19] or by exchange with Na^+ through the Na^+/Ca^{2+} exchangers [20]. Human platelets exhibit Na^+/Ca^{2+}exchange, as well as, K^+-dependent Na^+/Ca^{2+}exchange activity [21].

Significant inherent differences between mouse and human platelets have been described both at the level of platelet count and size and at the expression of certain Ca^{2+}-handling proteins. While these differences do not preclude the use of mouse models to investigate platelet Ca^{2+} homeostasis, the potential limitations suggest that caution must be exercised in the extrapolation of the data from mouse to human

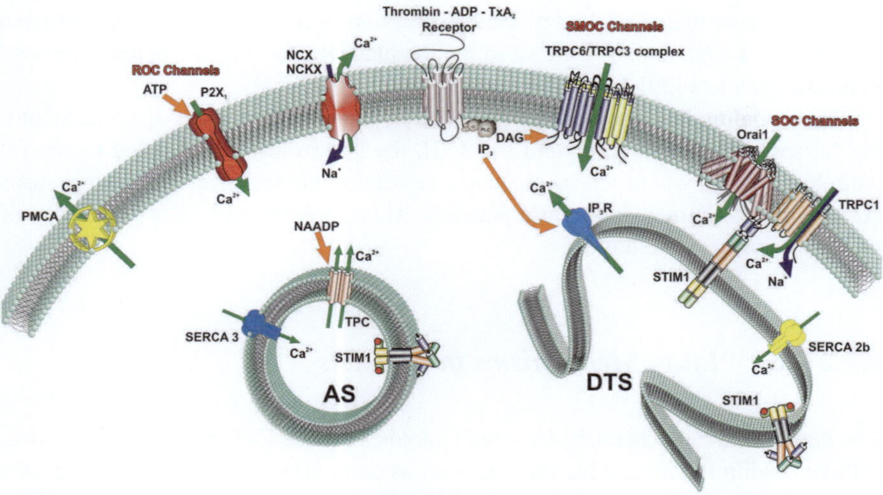

Fig. 8.1 Calcium tuning in human platelets: Cytosolic Ca^{2+} may be increased by three different main routes, which work together under physiological stimuli. **Receptor Operated Calcium** (*ROC*) channels, as the $P2X_1$ receptor: its activation by an agonist induces conformational changes allowing the pass of Ca^{2+} ions through them. **Second Messenger Operated Calcium** (*SMOC*) channels, like the complex *TRPC6/TRPC3*: as its name indicates, these channels are activated by second messengers, for instance diacylglycerol (*DAG*). **Store Operated Calcium** (*SOC*) **channels**, this calcium entry is activated by the depletion of the dense tubular system (*DTS*) and the acidic stores (*AS*), the intracellular calcium stores in human platelets, by the production of inositol triphosphate (IP_3) and nicotinic acid adenine dinucleotide phosphate (*NAADP*) upon agonist stimulation. Intraluminal calcium depletion is sensed by STIM1, located in both stores, and activates Orai and TRPC1 channels in the plasma membrane allowing calcium entry. Platelets present as well mechanisms to return the $[Ca^{2+}]_c$ to basal levels, which will prepare the cell for the response to new stimuli. Two members of the Sarcoplasmic Reticulum Ca^{2+} ATPase (*SERCA*) are expressed in human platelets, SERCA2b in the DTS and SERCA3 in the AS, accumulating high levels of Ca^{2+} in the stores. On the other hand, located on the plasma membrane, human platelets express proteins that extrude Ca^{2+} to the extracellular medium: the Plasma Membrane Calcium ATPase (*PMCA*), the Sodium Calcium Exchanger (*NCX*) and the Potassium dependent Sodium Calcium Exchanger (*NCKX*). *Abbreviations*: *ADP* Adenosine diphosphate, *AS* Acidic Stores, *ATP* Adenosine triphosphate, Ca^{2+} Calcium, *DAG* Diacylglycerol, *DTS* Dense tubular system, IP_3 Inositol 1,4,5-trisphosphate, IP_3R Inositol 1,4,5-trisphosphate Receptor, *NAADP* Nicotinic acid adenine dinucleotide phosphate, *NCKX* Potassium-dependent Sodium/Calcium Exchanger, *NCX* Sodium/Calcium Exchanger, $P2X_1$ ATP receptor, *PMCA* Plasma Membrane Calcium ATPase, *ROC Channels* Receptor Operated Ca^{2+} Channels, *SERCA 2b* Sarco Endoplasmic Calcium ATPase 2b, *SERCA3* Sarco Endoplasmic Calcium ATPase 3, *SMOC Channels* Second Messenger Operated Ca^{2+} Channels, *SOC Channels* Store Operated Ca^{2+} Channels, *TPC* Two Pore Channels, *TxA2* Tromboxane A2

platelets. Concerning Ca^{2+} entry, mouse platelets have been reported to express STIM1 [22] and STIM2 [23], as well as Orai1 [4], which is the predominant Orai member since low levels of *Orai2* and *Orai3* transcripts have been detected in these cells [24]. Mouse platelets express TRPC1, TRPC3 and TRPC6 [25–27], while other members of the TRPC subfamily have not been detected either at the transcript or protein level. Despite TRPM1, TRPM2 and TRPM7 have been reported in primary murine megakaryocytes, their expression in mouse platelets has not been demonstrated [28], and, in contrast to human platelets, the TRPV1 is not expressed in mouse platelets [29].

Ca^{2+} signaling has also been investigated in platelets from other species, including horses [30], rabbits [31], and rats [32], the latter mostly associated to several pathologies [33], where the mechanisms regulating intracellular Ca^{2+} homeostasis are common to human and mouse platelets. Therefore, we will refer to the cells of these two species hereinafter.

8.1.2 Ca^{2+} Entry Mechanisms in Platelets

The nature of the mechanisms by which agonists induce Ca^{2+} entry from the extracellular medium in human platelets, as well as other non-excitable cells, has been a matter of intense investigation in the last decades due to its relevance in cell physiology. In 1983, De Clerck and Van Nueten reported that flunarizine, a Ca^{2+} entry blocker, attenuated the release of thromboxane B2 and serotonin, a relevant mechanism for platelet-platelet interaction as well as for the communication of platelets with vascular cells [34]. Platelet stimulation with agonists results in an increase in the cytosolic free Ca^{2+} concentration ($[Ca^{2+}]_c$), which mainly consists of two components: the release of Ca^{2+} from intracellular stores, and Ca^{2+} entry across the plasma membrane [35]. Ca^{2+} release from finite intracellular stores induces a transient increase in $[Ca^{2+}]_c$; however, for full activation of platelet processes, as well as the refilling of the Ca^{2+} stores, Ca^{2+} entry plays a relevant role.

The mechanisms by which agonists induce Ca^{2+} entry in human platelets is not fully understood yet but three general routes for Ca^{2+} influx have been described, named receptor-operated Ca^{2+} entry (ROCE), second messenger-operated Ca^{2+} entry (SMOCE) and store-operated Ca^{2+} entry (SOCE) (Fig. 8.1). ROCE is the mechanism for Ca^{2+} influx directly activated by receptor occupation. This Ca^{2+} influx pathway occurs through receptor-operated channels (ROCs) gated by agonist-receptor binding. A good example for ROCE in platelets is the activation of the purinergic $P2X_1$ receptor, a ligand-gated non-selective cation channel activated by ATP [36] that induces a rapid and transient increase in $[Ca^{2+}]_c$. Despite this receptor is rapidly desensitized by purines in vitro studies, experimental maneuvers using high concentrations of apyrase to scavenge the ADP and ATP released have revealed that the activation of $P2X_1$ receptors is involved in a number of platelet functions, including reversible shape change [37], modulation of collagen, adrenaline, thromboxane A2 and thrombin responses [38–41]. In addition, activation of $P2X_1$ receptors in human platelets potentiates the responses induced by occupation of P2Y receptors by ADP [42]. Furthermore, in vivo studies using $P2X_1$ deficient mice have revealed that the $P2X_1$ receptor is involved in thrombus formation although deficient animals display no significant changes in tail bleeding, platelet count or the surface expression of adhesion receptors (see [43]).

SMOCE, also known as non-capacitative Ca^{2+} entry or store-independent Ca^{2+} entry, is a mechanism for Ca^{2+} entry described in platelets and other cells and activated by second messengers generated following G protein-mediated activation of phospholipase C (PLC). PLC hydrolyzes the membrane constituent phosphatidylinositol 4,5-bisphosphate (PIP_2) into IP_3, which, in turn, releases Ca^{2+} from the intracellular stores, and diacylglycerol (DAG). Besides activating protein kinase C (PKC), DAG is a well known direct activator of non-capacitative Ca^{2+} channels. In human and mouse platelets Ca^{2+} entry induced by DAG has been demonstrated using the membrane-permeable analog 1-oleoyl-2-acetyl-sn-glycerol (OAG), which revealed the presence of DAG-sensitive Ca^{2+} permeable channels in these cells [44–46]. In addition, a PKC-dependent DAG-induced route for Ca^{2+} influx has been described in human platelets, which plays a relevant role in thrombin-induced platelet Ca^{2+} signals [47]. This mechanism has been revealed to involve the activation of the isoform 3 of the Na^+-Ca^{2+} exchanger followed by the release of dense granule content, including ATP, and subsequent $P2X_1$ receptor activation [48].

Among the second messenger-operated channels (SMOCs) there is a body of evidence supporting the role of TRPC6 in platelets. Platelets from TRPC6-deficient mice show a reduced or undetectable (depending on the mouse strain) Ca^{2+} influx in response to OAG [27, 45, 46]. The involvement of TRPC6 in store-independent Ca^{2+} entry in human platelets was highlighted by Hassock et al. in 2002 [49]. Furthermore, platelet treatment with neutralizing antibodies against TRPC6 resulted in attenuated OAG-evoked Ca^{2+} influx [44]. In the plasma membrane, TRPC6 has been found associated with TRPC3 in membrane microdomains independent of the lipid rafts [8]. Gating of TRPC6 has been found to require a tyrosine phosphorylation step since it necessitates the activation of members of the Src family of tyrosine kinases and is enhanced by inhibitors of tyrosine phosphatases [50]. We have also observed that TRPC6 channels might also been involved in the conduction of SOCE in human platelets; thus, platelet stimulation with thrombin or pharmacological induction of extensive Ca^{2+} store depletion results in the interaction between TRPC6, TRPC1 and the SOCE proteins Orai1 and STIM1, while, cell stimulation with OAG displaces TRPC6 from the SOCE signalplex and enhances Ca^{2+}-dependent association between TRPC6 and TRPC3 [51]. These findings are consistent with a more recent analysis of the protein complexes generated upon the activation of SMOCE and SOCE, which highlighted the dynamism of the Ca^{2+} entry complexes upon Ca^{2+} store depletion or the activation by OAG. In human platelets, Ca^{2+} store discharge leads to rapid and Ca^{2+} independent association of STIM1, STIM2, Orai1, Orai2, TRPC6 and TRPC1. In contrast, platelet treatment with OAG results in the association of Orai3 with TRPC3 [7].

Recent studies have reported the expression of TMEM16F in platelets, a Ca^{2+}-activated nonselective cation channel permeable to monovalent and divalent cations that plays an important role in the regulation of lipid scrambling and pro-coagulant activity of platelets [52, 53]. TMEM16F has been identified as the gene altered in patients with Scott Syndrome, a rare congenital bleeding disorder caused by impaired surface exposure of phosphatidylserine and platelet pro-coagulant activity [54].

In addition to Ca^{2+} influx conducted by ROCs and SMOCs, a mechanism for Ca^{2+} entry regulated by the intracellular Ca^{2+} stores has been described in platelets, as well as in other cells investigated. The so called store-operated Ca^{2+} entry (SOCE), also known as capacitative Ca^{2+} entry or store-mediated Ca^{2+} entry, was first described in platelets soon after the discovery of this mechanism by Putney [55]. In 1989, Sage and coworkers demonstrated the presence of SOCE in human platelets using Mn^{2+} as a surrogate for Ca^{2+} and analyzing its quenching effect of Fura-2 at the Ca^{2+} isosbestic point (360 nM), an excitation wavelength at which Fura-2 fluorescence properties are independent of changes in $[Ca^{2+}]_c$. The authors used Mn^{2+} to avoid interference with other Ca^{2+} transport mechanisms [56]. SOCE has been reported to be induced in platelets by physiological agonists, such as thrombin [56] or ADP [57], as well as by pharmacological agents that induce a decrease in the intraluminal Ca^{2+} concentration, such as the SERCA inhibitors, thapsigargin, 2,5-di-(tert-butyl)-1,4-benzohydroquinone and cyclopiazonic acid [58, 59], or the intraluminal Ca^{2+} chelator N,N,N′,N′-tetrakis(2-pyridylmethyl)ethylenediamine (TPEN) [51].

The initial studies of the mechanism underlying the communication between the filling state of the Ca^{2+} stores and the plasma membrane Ca^{2+} channels in human platelets revealed the involvement of different intracellular messengers and signaling pathways including a protein tyrosine phosphorylation-dependent step [60, 61], an arachidonic acid derivative [62], Ca^{2+}-independent phospholipase A2 [63] or the participation of cyclic nucleotides [32, 64]. The activation of SOCE in human platelets was found to be finely regulated by the remodeling of the actin cytoskeleton, which was suggested as a target of the previously mentioned signaling pathways. The cytosolic actin network supports the intracellular trafficking that allows the approach of portions of the Ca^{2+} stores to the cell membrane, while the cortical actin filament network acts as a negative clamp preventing constitutive activation of SOCE [65, 66] in a model that allows reversible interaction between elements in the membrane of the Ca^{2+} stores and the plasma membrane. According to this model, we have found in human platelets that the SNARE protein SNAP-25 is required for full SOCE activation, probably by leading the establishment of a tight contact between the plasma membrane and the Ca^{2+} stores [67]. According to this, a role for tyrosine kinases, Ras family proteins, cAMP as well as cGMP-dependent protein kinases as well as the cytochrome P450 metabolite 5,6-epoxyeicosatrienoic acid in the regulation of platelet actin cytoskeleton, and, subsequently SOCE, has been described, revealing the regulatory role of the actin cytoskeleton in the activation and maintenance of Ca^{2+} entry [68–74]. In addition, in human platelets, where the existence of two separate agonist-sensitive Ca^{2+} stores have been described: the dense tubular system and the acidic organelles [2, 3], two pathways for SOCE, activated by each Ca^{2+} store and differentially modulated by the actin cytoskeleton have been identified. First, the actin network plays a dual role in the regulation of SOCE mediated by depletion of the dense tubular system, as described above, both supporting the transport of portions of the Ca^{2+} store to the proximity of the plasma membrane to allow the interaction to occur, but preventing constitutive activation of Ca^{2+} entry in the absence of store depletion through the cortical cytoskeleton. On the other hand, SOCE mediated by depletion of the acidic stores is solely regulated by the cortical cytoskeleton, which acts as a physical barrier to prevent coupling between the Ca^{2+} stores and the plasma membrane under resting conditions. For both pathways, the cortical actin cytoskeleton must be reorganized when the stores are depleted to facilitate the interaction between the Ca^{2+} sensor in the stores and the store-operated channels (SOCs). By contrast, only the pathway activated by the dense tubular system requires reorganization of the cytosolic actin network [75]. The reason of this difference might be attributed to the cellular location of the Ca^{2+} stores in relation to the plasma membrane or the open canalicular system.

Following the identification of STIM1 as the ER Ca^{2+} sensor and a key player for the activation of SOCE in *Drosophila* S2 and different mammalian cell types [76–79], in 2006 we revealed the expression and involvement of the protein STIM1 in the activation of SOCE in human platelets [1]. The identification of STIM proteins, as well as the Ca^{2+}-permeable channels of the Orai and TRPC families represented a significant advance in the characterization of the mechanism underlying SOCE and raised the necessary reinterpretation of the previous data.

8.1.3 Role of STIM, Orai and TRPC Channels

STIM1 has been recognized as the Ca^{2+} sensor in the intracellular Ca^{2+} stores [76–78, 80]. Discharge of the Ca^{2+} stores by agonists via the generation of diffusible messengers, such as IP_3, is sensed by STIM1 via its EF-hand domain. STIM1 becomes activated as the intraluminal Ca^{2+} concentration drops, resulting in the formation of clusters of STIM1-SOCs complexes called "puncta" in regions where the membrane of the Ca^{2+} stores is in close contact with the plasma membrane. As described in detail in Chaps. 2 and 3, STIM1 is a type I transmembrane protein that exhibits intraluminal EF-hand and sterile α-motif (SAM) domains that, in response to a decrease in the intraluminal Ca^{2+} concentration, undergo oligomerization and conformational change. The cytosolic structure of STIM1 is characterized by the presence of three putative coiled-coil segments and a C-terminal lysine-rich region. The coiled-coil segments contain the region of STIM1 that is necessary for the activation of Orai1, identified as STIM–Orai-activating region (SOAR; residues 344–442) [81], Orai-activating STIM fragment (OASF; residues 233–450/474) [82], CRAC-activating domain (CAD, residues 342–448) [83] and CC boundary nine fragment (residues 339–446) [84]. A modulatory domain between amino acids 474 and 485 has been reported to provide a negative feedback signal to Ca^{2+} entry by triggering fast Ca^{2+}-dependent inactivation of Orai channels [85]. Finally, a more distal region between amino acids 448 and 530 has been identified as a C-terminal inhibitory domain (CTID), which shows two lobes, the STIM1(448–490) lobe restricted the association of the regulatory protein SARAF to the STIM1 SOAR region, whereas the STIM1(490–530) lobe directed this association, thus modulating the slow Ca^{2+} -dependent inactivation of Orai channels [86].

Although the role of STIM1 as the intraluminal Ca^{2+} sensor in platelets and other cells is widely accepted, the nature and configuration of the SOC channels is still a matter of intense investigation. Current evidence in different cell types indicates that the channel Orai1 is involved in the conduction of the store-operated Ca^{2+} selective I_{CRAC} [87–89], the first identified store operated Ca^{2+} current [90]. In addition, there is a body of evidence supporting a role for different members of the TRPC subfamily in the conduction of the store-operated cationic current I_{SOC} [6, 91–95]. In 2006, Orai1 was identified as a SOC channel through whole-genome screening of *Drosophila* S2 cells and gene mapping in patients with hereditary severe combined immune deficiency (SCID) syndrome attributed to I_{CRAC} dysfunction [87, 96, 97]. The Orai family includes three human homologs, Orai1, Orai2 and Orai3 (see previous chapters for an extensive review of the architecture of the channel). The crystal structure of Orai from *Drosophila melanogaster* has revealed that the Ca^{2+} channel is composed of an hexameric assembly of Orai subunits arranged around a central ion pore delimited by the first transmembrane region (M1). According to this model, STIM1 interacts with the first and fourth transmembrane domains widening the pore by outward dilation of the M1 region and allowing Ca^{2+} to move [98].

The second type of channels that has been found to be involved in SOCE belongs to the TRPC subfamily. TRP proteins form cation channels identified in 1989 in the *trp* mutant of *Drosophila* [99], which exhibits a transient, rather than sustained, light-sensitive photoreceptor potential due to Na^+ and Ca^{2+} influx through the trp and trpl channels [100]. In 1995, the first mammalian TRP protein was identified, the canonical transient receptor potential protein-1 (TRPC1), both in human [101, 102] and mouse [103]. TRP proteins are classified into seven subfamilies, four groups closely related to *Drosophila* TRP (TRPC, TRPV, TRPA and TRPM), two subfamilies more distantly related (TRPP and TRPML), and the TRPN group that is only expressed in fish, flies and worms [104]. The TRPC subfamily comprises seven members (TRPC1-TRPC7) and, since the identification of mammalian TRPC1, the TRPC channels have been presented as SOC candidates in a variety of cell types, including human platelets [6, 91–95, 105], although their role in SOCE has not been demonstrated in all the cell types and cellular models investigated [106]. The general structure of TRPC channels has been thoroughly described in previous chapters of the present book. Among other mechanisms of gating, TRPC channels have been shown to be activated by store-depletion via STIM1. Mutational analyses have revealed that the C-terminal STIM1 lysine-rich domain directly gates TRPC channels by interacting with two conserved C-terminal negative charged residues [107]. Neutralization or reversing the negative charges D639 and D640 in TRPC1 and D697 and D698 in TRPC3 resulted in inactive channels. Similarly, STIM1 mutants where any positive charge of the polylysine domain, especially K684 and K685, were neutralized inhibited the activity of TRPC1 [108]. Further studies have revealed that the interaction of STIM1 with TRPC1 and TRPC3 also involves the interaction between the SOAR region of STIM1 and the N- and C-terminal coiled-coil domains of the TRPC channels [109]. Therefore, both STIM1 regions, the SOAR domain and the polylysine-rich motif, are required for the interaction and gating of TRPC channels by STIM1.

As mentioned above, STIM1 was described in human platelets in 2006 [1] and, 1 year later, its murine homolog was reported in mouse platelets [22]. By contrast, STIM2 was first reported in mouse platelets in 2010 [23] and the first analysis of the expression of STIM2 in human platelets was reported in 2011 [2]. The functional role of STIM1 in platelet SOCE has been demonstrated by different means. In mice with STIM1-deficient platelets a marked defect in SOCE and agonist-stimulated Ca^{2+} mobilization has been described, together with an impaired platelet activation and thrombus formation under flow in vitro [110]. The evidence supporting a role for STIM1 in SOCE in human platelets are more indirect mostly due to obvious technical limitations. STIM1 has been found to be highly expressed in human platelets and their precursors [22]. Furthermore, interference with STIM1 function by introduction of an antibody directed toward the N-termini of STIM1 has been found to attenuate SOCE in these cells [1]. These findings cannot be attributed to side effects due to the preservatives present in the antibody solution since introduction of a non-specific IgG of the same origin of the anti-STIM1 antibody and containing the same preservatives did not reproduce the effects observed with the anti-STIM1 antibody [111].

In human platelets, STIM1 has been found to be expressed both in the membranes of the dense tubular system and the acidic Ca^{2+} stores [1, 2] as well as in the plasma membrane [112]. STIM1 associates with endogenously expressed Orai1 as well as TRPC1 and TRPC6 upon Ca^{2+} store depletion in human platelets in a Ca^{2+}-independent manner [1, 6, 51]. A major role for Orai1 in SOCE has been reported in human platelets, as well as human megakaryocytes and human megakaryoblastic cell lines on the base of the expression level, as compared to other Ca^{2+}-permeable channels such as TRPC1 or TRPC6, and the electrophysiological recordings of I_{CRAC} in megakaryocytes [4]. Orai1 is also strongly expressed in mouse platelets and further studies in Orai1-deficient mice reported that platelets from Orai1$^{-/-}$ mice show impaired SOCE and defective thrombus formation [24]. In addition, a knock-in mice expressing a loss-of-function mutant of Orai1 (Orai1^{R93W}) revealed the functional role of Orai1 in the activation of SOCE and agonist-induced Ca^{2+} entry in mouse platelets [113].

As reported in different cell types [114–116], in human platelets we have found that STIM1 interacts with Orai1 and members of the TRPC subfamily, such as TRPC1 and TRPC3, forming a SOCE signalplex associated to lipid raft domains [6, 51, 105]. Thus, disruption of the association between Orai1 and TRPC1 results in displacement of TRPC1 from the STIM1-Orai1 complex and loss of responsiveness to store depletion [6].

The involvement of TRPC1 in SOCE was first suggested in human platelets in 2000, when we found that endogenously expressed TRPC1 co-immunoprecipitates with the type II IP_3R upon Ca^{2+} store depletion, but not in resting cells. The role of TRPC1 in SOCE in human platelets was supported by studies where interference with TRPC1 function by treatment with a specific antibody which recognizes the sequence TRPC1$^{557-571}$, located extracellularly in the pore-forming region, significantly inhibits SOCE in human platelets [91]. The role of TRPC1 in SOCE and platelet function has been challenged by studies reporting that platelets from

TRPC1$^{-/-}$ mice show fully intact SOCE compared to wild-type animals, and normal platelet function in vitro and in vivo [25]. The controversy between the data observed in human and mouse platelets deserves further studies. This might be attributed to interspecific differences or to a redundant function of TRPC channels in platelet physiology. More recently, we have reported that TRPC1 plays an important role in the activation of SOCE in these cells in the human megakaryoblastic cell line MEG-01, where TRPC1 expression was silenced using shRNA, an approach that avoids the development of compensatory mechanisms that might interfere with the observations in knock-out animal models [117].

The putative role of TRPC1 in SOCE in human platelets still remains to be clarified. Following the identification of the coupling between TRPC1 and the type II IP_3R, we found many characteristics of the coupling that shows parallelism with SOCE: (1) the interaction between both proteins was activated by depletion of the intracellular Ca^{2+} stores, and was reversed by refilling of the stores [118]; (2) both SOCE and the coupling between TRPC1 and the type II IP_3R were impaired by treatment of platelets with the IP_3R antagonist xestospongin C [118]; (3) stabilization of the membrane cytoskeleton using jasplakinolide prevented the coupling and impaired SOCE, indicating, as mentioned above, that the actin filaments at the cell periphery act as a negative clamp for both events. In addition, the cytosolic actin network plays a positive role, since disruption of the actin network inhibited the coupling and SOCE [65, 118]; and (4) immunophilins, such as FKBP52, are required both for the coupling between TRPC1 and the type II IP_3R and the activation of SOCE in human platelets. The parallelism between both phenomena suggested that a de novo association between TRPC1 in the plasma membrane and the IP_3R type II in the intracellular stores might play a role in SOCE in these cells. To further test this hypothesis we introduced into platelets a peptide corresponding to type II $IP_3R^{317-334}$, which impaired the association of the IP_3R with TRPC1. This maneuver revealed that the association between TRPC1 and the type II IP_3R is important for the maintenance, but not the initial stage, of SOCE [119]. Whether TRPC1, through its association with STIM1 and Orai1, is also involved in early stages of the activation of SOCE in human platelets deserves further analysis.

The role of other TRPC proteins, such as TRPC6, in platelet SOCE, has also been demonstrated. TRPC6 has been reported to participate in SOCE in human platelets by using a neutralizing antibody against the C-terminal region of TRPC6, which reduced both TG-induced Ca^{2+} and Mn^{2+} influx, as well as OAG-stimulated Ca^{2+} entry [44]. The role of TRPC6 in Ca^{2+} homeostasis in platelets is described in more detail in Chap. 10.

It is noteworthy to mention that a pathway for Ca^{2+} entry secondary to agonist stimulation has been reported in human platelets. Harper and coworkers have found that Na^+ entry through TRPC channels results in the activation of the Na^+/Ca^{2+} exchanger working in reverse mode [120]. The Na^+/Ca^{2+} exchanger-3 has been shown to be the predominant isoform in human platelets by using a proteomic screen of the platelet membrane [121] and Western blotting [48]. Ca^{2+} entry via Na^+/Ca^{2+} exchange has been found to potentiate both SOCE and $P2X_1$-mediated Ca^{2+} entry in human platelets [48, 120, 122], a mechanism that involves the increase in the concentration of Ca^{2+} in the pericellular region, which is recycled back into the cytosol to support cellular events such as dense granule secretion [122, 123].

8.2 Role of SOCE in Thrombosis and Hemostasis

8.2.1 The Importance of Ca^{2+} in Thrombosis and Hemostasis

Hemostasis refers to the property of the vascular system to stop blood loss after vascular damage. Besides platelet function, the early phase comprises the activation of the vascular endothelial cell layer located at sites of injury, local vessel wall contraction, the aggregation of platelets and other blood cells to form a blood clot and seal the broken area, and the participation of many plasma proteins known as the coagulation cascade, which act as thrombotic factors to stimulate and synergize the process (Fig. 8.2). The blood clot formed inside a vessel during this process is known as thrombus [124]. Once sealed the broken area, later phases of hemostasis take place to reestablish the initial conditions, for instance wound healing, thrombus clearance and the final reestablishment of the blood flow at the repaired area [125]. Non-mechanical agents such as chemical, pathological and inflammation are also able to initiate hemostasis and therefore thrombus formation without compromising the integrity of the vessel [126]. Interestingly, the early phase of hemostasis is constantly counter-regulated by many inhibitory agents and anti-thrombotic factors in order to avoid the aberrant and uncontrolled development of the process [127].

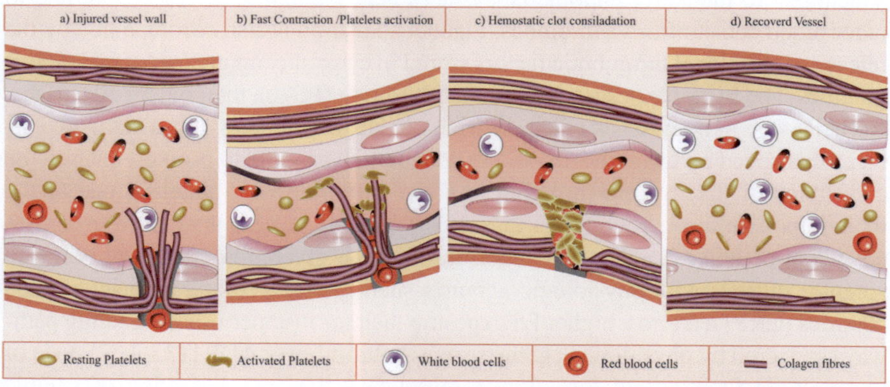

Fig. 8.2 Role of platelets in hemostasis: (**a**) Injuries in the walls of blood vessels lead to the exposition of collagen fibres in the exposed wall and the release of thromboregulatory molecules from the vascular cells, which will start platelets activation. (**b**) Fast contraction of the damaged vessel reduces the blood flow, thus, blood loss. Collagen-activated platelets will recruit and trigger new platelet activation to form, in addition to red blood cells and collagen fibres, a network, the clot, whose function is to stop bleeding. (**c**) When the hemostatic clot is consolidated and the bleeding is stopped, new vascular endothelial cells will be formed, and the clot components are reabsorbed. (**d**) The vessel recovers its original diameter and the normal blood flow is restored

These anti-thrombotic factors are also constantly released to the blood stream to prevent an unwanted spontaneous initiation of the hemostatic process in the intact vascular system. However, an abnormal uncontrolled thrombus formation can lead to a pathologic condition known as thrombosis. Thrombosis (-osis [-ō-sis], from greek, "pathologic process") is medically defined as the partial or complete occlusion of a blood vessel by a thrombus generated at the site of injury, also referred to as thromboembolism when the thrombus was generated elsewhere. Thus, the concept of thrombosis also extends to the pathologic initiation of the otherwise normal processes triggered during the early phase of hemostasis. The importance of thrombosis lies in how frequent it appears in many cardiovascular diseases, and depending on the occluded vessel and the affected organ or area, how easily it spontaneously develops into a life-threatening situation due to a dangerous reduction of the blood flow [126].

Blood platelets are considered the central cellular player of arterial thrombosis and hemostasis. Their decreased number or absence in the circulation leads to uncontrolled hemorrhage [128, 129], while abnormal hyperaggregability observed in many diseases, for instance diabetes mellitus, often leads to thrombosis [18, 126, 130]. In contrast, they seem to be less important in cases of deep venous thrombosis [131]. Circulating small rounded platelets are "inactive" and do not show adherent properties in the absence of damage. They circulate freely along the vascular system, detecting spontaneously released thrombotic and anti-thrombotic factors present in the blood plasma, and analyzing their activatory and inhibitory signals. Thus, platelets integrate these signals and initiate or not their activation if the balance of thrombotic signals exceeds the anti-thrombotic ones [132–134]. Platelet activation comprises cell shape changes and increased adhesive properties, promoting their attachment to the active endothelial surfaces near the damaged vessel wall and to other blood cells to form a thrombus. Platelet activation is also accompanied by the release of additional thrombotic factors stored in cytosolic organelles (α- and dense-granules) and by the exposure of phosphatidylserine (PS) on their outer cell surface, a crucial step in hemostasis [135]. PS-rich platelet membranes act as catalytic surfaces that speed up the activation of the coagulation cascade to promote mainly the conversion of inactive pro-thrombin to active thrombin near platelets, one of the most important thrombotic factor in hemostasis. This process is referred to as procoagulant activity. Platelet stimulation through the glycoprotein (GP)VI collagen receptor (GPVI) with physiologic agonists such as collagen and collagen-related peptides (CRP) leads to a higher PS exposure than for instance other signaling pathways triggered by physiologic agonists such as thrombin or ADP [135–137]. One of the multiple functions of thrombin is to catalyze the polymerization of soluble fibrinogen to insoluble fibrin fibers and thus stabilizing the growing thrombus [135].

New platelet functions in physiology are constantly reported [125, 138], and different platelet subpopulations displaying different specialized functions have been described [139]. However, the function of SOCE in platelets has mainly been studied during hemostasis so far.

The cellular machinery involved in platelet activation comprises Ca^{2+}-dependent and Ca^{2+}-independent signaling cascades. Raises in $[Ca^{2+}]_c$ are considered a central event in platelet physiology [140–142] as they experiment changes in $[Ca^{2+}]_c$ after attachment to the injured area and during the process of aggregation and thrombus formation [143]. In fact, removal of extracellular Ca^{2+} by chelating agents such as sodium citrate, potassium oxalate or ethylenediaminetetraacetic acid (EDTA) is a classical way to impair platelet function and prevent spontaneous aggregation [144, 145]. A growing body of evidence sustains the idea that SOCE and ROCE are the main pathways of Ca^{2+} entry into platelets, and that they might have a prominent function in their procoagulant activity during thrombosis and hemostasis [23, 24, 26, 45, 110, 113, 146, 147].

8.2.2 Role of STIM and Orai in Thrombosis and Hemostasis

As mentioned above, members of the STIM and Orai family are considered the main molecular components of SOCE. In the case of STIM proteins, there is no available data about STIM2 function for instance in humans, but the analysis of genetically modified murine models suggested a redundant or unclear role in platelet physiology and hemostasis [23, 148]. In contrast, recent evidences revealed that STIM1 plays a determinant role in human platelet physiology and hemostasis when STIM1 function is abnormally increased, while its absence has only a mild impact in hemostasis. Thus, it has been reported that STIM1 mutant proteins exhibiting constitutive or enhanced activity are probably responsible of the Stormorken syndrome in humans, causing an excessive Ca^{2+} entry into platelets and leading to serious problems, such as premature platelet activation, constitutive PS exposure in their outer cell surface, thrombocytopenia and bleeding disorders [149–151]. The heterozygous expression of similar STIM1 mutated forms has also been associated to the York Platelet syndrome (YPS) in humans, characterized by thrombocytopenia, defective Ca^{2+} storage inside platelet granules and aberrant morphology of their intracellular organelles [152]. In contrast, the lack of STIM1 in humans leads to a mild bleeding prolongation despite the decreased SOCE observed in their platelets, or their defect in α- and dense-granule secretion upon thrombin stimulation, or the slightly impaired platelet aggregation in the presence of epinephrine [153]. Interestingly, the analysis of mice lacking STIM1 only in blood cells suggested a more prominent role in arterial thrombosis than in hemostasis, since these animals do not develop vessel occlusion in the absence of STIM1 when they are challenged to develop thrombosis [110, 147]. As in humans, these animals show mild, almost unaltered bleeding times after mechanical injury, indicative of a redundant role in hemostasis. However, the induction of collagen- or thrombin-dependent thrombosis by $FeCl_3^-$ or laser-induced arterial damage in these animals generates unstable micro-thrombi, which are unable to firmly aggregate and form a single stable occluding thrombus. As a consequence, these genetically modified mice are protected from arterial thrombosis and ischemic brain infarction when STIM1 function is absent despite their normal hemostasis [110, 147]. Similar resistance to thrombosis has been reported in mouse bearing a constitutively active STIM1 mutant [22].

What is therefore the molecular mechanism underlying such different roles of STIM1 in thrombosis or hemostasis? Unfortunately, it is controversial, in part for the not completely understood models of thrombosis used in these studies [154, 155], or the genetic variability existing between all mouse strains used [110, 147]. What is clear is that most of the important physiological agonists, such as thrombin, ADP and collagen, trigger SOCE to mediate platelet function and aggregation [142]. STIM1-deficient platelets isolated from plasma show abrogated SOCE in fact, but surprisingly they respond normally to most physiologic agonists except to collagen [110, 147], probably due to a defective GPVI-dependent Ca^{2+} signaling [22, 113]. The reason of these different platelet responses to agonists in the absence of STIM1 is not clear, but might be explained in part by the different PLC isoforms activated by them [156, 157]. As a consequence to their defective collagen responses, STIM1-deficient platelets show diminished GPVI-induced activation of the fibrinogen and fibrin receptor $\alpha_{IIb}\beta_3$ integrin, defective PS exposure on their outer cell surface as well as in GPVI-induced procoagulant activity (thrombin generation near platelets). As a result, isolated platelets lacking STIM1 do not properly aggregate in response to collagen or CRP. In contrast, it is not clear whether they are able to attach and form stable thrombi in response to collagen [23, 110]. Conditional transgenic mice lacking STIM1 in platelets also display a delayed and reduced fibrin generation in a laser injury thrombosis model, suggesting also a defective fibrin generation as a consequence of their defective PS exposure [147]. In summary, STIM1-dependent SOCE seems to play a major role in GPVI/collagen-induced Ca^{2+} signaling, collagen-induced PS exposure and GPVI-dependent thrombin generation. But, if STIM1-deficient platelets are defective in GPVI-dependent signaling and do not respond properly to collagen, why do mice display normal hemostasis in the absence of STIM1? Interestingly, studies carried by Gilio et al. suggested that the presence of the coagulation cascade or thrombin compensates these GPVI-dependent signaling deficiencies observed in STIM1-deficient platelets and restores platelet function [23], probably by the compensation of their deficient SOCE through an activation of additional members of the TRPC family [23, 26, 158]. Therefore, the combined activity of both thrombin and collagen might compensate the defective function observed in STIM1-deficient platelets challenged only with collagen. Since tissue factor is one of the main initiators of the coagulation cascade [135], Gilio et al. suggested a mechanism that might explain the phenotype observed in mice lacking STIM1 in blood cells. These mice are protected from collagen-induced arterial thrombosis because the absence of STIM1 abrogates platelet adhesion to collagen, PS exposure and GPVI-induced thrombin generation, but cannot be compensated by the coagulation due to the limited tissue factor present in these vessels and the subsequent minimal amount of thrombin generated. However, tissue factor is abundantly exposed in wounds, and can compensate the deficient GPVI-induced signaling observed in STIM1-deficient platelets leading to an almost normal hemostasis [23].

However, a more recent study carried by Ahmad et al. in mice lacking STIM1 in platelets argued against this hypothesis, since they observed impaired thrombi formation and vessel occlusion also in a model of thrombin-dependent thrombosis by laser injury [147], in the presence of tissue factor-dependent thrombin generation

[159]. They also observed deficient PS exposure, but normal thrombus formation in response to collagen, in contrast to Gilio et al. Thus, Ahmad et al. concluded that their reported phenotype cannot be explained by a defective collagen adhesion, and it must be attributable to a defective PS exposure and limited thrombin generation in response to collagen, even in the presence of coagulation. Certainly, Ahmad et al. showed that the presence of an specific agonist of the thrombin PAR4 receptor, PAR4p, cannot compensate the defective PS exposure observed in STIM1-deficient platelets challenged by collagen [147], concluding that coagulation might not compensate the absence of STIM1 in platelet function as Gilio et al. proposed [23]. However, Ahmad et al. used a specific agonist of PAR4 [160] for that study, in contrast to the thrombin used by Gilio et al., which has a broader activity than PAR4p. Thus, the comparability between both assays and the conclusion taken must be carefully considered. In summary, Ahmad et al. proposed a more prominent function of STIM1 in PS exposure and procoagulant activity in platelets than in platelet adhesion as Gilio et al. suggested [23]. The appearance of later studies questioning the idea of $FeCl_3$-induced arterial injury as a collagen-induced model of thrombosis increases this discrepancy [154, 155]. Certainly, these data obtained in mice must be carefully considered due to the differences observed, as compared to humans or among mouse strains, in platelet aggregation in response to collagen or granule secretion in response to thrombin [23, 110, 147, 153], indicating possible interspecific differences in certain aspects of platelet function and hemostasis.

In the case of members of the Orai family, it has been reported that Orai1 mutant proteins displaying enhanced activity lead to Stormorken-like phenotype in humans, but interestingly they lack thrombosis or altered hemostasis [150]. Humans carrying inactive Orai1 mutants suffer from thrombocytopenia, but it seems to be more an autoimmune problem than a defect in hemostatic mechanisms, indicating the overall data a minor function of Orai1 in human hemostasis [161]. In contrast, similar alterations in thrombosis and hemostasis to those found in STIM1-deficient mice have been observed in mice lacking Orai1 in blood cells, indicating that those molecules work together in same signaling pathways as expected, and suggesting that SOCE might play a more prominent role in thrombosis than in hemostasis [23, 24, 113]. The role of other Orai isoforms expressed in platelets [7, 24] in thrombosis and hemostasis has not been addressed so far. Interestingly, the fact that both STIM1 and Orai1, and thus SOCE, display such prominent function in thrombus formation and stabilization but not in hemostasis after hemorrhage in mice, suggests that therapies directed against SOCE and these molecules during ischemic cardiovascular and cerebrovascular events could significantly improve traditional anti-thrombotic therapies [162].

8.2.3 Role of TRPC Channels in Thrombosis and Hemostasis

The available data about TRPC function has been obtained from murine models lacking these proteins, since there is no study until date about thrombosis and hemostasis performed in human patients carrying mutated versions of TRPC proteins.

However, there is evidence that genetically-based increase in TRPC6 expression is associated to idiopathic pulmonary arterial hypertension, a disease that is caused by thrombosis among other factors [163]. Thus, increase in TRPC6 expression might be associated to thrombosis in humans. In mice, while the function of TRPC1 seems to be redundant [25], the role of TRPC6 in thrombosis and hemostasis is controversial. Studies reported both normal [45] or prolonged bleedings after mechanical vessel injury and altered thrombus formation and hemostasis in TRPC6-deficient mice [164]. A recent report suggested that, in contrast to previously thought, TRPC3 is expressed in murine platelets and cooperates synergistically with TRPC6 to regulate PS exposure in platelets, and thus their procoagulant activity [26]. The role of TRPC6 in platelet Ca^{2+} homeostasis is also unclear, since it has been reported none, mild or severe reduction of ROCE in platelets isolated from different mouse strains lacking TRPC6 and stimulated with OAG [26, 45, 46]. On the other hand, the sensitivity of TRPC6 [46] and Orai1 [165] to pH might explain the decreased SOCE and the impaired platelet aggregation observed in acidic conditions [166] and in pathological situations where the local pH is very low, for instance during internal hemorrhages along the digestive tract [167] or in cases of massive prolonged bleedings [168]. Finally, it has been suggested that the increased activity of TRPC-dependent ROCE could be an important mechanism for the hyperreactive platelet response and the enhanced thrombosis observed in type 2 diabetic patients. Increased TRPC6-dependent ROCE but decreased Orai1/STIM1-dependent SOCE has been found in platelets isolated from these patients, despite the increased expression of Orai1/STIM1 [169–171], which might be explained because the TRPC6/phosphatidylinositol 3-kinase interaction seems to be favored, while STIM1/Orai1 or STIM1/TRPC interaction seems to be decreased under this pathologic condition. As a consequence of these interesting results, some research groups are considering the importance of TRPC agonists and blockers as useful pharmacological tools for antithrombotic therapies [172].

Acknowledgements This work was supported by MINECO (Grants BFU2013-45564-C2-1-P and BFU2013-45564-C2-2-P) and Junta de Extremadura-FEDER (GR15029). A. B-E is supported from UPF "MINECO – Plan Nacional de I+D+i(2008-2012) – SAF 2012-38140".

References

1. Lopez JJ, Salido GM, Pariente JA, Rosado JA (2006) Interaction of STIM1 with endogenously expressed human canonical TRP1 upon depletion of intracellular Ca^{2+} stores. J Biol Chem 281(38):28254–28264
2. Zbidi H, Jardin I, Woodard GE, Lopez JJ, Berna-Erro A, Salido GM, Rosado JA (2011) STIM1 and STIM2 are located in the acidic Ca^{2+} stores and associates with Orai1 upon depletion of the acidic stores in human platelets. J Biol Chem 286(14):12257–12270
3. Lopez JJ, Camello-Almaraz C, Pariente JA, Salido GM, Rosado JA (2005) Ca^{2+} accumulation into acidic organelles mediated by Ca^{2+}- and vacuolar H^+-ATPases in human platelets. Biochem J 390(Pt 1):243–252

4. Tolhurst G, Carter RN, Amisten S, Holdich JP, Erlinge D, Mahaut-Smith MP (2008) Expression profiling and electrophysiological studies suggest a major role for Orai1 in the store-operated Ca2+ influx pathway of platelets and megakaryocytes. Platelets 19(4):308–313
5. Redondo PC, Jardin I, Lopez JJ, Salido GM, Rosado JA (2008) Intracellular Ca^{2+} store depletion induces the formation of macromolecular complexes involving hTRPC1, hTRPC6, the type II IP_3 receptor and SERCA3 in human platelets. Biochim Biophys Acta 1783(6):1163–1176
6. Jardin I, Lopez JJ, Salido GM, Rosado JA (2008) Orai1 mediates the interaction between STIM1 and hTRPC1 and regulates the mode of activation of hTRPC1-forming Ca^{2+} channels. J Biol Chem 283(37):25296–25304
7. Berna-Erro A, Galan C, Dionisio N, Gomez LJ, Salido GM, Rosado JA (2012) Capacitative and non-capacitative signaling complexes in human platelets. Biochim Biophys Acta 1823(8):1242–1251
8. Brownlow SL, Sage SO (2005) Transient receptor potential protein subunit assembly and membrane distribution in human platelets. Thromb Haemost 94(4):839–845
9. Harper AG, Brownlow SL, Sage SO (2008) A role for TRPV1 in agonist-evoked activation of human platelets. J Thromb Haemost 7(2):330–338
10. Albarran L, Lopez JJ, Dionisio N, Smani T, Salido GM, Rosado JA (2013) Transient receptor potential ankyrin-1 (TRPA1) modulates store-operated Ca(2+) entry by regulation of STIM1-Orai1 association. Biochim Biophys Acta 1833(12):3025–3034
11. Brass LF, Joseph SK (1985) A role for inositol triphosphate in intracellular Ca2+ mobilization and granule secretion in platelets. J Biol Chem 260(28):15172–15179
12. Lopez JJ, Redondo PC, Salido GM, Pariente JA, Rosado JA (2006) Two distinct Ca^{2+} compartments show differential sensitivity to thrombin, ADP and vasopressin in human platelets. Cell Signal 18(3):373–381
13. El-Daher SS, Patel Y, Siddiqua A, Hassock S, Edmunds S, Maddison B, Patel G, Goulding D, Lupu F, Wojcikiewicz RJ, Authi KS (2000) Distinct localization and function of (1,4,5) IP(3) receptor subtypes and the (1,3,4,5)IP(4) receptor GAP1(IP4BP) in highly purified human platelet membranes. Blood 95(11):3412–3422
14. Dionisio N, Albarran L, Lopez JJ, Berna-Erro A, Salido GM, Bobe R, Rosado JA (2011) Acidic NAADP-releasable Ca(2+) compartments in the megakaryoblastic cell line MEG01. Biochim Biophys Acta 1813(8):1483–1494
15. Lopez JJ, Dionisio N, Berna-Erro A, Galan C, Salido GM, Rosado JA (2012) Two-pore channel 2 (TPC2) modulates store-operated Ca(2+) entry. Biochim Biophys Acta 1823(10):1976–1983
16. Bobe R, Bredoux R, Wuytack F, Quarck R, Kovacs T, Papp B, Corvazier E, Magnier C, Enouf J (1994) The rat platelet 97-kDa Ca^{2+}ATPase isoform is the sarcoendoplasmic reticulum Ca^{2+}ATPase 3 protein. J Biol Chem 269(2):1417–1424
17. Dean WL, Chen D, Brandt PC, Vanaman TC (1997) Regulation of platelet plasma membrane Ca^{2+}-ATPase by cAMP-dependent and tyrosine phosphorylation. J Biol Chem 272(24):15113–15119
18. Rosado JA, Saavedra FR, Redondo PC, Hernandez-Cruz JM, Salido GM, Pariente JA (2004) Reduced plasma membrane Ca^{2+}-ATPase function in platelets from patients with non-insulin-dependent diabetes mellitus. Haematologica 89(9):1142–1144
19. Bobe R, Bredoux R, Corvazier E, Lacabaratz-Porret C, Martin V, Kovacs T, Enouf J (2005) How many Ca^{2+}-ATPase isoforms are expressed in a cell type? A growing family of membrane proteins illustrated by studies in platelets. Platelets 16(3-4):133–150
20. Valant PA, Adjei PN, Haynes DH (1992) Rapid Ca2+ extrusion via the Na+/Ca2+ exchanger of the human platelet. J Membr Biol 130(1):63–82
21. Kimura M, Aviv A, Reeves JP (1993) K(+)-dependent Na+/Ca2+ exchange in human platelets. J Biol Chem 268(10):6874–6877

22. Grosse J, Braun A, Varga-Szabo D, Beyersdorf N, Schneider B, Zeitlmann L, Hanke P, Schropp P, Muhlstedt S, Zorn C, Huber M, Schmittwolf C, Jagla W, Yu P, Kerkau T, Schulze H, Nehls M, Nieswandt B (2007) An EF hand mutation in Stim1 causes premature platelet activation and bleeding in mice. J Clin Invest 117(11):3540–3550
23. Gilio K, van Kruchten R, Braun A, Berna-Erro A, Feijge MA, Stegner D, van der Meijden PE, Kuijpers MJ, Varga-Szabo D, Heemskerk JW, Nieswandt B (2010) Roles of platelet STIM1 and Orai1 in glycoprotein VI- and thrombin-dependent procoagulant activity and thrombus formation. J Biol Chem 285:23629–23638
24. Braun A, Varga-Szabo D, Kleinschnitz C, Pleines I, Bender M, Austinat M, Bosl M, Stoll G, Nieswandt B (2009) Orai1 (CRACM1) is the platelet SOC channel and essential for pathological thrombus formation. Blood 113(9):2056–2063
25. Varga-Szabo D, Authi KS, Braun A, Bender M, Ambily A, Hassock SR, Gudermann T, Dietrich A, Nieswandt B (2008) Store-operated Ca^{2+} entry in platelets occurs independently of transient receptor potential (TRP) C1. Pflugers Arch 457:377–387
26. Harper MT, Londono JE, Quick K, Londono JC, Flockerzi V, Philipp SE, Birnbaumer L, Freichel M, Poole AW (2013) Transient receptor potential channels function as a coincidence signal detector mediating phosphatidylserine exposure. Sci Signal 6(281):ra50
27. Albarran L, Berna-Erro A, Dionisio N, Redondo PC, Lopez E, Lopez JJ, Salido GM, Brull Sabate JM, Rosado JA (2014) TRPC6 participates in the regulation of cytosolic basal calcium concentration in murine resting platelets. Biochim Biophys Acta 1843(4):789–796
28. Carter RN, Tolhurst G, Walmsley G, Vizuete-Forster M, Miller N, Mahaut-Smith MP (2006) Molecular and electrophysiological characterization of transient receptor potential ion channels in the primary murine megakaryocyte. J Physiol 576(Pt 1):151–162
29. Sage SO, Jarvis GE, Jardin I, Rosado JA, Harper AG (2014) The TRPV1 ion channel is expressed in human but not mouse platelets. Platelets 25(5):390–392
30. Poole AW, Heath MF, Sage SO, Evans RJ (1993) A method for loading equine platelets with the fluorescent calcium indicator Fura-2: ADP induces a rise in the cytosolic free calcium ion concentration. Equine Vet J 25(1):45–48
31. Li BY, Bai Y, Li WH (1998) Enhancement of ADP-induced aggregation by 5-HT in rabbit platelets. Zhong Yao Li Xue Bao 19(1):58–62
32. Heemskerk JW, Feijge MA, Sage SO, Walter U (1994) Indirect regulation of Ca2+ entry by cAMP-dependent and cGMP-dependent protein kinases and phospholipase C in rat platelets. Eur J Biochem 223(2):543–551
33. Atucha NM, Iyu D, Alcaraz A, Rosa V, Martinez-Prieto C, Ortiz MC, Rosado JA, Garcia-Estan J (2007) Altered calcium signalling in platelets from bile-duct-ligated rats. Clin Sci (Lond) 112(3):167–174
34. De Clerck F, Van Nueten JM (1983) Platelet-mediated vascular contractions. Inhibition by flunarizine, a calcium-entry blocker. Biochem Pharmacol 32(5):765–771
35. Salido GM, Sage SO, Rosado JA (2009) Biochemical and functional properties of the store-operated Ca2+ channels. Cell Signal 21(4):457–461
36. Mahaut-Smith MP, Ennion SJ, Rolf MG, Evans RJ (2000) ADP is not an agonist at P2X(1) receptors: evidence for separate receptors stimulated by ATP and ADP on human platelets. Br J Pharmacol 131(1):108–114
37. Rolf MG, Brearley CA, Mahaut-Smith MP (2001) Platelet shape change evoked by selective activation of P2X1 purinoceptors with alpha, beta-methylene ATP. Thromb Haemost 85(2):303–308
38. Oury C, Toth-Zsamboki E, Thys C, Tytgat J, Vermylen J, Hoylaerts MF (2001) The ATP-gated P2X1 ion channel acts as a positive regulator of platelet responses to collagen. Thromb Haemost 86(5):1264–1271
39. Erhardt JA, Pillarisetti K, Toomey JR (2003) Potentiation of platelet activation through the stimulation of P2X1 receptors. J Thromb Haemost 1(12):2626–2635
40. Erhardt JA, Toomey JR, Douglas SA, Johns DG (2006) P2X1 stimulation promotes thrombin receptor-mediated platelet aggregation. J Thromb Haemost 4(4):882–890
41. Fung CY, Cendana C, Farndale RW, Mahaut-Smith MP (2007) Primary and secondary agonists can use P2X(1) receptors as a major pathway to increase intracellular Ca(2+) in the human platelet. J Thromb Haemost 5(5):910–917

42. Vial C, Rolf MG, Mahaut-Smith MP, Evans RJ (2002) A study of P2X1 receptor function in murine megakaryocytes and human platelets reveals synergy with P2Y receptors. Br J Pharmacol 135(2):363–372
43. Mahaut-Smith MP, Jones S, Evans RJ (2011) The P2X1 receptor and platelet function. Purinergic Signal 7(3):341–356
44. Jardin I, Redondo PC, Salido GM, Rosado JA (2008) Phosphatidylinositol 4,5-bisphosphate enhances store-operated calcium entry through hTRPC6 channel in human platelets. Biochim Biophys Acta 1783(1):84–97
45. Ramanathan G, Gupta S, Thielmann I, Pleines I, Varga-Szabo D, May F, Mannhalter C, Dietrich A, Nieswandt B, Braun A (2012) Defective diacylglycerol-induced Ca2+ entry but normal agonist-induced activation responses in TRPC6-deficient mouse platelets. J Thromb Haemost 10(3):419–429
46. Berna-Erro A, Albarran L, Dionisio N, Redondo PC, Alonso N, Gomez LJ, Salido GM, Rosado JA (2014) The canonical transient receptor potential 6 (TRPC6) channel is sensitive to extracellular pH in mouse platelets. Blood Cells Mol Dis 52(2–3):108–115
47. Rosado JA, Sage SO (2000) Protein kinase C activates non-capacitative calcium entry in human platelets. J Physiol 529(Pt 1):159–169
48. Harper MT, Mason MJ, Sage SO, Harper AG (2010) Phorbol ester-evoked Ca2+ signaling in human platelets is via autocrine activation of P(2X1) receptors, not a novel non-capacitative Ca2+ entry. J Thromb Haemost 8(7):1604–1613
49. Hassock SR, Zhu MX, Trost C, Flockerzi V, Authi KS (2002) Expression and role of TRPC proteins in human platelets: evidence that TRPC6 forms the store-independent calcium entry channel. Blood 100(8):2801–2811
50. Harper MT, Sage SO (2010) Src family tyrosine kinases activate thrombin-induced non-capacitative cation entry in human platelets. Platelets 21(6):445–450
51. Jardin I, Gomez LJ, Salido GM, Rosado JA (2009) Dynamic interaction of hTRPC6 with the Orai1/STIM1 complex or hTRPC3 mediates its role in capacitative or non-capacitative Ca^{2+} entry pathways. Biochem J 420:267–276
52. Yang H, Kim A, David T, Palmer D, Jin T, Tien J, Huang F, Cheng T, Coughlin SR, Jan YN, Jan LY (2012) TMEM16F forms a Ca2+ -activated cation channel required for lipid scrambling in platelets during blood coagulation. Cell 151(1):111–122
53. Suzuki J, Fujii T, Imao T, Ishihara K, Kuba H, Nagata S (2013) Calcium-dependent phospholipid scramblase activity of TMEM16 protein family members. J Biol Chem 288(19):13305–13316
54. Suzuki J, Umeda M, Sims PJ, Nagata S (2010) Calcium-dependent phospholipid scrambling by TMEM16F. Nature 468(7325):834–838
55. Putney JW Jr (1986) A model for receptor-regulated calcium entry. Cell Calcium 7(1):1–12
56. Sage SO, Merritt JE, Hallam TJ, Rink TJ (1989) Receptor-mediated calcium entry in fura-2-loaded human platelets stimulated with ADP and thrombin. Dual-wavelengths studies with Mn^{2+}. Biochem J 258(3):923–926
57. Sage SO, Reast R, Rink TJ (1990) ADP evokes biphasic Ca2+ influx in fura-2-loaded human platelets. Evidence for Ca2+ entry regulated by the intracellular Ca2+ store. Biochem J 265(3):675–680
58. Huang SJ, Kwan CY (1998) Cyclopiazonic acid and thapsigargin induce platelet aggregation resulting from Ca2+ influx through Ca2+ store-activated Ca2+ -channels. Eur J Pharmacol 341(2–3):343–347
59. Cavallini L, Coassin M, Alexandre A (1995) Two classes of agonist-sensitive Ca^{2+} stores in platelets, as identified by their differential sensitivity to 2,5-di-(tert-butyl)-1,4-benzohydroquinone and thapsigargin. Biochem J 310(Pt 2):449–452
60. Sargeant P, Farndale RW, Sage SO (1993) The tyrosine kinase inhibitors methyl 2,5-dihydroxycinnamate and genistein reduce thrombin-evoked tyrosine phosphorylation and Ca2+ entry in human platelets. FEBS Lett 315(3):242–246
61. Sargeant P, Farndale RW, Sage SO (1993) ADP- and thapsigargin-evoked Ca^{2+} entry and protein-tyrosine phosphorylation are inhibited by the tyrosine kinase inhibitors genistein and methyl-2,5-dihydroxycinnamate in fura-2-loaded human platelets. J Biol Chem 268(24):18151–18156

62. Alonso MT, Sanchez A, Garcia-Sancho J (1990) Arachidonic acid-induced calcium influx in human platelets. Comparison with the effect of thrombin. Biochem J 272(2):435–443
63. Smani T, Zakharov SI, Leno E, Csutora P, Trepakova ES, Bolotina VM (2003) Ca2+ -independent phospholipase A2 is a novel determinant of store-operated Ca2+ entry. J Biol Chem 278(14):11909–11915
64. Geiger J, Nolte C, Butt E, Sage SO, Walter U (1992) Role of cGMP and cGMP-dependent protein kinase in nitrovasodilator inhibition of agonist-evoked calcium elevation in human platelets. Proc Natl Acad Sci U S A 89(3):1031–1035
65. Rosado JA, Jenner S, Sage SO (2000) A role for the actin cytoskeleton in the initiation and maintenance of store-mediated calcium entry in human platelets. Evidence for conformational coupling. J Biol Chem 275(11):7527–7533
66. Rosado JA, Sage SO (2000) The actin cytoskeleton in store-mediated calcium entry. J Physiol 526(Pt 2):221–229
67. Redondo PC, Harper AG, Salido GM, Pariente JA, Sage SO, Rosado JA (2004) A role for SNAP-25 but not VAMPs in store-mediated Ca2+ entry in human platelets. J Physiol 558(Pt 1):99–109
68. Rosado JA, Graves D, Sage SO (2000) Tyrosine kinases activate store-mediated Ca2+ entry in human platelets through the reorganization of the actin cytoskeleton. Biochem J 351(Pt 2):429–437
69. Rosado JA, Sage SO (2000) Farnesylcysteine analogues inhibit store-regulated Ca^{2+} entry in human platelets: evidence for involvement of small GTP-binding proteins and actin cytoskeleton. Biochem J 347(Pt 1):183–192
70. Rosado JA, Porras T, Conde M, Sage SO (2001) Cyclic nucleotides modulate store-mediated calcium entry through the activation of protein-tyrosine phosphatases and altered actin polymerization in human platelets. J Biol Chem 276(19):15666–15675
71. Redondo PC, Ben-Amor N, Salido GM, Bartegi A, Pariente JA, Rosado JA (2005) Ca2+ -independent activation of Bruton's tyrosine kinase is required for store-mediated Ca2+ entry in human platelets. Cell Signal 17(8):1011–1021
72. Ben Amor N, Redondo PC, Bartegi A, Pariente JA, Salido GM, Rosado JA (2006) A role for 5,6-epoxyeicosatrienoic acid in calcium entry by de novo conformational coupling in human platelets. J Physiol 570(Pt 2):309–323
73. Redondo PC, Harper MT, Rosado JA, Sage SO (2006) A role for cofilin in the activation of store-operated calcium entry by de novo conformational coupling in human platelets. Blood 107(3):973–979
74. Harper AG, Sage SO (2007) A role for the intracellular protease calpain in the activation of store-operated calcium entry in human platelets. Cell Calcium 41(2):169–178
75. Rosado JA, Lopez JJ, Harper AG, Harper MT, Redondo PC, Pariente JA, Sage SO, Salido GM (2004) Two pathways for store-mediated calcium entry differentially dependent on the actin cytoskeleton in human platelets. J Biol Chem 279(28):29231–29235
76. Roos J, DiGregorio PJ, Yeromin AV, Ohlsen K, Lioudyno M, Zhang S, Safrina O, Kozak JA, Wagner SL, Cahalan MD, Velicelebi G, Stauderman KA (2005) STIM1, an essential and conserved component of store-operated Ca^{2+} channel function. J Cell Biol 169(3):435–445
77. Zhang SL, Yu Y, Roos J, Kozak JA, Deerinck TJ, Ellisman MH, Stauderman KA, Cahalan MD (2005) STIM1 is a Ca^{2+} sensor that activates CRAC channels and migrates from the Ca^{2+} store to the plasma membrane. Nature 437(7060):902–905
78. Huang GN, Zeng W, Kim JY, Yuan JP, Han L, Muallem S, Worley PF (2006) STIM1 carboxyl-terminus activates native SOC, I(crac) and TRPC1 channels. Nat Cell Biol 8(9):1003–1010
79. Baba Y, Hayashi K, Fujii Y, Mizushima A, Watarai H, Wakamori M, Numaga T, Mori Y, Iino M, Hikida M, Kurosaki T (2006) Coupling of STIM1 to store-operated Ca^{2+} entry through its constitutive and inducible movement in the endoplasmic reticulum. Proc Natl Acad Sci U S A 103(45):16704–16709
80. Kurosaki T, Baba Y (2010) Ca2+ signaling and STIM1. Prog Biophys Mol Biol 103(1):51–58

81. Yuan JP, Zeng W, Dorwart MR, Choi YJ, Worley PF, Muallem S (2009) SOAR and the polybasic STIM1 domains gate and regulate Orai channels. Nat Cell Biol 11(3):337–343
82. Muik M, Fahrner M, Derler I, Schindl R, Bergsmann J, Frischauf I, Groschner K, Romanin C (2009) A cytosolic homomerization and a modulatory domain within STIM1 C terminus determine coupling to ORAI1 channels. J Biol Chem 284(13):8421–8426
83. Park CY, Hoover PJ, Mullins FM, Bachhawat P, Covington ED, Raunser S, Walz T, Garcia KC, Dolmetsch RE, Lewis RS (2009) STIM1 clusters and activates CRAC channels via direct binding of a cytosolic domain to Orai1. Cell 136(5):876–890
84. Kawasaki T, Lange I, Feske S (2009) A minimal regulatory domain in the C terminus of STIM1 binds to and activates ORAI1 CRAC channels. Biochem Biophys Res Commun 385(1):49–54
85. Derler I, Fahrner M, Muik M, Lackner B, Schindl R, Groschner K, Romanin C (2009) A Ca2+ Release-activated Ca2+ (CRAC) Modulatory Domain (CMD) within STIM1 mediates fast Ca2+ -dependent inactivation of ORAI1 channels. J Biol Chem 284(37):24933–24938
86. Jha A, Ahuja M, Maleth J, Moreno CM, Yuan JP, Kim MS, Muallem S (2013) The STIM1 CTID domain determines access of SARAF to SOAR to regulate Orai1 channel function. J Cell Biol 202(1):71–79
87. Feske S, Gwack Y, Prakriya M, Srikanth S, Puppel SH, Tanasa B, Hogan PG, Lewis RS, Daly M, Rao A (2006) A mutation in Orai1 causes immune deficiency by abrogating CRAC channel function. Nature 441(7090):179–185
88. Peinelt C, Vig M, Koomoa DL, Beck A, Nadler MJ, Koblan-Huberson M, Lis A, Fleig A, Penner R, Kinet JP (2006) Amplification of CRAC current by STIM1 and CRACM1 (Orai1). Nat Cell Biol 8(7):771–773
89. Prakriya M, Feske S, Gwack Y, Srikanth S, Rao A, Hogan PG (2006) Orai1 is an essential pore subunit of the CRAC channel. Nature 443(7108):230–233
90. Hoth M, Penner R (1992) Depletion of intracellular calcium stores activates a calcium current in mast cells. Nature 355(6358):353–356
91. Rosado JA, Brownlow SL, Sage SO (2002) Endogenously expressed Trp1 is involved in store-mediated Ca^{2+} entry by conformational coupling in human platelets. J Biol Chem 277(44):42157–42163
92. Liu X, Singh BB, Ambudkar IS (2003) TRPC1 is required for functional store-operated Ca2+ channels. Role of acidic amino acid residues in the S5-S6 region. J Biol Chem 278(13):11337–11343
93. Ambudkar IS, Ong HL, Liu X, Bandyopadhyay B, Cheng KT (2007) TRPC1: the link between functionally distinct store-operated calcium channels. Cell Calcium 42(2):213–223
94. Ong HL, Cheng KT, Liu X, Bandyopadhyay BC, Paria BC, Soboloff J, Pani B, Gwack Y, Srikanth S, Singh BB, Gill DL, Ambudkar IS (2007) Dynamic assembly of TRPC1-STIM1-Orai1 ternary complex is involved in store-operated calcium influx. Evidence for similarities in store-operated and calcium release-activated calcium channel components. J Biol Chem 282(12):9105–9116
95. Sampieri A, Zepeda A, Saldaña C, Salgado A, Vaca L (2008) STIM1 converts TRPC1 from a receptor-operated to a store-operated channel: moving TRPC1 in and out of lipid rafts. Cell Calcium 44(5):479–491
96. Vig M, Peinelt C, Beck A, Koomoa DL, Rabah D, Koblan-Huberson M, Kraft S, Turner H, Fleig A, Penner R, Kinet JP (2006) CRACM1 is a plasma membrane protein essential for store-operated Ca^{2+} entry. Science 312(5777):1220–1223
97. Zhang SL, Yeromin AV, Zhang XH, Yu Y, Safrina O, Penna A, Roos J, Stauderman KA, Cahalan MD (2006) Genome-wide RNAi screen of Ca^{2+} influx identifies genes that regulate Ca^{2+} release-activated Ca^{2+} channel activity. Proc Natl Acad Sci U S A 103(24):9357–9362
98. Hou X, Pedi L, Diver MM, Long SB (2012) Crystal structure of the calcium release-activated calcium channel Orai. Science 338(6112):1308–1313
99. Montell C, Rubin GM (1989) Molecular characterization of the Drosophila trp locus: a putative integral membrane protein required for phototransduction. Neuron 2(4):1313–1323

100. Hardie RC, Minke B (1992) The trp gene is essential for a light-activated Ca^{2+} channel in Drosophila photoreceptors. Neuron 8(4):643–651
101. Wes PD, Chevesich J, Jeromin A, Rosenberg C, Stetten G, Montell C (1995) TRPC1, a human homolog of a Drosophila store-operated channel. Proc Natl Acad Sci U S A 92(21):9652–9656
102. Zhu X, Chu PB, Peyton M, Birnbaumer L (1995) Molecular cloning of a widely expressed human homologue for the Drosophila trp gene. FEBS Lett 373(3):193–198
103. Petersen CC, Berridge MJ, Borgese MF, Bennett DL (1995) Putative capacitative calcium entry channels: expression of Drosophila trp and evidence for the existence of vertebrate homologues. Biochem J 311(Pt 1):41–44
104. Montell C, Birnbaumer L, Flockerzi V, Bindels RJ, Bruford EA, Caterina MJ, Clapham DE, Harteneck C, Heller S, Julius D, Kojima I, Mori Y, Penner R, Prawitt D, Scharenberg AM, Schultz G, Shimizu N, Zhu MX (2002) A unified nomenclature for the superfamily of TRP cation channels. Mol Cell 9(2):229–231
105. Jardin I, Salido GM, Rosado JA (2008) Role of lipid rafts in the interaction between hTRPC1, Orai1 and STIM1. Channels (Austin) 2(6):401–403
106. DeHaven WI, Jones BF, Petranka JG, Smyth JT, Tomita T, Bird GS, Putney JW Jr (2009) TRPC channels function independently of STIM1 and Orai1. J Physiol 587(Pt 10):2275–2298
107. Zeng W, Yuan JP, Kim MS, Choi YJ, Huang GN, Worley PF, Muallem S (2008) STIM1 gates TRPC channels, but not Orai1, by electrostatic interaction. Mol Cell 32(3):439–448
108. Yuan JP, Kim MS, Zeng W, Shin DM, Huang G, Worley PF, Muallem S (2009) TRPC channels as STIM1-regulated SOCs. Channels (Austin) 3(4):221–225
109. Lee KP, Choi S, Hong JH, Ahuja M, Graham S, Ma R, So I, Shin DM, Muallem S, Yuan JP (2014) Molecular determinants mediating gating of Transient Receptor Potential Canonical (TRPC) channels by stromal interaction molecule 1 (STIM1). J Biol Chem 289(10):6372–6382
110. Varga-Szabo D, Braun A, Kleinschnitz C, Bender M, Pleines I, Pham M, Renne T, Stoll G, Nieswandt B (2008) The calcium sensor STIM1 is an essential mediator of arterial thrombosis and ischemic brain infarction. J Exp Med 205(7):1583–1591
111. Lopez JJ, Salido G, Rosado JA (2012) SOCE and Ca2+ handling in platelet dysfunction. In: Groschner K, Graier WF, Romanin C (eds) Store-operated Ca2+ entry (SOCE) pathways. Springer, New York, pp 377–396
112. Jardin I, Lopez JJ, Redondo PC, Salido GM, Rosado JA (2009) Store-operated Ca^{2+} entry is sensitive to the extracellular Ca^{2+} concentration through plasma membrane STIM1. Biochim Biophys Acta 1793(10):1614–1622
113. Bergmeier W, Oh-Hora M, McCarl CA, Roden RC, Bray PF, Feske S (2009) R93W mutation in Orai1 causes impaired calcium influx in platelets. Blood 113(3):675–678
114. Kim MS, Zeng W, Yuan JP, Shin DM, Worley PF, Muallem S (2009) Native store-operated Ca2+ influx requires the channel function of Orai1 and TRPC1. J Biol Chem 284(15):9733–9741
115. Lee KP, Yuan JP, Hong JH, So I, Worley PF, Muallem S (2009) An endoplasmic reticulum/ plasma membrane junction: STIM1/Orai1/TRPCs. FEBS Lett 584:2022–2027
116. Liao Y, Erxleben C, Yildirim E, Abramowitz J, Armstrong DL, Birnbaumer L (2007) Orai proteins interact with TRPC channels and confer responsiveness to store depletion. Proc Natl Acad Sci U S A 104(11):4682–4687
117. Lopez E, Berna-Erro A, Salido GM, Rosado JA, Redondo PC (2013) FKBP52 is involved in the regulation of SOCE channels in the human platelets and MEG 01 cells. Biochim Biophys Acta 1833(3):652–662
118. Rosado JA, Sage SO (2001) Activation of store-mediated calcium entry by secretion-like coupling between the inositol 1,4,5-trisphosphate receptor type II and human transient receptor potential (hTrp1) channels in human platelets. Biochem J 356(Pt 1):191–198

119. Jardin I, Lopez JJ, Salido GM, Rosado JA (2008) Functional relevance of the de novo coupling between hTRPC1 and type II IP_3 receptor in store-operated Ca^{2+} entry in human platelets. Cell Signal 20(4):737–747
120. Harper AG, Sage SO (2007) A key role for reverse Na^+/Ca^{2+} exchange influenced by the actin cytoskeleton in store-operated Ca^{2+} entry in human platelets: evidence against the de novo conformational coupling hypothesis. Cell Calcium 42(6):606–617
121. Lewandrowski U, Wortelkamp S, Lohrig K, Zahedi RP, Wolters DA, Walter U, Sickmann A (2009) Platelet membrane proteomics: a novel repository for functional research. Blood 114(1):e10–e19
122. Harper AG, Mason MJ, Sage SO (2009) A key role for dense granule secretion in potentiation of the Ca2+ signal arising from store-operated calcium entry in human platelets. Cell Calcium 45(5):413–420
123. Sage SO, Pugh N, Farndale RW, Harper AG (2013) Pericellular Ca(2+) recycling potentiates thrombin-evoked Ca(2+) signals in human platelets. Physiol Rep 1(5):e00085
124. McFadyen JD, Jackson SP (2013) Differentiating haemostasis from thrombosis for therapeutic benefit. Thromb Haemost 110(5):859–867
125. Nurden AT (2011) Platelets, inflammation and tissue regeneration. Thromb Haemost 105(Suppl 1):S13–S33
126. Furie B, Furie BC (2008) Mechanisms of thrombus formation. N Engl J Med 359(9):938–949
127. Brass LF, Wannemacher KM, Ma P, Stalker TJ (2011) Regulating thrombus growth and stability to achieve an optimal response to injury. J Thromb Haemost 9(Suppl 1):66–75
128. Shivdasani RA, Rosenblatt MF, Zucker-Franklin D, Jackson CW, Hunt P, Saris CJ, Orkin SH (1995) Transcription factor NF-E2 is required for platelet formation independent of the actions of thrombopoietin/MGDF in megakaryocyte development. Cell 81(5):695–704
129. Flores A, Buchanan GR (2013) Occult hemorrhage in immune thrombocytopenia. Semin Hematol 50(Suppl 1):S26–S30
130. Saavedra FR, Redondo PC, Hernandez-Cruz JM, Salido GM, Pariente JA, Rosado JA (2004) Store-operated Ca^{2+} entry and tyrosine kinase $pp60^{src}$ hyperactivity are modulated by hyperglycemia in platelets from patients with non insulin-dependent diabetes mellitus. Arch Biochem Biophys 432(2):261–268
131. Croce KJ, Sakuma M, Simon DI (2007) Platelet–leukocyte–endothelium cross talk. In: Gresele P, Fuster V, Lopez JA, Page CP, Vermylen J (eds) Platelets in hematologic and cardiovascular disorders: a clinical handbook. Cambridge University Press, Cambridge, pp 106–123
132. Crittenden JR, Bergmeier W, Zhang Y, Piffath CL, Liang Y, Wagner DD, Housman DE, Graybiel AM (2004) CalDAG-GEFI integrates signaling for platelet aggregation and thrombus formation. Nat Med 10(9):982–986
133. Stegner D, Nieswandt B (2011) Platelet receptor signaling in thrombus formation. J Mol Med (Berl) 89(2):109–121
134. Rosado JA, Meijer EM, Hamulyak K, Novakova I, Heemskerk JW, Sage SO (2001) Fibrinogen binding to the integrin alpha(IIb)beta(3) modulates store-mediated calcium entry in human platelets. Blood 97(9):2648–2656
135. Heemskerk JW, Bevers EM, Lindhout T (2002) Platelet activation and blood coagulation. Thromb Haemost 88(2):186–193
136. van der Meijden PE, Feijge MA, Giesen PL, Huijberts M, van Raak LP, Heemskerk JW (2005) Platelet P2Y12 receptors enhance signalling towards procoagulant activity and thrombin generation. A study with healthy subjects and patients at thrombotic risk. Thromb Haemost 93(6):1128–1136
137. Leon C, Ravanat C, Freund M, Cazenave JP, Gachet C (2003) Differential involvement of the P2Y1 and P2Y12 receptors in platelet procoagulant activity. Arterioscler Thromb Vasc Biol 23(10):1941–1947

138. Morrell CN, Maggirwar SB (2011) Recently recognized platelet agonists. Curr Opin Hematol 18(5):309–314
139. Heemskerk JW, Mattheij NJ, Cosemans JM (2013) Platelet-based coagulation: different populations, different functions. J Thromb Haemost 11(1):2–16
140. Davies TA, Drotts DL, Weil GJ, Simons ER (1989) Cytoplasmic Ca2+ is necessary for thrombin-induced platelet activation. J Biol Chem 264(33):19600–19606
141. Mahaut-Smith MP (2012) The unique contribution of ion channels to platelet and megakaryocyte function. J Thromb Haemost 10(9):1722–1732
142. Rosado JA (2011) Acidic Ca(2+) stores in platelets. Cell Calcium 50(2):168–174
143. Kuwahara M, Sugimoto M, Tsuji S, Miyata S, Yoshioka A (1999) Cytosolic calcium changes in a process of platelet adhesion and cohesion on a von Willebrand factor-coated surface under flow conditions. Blood 94(4):1149–1155
144. Tew JG, Roberts RR, Donaldson DM (1974) Release of beta-lysin from platelets by thrombin and by a factor produced in heparinized blood. Infect Immun 9(1):179–186
145. Eastham RD (1964) Rapid adhesive platelet count in whole blood. J Clin Pathol 17:45–46
146. Harper MT, Poole AW (2011) Store-operated calcium entry and non-capacitative calcium entry have distinct roles in thrombin-induced calcium signalling in human platelets. Cell Calcium 50(4):351–358
147. Ahmad F, Boulaftali Y, Greene TK, Ouellette TD, Poncz M, Feske S, Bergmeier W (2011) Relative contributions of stromal interaction molecule 1 and CalDAG-GEFI to calcium-dependent platelet activation and thrombosis. J Thromb Haemost 9(10):2077–2086
148. Berna-Erro A, Braun A, Kraft R, Kleinschnitz C, Schuhmann MK, Stegner D, Wultsch T, Eilers J, Meuth SG, Stoll G, Nieswandt B (2009) STIM2 regulates capacitive Ca2+ entry in neurons and plays a key role in hypoxic neuronal cell death. Sci Signal 2(93):ra67
149. Misceo D, Holmgren A, Louch WE, Holme PA, Mizobuchi M, Morales RJ, De Paula AM, Stray-Pedersen A, Lyle R, Dalhus B, Christensen G, Stormorken H, Tjonnfjord GE, Frengen E (2014) A dominant STIM1 mutation causes Stormorken syndrome. Hum Mutat 35(5):556–564
150. Nesin V, Wiley G, Kousi M, Ong EC, Lehmann T, Nicholl DJ, Suri M, Shahrizaila N, Katsanis N, Gaffney PM, Wierenga KJ, Tsiokas L (2014) Activating mutations in STIM1 and ORAI1 cause overlapping syndromes of tubular myopathy and congenital miosis. Proc Natl Acad Sci U S A 111(11):4197–4202
151. Morin G, Bruechle NO, Singh AR, Knopp C, Jedraszak G, Elbracht M, Bremond-Gignac D, Hartmann K, Sevestre H, Deutz P, Herent D, Nurnberg P, Romeo B, Konrad K, Mathieu-Dramard M, Oldenburg J, Bourges-Petit E, Shen Y, Zerres K, Ouadid-Ahidouch H, Rochette J (2014) Gain-of-function mutation in STIM1 (p.R304W) is associated with Stormorken Syndrome. Hum Mutat 35(10):1221–1232
152. Markello T, Chen D, Kwan JY, Horkayne-Szakaly I, Morrison A, Simakova O, Maric I, Lozier J, Cullinane AR, Kilo T, Meister L, Pakzad K, Bone W, Chainani S, Lee E, Links A, Boerkoel C, Fischer R, Toro C, White JG, Gahl WA, Gunay-Aygun M (2014) York platelet syndrome is a CRAC channelopathy due to gain-of-function mutations in STIM1. Mol Genet Metab 114(3):474–482
153. Nakamura L, Sandrock-Lang K, Speckmann C, Vraetz T, Buhrlen M, Ehl S, Heemskerk JW, Zieger B (2013) Platelet secretion defect in a patient with stromal interaction molecule 1 deficiency. Blood 122(22):3696–3698
154. Barr JD, Chauhan AK, Schaeffer GV, Hansen JK, Motto DG (2013) Red blood cells mediate the onset of thrombosis in the ferric chloride murine model. Blood 121(18):3733–3741
155. Eckly A, Hechler B, Freund M, Zerr M, Cazenave JP, Lanza F, Mangin PH, Gachet C (2011) Mechanisms underlying FeCl3-induced arterial thrombosis. J Thromb Haemost 9(4):779–789
156. Watson SP, Asazuma N, Atkinson B, Berlanga O, Best D, Bobe R, Jarvis G, Marshall S, Snell D, Stafford M, Tulasne D, Wilde J, Wonerow P, Frampton J (2001) The role of ITAM- and ITIM-coupled receptors in platelet activation by collagen. Thromb Haemost 86(1):276–288

157. Offermanns S (2006) Activation of platelet function through G protein-coupled receptors. Circ Res 99(12):1293–1304
158. Mahaut-Smith MP (2013) A role for platelet TRPC channels in the Ca2+ response that induces procoagulant activity. Sci Signal 6(281):pe23
159. Dubois C, Panicot-Dubois L, Gainor JF, Furie BC, Furie B (2007) Thrombin-initiated platelet activation in vivo is vWF independent during thrombus formation in a laser injury model. J Clin Invest 117(4):953–960
160. Cifuni SM, Wagner DD, Bergmeier W (2008) CalDAG-GEFI and protein kinase C represent alternative pathways leading to activation of integrin alphaIIbbeta3 in platelets. Blood 112(5):1696–1703
161. McCarl CA, Picard C, Khalil S, Kawasaki T, Rother J, Papolos A, Kutok J, Hivroz C, Ledeist F, Plogmann K, Ehl S, Notheis G, Albert MH, Belohradsky BH, Kirschner J, Rao A, Fischer A, Feske S (2009) ORAI1 deficiency and lack of store-operated Ca2+ entry cause immunodeficiency, myopathy, and ectodermal dysplasia. J Allergy Clin Immunol 124(6):1311–1318, e1317
162. van Kruchten R, Braun A, Feijge MA, Kuijpers MJ, Rivera-Galdos R, Kraft P, Stoll G, Kleinschnitz C, Bevers EM, Nieswandt B, Heemskerk JW (2012) Antithrombotic potential of blockers of store-operated calcium channels in platelets. Arterioscler Thromb Vasc Biol 32(7):1717–1723
163. Yu Y, Keller SH, Remillard CV, Safrina O, Nicholson A, Zhang SL, Jiang W, Vangala N, Landsberg JW, Wang JY, Thistlethwaite PA, Channick RN, Robbins IM, Loyd JE, Ghofrani HA, Grimminger F, Schermuly RT, Cahalan MD, Rubin LJ, Yuan JX (2009) A functional single-nucleotide polymorphism in the TRPC6 gene promoter associated with idiopathic pulmonary arterial hypertension. Circulation 119(17):2313–2322
164. Paez Espinosa EV, Murad JP, Ting HJ, Khasawneh FT (2012) Mouse transient receptor potential channel 6: role in hemostasis and thrombogenesis. Biochem Biophys Res Commun 417(2):853–856
165. Scrimgeour NR, Wilson DP, Rychkov GY (2012) Glu(1)(0)(6) in the Orai1 pore contributes to fast Ca(2)(+)-dependent inactivation and pH dependence of Ca(2)(+) release-activated Ca(2)(+) (CRAC) current. Biochem J 441(2):743–753
166. Marumo M, Suehiro A, Kakishita E, Groschner K, Wakabayashi I (2001) Extracellular pH affects platelet aggregation associated with modulation of store-operated Ca(2+) entry. Thromb Res 104(5):353–360
167. Green FW Jr, Kaplan MM, Curtis LE, Levine PH (1978) Effect of acid and pepsin on blood coagulation and platelet aggregation. A possible contributor prolonged gastroduodenal mucosal hemorrhage. Gastroenterology 74(1):38–43
168. Potzsch B, Ivaskevicius V (2011) Haemostasis management of massive bleeding. Hamostaseologie 31(1):15–20
169. Zbidi H, Lopez JJ, Amor NB, Bartegi A, Salido GM, Rosado JA (2009) Enhanced expression of STIM1/Orai1 and TRPC3 in platelets from patients with type 2 diabetes mellitus. Blood Cells Mol Dis 43(2):211–213
170. Liu D, Maier A, Scholze A, Rauch U, Boltzen U, Zhao Z, Zhu Z, Tepel M (2008) High glucose enhances transient receptor potential channel canonical type 6-dependent calcium influx in human platelets via phosphatidylinositol 3-kinase-dependent pathway. Arterioscler Thromb Vasc Biol 28(4):746–751
171. Jardin I, Lopez JJ, Zbidi H, Bartegi A, Salido GM, Rosado JA (2011) Attenuated store-operated divalent cation entry and association between STIM1, Orai1, hTRPC1 and hTRPC6 in platelets from type 2 diabetic patients. Blood Cells Mol Dis 46(3):252–260
172. Bon RS, Beech DJ (2013) In pursuit of small molecule chemistry for calcium-permeable non-selective TRPC channels – mirage or pot of gold? Br J Pharmacol 170(3):459–474

Part II
Non-capacitative Ca^{2+} Entry Pathways

Chapter 9
On the Roles of the Transient Receptor Potential Canonical 3 (TRPC3) Channel in Endothelium and Macrophages: Implications in Atherosclerosis

Guillermo Vazquez, Sumeet Solanki, Prabhatachandra Dube, Kathryn Smedlund, and Prince Ampem

Abstract In the cardiovascular and hematopoietic systems the Transient Receptor Potential Canonical 3 (TRPC3) channel has a well-recognized role in a number of signaling mechanisms that impact the function of diverse cells and tissues in physiology and disease. The latter includes, but is not limited to, molecular and cellular mechanisms associated to the pathogenesis of cardiac hypertrophy, hypertension and endothelial dysfunction. Despite several of these functions being closely related to atherorelevant mechanisms, the potential roles of TRPC3 in atherosclerosis, the major cause of coronary artery disease, have remained largely unexplored. Over recent years, a series of studies from the authors' laboratory revealed novel functions of TRPC3 in mechanisms related to endothelial inflammation, monocyte adhesion to endothelium and survival and apoptosis of macrophages. The relevance of these new TRPC3 functions to atherogenesis has recently began to receive validation through studies in mouse models of atherosclerosis with conditional gain or loss of TRPC3 function. This chapter summarizes these novel findings and provides a discussion of their impact in the context of atherosclerosis, in an attempt to delineate a framework for further exploration of this *terra incognita* in the TRPC field.

Keywords TRPC3 • Macrophages • Atherosclerosis • Endothelial cells • Calcium entry

G. Vazquez (✉) • S. Solanki • P. Dube • K. Smedlund • P. Ampem
Department of Physiology and Pharmacology, and Center for Hypertension and Personalized Medicine, University of Toledo College of Medicine and Life Sciences, University of Toledo Health Science Campus, 3000 Transverse Dr., UTHSC Mail stop 1008, Toledo, OH 43614, USA
e-mail: Guillermo.Vazquez@utoledo.edu

J.A. Rosado (ed.), *Calcium Entry Pathways in Non-excitable Cells*, Advances in Experimental Medicine and Biology 898, DOI 10.1007/978-3-319-26974-0_9

9.1 Introduction

In western societies, coronary artery disease (CAD) is the leading cause of morbidity and mortality due to cardiovascular diseases [1]. The major causative pathological entity behind CAD is atherosclerosis, a disease that affects medium- and large-sized arteries and whose initiation and progression have all the features of a chronic disease ruled by a maladaptive inflammatory response. Typical clinical manifestations of atherosclerosis are the appearance of ischemic symptoms subsequent to lesions causing critical stenosis of the arterial lumen, and/or acute thromboembolic events that often follow the rupture of unstable plaques [2, 3]. Atherosclerotic lesions develop through a complex series of interactions between pro-atherogenic humoral factors – v.g., apo-B containing lipoproteins, inflammatory cytokines–, cellular components of the arterial wall – i.e., endothelial and smooth muscle cells, macrophages-, and immune cells recruited from circulation – v.g., monocytes, lymphocytes–. The earliest molecular and cellular events in atherogenesis are dominated by endothelial cell activation, inflammation and recruitment of inflammatory monocytes to the arterial subintima. As the disease progresses, the added effect of cytokines and pro-inflammatory mediators that co-exist in the lesion environment, as well as the critical impact of macrophage death on plaque composition and stability, complete a maladaptive inflammatory vicious cycle that drives the lesion towards critically advanced and clinically relevant stages. A key therapeutic goal in atherosclerosis is to identify molecular targets that can be used to prevent or reduce endothelial inflammation and monocyte recruitment, and/or to improve plaque stability to reduce the risk of acute thromboembolic events. This goal has, so far, been unsatisfactorily met. Therefore, improving current knowledge of signaling mechanisms that mediate endothelial activation and macrophage functions in the plaque is an ongoing effort in the field and key to reach the above mentioned therapeutic objectives.

Transient Receptor Potential Canonical (TRPC) channels (TRPC1-7) are non-voltage gated, non-selective cation channels whose diverse functions in the cardiovascular and hematopoietic systems are now well-recognized (reviewed by us in [4]; see also [5]). And it is by virtue of their diverse functions in molecular and cellular mechanisms associated to the pathogenesis of cardiovascular and hematopoietic diseases that most TRPC proteins have been postulated as potential therapeutic targets [4, 5]. In the particular case of TRPC3, its expression and function have been shown to have implications in hypertension, endothelial dysfunction and, with solid support from in vivo studies, in cardiac hypertrophy [6–8]. Curiously, the potential roles of TRPC3 – and of any other TRPC for that matter – in atherosclerosis have remained largely unexplored. A series of recent studies from our laboratory have pioneered the field by providing experimental evidence revealing previously unforeseen roles of endothelial and macrophage TRPC3 in molecular and cellular mechanisms of relevance to the development of atherosclerotic lesions. The mode of regulation of TRPC3 in these and in most other cells where this channel is expressed, is still incompletely defined (for discussions on this topic, see [9–12]).

Nevertheless, understanding the functions of this and other TRPCs in cardiovascular pathology demands at the very least identification of cellular/molecular events that may be affected by channel expression/function, regardless of what the channel's activating/regulatory mechanism is. Most importantly, the generation of mouse models of atherosclerosis with conditional gain or loss of TRPC3 function has been instrumental in validating these findings and in paving the road for further exploration of this *terra incognita*. The sections below are aimed at providing a compilation of these novel findings with discussions of their implications in the context of atherosclerosis.

9.2 Structural and Topological Features of TRPC3

TRPC3 belongs to the TRPC family of channel forming proteins, which are part of the larger TRP superfamily [12, 13]. The TRPC group is subdivided into TRPC1, TRPC2 – a pseudogene in humans, old monkeys and apes, but a functional channel in rodents-, TRPC3/6/7 and TRPC 4/5, mostly on the basis of structural, functional and pharmacological properties [12]. All TRPC proteins form non-voltage gated, Ca^{2+}-permeable channels that are activated, at least under physiological conditions, downstream of receptor-stimulated phospholipases [4, 11]. Besides this receptor-regulated mode, TRPC3 channels exhibit significant constitutive function in both overexpression and native systems [10, 14]. Similar to the other members of the TRPC group, TRPC3 most salient topological features include six transmembrane domains (TM1-TM6) that separate cytoplasmic N- and C-ends, with a pore region framed in between TM5 and TM6 [15, 16]. The N-terminus contains three to four ankyrin repeats, a coiled-coil region and a caveolin binding domain. The ankyrin repeats form a protein binding interface that allows TRPC3 to interact with other proteins and it also affects trafficking of the assembled channel to the plasma membrane [17]. In fact, deletion of the ankyrin repeats leads to accumulation of TRPC3 in intracellular compartments [17]. The C-terminus contains the typical TRP signature motif (EWKFAR), a highly conserved proline rich motif, the CIRB (calmodulin/IP_3 receptor binding) domain and a predicted coiled-coil region. The coiled-coil regions participate in oligomerization of the channel subunits and contribute not only to the final assembly of a functional channel but also to the association of TRPC3 with other proteins [12]. The proline-rich motif and CIRB region have also been found to interact with a number of signaling molecules [12]. In both native and overexpression systems functional TRPC3 channels are made of homo- or hetero-tetrameric arrangements of four TRPC subunits [12, 15, 16]. When heteromerization occurs, a preference exists for association with members within the TRPC3/6/7 group, although interactions with other members of the TRPC family do take place [4, 12, 18]. The lack of correlation between some functional and pharmacological properties of overexpressed vs. native TRPC3 channels – v.g., sensitivity to low micromolar concentrations of gadolinium, responsiveness to synthetic diacylglycerols- have been traditionally attributed to differences in the final tetrameric buildup of the channel. However, definitive conclusions on this matter have not been reached.

9.3 Atherorelevant Functions of Endothelial TRPC3

9.3.1 TRPC3 and Regulation of Vascular Cell Adhesion Molecule-1 Expression in Coronary Artery Endothelial Cells

Endothelial inflammatory signaling is a major determinant of monocyte recruitment to the arterial subintima, and although present throughout all stages of atherosclerotic lesion development, it has a particular impact in the early phases of plaque formation (see [19] and references therein). Whereas several endothelial cell adhesion molecules (CAMs) participate in this process, intercellular adhesion molecule-1 (ICAM-1) and vascular cell adhesion molecule-1 (VCAM-1), which belong to the immunoglobulin superfamily [20, 21], define the firm attachment of the monocyte to the endothelium and its subsequent transmigration to the subintima. Perhaps the most important characteristic of VCAM-1 in early atherogenesis is the fact that whereas most other CAMs are constitutively expressed in the non-activated endothelium, VCAM-1 is nearly absent, but its expression increases dramatically when the endothelial cell is exposed to pro-inflammatory stimuli [21]. It is thus not surprising that a significant amount of research effort has been devoted at elucidating the mechanisms underlying VCAM-1 expression, with the expectation of identifying molecular targets to abrogate or at least reduce the recruitment of monocytes mediated by this CAM. Studies from different laboratories coincided in the observation that calcium, at least from intracellular sources, was a common factor in the mechanism by which a number of inflammatory mediators – v.g., substance P, ATP, lipoprotein a – induced VCAM-1 expression in the endothelium [22–24]. Until the publication of our first series of studies on the roles of TRPC3 in regulated expression of VCAM-1 in human coronary artery endothelial cells (HCAECs) [25], it was uncertain whether calcium influx had any specific role in regulation of VCAM-1 expression. Using a combined pharmacological and molecular approach, we found that, despite HCAECs expressing all members of the TRPC family, only TRPC3 formed endogenous calcium permeable channels in these cells [25]. Moreover, TRPC3 contributed, to a great extent, to the ATP-induced calcium influx and, most remarkably, accounted for most of the constitutive calcium influx in these cells. Notably, TRPC3 expression and function were found to be required for maximal induction of VCAM-1 and monocyte adhesion to HCAECs [25]. In most endothelial cells, regardless of their location in the vascular tree, VCAM-1 expression is regulated by the Nuclear Factor kappa B (NFkB) pathway [26]. In a follow up study, it was found that in HCAECs the constitutive function of TRPC3 had an obligatory role within the inflammatory signaling that drives VCAM-1 expression under pro-inflammatory/pro-atherogenic conditions [27]. In this mechanism TRPC3-mediated constitutive calcium entry was coupled to activation of an IKKβ/IkBα axis in a calmodulin-dependent protein kinase II (CAMKII) manner [27]. Perhaps the most remarkable observation was that the same pro-inflammatory/pro-atherogenic factors that induce VCAM-1 in HCAECs, are also strong inducers of TRPC3 expression in these cells, and this results in enhanced inflammatory signaling due to the augmented number of channels operating in constitutive mode [27]. These findings represented the first piece of experimental evidence suggesting a role of TRPC3 in molecular and cellular events that are critical in atherosclerosis. The in vivo relevance of these in vitro findings has been validated by our recent studies in Apoe knockout mice with endothelial-specific overexpression of TRPC3 (see Sect. 5.2 below).

9.3.2 *TRPC3 and Endoplasmic Reticulum Stress-Induced Apoptosis in Endothelial Cells*

Endoplasmic reticulum (ER) stress has emerged as a major mechanism of cell apoptosis in atherosclerotic plaques throughout all stages of the disease [28, 29]. A number of circulating atherorelevant stressors – v.g., low density lipoprotein, cytokines-can disturb endothelial ER homeostasis and trigger the unfolded protein response (UPR) in an attempt to restore normal ER function; however, if ER stress is sustained the UPR leads to activation of pro-apoptotic mechanisms [28, 30]. In a number of cell types persistent and/or exacerbated calcium influx can promote or contribute to mechanisms linked to ER stress [31, 32]. It is also known that in some cells and tissues a crosstalk exists between NFkB-dependent inflammatory signaling and the UPR, where ER stress can either activate or enhance NFkB signaling [33, 34]. In HCAECs augmented TRPC3 expression in response to pro-inflammatory factors results in augmented NFkB-signaling ([27] and Sect. 3.1 above). Interestingly, TRPC3 function has been linked to activation of the UPR in epithelial cells [31]. The question then raises whether in endothelial cells TRPC3 can also contribute to initiation and/or maintenance of UPR signaling, integrating the UPR response and inflammation. Recent work in our laboratory shows that in HCAECs inhibition of native TRPC3 with the selective TRPC3 blocker pyrazole-10 [35] results in a marked reduction in both UPR activation and the susceptibility of these cells to ER stress-induced apoptosis (Ampem, Smedlund and Vazquez, unpublished observations). Notably, the ER stress conditions that promote apoptosis of HCAECs also induce activation of NFkB in these cells. These findings do indeed suggest that TRPC3 may couple mechanisms of ER stress-induced apoptosis to inflammatory signaling in endothelial cells.

9.4 Atherorelevant Functions of Macrophage TRPC3

9.4.1 TRPC3 and Survival of Non-polarized Macrophages

Macrophage apoptosis has a prominent effect on the characteristics of atherosclerotic lesions [36]. One of the major pro-apoptotic mechanisms in plaque macrophages is mediated by activation of the UPR that follows chronic, unresolved, ER stress [29, 37]. Accumulation of apoptotic macrophages has a differential effect on early vs. advanced lesions [29, 38]. In early atherosclerotic plaques the clearance of apoptotic macrophages or efferocytosis, is efficient, and apoptotic cells barely accumulate, thus contributing minimally to lesion cellularity [38, 39]. However, in the more advanced lesion setting, increased rates of macrophage apoptosis combined with deficient clearance, lead to in situ cell death and secondary necrosis, major contributors to the enlargement of the plaque necrotic core [38, 40]. There is significant effort in the field to identify molecular components of mechanisms governing macrophage survival/apoptosis in the hope that this can lead to discovery of potential therapeutic targets. The phosphatidylinositol-3-kinase (PI3K)/AKT and NFkB pathways are typical mediators of survival mechanisms in macrophages. For instance, AKT-mediated phosphorylation of the pro-apoptotic Bcl-2 family member BAD prevents the association of BAD with the anti-apoptotic Bcl-2 family members Bcl-2 and Bcl-Xl, reducing apoptosis [41]. In addition, the expression level of these pro-survival genes is to some extent modulated by NFkB. Both pathways depend, directly or indirectly, upon calcium influx into the cell. Studies from our laboratory in both human THP-1-derived macrophages and mouse bone marrow-derived macrophages showed that constitutive calcium influx is required for optimal operation of AKT and NFkB dependent survival signaling [42, 43]. In fact, pharmacological maneuvers that lead to abrogation of constitutive calcium influx or of CAMKII function, impaired the activation of AKT or NFkB and a drastic increase in macrophage apoptosis [42, 43]. These findings indicated that the CAM/CAMKII/AKT and CAM/CAMKII/NFkB axes are common elements in coupling constitutive calcium influx to survival in both human and murine macrophages. The molecular nature of the channels mediating such constitutive calcium influx was not known at the time. In a follow up study we showed that bone marrow-derived macrophages prepared from mice with global deficiency of TRPC3 exhibited an almost complete loss of constitutive calcium influx, impaired survival signaling and increased apoptosis, compared to wild-type cells [44]. These findings represented the first experimental evidence identifying the molecular nature of a channel whose expression and constitutive function had an obligatory role in macrophage survival. Most importantly, the above observations led to the speculation that in the setting of atherosclerosis macrophage deficiency of TRPC3 may lead to accumulation of apoptotic cells and increased necrosis in advanced plaques. Contrarily to this prediction, studies using a bone marrow transplantation strategy in which Apoe knockout mice received TRPC3 deficient bone marrow, revealed unexpected findings ([45] and see Sect. 5.1 below for details). Most strikingly, advanced lesions in the aortic root of these mice showed a significant reduction in the number of apoptotic macrophages [45]. The speculation was then made that the characteristics of advanced plaques from mice with bone marrow deletion of TRPC3 may be due to the fact that polarized, rather than non-polarized macrophages predominate, and that the contribution of TRPC3 to mechanisms of macrophage apoptosis may depend on the differentiation status.

9.4.2 TRPC3 and Apoptosis of Polarized Macrophages

In lesions from both human and animal models of atherosclerosis, macrophages constitute a heterogenous population that includes a few distinguishable and relatively well-characterized phenotypes [46–48]. In particular, M1 or inflammatory or classically activated macrophages, and M2 or anti-inflammatory or alternatively activated macrophages, probably representing extremes of a wide spectrum of phenotypes, dominate in the setting of the atherosclerotic plaque. The relative abundance of these two subsets – the M1/M2 ratio – varies with lesion stage, and whereas this ratio at a given time point is influenced by a number of factors – i.e., lesion complexity, genetic background of the animal model – in general tends to increase as the lesion progresses. Nevertheless, both M1 and M2 types co-exist throughout plaque progression altogether with intermediate, ill-defined phenotypes [49, 50]. In line with the interpretation above that the effects of TRPC3 on macrophage apoptosis may be different in non-polarized vs. polarized macrophages, M1 cells prepared from mice with TRPC3 deficient bone marrow are notoriously less sensitive to ER stress-induced apoptosis than TRPC3 expressing M1 macrophages [45]. Notably, deletion of TRPC3 does not alter the apoptotic rates of M2 macrophages. In macrophages as in other eukaryotic cells, ER stress leads to activation of the UPR, which signals through the UPR sensors inositol requiring enzyme-1α (IRE-1α), protein kinase R-like endoplasmic reticulum kinase (PERK) and activating transcription factor-6 (ATF6) [51]. The observation that M1, but not M2 macrophages with deletion of TRPC3 had a reduced susceptibility to ER stress-induced apoptosis raised the question whether this was due to a specific effect of TRPC3 on the UPR signaling. This was addressed in a recent follow up study where the effect of TRPC3 on UPR and ER stress-induced apoptosis was examined in polarized M1 and M2 macrophages prepared from wild-type mice or from mice with macrophage-specific deletion of TRPC3 [52]. A number of interesting observations were made and are summarized as follows. Compared to controls, apoptosis induced by thapsigargin, tunicamycin or free-cholesterol loading – conditions that promote sustained, irreversible ER stress – was markedly reduced in TRPC3 deficient M1, but not M2 macrophages. In line with this, expression levels of CCAAT/enhancer binding protein homologous protein (CHOP), spliced X-box binding protein 1 – a proximal indicator of IRE1α activation-, PERK and ER oxidoreductase-1α – a transcriptional target of CHOP – were all decreased in TRPC3 deficient M1, but not M2 macrophages, evidencing impaired UPR activation in these cells [52]. ER stress-dependent activation of CAMKII and STAT1, mediators of UPR-dependent apoptosis in macrophages [53, 54], was severely impaired again in TRPC3 deficient M1 but not in M2 macrophages. Notably, in wild-type M1 macrophages basal activation of CAMKII was abrogated by genetic or pharmacological inhibition of TRPC3, suggesting that in M1 macrophages, similar to the findings in human and mouse non-polarized macrophages (Sect. 4.1 above), TRPC3 function is coupled to activation of this kinase [52]. This is interesting, as it suggests that despite naïve and M1 macrophages showing similar coupling between TRPC3 and CAMKII, modulation of downstream signaling pathways is distinctively affected by the macrophage phenotype.

Overall, the findings discussed in this section indicate that TRPC3 exerts a differential effect on apoptosis of M1 vs. M2 macrophages, with a clear contribution to mechanisms of ER stress-induced apoptosis in the M1, but not in the M2 cells. At this time, the nature of such selectivity remains unknown. It is possible that differences exist between M1 and M2 macrophages in terms of channel arrangement (i.e., homo- vs. hetero-tetrameric) or in their specific repertoire of potential signaling or adaptor proteins coupling TRPC3 to intracellular pathways.

9.4.3 Impact of TRPC3 Expression/Function on Efferocytic Properties of Macrophages

As described above (Sect. 4.1) the clearance of apoptotic macrophages from the lesion site by resident phagocytes – efferocytosis – is a process that shapes the lesion and has key consequences on its cellularity and volume. The observation that TRPC3 deficiency impairs survival and increases apoptosis of non-polarized macrophages led to the question whether the phagocytic capacity of these cells would also be affected by their TRPC3 expression status. By means of an in vitro efferocytosis assay in which both TRPC3 expressing and TRPC3 deficient macrophages were used either as phagocytes or apoptotic cells, it was found that phagocytes lacking TRPC3 had impaired efferocytic function when compared to wild-type cells [44]. In addition, deletion of TRPC3 resulted in apoptotic macrophages being poor substrates for efferocytosis, regardless of the TRPC3 expression status of the acting phagocyte. These findings suggested a mandatory requirement for TRPC3 both in phagocytosis mechanisms and in cell-cell recognition processes. Similar to the distinctive impact of TRPC3 on apoptosis of non-polarized vs. polarized macrophages (Sects. 4.1 and 4.2 above), the behavior of M1 and M2 phagocytes with deletion of TRPC3 was opposite to that of naïve phagocytes. In fact, TRPC3 deficient M1 and M2 macrophages exhibited higher efferocytic capacity compared to TRPC3 expressing cells [45]. Interestingly, protein levels of MERTK were higher in macrophages lacking TRPC3 compared to control cells [45], which can account, at least in part, for the higher efferocytic capacity of the TRPC3 deficient macrophages. The in vivo relevance of these findings has not yet been assessed.

9.5 Mouse Models of Atherosclerosis with Conditional Gain or Loss of TRPC3 Function

The in vitro findings described in Sects. 3 and 4 above, provided important evidence supporting a role of TRPC3 in molecular and cellular mechanisms of relevance to atherosclerosis. However, definitive associations between changes in TRPC3 expression and/or function and atherogenesis could not be established on the sole basis of in vitro studies. In order to evaluate the contribution of endothelial or macrophage TRPC3 to mechanisms of plaque development in vivo it was thus necessary to generate mouse models of atherosclerosis in which TRPC3 expression could be manipulated – i.e., conditional transgenic or knockout animals for TRPC3-. The sections below discuss some of the most recent findings in mice with hematopoietic deficiency of TRPC3 and mice with endothelial-specific overexpression of TRPC3.

9.5.1 Effects of Bone Marrow-Specific Deletion of TRPC3

The findings discussed in Sect. 4.1 regarding the impaired survival of TRPC3-deficient non-polarized macrophages led to the speculation that, in the setting of the advanced atherosclerotic plaque, macrophage deficiency of TRPC3 would result in accumulation of apoptotic cells and increased necrosis. This prediction was based on the fact that apoptotic macrophages contribute to expansion of the necrotic core in the advanced lesion (discussed in Sect. 4.1). The use of bone marrow transplantation models in gene-targeted mice is a powerful technique to evaluate the contribution to atherogenesis of a particular gene expressed by macrophages [39, 55, 56]. As a first approach to reveal the true impact of macrophage deficiency of TRPC3 in atherosclerotic lesions, we generated Apoe knockout mice chimeric for macrophage TRPC3 expression by means of bone marrow transplantation [45]. Contrarily to our predictions, we found that some typical undesirable features of advanced plaques were actually alleviated by the bone marrow deficiency of TRPC3 [45]. Compared to control mice, the advanced lesions in the aortic root of mice with bone marrow deletion of TRPC3 showed a marked reduction in necrosis, higher collagen content and larger fibrous cap thickness. In addition, despite no changes in lesion size or macrophage content, plaques from these mice had a twofold reduction in the number of apoptotic macrophages [45]. In the light of the previous in vitro findings in non-polarized macrophages, the phenotypic features of these plaques led us to speculate that the contribution of TRPC3 to mechanisms of macrophage apoptosis may actually depend on the differentiation status of these cells. This interpretation was later on supported by in vitro studies in M1 and M2 macrophages (discussed in Sect. 3.2 above).

Among the immune cells found in mouse lesions, lymphocytes, neutrophils and dendritic cells do not express TRPC3 [57–61], favoring the notion that the plaque phenotype in the bone marrow transplantation model presumably reflects the effects of TRPC3 on macrophage functions. However, TRPC3 is also expressed in the other two major plaque cell types namely, smooth muscle (SMC) and endothelial (ECs) cells [4, 5]. Atherorelevant functions of TRPC3 in SMCs, if any, are not known and recent work from our laboratory shows that in vivo, TRPC3 supports inflammation in endothelial cells ([62] and Sect. 5.2 below). Bone marrow-derived EC and SMC progenitors can contribute to replenishment of lesional cells [63, 64]. Therefore, it remains uncertain if and to what extent this could have contributed to the phenotype of mice with bone marrow deletion of TRPC3. Hence, establishing a more definitive relationship between macrophage TRPC3 and atherosclerosis awaits the characterization of plaques in the recently generated mice with macrophage-specific loss of TRPC3 function [52].

9.5.2 Effects of Endothelial-Specific Overexpression of TRPC3

In Sect. 3.1 above, we discussed the evidence that demonstrated that in HCAECs the endogenous TRPC3 channels mediate constitutive calcium influx and are obligatory for proper operation of inflammatory signaling through the NFkB pathway [25, 27]. The in vivo relevance of these findings was validated in recent studies using Apoe knockout mice with endothelial-specific overexpression of human TRPC3 [62]. Whereas early lesions – 10 weeks on high fat diet- in the aortic root were not different from control animals, advanced plaques – 16 weeks on high fat diet- in the transgenic mice were larger and of increased cellularity compared to non-transgenic mice. As predicted by the in vitro studies, there was increased endothelial inflammation in lesions of the TRPC3 transgenic mice, and this was already evident in the early lesions, long before differences in plaque size and cellularity became evident.

9.6 Targeting TRPC3: Myth or Reality?

The studies discussed in the preceding sections on the effects of TRPC3 on endothelial inflammation [25, 27] and macrophage apoptosis [45, 52], as well as the validation of these findings in mouse models of atherosclerosis with conditional gain or loss of TRPC3 function [45, 62], constitute the first available evidence demonstrating a pro-atherogenic role of a member of the TRPC family in atherosclerosis. The distinctive effects of TRPC3 on the characteristics of atherosclerotic plaques are consequence of the channel's unique functions in these two atherorelevant cell types. These observations are endowed with a therapeutic timbre that merits few considerations. First, the selective contribution of TRPC3 to mechanisms of UPR-related apoptosis in M1 macrophages with no effect on M2 cells provokes the notion of TRPC3 as a potential target for manipulation of macrophage functions in a phenotype selective manner. This is of most importance, as the identification of signaling molecules that may specifically affect a particular macrophage type but not others is, as of this writing, an as yet unmet goal. As discussed above ER stress is a major pro-apoptotic mechanism in lesional macrophages. Notably, macrophage differentiation towards the M2 phenotype seems to benefit from ER stress, to the extent that ameliorating ER stress shifts M2 differentiation towards the M1 type [65]. This suggests that targeting molecular components of the UPR as a strategy to decrease macrophage apoptosis with no discernment of phenotypes, may ultimately result in a deleterious effect on necrosis of the advanced plaque, as such maneuver may favor M2-to-M1 conversion. In this context, the findings discussed in Sects. 4.2 and 5.1 above suggest that targeting TRPC3, rather than its downstream effectors, may be a better strategy to reduce apoptosis of M1 macrophages and prevent or delay necrotic core enlargement in the advanced lesion. Second, the pro-atherogenic effect of endothelial TRPC3 suggests that molecular or pharmacological targeting of this channel at the endothelial level could also be beneficial in ameliorating endothelial cell activation, inflammation and recruitment of circulating inflammatory cells to the subintima.

Historically, L-type channels have been the only type of calcium permeable channels whose pharmacological targeting seems to be of some beneficial effect in atherosclerosis (discussed by us in [66]). However, these anti-atherogenic effects are for the most part inconsistent, mechanistically uncertain and often unrelated to the channel blocking properties of CCBs, raising the question whether these are manifestations of off-target effects or secondary to their beneficial impact on atherosclerosis risk factors. In this context, TRPC3 targeting is more likely to efficiently suppress signaling effectors downstream TRPC3 without interfering with pathways often affected by non-channel related actions of traditional CCBs [66]. Importantly, whereas the ubiquitous expression of TRPC3 may be seen as a limitation for selective targeting – a problem with most signaling molecules so far postulated as potential targets in atherosclerosis – endothelial – or macrophage-targeted drug delivery is now possible through nanoparticle delivery systems [67, 68]. It is our hope that the novel discoveries on the roles of TRPC3 in atherosclerosis will provoke additional efforts in the field to further our understanding of the functions of this and other TRPC proteins in atherorelevant cells and processes. With no doubt, these efforts should lead to development of a new dimension in experimental therapeutics of this disease.

Acknowledgments This work was supported by NIH grant R01HL111877-04 (to G.V.),

References

1. Mozaffarian D, Benjamin EJ, Go AS, Arnett DK, Blaha MJ, Cushman M, de Ferranti S, Després J-P, Fullerton HJ, Howard VJ, Huffman MD, Judd SE, Kissela BM, Lackland DT, Lichtman JH, Lisabeth LD, Liu S, Mackey RH, Matchar DB, McGuire DK, Mohler ER, Moy CS, Muntner P, Mussolino ME, Nasir K, Neumar RW, Nichol G, Palaniappan L, Pandey DK, Reeves MJ, Rodriguez CJ, Sorlie PD, Stein J, Towfighi A, Turan TN, Virani SS, Willey JZ,

Woo D, Yeh RW, Turner MB (2015) Heart disease and stroke statistics—2015 update: a report from the American Heart Association. Circulation 131(4):e29–e322
2. Falk E, Nakano M, Bentzon JF, Finn AV, Virmani R (2013) Update on acute coronary syndromes: the pathologists' view. Eur Heart J 34(10):719–728
3. Schwartz SM, Galis ZS, Rosenfeld ME, Falk E (2007) Plaque rupture in humans and mice. Arterioscler Thromb Vasc Biol 27(4):705–713
4. Smedlund K, Bah M, Vazquez G (2012) On the role of endothelial TRPC3 channels in endothelial dysfunction and cardiovascular disease. Cardiovasc Hematol Agents Med Chem 10(3):265–274
5. Abramowitz J, Birnbaumer L (2009) Physiology and pathophysiology of canonical transient receptor potential channels. FASEB J Off Publ Fed Am Soc Exp Biol 23(2):297–328
6. Bush EW, Hood DB, Papst PJ, Chapo JA, Minobe W, Bristow MR, Olson EN, McKinsey TA (2006) Canonical transient receptor potential channels promote cardiomyocyte hypertrophy through activation of calcineurin signaling. J Biol Chem 281(44):33487–33496
7. Nakayama H, Wilkin BJ, Bodi I, Molkentin JD (2006) Calcineurin-dependent cardiomyopathy is activated by TRPC in the adult mouse heart. FASEB J Off Publ Fed Am Soc Exp Biol 20(10):1660–1670
8. Shan D, Marchase RB, Chatham JC (2008) Overexpression of TRPC3 increases apoptosis but not necrosis in response to ischemia-reperfusion in adult mouse cardiomyocytes. Am J Physiol Cell Physiol 294(3):C833–C841
9. Smyth JT, DeHaven WI, Jones BF, Mercer JC, Trebak M, Vazquez G, Putney JJW (2006) Emerging perspectives in store-operated Ca^{2+} entry: roles of Orai, Stim and TRP. Biochim Biophys Acta (BBA) Mol Cell Res 1763(11):1147–1160
10. Vazquez G, Tano JY, Smedlund K (2010) On the potential role of source and species of diacylglycerol in phospholipase-dependent regulation of TRPC3 channels. Channels (Austin Tex) 4(3):232–240
11. Liao Y, Abramowitz J, Birnbaumer L (2014) The TRPC family of TRP channels: roles inferred (mostly) from knockout mice and relationship to ORAI proteins. Handb Exp Pharmacol 223:1055–1075
12. Vazquez G, Wedel BJ, Aziz O, Trebak M, Putney JW Jr (2004) The mammalian TRPC cation channels. Biochim Biophys Acta 1742(1–3):21–36
13. Birnbaumer L (2009) The TRPC class of ion channels: a critical review of their roles in slow, sustained increases in intracellular Ca(2+) concentrations. Annu Rev Pharmacol Toxicol 49:395–426
14. Trebak M, Vazquez G, Bird GS, Putney JW Jr (2003) The TRPC3/6/7 subfamily of cation channels. Cell Calcium 33(5–6):451–461
15. Birnbaumer L, Zhu X, Jiang M, Boulay G, Peyton M, Vannier B, Brown D, Platano D, Sadeghi H, Stefani E, Birnbaumer M (1996) On the molecular basis and regulation of cellular capacitative calcium entry: roles for Trp proteins. Proc Natl Acad Sci U S A 93(26):15195–15202
16. Vannier B, Zhu X, Brown D, Birnbaumer L (1998) The membrane topology of human transient receptor potential 3 as inferred from glycosylation-scanning mutagenesis and epitope immunocytochemistry. J Biol Chem 273(15):8675–8679
17. Wedel BJ, Vazquez G, McKay RR, St. J. Bird G, Putney JW Jr (2003) A calmodulin/inositol 1,4,5-trisphosphate (IP_3) receptor-binding region targets TRPC3 to the plasma membrane in a calmodulin/IP_3 receptor-independent process. J Biol Chem 278(28):25758–25765
18. Poteser M, Graziani A, Rosker C, Eder P, Derler I, Kahr H, Zhu MX, Romanin C, Groschner K (2006) TRPC3 and TRPC4 associate to form a redox-sensitive cation channel: evidence for expression of native TRPC3-TRPC4 heteromeric channels in endothelial cells. J Biol Chem 281(19):13588–13595
19. Tano J-Y, Smedlund K, Vazquez G (2010) Endothelial TRPC3/6/7 proteins at the edge of cardiovascular disease. Cardiovasc Hematol Agents Med Chem (Formerly Curr Med Chem Cardiovasc Hematol Agents) 8:76–86

20. Quehenberger O (2005) Molecular mechanisms regulating monocyte recruitment in atherosclerosis. J Lipid Res 46:1582–1590
21. Galkina E, Ley K (2007) Vascular adhesion molecules in atherosclerosis. Arterioscler Thromb Vasc Biol 11:2292–2301
22. Allen S, Khan S, S-P T, Koschinsky M, Taylor P, Yacoub M (1998) Expression of adhesion molecules by Lp(a): a potential novel mechanism for its atherogenicity. FASEB J 12(15):1765–1776
23. Seye CI, Yu N, Jain R, Kong Q, Minor T, Newton J, Erb L, Gonzalez FA, Weisman GA (2003) The P2Y2 nucleotide receptor mediates UTP-induced vascular cell adhesion molecule-1 expression in coronary artery endothelial cells. J Biol Chem 278(27):24960–24965
24. Quinlan KL, Naik SM, Cannon G, Armstrong CA, Bunnett NW, Ansel JC, Caughman SW (1999) Substance P activates coincident NF-AT- and NF-kB-dependent adhesion molecule gene expression in microvascular endothelial cells through intracellular calcium mobilization. J Immunol 163(10):5656–5665
25. Smedlund K, Vazquez G (2008) Involvement of native TRPC3 proteins in ATP-dependent expression of VCAM-1 and monocyte adherence in coronary artery endothelial cells. Arterioscler Thromb Vasc Biol 28(11):2049–2055
26. Zhang C (2008) The role of inflammatory cytokines in endothelial dysfunction. Basic Res Cardiol 103(5):398–406
27. Smedlund K, Tano JY, Vazquez G (2010) The constitutive function of native TRPC3 channels modulates vascular cell adhesion molecule-1 expression in coronary endothelial cells through nuclear factor kappaB signaling. Circ Res 106(9):1479–1488
28. Chistiakov DA, Sobenin IA, Orekhov AN, Bobryshev YV (2014) Role of endoplasmic reticulum stress in atherosclerosis and diabetic macrovascular complications. Biomed Res Int 2014:610140
29. Tabas I (2010) The role of endoplasmic reticulum stress in the progression of atherosclerosis. Circ Res 107(7):839–850
30. Kadowaki H, Nishitoh H (2013) Signaling pathways from the endoplasmic reticulum and their roles in disease. Genes 4(3):306–333
31. Kim MS, Lee KP, Yang D, Shin DM, Abramowitz J, Kiyonaka S, Birnbaumer L, Mori Y, Muallem S (2011) Genetic and pharmacologic inhibition of the Ca^{2+} influx channel TRPC3 protects secretory epithelia from Ca^{2+}-dependent toxicity. Gastroenterology 140(7):2107–2115, 2115 e2101-2104
32. Nakano T, Watanabe H, Ozeki M, Asai M, Katoh H, Satoh H, Hayashi H (2006) Endoplasmic reticulum Ca^{2+} depletion induces endothelial cell apoptosis independently of caspase-12. Cardiovasc Res 69(4):908–915
33. Tam AB, Mercado EL, Hoffmann A, Niwa M (2012) ER stress activates NF-kappaB by integrating functions of basal IKK activity, IRE1 and PERK. PLoS ONE 7(10):e45078
34. Prell T, Lautenschlager J, Weidemann L, Ruhmer J, Witte OW, Grosskreutz J (2014) Endoplasmic reticulum stress is accompanied by activation of NF-kappaB in amyotrophic lateral sclerosis. J Neuroimmunol 270(1–2):29–36
35. Schleifer H, Doleschal B, Lichtenegger M, Oppenrieder R, Derler I, Frischauf I, Glasnov TN, Kappe CO, Romanin C, Groschner K (2012) Novel pyrazole compounds for pharmacological discrimination between receptor-operated and store-operated Ca^{2+} entry pathways. Br J Pharmacol 167(8):1712–1722
36. Moore KJ, Tabas I (2011) Macrophages in the pathogenesis of atherosclerosis. Cell 145(3):341–355
37. Tabas I (2009) Macrophage apoptosis in atherosclerosis: consequences on plaque progression and the role of endoplasmic reticulum stress. Antioxid Redox Signal 11(9):2333–2339
38. Gautier EL, Huby T, Witztum JL, Ouzilleau B, Miller ER, Saint-Charles F, Aucouturier P, Chapman MJ, Lesnik P (2009) Macrophage apoptosis exerts divergent effects on atherogenesis as a function of lesion stage. Circulation 119(13):1795–1804

39. Babaev VR, Chew JD, Ding L, Davis S, Breyer MD, Breyer RM, Oates JA, Fazio S, Linton MF (2008) Macrophage EP4 deficiency increases apoptosis and suppresses early atherosclerosis. Cell Metab 8(6):492–501
40. Schrijvers DM, De Meyer GR, Kockx MM, Herman AG, Martinet W (2005) Phagocytosis of apoptotic cells by macrophages is impaired in atherosclerosis. Arterioscler Thromb Vasc Biol 25(6):1256–1261
41. Datta SR, Brunet A, Greenberg ME (1999) Cellular survival: a play in three Akts. Genes Dev 13(22):2905–2927
42. Tano JY, Vazquez G (2011) Requirement for non-regulated, constitutive calcium influx in macrophage survival signaling. Biochem Biophys Res Commun 407(2):432–437
43. Tano JY, Lee RH, Vazquez G (2012) Involvement of calmodulin and calmodulin kinase II in tumor necrosis factor alpha-induced survival of bone marrow derived macrophages. Biochem Biophys Res Commun 427(1):178–184
44. Tano JY, Smedlund K, Lee R, Abramowitz J, Birnbaumer L, Vazquez G (2011) Impairment of survival signaling and efferocytosis in TRPC3-deficient macrophages. Biochem Biophys Res Commun 410(3):643–647
45. Tano JY, Solanki S, Lee RH, Smedlund K, Birnbaumer L, Vazquez G (2014) Bone marrow deficiency of TRPC3 channel reduces early lesion burden and necrotic core of advanced plaques in a mouse model of atherosclerosis. Cardiovasc Res 101(1):138–144
46. Stoger JL, Gijbels MJ, van der Velden S, Manca M, van der Loos CM, Biessen EA, Daemen MJ, Lutgens E, de Winther MP (2012) Distribution of macrophage polarization markers in human atherosclerosis. Atherosclerosis 225(2):461–468
47. Leitinger N, Schulman IG (2013) Phenotypic polarization of macrophages in atherosclerosis. Arterioscler Thromb Vasc Biol 33(6):1120–1126
48. Wolfs IM, Donners MM, de Winther MP (2011) Differentiation factors and cytokines in the atherosclerotic plaque micro-environment as a trigger for macrophage polarisation. Thromb Haemost 106(5):763–771
49. Gordon S, Mantovani A (2011) Diversity and plasticity of mononuclear phagocytes. Eur J Immunol 41(9):2470–2472
50. Sica A, Mantovani A (2012) Macrophage plasticity and polarization: in vivo veritas. J Clin Invest 122(3):787–795
51. Xu C, Bailly-Maitre B, Reed JC (2005) Endoplasmic reticulum stress: cell life and death decisions. J Clin Invest 115(10):2656–2664
52. Solanki S, Dube PR, Tano JY, Birnbaumer L, Vazquez G (2014) Reduced endoplasmic reticulum stress-induced apoptosis and impaired unfolded protein response in TRPC3-deficient M1 macrophages. Am J Physiol Cell Physiol 307(6):C521–C531
53. Lim WS, Timmins JM, Seimon TA, Sadler A, Kolodgie FD, Virmani R, Tabas I (2008) Signal transducer and activator of transcription-1 is critical for apoptosis in macrophages subjected to endoplasmic reticulum stress in vitro and in advanced atherosclerotic lesions in vivo. Circulation 117(7):940–951
54. Timmins JM, Ozcan L, Seimon TA, Li G, Malagelada C, Backs J, Backs T, Bassel-Duby R, Olson EN, Anderson ME, Tabas I (2009) Calcium/calmodulin-dependent protein kinase II links ER stress with Fas and mitochondrial apoptosis pathways. J Clin Invest 119(10):2925–2941
55. Liu J, Thewke DP, Su YR, Linton MF, Fazio S, Sinensky MS (2005) Reduced macrophage apoptosis is associated with accelerated atherosclerosis in low-density lipoprotein receptor-null mice. Arterioscler Thromb Vasc Biol 25(1):174–179
56. Yancey PG, Blakemore J, Ding L, Fan D, Overton CD, Zhang Y, Linton MF, Fazio S (2010) Macrophage LRP-1 controls plaque cellularity by regulating efferocytosis and Akt activation. Arterioscler Thromb Vasc Biol 30(4):787–795
57. Damann N, Owsianik G, Li S, Poll C, Nilius B (2009) The calcium-conducting ion channel transient receptor potential canonical 6 is involved in macrophage inflammatory protein-2-induced migration of mouse neutrophils*. Acta Physiol 195(1):3–11

58. Inada H, Iida T, Tominaga M (2006) Different expression patterns of TRP genes in murine B and T lymphocytes. Biochem Biophys Res Commun 350(3):762–767
59. Heiner I, Eisfeld J, Halaszovich CR, Wehage E, Jungling E, Zitt C, Luckhoff A (2003) Expression profile of the transient receptor potential (TRP) family in neutrophil granulocytes: evidence for currents through long TRP channel 2 induced by ADP-ribose and NAD. Biochem J 371(Pt 3):1045–1053
60. Itagaki K, Barton BE, Murphy TF, Taheri S, Shu P, Huang H, Jordan ML (n.d.) Eicosanoid-induced store-operated calcium entry in dendritic cells. J Surg Res 169(2):301–310
61. Pantaler E, Luckhoff A (2009) Inhibitors of TRP channels reveal stimulus-dependent differential activation of Ca^{2+} influx pathways in human neutrophil granulocytes. Naunyn Schmiedeberg's Arch Pharmacol 380(6):497–507
62. Smedlund K, Birnbaumer L, Vazquez G (2015) Increased size and cellularity of advanced atherosclerotic lesions in mice with endothelial overexpression of the human TRPC3 channel Proc Natl Acad Sci U S A 112(17): E2201–E2206
63. Du F, Zhou J, Gong R, Huang X, Pansuria M, Virtue A, Li X, Wang H, Yang XF (2012) Endothelial progenitor cells in atherosclerosis. Front Biosci (Landmark Ed) 17:2327–2349
64. Fledderus JO, van Oostrom O, de Kleijn DP, den Ouden K, Penders AF, Gremmels H, de Bree P, Verhaar MC (2013) Increased amount of bone marrow-derived smooth muscle-like cells and accelerated atherosclerosis in diabetic apoE-deficient mice. Atherosclerosis 226(2):341–347
65. Oh J, Riek AE, Weng S, Petty M, Kim D, Colonna M, Cella M, Bernal-Mizrachi C (2012) Endoplasmic reticulum stress controls M2 macrophage differentiation and foam cell formation. J Biol Chem 287(15):11629–11641
66. Vazquez G (2012) TRPC channels as prospective targets in atherosclerosis: terra incognita. Front Biosci (Schol Ed) 4:157–166
67. Jain S, Amiji M (2012) Macrophage-targeted nanoparticle delivery systems. In: Svenson S, Prud'homme RK (eds) Multifunctional nanoparticles for drug delivery applications. Nanostructure science and technology. Springer, New York, pp 47–83
68. Hood ED, Chorny M, Greineder CF, SA I, Levy RJ, Muzykantov VR (2014) Endothelial targeting of nanocarriers loaded with antioxidant enzymes for protection against vascular oxidative stress and inflammation. Biomaterials 35(11):3708–3715

Chapter 10
Second Messenger-Operated Calcium Entry Through TRPC6

Alexandre Bouron, Sylvain Chauvet, Stuart Dryer, and Juan A. Rosado

Abstract Canonical transient receptor potential 6 (TRPC6) proteins assemble into heteromultimeric structures forming non-selective cation channels. In addition, many TRPC6-interacting proteins have been identified like some enzymes, channels, pumps, cytoskeleton-associated proteins, immunophilins, or cholesterol-binding proteins, indicating that TRPC6 are engaged into macromolecular complexes. Depending on the cell type and the experimental conditions used, TRPC6 activity has been reported to be controlled by diverse modalities. For instance, the second messenger diacylglycerol, store-depletion, the plant extract hyperforin or H_2O_2 have all been shown to trigger the opening of TRPC6 channels. A well-characterized consequence of TRPC6 activation is the elevation of the cytosolic concentration of Ca^{2+}. This latter response can reflect the entry of Ca^{2+} through open TRPC6 channels but it can also be due to the Na^+/Ca^{2+} exchanger (operating in its reverse mode) or voltage-gated Ca^{2+} channels (recruited in response to a TRPC6-mediated depolarization). Although TRPC6 controls a diverse array of biological functions in many tissues and cell types, its pathophysiological functions are far from being fully understood. This chapter covers some key features of TRPC6, with a special emphasis on their biological significance in kidney and blood cells.

Keywords TRPC6 • Ca^{2+} entry • DAG • Kidney • Blood cells

A. Bouron (✉) • S. Chauvet
Université Grenoble Alpes, 38000 Grenoble, France

CNRS, iRTSV-LCBM, 38000 Grenoble, France
e-mail: alexandre.bouron@cea.fr

S. Dryer
University of Houston, Houston, TX, USA

Baylor College of Medicine, Houston, TX, USA

J.A. Rosado
Departamento de Fisiología, University of Extremadura, Cáceres, Spain
e-mail: jarosado@unex.es

J.A. Rosado (ed.), *Calcium Entry Pathways in Non-excitable Cells*, Advances in Experimental Medicine and Biology 898, DOI 10.1007/978-3-319-26974-0_10

10.1 Introduction

In mammals, the super-family of transient receptor potential (TRP) ion channels gathers six families named TRPA, TRPC, TRPM, TRPML, TRPP, and TRPV [1]. Among these, members of the TRPC family, where C stands for "*classical*" or "*canonical*", have the closest homology with TRP and TRPL, the founding TRP members, which were identified in Drosophila [2]. The first TRPC was cloned in 1990s [3]. Since this pioneering work, seven members (TRPC1-TRPC7) are known to date, making the TRPC family an important class of cation channels [4] playing multiple roles in health and disease [5]. Within the TRPC family several groups have been identified: TRPC1, TRPC2, TRPC4/5, and TRPC3/TRPC6/TRPC7 [6]. Depending on the authors, TRPC1 is included into the TRPC4/5 subfamily [4].

In mice, the gene encoding TRPC6 (mTRPC6) has 13 exons and is located on chromosome 9 and full-length cDNA of mTRPC6 was isolated from the brain [7]. In human, the gene encoding TRPC6 (hTRPC6) is localized on chromosome 11q21–q22 and has 13 exons [8], and the cDNA was cloned from the placenta [9]. The mTRPC6 cDNA exhibits a similarity (and identity) of 85.1 % (and 74.1 %) with the mTRPC3 cDNA [7]. Mouse TRPC7 has 75 % identity with mTRPC6 and 84 % similarities [10]. Overall, TRPC6 has 70–80 % amino acid (a.a.) identity with TRPC3 and TRPC7.

The murine and human TRPC6 proteins have 930 and 931 amino acids, respectively, and their amino acid identity is 93 % [7, 9]. Several TRPC6 isoforms are known [11]. They result from the alternative splicing of the TRPC6 gene. However, some of these variants are not functional [12]. For instance, in the rat 3 isoforms were found: rTRPC6A, rTRPC6B, and rTRPC6C, having 930, 876 and 808 a.a., respectively [11]. rTRPC6B has 54 a.a. less than the A isoform, these missing a.a. are in the N-terminus whereas the C isoform lacks 68 a.a. near its C-terminus. rTRPC6A has 88.38 % and 94.41 % nucleic acid identity with the human and mouse TRPC6. Only the isoforms A and B are functional [11].

Nearly 20 years after the cloning of TRPC6, its expression and functional properties are much better characterized. Some of its hallmark features is the fact that TRPC6 is a lipid-dependent membrane protein forming non-selective cation channels conducting Na^+ and Ca^{2+}, but the functional significance of this Ca^{2+}-dependent transport system in health and disease is not yet completely elucidated.

10.2 Structure

Voltage-gated K^+, Na^+ and Ca^{2+} channels have served as structural models to establish the putative molecular architecture of all TRPC channels. They contain six transmembrane domains (named S1–S6), and intracellular N- and C-termini [13, 14]. The current model, which also applies to TRPC6, proposes that a TRPC channel would result from the association of four TRPC proteins. TRPC channels are thus regarded as tetrameric structures where S5 and S6 contribute to the pore function [15]. TRPC6 subunits can associate with TRPC6 and non-TRPC6 subunits like TRPC1, TRPC3 and TRPC7 subunits (see next Chapter), forming homo- or hetero-tetrameric structures. Within the large intracellular NH_2- and COOH-terminals, several important domains have been identified.

In the NH_2-Terminal

- The ankyrin (ANK) repeat, a relatively conserved structural motif of 33 residues, is a protein module with high affinity for other ANK repeats. These modules are present abundantly in a multitude of proteins and are regarded as ubiquitous scaffolds mediating protein-protein interactions [16] and involved in many processes like signal transduction, cell cycle regulation, vesicular trafficking and cytoskeleton integrity [17]. Members of the super-family of TRP proteins, including TRPCs, possess ANK repeats [18]. For instance, three ANK-like repeats located in the N-terminus were identified in the three rat TRPC6 isoforms [11].
- A coiled-coil-domain (close to the intracellular side of the transmembrane domain 1). This type of domain, thought to control oligomerisation, could thus participate in the oligo- or heteromerization of TRPC6 channels. It could also participate in the association of TRPC6 with other proteins.

In the COOH-Terminal

- TRP domain: it is a motif of 25 a.a. that is highly conserved throughout TRPC channels. This domain is located on the COOH-terminal part, near the sixth transmembrane domain. It contains a group of 6 invariant a.a. that are forming the TRP box (though to be involved in channel gating). This domain would be necessary for the binding of phosphoinositol phosphates (like PIP_2) (see later, section "Activation").
- Prolin-rich motif: a prolin-rich motif (found in all TRPCs) is present downstream to the TRP domain; it would participate in the interaction with immunophilins.
- A part of the COOH-terminal domain of TRPC6, containing several consensus calmodulin (CaM) binding motifs and also interacting with inositol 1,4,5-trisphosphate (IP_3) receptors (IP_3R), is described as the CaM and IP_3R binding domain (CIRB) which is conserved among all TRPC members [19]. The CaM-binding sequences of all TRPCs are known [19]. For mTRPC6 a motif lying from a.a. 838 to 872 has been identified. The binding of CaM to the CIRB motif of TRPCs inhibits their activity. The CIRB motif of mTRPC6 interacts with the COOH-terminal lobe of CaM [19].
- A coiled-coil-domain is present in the COOH-terminal domain.

Two sites participating in TRPC interaction are named *assembly domains* AD1 and AD2 [20]. They would stabilize the interactions between TRPCs. AD1 is located within the NH_2-terminus and overlaps the ankyrin and the coiled-coil domains. It is necessary for self-association of the NH_2-termini of adjacent TRPCs. AD2, which overlaps the pore region and the NH_2-terminal tail, is involved in the interaction with the NH_2- and COOH-terminal parts of adjacent TRPCs. Based on these results, a model was proposed by [20] with interactions between (i) the NH_2-termini and involving two regions (ANK repeats and coiled-coil), and (ii) between the NH_2- and the COOH-termini of an adjacent TRPC.

10.3 Activation

Understanding the physiological mechanism of activation of TRPCs, including TRPC6, has always been a difficult task. This part of the literature has generated discordant findings [21]. Indeed, multiple modes of TRPC activation have been documented and it has been proposed that their native environment would be essential in defining their recruitment under physiological conditions [21]. Similar remarks apply to TRPC6 which were reported to respond to different types of stimuli (chemical and mechanical) (see below). After its cloning, the analysis of the functional properties of mTRPC6 indicated that it is a non-selective Ca^{2+}-conducting channel blocked by La^{3+} and SKF-96365 and activated downstream to G-protein coupled receptors (GPCR), and not responding to Ca^{2+} store depletion [7]. A subsequent work showed that the gating of TRPC6 (and also TRPC3) channels can be directly controlled by diacylglycerol (DAG) without the need to deplete intracellular Ca^{2+} stores. TRPC6 channels were thus described as *second-messenger operated channels* that open downstream to receptors coupled to the hydrolysis of phosphatidyl inositol 4,5-bisphosphate (PIP_2) [9]. According to a generally held opinion, TRPC6 channels are activated downstream to GPCR or receptor tyrosine kinases via a PLC-dependent signaling process involving PLCβ or PLCγ and leading to the production of DAG.

10.3.1 DAG, Phosphoinositides

There is ample evidence that TRPC6 channels can function as DAG-activated channels and it is now widely accepted that DAG is a physiological activator, at least in some cell types. Experimentally, DAG analogues such as OAG (1-oleoyl-2-acetyl-*sn*-glycerol) are commonly used to activate TRPC6. [11] identified a stretch of a.a. (3–56) located in the NH_2-terminal part of TRPC6 that seemed to confer DAG sensitivity. However, conflicting data have been reported concerning the contribution of this motif of 54 a.a. on conferring the sensitivity to DAG-analogues like OAG. For instance, other OAG-sensitive channels like TRPC3 and TRPC7 lack this motif. In addition, an electrophysiological study conducted on the rTRPC6B isoform (which does not have this motif) and expressed in the T-REx-r6 cell line showed that it could be activated by OAG [22]. A similar finding has been reported for truncated TRPC3 channels, further questioning the role of this 54-a.a. motif [23]. Although, the exact mechanism of action of DAG on TRPC6 is still unresolved, a dissociation constant of DAG binding to TRPC6 of 20–40 μM has recently been calculated [24].

This latter study showed a parallel response between DAG production and TRPC6 activation downstream to receptor stimulation, supporting the view that the channel is DAG-sensitive.

Besides DAG, TRPC6 can be activated by DAG-containing arachidonic acids like SAG (1-stearoyl-2-arachidonyl-*sn*-glycerol) and PAG (1-palmytoyl-2-arachidonic acid-*sn*-glycerol), and by DAG-containing docosahexaenoic acid like SDG (1-stearoyl-2-docosahexaenoyl-*sn*-glycerol) and DOG (1,2-dioctanoyl-*sn*-glycerol) [25]. It was proposed that the arachidonic acid present at the *sn*-2 position is an important molecular determinant controlling TRPC6 activation [25]. By using DAG lipase inhibitors, these authors revealed that the activation of the channels was not due to the metabolism of SAG. Thus, the action of SAG is not mediated by free arachidonic acid or phosphatidic acid. Overall, SAG is the most effective DAG analogue in activating TRPC6 when compared to OAG, DOG and SLG (1-stearoyl-*sn*-glycerol). The SAG-dependent activation of TRPC6 channels requires the integrity of lipid rafts because their disruption with methyl-β-cyclodextrin (MβCD) prevents TRPC6 activation. This is also seen after knockdown of certain raft organizing proteins [26]. In addition, tyrosine phosphorylation would be a necessary step for the activation of the channels [25]. On the other hand, TRPC6 channels do not respond to the application of monoacylglycerols (e.g. 1-oleoyl-glycerol and 2-oleoyl-glycerol) [9]. Besides DAG and DAG analogues, TRPC6 can open in response to the application of DAG lipase inhibitors, which elevate DAG levels [9, 27, 28].

20-HETE (20-hydroxyeicosatetraenoic acid) is a metabolite of arachidonic acid. The external application of 20-HETE on HEK cells over-expressing mTRPC6 generates non-selective inward currents [29]. Arachidonic acid was also able to activate mTRPC6 channels but not the non metabolizable arachidonic acid analog ETYA, suggesting that TRPC6 channels are sensitive to metabolites of arachidonic acid. The authors propose that 20-HETE would be this natural activator [29].

The influence of lipids on TRPC6 activity has been further documented by [30] who showed that the lipid lysophosphatidylcholine is able to activate hTRPC6 channels over-expressed in HEK cells and native TRPC6 channels from bovine aortic endothelial cells in culture. A parallel increase in tyrosine phosphorylation of TRPC6 was noted after a 1 h treatment with lysophosphatidylcholine [30].

As mentioned above, TRPC6 are described as channels activated by the second messenger DAG which is produced in response to the recruitment of receptors coupled to PLCβ or PLCγ. Several phosphoinositides (PIs), like PIP_2, were shown to regulate TRPC6 activity. PIP_2 is more than a precursor of second messengers, it is itself a signaling molecule regulating the activity of many target proteins including ion channels [31, 32]. Contrasting results have been collected regarding the effect of PIP_2 on TRPC6. PIP_2 was shown to exert a positive effect on the activity of hTRPC6 channels [33]. Indeed, the COOH-terminal region of TRPC6 can bind to various PIs such as phosphatidylinositol-3,4,5-tisphosphate (PIP_3) and PIP_2. In fact, the COOH-terminal part of TRPC6 directly binds PIs, its affinity being higher for PIP_3 (compared to PIP_2) [34]. These PIP_3-activated TRPC6 Ca^{2+} responses were sensitive to Gd^{3+} (50 μM) and SKF-96365 (10 μM) but exposure of cells to other phosphoinositides like PI and PIP_2 failed to generate any Ca^{2+} signal [34]. The PI-binding site overlaps with the CaM-binding site but is not identical to it. It stretches from the residues 842 and 868, with three a.a. (arginine 853, lysine 860, and arginine 861) playing important roles in the binding of CaM and PI. The PI-dependent displacement of the CaM binding causes an enhancement of TRPC6-mediated currents [33]. TRPC6 channels (and also its close relatives TRPC3 and TRPC7) stably over-expressed in HEK cells were shown to be directly activated by PIP_2 [35]. However, in vascular smooth muscle cells the DAG-dependent activation of endogenous TRPC6 channels is antagonized by PIP_2, which is regarded as a physiological TRPC6 antagonist [36]. Based on results obtained with FRET experiments, it was concluded that PIP_2 controls TRPC6 inactivation [24].

Two important intracellular messengers are produced downstream to PIP_2: IP_3 and DAG. If the effect of the later one on TRPC6 activity has been well documented, much less is known about IP_3. It seems however that TRPC6 is not activated by IP_3 [9, 27, 37–39] that could act in a synergistic manner with DAG [39].

10.3.2 Store Depletion

The release of Ca^{2+} from the endoplasmic reticulum (ER) activates a plasma membrane conductance allowing the entry of Ca^{2+} into cells. This Ca^{2+} pathway, also named *capacitative* Ca^{2+}*entry* (CCE) [40], would reflect a *store-operated* Ca^{2+}*entry* (SOCE) via *store-operated channels* (SOCs). The activation of a PIP_2-dependent signaling pathway leading to DAG production also generates IP_3, which is followed by the recruitment of IP_3-sensitive intracellular channels (IP_3R) permitting the release of Ca^{2+} into the cytosolic compartment. Therefore, by depleting intracellular Ca^{2+} pools, IP_3 production can indirectly activate SOCs.

So far, TRPC6 channels have been predominantly described as contributing to a Ca^{2+} pathway not operated by Ca^{2+} stores. TRPC6 was indeed originally defined as non-activated by store depletion when over-expressed in HEK or CHO cells [7, 9, 27, 41]. In cardiac fibroblasts [42] and human platelets [43], TRPC6 was also reported to function as receptor-activated channels but not as store-operated channels. However, although the vast majority of studies published so far indicate that TRPC6 is a DAG-sensitive channel, several groups have challenged this view by showing that TRPC6 can be store-operated. This is for instance the case of rTRPC6 where its over-expression in COS cells enhances the thapsigargin (TG)-stimulated Ca^{2+} entry, indicating that it would function as a SOC [15]. A similar conclusion has been reached in pulmonary smooth muscle cells [44] and in the hepatoma cell line Huh-7 where TRPC6 contributes to the SOCE but not to the OAG- or receptor-operated Ca^{2+} entry [45].

Deciphering the molecular identity of SOCs has been a major challenge that generated controversies and lively debates [46]. TRPC proteins have long been thought to be the components of SOCs. However, a series of seminal articles published in 2006 provided strong experimental evidence for a central role of Stim and Orai proteins as important components of SOCs [47–50]. It is now well accepted that the release of Ca^{2+} from the ER influences the spatial distribution of Stim, a ER Ca^{2+} sensor, that relocalizes to regions near the plasma membrane [51]. If the contribution of Stim proteins in the control of SOCs seems consensual, the next step (namely, how Stim activates a plasma membrane conductance allowing the entry of Ca^{2+}?) is more disputed. Schematically, two models emerge. According to the first one, the translocation of Stim that follows the emptying of the Ca^{2+} stores permits the recruitment of Orai proteins. These are plasma membrane Ca^{2+} channels for which three isoforms are known (Orai1-3). This scenario implies that Stim and Orai are the central molecular components of SOCs able to generate an influx of Ca^{2+} (so-called SOCE) without the need of TRPC channels. This picture has however been challenged. Indeed, Stim but also Orai can interact with TRPC [52]. The second model thus proposes the existence of a more complex and dynamic picture with a tripartite complex involving Stim/Orai/TRPC as components of SOCs [53]. HL-60 cells expressed various TRPCs, with TRPC3 and TRPC6 being the predominant isoforms found [54]. By means of pharmacological agents acting on Ca^{2+}-conducting channels, it was concluded that TRPC6 would contribute (with TRPC1 and Orai1) to SOC in HL-60 cells whereas TRPC3 would contribute to receptor-operated channels [54].

In platelets, an even more dynamic picture has been proposed. Indeed, in these cells TRPC6 seem to be engaged in distinct macromolecular complexes and participate in two different Ca^{2+} routes: one (a SOCE) is under the control of intracellular Ca^{2+} stores whereas the second (non SOCE) functions independently of the Ca^{2+} stores [55]. According to these authors, TRPC6 could form a complex with Orai1 and Stim1 responding to Ca^{2+} store-depletion whereas applications of DAG analogues displaces TRPC6 from this complex, favoring its association with TRPC3. In this latter case, TRPC6 would function as a store-independent channel [55]. This view has however been recently challenged: Ca^{2+} imaging experiments on platelets from WT and TRPC6$^{-/-}$ mice indicated that in these cells TRPC6 does not contribute to SOCE but function as a receptor-operated channel [56].

10.3.3 Mechano-Sensation (or Stretch Activation)

Studies conducted on hTRPC6 channels expressed in HEK and CHO cells concluded that they could be sensors of mechanical stress: application of a hypo-osmotic medium activates TRPC6 in a PLC-independent manner [57]. The biophysical properties of this stretch-activated current are quite similar to the ones previously reported for TRPC6 with a single-channel conductance of ~25 pS. Furthermore, OAG and mechanical stress generate non additive responses. The stretch-induced TRPC6 activation is sensitive to the peptide GsMTx-4, an inhibitor of mechano-sensitive channels. GsMTx-4 also blocks OAG-activated TRPC6 currents further suggesting that stretch and OAG recruit the same cation channel. In their work [57], present TRPC6 channels as sensors of membrane stretch (whether the tension is induced physically or osmotically), and playing a role in regulating vascular tone. This conclusion was questioned in a subsequent study on hTRPC6 channels transiently over-expressed in CHO and COS-7 cells [58]. In this work, the authors argue against the idea that TRPC6 is a mechano-sensitive protein. Indeed, they found little or even no evidence for a role of TRPC6 as a mechano-sensitive protein. According to [58] analyzing the mechano-sensitivity of a channel over-expressed in a heterologous system is problematic and is not necessarily the best way to precise its mechano-sensitivity. They suggest that is it not possible to conclude from studies performed on heterologous systems that TRPC6 is a stretch-activated channel. A similar conclusion was reached by [59] who found that TRPC6 cannot be regarded as a mechano-sensitive channel. In fact, a mechanical stimulus seems capable of activating Gq-coupled GPCR. According to this scenario, TRPC6 channels are activated downstream to GqPCR that would be the real sensors of a mechanical stress in smooth muscle cells [59]. More recently, [60] show that in mesenteric and renal arteries the angiotensin II receptor of type 1a is able to function as a stretch-activated receptor without the need of its ligand. Furthermore, they showed that TRPC6 plays no role in the downstream response. These conclusions are not supported by all of the studies on vascular smooth muscle [61] or heterologous expression systems [62, 63]. In dorsal root ganglion neurons, TRPC6 channels may closely interact with TRPV4 channels to form a larger complex that can detect mechanical stimuli [64]. Similarly, recent studies of cochlear hair cells and sensory neuron cell lines have suggested that TRPC3/TRPC6 heteromers contribute to low-sensitivity and very short latency mechano-sensitivity [63]. These workers also found that the presence of TRPC3 subunits markedly enhances mechano-sensitivity of TRPC6, although it is significant that this was only observed in certain cell lines [63]. TRPC6 channels in ventricular myocytes [65] and podocytes are activated by a variety of mechanical stimuli, and these responses are also blocked by GsMTx4. Mechanical activation in podocytes persists when G protein signaling is completely inhibited [66] and clearly cannot be attributed to stretch activation of any GPCR in this cell type. Moreover, mechanical activation of TRPC6 in podocytes is enhanced by podocin knockdown, whereas responses to OAG or various GPCRs are enhanced by podocin knockdown [66–68]. It is possible that mechanical activation in this cell type is membrane delimited and requires other proteins to be present (e.g. TRPC3 or TRPC1), but the responses are completely abolished by TRPC6 knockdown. In summary, mechano-sensitivity of TRPC6-containing channels can occur through G protein mediated signaling, may be conferred by other proteins within a larger complex, but it also remains possible that intrinsic or membrane-delimited mechano-sensitivity of TRPC6 is suppressed by interacting proteins in certain cellular contexts.

10.3.4 Hyperforin

It is a molecule extracted from the medicinal plant *Hypericum perforatum*. It exhibits the interesting property to activate TRPC6 channels without influencing the activity of any other TRPC [69]. There is a growing interest for this pharmacological tool because it can activate native [70–72] and heterologously expressed TRPC6 channels [69]. Leuner and colleagues tried to gain a better molecular understanding of the effects of hyperforin by carrying out a structure-function study of various hyperforin derivatives. They concluded that 2,4-diacylphloroglucinol is a likely important structural part bearing the TRPC6 selectivity [73]. Of note, conflicting results have recently been published and showing that hyperforin controls a proton-conducting pathway unrelated to TRPC6 [74].

10.3.5 Flufenamic Acid (FFA)

FFA is a non-steroidal anti-inflammatory agent that blocks TRPC3- and TRPC7-mediated responses whereas it potentiates TRPC6-mediated currents [27, 75]. It was subsequently shown to directly activate native and heterogously expressed TRPC6 channels [76]. It was proposed to use this agent as a TRPC6 agonist in human podocytes [76]. However, caution is needed when using FFA. Indeed, it depresses TRPC6-mediated Ca^{2+} responses in PC12 cells [77]. In addition, it collapses the mitochondrial membrane potential and releases Ca^{2+} from mitochondria [78]. Investigations by one of us (Stuart Dryer unpublished observation) suggest that FFA acts as a partial agonist of TRPC6.

10.3.6 AlF_4- and Trimeric G Proteins

AlF4- has been shown to activate trimeric G proteins. To verify the role of these signaling molecules in TRPC6 activation, AlF4- was employed in some electrophysiological studies. For instance, AlF_4- (30 μM) was added to the pipette solution to trigger hTRPC6-mediated currents in HEK cells [9, 79]. Application of GTPγS can directly trigger the opening of TRPC6 channels [27]. Conversely application of GDPβS can prevent TRPC6 activation by GPCR [67, 68].

10.3.7 H_2O_2

A role of reactive oxygen species (ROS) in TRPC6 activation has been provided by [72, 80]. In these papers, the authors showed that a low concentration of H_2O_2 (e.g. 10 μM) can directly activate TRPC6 channels expressed in HEK cells and native TRPC6 channels in vascular smooth muscle cells. This H_2O_2-dependent TRPC6 activation does not occur through direct oxidation of the channel but requires thiol oxidation of a cytosolic component. In addition, OAG and H_2O_2 have synergistic effects [80]. This also occurs in a podocyte cell line [81]. Of note, when chronically applied to human mesangial cells for 2–24 h, H_2O_2 down-regulates TRPC6 expression in a concentration-dependent manner via a PKC mechanism [82].

10.3.8 Plasma Membrane Insertion of TRPC6 Upon Activation

The regulation of the plasma membrane abundance of TRPC6 is an important parameter influencing its activity. Experiments conducted on HEK cells over-expressing mTRPC6 revealed that the control of the trafficking of the channel and its insertion into the plasma membrane is a crucial regulatory mechanism. This can be induced by OAG and also by depletion of Ca^{2+} stores via a process that does not require any increase in cytosolic Ca^{2+}. This translocation of the channels leads to larger TRPC6-mediated currents [83]. A similar observation was made in PC12 cells where the application of an agonist recruiting GPCRs is accompanied by the insertion of TRPC6 channels into the plasma membrane; a process associated with an enhanced Ca^{2+} influx [84]. Interestingly, the trafficking of TRPC6 is negatively regulated by the membrane protein Klotho which prevents the exocytosis of TRPC6 by blocking the phosphoinositide-3-kinase. The Klotho-dependent regulation of TRPC6 is associated with depressed TRPC6-mediated Ca^{2+} responses [85]. H_2O_2 can stimulate TRPC6 translocation and its insertion into the plasma membrane but H_2O_2 is also able to influence TRPC6 activity even after blocking its trafficking process indicating that H_2O_2 may act at multiple sites [80]. Lysophosphatidylcholine triggers the translocation of TRPC6 and its insertion into the plasma membrane. This response was observed on hTRPC6 expressed in HEK cells as well as on native TRPC6 from bovine aortic endothelial cells [30]. The TRPC6 trafficking elicited by lysophosphatidylcholine occurs even when chelating intracellular Ca^{2+} or when incubating cells in a Ca^{2+}-free medium. Interestingly, this lysophosphatidylcholine-induced translocation of TRPC6 triggers the plasma membrane insertion of TRPC5 [30].

10.4 TRPC6 Interacting Partners

Many TRPC6-interacting partners have been identified (see for instance the TRIP Database: http://www.trpchannel.org [86]). Schematically, they can be classified as belonging to: channels and pumps, enzymes and intracellular signaling molecules, cytoskeleton-associated proteins, proteins involved in the trafficking, immunophilins, and Stim/Orai proteins (Fig. 10.1).

10.4.1 Interactions with Channels and Pumps

TRP Channels although TRPC6 can form homomeric channels, at least in some cell types like mesenteric artery myocytes [87], it is however able to interact with other TRPCs like TRPC1 [88], TRPC2 [89], TRPC3 [90–93], or TRPC7 [91, 92]. On the other hand, it does not seem to associate with TRPC4 [20, 88] or TRPC5 [30, 94].

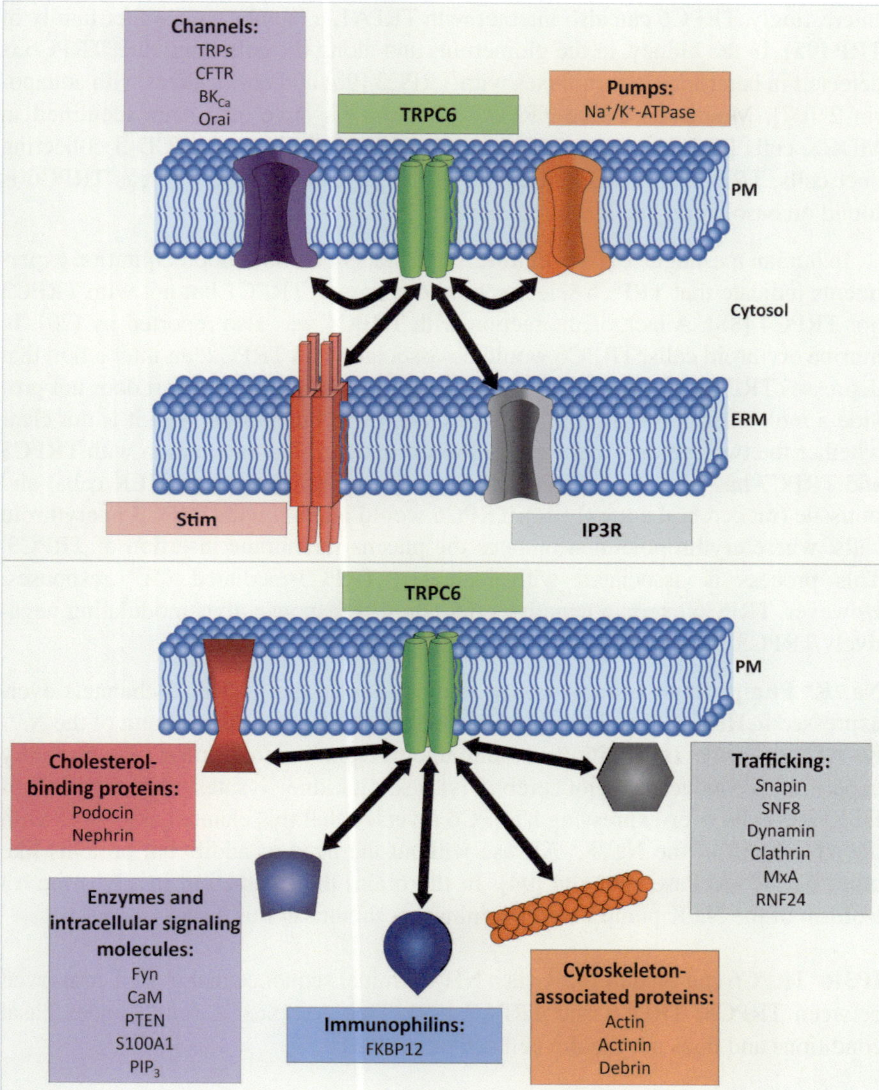

Fig. 10.1 TRPC6-interacting partners. Schematic representation of TRPC6 and its interacting partners. *Top panel*: TRPC6 has been shown to interact with channels and pumps of the plasma membrane (PM) as well as with actors located in the membrane of the endoplasmic reticulum (ERM). *Lower panel*: various groups of proteins have been proposed to interact with TRPC6 like cholesterol-binding proteins, enzymes and intracellular signaling molecules, cytoskeleton-associated proteins, and proteins involved in intracellular trafficking. See text for further details. TRPC6 is represented as a tetrameric structure, and for the sake of clarity the C- and N-termini have been omitted

Interestingly, TRPC6 can also interact with TRPA1, a member of another family of TRP [95]. In the kidney, in the glomerulus and along the collecting duct TRPC6 is detected in heteromeric complexes with TRPC3 [96] and co-localizes with aquaporin 2 [97]. Moreover, native TRPC3/6 heteromers have also been identified in MDCK cells [90]. However, in polarized cultures of M1 and IMCD-3 collecting duct cells, TRPC3 localizes exclusively to the apical domain, whereas TRPC6 is found on basolateral and apical membranes [98].

In human mesangial cells in culture, the results of coimmunoprecipitation experiments indicate that TRPC6 selectively interacts with TRPC1 but not with TRPC3 nor TRPC4 [88]. A lack of interaction with TRPC4 was also reported by [20]. In murine erythroid cells, TRPC6 would co-associate with TRPC2, an interaction that depresses TRPC2-mediated Ca^{2+} responses [89]. However, this report does not provide a molecular description of the TRPC2-TRPC6 interaction. And it is not clear whether the two proteins associate to form a channel. The interaction with TRPC3 and TRPC7 has been shown in a heterologous expression system (HEK cells) and in tissue (rat cerebral cortex) [92]. TRPC6 would interact with TRPC3 in erythroid cells, where erythropoietin stimulates the plasma membrane insertion of TRPC3. This process is associated with increased TRPC3-mediated Ca^{2+} responses. However, TRPC6 exerts a negative effect on this response, thus modulating negatively TRPC3 activity [93].

Na^+/K^+ Pump Immunofluorescence studies showed that hTRPC6 channels overexpressed in HEK-293 cells co-localize with the endogenous α1-subunit of the Na^+/K^+-ATPase [94]. In addition, co-immunoprecipitation and mass spectrometry experiments conducted on rat cerebral lysates, rat kidney lysates, and on lysates of HEK-293 cells over-expressing hTRPC6 revealed that this channel associates with the α1-subunit of the Na^+/K^+-ATPase without the need of additional proteins like other Na^+/K^+-ATPase subunits [94]. In the brain, the interaction involves the α3 isoform of the Na/K pump, the predominant α subunit in this organ.

IP3R TRPC6 can bind to IP3R via a NH_2-terminal sequence that is well conserved between TRPC3, TRPC6 and TRPC7 [99]. This complex is found under basal conditions and does not require cell activation [83].

CFTR or cystic fibrosis transmembrane conductance regulator, is a membrane protein controlling fluxes of chloride ions that plays a central role in the pathogenesis of cystic fibrosis. Experiments conducted on various epithelial cell lines and on freshly ciliated human bronchial epithelial cells showed that CFTR and TRPC6 interact each other and form a multiprotein complex [100].

BK_{Ca} TRPC6 binds directly to large-conductance Ca^{2+}-activated K^+ channels in podocytes, which are non-excitable visceral epithelial cells in kidney glomeruli [101]. This interaction may help to maintain driving force for Ca^{2+} influx during sustained TRPC6 activation in non-excitable cells [28].

Interactions with Stim/Orai Proteins

- Stim1: TRPC6 channels over-expressed in HEK cells and endogenous TRPC6 channels of the A10 smooth muscle cell line are not regulated by Stim1 [102]. These authors found no evidence for any functional regulation. TRPC6 is unable to directly interact with STIM proteins, but can indirectly interact with them in heterotrimeric complexes with TRPC1 [103]. STIM1 regulates the expression of TRPC6 in the plasma membrane and evokes its translocation to the ER [104].
- Interaction with Orai proteins: The question of the contribution of TRPC proteins to the entry of Ca^{2+} through SOCs is still under current investigation and this issue has been the subject of contrasting results. [52] found that COOH- and NH_2-terminal parts of TRPC6 (but also TRPC3) can interact with Orai proteins (Orai 1, 2 and 3). Two reports suggested a TRPC6-Orai interaction in a heterologous expression system (HEK293 cells) stably expressing TRPC6 by observing increased SOC currents [105, 106]. This Orai-TRPC interaction confers the sensitivity to Ca^{2+} store depletion via a Stim1-dependent process. In the hepatoma cell line Huh-7, TRPC6 would be engaged in a complex comprising Stim1 and Orai1 where they would participate in the SOC pathway [45]. A store-operated TRPC6-Stim1-Orai complex might exist in human platelets [55, 107].

10.4.2 Interactions with Enzymes and Intracellular Signaling Molecules

Calmodulin (CaM) TRPC members seem to possess several CaM binding sites on their COOH-terminal part but only one has been identified in the COOH-terminal part of TRPC6 [33, 108]. This CaM-binding site overlaps with the IP_3-binding site [109]. The interaction CaM-mTRPC6 is abolished in the presence of CaM inhibitors like calmidazolium or trifluoperazine [110].

The Tyrosine Kinase Fyn a binding site for Fyn located in the NH_2-terminal of mTRPC6 channels has been identified by [111]. Furthermore, mTRPC6 channels can be tyrosine phosphorylated by Fyn and Src. The Fyn-mTRPC6 interaction is not dependent on the tyrosine phosphorylation status of mTRPC6 [111].

PTEN is a phosphatase that exerts a positive role on TRPC6 activity. It interacts with TRPC6 but not with TRPC1, TRPC3 or TRPC4. This interaction facilitates the plasma membrane insertion of TRPC6 channels but the enzymatic activity of PTEN is not required for this regulatory process to develop [112].

A complex molecular rearrangement leading to the formation of a multiprotein complex centered on TRPC6 has been uncovered by [113]. They carried out their work on endogenous TRPC6 of PC12D cells (which express the TRPC6B isoform). In this cell model, the activation of muscarinic receptors triggers the association between these receptors and TRPC6 channels, a process that induces the recruitment of PKC. This causes the phosphorylation of TRPC6 in a PKC-dependent manner. This phosphorylation depends upon the activation of the muscarinic receptors. Two serine residues (ser768/714) seem to play a central role in this PKC-dependent phosphorylation. The formation of the tripartite complex muscarinic receptor-TRPC6-PKC allows the recruitment of the immunophilin FKBP12. Finally CaM and calcineurin are recruited to this complex, leading to the dephosphorylation of TRPC6 which correlates with the dissociation of the muscarinic receptor from TRPC6 [113].

S100A1 S100 proteins form a large family of Ca^{2+}-binding proteins participating in many cellular functions [114]. S100A, a protein mainly present in the heart where it controls cardiac performance, can interact with TRPC6 via its COOH-tail (801–878) [115].

PIP_3 Binding of PIP_3 was demonstrated to be at the CaM binding site [33] originally identified as CIRB site in TRPC3 [116]. Mutations decreasing the affinity of PIP_3 in this binding site enhanced CaM binding and reduced TRPC6-dependent current. Vice versa, mutations resulting in increased PIP_3 binding led to a reduced affinity of calmodulin and enhanced TRPC6 currents, while a triple amino acid substitution (R853Q/K860Q/R861Q in human TRPC6) resulted in reduced binding of both interaction partners [33].

10.4.3 Interactions with Cytoskeleton-Associated Proteins

Actin, Actinin and Debrin TRPC6 interacting partners like actin, actinin and debrin were found in rat brain lysates [94]. In HEK cells, the over-expression of mTRPC6 enhances the protein levels of ezrin and cofilin-1, two actin-associated proteins, and affects the actin cytoskeleton [117].

10.4.4 Interactions with Proteins Involved in the Trafficking

Snapin is a synaptic vesicle associated protein also known as a SNAP-25 binding protein of the SNARE complex. A motif of 24 a.a. located in the COOH-terminal part of the alpha1A-adrenergic receptor interacts with Snapin [84]. This interaction promotes the association with TRPC6. Furthermore, in the rat cortex, native TRPC6 associates with Snapin. The activation of alpha1A-adrenergic receptor promotes the plasma membrane insertion of TRPC6 via the formation of tripartite complex involving alpha1A-adrenergic receptor/Snapin/TRPC6 [84].

SNF8 a component of the ESCRT-II endosomal trafficking complex, interacts with TRPC6. The a.a. sequence 1–107 seems necessary and sufficient for this interaction to occur [118]. Co-expressing SNF8 and TRPC6 increases the current through TRPC6 without changing its surface expression. However, the mechanism by which SNF8 influences TRPC6 activity is not completely understood. SNF8 could regulate TRPC6 activity by influencing a TRPC6-associated protein [118].

Dynamin, Clathrin co-immunoprecipitation experiments indicate that dynamin and clathrin interact with TRPC6 (in rat brain) [94].

MxA is a member of the dynamin GTPase superfamily. Its COOH-terminal part interacts with the second ankyrin repeat of TRPC6 [41]. Furthermore, MxA upregulates TRPC6 activity without changing its amount at the plasma membrane, a regulatory process that requires the binding of GTP to MxA [41].

RNF24 or ring finger protein 24 is a zinc-binding membrane protein. It is a TRPC-interacting protein that controls the intracellular retention of TRPC6. The ANK-like repeats 1 and 2 of TRPC6 are important in mediating the interaction with RNF24 that seems to occur in the Golgi apparatus. It controls the intracellular trafficking of TRPC6 and its insertion into the plasma membrane but has no effect on the activity of the channel [119].

10.4.5 Immunophilins

FKBP12 or FK505-binding protein of 12 kDa is one of the best characterized immunophilins. These targets of immunosuppressant drugs have the ability to influence cellular Ca^{2+} signaling by altering the activity of Ca^{2+} channels. For instance the drosophila TRPL channels are regulated by the immunophilin FKBP59 [120]. Co-immunoprecipitation experiments indicated the existence of a constitutive association between FKBP12 and TRPC6 involving a stretch of 9 a.a (from a.a. 759 to a.a. 767) at the COOH-terminal of TRPC6A. The association could be disrupted by a low concentration of FK506, a drug that displaces immunophilins from their target proteins. In addition, FK506 inhibits TRPC6-mediated currents [121]. This observation stands in contrast with another report conducted with PC12D cells, and where FKBP12 does not associate with TRPC6 under basal (unstimulated) conditions. In this cell model, the FKBP12-TRPC6 association was only seen after the activation of muscarinic cholinergic receptors [113].

In addition to all these distinct types of TRPC6-interacting partners, TRPC6 seem able to interact with the cholesterol-binding proteins podocin and nephrin [122]. These interactions will be discussed separately further below.

10.5 Expression/Distribution

TRPC6 proteins are found throughout the body [5] (see Table 10.1), but they are mainly expressed in the pulmonary system, lung and brain. The expression of murine TRP channels was studied in various tissues by [123]. They found a special pattern of expression of TRPC6 in adult C57Bl/10SC mice with a high mRNA expression level in the brain and in tissues like the lung and ovary [123]. Of interest, they noted differences between mouse strains with, for instance, a higher mRNA TRPC6 expression in the heart of NOD mice when compared to the heart of C57Bl/10SC or Balbc mice [123]. When analyzing the relative mRNA expression of 22 TRP channels in various mouse organs, Jang et al. also found that mTRPC6 was predominantly expressed in the lung and, to a lower degree, in the brain [124]. TRPC6 expression has also been noted in various cell lines (See Table 10.2 for a non-exhaustive list of TRPC6-expressing cell lines).

At the cellular level, TRPC6 is predominantly found at the plasma membrane. Indeed, biochemical studies indicated the presence of two putative glycosylation sites (a.a. 472 and 560) on TRPC6, suggesting that it is a plasma membrane protein [7]. Since this initial report, many groups have also found that native and heterologously expressed TRPC6 proteins are glycosylated [44, 80, 89, 111, 119].

In polarized epithelial cells, TRPC6 is found in the luminal and basolateral regions [90]. In human platelets [43] found a specific location at the plasma membrane but no TRPC6 expression in inner membranes. On the other hand, in human

Table 10.1 TRPC6 expression in the body

Tissue	References
Adrenal	[15, 258]
Brain	[7, 15, 258]
Epididymis	[123]
Heart	[15, 42, 123, 258]
Intestine	[9, 259]
kidney	[15, 94, 124, 260]
Liver	[45, 124, 260]
Lung	[7, 9, 15, 123, 124, 258, 259]
Lymph nodes	[260]
Ovary	[9, 123, 258]
Pituitary gland	[259]
Placenta	[9, 259]
Prostate	[44]
Skeletal muscles	[259]
Skin	[260]
Stomach	[259]
Spleen	[9, 56, 123, 259, 260]
Testis	[123, 124, 258]
Thymus	[260]
Uterus	[261]

Table 10.2 Expression of TRPC6 in cell lines

Cell type	References
C2C12	[123]
CF-KM4	[100]
CHRF-288	[129]
ciPod	[76]
COS-1	[259]
COS-7	[259]
Dami	[129]
DU145	[26]
HaCaT	[70]
HEK	
HeLa	[259]
Hep G2	[45]
HL-60	[54]
Huh-7	[45]
J82	[26]
LD611	[26]
MCF-7	[262]
MEG01	[95]
MM39	[100]
MPC-5	[81]
PB-3c	[176]
PC3	[26]
RT4	[26]
SHSY5Y	[259]
V2D1	[176]

This table provides a non-exhaustive list of cell lines expressing TRPC6 channels. It is worth adding that conflicting results have been published regarding the expression of TRPC6 in some cell lines since for instance [139] and [263] failed to detect this channel in HaCaT and HL-60 cells, respectively

mesangial cells in culture, TRPC6 expression was primarily found to be localized to the nucleus [88].

Of particular interest is the association of TRPC6 with lipid rafts/caveolae, which are cholesterol-rich specialized microdomains located underneath the plasma membrane and playing crucial roles in calcium signaling [125]. Although one report found no evidence for the presence of TRPC6 in lipid rafts/caveolae in human platelets [126], these channels are commonly described as present in such microdomains [127, 128]. For instance, mTRPC6 expressed in HEK cells were shown to be located in domains of the plasma membrane enriched in caveolin-1/2 [83].

Like any other protein, TRPC6 mRNA and protein levels are subject to regulation. For instance, in human megakaryocytes, there is an up-regulation of the expression of the TRPC6 mRNA during the maturation process [129]. In pulmonary artery smooth muscle cells, PDGF and c-Jun increased mRNA and protein levels of TRPC6 [44]. In rat cardiac fibroblasts, TRPC6 expression is stimulated by $G\alpha_{12/13}$ [42] whereas forskolin, a PKA activator, inhibits it. A similar up-regulation was seen with angiotensin-II and endothelin-1. This ET-1-dependent regulation of TRPC6 expression is mediated via a signaling pathway involving $G\alpha_{12/13}$, the production of ROS and JNK activation [42]. This influences the transformation of cardiac fibroblasts into myofibroblasts. In Huh-7 cells, a hepatoma cell line, TRPC6 expression is strongly augmented by growth factors such as endothelial growth factor and hepatocyte growth factor [45]. In HL-60 cells, a human promyelocytic cell line, mTRPC6 expression is strongly up-regulated during the differentiation [54]. In human primary keratinocytes (hPK), Ca^{2+} and hyperforin, which induce the differentiation of the cells, stimulate the expression of TRPC6 [70]. A similar observation was made on cultured human skin explants.

TRPC6 seem to play a role in the control of the proliferation of cells, including tumor cells. Of interest, the gallium complex GaQ3 is a new anti-neoplastic agent regulating TRPC6 expression in cancer cells via a p53-dependent process. The trpc6 gene would be a transcriptional target of p53 [130]. Other metals influence TRPC6 expression like Co^{2+} [131] and Gd^{3+} [132]. In fact, many factors can positively or negatively affect TRPC6 expression like hypoxic conditions [133], the mammalian target of rapamycin (mTOR) complex 2 (mTORC2) [134, 135], a chronic treatment of mice with hyperforin [136], the bone morphogenetic protein 4 (BMP4) [137], the tumor necrosis factor alpha (TNF-α), ROS via an intracellular signaling cascade involving PKC [82] and the transcription factor NF-κB [138]. This ROS-dependent regulation of TRPC6 expression controlled by NF-κB [138] recruits the histone deacetylase HDAC2. Of interest, a recent work further highlighted the importance of deacetylases in regulating trpc6 gene expression. Indeed, the histone variant macroH2A negatively regulates the expression of TRPC6 (and TRPC3) without influencing the other TRPC channels. In bladder cells, the expression of macroH2A and TRPC6 is inversely correlated: cells with low levels of macroH2A expression have a high expression level of TRPC6 whereas in cells exhibiting higher expressions levels of macroH2A, TRPC6 was found less abundantly. In fact, macroH2A physically interacts with the histone deacetylases HDAC1 and HDAC2 which, in turns, control TRPC6 (and TRPC3) expression by silencing their genes. MacroH2A appears as a transcriptional regulator suppressing *trpc6* gene expression via a chromatin remodeling process involving deacetylases [26].

The presence of TRPC6 in HaCaT cells has been controversial: the TRPC6 mRNA was not found by [139] whereas [70] found evidence for the presence of TRPC6 mRNA in these cells. In human gingival keratinocytes, external Ca^{2+}, which stimulates keratinocyte differentiation, also stimulated the expression of TRPC6 mRNA [140]. Western blots show the presence of TRPC6 proteins and immunohistochemical experiments revealed that TRPC6 was present in all layers of the oral gingival epithelium [140]. The authors conclude that TRPC6 is present in undifferentiated and differentiated keratinocytes and TRPC6 may participate in the Ca^{2+}-dependent differentiation of these cells. In erythroid cells, TRPC6 expression decreases during erythroid differentiation [93]. In idiopathic pulmonary arterial hypertension (IPAH) patients, there is an up-regulation of the expression TRPC6 mRNA and protein in pulmonary artery smooth muscle cells, when compared to normal subjects. This indicates that TRPC6 may participate in the development of this disease.

Glucose seems to regulate TRPC6 expression but contrasting results have so far been reported. In glomerular mesengial cells high glucose diminishes TRPC6 expression [141] whereas in podocytes and platelets, high glucose enhances TRPC6 expression [142]. These effects are likely secondary to generation of ROS. Similar effects occur in response to hyperhomocysteinemia [143].

The promoter of the TRPC6 gene contains two conserved sites for the transcription factor NFAT. Furthermore, the calcineurin-dependent activation of NFAT enhances the transcription of the trpc6 gene. Additionally, increasing TRPC6 expression levels activates the calcineurin/NFAT-dependent pathway [144]. This dual regulation has been shown to be involved in the remodeling of cardiac cells under pathologic conditions [144]. A similar mode of regulation also occurs in podocytes [145, 146].

10.6 Regulation and Pharmacological Properties

TRPC6 form channels permeable to a wide range of cations like Na^+, Ca^{2+}, Cs^+, K^+, Ba^{2+} [9, 28]. But they are also permeable to non-canonical cations like iron [147], zinc [117, 148], and even N-methyl D glucamine [28]. TRPC6 opening generates inward and outward rectifying currents (dual rectification) with a reversal potential close to 0 mV [22]. The unitary conductance is in the order of 28–35 pS [9, 27]. Depending on the cell type, the relative permeability ratio is in the order of 5–6 [9, 38, 149]. A well-established consequence of TRPC6 activation is the elevation of the cytosolic concentration of Ca^{2+}. However, the mechanism by which this response occurs is not as trivial at it seems. The TRPC6-dependent Ca^{2+} rise could in fact have three origins: (i) a direct Ca^{2+} entry into cells via opened TRPC6 channels; (ii) it can reflect the contribution of voltage-gated Ca^{2+} channels recruited in response to a TRPC6-mediated depolarization; (iii) the elevation of Ca^{2+} could reflect the contribution of the Na^+/Ca^{2+} exchanger operating in its reverse mode. These pathways are not mutually exclusive. hTRPC6 over-expressed in HEK cells were found to have a low conductance for Ca^{2+}. Under physiological condition, it would rather be a Na^+ conducting channel [28]. Consequently, a Ca^{2+} rise would only be seen when the membrane potential is negative, indicating that TRPC6 functions primarily as a monovalent channel. According to this model, the TRPC6-dependent Ca^{2+} rise is the consequence of the depolarization and does not reflect the influx of Ca^{2+} through TRPC6 channels. The work of van Bremen et al. provided experimental evidence for the existence of a functional coupling between TRPC6 channels and the Na^+/Ca^{2+} exchanger: the entry of Na^+ through TRPC6 channels would promote the entry of Ca^{2+} by reversal of the Na^+/Ca^{2+} exchanger [150–152]. A similar finding has been noted in cultured cortical neurons where OAG stimulation provokes an intracellular Na^+ rise promoting the entry of Ca^{2+} [153]. The existence of such a functional coupling between the Na^+/Ca^{2+} exchanger and TRPC6 channels has also been suggested in the neuroblastoma cell line IMR-32 [154].

Like any other channel, the activity of TRPC6 can be regulated by various means. For instance, TRPC6-mediated responses can be finely tuned by phosphorylation/dephosphorylation reactions. Consensus sequences for various kinases such as PKA, PKC, PKG, casein kinase II and tyrosine kinases are known [43]. The following section will address the regulatory processes influencing TRPC6 activity.

CaM is generally viewed as a negative regulator of TRPCs, but it seems to exert opposite effects on TRPC6. CaM binding to mTRPC6 regulates its activity because inhibitors of CaM (calmidazolium, trifluoperazine) depress Ca^{2+} entry by abolishing the interaction CaM-mTRPC6 [110]. [38] reported that calmidazolium or an inhibitor of CaMK-II depresses mTRPC6 currents, indicating that phosphorylation of the channel by CaMKII is necessary for its activity.

Ca^{2+} the activity of TRPC6 channels is regulated by Ca^{2+} in a complex manner: chelating Ca^{2+}_{in} potentiates TRPC6-mediated currents whereas elevating Ca^{2+}_{in} has an opposite effect [37]. A role of external Ca^{2+} was documented for mTRPC6 channels over-expressed in HEK cells [38]: in this case, external Ca^{2+} exerts a complex regulatory action on TRPC6 with positive and negative effects on TPC6 activity. Maintaining human mesangial cells in a Ca^{2+}-enriched (5 mM) culture medium for 24 h augments their TRPC6 protein levels (when compared to cells kept in a culture medium containing 1 mM Ca^{2+}). This response, which is thought to involve the Ca^{2+}-sensing receptor, stimulates their proliferation [155].

Phosphorylation by Tyrosine Kinases Contradictory findings have been published regarding the regulation of TRPC6 channels by tyrosine phosphorylation. In human platelets, Hassock et al. found that TRPC6 does not undergo tyrosine phosphorylation [43]. However, in HEK cells, TRPC6 was reported to be tyrosine phosphorylated in a *Fyn*-dependent manner [111]. *Fyn*, a *src*-family tyrosine kinase, is a TRPC6-interacting protein (binding on the NH_2-terminal part of the channel). This Fyn-dependent phosphorylation of TRPC6 enhances TRPC6 channel activity. Experiments conducted on mTRPC6 expressed in embryonic fibroblasts gave contrasting results [156]: it was first shown that, in GST pull-down tests, *src* interacts with the NH_2-terminal of TRPC6. However, expressing TRPC6 in fibroblasts lacking *src*, *Fyn* and *yes* tyrosine kinases failed to affect TRPC6-mediated Ca^{2+} signals (when activated in response to the activation of GPCR with arginine vasopressin or carbachol). Furthermore, TRPC6-medicated Ca^{2+} signals are insensitive to genistein, a non-selective tyrosine kinase inhibitor. According to [156], tyrosine kinases are not necessary for the activity of TRPC6 channels. Experiments conducted with inhibitors of tyrosine kinases (PP2, SU6656) on mTRPC6 over expressed in HEK cells showed that tyrosine phosphorylation would be a necessary step for the activation of the channels [25]. An indirect evidence for the involvement of *src* in the modulation of TRPC6 activity in human leukocytes was provided by [157]. Of note, lysophosphatidylcholine, which induces the externalization of native and over-expressed TRPC6 channels, stimulates its tyrosine phosphorylation. But it is unclear whether this phosphorylation process is required to control the trafficking of TRPC6 [30].

Phosphorylation by PKC TRPC6 do not seem to open in response to PKC activation [27, 37] but they can be phosphorylated upon the application of the phorbol ester PMA, a well-known PKC activator, indicating that these channels can be directly phosphorylated by PKC [113]. In HEK cells expressing mTRPC6, TRPC6-mediated currents are inhibited by PKC agonists, indicating that this kinase regulates negatively TRPC6 channel activity [38]. However, in HEK cells expressing hTRPC6, PKC activation with PMA has no effect on OAG-activated TRPC6 channels whereas this protocol strongly depresses carbachol-activated TRPC6-mediated currents [28]. Biochemical studies found that mTRPC6 expressed in HEK cells are weakly phosphorylated under basal conditions but application of carbachol stimulates TRPC6 phosphorylation via a PKC-dependent process [158]. Furthermore, PKC stimulation depresses TRPC6-mediated Ca^{2+} responses. Mutagenesis studies indicated that none of the 12 putative PKC phosphorylation site seems to be involved. Instead, the authors identified a non-canonical phosphorylation site (Ser448) as a target for PKCδ. [158] propose that PKC acts as a negative feedback regulator of TRPC6 activity.

Regulation by PKG Protein sequence alignments revealed the existence of two putative PKG phosphorylation sites on TRPC6 [159]. Studies conducted on mouse TRPC6 channels expressed in HEK cells concluded that PKG activation by means of the NO donor SNAP or the PKG activator 8Br-cGMP depressed TRPC6-mediated currents [160]. This occurs without affecting the cell surface expression of TRPC6, indicating that PKG influences the gating of the channels rather than their plasma membrane abundance. In addition, this PKG-dependent depression of TRPC6-mediated responses critically depends on a phosphorylation site (Thr69) located in the COOH-terminal part of TRPC6 [160].

Regulation by PKA Studies focusing on the role of the cAMP-dependent protein kinase A (PKA) on TRPC6 activity have led to conflicting results. mTRPC6 and hTRPC6 can be phosphorylated by PKA, a process associated with reduced TRPC6-mediated Ca^{2+} responses [161, 162] that does not influence the cellular distribution of the channel [162]. This conclusion stands in contrast with other reports who observed that PKA exerts no modulatory action on TRPC6 activity [43, 163]. In glomerular mesangial cells, cAMP regulates TRPC6 activity but in a PKA-independent manner via a signaling cascade leading to the sequential activation of the following kinases: PIK3, MEK, and ERK1/2 kinase [163]. The existence of an interplay between TRPC6 and ERK has been recently confirmed in a study where the over-expression in HEK cells of gain-of-function TRPC6 mutants activates ERK [161].

Regulation by MLCK Based on their results [164] conclude that MLCK plays no (or very little) regulatory role on mTRPC6 channel activity.

Regulation by Podocin Podocin is a cholesterol-binding protein of the stomatin family expressed primarily in podocytes. The COOH-terminus of podocin interacts with the carboxyl-terminus of TRPC6 [66]. It regulates TRPC6 activity in a cholesterol-dependent manner [128]. Podocin enhances TRPC6 activation by DAG and through GPCRs [66–68] but reduces TRPC6 activation in podocytes by mechanical stimuli [66]. A chronic treatment of human B lymphoma Daudi cells with cholesterol or lovastatin, a statin known to inhibit cholesterol synthesis, respectively up-regulates or depresses TRPC6 expression and activity [165].

Glycosylation the original work on mTRPC6 proposed the presence of two putative glycosylation sites (a.a. 472 and 560) [7]. In the rat, the isoform A was mainly found in the (mature) glycosylated form whereas rTRPC6B was found in the (mature) glycosylated and (immature) non glycosylated forms. On the other hand, rTRPC6C is not glycosylated [11]. Since then, the glycosylation status of TRPC6 has been noted by several groups [44, 89]. This glycosylation has a strong impact on the basal (or constitutive) activity of TRPC6. By studying two related TRPC channels (TRPC3 and TRPC6) [149] could show that hTRPC3 and hTRPC6 channels over-expressed in HEK cells exhibit distinct features: TRPC3 is a mono-glycosylated channel displaying considerable basal activity in contrast to TRPC6, dually glycosylated (in the first and second extracellular loops), that has a low level of basal activity [37, 149]. It appears that the glycosylation status of TRPC6 plays important role in controlling its constitutive activity: the removal of one of these two sites strongly augments TRPC6 basal activity. On the other hand, mutant TRPC3 channels with two glycosylation sites become quiescent. The pattern of glycosylation has thus a strong effect on the basal activity of TRPC channels [149].

Regulation of TRPC6 Activity by Interacting Partners As pointed out, CFTR and TRPC6 are thought to be engaged in a macromolecular complex [100]. These two proteins seem to be functionally coupled. Indeed, suppression of CFTR expression up-regulates TRPC6 mediated Ca^{2+} responses. CFTR has been proposed to be a negative regulator of TRPC6 channels in bronchial cells [100].

Pharmacological Inhibitors Various compounds have been used to block the transport of ions through TRPC6 channels. One major challenge when studying the properties of TRPC channels is the lack of selective inhibitors. Therefore, most (if not all) compounds used to depress TRPC6-mediated responses are also inhibiting other members of the TRPC family. The blockers of cation channels 2-APB, SKF-96365, La^{3+} and Gd^{3+} are the more commonly used TRPC6 antagonists. For instance, the inhibitory effect of La^{3+} and Gd^{3+} on rTRPC6B channels has been defined [22]. These two cations are amongst the most potent TRPC6 antagonists since mTRPC6 expressed in HEK cells are blocked by La^{3+} with an IC_{50} of 4–6 μM and by Gd^{3+} with an IC_{50} of 1.9 μM [27]. These authors determined the IC_{50} values of various

Table 10.3 Pharmacological inhibition of TRPC6 channels

Blockers	Cd^{2+}	La^{3+}	Gd^{3+}	SKF-96365	Amiloride
IC_{50} (μM)	253	4.0	1.9	4.2	129

The table gives the IC_{50} values of several TRPC6 channel inhibitors (Data from Inoue et al. [27])

inhibitors (see Table 10.3). Besides, several other molecules were shown to affect TRPC6 channels like clotrimazole, an azole compound clinically used as antifungal drug, that was first described as a specific TRPM2 blocker [166] but in fact blocks TRPC6, TRPM3 and TRPV4 [167]. Pyrazole compounds like Pyr2 exert inhibitory properties on over-expressed TRPC6 channels [168]. In addition, U73122, an inhibitor of phospholipase C, has been widely used to analyze the contribution of this enzyme in the signaling pathway leading to TRPC6 activation. However, U73122 should not be used for this purpose because it exerts a direct inhibitory effect on TRPC6 channels [164, 169]. For instance, in HEK cells over-expressing mTRPC6, TRPC6-mediated currents activated by carbachol are inhibited by U73122 and by its inactive analogue U73343 [28]. Activation of native TRPC6 channels in podocytes by DAG analogs is also blocked by U73122 [68]. These observations call into question a fairly substantial literature in which these drugs have been used to probe Ca^{2+} dynamics under the supposition that they are specific inhibitors of phospholipases. When analyzing the putative regulatory action of the myosin light chain kinase (MLCK) [164] found that several MLCK inhibitors like ML-9, ML-7, and W7 reversibly depressed in a dose-dependent manner hTRPC6-mediated Ca^{2+} currents recorded in HEK cells. This occurs independently of their action on MLCK. They found IC_{50s} of 7.8 μM for ML-9, 10 μM for ML-7 and 50 μM for W7 [164]. ACA (N-(p-amylcinnamoyl) anthranilic acid) is a broad spectrum PLA2 inhibitor. It blocks various TRP channels, including hTRPC6 (expressed in HEK cells) with an IC_{50} of 2.3 μM [79]. It also blocks native TRPC6 channels of podocytes at micromolar concentrations (SED unpublished observations). On the other hand, another PLA2 inhibitor, p-bromophenacyl bromide (BPB) has no effect on hTRPC6-dependent currents when tested at the concentration of 100 μM [79]. In some studies LOE908 has been presented as a non-specific inhibitor of TRPC6 and receptor-operated non-capacitative calcium entry [170, 171]. In addition to these a number of proprietary inhibitors have been developed by pharmaceutical companies, although to date their specificity is not known.

10.7 Roles of TRPC6 Channels in Non-excitable Cells in Health and Disease

TRPC6 channels support many key physiological functions like the control of the vascular smooth muscle as revealed by experiments conducted on *Trpc6*$^{-/-}$ mice [172]. One key important TRPC6 contribution is its involvement in the cellular Ca^{2+} signaling. By allowing intracellular Ca^{2+} rises, changes in TRPC6 activity most likely influences a wide number of Ca^{2+}-dependent processes like the proliferation or differentiation of cells. For instance, in human keratinocytes, TRPC6 is sufficient to induce differentiation. Thus, TRPC6 activation by hyperforin triggers, while TRPC6-specific siRNAs inhibit, keratinocyte differentiation [70]. Moreover, triterpenes which also promote keratinocyte differentiation are able to increase TRPC6 expression in human keratinocytes [173]. Fibroblast transdifferentiation into myofibroblasts is an essential step in wound healing and tissue remodeling. In a genome-wide screen, TRPC6 was identified as sufficient for myofibroblast transformation. *Trpc6*$^{-/-}$ mice showed impaired dermal and cardiac wound healing because their fibroblasts were not able to initiate TGFβ- and AngII-induced transdifferentiation [174].

Another important physiological function of TRPC6 is its contribution to the control of the proliferation of normal and malignant cells. Indeed, a positive correlation between TRPC6 expression and the proliferation of tumor cells has been reported in cancers like lung, breast, prostate, ovarian, gastric, liver cancers and also in glioblastoma [175]. On the other hand, mTRPC6 expression was found to be down-regulated in a tumor cell line (mast cells) [176]. In addition, a role of TRPC6 in VEGF-induced angiogenesis has also been documented [177].

The above reported properties may be linked to the fact that TRPC6 channels influence the actin-based cytoskeleton and its dynamics [117, 169, 178, 179]. For instance, TRPC6 is involved in cytoskeletal rearrangements during neutrophil migration, and migration of *Trpc6*$^{-/-}$ neutrophils in response to Macrophage Inflammatory Protein-2 (MIP2 also known as CXCL2) is reduced compared to WT neutrophils [180]. TRPC6 channels are essential for CXR2-mediated intermediary chemotaxis but not for fMLP receptor-mediated end-target chemotaxis of neutrophils, indicating that not all receptors in neutrophils are in signaling complexes with TRPC6 [181].

More recently, some studies unveil additional TRPC6 functions like its contribution in the host-parasite response. Parasites like *Plasmodium falciparum* and *Toxoplasma gondii* use cellular signal transduction cascades for egress from the host cell. It was demonstrated that Ca^{2+} influx through TRPC6 activates host calpain to proteolyse the host cytoskeleton, allowing the release of parasites [182]. In addition, its involvement in the allergic response has been highlighted since ovalbumin-challenged *Trpc6*$^{-/-}$ mice exhibited reduced allergic responses as evidenced by a decrease in airway eosinophilia and blood IgE levels as well as levels of T-helper type 2 cytokines (IL-5, IL-13) in the bronchoalveolar lavage [183]. TRPC6 was also identified in other immune cells of the lungs like alveolar macrophages, and TRPC6 mRNA expression was significantly increased in macrophages obtained from COPD patients compared to healthy controls, while TRPC3 and TRPC7 levels remained unchanged [184]. Cell shrinkage and phospholipid scrambling, two hallmarks of eryptosis (erythrocytes programmed cell death triggered by Ca^{2+} influx), were significantly lower in Cl^--depleted erythrocytes from *Trpc6*$^{-/-}$ mice than from WT mice [185]. Moreover, Ca^{2+} entry in these cells was inhibited after pre-incubation with specific TRPC6 antibodies, but not with antibodies directed against TRPC3 or TRPM2 [185].

Below, the following chapters address the patho-physiological significance of TRPC6 in renal and blood cells where its expression and functions have been particularly studied.

10.7.1 TRPC6 in the Kidney

10.7.1.1 TRPC6 Dysfunction in Chronic Kidney Disease

In 2005 Michelle Winn and her co-workers documented a mutation (P112Q) in the first ANK repeat domain of TRPC6 in a large extended family in which several members comprising at least three generations had an aggressive form of focal and segmental glomerulosclerosis (FSGS) and high-grade proteinuria [186]. The disease in this family exhibited an autosomal dominant form of inheritance and occurred in adulthood (at a mean age of 33 years). The affected members of this family progressed to end-stage renal disease (ESRD) in 10 years or less after initial presentation [187]. The P112Q mutation causes a gain of function when these channels are expressed in HEK293 cells, resulting in markedly increased cationic currents and Ca^{2+} influx after application of DAG analogs or activation of GPCRs [186]. The biophysical basis for the gain of function in this allele is not entirely understood, but the initial report presented evidence for altered trafficking resulting in increased steady-state expression on the cell surface [186]. A number of other TRPC6 mutations were quickly identified in familial forms of FSGS [122]. These alleles also caused an aggressive disease with an adult onset and autosomal mode of inheritance, and most of the alleles encoded channels with at least a modest gain of function when expressed in heterologous expression systems. We should note that some of these mutations were identified from a limited number of patients, and the evidence is not equally strong in all cases that the TRPC6 allele is the cause of the disease, such as for the highly conservative S270T missense mutation, which does not appear to show any gain of function [122, 145]. Since these original reports, several more mutant alleles have been found in affected families [188–194]. At present, 21 different alleles have been reported. Among these, there is at least some evidence suggesting a gain of function for ten, whereas two do not appear to have a gain of function, and nine have not been characterized functionally. A mutation of particular interest (M132T) exhibited the largest gain of function seen to date, with current densities six to sevenfold greater than wild-type channels [195]. Notably, the M132T mutation was identified in a 9-year old girl with a severe nephrotic syndrome. It has been hypothesized on this basis that the age of onset of the disease is related to the extent of the overall gain of function as a result of a long-term "Ca^{2+} dose" effect. More recently, novel TRPC6 mutations have been observed in a 2-year old child (R895L) and in a 4-year old child (N125S) with severe steroid-resistant nephrotic syndrome [189]. Unfortunately, these later alleles have not been characterized using voltage-clamp methods. Most of the TRPC6 mutations identified in patients with FSGS have been observed in ankyrin repeat domains (such as P112Q and M132T) or close to the carboxyl terminus (such as R895C or E897K).

Nephrotic syndromes caused by TRPC6 mutations are relatively rare. By contrast, non-genetic "acquired" forms of FSGS are quite common, as are other severe glomerular diseases such as diabetic nephropathy, minimal change disease (MCD), and various inflammatory conditions associated with localized fixation of complement within glomerular capillaries. Overall TRPC6 expression appears to be increased in several of these conditions, as well as in rodent models of chronic kidney disease [196]. Moreover, selective over-expression of TRPC6 in podocytes in vivo induces modest proteinuria and glomerulosclerosis [197]. Therefore, it is possible that over-expression of TRPC6 could contribute to several forms of kidney disease, although this may serve initially as a mechanism to compensate for increased pressure within remaining glomerular capillaries that occurs whenever there is a loss of functional nephrons [81, 198].

10.7.1.2 Distribution of TRPC6 in the Kidney

TRPC6 channels, expressed by renal cells (see Table 10.4), are found at highest density within glomeruli, where they are expressed in podocytes and mesangial cells [97, 122, 186]. This distribution can explain why TRPC6 mutations induce glomerular diseases. Within podocytes, TRPC6 channels appear to be expressed at a high density within the slit diaphragm domains of foot processes [128], where they probably form complexes with signaling and scaffolding proteins including nephrin [122] and podocin [66, 128], and NADPH oxidases [199]. These membrane domains are characterized by unusually high cholesterol concentration [200]. TRPC6 can also be found along the major processes of podocytes and within its cell body [201] where nephrin and podocin are generally not expressed on the cell surface. While TRPC6 channels within glomeruli have been the most extensively studied, TRPC6 channels are also detected in aquaporin-2-containing collecting duct cells in cortex and medulla [97]. The function of those channels is not known, but based on their location, it is possible that they contribute to the formation of a concentrated urine. Moreover, based on molecular weight, it is possible that different splice variants of TRPC6 predominate in glomeruli compared to tubules [97]. TRPC3 is detected in the same renal cells as TRPC6, and at least some of the renal TRPC6 protein occurs in heteromers that contain TRPC3 subunits. However, in collecting duct, the TRPC6 appears to be preferentially localized in apical membranes, whereas TRPC3 is in the basolateral membrane [97].

10.7.1.3 Regulation of TRPC6 Gating and Trafficking in Podocytes and Mesangial Cells

The TRPC6 channels in podocytes are of particular interest since those cells are principally affected in patients with TRPC6 mutations, as well as in the early stages of primary FSGS. The TRPC6 channels of podocytes become active in response to

Table 10.4 TRPC6 expression in kidney

	References
Glomeruli	[88, 97, 122, 186]
Mesangial cells	[88, 122, 141]
Podocytes	[68, 122, 128, 264]
Tubules	[97, 122, 186]

three main classes of stimuli: (1) Canonical lipid activators such as DAG and 20-HETE, and their mimics such as hyperforin. (2) Stimuli that liberate ROS. (3) Mechanical deformation of the plasma membrane. Thus, simply perfusing 100 μM OAG, a membrane permeable analog of DAG causes a relatively rapid activation of TRPC6 channels in immortalized mouse podocyte cell lines [66, 81]. A similar effect is seen in response to 10 μM 20-HETE (SED unpublished data). Oxidative activation occurs with many members of the TRP family of channels, but was first observed for TRPC6 in HEK-293 cells, where H_2O_2 evoked increases in surface expression as well as direct activation of gating [80]. Similar effects are seen for native TRPC6 channels in podocytes, as application of 500 μM H_2O_2 causes rapid activation of podocyte TRPC6 accompanied by an increase in steady-state surface expression of these channels [81]. Native podocyte TRPC6 channels co-immunoprecipitate with NADPH oxidases, including NOX2 [199] and NOX4 (SED, unpublished data), whose primary function is to produce the ROS superoxide and H_2O_2 for use in normal physiological processes. Formation of NOX2-TRPC6 complexes only occurs in the presence of podocin, and therefore the interaction between these proteins is indirect [199]. Notably, ROS pathways appear to amplify responses to canonical lipids in podocytes, as TRPC6 activation by DAG analogs is markedly attenuated in the presence of ROS quenchers such as TEMPOL or MnTBAP. Moreover, DAG stimulates assembly of active NOX2 complexes on the cell surface [199]. Disruption of these complexes may explain why loss of podocin markedly decreases or eliminates responses to canonical lipids in podocytes. In this regard, inhibition or knockout of NOX enzymes, and ROS quenching, eliminates TRPC6 activation and mobilization evoked by angiotensin II (Ang II) [67, 202], P2Y receptor agonists [68], and insulin [81]. ROS generation also underlies increased TRPC6 activation in the presence of high glucose [203]. TRPC6 channels are also upregulated by other signals such as VEGF [204] but are down regulated by vitamin D [205]. Finally, we note that expression of syndecan-4 can lead to up-regulation of TRPC6 and an increase in their basal activation in podocytes [206]. In this regard, syndecan-4 knockout mice do not have detectable TRPC6 within glomeruli or in podocyte cell lines, although TRPC3 continues to be present [206].

The TRPC6 channels of mouse and rat podocytes are activated by mechanical stimuli that cause deformation of the plasma membrane in either direction [66]. This can be seen in response to hypo-osmotic stretch, inward indentation of the plasma membrane using a fire-polished glass probe, or by using hydrostatic pressure pulses delivered from a micropipette placed close to the cell surface [67]. These responses are seen in immortalized podocyte cell lines, as well as in podocytes in acutely isolated glomeruli still attached to the GBM [66]. These currents are not observed after TRPC6 knockdown, in the presence of 50 μM La^{3+} or SKF-96365, when the recording pipettes contain GDP-βS [66], by manipulations that eliminate responses to Ang II, ATP or DAG analogs in the same cells [66–68]. In addition, mechanical activation of podocyte TRPC6 channels is blocked by GsMTx4 [66] and also by proprietary TRPC6 inhibitors (SED, unpublished observations). The observation with GDP-βS suggests that the mechanical activation of TRPC6 occurs by signaling pathways independent of any G proteins, in marked contrast to the mechanical activation of TRPC6 in smooth muscle. In addition, knockdown of podocin markedly increases mechanical activation of podocyte TRPC6 channels, while simultaneously reducing or eliminating responses to canonical lipid signals [66] or responses mediated by G protein-dependent signaling pathways [67, 68]. This is of interest given that mutations in the *NPHS2* gene encoding podocin give rise to a common autosomal recessive form of FSGS primarily seen in children [207].

TRPC6 channels are also expressed in mesangial cells, along with TRPC3 and TRPC1 [97]. These channels have been less extensively studied than their counterparts in podocytes, and it appears that they are regulated differently in certain respects. TRPC6 channels in mesangial cells can be activated by signals through Ca^{2+}-sensing receptors [155] and by glucagon [163] through G protein-coupled pathways that may entail Erk and phosphatidyl inositol-3 kinase signaling. It is especially interesting that ROS and high glucose suppress TRPC6 activation in mesangial cells [82, 141, 208], because the same stimuli increase TRPC6 activation in podocytes [81, 203, 209]. Mesangial cells are excitable contractile cells located on the interior of the glomerular capillary. They make numerous connections with the inner face of the glomerular basement membrane (GBM), and contraction of mesangial cells is thought to prevent excessive expansion of the glomerular capillary in response to elevated transmural pressures [210]. By contrast, podocyte foot processes line the entire outer face of the GBM. Specialized connections between adjacent foot processes, known as slit diaphragms, form a porous matrix that plays a major role in determining the overall permselectivity of the glomerular filter [211]. The arrangement of mesangial cells and podocytes suggests that a modest mesangial relaxation accompanied by contraction of podocyte foot processes could serve to protect delicate slit diaphragms from sustained expansile forces caused by glomerular capillary pressures. Increased transmural glomerular capillary pressures occur, for example, in response to transient elevations in blood glucose [212] and also whenever there is substantial loss in the number of functional nephrons [213]. However, it is likely that a more severe abrogation of mesangial cell contractility would compromise glomerular function [208].

An important feature of TRPC6 channels in podocytes is that their sustained activation can activate a positive feedback loop through calcineurin-NFATc pathways that lead to increased transcription of the trpc6 gene [145, 146]. A similar pathway had earlier been described in the heart, where it is thought to contribute to cardiac hypertrophy [144]. This feedback loop may explain the increase in TRPC6 expression observed in chronic kidney diseases, in which there is an increase in glomerular capillary pressure [196]. In this regard, TRPC6 knockout mice or mice with reduced TRPC6 expression are protected in certain models of kidney disease, including proteinuria evoked by chronic Ang II infusion [214] or by intraperitoneal injection of high doses of bovine albumin [206]. At the same time, at least in the initial stages of glomerular disease, TRPC6 could play a protective role, as has been suggested in mouse models of inflammatory kidney diseases in which there is complement fixation within the kidney [198].

10.7.1.4 TRPC6-Interacting Proteins in Renal Cells

TRPC6 proteins are concentrated in the slit diaphragm domains of the podocyte foot process plasma membrane, and therefore it is not surprising that they interact with other proteins concentrated there. It bears noting that the slit diaphragm domain has a high concentration of sterols than other regions of the foot process plasma membrane [200]. Because proteins that partition within the same raft domain can co-immunoprecipitate, care has to be used in interpreting results. The strongest evidence for a direct interaction with podocyte TRPC6 channels is with podocin, because it can be detected by co-immunoprecipitation [66, 122, 128], and also using GST pull-down assays using soluble fragments of the proteins [66]. Podocin is a hairpin-loop intrinsic membrane protein in a family that also contains stomatin-like proteins, prohibitins, and flotillins, as well as the MEC-2 protein of *C. elegans* [215]. All of these proteins share a so-called PHB domain that binds cholesterol and usually localizes these proteins to lipid rafts. Proteins in this family also tend to form higher-order oligomers, and it is worth noting that several, including the stomatin-like proteins [216, 217] and MEC-2 [218], regulate (and usually facilitate) the gating of mechanosensitive channels. As already noted, podocin suppresses mechanically-induced activation of podocyte TRPC6 channels [66]. It also enhances activation by lipids or through GPCRs [66–68, 128] at least in part by facilitating co-localization with NOX2 [199]. The interactions between TRPC6 and podocin occur between their respective intracellular carboxyl terminals and do not require the transmembrane domains of either protein [66]. Podocyte TRPC6 channels also co-immunoprecipitate with nephrin [122], although it is not known if this interaction is direct. A wide range of proteins interact with the intracellular domains of nephrin forming so-called pleomorphic ensembles [219]. Among these protein are protein kinases such as *Fyn*, phosphoinositide 3-kinase, and podocin [220], which may therefore allow formation of signaling complexes that induce tyrosine phosphorylation of TRPC6 [221].

The striking deviation of TRPC6 channels from constant-field behavior does not allow for efficient Ca^{2+} flux as cells depolarize, and this effect is even greater than one would expect simply from a reduction in driving force [28]. In excitable cells, the TRPC6-dependent depolarization is sufficient to cause activation of voltage-dependent Ca^{2+} channels, which then allow for substantial Ca^{2+} flux. However, podocytes are non-excitable cells and do not express voltage-activated Ca^{2+} channels. Since TRPC6 activation should cause marked depolarization, the question arises as to how they can mediate sustained Ca^{2+} flux. One possibility is that TRPC6 channels in podocytes interact with large-conductance (BK-type) Ca^{2+}-activated K^+ channels, which raises the possibility that they are activated coordinately to prevent excessive depolarization in the face of continued TRPC6 activation [101].

10.7.2 TRPC6 in the Blood

10.7.2.1 TRPC6 in the Megakaryoblastic Lineage

The first evidence of the expression of TRPC6 in platelets was provided by Authi and coworkers in 2002 [43], reporting the expression of TRPC6 in the plasma membrane of human platelets and suggesting a role for TRPC6 in a non-capacitative, cAMP-dependent protein kinase-regulated, Ca^{2+} entry pathway. The location of TRPC6 in the plasma membrane of human platelets has been confirmed by a number of studies, i.e. [126, 222], and this channel has been reported to be mostly associated to membrane domains independent of lipid rafts [126], which is consistent with its initial putative role as a non-capacitative channel. Recent studies have provided evidence for a role of the protein STIM1 in the cellular location of TRPC6. Thus, in the NG115-401L cell line, expression of STIM1 results in translocation of a pool of TRPC6 from the plasma membrane to the endoplasmic reticulum, presumably to act as a Ca^{2+} leak channel [104]. The expression of TRPC6 has also been confirmed in murine platelets and megakaryocytes [129, 223, 224] and the megakaryoblastic cell line MEG01 [95] (see also Tables 10.2 and 10.5).

10.7.2.2 TRPC6-Interacting Proteins in Blood Cells

TRPC6 has been reported to interact with a number of intracellular proteins in platelets, including kinases, such as the cAMP-dependent protein kinase [43] and *Src* family tyrosine kinases [225], TRP channels like TRPC3 [55, 126], TRPC1 [226] and TRPA1 [95], the store-operated Orai1/STIM1 complex [55, 227], the Ca^{2+} channel Orai2 [107], and CaM as well as the type II IP_3R, through the CIRB site [228]. The association of TRPC6 with these proteins plays an important regulatory role on the function and behavior of the TRPC6-forming channels. For instance, the CIRB site allows a dual regulation of TRPC6 channel function by Ca^{2+} and IP_3 levels. Hence, elevation of cytosolic Ca^{2+} concentration in human platelets either by stimulation with physiological agonists, such as thrombin, or by treatment with pharmacological compounds, results in Ca^{2+}-dependent binding of CaM to TRPC6 channels and, thus, displacement of IP_3 receptors. Conversely, IP_3 generation increases the affinity of the IP_3R for the CIRB site of TRPC6, attenuating the displacement of IP_3R from this site and, as a result, attenuating the regulatory effect of CaM. Therefore, the CIRB site of TRPC6 dynamically regulates both Ca^{2+} influx and function stimulated by physiological agonists in human platelets [228].

Table 10.5 TRPC6 expression in blood cells

Cell type	References
Platelets	[43, 56, 129]
Megakaryocytes	[129, 223]
Polymorphonuclear leukocytes	[157]
Neutrophils	[54, 263]
T lymphocytes	[260, 265]
B lymphocytes	[260, 266]
Jurkat T cells	[34, 258, 265]
Erythroid cells	[89, 93]
Erythroid progenitors	[93, 185]
Erythrocytes	[185]

Furthermore, TRPC6 tyrosine phosphorylation by *Fyn*, a member of the Src family of protein tyrosine kinases, results in an increase in the channel activity [111], which is consistent with the finding that *Src* family kinase activation is required for store-independent Ca^{2+} entry in human platelets [225]. More interesting is the functional role of the association of TRPC6 with other channels also described in platelets, such as TRPC3, TRPC1 or Orai family members, which is associated to the behavior of TRPC6 channels as store-dependent or -independent channels.

10.7.2.3 Involvement of TRPC6 in Ca^{2+} Influx in Human Platelets

TRPC6 channels have been classically associated to store-independent, or non-capacitative, Ca^{2+} entry in a number of cell types as described above, activated by receptor occupation via DAG or changes in the phosphoinositide composition, or by physical stimuli, such as mechanical stretch. Activation of TRPC6 by DAG, or its membrane permeable analog OAG, has long been demonstrated in human [222] and mouse platelets [56, 227, 229]. In addition, there is a body of evidence supporting that TRPC6 might be sensitive to PIP_3, a substrate of phosphatidylinositol 3-kinase (PI3K), in platelets, Jurkat T cells, and RBL-2H3 mast cells. First of all, fluorometric analysis has revealed the ability of PIP_3 to selectively stimulate Ca^{2+} entry in TRPC6-expressing cells, a response that was altered by TRPC6 overexpression or down-regulation in Jurkat T cells; and, second, specific association of PIP_3 with TRPC6 has been demonstrated by pull-down studies. The activation of TRPC6-dependent non-capacitative Ca^{2+} entry by PIP_3 was found to be reminiscent to that evoked by OAG. However, other phosphoinositides failed to induce Ca^{2+} influx [34]. Furthermore, TRPC6 activation in human platelets has been found to be sensitive to the cellular level of PIP_2 [222], which might interact with the C-terminal region of TRPC6 stabilizing the channel in an open channel configuration, as well as disrupting the interaction of inhibitory molecules with the channel [230]. In human platelets, we hypothesize that activation of membrane receptors by physiological agonists might play a dual role on TRPC6 channel activity. First, PLC activation would result in IP_3 and DAG generation, leading to the activation of Ca^{2+} entry via TRPC6 and other channels, a process that is positively regulated by PIP_2, but, soon after receptor occupation, PLC-dependent PIP_2 hydrolysis would lead to inactivation of TRPC6, thereby limiting its activity to the initial step of Ca^{2+} influx as described for other PIP_2-regulated TRP channels [231].

In addition to the role of TRPC6 in the store-independent Ca^{2+} pathway, a role for TRPC6 in the conduction of SOCE has also been suggested in human platelets. SOCE was first described in platelets by Sage and coworkers in 1989 by using Mn^{2+} as a surrogate for Ca^{2+} influx, and analyzing its quenching effect on the Ca^{2+}-insensitive fura-2 fluorescence, to avoid contamination with other Ca^{2+} entry routes [232]. The mechanism underlying the communication between the filling state of the agonist-sensitive Ca^{2+} stores to the plasma membrane Ca^{2+} channels in these cells was revealed to involve the protein STIM1 [233], following its identification as the ER Ca^{2+} sensor for the activation of SOCE in *Drosophila* S2 and different mammalian cell types [234–237]. STIM1 and its homologue STIM2 are expressed both in the dense tubular system, the analog of the ER in platelets, and the agonist-releasable acidic Ca^{2+} stores, mostly including lysosomes and lysosome-like organelles [238]. Although the role of STIM1 as the intraluminal Ca^{2+} sensor is widely accepted, the nature of the SOC channels still remains as a matter of intense debate and investigation. Current evidence supports the role of the Ca^{2+} selective channel Orai1 in the conduction of the Ca^{2+} selective I_{CRAC}, the store operated Ca^{2+} release activated Ca^{2+} (CRAC) current, as well as TRPC proteins as candidates for the conduction of the non-selective and store-dependent I_{SOC} current [49, 239–241]. Since the identification of mammalian TRPC1, this channel has been proposed as a SOC candidate in different cell types, including human platelets [242, 243]. The role of TRPC6 in SOCE in human platelets was first suggested in 2008 by using a neutralizing antibody against the C-terminal region of TRPC6, which reduced both thapsigargin (TG)-induced and OAG-evoked Ca^{2+} influx, as well as TG-stimulated Mn^{2+} entry [222]. In addition, treatment with the PLA_2 and TRPC6 inhibitor ACA [244] attenuated TG-induced cation entry in human platelets [222]. Further evidence supporting a role for TRPC6 in SOCE in human platelets comes from studies performed using TPEN [N,N,N′,N′-tetrakis-(2-pyridylmethyl)-ethylenediamine] to decrease the intraluminal Ca^{2+} concentration avoiding the indirect effect induced by the amount of Ca^{2+} released into the cytosol. Introduction of a specific anti-TRPC6 antibody into platelets by electrotransjection resulted in attenuation of SOCE induced by treatment with TPEN [55]. The effect observed after incorporation of the antibody was not due to side-effects of the preservatives present in the antibody solution, which was tested in every set of experiments performed, for a reference see [245]. To further support the role of TRPC6 in SOCE (in platelets?), analysis of the signaling complexes generated upon induction of store-operated and store-independent Ca^{2+} influx revealed that Ca^{2+} store discharge results in the formation of a protein complex involving, among others, STIM1, STIM2, Orai1, TRPC6 and TRPC1 [107, 226], by contrast, stimulation of store-independent Ca^{2+} entry by OAG results in the association of TRPC3 with Orai3 [107]. Cell stimulation with thrombin or Ca^{2+}-store depletion using TG enhanced the interaction between TRPC6 with Orai1 and STIM1. In contrast, stimulation with OAG displaces TRPC6 from the Orai1/STIM1 complex and promotes the association of TRPC6 with the non-capacitative channel TRPC3. The interaction between TRPC6 and TRPC3 was abolished by cell loading with the intracellular Ca^{2+} chelator dimethyl-BAPTA, thus indicating that this phenomenon is Ca^{2+}-dependent [55]. Altogether, these findings support the hypothesis that TRPC6 might be a point of convergence between store-operated and store-independent Ca^{2+} influx pathways, participating in both mechanisms through its interaction with the Orai1/STIM1 complex or TRPC3/Orai3, respectively.

10.7.2.4 Functional Role of TRPC6 in Platelets

The role of TRPC6 in platelet function has been investigated using different murine transgenic models lacking TRPC6 [56, 246–248]. However, the observations reported have been contradictory and the function of TRPC6 in platelets is still uncertain. Ramanathan and coworkers found that platelets from TRPC6 deficient mice were insensitive to DAG while SOCE and agonist-induced Ca^{2+} entry were unaltered, as well as platelet function in vivo and in vitro [56]. Furthermore, this study showed a non-statistically significant decrease in the resting cytosolic Ca^{2+} concentration in platelets isolated from TRPC6-deficient mouse [56]. Paez-Espinosa and coworkers reported that TRPC6 knockout mice show prolonged bleeding time, detected using the tail bleeding time test, as well as an increased time for occlusion of the injured carotid artery as compared to wild type mice [246]. Harper and collaborators reported that Na^{+} entry via TRPC6 is required for full activation of phosphatidylserine exposure in mice platelets, a phenomenon that is essential for the generation of additional thrombin, thus supporting the mechanism of thrombosis and haemostasis [247]. Finally, we have recently found a significant reduction in the resting cytosolic Ca^{2+} concentration in platelets from mice lacking TRPC6 [248], despite we did not detect any significant alteration in thrombin-stimulated Ca^{2+} mobilization or platelet aggregation [229]. The reason of the discrepancies reported by the different groups that explored the function of TRPC6 in mouse platelets might be attributed to the different function in vivo protocols used, the distinct procedure used for platelet isolation and analysis, and/or the intrinsic variability generated by the different genetic background of the murine strains used. For instance, models of TRPC6-deficient mice were generated by inserting a selection cassette that replaced exon 4 or exon 7.

10.7.2.5 TRPC6 and Diabetes Mellitus

Finally, in the last part of this chapter, we would like to discuss the relationships between TRPC6 and diabetes mellitus. Cardiovascular complications are a frequent cause of morbidity and mortality in patients with diabetes mellitus and platelets have been proposed to contribute to diabetic micro and macroangiopathy [249,

250]. Platelet hyperactivity and hyper-aggregability has been associated to abnormal intracellular Ca^{2+} homeostasis, which includes enhanced Ca^{2+} influx and impairment of Ca^{2+} clearance mechanisms probably due to the altered redox state [251–255]. We have analyzed the expression of different Ca^{2+}-permeable channels in platelets from type 2 diabetic patients, including TRPC1, TRPC3, TRPC6 and Orai1, as well as that of STIM1 by Western blotting and found that the expression of TRPC1 in platelets from diabetics was similar to that in healthy donors, while the expression of TRPC3, Orai1 and STIM1 was found to be enhanced in platelets from diabetic patients as compared to controls and the expression of TRPC6 is reduced in diabetics [256]. Further studies by Liu and coworkers reported that direct platelet stimulation with glucose rapidly enhanced TRPC6 expression in the plasma membrane [142], thus suggesting that the reduced expression of TRPC6 at the protein level might be compensated by the increased surface expression of the channel. These observations are consistent with more recent findings from our laboratory testing the ability of platelets from type 2 diabetic donors to activate store-operated and store-independent Ca^{2+} influx mechanisms. For this set of experiments SOCE was stimulated by TG in the presence of specific purinergic and serotoninergic receptor antagonists, to avoid secondary activation of these receptors due to the increase in cytosolic Ca^{2+} concentration. In addition, SOCE was detected using Mn^{2+} as a surrogate for Ca^{2+} to prevent interference with other Ca^{2+} influx mechanisms. We found that store-operated Mn^{2+} entry induced by TG was reduced in platelets from diabetic donors as compared to healthy controls, probably as a result of the impaired association between STIM1 and Orai1, TRPC1 and TRPC6. These findings suggest that store-independent Ca^{2+} entry mediated by TRPC3, TRPC6 and/or Orai3, might be responsible for the enhanced Ca^{2+} influx observed in platelets from diabetic patients upon stimulation with physiological agonists [257].

Acknowledgements This work was supported by CNRS, l'Agence Nationale de la Recherche (ANR 13-NEUR-0003-02) and MINECO (Grant BFU2013-45564-C2-1-P).

References

1. Clapham DE, Julius D, Montell C, Schultz G (2005) International Union of Pharmacology. XLIX. Nomenclature and structure-function relationships of transient receptor potential channels. Pharmacol Rev 57(4):427–450
2. Minke B (2010) The history of the Drosophila TRP channel: the birth of a new channel superfamily. J Neurogenet 24(4):216–233
3. Zhu X, Jiang M, Peyton M, Boulay G, Hurst R, Stefani E, Birnbaumer L (1996) trp, a novel mammalian gene family essential for agonist-activated capacitative Ca2+ entry. Cell 85(5):661–671
4. Vazquez G, Wedel BJ, Aziz O, Trebak M, Putney JW Jr (2004) The mammalian TRPC cation channels. Biochim Biophys Acta 1742(1–3):21–36
5. Abramowitz J, Birnbaumer L (2009) Physiology and pathophysiology of canonical transient receptor potential channels. FASEB J 23(2):297–328

6. Montell C, Birnbaumer L, Flockerzi V (2002) The TRP channels, a remarkably functional family. Cell 108(5):595–598
7. Boulay G, Zhu X, Peyton M, Jiang M, Hurst R, Stefani E, Birnbaumer L (1997) Cloning and expression of a novel mammalian homolog of Drosophila transient receptor potential (Trp) involved in calcium entry secondary to activation of receptors coupled by the Gq class of G protein. J Biol Chem 272(47):29672–29680
8. D'Esposito M, Strazzullo M, Cuccurese M, Spalluto C, Rocchi M, D'Urso M, Ciccodicola A (1998) Identification and assignment of the human transient receptor potential channel 6 gene TRPC6 to chromosome 11q21-->q22. Cytogenet Cell Genet 83(1–2):46–47
9. Hofmann T, Obukhov AG, Schaefer M, Harteneck C, Gudermann T, Schultz G (1999) Direct activation of human TRPC6 and TRPC3 channels by diacylglycerol. Nature 397(6716):259–263
10. Okada T, Inoue R, Yamazaki K, Maeda A, Kurosaki T, Yamakuni T, Tanaka I, Shimizu S, Ikenaka K, Imoto K, Mori Y (1999) Molecular and functional characterization of a novel mouse transient receptor potential protein homologue TRP7. Ca(2+)-permeable cation channel that is constitutively activated and enhanced by stimulation of G protein-coupled receptor. J Biol Chem 274(39):27359–27370
11. Zhang L, Saffen D (2001) Muscarinic acetylcholine receptor regulation of TRP6 Ca2+ channel isoforms. Molecular structures and functional characterization. J Biol Chem 276(16):13331–13339
12. Vazquez E, Valverde MA (2006) A review of TRP channels splicing. Semin Cell Dev Biol 17(6):607–617
13. Clapham DE, Runnels LW, Strubing C (2001) The TRP ion channel family. Nat Rev Neurosci 2(6):387–396
14. Pedersen SF, Owsianik G, Nilius B (2005) TRP channels: an overview. Cell Calcium 38(3–4):233–252
15. Mizuno N, Kitayama S, Saishin Y, Shimada S, Morita K, Mitsuhata C, Kurihara H, Dohi T (1999) Molecular cloning and characterization of rat trp homologues from brain. Brain Res Mol Brain Res 64(1):41–51
16. Sedgwick SG, Smerdon SJ (1999) The ankyrin repeat: a diversity of interactions on a common structural framework. Trends Biochem Sci 24(8):311–316
17. Mosavi LK, Cammett TJ, Desrosiers DC, Peng ZY (2004) The ankyrin repeat as molecular architecture for protein recognition. Protein Sci 13(6):1435–1448
18. Gaudet R (2008) A primer on ankyrin repeat function in TRP channels and beyond. Mol BioSyst 4(5):372–379
19. Zhu MX (2005) Multiple roles of calmodulin and other Ca(2+)-binding proteins in the functional regulation of TRP channels. Pflugers Arch 451(1):105–115
20. Lepage PK, Lussier MP, Barajas-Martinez H, Bousquet SM, Blanchard AP, Francoeur N, Dumaine R, Boulay G (2006) Identification of two domains involved in the assembly of transient receptor potential canonical channels. J Biol Chem 281(41):30356–30364
21. Putney JW Jr (2004) The enigmatic TRPCs: multifunctional cation channels. Trends Cell Biol 14(6):282–286
22. Jung S, Muhle A, Schaefer M, Strotmann R, Schultz G, Plant TD (2003) Lanthanides potentiate TRPC5 currents by an action at extracellular sites close to the pore mouth. J Biol Chem 278(6):3562–3571
23. Wedel BJ, Vazquez G, McKay RR, St JBG, Putney JW Jr (2003) A calmodulin/inositol 1,4,5-trisphosphate (IP3) receptor-binding region targets TRPC3 to the plasma membrane in a calmodulin/IP3 receptor-independent process. J Biol Chem 278(28):25758–25765
24. Itsuki K, Imai Y, Hase H, Okamura Y, Inoue R, Mori MX (2014) PLC-mediated PI(4,5)P2 hydrolysis regulates activation and inactivation of TRPC6/7 channels. J Gen Physiol 143(2):183–201

25. Aires V, Hichami A, Boulay G, Khan NA (2007) Activation of TRPC6 calcium channels by diacylglycerol (DAG)-containing arachidonic acid: a comparative study with DAG-containing docosahexaenoic acid. Biochimie 89(8):926–937
26. Kim JM, Heo K, Choi J, Kim K, An W (2013) The histone variant MacroH2A regulates Ca(2+) influx through TRPC3 and TRPC6 channels. Oncogenesis 2, e77
27. Inoue R, Okada T, Onoue H, Hara Y, Shimizu S, Naitoh S, Ito Y, Mori Y (2001) The transient receptor potential protein homologue TRP6 is the essential component of vascular {{alpha}}1-adrenoceptor-activated Ca2+-permeable cation channel. Circ Res 88(3):325–332
28. Estacion M, Sinkins WG, Jones SW, Applegate MA, Schilling WP (2006) Human TRPC6 expressed in HEK 293 cells forms non-selective cation channels with limited Ca2+ permeability. J Physiol 572(Pt 2):359–377
29. Basora N, Boulay G, Bilodeau L, Rousseau E, Payet MD (2003) 20-hydroxyeicosatetraenoic acid (20-HETE) activates mouse TRPC6 channels expressed in HEK293 cells. J Biol Chem 278(34):31709–31716
30. Chaudhuri P, Colles SM, Bhat M, Van Wagoner DR, Birnbaumer L, Graham LM (2008) Elucidation of a TRPC6-TRPC5 channel cascade that restricts endothelial cell movement. Mol Biol Cell 19(8):3203–3211
31. Hardie RC (2003) Regulation of TRP channels via lipid second messengers. Annu Rev Physiol 65:735–759
32. Suh BC, Hille B (2008) PIP2 is a necessary cofactor for ion channel function: how and why? Annu Rev Biophys 37:175–195
33. Kwon Y, Hofmann T, Montell C (2007) Integration of phosphoinositide- and calmodulin-mediated regulation of TRPC6. Mol Cell 25(4):491–503
34. Tseng PH, Lin HP, Hu H, Wang C, Zhu MX, Chen CS (2004) The canonical transient receptor potential 6 channel as a putative phosphatidylinositol 3,4,5-trisphosphate-sensitive calcium entry system. Biochemistry 43(37):11701–11708
35. Lemonnier L, Trebak M, Putney JW Jr (2008) Complex regulation of the TRPC3, 6 and 7 channel subfamily by diacylglycerol and phosphatidylinositol-4,5-bisphosphate. Cell Calcium 43(5):506–514
36. Albert AP (2011) Gating mechanisms of canonical transient receptor potential channel proteins: role of phosphoinositols and diacylglycerol. Adv Exp Med Biol 704:391–411
37. Estacion M, Li S, Sinkins WG, Gosling M, Bahra P, Poll C, Westwick J, Schilling WP (2004) Activation of human TRPC6 channels by receptor stimulation. J Biol Chem 279(21):22047–22056
38. Shi J, Mori E, Mori Y, Mori M, Li J, Ito Y, Inoue R (2004) Multiple regulation by calcium of murine homologues of transient receptor potential proteins TRPC6 and TRPC7 expressed in HEK293 cells. J Physiol 561(Pt 2):415–432
39. Albert AP, Large WA (2003) Synergism between inositol phosphates and diacylglycerol on native TRPC6-like channels in rabbit portal vein myocytes. J Physiol 552(Pt 3):789–795
40. Putney JW (1997) Capacitative calcium entry. Landes Biochemichal Publishing, Austin
41. Lussier MP, Cayouette S, Lepage PK, Bernier CL, Francoeur N, St-Hilaire M, Pinard M, Boulay G (2005) MxA, a member of the dynamin superfamily, interacts with the ankyrin-like repeat domain of TRPC. J Biol Chem 280(19):19393–19400
42. Nishida M, Onohara N, Sato Y, Suda R, Ogushi M, Tanabe S, Inoue R, Mori Y, Kurose H (2007) Galpha12/13-mediated up-regulation of TRPC6 negatively regulates endothelin-1-induced cardiac myofibroblast formation and collagen synthesis through nuclear factor of activated T cells activation. J Biol Chem 282(32):23117–23128
43. Hassock SR, Zhu MX, Trost C, Flockerzi V, Authi KS (2002) Expression and role of TRPC proteins in human platelets: evidence that TRPC6 forms the store-independent calcium entry channel. Blood 100(8):2801–2811
44. Yu Y, Sweeney M, Zhang S, Platoshyn O, Landsberg J, Rothman A, Yuan JX (2003) PDGF stimulates pulmonary vascular smooth muscle cell proliferation by upregulating TRPC6 expression. Am J Physiol Cell Physiol 284(2):C316–C330

45. El Boustany C, Bidaux G, Enfissi A, Delcourt P, Prevarskaya N, Capiod T (2008) Capacitative calcium entry and transient receptor potential canonical 6 expression control human hepatoma cell proliferation. Hepatology 47(6):2068–2077
46. Nilius B (2003) From TRPs to SOCs, CCEs, and CRACs: consensus and controversies. Cell Calcium 33(5–6):293–298
47. Vig M, Peinelt C, Beck A, Koomoa DL, Rabah D, Koblan-Huberson M, Kraft S, Turner H, Fleig A, Penner R, Kinet JP (2006) CRACM1 is a plasma membrane protein essential for store-operated Ca2+ entry. Science 312(5777):1220–1223
48. Prakriya M, Feske S, Gwack Y, Srikanth S, Rao A, Hogan PG (2006) Orai1 is an essential pore subunit of the CRAC channel. Nature 443:230–233
49. Feske S, Gwack Y, Prakriya M, Srikanth S, Puppel SH, Tanasa B, Hogan PG, Lewis RS, Daly M, Rao A (2006) A mutation in Orai1 causes immune deficiency by abrogating CRAC channel function. Nature 441(7090):179–185
50. Zhang SL, Yeromin AV, Zhang XH, Yu Y, Safrina O, Penna A, Roos J, Stauderman KA, Cahalan MD (2006) Genome-wide RNAi screen of Ca(2+) influx identifies genes that regulate Ca(2+) release-activated Ca(2+) channel activity. Proc Natl Acad Sci U S A 103(24):9357–9362
51. Soboloff J, Rothberg BS, Madesh M, Gill DL (2012) STIM proteins: dynamic calcium signal transducers. Nat Rev Mol Cell Biol 13(9):549–565
52. Liao Y, Erxleben C, Yildirim E, Abramowitz J, Armstrong DL, Birnbaumer L (2007) Orai proteins interact with TRPC channels and confer responsiveness to store depletion. Proc Natl Acad Sci U S A 104(11):4682–4687
53. Salido GM, Sage SO, Rosado JA (2009) Biochemical and functional properties of the store-operated Ca(2+) channels. Cell Signal 21(4):457–461
54. Brechard S, Melchior C, Plancon S, Schenten V, Tschirhart EJ (2008) Store-operated Ca2+ channels formed by TRPC1, TRPC6 and Orai1 and non-store-operated channels formed by TRPC3 are involved in the regulation of NADPH oxidase in HL-60 granulocytes. Cell Calcium 44(5):492–506
55. Jardin I, Gomez LJ, Salido GM, Rosado JA (2009) Dynamic interaction of hTRPC6 with the Orai1-STIM1 complex or hTRPC3 mediates its role in capacitative or non-capacitative Ca(2+) entry pathways. Biochem J 420(2):267–276
56. Ramanathan G, Gupta S, Thielmann I, Pleines I, Varga-Szabo D, May F, Mannhalter C, Dietrich A, Nieswandt B, Braun A (2012) Defective diacylglycerol-induced Ca2+ entry but normal agonist-induced activation responses in TRPC6-deficient mouse platelets. J Thromb Haemost 10(3):419–429
57. Spassova MA, Hewavitharana T, Xu W, Soboloff J, Gill DL (2006) A common mechanism underlies stretch activation and receptor activation of TRPC6 channels. Proc Natl Acad Sci U S A 103(44):16586–16591
58. Gottlieb P, Folgering J, Maroto R, Raso A, Wood TG, Kurosky A, Bowman C, Bichet D, Patel A, Sachs F, Martinac B, Hamill OP, Honore E (2008) Revisiting TRPC1 and TRPC6 mechanosensitivity. Pflugers Arch 455(6):1097–1103
59. Mederos y Schnitzler M, Storch U, Meibers S, Nurwakagari P, Breit A, Essin K, Gollasch M, Gudermann T (2008) Gq-coupled receptors as mechanosensors mediating myogenic vasoconstriction. EMBO J 27(23):3092–3103
60. Schleifenbaum J, Kassmann M, Szijarto IA, Hercule HC, Tano JY, Weinert S, Heidenreich M, Pathan AR, Anistan YM, Alenina N, Rusch NJ, Bader M, Jentsch TJ, Gollasch M (2014) Stretch-activation of angiotensin II type 1a receptors contributes to the myogenic response of mouse mesenteric and renal arteries. Circ Res 115(2):263–272
61. Anfinogenova Y, Brett SE, Walsh MP, Harraz OF, Welsh DG (2011) Do TRPC-like currents and G protein-coupled receptors interact to facilitate myogenic tone development? Am J Physiol Heart Circ Physiol 301(4):H1378–H1388
62. Wilson C, Dryer SE (2014) A mutation in TRPC6 channels abolishes their activation by hypoosmotic stretch but does not affect activation by diacylglycerol or G protein signaling cascades. Am J Physiol Renal Physiol 306(9):F1018–F1025

63. Quick K, Zhao J, Eijkelkamp N, Linley JE, Rugiero F, Cox JJ, Raouf R, Gringhuis M, Sexton JE, Abramowitz J, Taylor R, Forge A, Ashmore J, Kirkwood N, Kros CJ, Richardson GP, Freichel M, Flockerzi V, Birnbaumer L, Wood JN (2012) TRPC3 and TRPC6 are essential for normal mechanotransduction in subsets of sensory neurons and cochlear hair cells. Open Biol 2(5):120068
64. Alessandri-Haber N, Dina OA, Chen X, Levine JD (2009) TRPC1 and TRPC6 channels cooperate with TRPV4 to mediate mechanical hyperalgesia and nociceptor sensitization. J Neurosci 29(19):6217–6228
65. Dyachenko V, Husse B, Rueckschloss U, Isenberg G (2009) Mechanical deformation of ventricular myocytes modulates both TRPC6 and Kir2.3 channels. Cell Calcium 45(1):38–54
66. Anderson M, Kim EY, Hagmann H, Benzing T, Dryer SE (2013) Opposing effects of podocin on the gating of podocyte TRPC6 channels evoked by membrane stretch or diacylglycerol. Am J Physiol Cell Physiol 305(3):C276–C289
67. Anderson M, Roshanravan H, Khine J, Dryer SE (2014) Angiotensin II activation of TRPC6 channels in rat podocytes requires generation of reactive oxygen species. J Cell Physiol 229(4):434–442
68. Roshanravan H, Dryer SE (2014) ATP acting through P2Y receptors causes activation of podocyte TRPC6 channels: role of podocin and reactive oxygen species. Am J Physiol Renal Physiol 306(9):F1088–F1097
69. Leuner K, Kazanski V, Muller M, Essin K, Henke B, Gollasch M, Harteneck C, Muller WE (2007) Hyperforin – a key constituent of St. John's wort specifically activates TRPC6 channels. FASEB J 21(14):4101–4111
70. Muller M, Essin K, Hill K, Beschmann H, Rubant S, Schempp CM, Gollasch M, Boehncke WH, Harteneck C, Muller WE, Leuner K (2008) Specific TRPC6 channel activation, a novel approach to stimulate keratinocyte differentiation. J Biol Chem 283(49):33942–33954
71. Leuner K, Li W, Amaral MD, Rudolph S, Calfa G, Schuwald AM, Harteneck C, Inoue T, Pozzo-Miller L (2013) Hyperforin modulates dendritic spine morphology in hippocampal pyramidal neurons by activating Ca(2+) -permeable TRPC6 channels. Hippocampus 23(1):40–52
72. Ding Y, Winters A, Ding M, Graham S, Akopova I, Muallem S, Wang Y, Hong JH, Gryczynski Z, Yang SH, Birnbaumer L, Ma R (2011) Reactive oxygen species-mediated TRPC6 protein activation in vascular myocytes, a mechanism for vasoconstrictor-regulated vascular tone. J Biol Chem 286(36):31799–31809
73. Leuner K, Heiser JH, Derksen S, Mladenov MI, Fehske CJ, Schubert R, Gollasch M, Schneider G, Harteneck C, Chatterjee SS, Muller WE (2010) Simple 2,4-diacylphloroglucinols as classic transient receptor potential-6 activators – identification of a novel pharmacophore. Mol Pharmacol 77(3):368–377
74. Sell TS, Belkacemi T, Flockerzi V, Beck A (2014) Protonophore properties of hyperforin are essential for its pharmacological activity. Sci Rep 4:7500
75. Jung S, Strotmann R, Schultz G, Plant TD (2002) TRPC6 is a candidate channel involved in receptor-stimulated cation currents in A7r5 smooth muscle cells. Am J Physiol Cell Physiol 282(2):C347–C359
76. Foster RR, Zadeh MA, Welsh GI, Satchell SC, Ye Y, Mathieson PW, Bates DO, Saleem MA (2009) Flufenamic acid is a tool for investigating TRPC6-mediated calcium signalling in human conditionally immortalised podocytes and HEK293 cells. Cell Calcium 45(4):384–390
77. Tesfai Y, Brereton HM, Barritt GJ (2001) A diacylglycerol-activated Ca2+ channel in PC12 cells (an adrenal chromaffin cell line) correlates with expression of the TRP-6 (transient receptor potential) protein. Biochem J 358(Pt 3):717–726
78. Tu P, Brandolin G, Bouron A (2009) The anti-inflammatory agent flufenamic acid depresses store-operated channels by altering mitochondrial calcium homeostasis. Neuropharmacology 56:1010–1016

79. Kraft R, Grimm C, Frenzel H, Harteneck C (2006) Inhibition of TRPM2 cation channels by N-(p-amylcinnamoyl)anthranilic acid. Br J Pharmacol 148(3):264–273
80. Graham S, Ding M, Ding Y, Sours-Brothers S, Luchowski R, Gryczynski Z, Yorio T, Ma H, Ma R (2010) Canonical transient receptor potential 6 (TRPC6), a redox-regulated cation channel. J Biol Chem 285(30):23466–23476
81. Kim EY, Anderson M, Dryer SE (2012) Sustained activation of N-methyl-D-aspartate receptors in podoctyes leads to oxidative stress, mobilization of transient receptor potential canonical 6 channels, nuclear factor of activated T cells activation, and apoptotic cell death. Mol Pharmacol 82(4):728–737
82. Graham S, Gorin Y, Abboud HE, Ding M, Lee DY, Shi H, Ding Y, Ma R (2011) Abundance of TRPC6 protein in glomerular mesangial cells is decreased by ROS and PKC in diabetes. Am J Physiol Cell Physiol 301(2):C304–C315
83. Cayouette S, Lussier MP, Mathieu E-L, Bousquet SM, Boulay G (2004) Exocytotic insertion of TRPC6 channel into the plasma membrane upon Gq protein-coupled receptor activation. J Biol Chem 279(8):7241–7246
84. Suzuki F, Morishima S, Tanaka T, Muramatsu I (2007) Snapin, a new regulator of receptor signaling, augments {alpha}1A-adrenoceptor-operated calcium influx through TRPC6. J Biol Chem 282(40):29563–29573. doi:10.1074/jbc.M702063200
85. Xie J, Cha SK, An SW, Kuro OM, Birnbaumer L, Huang CL (2012) Cardioprotection by Klotho through downregulation of TRPC6 channels in the mouse heart. Nat Commun 3:1238
86. Shin YC, Shin SY, So I, Kwon D, Jeon JH (2011) TRIP Database: a manually curated database of protein-protein interactions for mammalian TRP channels. Nucleic Acids Res 39(Database issue):D356–D361
87. Ju M, Shi J, Saleh SN, Albert AP, Large WA (2010) Ins(1,4,5)P3 interacts with PIP2 to regulate activation of TRPC6/C7 channels by diacylglycerol in native vascular myocytes. J Physiol 588(Pt 9):1419–1433
88. Sours S, Du J, Chu S, Ding M, Zhou XJ, Ma R (2006) Expression of canonical transient receptor potential (TRPC) proteins in human glomerular mesangial cells. Am J Physiol Renal Physiol 290(6):F1507–F1515
89. Chu X, Tong Q, Cheung JY, Wozney J, Conrad K, Mazack V, Zhang W, Stahl R, Barber DL, Miller BA (2004) Interaction of TRPC2 and TRPC6 in erythropoietin modulation of calcium influx. J Biol Chem 279(11):10514–10522
90. Bandyopadhyay BC, Swaim WD, Liu X, Redman RS, Patterson RL, Ambudkar IS (2005) Apical localization of a functional TRPC3/TRPC6-Ca2+-signaling complex in polarized epithelial cells. Role in apical Ca2+ influx. J Biol Chem 280(13):12908–12916
91. Hofmann T, Schaefer M, Schultz G, Gudermann T (2002) Subunit composition of mammalian transient receptor potential channels in living cells. Proc Natl Acad Sci U S A 99(11):7461–7466
92. Goel M, Sinkins WG, Schilling WP (2002) Selective association of TRPC channel subunits in rat brain synaptosomes. J Biol Chem 277:48303–48310
93. Hirschler-Laszkiewicz I, Tong Q, Conrad K, Zhang W, Flint WW, Barber AJ, Barber DL, Cheung JY, Miller BA (2009) TRPC3 activation by erythropoietin is modulated by TRPC6. J Biol Chem 284(7):4567–4581
94. Goel M, Sinkins W, Keightley A, Kinter M, Schilling WP (2005) Proteomic analysis of TRPC5- and TRPC6-binding partners reveals interaction with the plasmalemmal Na(+)/K(+)-ATPase. Pflugers Arch 451(1):87–98
95. Albarran L, Lopez JJ, Dionisio N, Smani T, Salido GM, Rosado JA (2013) Transient receptor potential ankyrin-1 (TRPA1) modulates store-operated Ca(2+) entry by regulation of STIM1-Orai1 association. Biochim Biophys Acta 1833(12):3025–3034
96. Hsu YJ, Hoenderop JG, Bindels RJ (2007) TRP channels in kidney disease. Biochim Biophys Acta 1772(8):928–936
97. Goel M, Sinkins WG, Zuo CD, Estacion M, Schilling WP (2006) Identification and localization of TRPC channels in the rat kidney. Am J Physiol Renal Physiol 290(5):F1241–F1252

98. Goel M, Zuo CD, Sinkins WG, Schilling WP (2007) TRPC3 channels colocalize with Na+/Ca2+ exchanger and Na+ pump in axial component of transverse-axial tubular system of rat ventricle. Am J Physiol Heart Circ Physiol 292(2):H874–H883
99. Boulay G, Brown DM, Qin N, Jiang M, Dietrich A, Zhu MX, Chen Z, Birnbaumer M, Mikoshiba K, Birnbaumer L (1999) Modulation of Ca(2+) entry by polypeptides of the inositol 1,4, 5-trisphosphate receptor (IP3R) that bind transient receptor potential (TRP): evidence for roles of TRP and IP3R in store depletion-activated Ca(2+) entry. Proc Natl Acad Sci U S A 96(26):14955–14960
100. Antigny F, Norez C, Dannhoffer L, Bertrand J, Raveau D, Corbi P, Jayle C, Becq F, Vandebrouck C (2011) Transient receptor potential canonical channel 6 links Ca2+ mishandling to cystic fibrosis transmembrane conductance regulator channel dysfunction in cystic fibrosis. Am J Respir Cell Mol Biol 44(1):83–90
101. Kim EY, Alvarez-Baron CP, Dryer SE (2009) Canonical transient receptor potential channel (TRPC)3 and TRPC6 associate with large-conductance Ca2+-activated K+ (BKCa) channels: role in BKCa trafficking to the surface of cultured podocytes. Mol Pharmacol 75(3):466–477
102. DeHaven WI, Jones BF, Petranka JG, Smyth JT, Tomita T, Bird GS, Putney JW Jr (2009) TRPC channels function independently of STIM1 and Orai1. J Physiol 587(Pt 10):2275–2298
103. Yuan JP, Zeng W, Huang GN, Worley PF, Muallem S (2007) STIM1 heteromultimerizes TRPC channels to determine their function as store-operated channels. Nat Cell Biol 9(6):636–645
104. Albarran L, Dionisio N, Lopez E, Salido GM, Redondo PC, Rosado JA (2014) STIM1 regulates TRPC6 heteromultimerization and subcellular location. Biochem J 463(3):373–381
105. Liao Y, Erxleben C, Abramowitz J, Flockerzi V, Zhu MX, Armstrong DL, Birnbaumer L (2008) Functional interactions among Orai1, TRPCs, and STIM1 suggest a STIM-regulated heteromeric Orai/TRPC model for SOCE/Icrac channels. Proc Natl Acad Sci U S A 105(8):2895–2900
106. Liao Y, Plummer NW, George MD, Abramowitz J, Zhu MX, Birnbaumer L (2009) A role for Orai in TRPC-mediated Ca2+ entry suggests that a TRPC:Orai complex may mediate store and receptor operated Ca2+ entry. Proc Natl Acad Sci U S A 106(9):3202–3206
107. Berna-Erro A, Galan C, Dionisio N, Gomez LJ, Salido GM, Rosado JA (2012) Capacitative and non-capacitative signaling complexes in human platelets. Biochim Biophys Acta 1823(8):1242–1251
108. Friedlova E, Grycova L, Holakovska B, Silhan J, Janouskova H, Sulc M, Obsilova V, Obsil T, Teisinger J (2010) The interactions of the C-terminal region of the TRPC6 channel with calmodulin. Neurochem Int 56(2):363–366
109. Tang J, Lin Y, Zhang Z, Tikunova S, Birnbaumer L, Zhu MX (2001) Identification of common binding sites for calmodulin and inositol 1,4,5-trisphosphate receptors on the carboxyl termini of trp channels. J Biol Chem 276(24):21303–21310
110. Boulay G (2002) Ca(2+)-calmodulin regulates receptor-operated Ca(2+) entry activity of TRPC6 in HEK-293 cells. Cell Calcium 32(4):201–207
111. Hisatsune C, Kuroda Y, Nakamura K, Inoue T, Nakamura T, Michikawa T, Mizutani A, Mikoshiba K (2004) Regulation of TRPC6 channel activity by tyrosine phosphorylation. J Biol Chem 279(18):18887–18894
112. Kini V, Chavez A, Mehta D (2010) A new role for PTEN in regulating transient receptor potential canonical channel 6-mediated Ca2+ entry, endothelial permeability, and angiogenesis. J Biol Chem 285(43):33082–33091
113. Kim JY, Saffen D (2005) Activation of M1 muscarinic acetylcholine receptors stimulates the formation of a multiprotein complex centered on TRPC6 channels. J Biol Chem 280(36):32035–32047
114. Leclerc E, Heizmann CW (2011) The importance of Ca2+/Zn2+ signaling S100 proteins and RAGE in translational medicine. Front Biosci 3:1232–1262

115. Bily J, Grycova L, Holendova B, Jirku M, Janouskova H, Bousova K, Teisinger J (2013) Characterization of the S100A1 protein binding site on TRPC6 C-terminus. PLoS ONE 8(5), e62677
116. Zhang Z, Tang J, Tikunova S, Johnson JD, Chen Z, Qin N, Dietrich A, Stefani E, Birnbaumer L, Zhu MX (2001) Activation of Trp3 by inositol 1,4,5-trisphosphate receptors through displacement of inhibitory calmodulin from a common binding domain. Proc Natl Acad Sci U S A 98(6):3168–3173
117. Chevallet M, Jarvis L, Harel A, Luche S, Degot S, Chapuis V, Boulay G, Rabilloud T, Bouron A (2014) Functional consequences of the over-expression of TRPC6 channels in HEK cells: impact on the homeostasis of zinc. Metallomics 6(7):1269–1276
118. Carrasquillo R, Tian D, Krishna S, Pollak MR, Greka A, Schlondorff J (2012) SNF8, a member of the ESCRT-II complex, interacts with TRPC6 and enhances its channel activity. BMC Cell Biol 13:33
119. Lussier MP, Lepage PK, Bousquet SM, Boulay G (2008) RNF24, a new TRPC interacting protein, causes the intracellular retention of TRPC. Cell Calcium 43(5):432–443
120. Goel M, Garcia R, Estacion M, Schilling WP (2001) Regulation of Drosophila TRPL channels by immunophilin FKBP59. J Biol Chem 276(42):38762–38773
121. Sinkins WG, Goel M, Estacion M, Schilling WP (2004) Association of immunophilins with mammalian TRPC channels. J Biol Chem 279(33):34521–34529
122. Reiser J, Polu KR, Moller CC, Kenlan P, Altintas MM, Wei C, Faul C, Herbert S, Villegas I, Avila-Casado C, McGee M, Sugimoto H, Brown D, Kalluri R, Mundel P, Smith PL, Clapham DE, Pollak MR (2005) TRPC6 is a glomerular slit diaphragm-associated channel required for normal renal function. Nat Genet 37(7):739–744
123. Kunert-Keil C, Bisping F, Kruger J, Brinkmeier H (2006) Tissue-specific expression of TRP channel genes in the mouse and its variation in three different mouse strains. BMC Genomics 7:159
124. Jang Y, Lee Y, Kim SM, Yang YD, Jung J, Oh U (2012) Quantitative analysis of TRP channel genes in mouse organs. Arch Pharm Res 35(10):1823–1830
125. Pani B, Singh BB (2009) Lipid rafts/caveolae as microdomains of calcium signaling. Cell Calcium 45(6):625–633
126. Brownlow SL, Sage SO (2005) Transient receptor potential protein subunit assembly and membrane distribution in human platelets. Thromb Haemost 94(4):839–845
127. Lei L, Lu S, Wang Y, Kim T, Mehta D, Wang Y (2014) The role of mechanical tension on lipid raft dependent PDGF-induced TRPC6 activation. Biomaterials 35(9):2868–2877
128. Huber TB, Schermer B, Muller RU, Hohne M, Bartram M, Calixto A, Hagmann H, Reinhardt C, Koos F, Kunzelmann K, Shirokova E, Krautwurst D, Harteneck C, Simons M, Pavenstadt H, Kerjaschki D, Thiele C, Walz G, Chalfie M, Benzing T (2006) Podocin and MEC-2 bind cholesterol to regulate the activity of associated ion channels. Proc Natl Acad Sci U S A 103(46):17079–17086
129. den Dekker E, Molin DG, Breikers G, van Oerle R, Akkerman JW, van Eys GJ, Heemskerk JW (2001) Expression of transient receptor potential mRNA isoforms and Ca(2+) influx in differentiating human stem cells and platelets. Biochim Biophys Acta 1539(3):243–255
130. Madan E, Gogna R, Keppler B, Pati U (2013) p53 increases intra-cellular calcium release by transcriptional regulation of calcium channel TRPC6 in GaQ3-treated cancer cells. PLoS ONE 8(8), e71016
131. Chigurupati S, Venkataraman R, Barrera D, Naganathan A, Madan M, Paul L, Pattisapu JV, Kyriazis GA, Sugaya K, Bushnev S, Lathia JD, Rich JN, Chan SL (2010) Receptor channel TRPC6 is a key mediator of Notch-driven glioblastoma growth and invasiveness. Cancer Res 70(1):418–427
132. Sun YH, Li YQ, Feng SL, Li BX, Pan ZW, Xu CQ, Li TT, Yang BF (2010) Calcium-sensing receptor activation contributed to apoptosis stimulates TRPC6 channel in rat neonatal ventricular myocytes. Biochem Biophys Res Commun 394(4):955–961
133. Lin MJ, Leung GP, Zhang WM, Yang XR, Yip KP, Tse CM, Sham JS (2004) Chronic hypoxia-induced upregulation of store-operated and receptor-operated Ca2+ channels in pulmonary

arterial smooth muscle cells: a novel mechanism of hypoxic pulmonary hypertension. Circ Res 95(5):496–505
134. Ding F, Zhang X, Li X, Zhang Y, Li B, Ding J (2014) Mammalian target of rapamycin complex 2 signaling pathway regulates transient receptor potential cation channel 6 in podocytes. PLoS ONE 9(11), e112972
135. Vollenbroker B, George B, Wolfgart M, Saleem MA, Pavenstadt H, Weide T (2009) mTOR regulates expression of slit diaphragm proteins and cytoskeleton structure in podocytes. Am J Physiol Renal Physiol 296(2):F418–F426
136. Gibon J, Deloulme JC, Chevallier T, Ladeveze E, Abrous DN, Bouron A (2013) The antidepressant hyperforin increases the phosphorylation of CREB and the expression of TrkB in a tissue-specific manner. Int J Neuropsychopharmacol 16(1):189–198
137. Lu W, Ran P, Zhang D, Lai N, Zhong N, Wang J (2010) Bone morphogenetic protein 4 enhances canonical transient receptor potential expression, store-operated Ca2+ entry, and basal [Ca2+]i in rat distal pulmonary arterial smooth muscle cells. Am J Physiol Cell Physiol 299(6):C1370–C1378
138. Wang Y, Ding M, Chaudhari S, Ding Y, Yuan J, Stankowska D, He S, Krishnamoorthy R, Cunningham JT, Ma R (2013) Nuclear factor kappaB mediates suppression of canonical transient receptor potential 6 expression by reactive oxygen species and protein kinase C in kidney cells. J Biol Chem 288(18):12852–12865
139. Beck B, Zholos A, Sydorenko V, Roudbaraki M, Lehen'kyi V, Bordat P, Prevarskaya N, Skryma R (2006) TRPC7 is a receptor-operated DAG-activated channel in human keratinocytes. J Invest Dermatol 126(9):1982–1993
140. Cai S, Fatherazi S, Presland RB, Belton CM, Izutsu KT (2005) TRPC channel expression during calcium-induced differentiation of human gingival keratinocytes. J Dermatol Sci 40(1):21–28
141. Graham S, Ding M, Sours-Brothers S, Yorio T, Ma JX, Ma R (2007) Downregulation of TRPC6 protein expression by high glucose, a possible mechanism for the impaired Ca2+ signaling in glomerular mesangial cells in diabetes. Am J Physiol Renal Physiol 293(4):F1381–F1390
142. Liu D, Maier A, Scholze A, Rauch U, Boltzen U, Zhao Z, Zhu Z, Tepel M (2008) High glucose enhances transient receptor potential channel canonical type 6-dependent calcium influx in human platelets via phosphatidylinositol 3-kinase-dependent pathway. Arterioscler Thromb Vasc Biol 28(4):746–751
143. Han H, Wang Y, Li X, Wang PA, Wei X, Liang W, Ding G, Yu X, Bao C, Zhang Y, Wang Z, Yi F (2013) Novel role of NOD2 in mediating Ca2+ signaling: evidence from NOD2-regulated podocyte TRPC6 channels in hyperhomocysteinemia. Hypertension 62(3):506–511
144. Kuwahara K, Wang Y, McAnally J, Richardson JA, Bassel-Duby R, Hill JA, Olson EN (2006) TRPC6 fulfills a calcineurin signaling circuit during pathologic cardiac remodeling. J Clin Invest 116(12):3114–3126
145. Schlondorff J, Del Camino D, Carrasquillo R, Lacey V, Pollak MR (2009) TRPC6 mutations associated with focal segmental glomerulosclerosis cause constitutive activation of NFAT-dependent transcription. Am J Physiol Cell Physiol 296(3):C558–C569
146. Nijenhuis T, Sloan AJ, Hoenderop JG, Flesche J, van Goor H, Kistler AD, Bakker M, Bindels RJ, de Boer RA, Moller CC, Hamming I, Navis G, Wetzels JF, Berden JH, Reiser J, Faul C, van der Vlag J (2011) Angiotensin II contributes to podocyte injury by increasing TRPC6 expression via an NFAT-mediated positive feedback signaling pathway. Am J Pathol 179(4):1719–1732
147. Mwanjewe J, Grover AK (2004) Role of transient receptor potential canonical 6 (TRPC6) in non-transferrin-bound iron uptake in neuronal phenotype PC12 cells. Biochem J 378(Pt 3):975–982

148. Gibon J, Tu P, Bohic S, Richaud P, Arnaud J, Zhu M, Boulay G, Bouron A (2011) The overexpression of TRPC6 channels in HEK-293 cells favours the intracellular accumulation of zinc. Biochim Biophys Acta 1808(12):2807–2818
149. Dietrich A, Mederos y Schnitzler M, Emmel J, Kalwa H, Hofmann T, Gudermann T (2003) N-linked protein glycosylation is a major determinant for basal TRPC3 and TRPC6 channel activity. J Biol Chem 278(48):47842–47852
150. Lemos VS, Poburko D, Liao CH, Cole WC, van Breemen C (2007) Na+ entry via TRPC6 causes Ca2+ entry via NCX reversal in ATP stimulated smooth muscle cells. Biochem Biophys Res Commun 352(1):130–134
151. Poburko D, Liao CH, Lemos VS, Lin E, Maruyama Y, Cole WC, van Breemen C (2007) Transient receptor potential channel 6-mediated, localized cytosolic [Na+] transients drive Na+/Ca2+ exchanger-mediated Ca2+ entry in purinergically stimulated aorta smooth muscle cells. Circ Res 101(10):1030–1038
152. Syyong HT, Poburko D, Fameli N, van Breemen C (2007) ATP promotes NCX-reversal in aortic smooth muscle cells by DAG-activated Na+ entry. Biochem Biophys Res Commun 357(4):1177–1182
153. Tu P, Kunert-Keil C, Lucke S, Brinkmeier H, Bouron A (2009) Diacylglycerol analogues activate second messenger-operated calcium channels exhibiting TRPC-like properties in cortical neurons. J Neurochem 108(1):126–138
154. Louhivuori LM, Jansson L, Nordstrom T, Bart G, Nasman J, Akerman KE (2010) Selective interference with TRPC3/6 channels disrupts OX1 receptor signalling via NCX and reveals a distinct calcium influx pathway. Cell Calcium 48(2–3):114–123
155. Meng K, Xu J, Zhang C, Zhang R, Yang H, Liao C, Jiao J (2014) Calcium sensing receptor modulates extracellular calcium entry and proliferation via TRPC3/6 channels in cultured human mesangial cells. PLoS ONE 9(6), e98777
156. Kawasaki BT, Liao Y, Birnbaumer L (2006) Role of Src in C3 transient receptor potential channel function and evidence for a heterogeneous makeup of receptor- and store-operated Ca2+ entry channels. Proc Natl Acad Sci U S A 103(2):335–340
157. McMeekin SR, Dransfield I, Rossi AG, Haslett C, Walker TR (2006) E-selectin permits communication between PAF receptors and TRPC channels in human neutrophils. Blood 107(12):4938–4945
158. Bousquet SM, Monet M, Boulay G (2010) Protein kinase C-dependent phosphorylation of transient receptor potential canonical 6 (TRPC6) on serine 448 causes channel inhibition. J Biol Chem 285:40534–40543
159. Kwan HY, Huang Y, Yao X (2004) Regulation of canonical transient receptor potential isoform 3 (TRPC3) channel by protein kinase G. Proc Natl Acad Sci U S A 101(8):2625–2630
160. Takahashi S, Lin H, Geshi N, Mori Y, Kawarabayashi Y, Takami N, Mori MX, Honda A, Inoue R (2008) Nitric oxide-cGMP-protein kinase G pathway negatively regulates vascular transient receptor potential channel TRPC6. J Physiol 586(Pt 17):4209–4223
161. Chiluiza D, Krishna S, Schumacher VA, Schlondorff J (2013) Gain-of-function mutations in transient receptor potential C6 (TRPC6) activate extracellular signal-regulated kinases 1/2 (ERK1/2). J Biol Chem 288(25):18407–18420
162. Horinouchi T, Higa T, Aoyagi H, Nishiya T, Terada K, Miwa S (2012) Adenylate cyclase/cAMP/protein kinase A signaling pathway inhibits endothelin type A receptor-operated Ca(2)(+) entry mediated via transient receptor potential canonical 6 channels. J Pharmacol Exp Ther 340(1):143–151
163. Shen B, Kwan HY, Ma X, Wong CO, Du J, Huang Y, Yao X (2011) cAMP activates TRPC6 channels via the phosphatidylinositol 3-kinase (PI3K)-protein kinase B (PKB)-mitogen-activated protein kinase kinase (MEK)-ERK1/2 signaling pathway. J Biol Chem 286(22):19439–19445
164. Shi J, Takahashi S, Jin XH, Li YQ, Ito Y, Mori Y, Inoue R (2007) Myosin light chain kinase-independent inhibition by ML-9 of murine TRPC6 channels expressed in HEK293 cells. Br J Pharmacol 152:122–131

165. Song X, Liu BC, Lu XY, Yang LL, Zhai YJ, Eaton AF, Thai TL, Eaton DC, Ma HP, Shen BZ (2014) Lovastatin inhibits human B lymphoma cell proliferation by reducing intracellular ROS and TRPC6 expression. Biochim Biophys Acta 1843(5):894–901
166. Hill K, McNulty S, Randall AD (2004) Inhibition of TRPM2 channels by the antifungal agents clotrimazole and econazole. Naunyn Schmiedeberg's Arch Pharmacol 370(4):227–237
167. Harteneck C, Klose C, Krautwurst D (2011) Synthetic modulators of TRP channel activity. Adv Exp Med Biol 704:87–106
168. Kiyonaka S, Kato K, Nishida M, Mio K, Numaga T, Sawaguchi Y, Yoshida T, Wakamori M, Mori E, Numata T, Ishii M, Takemoto H, Ojida A, Watanabe K, Uemura A, Kurose H, Morii T, Kobayashi T, Sato Y, Sato C, Hamachi I, Mori Y (2009) Selective and direct inhibition of TRPC3 channels underlies biological activities of a pyrazole compound. Proc Natl Acad Sci U S A 106(13):5400–5405
169. Jiang L, Ding J, Tsai H, Li L, Feng Q, Miao J, Fan Q (2011) Over-expressing transient receptor potential cation channel 6 in podocytes induces cytoskeleton rearrangement through increases of intracellular Ca2+ and RhoA activation. Exp Biol Med (Maywood) 236(2):184–193
170. Kassouf N, Ambily A, Watson S, Hassock S, Authi HS, Srivastava S, Watson SP, Authi KS (2015) Phosphatidylinositol-3,4,5-trisphosphate stimulates Ca elevation and Akt phosphorylation to constitute a major mechanism of thromboxane A formation in human platelets. Cell Signal 27(7):1488–1498
171. Krautwurst D, Hescheler J, Arndts D, Losel W, Hammer R, Schultz G (1993) Novel potent inhibitor of receptor-activated nonselective cation currents in HL-60 cells. Mol Pharmacol 43(5):655–659
172. Dietrich A, Mederos YSM, Gollasch M, Gross V, Storch U, Dubrovska G, Obst M, Yildirim E, Salanova B, Kalwa H, Essin K, Pinkenburg O, Luft FC, Gudermann T, Birnbaumer L (2005) Increased vascular smooth muscle contractility in TRPC6-/- mice. Mol Cell Biol 25(16):6980–6989
173. Woelfle U, Laszczyk MN, Kraus M, Leuner K, Kersten A, Simon-Haarhaus B, Scheffler A, Martin SF, Muller WE, Nashan D, Schempp CM (2010) Triterpenes promote keratinocyte differentiation in vitro, ex vivo and in vivo: a role for the transient receptor potential canonical (subtype) 6. J Invest Dermatol 130(1):113–123
174. Davis J, Burr AR, Davis GF, Birnbaumer L, Molkentin JD (2012) A TRPC6-dependent pathway for myofibroblast transdifferentiation and wound healing in vivo. Dev Cell 23(4):705–715
175. Chen J, Luan Y, Yu R, Zhang Z, Zhang J, Wang W (2014) Transient receptor potential (TRP) channels, promising potential diagnostic and therapeutic tools for cancer. Biosci Trends 8(1):1–10
176. Buess M, Engler O, Hirsch HH, Moroni C (1999) Search for oncogenic regulators in an autocrine tumor model using differential display PCR: identification of novel candidate genes including the calcium channel mtrp6. Oncogene 18(7):1487–1494
177. Ge R, Tai Y, Sun Y, Zhou K, Yang S, Cheng T, Zou Q, Shen F, Wang Y (2009) Critical role of TRPC6 channels in VEGF-mediated angiogenesis. Cancer Lett 283(1):43–51
178. Singh I, Knezevic N, Ahmmed GU, Kini V, Malik AB, Mehta D (2007) Galphaq-TRPC6-mediated Ca2+ entry induces RhoA activation and resultant endothelial cell shape change in response to thrombin. J Biol Chem 282(11):7833–7843
179. Tian D, Jacobo SM, Billing D, Rozkalne A, Gage SD, Anagnostou T, Pavenstadt H, Hsu HH, Schlondorff J, Ramos A, Greka A (2010) Antagonistic regulation of actin dynamics and cell motility by TRPC5 and TRPC6 channels. Sci Signal 3(145):ra77
180. Damann N, Owsianik G, Li S, Poll C, Nilius B (2009) The calcium-conducting ion channel transient receptor potential canonical 6 is involved in macrophage inflammatory protein-2-induced migration of mouse neutrophils. Acta Physiol 195(1):3–11

181. Lindemann O, Umlauf D, Frank S, Schimmelpfennig S, Bertrand J, Pap T, Hanley PJ, Fabian A, Dietrich A, Schwab A (2013) TRPC6 regulates CXCR2-mediated chemotaxis of murine neutrophils. J Immunol 190(11):5496–5505
182. Millholland MG, Mishra S, Dupont CD, Love MS, Patel B, Shilling D, Kazanietz MG, Foskett JK, Hunter CA, Sinnis P, Greenbaum DC (2013) A host GPCR signaling network required for the cytolysis of infected cells facilitates release of apicomplexan parasites. Cell Host Microbe 13(1):15–28
183. Sel S, Rost BR, Yildirim AO, Sel B, Kalwa H, Fehrenbach H, Renz H, Gudermann T, Dietrich A (2008) Loss of classical transient receptor potential 6 channel reduces allergic airway response. Clin Exp Allergy 38(9):1548–1558
184. Finney-Hayward TK, Popa MO, Bahra P, Li S, Poll CT, Gosling M, Nicholson AG, Russell RE, Kon OM, Jarai G, Westwick J, Barnes PJ, Donnelly LE (2010) Expression of transient receptor potential C6 channels in human lung macrophages. Am J Respir Cell Mol Biol 43(3):296–304
185. Foller M, Kasinathan RS, Koka S, Lang C, Shumilina E, Birnbaumer L, Lang F, Huber SM (2008) TRPC6 contributes to the Ca(2+) leak of human erythrocytes. Cell Physiol Biochem 21(1–3):183–192
186. Winn MP, Conlon PJ, Lynn KL, Farrington MK, Creazzo T, Hawkins AF, Daskalakis N, Kwan SY, Ebersviller S, Burchette JL, Pericak-Vance MA, Howell DN, Vance JM, Rosenberg PB (2005) A mutation in the TRPC6 cation channel causes familial focal segmental glomerulosclerosis. Science 308(5729):1801–1804
187. Winn MP, Conlon PJ, Lynn KL, Howell DN, Slotterbeck BD, Smith AH, Graham FL, Bembe M, Quarles LD, Pericak-Vance MA, Vance JM (1999) Linkage of a gene causing familial focal segmental glomerulosclerosis to chromosome 11 and further evidence of genetic heterogeneity. Genomics 58(2):113–120
188. Santin S, Ars E, Rossetti S, Salido E, Silva I, Garcia-Maset R, Gimenez I, Ruiz P, Mendizabal S, Luciano Nieto J, Pena A, Camacho JA, Fraga G, Cobo MA, Bernis C, Ortiz A, de Pablos AL, Sanchez-Moreno A, Pintos G, Mirapeix E, Fernandez-Llama P, Ballarin J, Torra R, Group FS, Zamora I, Lopez-Hellin J, Madrid A, Ventura C, Vilalta R, Espinosa L, Garcia C, Melgosa M, Navarro M, Gimenez A, Cots JV, Alexandra S, Caramelo C, Egido J, San Jose MD, de la Cerda F, Sala P, Raspall F, Vila A, Daza AM, Vazquez M, Ecija JL, Espinosa M, Justa ML, Poveda R, Aparicio C, Rosell J, Muley R, Montenegro J, Gonzalez D, Hidalgo E, de Frutos DB, Trillo E, Gracia S, de los Rios FJ (2009) TRPC6 mutational analysis in a large cohort of patients with focal segmental glomerulosclerosis. Nephrol Dial Transplant 24(10):3089–3096
189. Gigante M, Caridi G, Montemurno E, Soccio M, d'Apolito M, Cerullo G, Aucella F, Schirinzi A, Emma F, Massella L, Messina G, De Palo T, Ranieri E, Ghiggeri GM, Gesualdo L (2011) TRPC6 mutations in children with steroid-resistant nephrotic syndrome and atypical phenotype. Clin J Am Soc Nephrol 6(7):1626–1634
190. Buscher AK, Konrad M, Nagel M, Witzke O, Kribben A, Hoyer PF, Weber S (2012) Mutations in podocyte genes are a rare cause of primary FSGS associated with ESRD in adult patients. Clin Nephrol 78(1):47–53
191. Mir S, Yavascan O, Berdeli A, Sozeri B (2012) TRPC6 gene variants in Turkish children with steroid-resistant nephrotic syndrome. Nephrol Dial Transplant 27(1):205–209
192. Mottl AK, Lu M, Fine CA, Weck KE (2013) A novel TRPC6 mutation in a family with podocytopathy and clinical variability. BMC Nephrol 14:104
193. Hofstra JM, Lainez S, van Kuijk WH, Schoots J, Baltissen MP, Hoefsloot LH, Knoers NV, Berden JH, Bindels RJ, van der Vlag J, Hoenderop JG, Wetzels JF, Nijenhuis T (2013) New TRPC6 gain-of-function mutation in a non-consanguineous Dutch family with late-onset focal segmental glomerulosclerosis. Nephrol Dial Transplant 28(7):1830–1838
194. Zhu B, Chen N, Wang ZH, Pan XX, Ren H, Zhang W, Wang WM (2009) Identification and functional analysis of a novel TRPC6 mutation associated with late onset familial focal segmental glomerulosclerosis in Chinese patients. Mutat Res 664(1–2):84–90

195. Heeringa SF, Moller CC, Du J, Yue L, Hinkes B, Chernin G, Vlangos CN, Hoyer PF, Reiser J, Hildebrandt F (2009) A novel TRPC6 mutation that causes childhood FSGS. PLoS ONE 4(11), e7771
196. Moller CC, Wei C, Altintas MM, Li J, Greka A, Ohse T, Pippin JW, Rastaldi MP, Wawersik S, Schiavi S, Henger A, Kretzler M, Shankland SJ, Reiser J (2007) Induction of TRPC6 channel in acquired forms of proteinuric kidney disease. J Am Soc Nephrol 18(1):29–36
197. Krall P, Canales CP, Kairath P, Carmona-Mora P, Molina J, Carpio JD, Ruiz P, Mezzano SA, Li J, Wei C, Reiser J, Young JI, Walz K (2010) Podocyte-specific overexpression of wild type or mutant trpc6 in mice is sufficient to cause glomerular disease. PLoS ONE 5(9), e12859
198. Kistler AD, Singh G, Altintas MM, Yu H, Fernandez IC, Gu C, Wilson C, Srivastava SK, Dietrich A, Walz K, Kerjaschki D, Ruiz P, Dryer S, Sever S, Dinda AK, Faul C, Reiser J (2013) Transient receptor potential channel 6 (TRPC6) protects podocytes during complement-mediated glomerular disease. J Biol Chem 288(51):36598–36609
199. Kim EY, Anderson M, Wilson C, Hagmann H, Benzing T, Dryer SE (2013) NOX2 interacts with podocyte TRPC6 channels and contributes to their activation by diacylglycerol: essential role of podocin in formation of this complex. Am J Physiol Cell Physiol 305(9):C960–C971
200. Orci L, Singh A, Amherdt M, Brown D, Perrelet A (1981) Microheterogeneity of protein and sterol content in kidney podocyte membrane. Nature 293(5834):646–647
201. Dryer SE, Reiser J (2010) TRPC6 channels and their binding partners in podocytes: role in glomerular filtration and pathophysiology. Am J Physiol Renal Physiol 299(4):F689–F701
202. Ilatovskaya DV, Palygin O, Chubinskiy-Nadezhdin V, Negulyaev YA, Ma R, Birnbaumer L, Staruschenko A (2014) Angiotensin II has acute effects on TRPC6 channels in podocytes of freshly isolated glomeruli. Kidney Int 86(3):506–514
203. Liu BC, Song X, Lu XY, Li DT, Eaton DC, Shen BZ, Li XQ, Ma HP (2013) High glucose induces podocyte apoptosis by stimulating TRPC6 via elevation of reactive oxygen species. Biochim Biophys Acta 1833(6):1434–1442
204. Thilo F, Liu Y, Loddenkemper C, Schuelein R, Schmidt A, Yan Z, Zhu Z, Zakrzewicz A, Gollasch M, Tepel M (2012) VEGF regulates TRPC6 channels in podocytes. Nephrol Dial Transplant 27(3):921–929
205. Sonneveld R, Ferre S, Hoenderop JG, Dijkman HB, Berden JH, Bindels RJ, Wetzels JF, van der Vlag J, Nijenhuis T (2013) Vitamin D down-regulates TRPC6 expression in podocyte injury and proteinuric glomerular disease. Am J Pathol 182(4):1196–1204
206. Liu Y, Echtermeyer F, Thilo F, Theilmeier G, Schmidt A, Schulein R, Jensen BL, Loddenkemper C, Jankowski V, Marcussen N, Gollasch M, Arendshorst WJ, Tepel M (2012) The proteoglycan syndecan 4 regulates transient receptor potential canonical 6 channels via RhoA/Rho-associated protein kinase signaling. Arterioscler Thromb Vasc Biol 32(2):378–385
207. Boute N, Gribouval O, Roselli S, Benessy F, Lee H, Fuchshuber A, Dahan K, Gubler MC, Niaudet P, Antignac C (2000) NPHS2, encoding the glomerular protein podocin, is mutated in autosomal recessive steroid-resistant nephrotic syndrome. Nat Genet 24(4):349–354
208. Sun L, Li W, Li W, Xiong L, Li G, Ma R (2014) Astragaloside IV prevents damage to human mesangial cells through the inhibition of the NADPH oxidase/ROS/Akt/NFkappaB pathway under high glucose conditions. Int J Mol Med 34(1):167–176
209. Sonneveld R, van der Vlag J, Baltissen MP, Verkaart SA, Wetzels JF, Berden JH, Hoenderop JG, Nijenhuis T (2014) Glucose specifically regulates TRPC6 expression in the podocyte in an AngII-dependent manner. Am J Pathol 184(6):1715–1726
210. Kriz W, Elger M, Mundel P, Lemley KV (1995) Structure-stabilizing forces in the glomerular tuft. J Am Soc Nephrol 5(10):1731–1739
211. Pavenstadt H, Kriz W, Kretzler M (2003) Cell biology of the glomerular podocyte. Physiol Rev 83(1):253–307

212. Vallon V, Richter K, Blantz RC, Thomson S, Osswald H (1999) Glomerular hyperfiltration in experimental diabetes mellitus: potential role of tubular reabsorption. J Am Soc Nephrol 10(12):2569–2576
213. Deen WM, Maddox DA, Robertson CR, Brenner BM (1974) Dynamics of glomerular ultrafiltration in the rat.VII. Response to reduced renal mass. Am J Physiol 227(3):556–562
214. Eckel J, Lavin PJ, Finch EA, Mukerji N, Burch J, Gbadegesin R, Wu G, Bowling B, Byrd A, Hall G, Sparks M, Zhang ZS, Homstad A, Barisoni L, Birbaumer L, Rosenberg P, Winn MP (2011) TRPC6 enhances angiotensin II-induced albuminuria. J Am Soc Nephrol 22(3):526–535
215. Lapatsina L, Brand J, Poole K, Daumke O, Lewin GR (2012) Stomatin-domain proteins. Eur J Cell Biol 91(4):240–245
216. Moshourab RA, Wetzel C, Martinez-Salgado C, Lewin GR (2013) Stomatin-domain protein interactions with acid-sensing ion channels modulate nociceptor mechanosensitivity. J Physiol 591(Pt 22):5555–5574
217. Wetzel C, Hu J, Riethmacher D, Benckendorff A, Harder L, Eilers A, Moshourab R, Kozlenkov A, Labuz D, Caspani O, Erdmann B, Machelska H, Heppenstall PA, Lewin GR (2007) A stomatin-domain protein essential for touch sensation in the mouse. Nature 445(7124):206–209
218. Goodman MB, Ernstrom GG, Chelur DS, O'Hagan R, Yao CA, Chalfie M (2002) MEC-2 regulates C. elegans DEG/ENaC channels needed for mechanosensation. Nature 415(6875):1039–1042
219. Falkenberg CV, Blinov ML, Loew LM (2013) Pleomorphic ensembles: formation of large clusters composed of weakly interacting multivalent molecules. Biophys J 105(11):2451–2460
220. Liu XL, Kilpelainen P, Hellman U, Sun Y, Wartiovaara J, Morgunova E, Pikkarainen T, Yan K, Jonsson AP, Tryggvason K (2005) Characterization of the interactions of the nephrin intracellular domain. FEBS J 272(1):228–243
221. Yu L, Lin Q, Liao H, Feng J, Dong X, Ye J (2010) TGF-beta1 induces podocyte injury through Smad3-ERK-NF-kappaB pathway and Fyn-dependent TRPC6 phosphorylation. Cell Physiol Biochem 26(6):869–878
222. Jardin I, Redondo PC, Salido GM, Rosado JA (2008) Phosphatidylinositol 4,5-bisphosphate enhances store-operated calcium entry through hTRPC6 channel in human platelets. Biochim Biophys Acta 1783(1):84–97
223. Carter RN, Tolhurst G, Walmsley G, Vizuete-Forster M, Miller N, Mahaut-Smith MP (2006) Molecular and electrophysiological characterization of transient receptor potential ion channels in the primary murine megakaryocyte. J Physiol 576(Pt 1):151–162
224. Tolhurst G, Carter RN, Amisten S, Holdich JP, Erlinge D, Mahaut-Smith MP (2008) Expression profiling and electrophysiological studies suggest a major role for Orai1 in the store-operated Ca2+ influx pathway of platelets and megakaryocytes. Platelets 19(4):308–313
225. Harper MT, Sage SO (2010) Src family tyrosine kinases activate thrombin-induced non-capacitative cation entry in human platelets. Platelets 21(6):445–450
226. Redondo PC, Jardin I, Lopez JJ, Salido GM, Rosado JA (2008) Intracellular Ca2+ store depletion induces the formation of macromolecular complexes involving hTRPC1, hTRPC6, the type II IP3 receptor and SERCA3 in human platelets. Biochim Biophys Acta 1783(6):1163–1176
227. Chen W, Thielmann I, Gupta S, Subramanian H, Stegner D, van Kruchten R, Dietrich A, Gambaryan S, Heemskerk JW, Hermanns HM, Nieswandt B, Braun A (2014) Orai1-induced store-operated Ca(2+) entry enhances phospholipase activity and modulates canonical transient receptor potential channel 6 function in murine platelets. J Thromb Haemost 12(4):528–539
228. Dionisio N, Albarran L, Berna-Erro A, Hernandez-Cruz JM, Salido GM, Rosado JA (2011) Functional role of the calmodulin- and inositol 1,4,5-trisphosphate receptor-binding (CIRB) site of TRPC6 in human platelet activation. Cell Signal 23(11):1850–1856

229. Berna-Erro A, Albarran L, Dionisio N, Redondo PC, Alonso N, Gomez LJ, Salido GM, Rosado JA (2014) The canonical transient receptor potential 6 (TRPC6) channel is sensitive to extracellular pH in mouse platelets. Blood Cells Mol Dis 52(2–3):108–115
230. Voets T, Nilius B (2007) Modulation of TRPs by PIPs. J Physiol 582(Pt 3):939–944
231. Nilius B, Mahieu F, Prenen J, Janssens A, Owsianik G, Vennekens R, Voets T (2006) The Ca2+-activated cation channel TRPM4 is regulated by phosphatidylinositol 4,5-biphosphate. EMBO J 25(3):467–478
232. Sage SO, Merritt JE, Hallam TJ, Rink TJ (1989) Receptor-mediated calcium entry in fura-2-loaded human platelets stimulated with ADP and thrombin. Dual-wavelengths studies with Mn^{2+}. Biochem J 258(3):923–926
233. Lopez JJ, Salido GM, Pariente JA, Rosado JA (2006) Interaction of STIM1 with endogenously expressed human canonical TRP1 upon depletion of intracellular Ca^{2+} stores. J Biol Chem 281(38):28254–28264
234. Roos J, DiGregorio PJ, Yeromin AV, Ohlsen K, Lioudyno M, Zhang S, Safrina O, Kozak JA, Wagner SL, Cahalan MD, Velicelebi G, Stauderman KA (2005) STIM1, an essential and conserved component of store-operated Ca2+ channel function. J Cell Biol 169(3):435–445
235. Zhang SL, Yu Y, Roos J, Kozak JA, Deerinck TJ, Ellisman MH, Stauderman KA, Cahalan MD (2005) STIM1 is a Ca2+ sensor that activates CRAC channels and migrates from the Ca2+ store to the plasma membrane. Nature 437(7060):902–905
236. Baba Y, Hayashi K, Fujii Y, Mizushima A, Watarai H, Wakamori M, Numaga T, Mori Y, Iino M, Hikida M, Kurosaki T (2006) Coupling of STIM1 to store-operated Ca^{2+} entry through its constitutive and inducible movement in the endoplasmic reticulum. Proc Natl Acad Sci U S A 103(45):16704–16709
237. Huang GN, Zeng W, Kim JY, Yuan JP, Han L, Muallem S, Worley PF (2006) STIM1 carboxyl-terminus activates native SOC, I(crac) and TRPC1 channels. Nat Cell Biol 8(9):1003–1010
238. Zbidi H, Jardin I, Woodard GE, Lopez JJ, Berna-Erro A, Salido GM, Rosado JA (2011) STIM1 and STIM2 are located in the acidic Ca^{2+} stores and associates with Orai1 upon depletion of the acidic stores in human platelets. J Biol Chem 286(14):12257–12270
239. Prakriya M, Feske S, Gwack Y, Srikanth S, Rao A, Hogan PG (2006) Orai1 is an essential pore subunit of the CRAC channel. Nature 443(7108):230–233
240. Cheng KT, Ong HL, Liu X, Ambudkar IS (2013) Contribution and regulation of TRPC channels in store-operated Ca2+ entry. Curr Top Membr 71:149–179
241. Dionisio N, Redondo PC, Jardin I, Rosado JA (2012) Transient receptor potential channels in human platelets: expression and functional role. Curr Mol Med 12(10):1319–1328
242. Salido GM, Sage SO, Rosado JA (2009) Biochemical and functional properties of the store-operated Ca^{2+} channels. Cell Signal 21(4):457–461
243. Cheng KT, Ong HL, Liu X, Ambudkar IS (2011) Contribution of TRPC1 and Orai1 to Ca(2+) entry activated by store depletion. Adv Exp Med Biol 704:435–449
244. Harteneck C, Frenzel H, Kraft R (2007) N-(p-amylcinnamoyl)anthranilic acid (ACA): a phospholipase A(2) inhibitor and TRP channel blocker. Cardiovasc Drug Rev 25(1):61–75
245. Lopez JJ, Salido G, Rosado JA (2012) SOCE and Ca2+ handling in platelet dysfunction. In: Groschner K, Graier WF, Romanin C (eds) Store-operated Ca2+ entry (SOCE) pathways. Springer, New York, pp 377–396
246. Paez Espinosa EV, Murad JP, Ting HJ, Khasawneh FT (2012) Mouse transient receptor potential channel 6: role in hemostasis and thrombogenesis. Biochem Biophys Res Commun 417(2):853–856
247. Harper MT, Londono JE, Quick K, Londono JC, Flockerzi V, Philipp SE, Birnbaumer L, Freichel M, Poole AW (2013) Transient receptor potential channels function as a coincidence signal detector mediating phosphatidylserine exposure. Sci Signal 6(281):ra50
248. Albarran L, Berna-Erro A, Dionisio N, Redondo PC, Lopez E, Lopez JJ, Salido GM, Brull Sabate JM, Rosado JA (2014) TRPC6 participates in the regulation of cytosolic basal calcium concentration in murine resting platelets. Biochim Biophys Acta 1843(4):789–796

249. Vinik AI, Erbas T, Park TS, Nolan R, Pittenger GL (2001) Platelet dysfunction in type 2 diabetes. Diabetes Care 24(8):1476–1485
250. Rosado JA, Saavedra FR, Redondo PC, Hernandez-Cruz JM, Salido GM, Pariente JA (2004) Reduced plasma membrane Ca^{2+}-ATPase function in platelets from patients with non-insulin-dependent diabetes mellitus. Haematologica 89(9):1142–1144
251. Leoncini G, Signorello MG, Piana A, Carrubba M, Armani U (1997) Hyperactivity and increased hydrogen peroxide formation in platelets of NIDDM patients. Thromb Res 86(2):153–160
252. Bose R, Li Y, Woo V (2001) Sodium-calcium exchange in platelets of diabetics. Proc West Pharmacol Soc 44:183–184
253. Saavedra FR, Redondo PC, Hernandez-Cruz JM, Salido GM, Pariente JA, Rosado JA (2004) Store-operated Ca^{2+} entry and tyrosine kinase $pp60^{src}$ hyperactivity are modulated by hyperglycemia in platelets from patients with non insulin-dependent diabetes mellitus. Arch Biochem Biophys 432(2):261–268
254. Redondo PC, Jardin I, Hernandez-Cruz JM, Pariente JA, Salido GM, Rosado JA (2005) Hydrogen peroxide and peroxynitrite enhance Ca2+ mobilization and aggregation in platelets from type 2 diabetic patients. Biochem Biophys Res Commun 333(3):794–802
255. Jardin I, Redondo PC, Salido GM, Pariente JA, Rosado JA (2006) Endogenously generated reactive oxygen species reduce PMCA activity in platelets from patients with non-insulin-dependent diabetes mellitus. Platelets 17(5):283–288
256. Zbidi H, Lopez JJ, Amor NB, Bartegi A, Salido GM, Rosado JA (2009) Enhanced expression of STIM1/Orai1 and TRPC3 in platelets from patients with type 2 diabetes mellitus. Blood Cells Mol Dis 43(2):211–213
257. Jardin I, Lopez JJ, Zbidi H, Bartegi A, Salido GM, Rosado JA (2011) Attenuated store-operated divalent cation entry and association between STIM1, Orai1, hTRPC1 and hTRPC6 in platelets from type 2 diabetic patients. Blood Cells Mol Dis 46(3):252–260
258. Garcia RL, Schilling WP (1997) Differential expression of mammalian TRP homologues across tissues and cell lines. Biochem Biophys Res Commun 239(1):279–283
259. Riccio A, Medhurst AD, Mattei C, Kelsell RE, Calver AR, Randall AD, Benham CD, Pangalos MN (2002) mRNA distribution analysis of human TRPC family in CNS and peripheral tissues. Brain Res Mol Brain Res 109(1–2):95–104
260. Inada H, Iida T, Tominaga M (2006) Different expression patterns of TRP genes in murine B and T lymphocytes. Biochem Biophys Res Commun 350(3):762–767
261. Dalrymple A, Slater DM, Beech D, Poston L, Tribe RM (2002) Molecular identification and localization of Trp homologues, putative calcium channels, in pregnant human uterus. Mol Hum Reprod 8(10):946–951
262. Guilbert A, Dhennin-Duthille I, Hiani YE, Haren N, Khorsi H, Sevestre H, Ahidouch A, Ouadid-Ahidouch H (2008) Expression of TRPC6 channels in human epithelial breast cancer cells. BMC Cancer 8:125
263. Heiner I, Eisfeld J, Halaszovich CR, Wehage E, Jungling E, Zitt C, Luckhoff A (2003) Expression profile of the transient receptor potential (TRP) family in neutrophil granulocytes: evidence for currents through long TRP channel 2 induced by ADP-ribose and NAD. Biochem J 371(Pt 3):1045–1053
264. Foster RR, Welsh GI, Satchell SC, Marlow RD, Wherlock MD, Pons D, Mathieson PW, Bates DO, Saleem MA (2010) Functional distinctions in cytosolic calcium regulation between cells of the glomerular filtration barrier. Cell Calcium 48(1):44–53
265. Gamberucci A, Giurisato E, Pizzo P, Tassi M, Giunti R, McIntosh DP, Benedetti A (2002) Diacylglycerol activates the influx of extracellular cations in T-lymphocytes independently of intracellular calcium-store depletion and possibly involving endogenous TRP6 gene products. Biochem J 364(Pt 1):245–254
266. Roedding AS, Li PP, Warsh JJ (2006) Characterization of the transient receptor potential channels mediating lysophosphatidic acid-stimulated calcium mobilization in B lymphoblasts. Life Sci 80(2):89–97

Chapter 11
Transient Receptor Potential Canonical 7 (TRPC7), a Calcium (Ca^{2+}) Permeable Non-selective Cation Channel

Xuexin Zhang, Amy M. Spinelli, Timothy Masiello, and Mohamed Trebak

Abstract Transient receptor potential canonical subfamily, member 7 (TRPC7) is the most recently identified member of the TRPC family of Ca^{2+}-permeable non-selective cation channels. The gene encoding the TRPC7 channel plasma membrane protein was first cloned from mouse brain. TRPC7 mRNA and protein have been detected in cell types derived from multiple organ systems from various species including humans. G_q-coupled protein receptor activation is the predominant mode of TRPC7 activation. Lipid metabolites involved in the phospholipase C (PLC) signaling pathway, including diacylglycerol (DAG) and its precursor the phosphatidylinositol-4,5-bisphosphate (PIP_2), have been shown to be direct regulators of TRPC7 channel. TRPC7 channels have been linked to the regulation of various cellular functions however, the depth of our understanding of TRPC7 channel function and regulation is limited in comparison to other TRP channel family members. This review takes a historical look at our current knowledge of TRPC7 mechanisms of activation and its role in cellular physiology and pathophysiology.

Keywords TRPC7 • Non-selective cation channel • Diacylglycerol • PIP_2 • Phospholipase C • Ca^{2+} signaling • Non-excitable cells

11.1 Gene and Expression

Transient receptor potential canonical subfamily (TRPC) channels are a family of non-selective cation plasma membrane channel proteins. Seven members have currently been identified in mammals (TRPC1-7) [1] while 6 TRPC channel proteins have been identified in humans (TRPC1, TRPC3-7) [2]. The most recently identified member of this family is TRPC7. Below is a brief timeline and discussion of the species specific characterization of the TRPC7 gene and expression patterns of this cation channel (Table 11.1) [3].

X. Zhang • A.M. Spinelli • T. Masiello • M. Trebak (✉)
Department of Cellular and Molecular Physiology, The Pennsylvania State University College of Medicine, Hershey, PA 17033, USA
e-mail: mtrebak@hmc.psu.edu

J.A. Rosado (ed.), *Calcium Entry Pathways in Non-excitable Cells*, Advances in Experimental Medicine and Biology 898, DOI 10.1007/978-3-319-26974-0_11

Table 11.1 Species specific TRPC7 expression

Species	mRNA/protein	Date of characterization
Mouse	**mRNA**:	
	Heart, lung, eye, brain, spleen, and testis (1)	1999
	Smooth muscle cells (2)	2001
	MC3T3 osteoblasts (3)	2009
	Embryonic brain and cortex of E13 C57BL6/J mice (4)	2009
	Melanopsin-expressing ganglion cells (5)	2011
	Brain, testis, lung, liver, heart, kidney, and DRG (6)	2012
	MnPO Glutamatergic Neurons (7)	2012
	Protein:	
	Retina (8)	2007
	Cerebellum and ventral respiratory group island (9)	2010
Rat	**mRNA**:	
	Hippocampal H19-7 cells (10, 11)	2004
	A7r5 vascular smooth muscle cells (12)	2006
	Failing myocardium of Dahl salt-sensitive rats (13)	2007
	Cultured rat microglia (14)	2009
	Hypocretin/orexin neurons (15)	2010
	Protein:	
	Brain synaptosomes (16)	2002
	Ganglion neurons (17)	2003
	H19-7 cells (10, 11)	2004
	GH4C1 pituitary cells (18)	2008
	A7r5 vascular smooth muscle cells (12)	2006
Human	**mRNA**:	
	Central nervous system and pituitary gland, kidney, intestine, prostate and cartilage (19)	2002
	Pregnant women myometrium and myometrial cell lines (20, 21)	2002
	Coronary artery endothelial cells (22)	2004
	HEK293 cells (23)	2005
	Undifferentiated gingival keratinocytes (24)	2005
	Differentiated IMR-32 neuroblastoma cells (25)	2006
	HaCaT keratinocyte (26)	2006
	Human breast cancer cell line (ZR-75-1, MCF7 and MDA-MB-231) and human breast epithelial cell line (hTERT-HME1) (27)	2014
	Lung tissue (28)	2010
	Protein:	
	HEK293 cells (29)	2005

(1) Okada et al. [4], (2) Walker et al. [5], (3) Abed et al. [10], (4) Boisseau et al. [7], (5) Perez-Leighton et al. [11], (6) Jang et al. [9], (7) Tabarean [12], (8) Sekaran et al. [6], (9) Ben-Mabrouk and Tryba [13], (10) Wu et al. [22], (11) Numaga et al. [21], (12) Maruyama et al. [23], (13) Satoh et al. [18], (14) Ohana et al. [19], (15) Cvetkovic-Lopes et al. [20], (16) Goel et al. [14], (17) Buniel et al. [15], (18) Lavender et al. [24], (19) Riccio et al. [29], (20) Yang et al. [31], (21) Dalrymple et al. [30], (22) Yip et al. [32], (23) Zagranichnaya et al. [34], (24) Cai et al. [33], (25) Nasman et al. [36], (26) Beck et al. [35], (27) Gogebakan et al. [38], (28) Finney-Hayward et al. [37], (29) Zagranichnaya et al. [34]

Mouse The gene encoding the TRPC7 channel protein was first isolated from mouse brain by Okada et al. in 1999. This gene is made up of 12 exons, and is located at chromosomal region 13 B2. TRPC7 mRNA is highly expressed in mouse eye, lung, and heart; expression was also detected in the brain, spleen and testis, although to a lesser extent [4]. In 2001, TRPC7 transcripts and splice variants were detected in smooth muscle derived from murine colon and jejunum [5]. In 2007, TRPC7 was identified in the mouse retina [6]. In 2009, Boisseau et al. examined the expression patterns of TRPC channels in the embryonic forebrain of C57BL6/J mice; at E13 TRPC7 mRNA expression was detected although to a lesser extent than other TRPC family members [7]. No TRPC7 mRNA expression was detected in skeletal muscle and inner ear organs from embryonic and early postnatal Swiss Webster mice [8, 9]; these studies highlight a development-dependent pattern of expression. TRPC7 mRNA expression was also observed in the mouse osteoblast cell line, MC3T3 in a 2009 study [10]. Recently major focus has centered on characterizing TRPC7 in neuronal cells after TRPC7 was identified in photosensitive retinal ganglion cells in 2011 [11] and glutamatergic preoptic neurons in 2012 [12]. In 2010, examination of mouse brain stem slices containing the respiratory-regulating neuronal bundle, the pre-Bötzinger complex, detected TRPC7 protein expression localized to this regulatory region [13].

Rat Rat TRPC7 genes are also made up of 12 exons, and have been mapped to chromosomal region 17p14. In rat, TRPC7 is predominantly expressed in the nervous system. In 2002, TRPC7 protein was isolated from brain synaptosomes [14]. In 2003, TRPC7 mRNA expression was detected in neurons throughout the ganglia [15]. TRPC7 protein was observed in rat striatal cholinergic interneurons [16], and had a strikingly high level of expression in the neuropil in the rat globus pallidus [17]. In 2007, TRPC7 expression was found to be up-regulated in the failing myocardium of Dahl salt-sensitive rats [18]. Characterization of TRPC7 mRNA expression has also been analyzed in various rat cell lines, including cultured rat microglia where expression was detected in 2009 [19] and in hypocretin/orexin neurons in 2010 [20]. In rat hippocampal H19-7 cells, the levels of mRNA and protein for TRPC7 are high in proliferating cells and decline upon differentiation [21, 22]. In rat vascular smooth muscle cell line A7r5 [23] and pituitary cell line GH_4C_1 [24] TRPC7 transcripts were also detected. TRPC7 mRNA was not detected in freshly isolated rat renal resistance vessels, glomeruli, and aorta [25], nor was it found in rat distal pulmonary arterial smooth muscle [26], and rat dorsal root ganglia (DRG) neurons [27].

Human In 2002, Riccio et al. cloned human TRPC7 gene from brain [28]. The gene encoding human TRPC7 is located at the chromosomal region 5q31.1, and similarly to the mouse and rat gene, has 12 exons. Both human and mouse TRPC7 genes consist of an open reading frame of 2,589 bp yielding a protein of 862 amino acids [4, 28]. Human TRPC7 has 98 % sequence homology with mouse TRPC7 [21]. Human TRPC7 mRNA expression is mostly restricted to the central nervous system however expression has been detected in some peripheral tissues including cartilage, intestine, kidney, pituitary gland and prostate [29]. TRPC7 expression has also been examined in various human cell lines. In 2002, human myometrial cells were found to express TRPC7 mRNA, and its expression was upregulated in myometrium obtained during active labor from full term pregnant women [30, 31]. In 2004, TRPC7 expression was found in human coronary artery endothelial cells, although not in vascular smooth muscle cells [32]. In 2005, TRPC7 mRNA was also detected in undifferentiated human gingival keratinocytes [33] and protein expression was observed in the human embryonic kidney (HEK) 293 cell line [34]. In 2006, detection was observed in human differentiated keratinocytes [35] and differentiated IMR-32 neuroblastoma cells [36]. In 2009, studies failed to detect TRPC7 mRNA in human osteoblast cell lines MG-63, SaOS, and U2 OS [10], while in 2010 expression was observed in lung tissue [37]. Recently TRPC7 expression was characterized in various breast cancer cell lines including ZR-75-1, MCF7 and MDA-MB-231 and the human breast epithelial cell line hTERT-HME1 [38].

11.2 Protein Structure and Channel Configuration

TRPC7 is composed of 862 amino acid residues [4]. TRPC channel proteins including TRPC7 possess a similar basic domain structures. These defining structures include a cytosolic N-terminal domain containing four ankyrin repeats, six transmembrane spanning domains, within transmembrane domain 5 and 6 lies a highly conserved hydrophobic α helix pore-loop motif and finally a cytosolic C-terminal region [39]. This carboxy terminal region is made up of the TRP domain containing an EWKFAR and proline rich domain [40]. In 2001, Tang et al. further characterized the TRPC family C- terminal domain by identifying a calmodulin and Inositol1,4,5-Trisphosphate (IP_3) receptor binding (CIRB) region [41]. Variable regions between TRPC family members exist within this C-terminal domain contributing to their differential modes of regulation [42].

TRPC family proteins are believed to assemble into tetramers. The stoichiometry of this assembly varies depending on both TRPC family member and cell type, as homo- and heterotetrameric assembly has been proposed [43]. Specifically, TRPC7 has been shown to selectively interact with TRPC3 and TRPC6. TRPC3/6/7 channels share 70–80 % structural homology and are members of a sub-family within the TRPC family [4, 39, 43]. Co-expression of these channels has been observed in various smooth muscle tissues [44] suggesting possible heterotertrameric assemblies. However, existence of cell specific expression of different TRPC channel isoforms does not necessarily mean heteromultimeric assembly.

A study examined native TRPC7 and TRPC6 function highlighting that both the assembly and function of these channels is cell type specific. Biochemical analysis in rabbit portal vein myocytes determined strong association between TRPC6 and TRPC7 and single channel patch recordings in cells treated with anti-TRPC6 and anti-TRPC7 antibodies concluded both of these subunits were necessary for channel activity. However, examination of myocytes from rabbit mesenteric arteries revealed that the biochemical and functional characteristics of these channel proteins were not the same [45]. More studies assessing native TRPC7 expression and function and the use of TRPC7 deficient animals are necessary to further understand in vivo the native assembly and function of these tetrameric channels [39]. While TRPC7 has been well characterized as a plasma membrane channel, one overexpression study examining localization of this channel protein in COS-7 cells observed a pool of intracellular TRPC7 protein that co-localized with the golgi network. Similar patterns of endogenous TRPC7 distribution were observed in the rat pituitary cell line, GH_4C_1. This localization within the golgi stack was thought to enhance cellular secretion either by mediating cargo transport or plasma membrane fusion [24].

11.3 Biophysical Properties of TRPC7 Channel

Okada et al. studied the permeability of TRPC7 channel using patch clamp recordings from TRPC7-expressing cells and showed constitutively activated and ATP-enhanced inward cation currents mediated by TRPC7 with permeability ratios $P_{Cs}:P_{Na}:P_{Ca}:P_{Ba}$ of 1:1:1.9:3.5 and 1:1.1:5.9:5.0, respectively [4]. Lievremont et al. used Fura2 imaging and observed that diacylglycerol analog OAG- or muscarinic receptor stimulation-induced Ba^{2+} entry that was significantly reduced in TRPC7-deficient DT40 cell, further confirming that TRPC7 channel permeates Ba^{2+} ions [46]. Shi et al. reported that the current-voltage (I-V) relationship obtained from HEK293 cells transfected with mouse or human TRPC7 gene is almost linear with a slight flattening around the reversal potentials; mouse TRPC7 single-channel conductance is 24.3 pS and 24.8 pS under bath solutions containing 10 mM Ca^{2+} and 1 mM Ca^{2+}, respectively. However, when cells were bathed in Ca^{2+} free solution, single-channel conductance was increased to 49.3 pS [42]. Okada et al. also showed that heterologously expressed mouse TRPC7 behaves as receptor-operated DAG-activated cation channel that is insensitive to store depletion. Subsequent studies on native TRPC7 in DT40 cells and on TRPC7 ectopically expressed in HEK293 cells further suggested that TRPC7 forms a store-independent receptor-operated channel [4, 46, 47]. Thus, TRPC7 is a receptor-regulated, store-independent, non-selective cation channel activated through phospholipase C (PLC)-mediated metabolism of phosphatidylinositol 4,5-bisphosphate (PIP_2) and production of DAG [4, 48]. Regarding the pharmacological profile of TRPC7, few inhibitors are known to interfere with TRPC7 activation and none of them are specific to TRPC7. Okada et al. showed that ATP-activated mouse TRPC7 ectopically expressed HEK293 cells is more efficiently inhibited with 100 μM lanthanum (La^{3+}) and 25 μM SKF96365 than 100 μM gadolinium (Gd^{3+}) [4]. Riccio et al. used 250 μM Gd^{3+}, 250 μM La^{3+}, and 25 μM SKF96365 to completely block the Ca^{2+} entry in HEK293 cells expressing human TRPC7 [28]. 300 μM Amiloride also inhibited constitutively activated TRPC7 currents, and 2-aminoethoxydiphenyl borate (2-APB) at concentrations of 10–100 μM partially inhibited TRPC3, TRPC6, and TRPC7 channels by a mechanism not involving the known inhibitory effects of 2-APB on IP_3 receptors [4, 49].

The Schaefer group has screened a chembionet library and found three compounds, 6228-0353, 6228-0473 and 2910-0498 showing strong inhibition towards TRPC7 channels [50].

11.4 TRPC7 Channel Regulation and Function

The exact regulatory mechanisms that control TRPC7 channel activity have been the subject of contention since TRPC7 was first cloned in 1999 with many discrepancies between different groups. TRPC7 channel activity has been shown to vary depending on the molecular reagents used and the expression levels of this channel protein in heterologous model systems [51]. The use of cell specific TRPC7 gene deficient transgenic animals will likely generate more accurate and reproducible characterization of TRPC7 mediated channel activity. The general consensus is that TRPC7 is a receptor-activated channel that depends on DAG production through the catalytic activity of PLC. Below is a brief description of the current understanding of the mediators that regulate TRPC7 channel activity.

11.4.1 Metabolites of the PLC Pathway as Regulators of TRPC7 Channel

G_q-coupled receptor mediated activation is the predominant mode of regulation of the TRPC7 channel. Receptor mediated activation of PLC leads to hydrolysis of PIP_2 and the production of the signaling molecules IP_3 and DAG. TRPC7 channel function was first characterized by ectopically expressing this protein in cultured HEK293 cells [4]; this study proposed DAG as the activator of TRPC7 channels based in the use of exogenous OAG. Addition of OAG to the bath solution during whole-cell patch clamp recordings activated TRPC7 currents in a PKC-independent manner. Subsequently, Beck et al. reported that in human keratinocytes, OAG evoked a TRPC7 channel cation current, and this OAG-induced current was decreased by TRPC7 knock down [35]. Itsuki et al. identified TRPC7 activity being mediated by DAG produced through the hydrolysis of PIP_2 by PLC and reported a requirement for PIP_2 in channel activation as TRPC7 current subsequently inactivated upon hydrolysis of PIP_2. Further, PIP_2 depletion by a heterologously expressed phosphatase inhibited TRPC7 activity independently of DAG [52]. Patch clamp recordings on TRPC7-expressing HEK293 cells using the cell-attached mode detected TRPC7 currents after addition of OAG, but OAG failed to activate TRPC7 channels in excised patches. This finding suggests that DAG does activate TRPC7 channels rather indirectly perhaps through cytosolic proteins or factors that are lost in excised patches [53].

PIP_2 has been shown to directly regulate several ion channels, including the broader TRP family of channels [3, 53]. The ability of PIP_2 to directly activate TRPC7 as well as members of its subfamily TRPC3 and TRPC6 was first observed in 1998 using the phosphoinositide lipid kinase inhibitor, LY294002 which causes PIP_2 depletion [54]. Treatment of HEK293 cells overexpressing TRPC7 with this drug prevented receptor- and OAG-mediated TRPC7 channel activation. In excised inside – out patches, direct application of either PIP_2 or ATP activated TRPC7 single channels while IP_3 application had no effect [53]. Imai et al. [55] used an ectopic expression system of the voltage-sensing PIP phosphatase (DrVSP) and showed that Carbachol, OAG or RHC80267 (a DAG lipase inhibitor that enhances endogenous DAG and thus activates TRPC7) -mediated channel activation was inhibited by DrVSP activation. The extent of phosphatase inhibition was TRPC isoform specific as TRPC7 currents were attenuated to a larger extent than either TRPC6 or TRPC3 currents. Ju et al. used antibodies to inhibit TRPC isoforms in inside-out patches and concluded that noradrenaline-activated native cationic currents in portal vein myocytes are mediated by TRPC6/TRPC7 heteromultimers. These authors reported that PIP_2 inhibited this OAG-activated TRPC6/TRPC7 channel, and that this PIP_2 inhibition is rescued by IP_3 through an IP_3 receptor-independent manner [45]. The reasons for the discrepancy between this study and those reporting PIP_2 requirement for TRPC7 channel activation are unclear.

11.4.2 TRPC7 and Store Operated Ca^{2+} Entry

TRPC7 along with other TRPC channels have been proposed to constitute the pore forming unit of store-operated calcium entry (SOCE) channels [56]; this notion however remains highly controversial. In stably expressing TRPC7 HEK-293 cells either pharmacological Ca^{2+} store depletion with thapsigargin or activation of the PLC signaling pathway was able to promote TRPC7 channel activation [57]. However, in a transient expression system only stimulation of PLC-coupled receptors induced TRPC7 activation. STIM1, the ER residing Ca^{2+} sensor [58, 59] and Orai1, the highly Ca^{2+}-selective plasma membrane ion channel subunit have been established as the *bona fide* molecular components of SOCE and the calcium release-activated calcium (CRAC) current [60]. A number of studies have examined the role of STIM1 and Orai1 in regulating TRPC7-mediated calcium entry. In stably expressing TRPC7 HEK293 cells, silencing of STIM1 and Orai1 proteins by RNA interference did not affect TRPC7 activity nor did the expression of a constitutively active STIM1 mutant [47]. Studies from one group however suggested that STIM1 is capable of regulating specific isoforms of TRPC channels. Huang et al. studied the interaction between STIM1 and TRPC channels, concluding that STIM1 ERM domain mediates the selective binding of STIM1 to TRPC1, TRPC2 and TRPC4, but not to TRPC3, TRPC6 or TRPC7 [61]. Subsequently, the same group reported that STIM1 binds TRPC1, TRPC4 and TRPC5 allowing them to function as store-operated channels. The authors also demonstrated that STIM1 indirectly regulates TRPC3 and TRPC6 channels by mediating the heteromultimerization of TRPC3 with TRPC1 and TRPC6 with TRPC4. In these studies, TRPC7 was the only TRPC family member that was not regulated by STIM1 [62].

11.4.3 Ca^{2+} and Calmodulin

Most Ca^{2+}-permeable channels including TRPC7 channels are themselves regulated by Ca^{2+}. Specifically studies have shown that TRPC7 channels can be negatively regulated by Ca^{2+} both directly and indirectly. Lemonnier et al. identified a negative feedback mechanism of TRPC7 channels in HEK293 cells, whereby Ca^{2+} entry through TRPC7 channel negatively regulates its own activity [63]. This Ca^{2+}-mediated negative feedback at the mouth of TRPC7 channels is more pronounced when the SERCA pump is inhibited by thapsigargin, suggesting that physiologically SERCA pumps that are closely associated with TRPC7 channels attenuate this negative feedback by buffering calcium into the endoplasmic reticulum. Indeed, application of the SERCA pump inhibitor thapsigargin or CPA prevented OAG-activated TRPC7 channel activation. The inhibitory effect of thapsigargin was reversed by inhibition of calmodulin and was recapitulated by pharmacological disruption of the actin cytoskeleton. Shi et al. used whole-cell and single-channel recordings to reveal that voltage-dependent inhibitory actions of extracellular Ca^{2+} on TRPC7 channel currents are likely mediated through Ca^{2+} interaction with an extracellular site capable of sensing the membrane potential. This group also reported a concentration-dependent inhibitory effect of intracellular Ca^{2+} on TRPC7 channel currents, which they suggested was mediated by calmodulin [42].

The variable C-terminal domain distinguishes members of the TRPC family from one another [42]. TRPC7 has the CIRB domain within this variable region that contains binding sequences for both calmodulin and the IP_3R. Direct interactions between these proteins and TRPC7 have been shown to regulate TRPC7 channel activity. As discussed above, Shi et al. used cell-attached mode single-channel recordings to show that cytosolic Ca^{2+}-mediated inhibition of TRPC7 channel activity was strongly attenuated by pretreatment with pharmacological agent calmidazolium (CMZ), a potent calmodulin (CaM) antagonist or coexpression of the calmodulin mutant, mutCaM. These findings suggests binding of calmodulin to the TRPC7 C-terminal CIRB domain inhibits channel activity [42]. In the avian B-cell line, DT40, using the cell-attached mode of patch clamp, OAG activated single channel activity was not observed in $TRPC7^{-/-}$ cells, nor in $IP_3R^{-/-}$ cells. Interestingly, exogenous expression of either TRPC7 in the $TRPC7^{-/-}$ cells or IP_3R in the $IP_3R^{-/-}$ cells rescued OAG-activated single channel activity to the same extent as wild type cells [51]. These findings support a native role for the TRPC7 channel protein in mediating DAG-activated currents that are dependent on IP_3R.

11.4.4 N-Terminal Binding Proteins

The N-terminal domain of TRPC7 is made up of four ankyrin-like repeats. These protein-protein interacting domains have been shown to regulate TRPC7 channel activity and localization. Co-immunoprecipitation studies revealed that the cytosolic cGMP/cGMP-dependent protein kinase (cGK) isoform, cGK-I α binds to the N-terminal ankyrin-like repeat domain of TRPC7. Yuasa et al. also observed cGK-I α phosphorylates TRPC7 on its threonine 15 site without any effect on TRPC3. This TRPC7 specific phosphorylation reduced carbachol-activated calcium signaling and phosphorylation of the transcription factor, CREB [64]. Lussier and colleagues used protein-protein interaction assays, including yeast two-hybrid screen, GST pull-down, and co-immunoprecipitation and demonstrated that MxA which is a member of the dynamin GTPase superfamily that interacts with the N-terminal second ankyrin-like repeat domain of TRPC1, TRPC3, TRPC4, TRPC5, TRPC6, and TRPC7 [65]. It was proposed that MxA enhances channel activity by regulating trafficking of the channel proteins to the plasma membrane. Although this functional role has not been tested for TRPC7, mutation studies involving subfamily members TRPC3 and TRPC6 have shown deletion of the ankyrin-like repeat domain leads to intracellular retention of channel protein, inhibition of plasma membrane localization and overall channel dysfunction [43, 66]. Future studies looking at TRPC7 channel specific protein trafficking are necessary to gain insights into TRPC7 regulation through membrane trafficking.

11.5 TRPC7 in Pathophysiology of Disease

Altered TRPC7 channel expression and activity are observed in various diseases including malignant breast cancer tumors. Regulation of multiple TRPC genes has been shown to be mediated by the Rho kinase pathway [67]. Rho kinase activity has been associated with metastasis of esophageal cancer cell lines and inhibition of this pathway using the pharmacological inhibitor Y-27632 prevented the growth and invasiveness of these cancer cells [68]. In human breast cancer cell lines ZR-75-1, MCF7, and MDA-MB-231 and the human breast cancer epithelial cell line hTERT-HME1 inhibition of the Rho-kinase pathway with Y-27632 increased the expression of TRPC7 in all cell lines as detected by quantitative real time PCR [38]. Interestingly the Rho-kinase inhibitors attenuated expression of TRPC1 and another TRP family member TRPV2. This study suggests a potential protective role of TRPC7 in cancer cell growth and progression.

TRPC7 expression has also been shown to be up-regulated in animal models of disease. In Dahl salt sensitive rats, TRPC7 mRNA was increased in the failing myocardium of these animals. This increased expression correlated with an increased level of apoptosis as assessed by TUNEL staining. Heart failure in these mice was suggested to be contributed by angiotensin II-induced Ca^{2+} entry through activation of TRPC7 and subsequent myocardial apoptosis [18].

In a mouse model of pilocarpine induced status epilepticus, TRPC7 was shown to mediate the initiation of acute seizures, as observed by the reduction in pilocarpine-induced gamma wave activity in TRPC7 knockout animals [69]. The mechanism by which TRPC7 mediates seizure induction was characterized in brain slices derived from the hippocampal region cornu ammonis CA3, a central region involved in seizure generation. TRPC7 was shown to generate spontaneous epileptiform burst firing, these signals of seizure induction were initiated at the synaptic level where TRPC7 was shown to be involved in the potentiation of these signals in the CA3 synapses and Schaffer collateral-CA1 synapses. This study using TRPC7 knockout animals was one of the first studies to provide insight into the in vivo function of this enigmatic channel protein.

11.6 Conclusion

Since the initial characterization of the mouse TRPC7 gene in 1999 [4], the following two decades of research studying this channel protein have shown that: (i) TRPC7 expression is present in both excitable and non-excitable cells; (ii) Biophysical and functional studies have described TRPC7 as a store-independent DAG-activated Ca^{2+}-permeable non-selective cation channels [47, 49]; (iii) The mode of TRPC7 channel regulation is likely complex, attracting much debate, including the role of PIP_2 in regulating TRPC7 function. Channel heteromultimerization with other TRPC isoforms and isoforms of the broader TRP superfamily are likely to enhance the diversity of signaling mechanisms through TRPC7. Thus far, TRPC7-associated diseases have rarely been reported. With increasing number of studies employing transgenic TRPC7 knockout animals, understanding of the various physiological functions controlled by this dynamic Ca^{2+}-permeable non-selective cation channel is likely increase.

References

1. Montell C, Birnbaumer L, Flockerzi V, Bindels RJ, Bruford EA, Caterina MJ, Clapham DE, Harteneck C, Heller S, Julius D, Kojima I, Mori Y, Penner R, Prawitt D, Scharenberg AM, Schultz G, Shimizu N, Zhu MX (2002) A unified nomenclature for the superfamily of TRP cation channels. Mol Cell 9(2):229–231
2. Wissenbach U, Schroth G, Philipp S, Flockerzi V (1998) Structure and mRNA expression of a bovine trp homologue related to mammalian trp2 transcripts. FEBS Lett 429(1):61–66
3. Trebak M, Lemonnier L, DeHaven WI, Wedel BJ, Bird GS, Putney JW Jr (2009) Complex functions of phosphatidylinositol 4,5-bisphosphate in regulation of TRPC5 cation channels. Pflugers Arch 457(4):757–769

4. Okada T, Inoue R, Yamazaki K, Maeda A, Kurosaki T, Yamakuni T, Tanaka I, Shimizu S, Ikenaka K, Imoto K, Mori Y (1999) Molecular and functional characterization of a novel mouse transient receptor potential protein homologue TRP7. Ca(2+)-permeable cation channel that is constitutively activated and enhanced by stimulation of G protein-coupled receptor. J Biol Chem 274(39):27359–27370
5. Walker RL, Hume JR, Horowitz B (2001) Differential expression and alternative splicing of TRP channel genes in smooth muscles. Am J Physiol Cell Physiol 280(5):C1184–C1192
6. Sekaran S, Lall GS, Ralphs KL, Wolstenholme AJ, Lucas RJ, Foster RG, Hankins MW (2007) 2-Aminoethoxydiphenylborane is an acute inhibitor of directly photosensitive retinal ganglion cell activity in vitro and in vivo. J Neurosci Off J Soc Neurosci 27(15):3981–3986
7. Boisseau S, Kunert-Keil C, Lucke S, Bouron A (2009) Heterogeneous distribution of TRPC proteins in the embryonic cortex. Histochem Cell Biol 131(3):355–363
8. Asai Y, Holt JR, Geleoc GS (2010) A quantitative analysis of the spatiotemporal pattern of transient receptor potential gene expression in the developing mouse cochlea. J Assoc Res Otolaryngol 11(1):27–37
9. Jang Y, Lee Y, Kim SM, Yang YD, Jung J, Oh U (2012) Quantitative analysis of TRP channel genes in mouse organs. Arch Pharm Res 35(10):1823–1830
10. Abed E, Labelle D, Martineau C, Loghin A, Moreau R (2009) Expression of transient receptor potential (TRP) channels in human and murine osteoblast-like cells. Mol Membr Biol 26(3):146–158
11. Perez-Leighton CE, Schmidt TM, Abramowitz J, Birnbaumer L, Kofuji P (2011) Intrinsic phototransduction persists in melanopsin-expressing ganglion cells lacking diacylglycerol-sensitive TRPC subunits. Eur J Neurosci 33(5):856–867
12. Tabarean IV (2012) Persistent histamine excitation of glutamatergic preoptic neurons. PLoS One 7(10):e47700
13. Ben-Mabrouk F, Tryba AK (2010) Substance P modulation of TRPC3/7 channels improves respiratory rhythm regularity and ICAN-dependent pacemaker activity. Eur J Neurosci 31(7):1219–1232
14. Goel M, Sinkins WG, Schilling WP (2002) Selective association of TRPC channel subunits in rat brain synaptosomes. J Biol Chem 277(50):48303–48310
15. Buniel MC, Schilling WP, Kunze DL (2003) Distribution of transient receptor potential channels in the rat carotid chemosensory pathway. J Comp Neurol 464(3):404–413
16. Berg AP, Sen N, Bayliss DA (2007) TrpC3/C7 and Slo2.1 are molecular targets for metabotropic glutamate receptor signaling in rat striatal cholinergic interneurons. J Neurosci Off J Soc Neurosci 27(33):8845–8856
17. Chung YH, Kim D, Moon NJ, Oh CS, Lee E, Shin DH, Kim SS, Lee WB, Lee JY, Cha CI (2007) Immunohistochemical study on the distribution of canonical transient receptor potential channels in rat basal ganglia. Neurosci Lett 422(1):18–23
18. Satoh S, Tanaka H, Ueda Y, Oyama J, Sugano M, Sumimoto H, Mori Y, Makino N (2007) Transient receptor potential (TRP) protein 7 acts as a G protein-activated Ca2+ channel mediating angiotensin II-induced myocardial apoptosis. Mol Cell Biochem 294(1–2):205–215
19. Ohana L, Newell EW, Stanley EF, Schlichter LC (2009) The Ca2+ release-activated Ca2+ current (I(CRAC)) mediates store-operated Ca2+ entry in rat microglia. Channels 3(2):129–139
20. Cvetkovic-Lopes V, Eggermann E, Uschakov A, Grivel J, Bayer L, Jones BE, Serafin M, Muhlethaler M (2010) Rat hypocretin/orexin neurons are maintained in a depolarized state by TRPC channels. PLoS One 5(12):e15673
21. Numaga T, Wakamori M, Mori Y (2007) Trpc7. In: Transient receptor potential (TRP) channels, vol 179, Handbook of experimental pharmacology. Springer, Berlin/Heidelberg, pp 143–151
22. Wu X, Zagranichnaya TK, Gurda GT, Eves EM, Villereal ML (2004) A TRPC1/TRPC3-mediated increase in store-operated calcium entry is required for differentiation of H19-7 hippocampal neuronal cells. J Biol Chem 279(42):43392–43402

23. Maruyama Y, Nakanishi Y, Walsh EJ, Wilson DP, Welsh DG, Cole WC (2006) Heteromultimeric TRPC6-TRPC7 channels contribute to arginine vasopressin-induced cation current of A7r5 vascular smooth muscle cells. Circ Res 98(12):1520–1527
24. Lavender V, Chong S, Ralphs K, Wolstenholme AJ, Reaves BJ (2008) Increasing the expression of calcium-permeable TRPC3 and TRPC7 channels enhances constitutive secretion. Biochem J 413(3):437–446
25. Facemire CS, Mohler PJ, Arendshorst WJ (2004) Expression and relative abundance of short transient receptor potential channels in the rat renal microcirculation. Am J Physiol Ren Physiol 286(3):F546–F551
26. Wang J, Shimoda LA, Sylvester JT (2004) Capacitative calcium entry and TRPC channel proteins are expressed in rat distal pulmonary arterial smooth muscle. Am J Physiol Lung Cell Mol Physiol 286(4):L848–L858
27. Wu D, Huang W, Richardson PM, Priestley JV, Liu M (2008) TRPC4 in rat dorsal root ganglion neurons is increased after nerve injury and is necessary for neurite outgrowth. J Biol Chem 283(1):416–426
28. Riccio A, Mattei C, Kelsell RE, Medhurst AD, Calver AR, Randall AD, Davis JB, Benham CD, Pangalos MN (2002) Cloning and functional expression of human short TRP7, a candidate protein for store-operated Ca2+ influx. J Biol Chem 277(14):12302–12309
29. Riccio A, Medhurst AD, Mattei C, Kelsell RE, Calver AR, Randall AD, Benham CD, Pangalos MN (2002) mRNA distribution analysis of human TRPC family in CNS and peripheral tissues. Brain Res Mol Brain Res 109(1–2):95–104
30. Dalrymple A, Slater DM, Beech D, Poston L, Tribe RM (2002) Molecular identification and localization of Trp homologues, putative calcium channels, in pregnant human uterus. Mol Hum Reprod 8(10):946–951
31. Yang M, Gupta A, Shlykov SG, Corrigan R, Tsujimoto S, Sanborn BM (2002) Multiple Trp isoforms implicated in capacitative calcium entry are expressed in human pregnant myometrium and myometrial cells. Biol Reprod 67(3):988–994
32. Yip H, Chan WY, Leung PC, Kwan HY, Liu C, Huang Y, Michel V, Yew DT, Yao X (2004) Expression of TRPC homologs in endothelial cells and smooth muscle layers of human arteries. Histochem Cell Biol 122(6):553–561
33. Cai S, Fatherazi S, Presland RB, Belton CM, Izutsu KT (2005) TRPC channel expression during calcium-induced differentiation of human gingival keratinocytes. J Dermatol Sci 40(1):21–28
34. Zagranichnaya TK, Wu X, Villereal ML (2005) Endogenous TRPC1, TRPC3, and TRPC7 proteins combine to form native store-operated channels in HEK-293 cells. J Biol Chem 280(33):29559–29569
35. Beck B, Zholos A, Sydorenko V, Roudbaraki M, Lehen'kyi V, Bordat P, Prevarskaya N, Skryma R (2006) TRPC7 is a receptor-operated DAG-activated channel in human keratinocytes. J Invest Dermatol 126(9):1982–1993
36. Nasman J, Bart G, Larsson K, Louhivuori L, Peltonen H, Akerman KE (2006) The orexin OX1 receptor regulates Ca2+ entry via diacylglycerol-activated channels in differentiated neuroblastoma cells. J Neurosci Off J Soc Neurosci 26(42):10658–10666
37. Finney-Hayward TK, Popa MO, Bahra P, Li S, Poll CT, Gosling M, Nicholson AG, Russell RE, Kon OM, Jarai G, Westwick J, Barnes PJ, Donnelly LE (2010) Expression of transient receptor potential C6 channels in human lung macrophages. Am J Respir Cell Mol Biol 43(3):296–304
38. Gogebakan B, Bayraktar R, Suner A, Balakan O, Ulasli M, Izmirli M, Oztuzcu S, Camci C (2014) Do fasudil and Y-27632 affect the level of transient receptor potential (TRP) gene expressions in breast cancer cell lines? Tumour Biol: J Int Soc Oncodevelopmental Biol Med 35(8):8033–8041
39. Dietrich A, Kalwa H, Rost BR, Gudermann T (2005) The diacylgylcerol-sensitive TRPC3/6/7 subfamily of cation channels: functional characterization and physiological relevance. Pflugers Arch 451(1):72–80

40. Venkatachalam K, Montell C (2007) TRP channels. Annu Rev Biochem 76:387–417
41. Tang J, Lin Y, Zhang Z, Tikunova S, Birnbaumer L, Zhu MX (2001) Identification of common binding sites for calmodulin and inositol 1,4,5-trisphosphate receptors on the carboxyl termini of trp channels. J Biol Chem 276(24):21303–21310
42. Shi J, Mori E, Mori Y, Mori M, Li J, Ito Y, Inoue R (2004) Multiple regulation by calcium of murine homologues of transient receptor potential proteins TRPC6 and TRPC7 expressed in HEK293 cells. J Physiol 561(Pt 2):415–432
43. Hofmann T, Schaefer M, Schultz G, Gudermann T (2002) Subunit composition of mammalian transient receptor potential channels in living cells. Proc Natl Acad Sci U S A 99(11):7461–7466
44. Beech DJ, Muraki K, Flemming R (2004) Non-selective cationic channels of smooth muscle and the mammalian homologues of Drosophila TRP. J Physiol 559(Pt 3):685–706
45. Ju M, Shi J, Saleh SN, Albert AP, Large WA (2010) Ins(1,4,5)P3 interacts with PIP2 to regulate activation of TRPC6/C7 channels by diacylglycerol in native vascular myocytes. J Physiol 588(Pt 9):1419–1433
46. Lievremont JP, Numaga T, Vazquez G, Lemonnier L, Hara Y, Mori E, Trebak M, Moss SE, Bird GS, Mori Y, Putney JW Jr (2005) The role of canonical transient receptor potential 7 in B-cell receptor-activated channels. J Biol Chem 280(42):35346–35351
47. DeHaven WI, Jones BF, Petranka JG, Smyth JT, Tomita T, Bird GS, Putney JW Jr (2009) TRPC channels function independently of STIM1 and Orai1. J Physiol 587(Pt 10):2275–2298
48. Trebak M, Vazquez G, Bird GS, Putney JW Jr (2003) The TRPC3/6/7 subfamily of cation channels. Cell Calcium 33(5–6):451–461
49. Lievremont JP, Bird GS, Putney JW Jr (2005) Mechanism of inhibition of TRPC cation channels by 2-aminoethoxydiphenylborane. Mol Pharmacol 68(3):758–762
50. Urban N, Hill K, Wang L, Kuebler WM, Schaefer M (2012) Novel pharmacological TRPC inhibitors block hypoxia-induced vasoconstriction. Cell Calcium 51(2):194–206
51. Vazquez G, Bird GS, Mori Y, Putney JW Jr (2006) Native TRPC7 channel activation by an inositol trisphosphate receptor-dependent mechanism. J Biol Chem 281(35):25250–25258
52. Itsuki K, Imai Y, Hase H, Okamura Y, Inoue R, Mori MX (2014) PLC-mediated PI(4,5)P2 hydrolysis regulates activation and inactivation of TRPC6/7 channels. J Gen Physiol 143(2):183–201
53. Lemonnier L, Trebak M, Putney JW Jr (2008) Complex regulation of the TRPC3, 6 and 7 channel subfamily by diacylglycerol and phosphatidylinositol-4,5-bisphosphate. Cell Calcium 43(5):506–514
54. Linseman DA, McEwen EL, Sorensen SD, Fisher SK (1998) Cytoskeletal and phosphoinositide requirements for muscarinic receptor signaling to focal adhesion kinase and paxillin. J Neurochem 70(3):940–950
55. Imai Y, Itsuki K, Okamura Y, Inoue R, Mori MX (2012) A self-limiting regulation of vasoconstrictor-activated TRPC3/C6/C7 channels coupled to PI(4,5)P(2)-diacylglycerol signalling. J Physiol 590(Pt 5):1101–1119
56. Worley PF, Zeng W, Huang GN, Yuan JP, Kim JY, Lee MG, Muallem S (2007) TRPC channels as STIM1-regulated store-operated channels. Cell Calcium 42(2):205–211
57. Lievremont JP, Bird GS, Putney JW Jr (2004) Canonical transient receptor potential TRPC7 can function as both a receptor- and store-operated channel in HEK-293 cells. Am J Physiol Cell Physiol 287(6):C1709–C1716
58. Liou J, Kim ML, Heo WD, Jones JT, Myers JW, Ferrell JE Jr, Meyer T (2005) STIM is a Ca2+ sensor essential for Ca2+-store-depletion-triggered Ca2+ influx. Curr Biol: CB 15(13):1235–1241
59. Roos J, DiGregorio PJ, Yeromin AV, Ohlsen K, Lioudyno M, Zhang S, Safrina O, Kozak JA, Wagner SL, Cahalan MD, Velicelebi G, Stauderman KA (2005) STIM1, an essential and conserved component of store-operated Ca^{2+} channel function. J Cell Biol 169(3):435–445

60. Potier M, Trebak M (2008) New developments in the signaling mechanisms of the store-operated calcium entry pathway. Pflugers Arch 457(2):405–415
61. Huang GN, Zeng W, Kim JY, Yuan JP, Han L, Muallem S, Worley PF (2006) STIM1 carboxyl-terminus activates native SOC, I(crac) and TRPC1 channels. Nat Cell Biol 8(9):1003–1010
62. Yuan JP, Zeng W, Huang GN, Worley PF, Muallem S (2007) STIM1 heteromultimerizes TRPC channels to determine their function as store-operated channels. Nat Cell Biol 9(6):636–645
63. Lemonnier L, Trebak M, Lievremont JP, Bird GS, Putney JW Jr (2006) Protection of TRPC7 cation channels from calcium inhibition by closely associated SERCA pumps. FASEB J: Off Publ Fed Am Soc Exp Biol 20(3):503–505
64. Yuasa K, Matsuda T, Tsuji A (2011) Functional regulation of transient receptor potential canonical 7 by cGMP-dependent protein kinase Ialpha. Cell Signal 23(7):1179–1187
65. Lussier MP, Cayouette S, Lepage PK, Bernier CL, Francoeur N, St-Hilaire M, Pinard M, Boulay G (2005) MxA, a member of the dynamin superfamily, interacts with the ankyrin-like repeat domain of TRPC. J Biol Chem 280(19):19393–19400
66. Wedel BJ, Vazquez G, McKay RR, St JBG, Putney JW Jr (2003) A calmodulin/inositol 1,4,5-trisphosphate (IP3) receptor-binding region targets TRPC3 to the plasma membrane in a calmodulin/IP3 receptor-independent process. J Biol Chem 278(28):25758–25765
67. Merico D, Isserlin R, Stueker O, Emili A, Bader GD (2010) Enrichment map: a network-based method for gene-set enrichment visualization and interpretation. PLoS One 5(11):e13984
68. Lawler K, Foran E, O'Sullivan G, Long A, Kenny D (2006) Mobility and invasiveness of metastatic esophageal cancer are potentiated by shear stress in a ROCK- and Ras-dependent manner. Am J Physiol Cell Physiol 291(4):C668–C677
69. Phelan KD, Shwe UT, Abramowitz J, Birnbaumer L, Zheng F (2014) Critical role of canonical transient receptor potential channel 7 in initiation of seizures. Proc Natl Acad Sci U S A 111(31):11533–11538

Chapter 12
Calcium Entry Through Thermosensory Channels

Francisco J. Taberner, Isabel Devesa, and Antonio Ferrer-Montiel

Abstract ThermoTRPs are unique channels that mediate Na^+ and Ca^{2+} currents in response to changes in ambient temperature. In combination with their activation by other physical and chemical stimuli, they are considered key integrators of environmental cues into neuronal excitability. Furthermore, roles of thermoTRPs in non-neuronal tissues are currently emerging such as insulin secretion in pancreatic β-cells, and links to cancer. Calcium permeability through thermoTRPs appears a central hallmark for their physiological and pathological activities. Moreover, it is currently being proposed that beyond working as a second messenger, Ca^{2+} can function locally by acting on protein complexes near the membrane. Interestingly, thermoTRPs can enhance and expand the inherent plasticity of signalplexes by conferring them temperature, pH and lipid regulation through Ca^{2+} signalling. Thus, unveiling the local role of Ca^{2+} fluxes induced by thermoTRPs on the dynamics of membrane-attached signalling complexes as well as their significance in cellular processes, are central issues that will expand the opportunities for therapeutic intervention in disorders involving dysfunction of thermoTRP channels.

Keywords TRP • ThermoTRP • Structure-function • Signaling • Pathology • Targets

12.1 Introduction

Transient receptor potential (TRP) channels comprehend a large family of ion channels that play a broad diversity of physiological functions [1]. These channels are widely expressed in a large number of tissues and genetic studies have linked mutations in them to human diseases [2, 3]. The majority of TRP channels are non-selective cation channels that permeate Na^+ and $K^{+,}$ and most of them display significant Ca^{2+} permeability. Ca^{2+}-influx through TRP channels is physiologically

F.J. Taberner (✉) • I. Devesa (✉) • A. Ferrer-Montiel
Instituto de Biología Molecular y Celular, Universidad Miguel Hernández,
Av. De la Universidad s/n, 03202 Elche, Spain
e-mail: ftaberner@umh.es; idevesa@umh.es

J.A. Rosado (ed.), *Calcium Entry Pathways in Non-excitable Cells*, Advances in Experimental Medicine and Biology 898, DOI 10.1007/978-3-319-26974-0_12

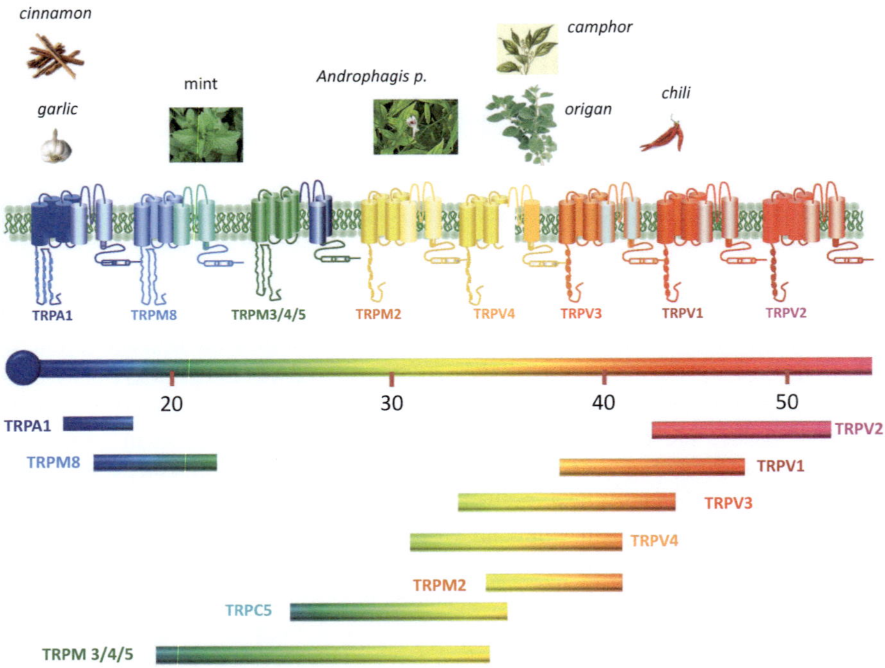

Fig. 12.1 ThermoTRP channels. Structurally thermoTRP tetramers and each subunit contains six transmembrane domains (S1–S6), a hydrophobic pore loop linking transmembrane S5 and S6, and large cytoplasmic *N*- and *C*-terminals (NB not drawn to scale). All thermoTRPs have a variable number of ankyrin repeat domains in the *N*-terminus, except *TRPM8* which has none and instead contains TRPM homology region. ThermoTRPs display distinct thermal thresholds from very noxious cold (TRPA1) to harmful hot (TRPV2). Each thermoTRP is also activated by specific natural or synthetic compounds, known to induce the relevant thermal and pain sensations in humans (Figure adapted from [4])

and pathophysiologically important due to the critical role regulating diverse cellular functions. In mammals, the TRP family consists of 28 different members grouped in 7 subfamilies: TRPC (classical or canonical), TRPV (vanilloid), TRPM (melastatin), TRPA (ankyrin-like), TRPP (polycysteine), TRPML (mucolipin) and the TRPN (no mechanoreceptor potential C; NOMPC), while the TRPY is present in yeast. For a detailed review, see the excellent book edited in 2014 by Nilius and Flockerzi.

Thermosensory channels, also named "thermoTRPs", define a subfamily of the TRP channels that are activated by changes in the environmental temperature, from noxious cold (<15 °C) to injurious heat (>42 °C) (Fig. 12.1). However, these channels are also activated by a wide range of other physical (voltage, pressure) and chemical stimuli [1]. Acting as integrators of several stimuli and signaling pathways, dysfunction of these channels contributes to diverse pathophysiological conditions such as chronic inflammation, neurological disorders, pain or cancer among others [2, 5–7]. For this reason, thermoTRPs have become promising drug targets, and the development of therapeutic compounds for pharmacological intervention is actively pursued by the academy and the industry [8, 9].

ThermoTRPs participate in Ca^{2+} homeostasis and signaling through different mechanisms: (i) most of them are Ca^{2+} permeable channels allowing extracellular Ca^{2+} entry; (ii) channel gating can be directly modulated by Ca^{2+}, but can also be indirectly regulated by Ca^{2+}-dependent proteins or signaling pathways; (iii) an increase of Ca^{2+} influx through TRP induces Ca^{2+}-dependent exocytosis promoting the release of mediators as well as trafficking of new thermoTRP receptors to the plasma membrane.

Here, we will describe the current knowledge in Ca^{2+}-dependent signaling and modulation of the classical thermosensory channels, namely TRPV1, TRPV2, TRPV4, TRPM8, and TRPA1, as well as the recently included TRPM2, TRPM3, TRPM4, TRPM5 and TRPC5. This book chapter will focus on the direct and indirect effects of Ca^{2+} on/ or through the thermoTRP channels (Table 12.1). The important role of these receptors has opened exciting research lines driven to the development of novel therapeutic approaches with successful clinical results in the treatment of several pathologies.

12.2 Structural Insights into Ion Selectivity and Permeation Through ThermoTRPs

The generally accepted functional TRP channel is an ensemble of four identical subunits. Akin to voltage gated potassium channels each subunit is composed of six membrane-spanning alpha-helices (S1–S6) flanked by the cytosolic N and C-termini [10–13]. By folding together, transmembrane helices S5 and S6 from each subunit delineate the ion-conducting pore whose opening is influenced by agonist-sensing modules in other regions of the protein [6, 12, 14]. Except for the TRPM family, the N-terminus of the channels contains a varying number of ankyrin repeat domains (ARDs) which are involved in channel regulation [15]. Agonist-driven activation requires the TRP domain in the C-terminus [16–18], a 25 amino acids region that is hallmark in TRPC, TRPV and TRPM ion channels. The region contiguous to the S6 plays a central role in the allosteric regulation of TRPV1 and TRPM8 by coupling the stimuli-induced change to open the channel's gate [14, 19, 20].

ThermoTRP channels permeate non-selectively mono and divalent cations, with a discrimination principally dictated by the selectivity filter (SF) located in the loop bridging S5–S6 transmembrane helices. The largest fractional current is due to Na^+ permeation with Ca^{2+} playing a minor contribution. Among the different TRPs, the sequence homology at the SF is limited. However, for members of the TRPV family, the SF signature GM(L)GD has been proposed. The range of selectivity for Ca^{2+} ions varies from channels impermeable to Ca^{2+} (TRPM4 and TRPM5) to channels highly selective (TRPV1, $P_{Ca^{2+}}/P_{Na^+} \sim 10$).

Table 12.1 Ca^{+2} -permeability and -modulation of thermoTRP

Channel	Temperature	Ca^{2+} permeability (estimated P_{Ca}/P_{Na})	Ca^{2+} dependent modulation
TRPV1	>42 °C	10	Desensitization by:
			Calcineurin activation
			High PLC activation
			Sensitization by:
			CaMKII phosphorylation
			Mild PLC activation
			Regulated exocytosis
TRPV2	>52 °C	1–3	Desensitization by:
			PLC activation
			Sensitization by:
			Membrane mobilization
TRPV3	>33 °C	10–11	Rectification due to extracellular Ca^{2+}
			Desensitization by:
			CaM binding
			Sensitization by:
			PLC activation
TRPV4	24–38 °C	6	Rectification due to extracellular Ca^{2+}
			Desensitization by:
			CaM competing with ATP potentiation*
			Potentiation by:
			CaM Binding*
			*Depends on [Ca^{2+}]
TRPM2	>35 °C	0.4–1.8	Potentiation depends on CaM-Ca^{2+}
TRPM3	>35 °C	1.6–10	CaM-Ca^{2+} isoform dependent regulation
TRPM4	15–35 °C	Non permeable	Activation by:
			Direct activation Ca^{2+}
			CaM-Ca^{2+}
			Desensitization by:
			Ca^{2+} dependent PIP2 depletion
TRPM5	>15 °C	Non permeable	Activation by:
			Ca^{2+} or unknown Ca^{2+} interacting protein
TRPM8	<22 °C	1–3.3	Inhibition by:
			Extracellular Ca^{2+}
			Desensitization by:
			Ca^{2+} dependent PLC
			CaM-Ca^{2+}

(continued)

Table 12.1 (continued)

Channel	Temperature	Ca^{2+} permeability (estimated P_{Ca}/P_{Na})	Ca^{2+} dependent modulation
TRPC5	25–37 °C	9–10	Inhibition by interaction with:
			CaM-Ca^{2+}
			CaBP1-Ca^{2+}
TRPA1	<10 °C species dependent	0.8–6	Potentiation by:
			Ca^{2+} entry
			Ca^{2+} dependent membrane recruitment of channels
			Desensitization by:
			Sustained Ca^{2+} entry

The major insights into the structural determinants of Ca^{2+} permeability derive from structure-function studies guided by sequence homology and aided by molecular dynamic simulations. Studies in TRPV5, a highly Ca^{2+} selective channel, indicate that selectivity and permeation properties are mainly determined by a ring of Asp residues within the pore [21], reminiscent to the ring of four negative residues present in voltage gated Ca^{2+} channels [21–24]. Exchange of residues in the S5-S6 loop of the warm-temperature activated Ca^{2+} impermeable TRPM4, with amino acids of the Ca^{2+} selective TRPV6, yielded a functional Ca^{2+} permeable channel that recapitulated the TRPV6 sensitivity to extracellular Ca^{2+} and Mg^{2+} blockade [25], substantiating the role of Asp residues in Ca^{2+} permeability. Studies in thermoTRPs further support the requirement of negatively charged residues in ion selectivity. Substitutions of Gln981 and Pro983 of the low Ca^{2+} selective human TRPM2 with equivalent residues Glu and Tyr present in more Ca^{2+} permeable TRPM2 channels from other organisms, increased the Ca^{2+} permeability by fourfold [26]. Conversely, neutralization of the negatively charged Asp546 and Glu682 in the TRPV1 and TRPV4 selectivity filter reduced the divalent permeability of the channels [27, 28]. Noteworthy, evidences in TRPM2 and TRPV4 indicate that residues outside the SF can also contribute to Ca^{2+} selectivity [28, 29].

12.2.1 Pore Dilation

The high resolution TRPV1 structure unveiled that, in contrast to the SF of Nav channels, TRPV1 lacks hydrogen bonding between adjacent pore helices. This feature results in a more flexible SF [12]. Molecular dynamic simulations further support the tenet of a TRPV1 SF highly dynamic, whose conformation is greatly influenced by the permeating ion [30]. The plausible dynamic nature of the SF of thermoTRP channels could underlie the pore dilation phenomenon by which some of them can permeate large organic cations [12]. For instance, long exposure of TRPV1 to capsaicin results in induced permeability to the large monovalent cation

N-methyl-D-glutamine accompanied with major changes in Ca^{2+} permeability [31]. Noteworthy, TRPV1 pore dilation can be influenced from the intracellular side of the channels through signaling cascades that tune channel function [12]. Analogously to the vanilloid receptor, TRPA1 exposure to agonists resulted in a 3 Å channel dilation with a concomitant increase in Ca^{2+} permeability and the fraction of current mediated by Ca^{2+} ions [32, 33]. Pore dilation also allows cells to uptake the dye Yo-Pro, a convenient tool to assess increased permeability to large cations. The inability of TRPM8 to support Yo-Pro entry [34], highlights significant differences in SF flexibility among different TRP channels.

The dynamic nature of the thermoTRP channel pore suggests that modifications of residues near the SF can greatly influence channels permeability. Indeed amino acid modification in the loop S5-S6 also influences cation permeation. Glycosylation of Asp604 in TRPV1, despite not impeding membrane trafficking of the channel, is necessary for sustained Ca^{2+} entry and for Yo-Pro uptake [35].

12.2.2 TRP Heteromers

Although not frequent, thermoTRP channels also form heteromers additionally contributing to modulate the Ca^{2+} permeability spectrum of TRP channels. The cold-activated TRPC5, form heterotetramers with TRPC1, TRPC4 and other TRPC members resulting in permeation properties differing from those of the homotetramers [36–39]. Similarly, TRPV1 has been proposed to form heteromers with TRPV2 and TRPV3 which varies their electrophysiological properties [40, 41]. The use of TRPV1-TRPA1 concatemers has evidenced the compatibility of TRPV1 and TRPA1 monomers to form functional channels with different regulation [42]. The heteromerization capability also extends to TRPV4, which functional heteromers with TRPP2 and TRPC1 are involved in thermal and mechanosensation in kidney cilium and in the store operated Ca^{2+} entry in endothelial cells [43–46].

12.3 ThermoTRP Regulation by Ca^{2+}

In addition to permeate through thermoTRP channels, Ca^{2+} can greatly affect channel function by directly or indirectly influencing posttranslational modifications or protein/lipid interactions. The reports describing a direct effect of Ca^{2+} ions in thermoTRP channel regulation are scarce. However, its action is determinant for TRPM4 and TRPM5 activity, which are impermeable to this cation. Ca^{2+} released from the internal stores activates TRPM4 and TRPM5 [47, 48]. Calmodulin (CaM) can greatly influence TRPM4 functions through its binding in the C-terminus. When CaM interaction is impaired, high concentrations of Ca^{2+} still activate the channel [49], suggesting a direct effect of Ca^{2+} on channel gating. Indeed, negative residues

Asp1049 and Glu1062 within the TRP domain are proposed to form a Ca^{2+} binding site that is essential for the normal TRPM4 Ca^{2+} response [50].

Other thermoTRP channels regulated by Ca^{2+} are TRPC5 and TRPM8. Intracellular increase of Ca^{2+} concentration also activates murine TRPC5 heterologously expressed in HEK cells independently from other regulators and agonists [51, 52]. In TRPM8, extracellular Ca^{2+} opposes its activation by agonists. High amounts of Ca^{2+} (or other divalent cations) shield the effective electrical field detected by the TRPM8 voltage sensor by neutralizing negative charges in the extracellular leaflet of the membrane therefore reducing channel response [53]. TRPA1 is also directly gated by Ca^{2+}. The ability to respond to intracellular Ca^{2+} elevations resides in the EF-Hand Ca^{+2}-binding domain at the N-terminus [54, 55]. Ca^{2+} also appears to modulate TRPV1 desensitization and tachyphylaxia through Ca^{2+}-dependent covalent modification.

By acting on signaling pathways or reshaping channel interactions, Ca^{2+} indirectly regulates thermoTRP function. This regulation is complex and in most cases channel dependent as variations in Ca^{2+} concentration can result in channel activation, potentiation of the basal current, or desensitization. Therefore, the understanding of this layer of regulation deserves a closer look into specific channel biology. We next expose the specific information accrued for thermoTRPs.

12.3.1 TRPV Family

All mammalian homologues of TRPVs are Ca^{2+}-permeable channels, with the thermosensors TRPV1–4 characterized as moderately Ca^{2+}-selective cationic channels [56–58], and the non-thermally activated TRPV5 and TRPV6 as highly Ca^{2+}-selective channels.

12.3.1.1 TRPV1

TRPV1, the founder member of the TRPV1 subfamily, is the most extensively studied thermoTRP channel. Since its cloning and description as the vanilloid receptor [59], TRPV1 has been involved in central processes in different organisms. Its ability to sense different pungent compounds and noxious heat plus its regulation by inflammatory mediators has entitled it as a central player in temperature sensing, body temperature regulation or pain transduction. Besides being activated by capsaicin, is also gated by other pungent compounds such as resiniferatoxin, gingerol or piperine [60]. Remarkably, venoms from different organism such as spiders or scorpions target TRPV1 contributing to the pain sensation [61]. Additionally, TRPV1 also responds to physical stimuli including highly depolarizing voltages and noxious heat [59, 62]. TRPV1 activity is highly temperature dependent (Q10 around 25 versus 1–4 from non-thermal-activated channels) and temperatures over 42 °C readily evoke inward ionic currents [59]. Conversely, it can be pharmacologically blocked by capsazepine, BCTC, TRPducins and other synthetic compounds [63–65].

Structure-function studies have isolated the binding sites of vanilloids as well as key determinants temperature sensing and function in TRPV1. Domain swapping among capsaicin sensitive and insensitive TRPV1 orthologues, unveiled the role of S2–S3 region in capsaicin and resiniferatoxin activation [66, 67]. Nonetheless, the region accounting for temperature sensing is not well delimited. Indeed, several regions of the protein may take part. In the N-terminus, the region adjacent to the S1 plays a critical role [68]. Mutations in the pore region enhance [69] or reduce [70] channel sensibility to heat and affect proton sensing and capsacin response as well. Transferring the C-terminus of the cold receptor TRPM8 into TRPV1, results in a chimera that responds to cooling [71]. Additionally, mutations in the TRP domain also affect heat, capsaicin and voltage activation [18] by affecting the allosteric coupling of the sensors and the gate [19]. All these findings ought to be explained by the recent high-resolution cryoelectron microscopy model of TRPV1.

The regulation of TRPV1 is complex. Long or repeated agonist stimulation desensitizes the channel, a process completely dependent on Ca^{2+}. In fact, Ca^{2+} dependent phosphatase calcineurin activation results in channel desensitization [72, 73]. Similarly, CaM appears to compete with ATP for binding to the ARD decreasing the channel response. More controversial is the role of phosphoinositides (PIs) on TRPV1. Some findings indicate that PIP_2 has an inhibitory role, and that the activation of phospholipase C (PLC) through G protein coupled receptor (GPCR) increases channel response. Other findings indicate that PIs are central for channel function and Ca^{2+} dependent PLC activity inhibits channel responsiveness [74]. On the other hand, TRPV1 response to heat or chemicals can be potentiated by pro-algesic agents through activation of different signaling cascades promoting post-traductional modification, usually through phosphorylation/dephosphorylation mechanisms by protein kinase A (PKA), protein kinase C (PKC), Src or Ca^{2+}-CaM kinase II (CaMKII). PKC phosphorylation increases the likelihood of channel opening at resting membrane potential. PKA and CaMKII phosphorylation contributes to recovery of desensitization [75, 76]. In addition to these mechanisms, pro-inflammatory mediators also promote the rapid exocytosis of a vesicle reservoir located near the plasma membrane. These vesicles contain new TRPV1 channels that are transported to the membrane through SNARE-dependent mechanism, which is highly sensitive to Ca^{2+} and sensible to botulinotoxin or to a botulinomimetic peptide treatment [77, 78].

In the peripheral nervous system, TRPV1 is highly expressed in dorsal root, trigeminal and nodose ganglia [59], mainly in small and medium peptidergic neurons, and to a lesser extent in non-peptidergic nociceptors [62, 79]. Gating of TRPV1 allows Ca^{2+} entry eliciting pain signaling, promotes Ca^{2+}-dependent release of pro-inflammatory neuropeptides calcitonin-gene related peptide (CGRP) and Substance P, and increases TRPV1 plasma membrane translocation. The enhancement of TRPV1 trafficking from large-dense core vesicles in response to exogenous stimuli facilitates rapid modulation of neuronal excitability [80]. The presence of TRPV1 in sensory nerves and in areas involved in detection, transmission and regulation of pain revealed its potential key role in nociception. Actually, TRPV1 turned to be a thermosensor able to transduce physical, chemical and thermal noxious stimuli. Pharmacological and knockout studies have demonstrated TRPV1 as an essential channel to pain signaling during inflammation, being responsible of the development and maintenance of thermal hyperalgesia. Indeed, TRPV1 is overexpressed in several chronic painful pathologies such as rheumatoid arthritis [81, 82], osteoarthritis [83], bone cancer-induced pain [84] and several neuropathies [85, 86] among others [87]. All these evidences rapidly promoted TRPV1 as an interesting pharmacological target to develop new analgesic treatments especially in diseases with an inflammatory component.

In the central nervous system, although with some initial controversy on its brain distribution, the role of TRPV1 in synaptic transmission, neurotransmitter release and plasticity is proposed by several groups [88–90]. Prolonged activation of TRPV1 allows Ca^{2+} overload inducing Ca^{2+}-dependent programmed cell death in cortical or hippocampal neurons [91–94]. A potential contribution in cognition, perception and neuropsychiatric disorders is also suggested [95]. Nevertheless, more studies are needed before the relevance of TRPV1 in brain activity can be clearly stated.

There is an extensive list of non-neuronal tissues expressing TRPV1 channel [96, 97], with a diverse range of functions. For instance, TRPV1 participates in many aspects of skin biology [97, 98]. In keratinocytes, TRPV1- mediated Ca^{2+} influx promotes the release of extracellular molecules that may affect the activity of surrounding cells, but also regulates directional migration [99] or contributes to metalloproteinase expression, which is critical for skin inflammation and aging [100]. TRPV1 function is necessary to keep skin homeostasis, as absence of TRPV1 function results in a striking increase in skin carcinogenesis [101]. Ca^{2+} entry through this thermoTRP channel participates in bone homeostasis as well as pathophysiology. In addition to its contributions in joint pain [102], TRPV1 promotes the differentiation of osteoblasts and osteoclasts, and its suppression protects against ovariectomy-induced bone loss [103, 104].

12.3.1.2 TRPV2

The second member of the TRPV family is also gated by heat. Actually, TRPV2 presents the highest temperature activation threshold requiring thermal stimuli over 52 °C. This high temperature requirement questions its relevance as a physiological thermosensor. Indeed, TRPV2 deficient mice do not show abnormalities in thermal sensing [105]. Additionally, TRPV2 responds to mechanical stress and to various ligands as for instance 2-aminoethoxydiphenyl borate (APB), probenecid, lysophospholypids and cannabinoids. Nevertheless, most of them are not specific and act in the micromolar range. These circumstances severely hamper the study of TRPV2, further hindered by the fact that some agonists show species dependent effect. Human TRPV2 insensitivity to 2-APB is a case in point. The pharmacological toolbox for channel blockade is scarce. Only three antagonists have been described so far, namely ruthenium red, SKF96365 and tranilast, and all of them are non-selective.

Findings on TRPV2 regulation are limited. PIP_2 interaction through the TRP domain is central for channel function since TRPV2 currents show desensitization by Ca^{2+}-dependent PIP_2 depletion [106]. In marked contrast to TRP members with ARDs, TRPV2 ARDs do not bind CaM. Indeed, it has been reported that CaM directly binds to TRPV2 C-terminus through a motif that overlaps the TRP domain. Strikingly, despite the plausible PIP_2/CaM competition, CaM addition does not result in channel desensitization. Consistent with the absence of ATP binding sites in the ARDs, the channel does not respond to intracellular changes in ATP concentration [106, 107]. Membrane mobilization of TRPV2 channels from intracellular membranes after PI3K stimulation (mainly mediated by IGF-1) has been considered an important part of the channel regulation [108]; however, this regulatory step is controversial [109].

In the peripheral nervous system, TRPV2 is expressed in many peptidergic neurons, and similar to TRPV1, TRPV2 activation allows Ca^{2+} entry and CGRP release contributing to the development of inflammatory pain [110]. Cytosolic Ca^{2+} increase through TRPV2 is important for axonal outgrowth during development [111]; while in intestinal sensory neurons regulates the intestinal motility [112].

Most of the thermosensitive TRP channels can function as multimodal receptors in pancreatic β-cells causing Ca^{2+} influx and membrane depolarization at physiological body temperature. In this context, glucose-induced insulin secretion is dependent of TRPV2-mediated intracellular Ca^{2+} increase [113], and intriguingly, insulin promotes a Ca^{2+}-dependent translocation of TRPV2 channel to the plasma membrane [114].

TRPV2 is involved in several physiological as well as pathological conditions in the immune system, where Ca^{2+} signaling through this channel plays an important role [115–117]. In response to high temperatures, as well as other physical and chemical stimuli, TRPV2 allows Ca^{2+} entry inducing pro-inflammatory degranulation of mast cells [118, 119]. In macrophages, sustained elevation of cytosolic Ca^{2+} by activation of PI3kinase pathway allows translocation of TRPV2 from the endoplasmic reticulum to the plasma membrane promoting macrophage migration [120]. In fact, disruption of TRPV2 membrane insertion reduces cytosolic Ca^{2+} increase, which critically regulates assembly of the podosome [120]. In this line, absence of TRPV2 expression reduces the recruitment of macrophages in experimental colitis [121], migration towards injured cardiomyocytes [122] and, attenuates phagocytosis [123]. Interestingly, TRPV2 shows a restricted expression in normal immune cells, in contrast, TRPV2 is widely expressed in the myeloid and lymphoid leukemias, with a very peculiar expression of this channel in different types of B cell lymphomas and multiple myeloma [124].

TRPV2 is a very promising target for early diagnosis or therapy of different human cancers [125]. For instance, TRPV2 negatively controls cell survival and proliferation of human glioma [126, 127] and bladder cancer [128, 129] by induction of apoptotic cell death. Remarkably, activation of TRPV2 synergizes cytotoxic effects of chemotherapies in human gliomas [130]. In contrast, TRPV2 might be a novel prognostic marker of patients with hepatocarcinogenesis [131], resected esophageal squamous carcinoma [132] or prostate cancer, where TRPV2 overexpression is present [133]. In particular, TRPV2 promotes prostate cancer cell migration and invasive phenotype influencing resting intracellular Ca^{2+} levels [133, 134].

Critical for the maintenance of cardiac structure and function [135], TRPV2 seems a potential target to treat cardiovascular diseases. Under physiological conditions, TRPV2 modulates Ca^{2+} transients and sarcoplasmic reticulum Ca^{2+} loading on myocytes that regulates cardiac contractility, while in smooth muscle contributes to global Ca^{2+} entry helping constriction [136]. TRPV2 deletion results in an impairment in Ca^{2+} handling together with defects on myocardial conduction, among other cardiac features [137]. In muscular dystrophic animals, TRPV2 cell surface expression is increased causing Ca^{2+} overload and cell dead [138]. Noteworthy, attenuation of TRPV2 translocation to the plasma membrane reduces Ca^{2+} increase in muscle fibers attenuating the progression of dystrophic pathology [139]. In this line, abrogation of TRPV2 plasma membrane accumulation, inhibition of TRPV2-mediated Ca^{2+} entry or removal of sarcolemmal TRPV2 ameliorates contractile dysfunction [140, 141]. Consistently, TRPV2 is accumulated in sarcolemma of dilated cardiomyopathy patients and appears to have pathological impact on disease progression through excessive Ca^{2+} influx. In conclusion, TRPV2 contributes to defective cellular Ca^{2+} handling in dystrophic cardiomyopathy [142].

12.3.1.3 TRPV3

TRPV3 is a heat-activated receptor that gates in response to moderate heat (>33 °C). Temperature sensitivity is influenced by voltage as well as the rate at which heat is applied [143]. TRPV3 is also potently and specifically activated by the endogenous molecule farnesyl pyrophosphate from the biosynthetic pathway of steroid hormones [144]. The most used agonist for studying TRPV3 is 2-APB. This molecule as well as other activators such as carvacrol, eugenol, thymol, citral or camphor are not specific and require high concentrations.

A unique feature of TRPV3 is that contrary to other TRPs, currents are potentiated by repeated stimuli application. Noticeably, this property derives from Ca^{2+} regulation of channel activity at both inside and outside domains [145]. By interacting with acidic residues in the pore, extracellular Ca^{2+} channel inhibition is the origin on the double rectifying I–V curves. In the cytosol, CaM-Ca^{2+} complexes have an inhibitory action that is released with repeated application of 2-APB [145]. A CaM binding motif in the N-terminus mediates this inhibitory effect. Lysine residues within this domain are critical for Ca^{2+} dependent regulation as well as for ATP inhibition of TRPV3 currents [145, 146]. Interestingly, CaM and ATP modulatory effects are opposite to those described for TRPV1. On the contrary, TRPV3 regulation by PIP_2 mimics TRPV1 since PLC activation releases PIP_2 tonic inhibition through the TRP domain [54]. Less studied is the channel regulation by phosphorylation. Activation of PKC with a phorbol ester enhances channel function; however, there is no evidence that the potentiation arises by direct phosphorylation.

As TRPV3 is activated by innocuous warm and noxious hot temperatures, this channel was immediately proposed as thermosensor. Interestingly, in humans is activated by heat, but in lower vertebrates by cold [147]. Although the expression of TRPV3 in sensory neurons is not really high in rodents [148], first reports showed that TRPV3 deficiency was critical for thermal sensing of noxious and innocuous heat [149], but later studies restricted the role of TRPV3 on thermoregulation [150].

TRPV3 is a therapeutic target in dermatology, and TRPV3 antagonists might be useful tools in the management of a wide array of skin diseases. TRPV3 is mostly expressed in keratinocytes and it is involved in numerous cutaneous regulatory mechanisms [151, 152]. Increase in Ca^{2+} influx through TRPV3 activation directly inhibits keratinocyte proliferation, decreases hair loss, regulates differentiation, induces apoptosis and maintains cutaneous barrier function. Indirectly, TRPV3 function results in the release of numerous agents that can activate the surrounding cells, such as nociceptors, other keratinocytes or immune cells, inducing pain, itch or skin inflammation. In this regard, activation of TRPV3 upon thermal stimulation promotes a concomitant cytosolic Ca^{2+} increase in keratinocytes leading to the release of pro-inflammatory agents such as ATP, prostaglandin E_2, IL-1b, bradykinin or histamine that are able to activate and sensitize nociceptors [146, 153, 154]. Therefore, epidermal keratinocytes are indirect nociceptor transducers, since via TRPV3-coupled signal mechanism participate in thermal pain signaling. Consequently, molecules able to inhibit TRPV3 channel activity are promising analgesic therapies [9].

Genetic deletion of TRPV3 impairs epidermal barrier structure and hair morphology [155]. In contrast, the gain-of- function Glu573Ser mutation results in a sever dermatitis and itching in mice. Interestingly, identical mutation is described in Olmsted syndrome in humans, together with other two point mutations on TRPV3, Gly573Cys and Trp692Gly, causing constitutive activation of TRPV3 [156]. The enhanced basal activity is accompanied with an increase in Ca^{2+} entry diminishing cell growth and inducing cell death. This channelopathy characterized by a gain of function of the TRPV3 phenotype, causes dramatic effects on skin including hair loss, dermatitis, skin inflammation, keratoderma and deformity of digits. In fact, gain-of-function mutations of TRPV3 result in multiple problems of skin health, which further supports the potential value of TRPV3 antagonist in the treatment of skin diseases [151].

12.3.1.4 TRPV4

TRPV4 is activated by various stimuli ranging from physical stimuli to chemical activators, being considered as a mechano- or osmo-sensor and a moderate heat sensor [157, 158], with higher permeability to Ca^{2+} and Mg^{2+} than Na^{+}. TRPV4 responds to temperature in the range from 24 to 38 °C with strong temperature dependence as indicated by a Q10 between 10 and 20. Initial studies showed that TRPV4 only was heat sensitive in intact cells [159, 160], suggesting that the temperature sensing was dependent on an accessory factor. This tenet has been recently rebated with findings that demonstrate that TRPV4-PIP_2 interaction, lost in excised patches, is mandatory for temperature and hyposomotic activation [161]. Mutations in the PIP_2 binding domain [161] or in Tyr556 [162, 163] impaired activation by heat. The most extensively used chemical agonist is 4αPDD although it also responds to bisandrographolidea in the mid nanomolar range. The non-specific TRP channel blockers ruthenium red, gadolinium and lanthanum, inhibit TRPV4 currents. In a more targeted manner, the insect repellent citral and the synthetic compounds HC-067047 and GKS block the channel [161].

TRPV4 is subjected to dual Ca^{2+}-dependent regulation, with channel activity potentiated or inactivated during agonist-mediated activation [164]. Ca^{2+} regulation of TRPV4 is intricate. Ca^{2+} affects the channel from both sides of the plasma membrane. Outside the cell, Ca^{2+} accounts for the observed rectification in whole cell currents. Inside the cell, depending on the concentration, it can inhibit or potentiate the channel by a mechanism not yet understood. CaM also regulates TRPV4; while some findings indicate that CaM is a positive modulator [165] previous results show an inhibitory effect [146]. It has been reported a CaM binding site in the N-terminus, which, when complexed with ATP, results in channel potentiation [146]. A second CaM interacting motif also accounts for IP_3R potentiation. TRPV4 is desensitized by PIP_2 depletion since the extended conformation that enables channel response to temperature and hypotonicity is lost. Conversely, TRPV4 currents are potentiated by PKC, PKA and SRC [160, 166, 167].

TRPV4 gating can transduce external mechanical, thermal or osmotic stimuli into the form of elevated intracellular Ca^{2+} concentration, initiating multiple cellular responses [168]. Medical interest in TRPV4 has recently risen since mutations in TRPV4 gene results in several genetic disorders affecting the peripheral nervous and osteoarticular system [157, 169, 170]. Although, the knowledge about the exact underling mechanism is still limited, Ca^{2+}-influx mediated through TRPV4 is critical.

TRPV4 is essential for joint homeostasis and maintenance of musculoskeletal tissues normal function. Expressed in cartilage, bone, and synovium [171], TRPV4 mutation results in the development of a spectrum of skeletal dysplasias and arthropathies. Interestingly, skeletal dysplasias are mainly due to TRPV4 gain-of-function mutations, increasing channel permeability to resting Ca^{2+} as well as Ca^{2+} influx in response to a stimulus [172–174]. The increased TRPV4-mediated Ca^{2+} entry into the chondrocytes results in an improper endochondral ossification [175]. In contrast, arthropathy related TRPV4 mutations cause a decrease in channel function by preventing normal trafficking of TRPV4 channels to the membrane [176]. Actually, TRPV4 knockout mice present an age and sex-dependent development of osteoarthritis and a decreased osteoclast function [177, 178]. As a result, TRPV4 is a possible therapeutic target for osteoarthritis treatment and recently also a valuable tool for the development of tissue-engineered constructs promoting extracellular matrix biosynthesis [179].

TRPV4 is also a critical key player in hereditary neuropathies. TRPV4-axonal neuropathies present a predominantly motor axonal peripheral neuropathy or can be associated with sensory disturbances [157]. A proposed mechanism for motor neuron degeneration in TRPV4-dependent neuropathies is cell toxicity and increased cell death through Ca^{2+}-overload TRPV4 mutants [180, 181]; however, it is unlikely that Ca^{2+}-induced cell toxicity is a general pathophysiological mechanism. Other potential mechanisms that are altered are neuritogenesis or dysregulation of protein–protein interactions.

TRPV4 has been involved in vascular function, nociception, metabolism or bladder function among others. Abnormal TRPV4 function has been associated with chronic obstructive pulmonary disease, pancreatitis or high risk of hyponatremia [158].

12.3.2 *TRPM Family*

The TRPM family can be subdivided in different cation channels subgroups that are either highly permeable for Ca^{2+} (TRPM3/6/7), non-selective (TRPM2/8), or Ca^{2+} impermeable (TRPM4/5) [182].

12.3.2.1 TRPM2

TRPM2 is the closest phylogenetic channel to the TRPM8 receptor. Notwithstanding, it responds to warm temperatures (>35 °C) in heterologous expression systems and in pancreatic β-cells. Temperature-elicited currents are synergistically increased by the application of other agonists [183]. It also responds to ADP-ribose/Ca^{2+}, H_2O_2 as well as reactive oxygen and nitrogen species (ROS/NOS). Moreover, channel function is enhanced by NAADP and cyclic ADP-ribose. On the contrary, acidic pH and AMP attenuates its activity [184]. TRPM2 characteristics suggest a major role as cellular metabolic and oxidative stress detector. Noteworthy, several splicing variants of TRPM2 have been documented and some of them are physiologically relevant for tuning channel function.

Ca^{2+} is a central regulator of TRPM2, and this modulation is CaM-dependent. Indeed, different studies indicate that CaM interacts with the N-terminus of the channel [185, 186]. Particularly, the IQ-like CaM binding motif mediates Ca^{2+} activated TRPM2 currents [185]. This interaction is deemed to account for part of the Ca^{2+} dependent activation of the channel. In β-pancreatic cells, TRPM2 currents are potentiated by PKA activation [183]. Physiologically, incretin contribution to insulin secretion is believed to result from potentiation of TRPM2 currents through receptor mediated activation of PKA [187]. Silencing or inhibiting PKC activity reduces Ca^{2+} influx via TRPM2. Interestingly, PKC phosphorylation of the splicing variant TRPM2-S releases TRPM2 from the isoform negative-dominant effect [188]. Some reports evidence that Ca^{2+} entry can directly gates TRPM2 in a dose-dependent manner, although other studies have challenged this statement [185, 189].

As other thermoTRP channels, Ca^{2+} entry through TRPM2 promotes different intracellular signaling cascades and physiological processes. TRPM2 activation by H_2O_2 is being implicated in stress-related inflammatory, vascular and neurodegenerative conditions. The physiological and pathophysiological context of ROS-mediated events makes TRPM2 a promising target for the development of therapeutic tools of inflammatory and degenerative diseases. During inflammation, TRPM2 regulates innate and adaptive immunity, acting as an oxidative stress and metabolic sensor [190]. Through lysosomal Ca^{2+} release, TRPM2 participates in maturation of dendritic cells [191]. In contrast, in macrophages, TRPM2 in the cell surface mediates Ca^{2+} -dependent chemokine and cytokine release, inflammasome formation, cell infiltration and cell death [192, 193]. In addition, TRPM2 contributes to antigen-stimulated Ca^{2+} influx in mucosal mast cells and induces degranulation [194].

TRPM2 is expressed in numerous cell types [195] being embedded in the plasma membrane and in lysosomal compartments. This channel is highly expressed in the central as well as the peripheral nervous system. Due to its high permeability to Ca^{2+}, it exerts a crucial regulatory function, most notably in modulation of DRG neurons to painful signals [196, 197]. Altered intracellular Ca^{2+} homeostasis and oxidative stress are involved in the pathophysiology of bipolar disorder, and TRPM2 genetic variants are associated with this pathogenesis [198]. TRPM2 also spotlights as possible therapeutic target for neurodegenerative diseases, as TRPM2 mediated Ca^{2+}-influx drives neuronal death [199, 200]. Although with still controversy, targeting TRPM2 for potential stroke therapy requires a balance between effects on microglia and neurons, but further investigation is required in this area [201].

Interestingly, temperatures above 35 °C can directly activate TRPM2 and potentiate Ca^{2+} entry upon ADP-ribose stimulation of neuronal cells [202], as well as in pancreatic cells [183]. Here, TRPM2 function acts as lysosomal Ca^{2+} release channel [203], inducing insulin secretion through modulation of intracellular Ca^{2+} influx. Lack of TRPM2 impairs insulin secretion and glucose metabolisms showing lower Ca^{2+} signals in response to glucose [187] and prevents diet-induce obesity [204]. Therefore, TRPM2 also represent a potential new target for diabetes therapy [205].

12.3.2.2 TRPM3

TRPM3 channels are non-selective cationic channels activated by heat, but also by voltage, hypotonic cell swelling and chemical stimuli (pregnenolone sulfate, D-erythro-sphingosine). Nevertheless, less is known about the post-translational mechanisms regulating TRPM3 channel activity. There are only limited studies reporting some insights into the features and/or regulation of Ca^{2+}. Important domains for protein-protein interaction such as CaM and PIP_2 binding domains are present in the N-terminus [206, 207], but there is limited information on the functional implication of this association.

Several splice variants of TRPM3 are known which are divided in three groups depending on their first exon. TRPM3α (α1–α5) isoforms start with exon 1 and lack exon 2, while TRPM3β isoforms (β1–β17) start with exon 2. A third group is composed of isoforms starting with exon 3 [208]. This plethora of splice variants together with the microRNA seems to facilitate a fine tuning of the channel expression level. The TRPM3β variants seem to display a different kind of Ca^{2+}–dependent regulation [207]. Alternative splicing in exon 24, leads to two different pores and influences cation selectivity [209].

The physiological and/or pathological role of TRPM3 still remains to be deeply elucidated, and only the endogenously Ca^{2+}-permeable variants have been studied. TRPM3 expression is found in the brain, eyes, reproductive system, pituitary gland, and adipose tissue. In pancreatic β-cells, TRPM3 activation induces an intracellular signaling cascade, with a concomitant rise in cytosolic Ca^{2+} followed by an increase in insulin secretion [210]. Therefore, as other thermoTRPs, TRPM3 may influence main functions of pancreatic β-cells using a similar signaling cascade induced by glucose. The functional relevance of TRPM3 in contractile and proliferating vascular smooth muscle cells is via a Ca^{2+}-dependent mechanism [211]. However, TRPM3 has recently drawn much attention since it has been described as a heat sensing channel expressed in nociceptors [212]. Present in a large number of small-diameter DRG and TG neurons, TRPM3 detects noxious heat, is involved in heat-, but not cold-mediated thermal hyperalgesia, but does not regulate homeostasis of body temperature.

12.3.2.3 TRPM4

In marked contrast to the other TRP members, TRPM4 and TRPM5 do not permeate Ca^{2+}. Nevertheless, Ca^{2+} activates their currents [213], foretelling important Ca^{2+} regulation of channel function. Despite Ca^{2+} being able to directly activate TRPM4, it requires high intracellular Ca^{2+} levels. At more physiological Ca^{2+} concentrations, TRPM4 currents are likely to be triggered by CaM interaction with CaM-binding sequences. Actually, mutations on the three putative binding sites impaired current activation [214]. TRPM4 behaves as a heat receptor in the temperature range between 15 and 35 °C [215] in which, the temperature dependency Q10 value is 8.5. In addition, it shows weak voltage sensitivity. Pharmacologically, it can be activated by vanadate and BTP2 and blocked by compounds such as 9-phenanthrol, flufenamic acid, quinine and quinidine [216].

The activity of TRPM4 is dependent on PIP_2 as shown by the inhibition of the currents following PIP_2 depletion. Neutralization of Lys residues within two putative Pleckstrin homology domains (located in the C-terminus), results in low sensitivity to PIP_2 and rapid desensitization [217]. In addition to CaM, and PIP_2 binding motifs, TRPM4 also contains multiple ATP biding sites that greatly influence Ca^{2+} sensitivity. ATP modulation seems to depend on the binding site; it is considered that interaction with one ATP binding site has an inhibitory role while interaction with the other site facilitates Ca^{2+} activation [216]. The mechanism underlying this bimodal regulation is not yet understood.

Under certain conditions, TRPM4 function is also potentiated, by the activity of PKC on Ser1145 and Ser1152 in the C-terminus [49]. Activation of PKCdelta contributes to channel potentiation through increasing its surface expression [218].

Physiologically, TRPM4 is a key regulator of the driving force for Ca^{2+} entry being widely expressed in the body [216]. This thermoTRP channel plays an important role in the immune, cardiovascular and central nervous system. For instance, in dendritic cells, the absence of TRPM4 leads to a Ca^{2+} overload which impairs chemokine–dependent migration [219]; while in mast cells, decreases migration and increases histamine release aggravating anaphylactic reactions [220]. In both cases, due to the absence of the Ca^{2+}-activated depolarizing current mediated by TRPM4. Additionally, deletion of TRPM4 expression, impairs phagocytosis in macrophages [221], and increases Ca^{2+} influx in lymphocytes Th2 [222].

Excessive Na^+ influx through TRPM4 induced by Ca^{2+} overload is a critical mechanism in experimental autoimmune encephalomyelitis, multiple sclerosis or spinal cord injury. Rise of cytosolic Ca^{2+} activates TRPM4 in neurons contributing to neuronal cell death, which turned to aggravate autoimmune encephalomyelitis [223]. TRPM4 expression is upregulated in spinal cord injury and absence of TRPM4 function reduces secondary hemorrhages and improves behavioral performance [224].

Another example is how Ca^{2+}-dependent modulation of TRPM4 increases osteogenesis or decreases adipogenesis [225]. In fact, blockade of TRPM4 activity reduces glucose-induced Ca^{2+} signal and insulin secretion in β-pancreatic cells, while in α-cells knockdown of TRPM4 decreases Ca^{2+} responses and glucagon secretion [40, 226].

TRPM4 is involved in several aspects of cardiac rhythmicity, including cardiac conduction, pacemaking and action-potential repolarization [227]. The regulation of TRPM4 function in arterial smooth muscle cells is complex and involves release of Ca^{2+} from the sarcoplasmic reticulum through IP_3 receptors and translocation of TRPM4 channels to the plasma membrane [213, 220, 228] and is critical for pressure-induced cerebral arterial myocyte depolarization and myogenic vasoconstriction [229]. A specific mutation in the TRPM4 gene leads to gain-of function of the channel in a progressive Familial Heart Block Type I pathology [230, 231], but this mutation is not determinant as additional factors are required [232].

12.3.2.4 TRPM5

TRPM5 is activated by warm temperatures (>15 °C). Akin to TRPM4, TRPM5 channel does not permeate Ca^{2+} ions; nevertheless, it is activated by Ca^{2+} [233]. The binding of bitter, sweet or umami taste molecules to their cognate receptors activates the PLCβ2 signalling, that ultimately evokes Ca^{2+} release from intracellular stores and activation of TRPM5 [233]. A comparison with TRPM4 reveals that TRPM5 opening requires less Ca^{2+} [234, 235]. Remarkably, TRPM5 seems to lack the CaM-facilitated activation; therefore, it is proposed that either another Ca^{2+} binding protein or Ca^{2+} itself gates the channel [48]. With a total identity of 40 %, both channels share similar temperature dependence (potentiation over 15 °C) and PLC-mediated desensitization. However, ATP regulation is absent as a consequence of the lack of functional nucleotide binding sites. Pharmacologically, TRPM5 is blocked by quinine, a general inhibitor of ion channels, and triphenylphosphine oxide [236].

As mentioned, TRPM5 channel plays an important role in taste and thermal sensitivity and is highly expressed in the subset of taste receptor cells. TRPM5 contributes to temperature dependence of sweet sensation [215]. Heat is not sufficient to directly activate TRPM5, which at warm temperatures requires an increase of intracellular Ca^{2+}. Because of its role in normal taste signaling, drugs designed to modulate TRPM5 could be useful in controlling taste sensory signals and potentially developed as the anti-obesity therapies with limited drug distribution to the site of action [236]. In small intestine and stomach, as well as respiratory and olfactory epithelium, TRPM5 appears to be specialized in chemosensation [237, 238]. Recently, TRPM5 channels have been shown essential for normal insulin secretion and glucose tolerance [239, 240]. Noteworthy, a genetic variation of TRPM5 gene could contribute to propensity to develop diabetes [241], highlighting TRPM5 as a potential target to control diabetes.

12.3.2.5 TRPM8

In contrast to the above-described thermoTRPs, TRPM8 is considered as the main detector of cold temperatures in mammals. Indeed knock-out mouse lines lose the cold induced responses [242–244]. At the protein level, the channel not only responds to temperatures below 22 °C but it is also active after membrane depolarization and in the presence of compounds such as menthol or icilin [245, 246]. TRPM8 has multiple physiological roles such as cool temperature discrimination, noxious cold sensing, thermoregulation, cold-induced analgesia and an anti-inflammatory effect [242–244].

TRPM8 can be regulated by various intracellular secondary messengers and signalling pathways that allow a fine tuning of channel activity by the cellular environment [247]. Presence of external Ca^{2+} markedly desensitizes TRPM8-dependent currents [246], and recent evidence shows that promotes a displacement of voltage dependent activation of TRPM8 towards more positive potentials [53]. Additionally, TRPM8 acute desensitization during a single agonist application is a Ca^{2+}-dependent mechanism triggered by CaM [248].

Besides, Ca^{2+} influx through TRPM8 activates PLC isoforms which hydrolyze PIP_2 and decreases channel activity, being responsible of Ca^{2+}-dependent tachyphylaxia [249–252]. PIP_2 is a necessary cofactor for TRPM8 activity. The fast run down of the currents is reversed upon addition of PIP_2 analogues to the bath solution [250]. Furthermore, in the presence of extracellular Ca^{2+}, TRPM8 is desensitized as a result of the PIP_2 hydrolysis by a Ca^{2+} dependent PLC pathways [253].

TRPM8 is also positively regulated by PKA through phosphorylation of Ser9 and Thr17 in the N-terminus [254]. However, PKA-mediated phosphorylation has also been described to desensitize the channel [255]. In heterologous systems and in cold sensitive neurons, chemical activation of PKC inhibits TRPM8 currents [256–258]. Interestingly, inhibition does not result from PKC phosphorylation of the channel but from activation of calcineurin and subsequent dephosphorylation in a Ca^{2+} dependent manner [256].

TRPM8 has a relatively restricted expression in normal tissues, and its overexpression or increased activity in tumors offers an interesting tool for diagnosis or treatment [259]. TRPM8 has emerged as a putative therapeutic target for prostate cancer, as TRPM8 localization and activity are regulated depending on the differentiation and oncogenic status of prostate cancer cells [260]. Expressed in the plasma membrane, TRPM8 seems important for Ca^{2+} signaling involved in proliferation through a short and repeated Ca^{2+} entry. In contrast, activation of TRPM8 in the endoplasmic reticulum, acting as a Ca^{2+} release channel [261, 262], promotes Ca^{2+} stored depletion inducing growth arrest and apoptosis in prostate epithelial cells in advanced prostate cancer [260, 263]. Similarly, in melanoma cells, activation of TRPM8 induces a transient and sustained rise in intracellular free Ca^{2+} concentration that reduces cell viability [264, 265], while in neuroendocrine pancreatic tumor cells Ca^{2+} increase stimulates neurotensin secretion [266]. TRPM8 activation stimulates Ca^{2+} influx membrane current, and migration of human glioblastoma cells [267], but inhibites cell motility in pancreatic ductal adenocarcinoma [268]. Nowadays, an oral active TRPM8 agonist is on clinical evaluation for the treatment of refractory solid tumors (D-3263, Dendreon). The hypothesis is that TRPM8 activation will disrupt Ca^{2+} and Na^+ homeostasis allowing entry of both ions, which will specifically induce cell death only in TRPM8- expressing neurons.

For a long time, cooling or topical application of menthol has been used for their analgesic effects. This evidence, together with TRPM8 presence in sensory neurons [269] and fibers innervating the skin, mucosa or visceral organs, suggests a potential role of TRPM8 in thermal sensing and nociception [9, 246, 270–272]. In contrast to the pro-algesic activity of other thermoTRPs, TRPM8 mediates analgesia in neuropathic and inflammatory pain models. TRPM8 is a major integrator of cold allodynia, and it is required for behavioral responses to injury-evoked cold hypersensitivity, cooling-mediated analgesia, and oxaliplatin-induced cold pain [273–276].

Recent observations of TRPM8 expression in vagal neurons innervating bronchopulmonary tissue have brought up TRPM8 as a drug target for various respiratory disorders. Other diseases where TRPM8 is involved include dry-eye-syndrome, amyloidotic polyneuropathy, and diseases of the urogenital tract like overactive bladder syndrome and pain bladder syndrome [277–280].

12.3.3 TRPC5

TRPC5 is principally a receptor-operated channel depending on the activation of Gq, phospholipase C and/or Gi receptor to gate the channel. Nonetheless, it is also directly activated by Ca^{2+} [51], nitric oxide [281] and lysophospholipids [282]. Its role in temperature sensing involves channel activation at innocuous temperatures ranging from 25 to 37 °C [36]. The cytosolic C-terminus contains interaction sites for Ca^{2+} binding proteins that notably modulate channels function [52]. CaM can interact with TRPC5 at different sites [283]. One of these sites, namely CIRB, is additionally targeted by IP_3 receptors that are proposed to displace CaM, therefore releasing the inhibition [283]. Furthermore, the C-terminus can also accommodate Ca^{2+} binding protein 1 that inhibits intracellular Ca^{2+} activation of the channel [284]. In addition, TRPC5 is phosphorylated by PKA (Gs/cAMP/PKA) at residues Ser794, Ser796 resulting in an inhibition of the currents through the channel. Mechanistically, this phosphorylation in the PDZ domain may regulate functionally relevant interactions in a Ca^{2+} dependent manner [285]. The channel is also regulated by PKC phosphorylation that results in desensitization of the currents [283].

TRPC5 is broadly expressed acting as an integrator of extracellular and intracellular signals at the level of Ca^{2+} entry. TRPC5 expressed as homotetramer allows Ca^{2+} entry and causes an intracellular Ca^{2+} increase underlying specific cell functions; however, in complexes with other TRPC proteins can acts as a store operated channel. TRPC5 is involved in many physiological and pathological processes [286]. For instance, TRPC5 is predominantly expressed in central nervous system, participating in brain synapsis [287]. Here, this channel is mainly present in dendrites as well as cell nuclei, suggesting a role of TRPC5 in the nuclear Ca^{2+} signaling [288, 289]. TRPC5 participates in electrogenesis in neurons with a key role in axon guidance during brain development. In this context, TRPC5 is carried to newly forming growth cones and synapses where acts as regulator of hippocampal neurite length and growth cone morphology mediating Ca^{2+} influx and signaling [290]. This receptor is also essential in dendrite regulation through Ca^{2+}/CaM interaction [291]. In mouse sensory neurons, TRPC5 expression is progressively increased during embriogenesis, and may play a role in sensory physiology [36, 292]. In the peripheral organs, TRPC5 is present in liver, heart, vasculature, adipocytes, and kidney among others [286].

12.3.4 TRPA1

Besides being activated by noxious cold temperatures (<10 °C) and Ca^{2+}, TRPA1 is gated by a vast and ever-growing array of compounds among which irritants take the lead. Mustard oil, cinamaldehyde, allylisothiocyanate and icilin are classical TRPA1 activators [293]. Most of them share an electrophilic nature and activate the channel by covalently modifying Cys residues in the N-terminus [109, 294]. Noteworthy, the precise temperature of TRPA1 activation is a hot topic since several findings indicate that it is organism dependent. For instance, while mouse TRPA1 responds to low temperatures, TRPA1 from snakes and Drosophila detects warmth [295, 296]. Initial findings in human TRPA1 was not gated by cold [297]. However, it has been recently shown that human TRPA1 is intrinsically cold sensitive [298].

The activity of TRPA1 is potentiated by PLC and PKA pathways [299, 300]. Interestingly, the response of TRPA1 is also regulated by Ca^{2+}. Findings in the TRPA1^{Glu918Ala} mutant, whose Ca^{2+} permeability is reduced, indicates that the Ca^{2+} entry through the channel initially potentiates the currents and secondly mediates its inactivation [301]. Similar to TRPV1, TRPA1 activity is also potentiated by the recruitment of new channels to the cell membrane through vesicle fusion. The receptor increase on the cell surface requires Ca^{2+} influx through TRPA1 [302], and is at least partially dependent on SNARE-mediated vesicle-fusion [303]. Intriguingly, in sensory neurons TRPA1 is desensitized by the increase in intracellular Ca^{2+} through TRPV1 [304] pointing to a regulatory crosstalk among TRP channels.

TRPA1-positive C-fibers densely innervate the skin, airways and gastrointestinal tract. This location, along with the robust activation of TRPA1 by a long list of exogenous environmental compounds, chemicals as well as endogenous inflammatory mediators, has implicated TRPA1 in pain perception [305, 306]. Pharmacological as well as knockout studies have evidenced the functional role of TRPA1 as chemo-, mechano- and cold sensor in the peripheral nervous system [294]. Although with still some controversy [299], TRPA1 participates in cold allodynia induced by inflammation, nerve injury, diabetic neuropathy or chemotherapy-induced neuropathies; however, it seems that cold responses are independent of Ca^{2+} [307, 308], although Ca^{2+} is an important regulator of TRPA1-mediated responses to cold. The essential role of TRPA1 in pain perception is evidenced by the TRPA1 associated channelopathy. A gain-of-function mutation in TRPA1 (Asn885Ser) is related to an autosomal dominant familial episodic pain syndrome [309]. This mutation renders the channel hyperresponsive to stimuli altering its voltage dependence. Patients are in general healthy but suffer hyperalgesia upon application of TRPA1 agonists or cold. Hence, TRPA1 has become a potential pharmacological target for the treatment of chronic pain [310].

TRPA1 is expressed in both peptidergic and non-peptidergic sensory neurons [311, 312], and mostly found, but not exclusively, in a subpopulation of TRPV1-positive neurons. In addition to pain transduction, Ca^{2+}-entry through TRPA1 causes the release of the pro-inflammatory neuropeptides αCGRP and/or SP from sensory neurons, which enhances neurogenic inflammation, pain signaling, cerebral blood flow [313], cutaneous vasodilatation [314], or promotes and maintains colitis in mice [315]. For instance, TRPA1 acts as a primary vascular cold sensor, through an initially vasoconstriction followed by a subsequent restorative blood flow dependent on αCGRP release [316]. Interestingly, TRPA1 activators are well-known environmental migraine triggers able to induce the release of the pro-migraine peptide αCGRP [305, 317].

The functional relevance of TRPA1 in the central nervous system is still not deeply understood. Recent findings further support the role of TRPA1 on mechanical allodynia central sensitization in the spinal cord [318–320], but also disrupts spinal nociceptive transmission through a Ca^{2+}-dependent mechanism [321]. In astrocytes, TRPA1 channel-mediated fluxes contribute to basal Ca^{2+} levels and neuronal function via constitutive D-serine release [322].

In non-neuronal tissues, TRPA1 seems to trigger airway inflammation as part of the lung defense, but overstimulation leads to pathological states. TRPA1 participates in cough, bronchoconstriction, asthma or chronic obstructive pulmonary diseases [323–325]. Exposure to environmental and occupational respiratory irritants activates TRPA1 on vagal sensory afferents contributing to the pathophysiology of respiratory diseases [326]. Notably, exogenous activators of TRPA1 are able to promote cell survival of small cell lung cancer cells [327]. Therefore, TRPA1 antagonists are actively pursued as therapeutic agents for airway diseases.

Finally, in skin, TRPA1 is also involved in non-histaminergic itch [328], and participates in the integration of immune and neuronal mechanisms leading to chronic inflammatory responses and pruritus associated with contact dermatitis [329].

12.4 Conclusions

ThermoTRP channels enclose a group of structural related receptors activated by different temperatures with a wide range of physiological functions; however, in all of them, directly or indirectly, Ca^{2+} is a fundamental element on their function. Most of the thermoTRP channels permeate Ca^{2+} upon activation leading to different cellular responses such as action potentials, proliferation or cell migration, among others. Additionally, an increase of cytosolic Ca^{2+} through some thermoTRP promotes Ca^{2+}-dependent exocytosis allowing the release of mediators as well as trafficking of new thermoTRP receptors to the plasma membrane. Furthermore, Ca^{2+} also plays a critical role on the regulation of thermoTRP gating through a direct binding or indirectly through Ca^{2+}-dependent proteins or signaling pathways. However, while some of these receptors are desensitized by Ca^{2+} or Ca^{2+} signaling others are potentiated. The contribution of each thermoTRP channel to Ca^{2+} homeostasis and its physiological consequence is completely dependent on the subtype of receptor and the cellular context where it is expressed. The essential role of thermoTRP in maintenance of cellular homeostasis is evidenced by the dramatic consequences observed in patients suffering thermoTRP-associated channelopathies. Moreover, these receptors are also involved in the maintenance and development of numerous pathological conditions such as chronic pain, inflammation, cancer, diabetes, or cardiac disorders. In both cases, there is an unbalance of the cytosolic Ca^{2+} levels resulting in numerous and diverse pathological symptoms. Noticeable, in the last years, there is an increasing progress carried out to fully understand pathophysiological modulation of thermoTRP channels by Ca^{2+} and the implication of these receptors on Ca^{2+} homeostasis. Notwithstanding, although further research is still required, novel therapies are currently under evaluation targeting thermoTRPs with the aim to control their dysfunction.

Acknowledgments This work was supported by MINECO (Grants BFU2012-39092-C02-01 and CONSOLIDER-INGENIO 2010 CSD2008-00005) and by GVA (Grant PROMETEO/2014/011).

References

1. Nilius B, Flockerzi V (2014) Mammalian transient receptor potential (TRP) cation channels. Handbook of experimental pharmacology, vol 222. Springer, Berlin
2. Nilius B, Owsianik G, Voets T, Peters JA (2007) Transient receptor potential cation channels in disease. Physiol Rev 87(1):165–217
3. Venkatachalam K, Montell C (2007) TRP channels. Annu Rev Biochem 76:387–417
4. Baez D, Raddatz N, Ferreira G, Gonzalez C, Latorre R (2014) Gating of thermally activated channels. Curr Top Membr 74:51–87. doi:10.1016/B978-0-12-800181-3.00003-8
5. Brederson JD, Kym PR, Szallasi A (2013) Targeting TRP channels for pain relief. Eur J Pharmacol 716(1–3):61–76
6. Latorre R, Brauchi S, Orta G, Zaelzer C, Vargas G (2007) ThermoTRP channels as modular proteins with allosteric gating. Cell Calcium 42(4–5):427–438
7. Schepers RJ, Ringkamp M (2010) Thermoreceptors and thermosensitive afferents. Neurosci Biobehav Rev 34(2):177–184
8. Fernandez-Carvajal A, Fernandez-Ballester G, Devesa I, Gonzalez-Ros JM, Ferrer-Montiel A (2011) New strategies to develop novel pain therapies: addressing thermoreceptors from different points of view. Pharm (Basel) 5(1):16–48
9. Ferrer-Montiel A, Fernandez-Carvajal A, Planells-Cases R, Fernandez-Ballester G, Gonzalez-Ros JM, Messeguer A, Gonzalez-Muniz R (2012) Advances in modulating thermosensory TRP channels. Expert Opin Ther Pathol 22(9):999–1017
10. Cvetkov TL, Huynh KW, Cohen MR, Moiseenkova-Bell VY (2011) Molecular architecture and subunit organization of TRPA1 ion channel revealed by electron microscopy. J Biol Chem 286(44):38168–38176
11. Huynh KW, Cohen MR, Chakrapani S, Holdaway HA, Stewart PL, Moiseenkova-Bell VY (2014) Structural insight into the assembly of TRPV channels. Structure 22(2):260–268
12. Liao M, Cao E, Julius D, Cheng Y (2013) Structure of the TRPV1 ion channel determined by electron cryo-microscopy. Nature 504(7478):107–112
13. Moiseenkova-Bell VY, Stanciu LA, Serysheva II, Tobe BJ, Wensel TG (2008) Structure of TRPV1 channel revealed by electron cryomicroscopy. Proc Natl Acad Sci U S A 105(21):7451–7455
14. Cao E, Liao M, Cheng Y, Julius D (2013) TRPV1 structures in distinct conformations reveal activation mechanisms. Nature 504:113–118
15. Phelps CB, Huang RJ, Lishko PV, Wang RR, Gaudet R (2008) Structural analyses of the ankyrin repeat domain of TRPV6 and related TRPV ion channels. Biochemistry 47(8):2476–2484
16. Garcia-Sanz N, Fernandez-Carvajal A, Morenilla-Palao C, Planells-Cases R, Fajardo-Sanchez E, Fernandez-Ballester G, Ferrer-Montiel A (2004) Identification of a tetramerization domain in the C terminus of the vanilloid receptor. J Neurosci 24(23):5307–5314
17. Garcia-Sanz N, Valente P, Gomis A, Fernandez-Carvajal A, Fernandez-Ballester G, Viana F, Belmonte C, Ferrer-Montiel A (2007) A role of the transient receptor potential domain of vanilloid receptor I in channel gating. J Neurosci 27(43):11641–11650
18. Valente P, Garcia-Sanz N, Gomis A, Fernandez-Carvajal A, Fernandez-Ballester G, Viana F, Belmonte C, Ferrer-Montiel A (2008) Identification of molecular determinants of channel gating in the transient receptor potential box of vanilloid receptor I. FASEB J 22:3298–3309
19. Gregorio-Teruel L, Valente P, González-Ros JM, Fernández-Ballester G, Ferrer-Montiel A (2014) Mutation of I696 and W697 in the TRP box of vanilloid receptor subtype I modulates allosteric channel activation. J Gen Physiol 143:361–375
20. Taberner FJ, Lopez-Cordoba A, Fernandez-Ballester G, Korchev Y, Ferrer-Montiel A (2014) The region adjacent to the C-end of the inner gate in transient receptor potential melastatin 8 (TRPM8) channels plays a central role in allosteric channel activation. J Biol Chem 289(41):28579–28594

21. Nilius B, Vennekens R, Prenen J, Hoenderop JGJ, Droogmans G, Bindels RJM (2001) The single pore residue Asp542 determines Ca2+ permeation and Mg2+ block of the epithelial Ca2+ channel. J Biol Chem 276:1020–1025
22. Talavera K, Staes M, Janssens A, Klugbauer N, Droogmans G, Hofmann F, Nilius B (2001) Aspartate residues of the Glu-Glu-Asp-Asp (EEDD) pore locus control selectivity and permeation of the T-type Ca2+channel α1G. J Biol Chem 276:45628–45635
23. Tang L, Gamal El-Din TM, Payandeh J, Martinez GQ, Heard TM, Scheuer T, Zheng N, Catterall WA (2013) Structural basis for Ca2+ selectivity of a voltage-gated calcium channel. Nat Adv Online Publ 505(7481):56
24. Yang J, Elllnor PT, Sather WA, Zhang J-F, Tsien RW (1993) Molecular determinants of Ca2+ selectivity and ion permeation in L-type Ca2+ channels. Nature 366:158–161
25. Nilius B, Prenen J, Janssens A, Owsianik G, Wang C, Zhu MX, Voets T (2005) The selectivity filter of the cation channel TRPM4. J Biol Chem 280:22899–22906
26. Mederos y Schnitzler M, Wäring J, Gudermann T, Chubanov V (2008) Evolutionary determinants of divergent calcium selectivity of TRPM channels. FASEB J 22:1540–1551
27. García-Martínez C, Morenilla-Palao C, Planells-Cases R, Merino JM, Ferrer-Montiel A (2000) Identification of an aspartic residue in the P-loop of the vanilloid receptor that modulates pore properties. J Biol Chem 275:32552–32558
28. Voets T, Prenen J, Vriens J, Watanabe H, Janssens A, Wissenbach U, Bödding M, Droogmans G, Nilius B (2002) Molecular determinants of permeation through the cation channel TRPV4. J Biol Chem 277:33704–33710
29. Xia R, Mei Z-Z, Mao H-J, Yang W, Dong L, Bradley H, Beech DJ, Jiang L-H (2008) Identification of pore residues engaged in determining divalent cationic permeation in transient receptor potential melastatin subtype channel 2. J Biol Chem 283:27426–27432
30. Darré L, Furini S, Domene C (2015) Permeation and dynamics of an open-activated TRPV1 channel. J Mol Biol 427:537–549
31. Chung M-K, Güler AD, Caterina MJ (2008) TRPV1 shows dynamic ionic selectivity during agonist stimulation. Nat Neurosci 11:555–564
32. Banke TG, Chaplan SR, Wickenden AD (2010) Dynamic changes in the TRPA1 selectivity filter lead to progressive but reversible pore dilation. Am J Physiol Cell Physiol 298:C1457–C1468
33. Karashima Y, Prenen J, Talavera K, Janssens A, Voets T, Nilius B (2010) Agonist-induced changes in Ca2+ permeation through the nociceptor cation channel TRPA1. Biophys J 98:773–783
34. Chen J, Kim D, Bianchi BR, Cavanaugh EJ, Faltynek CR, Kym PR, Reilly RM (2009) Pore dilation occurs in TRPA1 but not in TRPM8 channels. Mol Pain 5:3
35. Veldhuis NA, Lew MJ, Abogadie FC, Poole DP, Jennings EA, Ivanusic JJ, Eilers H, Bunnett NW, McIntyre P (2012) N-glycosylation determines ionic permeability and desensitization of the TRPV1 capsaicin receptor. J Biol Chem 287:21765–21772
36. Zimmermann K, Lennerz JK, Hein A, Link AS, Kaczmarek JS, Delling M, Uysal S, Pfeifer JD, Riccio A, Clapham DE (2011) Transient receptor potential cation channel, subfamily C, member 5 (TRPC5) is a cold-transducer in the peripheral nervous system. Proc Natl Acad Sci U S A 108(44):18114–18119
37. Schilling WP, Goel M (2004) Mammalian TRPC channel subunit assembly. Novartis Found Symp 258:18–30; discussion 30–43, 98–102, 263–266
38. Strübing C, Krapivinsky G, Krapivinsky L, Clapham DE (2001) TRPC1 and TRPC5 form a novel cation channel in mammalian brain. Neuron 29:645–655
39. Strübing C, Krapivinsky G, Krapivinsky L, Clapham DE (2003) Formation of novel TRPC channels by complex subunit interactions in embryonic brain. J Biol Chem 278:39014–39019
40. Cheng H, Beck A, Launay P, Gross SA, Stokes AJ, Kinet JP, Fleig A, Penner R (2007) TRPM4 controls insulin secretion in pancreatic beta-cells. Cell Calcium 41(1):51–61

41. Hellwig N, Albrecht N, Harteneck C, Schultz G, Schaefer M (2005) Homo- and heteromeric assembly of TRPV channel subunits. J Cell Sci 118:917–928
42. Fischer MJM, Balasuriya D, Jeggle P, Goetze TA, McNaughton PA, Reeh PW, Edwardson JM (2014) Direct evidence for functional TRPV1/TRPA1 heteromers. Pflugers Arch – Eur J Physiol 466:2229–2241
43. Köttgen M, Buchholz B, Garcia-Gonzalez MA, Kotsis F, Fu X, Doerken M, Boehlke C, Steffl D, Tauber R, Wegierski T, Nitschke R, Suzuki M, Kramer-Zucker A, Germino GG, Watnick T, Prenen J, Nilius B, Kuehn EW, Walz G (2008) TRPP2 and TRPV4 form a polymodal sensory channel complex. J Cell Biol 182:437–447
44. Ma X, Nilius B, Wong JW-Y, Huang Y, Yao X (2011) Electrophysiological properties of heteromeric TRPV4–C1 channels. Biochim Biophys Acta Biomembr 1808:2789–2797
45. Ma X, Cheng K-T, Wong C-O, O'Neil RG, Birnbaumer L, Ambudkar IS, Yao X (2011) Heteromeric TRPV4-C1 channels contribute to store-operated Ca2+ entry in vascular endothelial cells. Cell Calcium 50:502–509
46. Stewart AP, Smith GD, Sandford RN, Edwardson JM (2010) Atomic force microscopy reveals the alternating subunit arrangement of the TRPP2-TRPV4 heterotetramer. Biophys J 99:790–797
47. Launay P, Fleig A, Perraud A-L, Scharenberg AM, Penner R, Kinet J-P (2002) TRPM4 is a Ca2+-activated nonselective cation channel mediating cell membrane depolarization. Cell 109:397–407
48. Hofmann T, Chubanov V, Gudermann T, Montell C (2003) TRPM5 is a voltage-modulated and Ca2+-activated monovalent selective cation channel. Curr Biol 13:1153–1158
49. Nilius B, Prenen J, Tang J, Wang C, Owsianik G, Janssens A, Voets T, Zhu MX (2005) Regulation of the Ca2+ sensitivity of the nonselective cation channel TRPM4. J Biol Chem 280:6423–6433
50. Yamaguchi S, Tanimoto A, K-i O, Hibino H, Ito S (2014) Negatively charged amino acids near and in transient receptor potential (TRP) domain of TRPM4 channel are one determinant of its Ca2+ sensitivity. J Biol Chem 289:35265–35282
51. Gross SA, Guzmán GA, Wissenbach U, Philipp SE, Zhu MX, Bruns D, Cavalié A (2009) TRPC5 is a Ca2+-activated channel functionally coupled to Ca2+-selective ion channels. J Biol Chem 284:34423–34432
52. Zholos AV (2014) TRPC5. In: Nilius B, Flockerzi V (eds) Mammalian transient receptor potential (TRP) cation channels. Handbook of experimental pharmacology. Springer, Berlin, pp 129–156
53. Mahieu F, Janssens A, Gees M, Talavera K, Nilius B, Voets T (2010) Modulation of the cold-activated cation channel TRPM8 by surface charge screening. J Physiol 588(Pt 2):315–324
54. Doerner JF, Hatt H, Ramsey IS (2011) Voltage- and temperature-dependent activation of TRPV3 channels is potentiated by receptor-mediated PI(4,5)P2 hydrolysis. J Gen Physiol 137:271–288
55. Zurborg S, Yurgionas B, Jira JA, Caspani O, Heppenstall PA (2007) Direct activation of the ion channel TRPA1 by Ca2+. Nat Neurosci 10:277–279
56. Nilius B, Voets T, Peters J (2005) TRP channels in disease. Sci STKE 2005(295):re8
57. O'Neil RG, Brown RC (2003) The vanilloid receptor family of calcium-permeable channels: molecular integrators of microenvironmental stimuli. News Physiol Sci 18:226–231
58. Benham CD, Davis JB, Randall AD (2002) Vanilloid and TRP channels: a family of lipid-gated cation channels. Neuropharmacology 42(7):873–888
59. Caterina MJ, Schumacher MA, Tominaga M, Rosen TA, Levine JD, Julius D (1997) The capsaicin receptor: a heat-activated ion channel in the pain pathway. Nature 389(6653):816–824
60. McNamara FN, Randall A, Gunthorpe MJ (2005) Effects of piperine, the pungent component of black pepper, at the human vanilloid receptor (TRPV1). Br J Pharmacol 144:781–790

61. Akopian AN, Ruparel NB, Jeske NA, Hargreaves KM (2007) Transient receptor potential TRPA1 channel desensitization in sensory neurons is agonist dependent and regulated by TRPV1-directed internalization. J Physiol 583:175–193
62. Tominaga M, Caterina MJ, Malmberg AB, Rosen TA, Gilbert H, Skinner K, Raumann BE, Basbaum AI, Julius D (1998) The cloned capsaicin receptor integrates multiple pain-producing stimuli. Neuron 21(3):531–543
63. Bevan S, Hothi S, Hughes G, James IF, Rang HP, Shah K, Walpole CS, Yeats JC (1992) Capsazepine: a competitive antagonist of the sensory neurone excitant capsaicin. Br J Pharmacol 107:544–552
64. Valente P, Fernández-Carvajal A, Camprubí-Robles M, Gomis A, Quirce S, Viana F, Fernández-Ballester G, González-Ros JM, Belmonte C, Planells-Cases R, Ferrer-Montiel A (2011) Membrane-tethered peptides patterned after the TRP domain (TRPducins) selectively inhibit TRPV1 channel activity. FASEB J 25:1628–1640
65. Valenzano KJ, Grant ER, Wu G, Hachicha M, Schmid L, Tafesse L, Sun Q, Rotshteyn Y, Francis J, Limberis J, Malik S, Whittemore ER, Hodges D (2003) N-(4-tertiarybutylphenyl)-4-(3-chloropyridin-2-yl)tetrahydropyrazine -1(2H)-carbox-amide (BCTC), a novel, orally effective vanilloid receptor 1 antagonist with analgesic properties: I. In vitro characterization and pharmacokinetic properties. J Pharmacol Exp Ther 306:377–386
66. Gavva NR, Klionsky L, Qu Y, Shi L, Tamir R, Edenson S, Zhang TJ, Viswanadhan VN, Toth A, Pearce LV, Vanderah TW, Porreca F, Blumberg PM, Lile J, Sun Y, Wild K, Louis J-C, Treanor JJS (2004) Molecular determinants of vanilloid sensitivity in TRPV1. J Biol Chem 279:20283–20295
67. Jordt S-E, Julius D (2002) Molecular basis for species-specific sensitivity to "hot" chili peppers. Cell 108:421–430
68. Yao J, Liu B, Qin F (2011) Modular thermal sensors in temperature-gated transient receptor potential (TRP) channels. Proc Natl Acad Sci 108:11109–11114
69. Myers BR, Bohlen CJ, Julius D (2008) A yeast genetic screen reveals a critical role for the pore helix domain in TRP channel gating. Neuron 58(3):362–373
70. Kim SE, Patapoutian A, Grandl J (2013) Single residues in the outer pore of TRPV1 and TRPV3 have temperature-dependent conformations. PLoS One 8(3):e59593
71. Brauchi S, Orta G, Salazar M, Rosenmann E, Latorre R (2006) A hot-sensing cold receptor: C-terminal domain determines thermosensation in transient receptor potential channels. J Neurosci 26:4835–4840
72. Docherty RJ, Yeats JC, Bevan S, Boddeke HW (1996) Inhibition of calcineurin inhibits the desensitization of capsaicin-evoked currents in cultured dorsal root ganglion neurones from adult rats. Pflugers Arch – Eur J Physiol 431(6):828–837
73. Koplas PA, Rosenberg RL, Oxford GS (1997) The role of calcium in the desensitization of capsaicin responses in rat dorsal root ganglion neurons. J Neurosci 17(10):3525–3537
74. Lukacs V, Rives J-M, Sun X, Zakharian E, Rohacs T (2013) Promiscuous activation of transient receptor potential vanilloid 1 (TRPV1) channels by negatively charged intracellular lipids. The key role of endogenous phosphoinositides in maintaining channel activity. J Biol Chem 288:35003–35013
75. Devesa I, Planells-Cases R, Fernandez-Ballester G, Gonzalez-Ros JM, Ferrer-Montiel A, Fernandez-Carvajal A (2011) Role of the transient receptor potential vanilloid 1 in inflammation and sepsis. J Inflamm Res 4:67–81
76. Planells-Cases R, Valente P, Ferrer-Montiel A, Qin F, Szallasi A (2011) Complex regulation of TRPV1 and related thermo-TRPs: implications for therapeutic intervention. Adv Exp Med Biol 704:491–515
77. Camprubi-Robles M, Planells-Cases R, Ferrer-Montiel A (2009) Differential contribution of SNARE-dependent exocytosis to inflammatory potentiation of TRPV1 in nociceptors. FASEB J 23(11):3722–3733
78. Morenilla-Palao C, Planells-Cases R, Garcia-Sanz N, Ferrer-Montiel A (2004) Regulated exocytosis contributes to protein kinase C potentiation of vanilloid receptor activity. J Biol Chem 279(24):25665–25672

79. Price TJ, Flores CM (2007) Critical evaluation of the colocalization between calcitonin gene-related peptide, substance P, transient receptor potential vanilloid subfamily type 1 immunoreactivities, and isolectin B4 binding in primary afferent neurons of the rat and mouse. J Pain 8(3):263–272
80. Devesa I, Ferrandiz-Huertas C, Mathivanan S, Wolf C, Lujan R, Changeux JP, Ferrer-Montiel A (2014) alphaCGRP is essential for algesic exocytotic mobilization of TRPV1 channels in peptidergic nociceptors. Proc Natl Acad Sci U S A 111(51):18345–18350
81. Barton NJ, McQueen DS, Thomson D, Gauldie SD, Wilson AW, Salter DM, Chessell IP (2006) Attenuation of experimental arthritis in TRPV1R knockout mice. Exp Mol Pathol 81(2):166–170
82. Keeble J, Russell F, Curtis B, Starr A, Pinter E, Brain SD (2005) Involvement of transient receptor potential vanilloid 1 in the vascular and hyperalgesic components of joint inflammation. Arthritis Rheum 52(10):3248–3256
83. Engler A, Aeschlimann A, Simmen BR, Michel BA, Gay RE, Gay S, Sprott H (2007) Expression of transient receptor potential vanilloid 1 (TRPV1) in synovial fibroblasts from patients with osteoarthritis and rheumatoid arthritis. Biochem Biophys Res Commun 359(4):884–888
84. Menendez L, Juarez L, Garcia E, Garcia-Suarez O, Hidalgo A, Baamonde A (2006) Analgesic effects of capsazepine and resiniferatoxin on bone cancer pain in mice. Neurosci Lett 393(1):70–73
85. Hong S, Wiley JW (2005) Early painful diabetic neuropathy is associated with differential changes in the expression and function of vanilloid receptor 1. J Biol Chem 280(1):618–627
86. Jhaveri MD, Elmes SJ, Kendall DA, Chapman V (2005) Inhibition of peripheral vanilloid TRPV1 receptors reduces noxious heat-evoked responses of dorsal horn neurons in naive, carrageenan-inflamed and neuropathic rats. Eur J Neurosci 22(2):361–370
87. Richardson JD, Vasko MR (2002) Cellular mechanisms of neurogenic inflammation. J Pharmacol Exp Ther 302(3):839–845
88. Edwards JG (2014) TRPV1 in the central nervous system: synaptic plasticity, function, and pharmacological implications. Prog Drug Res 68:77–104
89. Maione S, Cristino L, Migliozzi AL, Georgiou AL, Starowicz K, Salt TE, Di MV (2009) TRPV1 channels control synaptic plasticity in the developing superior colliculus. J Physiol 587(Pt 11):2521–2535
90. Matta JA, Ahern GP (2011) TRPV1 and synaptic transmission. Curr Pharm Biotechnol 12(1):95–101
91. Song J, Lee JH, Lee SH, Park KA, Lee WT, Lee JE (2013) TRPV1 activation in primary cortical neurons induces calcium-dependent programmed cell death. Exp Neurobiol 22(1):51–57
92. Shirakawa H, Yamaoka T, Sanpei K, Sasaoka H, Nakagawa T, Kaneko S (2008) TRPV1 stimulation triggers apoptotic cell death of rat cortical neurons. Biochem Biophys Res Commun 377(4):1211–1215
93. Ghazizadeh V, Naziroglu M (2014) Electromagnetic radiation (Wi-Fi) and epilepsy induce calcium entry and apoptosis through activation of TRPV1 channel in hippocampus and dorsal root ganglion of rats. Metab Brain Dis 29(3):787–799
94. Ovey IS, Naziroglu M (2015) Homocysteine and cytosolic GSH depletion induce apoptosis and oxidative toxicity through cytosolic calcium overload in the hippocampus of aged mice: involvement of TRPM2 and TRPV1 channels. Neuroscience 284:225–233
95. Starowicz K, Cristino L, Di MV (2008) TRPV1 receptors in the central nervous system: potential for previously unforeseen therapeutic applications. Curr Pharm Des 14(1):42–54
96. Alawi K, Keeble J (2010) The paradoxical role of the transient receptor potential vanilloid 1 receptor in inflammation. Pharmacol Ther 125(2):181–195
97. Fernandes ES, Fernandes MA, Keeble JE (2012) The functions of TRPA1 and TRPV1: moving away from sensory nerves. Br J Pharmacol 166(2):510–521

98. Denda M, Tsutsumi M (2011) Roles of transient receptor potential proteins (TRPs) in epidermal keratinocytes. Adv Exp Med Biol 704:847–860
99. Miyazaki A, Ohkubo T, Hatta M, Ishikawa H, Yamazaki J (2015) Integrin alpha6beta4 and TRPV1 channel coordinately regulate directional keratinocyte migration. Biochem Biophys Res Commun 458(1):161–167
100. Lee YM, Kang SM, Chung JH (2012) The role of TRPV1 channel in aged human skin. J Dermatol Sci 65(2):81–85
101. Li S, Bode AM, Zhu F, Liu K, Zhang J, Kim MO, Reddy K, Zykova T, Ma WY, Carper AL, Langfald AK, Dong Z (2011) TRPV1-antagonist AMG9810 promotes mouse skin tumorigenesis through EGFR/Akt signaling. Carcinogenesis 32(5):779–785
102. Lieben L, Carmeliet G (2012) The involvement of TRP channels in bone homeostasis. Front Endocrinol 3:99
103. Idris AI, Landao-Bassonga E, Ralston SH (2010) The TRPV1 ion channel antagonist capsazepine inhibits osteoclast and osteoblast differentiation in vitro and ovariectomy induced bone loss in vivo. Bone 46(4):1089–1099
104. Rossi F, Bellini G, Torella M, Tortora C, Manzo I, Giordano C, Guida F, Luongo L, Papale F, Rosso F, Nobili B, Maione S (2014) The genetic ablation or pharmacological inhibition of TRPV1 signalling is beneficial for the restoration of quiescent osteoclast activity in ovariectomized mice. Br J Pharmacol 171(10):2621–2630
105. Park U, Vastani N, Guan Y, Raja SN, Koltzenburg M, Caterina MJ (2011) TRP vanilloid 2 knock-out mice are susceptible to perinatal lethality but display normal thermal and mechanical nociception. J Neurosci 31(32):11425–11436
106. Mercado J, Gordon-Shaag A, Zagotta WN, Gordon SE (2010) Ca2+-dependent desensitization of TRPV2 channels is mediated by hydrolysis of phosphatidylinositol 4,5-bisphosphate. J Neurosci 30:13338–13347
107. Gordon-Shaag A, Zagotta WN, Gordon SE (2008) Mechanism of Ca2+-dependent desensitization in TRP channels. Channels 2:125–129
108. Kanzaki M, Zhang YQ, Mashima H, Li L, Shibata H, Kojima I (1999) Translocation of a calcium-permeable cation channel induced by insulin-like growth factor-I. Nat Cell Biol 1:165–170
109. Cohen MR, Huynh KW, Cawley D, Moiseenkova-Bell VY (2013) Understanding the cellular function of TRPV2 channel through generation of specific monoclonal antibodies. PLoS One 8:e85392
110. Qin N, Neeper MP, Liu Y, Hutchinson TL, Lubin ML, Flores CM (2008) TRPV2 is activated by cannabidiol and mediates CGRP release in cultured rat dorsal root ganglion neurons. J Neurosci 28(24):6231–6238
111. Shibasaki K, Murayama N, Ono K, Ishizaki Y, Tominaga M (2010) TRPV2 enhances axon outgrowth through its activation by membrane stretch in developing sensory and motor neurons. J Neurosci 30(13):4601–4612
112. Mihara H, Boudaka A, Shibasaki K, Yamanaka A, Sugiyama T, Tominaga M (2010) Involvement of TRPV2 activation in intestinal movement through nitric oxide production in mice. J Neurosci 30(49):16536–16544
113. Aoyagi K, Ohara-Imaizumi M, Nishiwaki C, Nakamichi Y, Nagamatsu S (2010) Insulin/phosphoinositide 3-kinase pathway accelerates the glucose-induced first-phase insulin secretion through TrpV2 recruitment in pancreatic beta-cells. Biochem J 432(2):375–386
114. Hisanaga E, Nagasawa M, Ueki K, Kulkarni RN, Mori M, Kojima I (2009) Regulation of calcium-permeable TRPV2 channel by insulin in pancreatic beta-cells. Diabetes 58(1):174–184
115. Pottosin I, Delgado-Enciso I, Bonales-Alatorre E, Nieto-Pescador MG, Moreno-Galindo EG, Dobrovinskaya O (2015) Mechanosensitive Ca(2)(+)-permeable channels in human leukemic cells: pharmacological and molecular evidence for TRPV2. Biochim Biophys Acta 1848(1 Pt A):51–59

116. Schwarz EC, Wolfs MJ, Tonner S, Wenning AS, Quintana A, Griesemer D, Hoth M (2007) TRP channels in lymphocytes. Handb Exp Pharmacol 179:445–456
117. Santoni G, Farfariello V, Liberati S, Morelli MB, Nabissi M, Santoni M, Amantini C (2013) The role of transient receptor potential vanilloid type-2 ion channels in innate and adaptive immune responses. Front Immunol 4:34
118. Zhang D, Spielmann A, Wang L, Ding G, Huang F, Gu Q, Schwarz W (2012) Mast-cell degranulation induced by physical stimuli involves the activation of transient-receptor-potential channel TRPV2. Physiol Res 61(1):113–124
119. Stokes AJ, Shimoda LM, Koblan-Huberson M, Adra CN, Turner H (2004) A TRPV2-PKA signaling module for transduction of physical stimuli in mast cells. J Exp Med 200(2):137–147
120. Nagasawa M, Nakagawa Y, Tanaka S, Kojima I (2007) Chemotactic peptide fMetLeuPhe induces translocation of the TRPV2 channel in macrophages. J Cell Physiol 210(3):692–702
121. Issa CM, Hambly BD, Wang Y, Maleki S, Wang W, Fei J, Bao S (2014) TRPV2 in the development of experimental colitis. Scand J Immunol 80(5):307–312
122. Entin-Meer M, Levy R, Goryainov P, Landa N, Barshack I, Avivi C, Semo J, Keren G (2014) The transient receptor potential vanilloid 2 cation channel is abundant in macrophages accumulating at the peri-infarct zone and may enhance their migration capacity towards injured cardiomyocytes following myocardial infarction. PLoS One 9(8):e105055
123. Link TM, Park U, Vonakis BM, Raben DM, Soloski MJ, Caterina MJ (2010) TRPV2 has a pivotal role in macrophage particle binding and phagocytosis. Nat Immunol 11(3):232–239
124. Morelli MB, Liberati S, Amantini C, Nabiss M, Santoni M, Farfariello V, Santoni G (2013) Expression and function of the transient receptor potential ion channel family in the hematologic malignancies. Curr Mol Pharmacol 6(3):137–148
125. Liberati S, Morelli MB, Amantini C, Santoni M, Nabissi M, Cardinali C, Santoni G (2014) Advances in transient receptor potential vanilloid-2 channel expression and function in tumor growth and progression. Curr Protein Pept Sci 15(7):732–737
126. Morelli MB, Nabissi M, Amantini C, Farfariello V, Ricci-Vitiani L, di Martino S, Pallini R, Larocca LM, Caprodossi S, Santoni M, De Maria R, Santoni G (2012) The transient receptor potential vanilloid-2 cation channel impairs glioblastoma stem-like cell proliferation and promotes differentiation. International journal of cancer. J Int Cancer 131(7):E1067–E1077
127. Nabissi M, Morelli MB, Amantini C, Farfariello V, Ricci-Vitiani L, Caprodossi S, Arcella A, Santoni M, Giangaspero F, De Maria R, Santoni G (2010) TRPV2 channel negatively controls glioma cell proliferation and resistance to Fas-induced apoptosis in ERK-dependent manner. Carcinogenesis 31(5):794–803
128. Yamada T, Ueda T, Shibata Y, Ikegami Y, Saito M, Ishida Y, Ugawa S, Kohri K, Shimada S (2010) TRPV2 activation induces apoptotic cell death in human T24 bladder cancer cells: a potential therapeutic target for bladder cancer. Urology 76(2):509 e501–509 e507
129. Liu Q, Wang X (2013) Effect of TRPV2 cation channels on the proliferation, migration and invasion of 5637 bladder cancer cells. Exp Ther Med 6(5):1277–1282
130. Nabissi M, Morelli MB, Santoni M, Santoni G (2013) Triggering of the TRPV2 channel by cannabidiol sensitizes glioblastoma cells to cytotoxic chemotherapeutic agents. Carcinogenesis 34(1):48–57
131. Liu G, Xie C, Sun F, Xu X, Yang Y, Zhang T, Deng Y, Wang D, Huang Z, Yang L, Huang S, Wang Q, Liu G, Zhong D, Miao X (2010) Clinical significance of transient receptor potential vanilloid 2 expression in human hepatocellular carcinoma. Cancer Genet Cytogenet 197(1):54–59
132. Zhou K, Zhang SS, Yan Y, Zhao S (2014) Overexpression of transient receptor potential vanilloid 2 is associated with poor prognosis in patients with esophageal squamous cell carcinoma. Med Oncol 31(7):17
133. Monet M, Lehen'kyi V, Gackiere F, Firlej V, Vandenberghe M, Roudbaraki M, Gkika D, Pourtier A, Bidaux G, Slomianny C, Delcourt P, Rassendren F, Bergerat JP, Ceraline J, Cabon

F, Humez S, Prevarskaya N (2010) Role of cationic channel TRPV2 in promoting prostate cancer migration and progression to androgen resistance. Cancer Res 70(3):1225–1235
134. Oulidi A, Bokhobza A, Gkika D, Vanden Abeele F, Lehen'kyi V, Ouafik L, Mauroy B, Prevarskaya N (2013) TRPV2 mediates adrenomedullin stimulation of prostate and urothelial cancer cell adhesion, migration and invasion. PLoS One 8(5):e64885
135. Katanosaka Y, Iwasaki K, Ujihara Y, Takatsu S, Nishitsuji K, Kanagawa M, Sudo A, Toda T, Katanosaka K, Mohri S, Naruse K (2014) TRPV2 is critical for the maintenance of cardiac structure and function in mice. Nat Commun 5:3932
136. Koch SE, Gao X, Haar L, Jiang M, Lasko VM, Robbins N, Cai W, Brokamp C, Varma P, Tranter M, Liu Y, Ren X, Lorenz JN, Wang HS, Jones WK, Rubinstein J (2012) Probenecid: novel use as a non-injurious positive inotrope acting via cardiac TRPV2 stimulation. J Mol Cell Cardiol 53(1):134–144
137. Rubinstein J, Lasko VM, Koch SE, Singh VP, Carreira V, Robbins N, Patel AR, Jiang M, Bidwell P, Kranias EG, Jones WK, Lorenz JN (2014) Novel role of transient receptor potential vanilloid 2 in the regulation of cardiac performance. Am J Physiol Heart Circ Physiol 306(4):H574–H584
138. Iwata Y, Katanosaka Y, Arai Y, Komamura K, Miyatake K, Shigekawa M (2003) A novel mechanism of myocyte degeneration involving the Ca2+-permeable growth factor-regulated channel. J Cell Biol 161(5):957–967
139. Iwata Y, Katanosaka Y, Arai Y, Shigekawa M, Wakabayashi S (2009) Dominant-negative inhibition of Ca2+ influx via TRPV2 ameliorates muscular dystrophy in animal models. Hum Mol Genet 18(5):824–834
140. Harisseh R, Chatelier A, Magaud C, Deliot N, Constantin B (2013) Involvement of TRPV2 and SOCE in calcium influx disorder in DMD primary human myotubes with a specific contribution of alpha1-syntrophin and PLC/PKC in SOCE regulation. Am J Physiol Cell Physiol 304(9):C881–C894
141. Iwata Y, Ohtake H, Suzuki O, Matsuda J, Komamura K, Wakabayashi S (2013) Blockade of sarcolemmal TRPV2 accumulation inhibits progression of dilated cardiomyopathy. Cardiovasc Res 99(4):760–768
142. Lorin C, Vogeli I, Niggli E (2015) Dystrophic cardiomyopathy: role of TRPV2 channels in stretch-induced cell damage. Cardiovasc Res 106:153–162
143. Xu H, Ramsey IS, Kotecha SA, Moran MM, Chong JA, Lawson D, Ge P, Lilly J, Silos-Santiago I, Xie Y, DiStefano PS, Curtis R, Clapham DE (2002) TRPV3 is a calcium-permeable temperature-sensitive cation channel. Nature 418:181–186
144. Bang S, Yoo S, Yang T-J, Cho H, Hwang SW (2010) Farnesyl pyrophosphate is a novel pain-producing molecule via specific activation of TRPV3. J Biol Chem 285:19362–19371
145. Xiao R, Tang J, Wang C, Colton CK, Tian J, Zhu MX (2008) Calcium plays a central role in the sensitization of TRPV3 channel to repetitive stimulations. J Biol Chem 283:6162–6174
146. Phelps CB, Wang RR, Choo SS, Gaudet R (2010) Differential regulation of TRPV1, TRPV3, and TRPV4 sensitivity through a conserved binding site on the ankyrin repeat domain. J Biol Chem 285(1):731–740
147. Saito S, Fukuta N, Shingai R, Tominaga M (2011) Evolution of vertebrate transient receptor potential vanilloid 3 channels: opposite temperature sensitivity between mammals and western clawed frogs. PLoS Genet 7(4):e1002041
148. Peier AM, Reeve AJ, Andersson DA, Moqrich A, Earley TJ, Hergarden AC, Story GM, Colley S, Hogenesch JB, McIntyre P, Bevan S, Patapoutian A (2002) A heat-sensitive TRP channel expressed in keratinocytes. Science 296(5575):2046–2049
149. Moqrich A, Hwang SW, Earley TJ, Petrus MJ, Murray AN, Spencer KS, Andahazy M, Story GM, Patapoutian A (2005) Impaired thermosensation in mice lacking TRPV3, a heat and camphor sensor in the skin. Science 307(5714):1468–1472
150. Huang SM, Li X, Yu Y, Wang J, Caterina MJ (2011) TRPV3 and TRPV4 ion channels are not major contributors to mouse heat sensation. Mol Pain 7:37

151. Nilius B, Biro T (2013) TRPV3: a 'more than skinny' channel. Exp Dermatol 22(7):447–452
152. Nilius B, Biro T, Owsianik G (2014) TRPV3: time to decipher a poorly understood family member! J Physiol 592(Pt 2):295–304
153. Huang SM, Lee H, Chung MK, Park U, Yu YY, Bradshaw HB, Coulombe PA, Walker JM, Caterina MJ (2008) Overexpressed transient receptor potential vanilloid 3 ion channels in skin keratinocytes modulate pain sensitivity via prostaglandin E2. J Neurosci 28(51):13727–13737
154. Masamoto Y, Kawabata F, Fushiki T (2009) Intragastric administration of TRPV1, TRPV3, TRPM8, and TRPA1 agonists modulates autonomic thermoregulation in different manners in mice. Biosci Biotechnol Biochem 73(5):1021–1027
155. Cheng X, Jin J, Hu L, Shen D, Dong XP, Samie MA, Knoff J, Eisinger B, Liu ML, Huang SM, Caterina MJ, Dempsey P, Michael LE, Dlugosz AA, Andrews NC, Clapham DE, Xu H (2010) TRP channel regulates EGFR signaling in hair morphogenesis and skin barrier formation. Cell 141(2):331–343
156. Xiao R, Tian J, Tang J, Zhu MX (2008) The TRPV3 mutation associated with the hairless phenotype in rodents is constitutively active. Cell Calcium 43(4):334–343
157. Nilius B, Voets T (2013) The puzzle of TRPV4 channelopathies. EMBO Rep 14(2):152–163
158. Garcia-Elias A, Mrkonjic S, Jung C, Pardo-Pastor C, Vicente R, Valverde MA (2014) The TRPV4 channel. Handb Exp Pharmacol 222:293–319
159. Güler AD, Lee H, Iida T, Shimizu I, Tominaga M, Caterina M (2002) Heat-evoked activation of the ion channel, TRPV4. J Neurosci Off J Soc Neurosci 22:6408–6414
160. Watanabe H, Vriens J, Suh SH, Benham CD, Droogmans G, Nilius B (2002) Heat-evoked activation of TRPV4 channels in a HEK293 cell expression system and in native mouse aorta endothelial cells. J Biol Chem 277:47044–47051
161. Garcia-Elias A, Mrkonjic S, Pardo-Pastor C, Inada H, Hellmich UA, Rubio-Moscardo F, Plata C, Gaudet R, Vicente R, Valverde MA (2013) Phosphatidylinositol-4,5-biphosphate-dependent rearrangement of TRPV4 cytosolic tails enables channel activation by physiological stimuli. Proc Natl Acad Sci U S A 110(23):9553–9558
162. Vriens J, Owsianik G, Janssens A, Voets T, Nilius B (2007) Determinants of 4 alpha-phorbol sensitivity in transmembrane domains 3 and 4 of the cation channel TRPV4. J Biol Chem 282:12796–12803
163. Vriens J, Watanabe H, Janssens A, Droogmans G, Voets T, Nilius B (2004) Cell swelling, heat, and chemical agonists use distinct pathways for the activation of the cation channel TRPV4. Proc Natl Acad Sci U S A 101:396–401
164. Watanabe H, Vriens J, Janssens A, Wondergem R, Droogmans G, Nilius B (2003) Modulation of TRPV4 gating by intra- and extracellular Ca2+. Cell Calcium 33(5–6):489–495
165. Strotmann R, Semtner M, Kepura F, Plant TD, Schöneberg T (2010) Interdomain interactions control Ca2+-dependent potentiation in the cation channel TRPV4. PLoS One 5:e10580
166. Alessandri-Haber N, Dina OA, Joseph EK, Reichling D, Levine JD (2006) A transient receptor potential vanilloid 4-dependent mechanism of hyperalgesia is engaged by concerted action of inflammatory mediators. J Neurosci Off J Soc Neurosci 26:3864–3874
167. Xu F, Satoh E, Iijima T (2003) Protein kinase C-mediated Ca2+ entry in HEK 293 cells transiently expressing human TRPV4. Br J Pharmacol 140:413–421
168. Everaerts W, Nilius B, Owsianik G (2010) The vanilloid transient receptor potential channel TRPV4: from structure to disease. Prog Biophys Mol Biol 103(1):2–17
169. Dai J, Cho TJ, Unger S, Lausch E, Nishimura G, Kim OH, Superti-Furga A, Ikegawa S (2010) TRPV4-pathy, a novel channelopathy affecting diverse systems. J Hum Genet 55(7):400–402
170. Verma P, Kumar A, Goswami C (2010) TRPV4-mediated channelopathies. Channels 4(4):319–328

171. McNulty AL, Leddy HA, Liedtke W, Guilak F (2015) TRPV4 as a therapeutic target for joint diseases. Naunyn-Schmiedeberg's Arch Pharmacol 388(4):437–450
172. Krakow D, Vriens J, Camacho N, Luong P, Deixler H, Funari TL, Bacino CA, Irons MB, Holm IA, Sadler L, Okenfuss EB, Janssens A, Voets T, Rimoin DL, Lachman RS, Nilius B, Cohn DH (2009) Mutations in the gene encoding the calcium-permeable ion channel TRPV4 produce spondylometaphyseal dysplasia, Kozlowski type and metatropic dysplasia. Am J Hum Genet 84(3):307–315
173. Camacho N, Krakow D, Johnykutty S, Katzman PJ, Pepkowitz S, Vriens J, Nilius B, Boyce BF, Cohn DH (2010) Dominant TRPV4 mutations in nonlethal and lethal metatropic dysplasia. Am J Med Genet Part A 152A(5):1169–1177
174. Rock MJ, Prenen J, Funari VA, Funari TL, Merriman B, Nelson SF, Lachman RS, Wilcox WR, Reyno S, Quadrelli R, Vaglio A, Owsianik G, Janssens A, Voets T, Ikegawa S, Nagai T, Rimoin DL, Nilius B, Cohn DH (2008) Gain-of-function mutations in TRPV4 cause autosomal dominant brachyolmia. Nat Genet 40(8):999–1003
175. Weinstein MM, Tompson SW, Chen Y, Lee B, Cohn DH (2014) Mice expressing mutant Trpv4 recapitulate the human TRPV4 disorders. J Bone Mineral Res 29(8):1815–1822
176. Lamande SR, Yuan Y, Gresshoff IL, Rowley L, Belluoccio D, Kaluarachchi K, Little CB, Botzenhart E, Zerres K, Amor DJ, Cole WG, Savarirayan R, McIntyre P, Bateman JF (2011) Mutations in TRPV4 cause an inherited arthropathy of hands and feet. Nat Genet 43(11):1142–1146
177. Clark AL, Votta BJ, Kumar S, Liedtke W, Guilak F (2010) Chondroprotective role of the osmotically sensitive ion channel transient receptor potential vanilloid 4: age- and sex-dependent progression of osteoarthritis in Trpv4-deficient mice. Arthritis Rheum 62(10):2973–2983
178. Masuyama R, Vriens J, Voets T, Karashima Y, Owsianik G, Vennekens R, Lieben L, Torrekens S, Moermans K, Vanden Bosch A, Bouillon R, Nilius B, Carmeliet G (2008) TRPV4-mediated calcium influx regulates terminal differentiation of osteoclasts. Cell Metab 8(3):257–265
179. O'Conor CJ, Leddy HA, Benefield HC, Liedtke WB, Guilak F (2014) TRPV4-mediated mechanotransduction regulates the metabolic response of chondrocytes to dynamic loading. Proc Natl Acad Sci U S A 111(4):1316–1321
180. Fecto F, Shi Y, Huda R, Martina M, Siddique T, Deng HX (2011) Mutant TRPV4-mediated toxicity is linked to increased constitutive function in axonal neuropathies. J Biol Chem 286(19):17281–17291
181. Klein CJ, Shi Y, Fecto F, Donaghy M, Nicholson G, McEntagart ME, Crosby AH, Wu Y, Lou H, McEvoy KM, Siddique T, Deng HX, Dyck PJ (2011) TRPV4 mutations and cytotoxic hypercalcemia in axonal Charcot-Marie-Tooth neuropathies. Neurology 76(10):887–894
182. Mederos y Schnitzler M, Waring J, Gudermann T, Chubanov V (2008) Evolutionary determinants of divergent calcium selectivity of TRPM channels. FASEB J 22(5):1540–1551
183. Togashi K, Hara Y, Tominaga T, Higashi T, Konishi Y, Mori Y, Tominaga M (2006) TRPM2 activation by cyclic ADP-ribose at body temperature is involved in insulin secretion. EMBO J 25(9):1804–1815
184. Faouzi M, Penner R (2014) Trpm2. Handb Exp Pharmacol 222:403–426
185. Du J, Xie J, Yue L (2009) Intracellular calcium activates TRPM2 and its alternative spliced isoforms. Proc Natl Acad Sci U S A 106(17):7239–7244
186. Tong Q, Zhang W, Conrad K, Mostoller K, Cheung JY, Peterson BZ, Miller BA (2006) Regulation of the transient receptor potential channel TRPM2 by the Ca2+ sensor calmodulin. J Biol Chem 281(14):9076–9085
187. Uchida K, Tominaga M (2011) TRPM2 modulates insulin secretion in pancreatic beta-cells. Islets 3(4):209–211
188. Hecquet CM, Malik AB (2009) Role of H(2)O(2)-activated TRPM2 calcium channel in oxidant-induced endothelial injury. Thromb Haemost 101(4):619–625

189. Starkus J, Beck A, Fleig A, Penner R (2007) Regulation of TRPM2 by extra- and intracellular calcium. J Gen Physiol 130(4):427–440
190. Knowles H, Li Y, Perraud AL (2013) The TRPM2 ion channel, an oxidative stress and metabolic sensor regulating innate immunity and inflammation. Immunol Res 55(1–3):241–248
191. Sumoza-Toledo A, Lange I, Cortado H, Bhagat H, Mori Y, Fleig A, Penner R, Partida-Sanchez S (2011) Dendritic cell maturation and chemotaxis is regulated by TRPM2-mediated lysosomal Ca2+ release. FASEB J 25(10):3529–3542
192. Takahashi N, Kozai D, Kobayashi R, Ebert M, Mori Y (2011) Roles of TRPM2 in oxidative stress. Cell Calcium 50(3):279–287
193. Yamamoto S, Shimizu S, Kiyonaka S, Takahashi N, Wajima T, Hara Y, Negoro T, Hiroi T, Kiuchi Y, Okada T, Kaneko S, Lange I, Fleig A, Penner R, Nishi M, Takeshima H, Mori Y (2008) TRPM2-mediated Ca2+influx induces chemokine production in monocytes that aggravates inflammatory neutrophil infiltration. Nat Med 14(7):738–747
194. Oda S, Uchida K, Wang X, Lee J, Shimada Y, Tominaga M, Kadowaki M (2013) TRPM2 contributes to antigen-stimulated Ca(2)(+) influx in mucosal mast cells. Pflugers Arch – Eur J Physiol 465(7):1023–1030
195. Fonfria E, Murdock PR, Cusdin FS, Benham CD, Kelsell RE, McNulty S (2006) Tissue distribution profiles of the human TRPM cation channel family. J Recept Signal Transduct Res 26(3):159–178
196. Vandewauw I, Owsianik G, Voets T (2013) Systematic and quantitative mRNA expression analysis of TRP channel genes at the single trigeminal and dorsal root ganglion level in mouse. BMC Neurosci 14:21
197. Naziroglu M, Ozgul C (2013) Vitamin E modulates oxidative stress and protein kinase C activator (PMA)-induced TRPM2 channel gate in dorsal root ganglion of rats. J Bioenerg Biomembr 45(6):541–549
198. Xu C, Macciardi F, Li PP, Yoon IS, Cooke RG, Hughes B, Parikh SV, McIntyre RS, Kennedy JL, Warsh JJ (2006) Association of the putative susceptibility gene, transient receptor potential protein melastatin type 2, with bipolar disorder. Am J Med Genet Part B Neuropsychiatr Genet 141B(1):36–43
199. Kaneko S, Kawakami S, Hara Y, Wakamori M, Itoh E, Minami T, Takada Y, Kume T, Katsuki H, Mori Y, Akaike A (2006) A critical role of TRPM2 in neuronal cell death by hydrogen peroxide. J Pharmacol Sci 101(1):66–76
200. Hermosura MC, Garruto RM (2007) TRPM7 and TRPM2-candidate susceptibility genes for Western Pacific ALS and PD? Biochim Biophys Acta 1772(8):822–835
201. Zhang E, Liao P (2015) Brain transient receptor potential channels and stroke. J Neurosci Res 93(8):1165–1183
202. Amina S, Hashii M, Ma WJ, Yokoyama S, Lopatina O, Liu HX, Islam MS, Higashida H (2010) Intracellular calcium elevation induced by extracellular application of cyclic-ADP-ribose or oxytocin is temperature-sensitive in rodent NG108-15 neuronal cells with or without exogenous expression of human oxytocin receptors. J Neuroendocrinol 22(5):460–466
203. Lange I, Yamamoto S, Partida-Sanchez S, Mori Y, Fleig A, Penner R (2009) TRPM2 functions as a lysosomal Ca2+-release channel in beta cells. Sci Signal 2(71):ra23
204. Zhang Z, Zhang W, Jung DY, Ko HJ, Lee Y, Friedline RH, Lee E, Jun J, Ma Z, Kim F, Tsitsilianos N, Chapman K, Morrison A, Cooper MP, Miller BA, Kim JK (2012) TRPM2 Ca2+ channel regulates energy balance and glucose metabolism. Am J Physiol Endocrinol Metab 302(7):E807–E816
205. Uchida K, Tominaga M (2014) The role of TRPM2 in pancreatic beta-cells and the development of diabetes. Cell Calcium 56(5):332–339
206. Oberwinkler J, Philipp SE (2014) Trpm3. Handb Exp Pharmacol 222:427–459
207. Holendova B, Grycova L, Jirku M, Teisinger J (2012) PtdIns(4,5)P2 interacts with CaM binding domains on TRPM3 N-terminus. Channels 6(6):479–482

208. Lee N, Chen J, Sun L, Wu S, Gray KR, Rich A, Huang M, Lin JH, Feder JN, Janovitz EB, Levesque PC, Blanar MA (2003) Expression and characterization of human transient receptor potential melastatin 3 (hTRPM3). J Biol Chem 278(23):20890–20897
209. Oberwinkler J, Lis A, Giehl KM, Flockerzi V, Philipp SE (2005) Alternative splicing switches the divalent cation selectivity of TRPM3 channels. J Biol Chem 280(23):22540–22548
210. Thiel G, Muller I, Rossler OG (2013) Signal transduction via TRPM3 channels in pancreatic beta-cells. J Mol Endocrinol 50(3):R75–R83
211. Naylor J, Li J, Milligan CJ, Zeng F, Sukumar P, Hou B, Sedo A, Yuldasheva N, Majeed Y, Beri D, Jiang S, Seymour VA, McKeown L, Kumar B, Harteneck C, O'Regan D, Wheatcroft SB, Kearney MT, Jones C, Porter KE, Beech DJ (2010) Pregnenolone sulphate- and cholesterol-regulated TRPM3 channels coupled to vascular smooth muscle secretion and contraction. Circ Res 106(9):1507–1515
212. Vriens J, Owsianik G, Hofmann T, Philipp SE, Stab J, Chen X, Benoit M, Xue F, Janssens A, Kerselaers S, Oberwinkler J, Vennekens R, Gudermann T, Nilius B, Voets T (2011) TRPM3 is a nociceptor channel involved in the detection of noxious heat. Neuron 70(3):482–494
213. Gonzales AL, Amberg GC, Earley S (2010) Ca2+ release from the sarcoplasmic reticulum is required for sustained TRPM4 activity in cerebral artery smooth muscle cells. Am J Physiol Cell Physiol 299(2):C279–C288
214. McHugh D, Flemming R, Xu S-Z, Perraud A-L, Beech DJ (2003) Critical intracellular Ca2+ dependence of transient receptor potential melastatin 2 (TRPM2) cation channel activation. J Biol Chem 278:11002–11006
215. Talavera K, Yasumatsu K, Voets T, Droogmans G, Shigemura N, Ninomiya Y, Margolskee RF, Nilius B (2005) Heat activation of TRPM5 underlies thermal sensitivity of sweet taste. Nature 438(7070):1022–1025
216. Mathar I, Jacobs G, Kecskes M, Menigoz A, Philippaert K, Vennekens R (2014) Trpm4. Handb Exp Pharmacol 222:461–487
217. Nilius B, Mahieu F, Prenen J, Janssens A, Owsianik G, Vennekens R, Voets T (2006) The Ca2+-activated cation channel TRPM4 is regulated by phosphatidylinositol 4,5-biphosphate. EMBO J 25(3):467–478
218. Crnich R, Amberg GC, Leo MD, Gonzales AL, Tamkun MM, Jaggar JH, Earley S (2010) Vasoconstriction resulting from dynamic membrane trafficking of TRPM4 in vascular smooth muscle cells. Am J Physiol Cell Physiol 299(3):C682–C694
219. Barbet G, Demion M, Moura IC, Serafini N, Leger T, Vrtovsnik F, Monteiro RC, Guinamard R, Kinet JP, Launay P (2008) The calcium-activated nonselective cation channel TRPM4 is essential for the migration but not the maturation of dendritic cells. Nat Immunol 9(10):1148–1156
220. Vennekens R, Olausson J, Meissner M, Bloch W, Mathar I, Philipp SE, Schmitz F, Weissgerber P, Nilius B, Flockerzi V, Freichel M (2007) Increased IgE-dependent mast cell activation and anaphylactic responses in mice lacking the calcium-activated nonselective cation channel TRPM4. Nat Immunol 8(3):312–320
221. Serafini N, Dahdah A, Barbet G, Demion M, Attout T, Gautier G, Arcos-Fajardo M, Souchet H, Jouvin MH, Vrtovsnik F, Kinet JP, Benhamou M, Monteiro RC, Launay P (2012) The TRPM4 channel controls monocyte and macrophage, but not neutrophil, function for survival in sepsis. J Immunol 189(7):3689–3699
222. Weber KS, Hildner K, Murphy KM, Allen PM (2010) Trpm4 differentially regulates Th1 and Th2 function by altering calcium signaling and NFAT localization. J Immunol 185(5):2836–2846
223. Schattling B, Steinbach K, Thies E, Kruse M, Menigoz A, Ufer F, Flockerzi V, Bruck W, Pongs O, Vennekens R, Kneussel M, Freichel M, Merkler D, Friese MA (2012) TRPM4 cation channel mediates axonal and neuronal degeneration in experimental autoimmune encephalomyelitis and multiple sclerosis. Nat Med 18(12):1805–1811

224. Gerzanich V, Woo SK, Vennekens R, Tsymbalyuk O, Ivanova S, Ivanov A, Geng Z, Chen Z, Nilius B, Flockerzi V, Freichel M, Simard JM (2009) De novo expression of Trpm4 initiates secondary hemorrhage in spinal cord injury. Nat Med 15(2):185–191
225. Nelson P, Ngoc Tran TD, Zhang H, Zolochevska O, Figueiredo M, Feng JM, Gutierrez DL, Xiao R, Yao S, Penn A, Yang LJ, Cheng H (2013) Transient receptor potential melastatin 4 channel controls calcium signals and dental follicle stem cell differentiation. Stem Cells 31(1):167–177
226. Nelson PL, Zolochevska O, Figueiredo ML, Soliman A, Hsu WH, Feng JM, Zhang H, Cheng H (2011) Regulation of Ca(2+)-entry in pancreatic alpha-cell line by transient receptor potential melastatin 4 plays a vital role in glucagon release. Mol Cell Endocrinol 335(2):126–134
227. Kruse M, Pongs O (2014) TRPM4 channels in the cardiovascular system. Curr Opin Pharmacol 15:68–73
228. Gonzales AL, Earley S (2012) Endogenous cytosolic Ca(2+) buffering is necessary for TRPM4 activity in cerebral artery smooth muscle cells. Cell Calcium 51(1):82–93
229. Earley S (2013) TRPM4 channels in smooth muscle function. Pflugers Arch – Eur J Physiol 465(9):1223–1231
230. Kruse M, Schulze-Bahr E, Corfield V, Beckmann A, Stallmeyer B, Kurtbay G, Ohmert I, Schulze-Bahr E, Brink P, Pongs O (2009) Impaired endocytosis of the ion channel TRPM4 is associated with human progressive familial heart block type I. J Clin Invest 119(9):2737–2744
231. Brink AJ (2009) Breaking new ground in research. Cardiovasc J Afr 20(6):335
232. Stallmeyer B, Zumhagen S, Denjoy I, Duthoit G, Hebert JL, Ferrer X, Maugenre S, Schmitz W, Kirchhefer U, Schulze-Bahr E, Guicheney P, Schulze-Bahr E (2012) Mutational spectrum in the Ca(2+) – activated cation channel gene TRPM4 in patients with cardiac conductance disturbances. Hum Mutat 33(1):109–117
233. Liu D, Liman ER (2003) Intracellular Ca2+ and the phospholipid PIP2 regulate the taste transduction ion channel TRPM5. Proc Natl Acad Sci U S A 100(25):15160–15165
234. Zhang Z, Zhao Z, Margolskee R, Liman E (2007) The transduction channel TRPM5 is gated by intracellular calcium in taste cells. J Neurosci Off J Soc Neurosci 27:5777–5786
235. Ullrich ND, Voets T, Prenen J, Vennekens R, Talavera K, Droogmans G, Nilius B (2005) Comparison of functional properties of the Ca2+-activated cation channels TRPM4 and TRPM5 from mice. Cell Calcium 37:267–278
236. Palmer RK, Lunn CA (2013) TRP channels as targets for therapeutic intervention in obesity: focus on TRPV1 and TRPM5. Curr Top Med Chem 13(3):247–257
237. Lin W, Margolskee R, Donnert G, Hell SW, Restrepo D (2007) Olfactory neurons expressing transient receptor potential channel M5 (TRPM5) are involved in sensing semiochemicals. Proc Natl Acad Sci U S A 104(7):2471–2476
238. Perez CA, Huang L, Rong M, Kozak JA, Preuss AK, Zhang H, Max M, Margolskee RF (2002) A transient receptor potential channel expressed in taste receptor cells. Nat Neurosci 5(11):1169–1176
239. Colsoul B, Schraenen A, Lemaire K, Quintens R, Van Lommel L, Segal A, Owsianik G, Talavera K, Voets T, Margolskee RF, Kokrashvili Z, Gilon P, Nilius B, Schuit FC, Vennekens R (2010) Loss of high-frequency glucose-induced Ca2+ oscillations in pancreatic islets correlates with impaired glucose tolerance in Trpm5-/- mice. Proc Natl Acad Sci U S A 107(11):5208–5213
240. Brixel LR, Monteilh-Zoller MK, Ingenbrandt CS, Fleig A, Penner R, Enklaar T, Zabel BU, Prawitt D (2010) TRPM5 regulates glucose-stimulated insulin secretion. Pflugers Arch – Eur J Physiol 460(1):69–76
241. Ketterer C, Mussig K, Heni M, Dudziak K, Randrianarisoa E, Wagner R, Machicao F, Stefan N, Holst JJ, Fritsche A, Haring HU, Staiger H (2011) Genetic variation within the TRPM5 locus associates with prediabetic phenotypes in subjects at increased risk for type 2 diabetes. Metab Clin Exp 60(9):1325–1333

242. Bautista DM, Siemens J, Glazer JM, Tsuruda PR, Basbaum AI, Stucky CL, Jordt SE, Julius D (2007) The menthol receptor TRPM8 is the principal detector of environmental cold. Nature 448(7150):204–208
242. Colburn RW, Lubin ML, Stone DJ Jr, Wang Y, Lawrence D, D'Andrea MR, Brandt MR, Liu Y, Flores CM, Qin N (2007) Attenuated cold sensitivity in TRPM8 null mice. Neuron 54(3):379–386
244. Dhaka A, Murray AN, Mathur J, Earley TJ, Petrus MJ, Patapoutian A (2007) TRPM8 is required for cold sensation in mice. Neuron 54(3):371–378
245. Peier AM, Moqrich A, Hergarden AC, Reeve AJ, Andersson DA, Story GM, Earley TJ, Dragoni I, McIntyre P, Bevan S, Patapoutian A (2002) A TRP channel that senses cold stimuli and menthol. Cell 108:705–715
246. McKemy DD, Neuhausser WM, Julius D (2002) Identification of a cold receptor reveals a general role for TRP channels in thermosensation. Nature 416(6876):52–58
247. Yudin Y, Rohacs T (2012) Regulation of TRPM8 channel activity. Mol Cell Endocrinol 353(1–2):68–74
248. Sarria I, Gu J (2010) Menthol response and adaptation in nociceptive-like and nonnociceptive-like neurons: role of protein kinases. Mol Pain 6:47
249. Sarria I, Ling J, Zhu MX, Gu JG (2011) TRPM8 acute desensitization is mediated by calmodulin and requires PIP(2): distinction from tachyphylaxis. J Neurophysiol 106(6):3056–3066
250. Rohacs T, Lopes CM, Michailidis I, Logothetis DE (2005) PI(4,5)P2 regulates the activation and desensitization of TRPM8 channels through the TRP domain. Nat Neurosci 8(5):626–634
251. Liu B, Qin F (2005) Functional control of cold- and menthol-sensitive TRPM8 ion channels by phosphatidylinositol 4,5-bisphosphate. J Neurosci 25(7):1674–1681
252. Daniels RL, Takashima Y, McKemy DD (2009) Activity of the neuronal cold sensor TRPM8 is regulated by phospholipase C via the phospholipid phosphoinositol 4,5-bisphosphate. J Biol Chem 284(3):1570–1582
253. Yudin Y, Lukacs V, Cao C, Rohacs T (2011) Decrease in phosphatidylinositol 4,5-bisphosphate levels mediates desensitization of the cold sensor TRPM8 channels. J Physiol 589:6007–6027
254. Bavencoffe A, Gkika D, Kondratskyi A, Beck B, Borowiec AS, Bidaux G, Busserolles J, Eschalier A, Shuba Y, Skryma R, Prevarskaya N (2010) The transient receptor potential channel TRPM8 is inhibited via the alpha 2A adrenoreceptor signaling pathway. J Biol Chem 285(13):9410–9419
255. De Petrocellis L, Starowicz K, Moriello AS, Vivese M, Orlando P, Di Marzo V (2007) Regulation of transient receptor potential channels of melastatin type 8 (TRPM8): effect of cAMP, cannabinoid CB1 receptors and endovanilloids. Exp Cell Res 313:1911–1920
256. Linte RM, Ciobanu C, Reid G, Babes A (2007) Desensitization of cold- and menthol-sensitive rat dorsal root ganglion neurones by inflammatory mediators. Exp Brain Res 178(1):89–98
257. Abe J, Hosokawa H, Sawada Y, Matsumura K, Kobayashi S (2006) Ca2+-dependent PKC activation mediates menthol-induced desensitization of transient receptor potential M8. Neurosci Lett 397(1–2):140–144
258. Premkumar LS, Raisinghani M, Pingle SC, Long C, Pimentel F (2005) Downregulation of transient receptor potential melastatin 8 by protein kinase C-mediated dephosphorylation. J Neurosci 25(49):11322–11329
259. Malkia A, Morenilla-Palao C, Viana F (2011) The emerging pharmacology of TRPM8 channels: hidden therapeutic potential underneath a cold surface. Curr Pharm Biotechnol 12(1):54–67
260. Bidaux G, Flourakis M, Thebault S, Zholos A, Beck B, Gkika D, Roudbaraki M, Bonnal JL, Mauroy B, Shuba Y, Skryma R, Prevarskaya N (2007) Prostate cell differentiation status determines transient receptor potential melastatin member 8 channel subcellular localization and function. J Clin Invest 117(6):1647–1657

261. Thebault S, Lemonnier L, Bidaux G, Flourakis M, Bavencoffe A, Gordienko D, Roudbaraki M, Delcourt P, Panchin Y, Shuba Y, Skryma R, Prevarskaya N (2005) Novel role of cold/menthol-sensitive transient receptor potential melastatine family member 8 (TRPM8) in the activation of store-operated channels in LNCaP human prostate cancer epithelial cells. J Biol Chem 280(47):39423–39435
262. Zhang L, Barritt GJ (2006) TRPM8 in prostate cancer cells: a potential diagnostic and prognostic marker with a secretory function? Endocr Relat Cancer 13(1):27–38
263. Flourakis M, Prevarskaya N (2009) Insights into Ca2+ homeostasis of advanced prostate cancer cells. Biochim Biophys Acta 1793(6):1105–1109
264. Mergler S, Derckx R, Reinach PS, Garreis F, Bohm A, Schmelzer L, Skosyrski S, Ramesh N, Abdelmessih S, Polat OK, Khajavi N, Riechardt AI (2014) Calcium regulation by temperature-sensitive transient receptor potential channels in human uveal melanoma cells. Cell Signal 26(1):56–69
265. Yamamura H, Ugawa S, Ueda T, Morita A, Shimada S (2008) TRPM8 activation suppresses cellular viability in human melanoma. Am J Physiol Cell Physiol 295(2):C296–C301
266. Mergler S, Strowski MZ, Kaiser S, Plath T, Giesecke Y, Neumann M, Hosokawa H, Kobayashi S, Langrehr J, Neuhaus P, Plockinger U, Wiedenmann B, Grotzinger C (2007) Transient receptor potential channel TRPM8 agonists stimulate calcium influx and neurotensin secretion in neuroendocrine tumor cells. Neuroendocrinology 85(2):81–92
267. Wondergem R, Ecay TW, Mahieu F, Owsianik G, Nilius B (2008) HGF/SF and menthol increase human glioblastoma cell calcium and migration. Biochem Biophys Res Commun 372(1):210–215
268. Cucu D, Chiritoiu G, Petrescu S, Babes A, Stanica L, Duda DG, Horii A, Dima SO, Popescu I (2014) Characterization of functional transient receptor potential melastatin 8 channels in human pancreatic ductal adenocarcinoma cells. Pancreas 43(5):795–800
269. Eid SR, Cortright DN (2009) Transient receptor potential channels on sensory nerves. Handb Exp Pharmacol 194:261–281
270. Levine JD, Alessandri-Haber N (2007) TRP channels: targets for the relief of pain. Biochim Biophys Acta 1772(8):989–1003
271. Chung MK, Jung SJ, Oh SB (2011) Role of TRP channels in pain sensation. Adv Exp Med Biol 704:615–636
272. Almaraz L, Manenschijn JA, de la Pena E, Viana F (2014) Trpm8. Handb Exp Pharmacol 222:547–579
273. Proudfoot CJ, Garry EM, Cottrell DF, Rosie R, Anderson H, Robertson DC, Fleetwood-Walker SM, Mitchell R (2006) Analgesia mediated by the TRPM8 cold receptor in chronic neuropathic pain. Curr Biol 16(16):1591–1605
274. Knowlton WM, Daniels RL, Palkar R, McCoy DD, McKemy DD (2011) Pharmacological blockade of TRPM8 ion channels alters cold and cold pain responses in mice. PLoS One 6(9):e25894
275. Zuo X, Ling JX, Xu GY, Gu JG (2013) Operant behavioral responses to orofacial cold stimuli in rats with chronic constrictive trigeminal nerve injury: effects of menthol and capsazepine. Mol Pain 9:28
276. Xing H, Chen M, Ling J, Tan W, Gu JG (2007) TRPM8 mechanism of cold allodynia after chronic nerve injury. J Neurosci 27(50):13680–13690
277. Kurose M, Meng ID (2013) Dry eye modifies the thermal and menthol responses in rat corneal primary afferent cool cells. J Neurophysiol 110(2):495–504
278. Parra A, Madrid R, Echevarria D, del Olmo S, Morenilla-Palao C, Acosta MC, Gallar J, Dhaka A, Viana F, Belmonte C (2010) Ocular surface wetness is regulated by TRPM8-dependent cold thermoreceptors of the cornea. Nat Med 16(12):1396–1399
279. Lashinger ES, Steiginga MS, Hieble JP, Leon LA, Gardner SD, Nagilla R, Davenport EA, Hoffman BE, Laping NJ, Su X (2008) AMTB, a TRPM8 channel blocker: evidence in rats for activity in overactive bladder and painful bladder syndrome. Am J Physiol Renal Physiol 295(3):F803–F810

280. Mukerji G, Yiangou Y, Corcoran SL, Selmer IS, Smith GD, Benham CD, Bountra C, Agarwal SK, Anand P (2006) Cool and menthol receptor TRPM8 in human urinary bladder disorders and clinical correlations. BMC Urol 6:6
281. Yoshida T, Inoue R, Morii T, Takahashi N, Yamamoto S, Hara Y, Tominaga M, Shimizu S, Sato Y, Mori Y (2006) Nitric oxide activates TRP channels by cysteine S-nitrosylation. Nat Chem Biol 2:596–607
282. Flemming PK, Dedman AM, Xu S-Z, Li J, Zeng F, Naylor J, Benham CD, Bateson AN, Muraki K, Beech DJ (2006) Sensing of lysophospholipids by TRPC5 calcium channel. J Biol Chem 281:4977–4982
283. Zhu MH, Chae M, Kim HJ, Lee YM, Kim MJ, Jin NG, Yang DK, So I, Kim KW (2005) Desensitization of canonical transient receptor potential channel 5 by protein kinase C. Am J Physiol Cell Physiol 289(3):C591–C600
284. Kinoshita-Kawada M, Tang J, Xiao R, Kaneko S, Foskett JK, Zhu MX (2005) Inhibition of TRPC5 channels by Ca2+-binding protein 1 in Xenopus oocytes. Pflügers Arch 450:345–354
285. Sung TS, Jeon JP, Kim BJ, Hong C, Kim SY, Kim J, Jeon JH, Kim HJ, Suh CK, Kim SJ, So I (2011) Molecular determinants of PKA-dependent inhibition of TRPC5 channel. Am J Physiol Cell Physiol 301:C823–C832
286. Zholos AV (2014) Trpc5. Handb Exp Pharmacol 222:129–156
287. Faber ES, Sedlak P, Vidovic M, Sah P (2006) Synaptic activation of transient receptor potential channels by metabotropic glutamate receptors in the lateral amygdala. Neuroscience 137(3):781–794
288. von Bohlen Und Halbach O, Hinz U, Unsicker K, Egorov AV (2005) Distribution of TRPC1 and TRPC5 in medial temporal lobe structures of mice. Cell Tissue Res 322(2):201–206
289. De March Z, Giampa C, Patassini S, Bernardi G, Fusco FR (2006) Cellular localization of TRPC5 in the substantia nigra of rat. Neurosci Lett 402(1–2):35–39
290. Greka A, Navarro B, Oancea E, Duggan A, Clapham DE (2003) TRPC5 is a regulator of hippocampal neurite length and growth cone morphology. Nat Neurosci 6(8):837–845
291. He Z, Jia C, Feng S, Zhou K, Tai Y, Bai X, Wang Y (2012) TRPC5 channel is the mediator of neurotrophin-3 in regulating dendritic growth via CaMKIIalpha in rat hippocampal neurons. J Neurosci 32(27):9383–9395
292. Gomis A, Soriano S, Belmonte C, Viana F (2008) Hypoosmotic- and pressure-induced membrane stretch activate TRPC5 channels. J Physiol 586(Pt 23):5633–5649
293. Min J-W, Liu W-H, He X-H, Peng B-W (2013) Different types of toxins targeting TRPV1 in pain. Toxicon 71:66–75
294. Zygmunt PM, Hogestatt ED (2014) Trpa1. Handb Exp Pharmacol 222:583–630
295. Hinman A, H-h C, Bautista DM, Julius D (2006) TRP channel activation by reversible covalent modification. Proc Natl Acad Sci 103:19564–19568
296. Macpherson LJ, Dubin AE, Evans MJ, Marr F, Schultz PG, Cravatt BF, Patapoutian A (2007) Noxious compounds activate TRPA1 ion channels through covalent modification of cysteines. Nature 445:541–545
297. Gracheva EO, Ingolia NT, Kelly YM, Cordero-Morales JF, Hollopeter G, Chesler AT, Sánchez EE, Perez JC, Weissman JS, Julius D (2010) Molecular basis of infrared detection by snakes. Nature 464:1006–1011
298. Moparthi L, Survery S, Kreir M, Simonsen C, Kjellbom P, Högestätt ED, Johanson U, Zygmunt PM (2014) Human TRPA1 is intrinsically cold- and chemosensitive with and without its N-terminal ankyrin repeat domain. Proc Natl Acad Sci 111:16901–16906
299. Chen J, Kang D, Xu J, Lake M, Hogan JO, Sun C, Walter K, Yao B, Kim D (2013) Species differences and molecular determinant of TRPA1 cold sensitivity. Nat Commun 4:2501
300. Hamada FN, Rosenzweig M, Kang K, Pulver SR, Ghezzi A, Jegla TJ, Garrity PA (2008) An internal thermal sensor controlling temperature preference in Drosophila. Nature 454:217–220

301. Dai Y, Wang S, Tominaga M, Yamamoto S, Fukuoka T, Higashi T, Kobayashi K, Obata K, Yamanaka H, Noguchi K (2007) Sensitization of TRPA1 by PAR2 contributes to the sensation of inflammatory pain. J Clin Investig 117:1979–1987
302. Wang S, Dai Y, Fukuoka T, Yamanaka H, Kobayashi K, Obata K, Cui X, Tominaga M, Noguchi K (2008) Phospholipase C and protein kinase A mediate bradykinin sensitization of TRPA1: a molecular mechanism of inflammatory pain. Brain 131:1241–1251
303. Schmidt M, Dubin AE, Petrus MJ, Earley TJ, Patapoutian A (2009) Nociceptive signals induce trafficking of TRPA1 to the plasma membrane. Neuron 64(4):498–509
304. Wang YY, Chang RB, Waters HN, McKemy DD, Liman ER (2008) The nociceptor Ion channel TRPA1 is potentiated and inactivated by permeating calcium ions. J Biol Chem 283:32691–32703
305. Nassini R, Materazzi S, Benemei S, Geppetti P (2014) The TRPA1 channel in inflammatory and neuropathic pain and migraine. Rev Physiol Biochem Pharmacol 167:1–43
306. Andrade EL, Meotti FC, Calixto JB (2012) TRPA1 antagonists as potential analgesic drugs. Pharmacol Ther 133(2):189–204
307. Sawada Y, Hosokawa H, Hori A, Matsumura K, Kobayashi S (2007) Cold sensitivity of recombinant TRPA1 channels. Brain Res 1160:39–46
308. Karashima Y, Talavera K, Everaerts W, Janssens A, Kwan KY, Vennekens R, Nilius B, Voets T (2009) TRPA1 acts as a cold sensor in vitro and in vivo. Proc Natl Acad Sci U S A 106(4):1273–1278
309. Kremeyer B, Lopera F, Cox JJ, Momin A, Rugiero F, Marsh S, Woods CG, Jones NG, Paterson KJ, Fricker FR, Villegas A, Acosta N, Pineda-Trujillo NG, Ramirez JD, Zea J, Burley MW, Bedoya G, Bennett DL, Wood JN, Ruiz-Linares A (2010) A gain-of-function mutation in TRPA1 causes familial episodic pain syndrome. Neuron 66(5):671–680
310. Patapoutian A, Macpherson L (2006) Channeling pain. Nat Med 12(5):506–507
311. Hjerling-Leffler J, Alqatari M, Ernfors P, Koltzenburg M (2007) Emergence of functional sensory subtypes as defined by transient receptor potential channel expression. J Neurosci 27(10):2435–2443
312. Kim YS, Son JY, Kim TH, Paik SK, Dai Y, Noguchi K, Ahn DK, Bae YC (2010) Expression of transient receptor potential ankyrin 1 (TRPA1) in the rat trigeminal sensory afferents and spinal dorsal horn. J Comp Neurol 518(5):687–698
313. Kunkler PE, Ballard CJ, Oxford GS, Hurley JH (2011) TRPA1 receptors mediate environmental irritant-induced meningeal vasodilatation. Pain 152(1):38–44
314. Pozsgai G, Bodkin JV, Graepel R, Bevan S, Andersson DA, Brain SD (2010) Evidence for the pathophysiological relevance of TRPA1 receptors in the cardiovascular system in vivo. Cardiovasc Res 87(4):760–768
315. Engel MA, Leffler A, Niedermirtl F, Babes A, Zimmermann K, Filipovic MR, Izydorczyk I, Eberhardt M, Kichko TI, Mueller-Tribbensee SM, Khalil M, Siklosi N, Nau C, Ivanovic-Burmazovic I, Neuhuber WL, Becker C, Neurath MF, Reeh PW (2011) TRPA1 and substance P mediate colitis in mice. Gastroenterology 141(4):1346–1358
316. Aubdool AA, Graepel R, Kodji X, Alawi KM, Bodkin JV, Srivastava S, Gentry C, Heads R, Grant AD, Fernandes ES, Bevan S, Brain SD (2014) TRPA1 is essential for the vascular response to environmental cold exposure. Nat Commun 5:5732
317. Benemei S, Fusi C, Trevisan G, Geppetti P (2014) The TRPA1 channel in migraine mechanism and treatment. Br J Pharmacol 171(10):2552–2567
318. Wei H, Koivisto A, Saarnilehto M, Chapman H, Kuokkanen K, Hao B, Huang JL, Wang YX, Pertovaara A (2011) Spinal transient receptor potential ankyrin 1 channel contributes to central pain hypersensitivity in various pathophysiological conditions in the rat. Pain 152(3):582–591
319. da Costa DS, Meotti FC, Andrade EL, Leal PC, Motta EM, Calixto JB (2010) The involvement of the transient receptor potential A1 (TRPA1) in the maintenance of mechanical and cold hyperalgesia in persistent inflammation. Pain 148(3):431–437

320. Miyakawa T, Terashima Y, Takebayashi T, Tanimoto K, Iwase T, Ogon I, Kobayashi T, Tohse N, Yamashita T (2014) Transient receptor potential ankyrin 1 in spinal cord dorsal horn is involved in neuropathic pain in nerve root constriction rats. Mol Pain 10:58
321. Andersson DA, Gentry C, Alenmyr L, Killander D, Lewis SE, Andersson A, Bucher B, Galzi JL, Sterner O, Bevan S, Hogestatt ED, Zygmunt PM (2011) TRPA1 mediates spinal antinociception induced by acetaminophen and the cannabinoid Delta(9)-tetrahydrocannabiorcol. Nat Commun 2:551
322. Shigetomi E, Tong X, Kwan KY, Corey DP, Khakh BS (2012) TRPA1 channels regulate astrocyte resting calcium and inhibitory synapse efficacy through GAT-3. Nat Neurosci 15(1):70–80
323. Belvisi MG, Dubuis E, Birrell MA (2011) Transient receptor potential A1 channels: insights into cough and airway inflammatory disease. Chest 140(4):1040–1047
324. Geppetti P, Patacchini R, Nassini R (2014) Transient receptor potential channels and occupational exposure. Curr Opin Allergy Clin Immunol 14(2):77–83
325. Grace MS, Baxter M, Dubuis E, Birrell MA, Belvisi MG (2014) Transient receptor potential (TRP) channels in the airway: role in airway disease. Br J Pharmacol 171(10):2593–2607
326. Grace MS, Belvisi MG (2011) TRPA1 receptors in cough. Pulm Pharmacol Ther 24(3):286–288
327. Schaefer EA, Stohr S, Meister M, Aigner A, Gudermann T, Buech TR (2013) Stimulation of the chemosensory TRPA1 cation channel by volatile toxic substances promotes cell survival of small cell lung cancer cells. Biochem Pharmacol 85(3):426–438
328. Wilson SR, Gerhold KA, Bifolck-Fisher A, Liu Q, Patel KN, Dong X, Bautista DM (2011) TRPA1 is required for histamine-independent, Mas-related G protein-coupled receptor-mediated itch. Nat Neurosci 14(5):595–602
329. Liu B, Escalera J, Balakrishna S, Fan L, Caceres AI, Robinson E, Sui A, McKay MC, McAlexander MA, Herrick CA, Jordt SE (2013) TRPA1 controls inflammation and pruritogen responses in allergic contact dermatitis. FASEB J 27(9):3549–3563

Chapter 13
Calcium Signalling through Ligand-Gated Ion Channels such as P2X1 Receptors in the Platelet and other Non-Excitable Cells

Martyn P. Mahaut-Smith, Kirk A. Taylor, and Richard J. Evans

Abstract Ligand-gated ion channels on the cell surface are directly activated by the binding of an agonist to their extracellular domain and often referred to as ionotropic receptors. P2X receptors are ligand-gated non-selective cation channels with significant permeability to Ca^{2+} whose principal physiological agonist is ATP. This chapter focuses on the mechanisms by which P2X1 receptors, a ubiquitously expressed member of the family of ATP-gated channels, can contribute to cellular responses in non-excitable cells. Much of the detailed information on the contribution of P2X1 to Ca^{2+} signalling and downstream functional events has been derived from the platelet. The underlying primary P2X1-generated signalling event in non-excitable cells is principally due to Ca^{2+} influx, although Na^+ entry will also occur along with membrane depolarization. P2X1 receptor stimulation can lead to additional Ca^{2+} mobilization via a range of routes such as amplification of G-protein-coupled receptor-dependent Ca^{2+} responses. This chapter also considers the mechanism by which cells generate extracellular ATP for autocrine or paracrine activation of P2X1 receptors. For example cytosolic ATP efflux can result from opening of pannexin anion-permeable channels or following damage to the cell membrane. Alternatively, ATP stored in specialised secretory vesicles can undergo quantal release via the process of exocytosis. Examples of physiological or pathophysiological roles of P2X1-dependent signalling in non-excitable cells are also discussed, such as thrombosis and immune responses.

Keywords P2X receptors • Ca^{2+} entry • P2X1 • Ligand-gated ion channels • ATP • Leukocyte • Platelets • Exocytosis • Pannexin • Ionotropic receptors • Intracellular Ca^{2+} • Thrombosis • Inflammation

---

M.P. Mahaut-Smith (✉) • R.J. Evans
Department of Molecular and Cell Biology, University of Leicester, Henry Wellcome Building, Lancaster Road, Leicester LE1 9HN, UK
e-mail: mpms1@le.ac.uk

K.A. Taylor
Department of Biomedical and Forensic Sciences, Anglia Ruskin University, Cambridge, UK

J.A. Rosado (ed.), *Calcium Entry Pathways in Non-excitable Cells*, Advances in Experimental Medicine and Biology 898, DOI 10.1007/978-3-319-26974-0_13

13.1 General Background on Ligand-Gated Ca^{2+}-Permeable Receptor Ion Channels

Surface-expressed, ligand-gated P2X receptors represent a unique mechanism whereby cytosolic Ca^{2+} can be elevated during cellular activation [1, 2]. The main difference between this class of ion channel and other Ca^{2+} mobilization pathways is the rapid coupling of ligand binding to pore opening; consequently membrane currents and calcium increases can be detected almost instantaneously following agonist application [3–5]. This compares with a latency of several hundreds of milliseconds for Ca^{2+} increases downstream of G-protein-coupled receptor (GPCR) activation [6, 7] due to the time required for the hydrolysis of phosphatidylinositol 4,5-bisphosphate and for released inositol 1,4,5 trisphosphate (IP_3) to diffuse to the IP_3 –dependent Ca^{2+}-permeable channels on intracellular stores. P2X receptors consist of a family of seven subunits which are widely expressed and can form a range of homotrimeric and heterotrimeric functional channels [2]. They are non-selective cation channels with significant Ca^{2+}-permeability (P_{Ca}/P_{Na} in the range 1.1–4 depending upon the receptor type and ionic conditions [2]). Although non-excitable cells have been reported to express several other types of ligand-gated cation channels (examples are glutamate-, serotonin- and acetylcholine-stimulated channels), their role in Ca^{2+} signalling is less clear. For example, platelets express a number of ionotropic glutamate receptor subunits of the AMPA and kainate family, however all evidence points towards functional channels that are selective for monovalent cations [8, 9]. For this reason, this chapter will consider only P2X receptors. Furthermore, discussion will concentrate on the P2X1 receptor, which is widely expressed in both excitable and non-excitable cells and has been extensively studied in terms of its role in Ca^{2+} signalling. Much of this information has been derived from experiments in the platelet (summarised in Figs. 13.1 and 13.2), where P2X1 is the only P2X receptor subtype expressed [10–12] and contributes to the responses to several agonists in vitro [13] and to thrombosis in vivo [14, 15]. We will also consider more recent evidence for a contribution by P2X1 receptors to physiological and pathophysiological responses via activation of T lymphocytes and neutrophils [16–19].

13.2 Structure, Activation and Regulation of P2X Cation Channels

13.2.1 Molecular Structure and Activation

The cloning of the first P2X receptor subunits over 20 years ago [20, 21] was a catalyst for understanding the role of extracellular ATP as a signalling molecule and the molecular diversity/properties of the receptors [22]. The seven mammalian P2X receptor subunits (P2X1-7) have intracellular amino and carboxyl termini, two

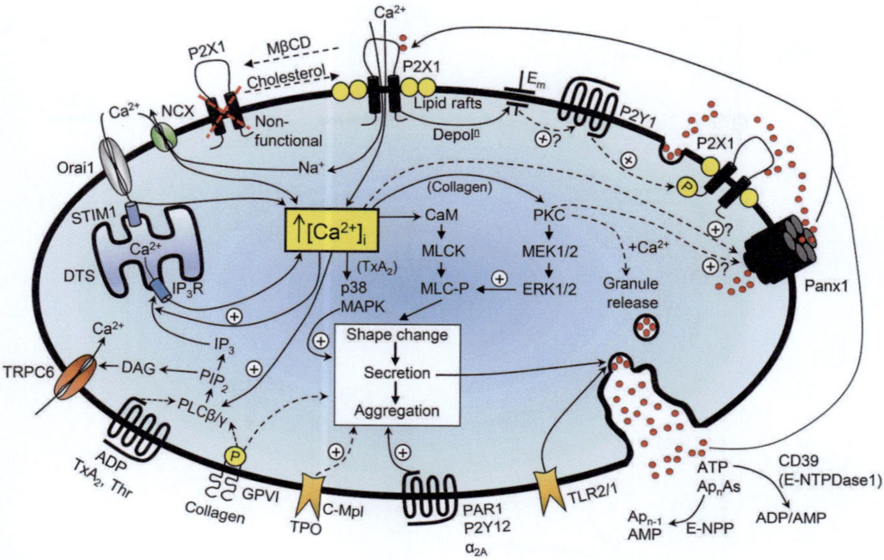

Fig. 13.1 Regulation of P2X1 receptors and interplay with other Ca^{2+}-mobilising pathways in the platelet: refer to text for full explanation. *Panx1* pannexin1, *TxA_2* thromboxane A_2, *MβCD* methyl-β-cyclodextrin, *PAR1* protease activated receptor 1, *thr* thrombin, *Depol^n* depolarization, *NCX* Na^+/Ca^{2+} exchanger, *C-Mpl* thrombopoietin receptor, *CaM* calmodulin, *TRPC6* canonical transient receptor potential channel 6, *TPO* thrombopoietin, *TLR1/2* Toll-like receptor1/2, *IP_3* inositol 1,4,5 trisphosphate, *PLC* phospholipase-C, *PIP_2* phosphatidylinositol 4,5-bisphosphate, *DAG* diacylglycerol, *PKC* protein kinase C, *Em* membrane potential, *Ap_nA* diadenosine polyphosphates, α_{2A} alpha$_{2A}$-adrenergic receptors, *GPVI* glycoprotein VI, *MLCK* myosin light chain kinase, *MLC-P* phosphorylated myosin light chain, *ERK2* extracellular receptor kinase 2, *p38 MAPK* p38 mitogen activated protein kinase, *MEK* MAPK/ERK kinase, *E-NTPDase1* ecto-nucleoside 5′-triphosphate diphosphohydrolase1, *E-NPP* ectonucleotide pyrophosphatase/phosphodiesterase (the enzyme products will be AMP + Ap_{n-1}, i.e. AMP + ATP for Ap_4A), *DTS* dense tubular system

transmembrane domains and a large extracellular ligand binding loop [23]. The subunits assemble to form a range of homo- and hetero-trimeric ATP-gated cation channels with properties dependent on the subunit composition [23]. The crystallization of the zebrafish P2X4 receptor in agonist-free and ATP-bound forms was a major advance, providing detailed structural insight into agonist binding and the gating of the receptor/channel [24, 25]. The crystal structures show the trimeric assembly of subunits with the ATP binding site(s) forming at the subunit interfaces and the second transmembrane domains lining the channel pore. ATP binding induces conformational change in the nucleotide binding pocket (≈40 Å from the pore-spanning transmembrane region) resulting in an iris-like expansion of the transmembrane helices and channel gating. A wide range of biochemical and electrophysiological studies have supported these conclusions and identified residues involved in the inter-subunit ATP binding site and formation of the channel pore [22, 26, 27].

Cysteine scanning mutagenesis has been used by several groups to investigate the pore-forming second transmembrane domain [28–30]. Silver and cadmium can

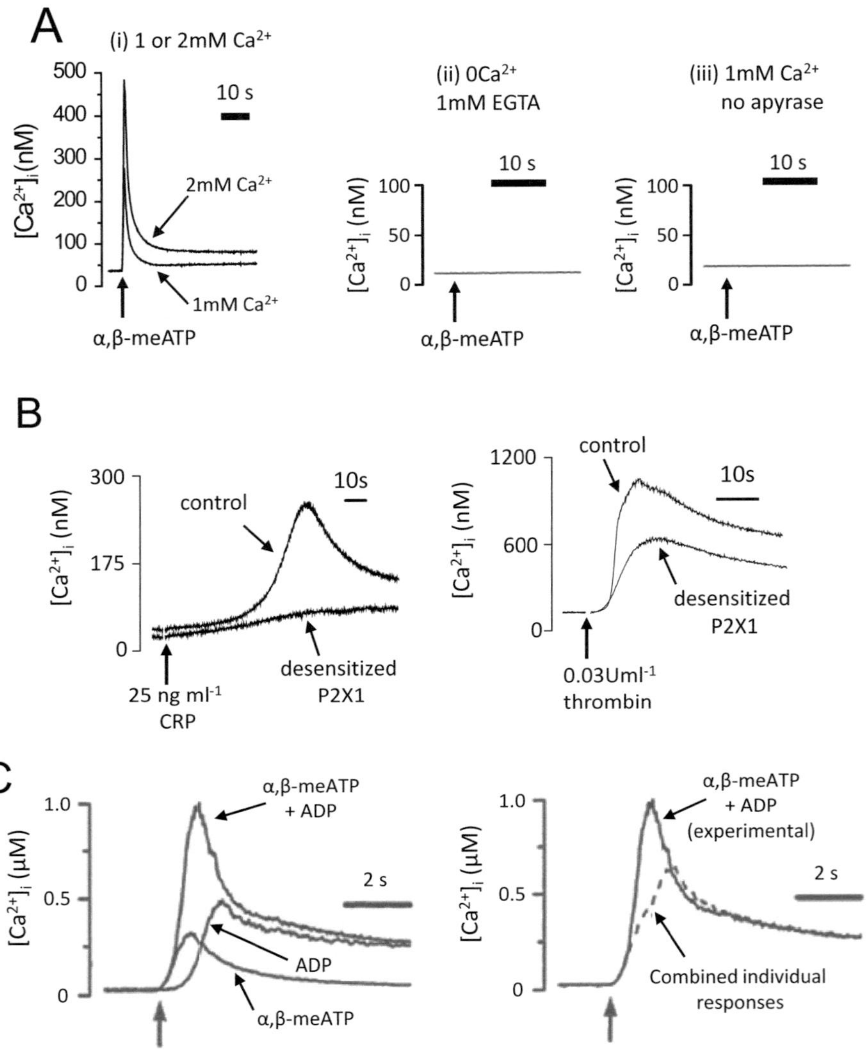

Fig. 13.2 P2X1 receptor-dependent intracellular Ca^{2+} responses in human platelets: average measurements from stirred suspensions of platelets loaded with the fluorescent indicator fura-2. **A** Selective activation of P2X1 receptors by a maximal concentration (10 μM) of the non-hydrolysable ATP analogue α,β-meATP leads to a rapid, transient increase in intracellular free Ca^{2+} in human platelets (i). An increase in extracellular Ca^{2+} results in a significant increase in peak response (i), whilst Ca^{2+} responses are abolished by removal of extracellular Ca^{2+} (ii) or omission of the ATP-degrading enzyme apyrase from the resuspension buffer due to receptor desensitization (iii). **B** Examples of the contribution of P2X1 receptors, activated by ATP released via exocytosis and/or Panx-1, to responses downstream of major platelet receptors; CRP (collagen receptor peptide) stimulates the TKR GPVI and thrombin co-stimulates the GPCRs PAR1 and PAR4; P2X1 function was prevented by desensitization with a low concentration of α,β-meATP prior to addition of extracellular Ca^{2+} (see [13]). **C** Synergy between P2X1 and P2Y receptors. The *left panel* shows responses to individual or combined addition of 10 μM α,β-meATP and 10 μM ADP. The *right panel* shows the experimental response to co-addition of these agonists (trace from the *left panel*) together with the response predicted by summation of the individual responses. α,β-meATP selectively activates P2X1 receptors whilst ADP generates Ca^{2+} responses primarily through P2Y1 receptors (Taken from [12, 13, 52, 148] with permission)

permeate the P2X receptor channel and the characterization of the rates of modification of cysteine mutants (in the absence or presence of ATP) by these metals highlights not only the residues that line the channel pore but also the location of the channel gate [30–32] and confirms findings from structural studies. P2X receptors show appreciable permeability to Ca^{2+} ([5]; see below). Mutagenesis studies on the P2X2 receptor have identified three polar pore-lining residues around the channel gate (Thr336, Thr339 and Ser340) that are important for high Ca^{2+} permeability [33], suggesting that these residues are forming the selectivity filter. Polar residues are found at two corresponding positions in the human P2X1 receptor (336-Ser335 and 339-Gly340, human P2X1 receptor), but at the position corresponding to 340 there is non-polar isoleucine (Iso341). However the contribution of these residues to Ca^{2+} permeability at the P2X1 receptor has not been determined and there does not appear to be a consensus of residues at this location within the P2X receptor family. For the rat P2X1 receptor, juxta-membrane acidic amino acids at the extracellular vestibule contribute to the high Ca^{2+} permeability of the receptor [34].

In myeloid cells such as platelets, megakaryocytes and neutrophils, the P2X1 receptor functions as a homomeric channel [12, 14, 19]. One of the characteristic features of the P2X1 receptor is its rapid desensitization; the channel closes with a time constant of ≈200–500 ms during the continued application of maximal concentrations of ATP. However, at lower concentrations of ATP the desensitization is slower, reflecting the cumulative activation of the receptors, and in this way P2X1 receptor activation can give rise to longer lasting, smaller rises in intracellular calcium [35]. Studies on chimeric P2X receptors and point mutants have shown that the molecular basis of the desensitizing phenotype is dominated by the intracellular domains and involves complex interactions with the transmembrane regions [36, 37] as well as structural changes around the ATP binding pocket [38].

13.2.2 Regulation of P2X1 Receptor Function by Cellular Signals and Agonist Removal

The activity of cell surface P2X1 receptors can be dynamically regulated by a range of mechanisms. Activation of Gαq-coupled receptors potentiates P2X1 receptor-mediated responses via a protein kinase C (PKC)-dependent mechanism [39, 40], likely via phosphorylation of a regulatory protein associated with the amino terminus [41]. Cholesterol depletion and disruption of lipid rafts inhibits P2X1 receptor-mediated responses [42, 43] and this is likely through an interaction with the pre-transmembrane 1 intracellular amino terminal region [44]. Purification of the P2X1 receptor from cell membranes has indicated interactions with the actin cytoskeleton and heat shock protein 90 [45, 46]. Disruption of the actin cytoskeleton inhibited P2X1 receptor mediated responses. Interestingly, stabilization of the actin cytoskeleton abolished the inhibitory effects of cholesterol depletion, suggesting that lipid rafts regulate the P2X1 receptor through stabilization of the cytoskeleton [45]. In platelets, approximately 85 % of receptors were not associated with lipid rafts, and this proportion are predicted to be non-functional [43]. Therefore, it is interesting to speculate that high cholesterol levels may enhance P2X1 receptor responses and thus thrombosis.

Another factor that regulates P2X receptor responses is the rate of removal of the agonist. Diffusion will contribute to the movement of ATP away from the cell surface in all cells, and blood flow will exert profound effects on the duration of agonist exposure to P2X receptors on endothelial cells, platelets, erythrocytes and leukocytes. In addition, enzymatic degradation by ectonucleotidases plays a key regulatory influence, particularly E-NTPDase1 (CD39) [47–49]. Within in vitro experiments on human platelets, a ten-fold increase in the concentration of extracellular apyrase (a form of NTPDase from potato that displays similar properties to human CD39 [50]) reduced the activation of P2X1 receptors following collagen exposure [51]. This may result from "buffering" and/or active degradation of released ATP. On the other hand, a basal level of ATP-degrading activity is essential to prevent P2X1 desensitization in vitro [52] and in vivo [53] (see Fig. 13.2Aiii). Platelets, leukocytes, endothelial cells and blood-borne microparticles all express CD39 [54–58] and plasma also has ATP degrading activity [59–61]. Ectonucleotidase activity on the surface of platelets is insufficient alone in preventing P2X1 receptor desensitization in these cells, as receptor activity is rapidly lost following resuspension in apyrase-free saline [52]. The relative importance of plasma, microparticle or cellular CD39 in protecting platelet P2X1 desensitization in vivo remains to be ascertained. Although never directly tested, it is likely that loss of NTPDase1 activity in vivo results in complete desensitization of P2X1 in platelets and other blood cells since NTPDase1$^{-/-}$ mice suffer from chronically desensitized platelet P2Y1 [62] and *vas deferens* smooth muscle P2X1 receptors [53]. It is worth noting that P2X1 receptors are more readily lost in cellular preparations compared to other Ca^{2+}-mobilising pathways [52]. In addition, since it is not common practice in many laboratories to include apyrase in cellular preparations, studies will often be conducted under conditions where this cation channel has become completely desensitized (due to ATP release during cell preparation/culture), leading to an underestimation of its overall contribution. Of relevance and importance for future work, P2X1 receptor signals are at least partially recoverable in preparations of both platelets and continuous cell lines by exposure to a nucleotide-degrading environment [61, 63].

13.3 Sources of Ligand for Activation of P2X1 Receptors

13.3.1 The Importance of ATP as the Physiological Agonist of P2X1 Receptors

The principal physiological stimulus for gating P2X1 receptors is an increase in extracellular ATP. Although early studies in both platelets and heterologous expression systems suggested that ADP is also able to gate the opening of P2X1 receptors

[3, 6, 21, 64, 65], it is now recognised that this activity is entirely due to the 1–2 % contamination of commercial samples with ATP [66, 67]. It is clear that platelets possess one ATP-gated receptor (P2X1) and two ADP-gated receptors (P2Y1 and P2Y12) [68, 69]. Whilst ATP can activate both these GPCRs at high levels of expression in heterologous systems [70–72], it is an antagonist rather than an agonist of P2Y receptors in the platelet due to the lower receptor density levels [73–75]. Therefore, ATP stimulates platelets entirely through its ability to activate P2X1 receptors. Other possible physiological agonists of P2X1 include the diadenosine polyphosphates (Ap_nAs), and adenosine polyphosphoguanosines (Ap_nGs) [76–81], which are secreted from platelets. Ap_4A and Ap_5A are the most abundant of these compounds in platelet dense granules, albeit at levels substantially lower than ATP [82–84]. It is also important to note that Ap_nAs and Ap_nGs are only partial agonists at P2X1 receptors in rat mesenteric arteries and have a potency ten-fold lower than ATP [35, 79]. However, it is possible that diadenosine polyphosphates may outlast ATP in the circulation as they are degraded more slowly than ATP [85–87], and thus may act as longer-range or longer duration P2X1 agonists. Other possible physiological P2X1 agonists are uridine adenosine tetraphosphate (Up_4A) and adenosine tetraphosphate (Ap_4) which are released from endothelial cells and reach plasma concentrations sufficient to induce potent P2X1-dependent vasoconstriction [88, 89]. However, at present ATP is the only agonist with a clear physiological role in platelet P2X1 receptor activation.

There are three well-established routes whereby cells can be exposed to an increase in extracellular ATP and thereby gate the opening of P2X receptors. Firstly, there is simple diffusion from the cytoplasm as a result of general damage to the cell membrane following injury or necrosis. Secondly, several cells including platelets store ATP at high concentrations in specialised vesicles that are released via exocytosis. Thirdly, the pannexin family of anion-permeable channels represent a route whereby cellular signals can promote ATP efflux from healthy cells or during apoptosis. Exocytosis and pannexins represent routes for ATP release following cellular stimulation, for example by agonists of GPCRs or tyrosine kinase-linked receptors (TKRs). In addition, both of these ATP efflux pathways may also contribute to the background (i.e. constitutive) release of ATP that has been reported in many cells [90–92].

13.3.2 Release of Cytoplasmic ATP Following Cellular Damage

Cytoplasmic ATP concentrations are in the order of a millimolar or more in a healthy cell as a consequence of the ubiquitous use of this nucleotide as a cellular energy source. The plasma membrane is not normally permeable to ATP, and extracellular ATP levels are generally kept very low with the aid of ectonucleotidases [47]. It is therefore easy to understand how damage to the plasma membrane can generate the micromolar extracellular concentrations of ATP that are required to stimulate P2X receptors (for example the EC_{50} for stimulation of P2X1 receptors is approximately 1 μM [2, 20, 21]). ATP levels in whole blood samples a few seconds after vascular injury have been reported to be in the range 0.2–2 μM depending upon the species, increasing to 20 μM ≈3–5 min after injury [93]. Damaged cells in the injured vessel walls were identified as the source of this ATP and it is likely that the ATP concentrations are much higher near the site of injury. Therefore, in theory this represents a potent signal to activate P2X receptors on platelets and other cells at the site of vascular injury, although the relative importance of this initial source of ATP within thrombus formation has not been clearly defined.

13.3.3 Exocytosis

Many non-excitable cells store ATP in secretory vesicles that can be released following regulated exocytosis [94]. For example, the dense granules of human platelets contain ≈0.5 M ATP, which is released via an exocytotic pathway involving SNARE (soluble N-ethylmaleimide-sensitive factor attachment protein receptor) family proteins, assisted by accessory factors [95, 96]. It is clear that this is the major mechanism of ATP release in many cells [95, 97, 98], the exception being mature human erythrocytes which do not contain secretory vesicles [99, 100]. As in excitable cells, an increase in intracellular Ca^{2+} is a key stimulus for secretion in non-excitable cells, although other signalling events will contribute [95, 101–104]. Bulk phase measurements in platelet suspensions show that ATP can reach 1–5 μM following activation by a number of stimuli [105–109], however a surface-attached luciferase shows a considerably higher peak of 15–20 μM after simulation by thrombin [98, 110]. The EC_{50} for activation of P2X1 by ATP is 1 μM and the EC_{90} is ≈10 μM, therefore the extracellular level of this nucleotide following exocytosis can achieve near-maximal stimulation of the channel [78]. Autocrine activation of P2X receptors in patch clamp studies occurs as a series of transient inward currents in a quantal manner resembling the fundamental nature of exocytosis. This also suggests that vesicular ATP release and channel activation occurs at several different sites across the surface of an individual cell [111–113]. Similar multiple secretory events from single platelets have also been observed using electrochemical measurements of serotonin, which is stored and released from dense granules [114, 115]. In these real time voltammetric measurements, a high concentration of Ca^{2+} ionophore stimulated many more secretory events per platelet than the estimated total number of dense granules (≈6–8) [98, 116–118], This suggests that each dense granule will release its content via multiple exocytotic events rather than in an all or nothing manner. Theoretically, this could allow repeated activation of P2X receptors; however given the rapid desensitization and slow recovery (minutes) of the P2X1 receptor, it is unclear to what extent repeated release from each granule contributes to the substantial autocrine/paracrine activation of the ionotropic receptor observed in platelets [13].

P2X receptor activation itself can lead to granular secretion, as occurs after P2X7 receptor stimulation in mast cells [119–121]. Another possibility whereby exocytosis can use P2X receptors to amplify responses is via insertion of additional channels from internal membranes into the plasma membrane. This has been demonstrated in alveolar type II epithelial cells, where P2X4 receptors on the lamellar bodies lead to "Fusion-activated Ca^{2+} entry" when these organelles are inserted into the plasma membrane [122].

13.3.4 Release through Pannexin Channels

Pannexins are anion-permeable channels which also allow passage of molecules up to a molecular weight of approximately 1 kDa and serve as a route for non-lytic and non-vesicular regulated release of cytosolic ATP [123, 124]. Three isoforms exist, pannexin-1, -2 and -3, which are structurally related to the connexin family that form gap junctions between adjacent cells. Connexins can also form ion channels in the plasma membrane of a solo cell, when they are referred to as hemichannels or connexons. Consequently, the term hemichannel has been ascribed to pannexin channels, however this is a misnomer since pannexins do not form gap junctions due to glycosylation of the extracellular domain [125–127]. Pannexin-1 (Panx1) is ubiquitously expressed whereas Panx2 is found in the brain and central nervous system and Panx3 is restricted to the bone and skin [128]. Panx1 is expressed on several non-excitable cell types, including macrophages, platelets, astrocytes, airway epithelial cells and erythrocytes (reviewed in [100, 129]). Panx1 channel opening has been reported in response to oxygen-glucose deprivation, caspase cleavage, mechanical stimulation and elevation of $[Ca^{2+}]_i$ [123, 130–133]. However, we are really only at an early stage in understanding the cytosolic signals and environmental cues that regulate pannexins. It is clear that they represent an important route for release of ATP in several non-excitable cell types, particularly in erythrocytes that lack the capacity for regulated exocytosis [99, 100]. It is worth noting that although other ion channels have been proposed as pathways for cytosolic ATP efflux, including the cystic fibrosis transmembrane regulator, the maxi-anion channel and connexin hemichannels, clear evidence only exists for pannexins in this role [100].

Several studies have reported interactions between Panx1-dependent ATP release and activation of purinergic P2 receptors across a number of cell types [99, 134–140]. Recently, it has been shown that agonist-dependent stimulation of platelets causes efflux of the anionic dye calcein (similar molecular weight to ATP) through Panx1 channels. This ATP efflux pathway amplified platelet functional responses (e.g. aggregation) at low concentrations of agonist via activation of P2X1 receptors [129]. Another recent study has confirmed the presence of Panx1 in human and murine platelets and highlights that the channel serves as a particularly important route for collagen-evoked ATP release and subsequent aggregation through activation of P2X1 receptors [141]. Furthermore, this study identified a single nucleotide polymorphism 400C>A, which causes a switch from glutamine to histidine within the N-terminal domain region and represents a gain in function mutation leading to increased ATP release and collagen-evoked responses. Panx1-dependent ATP release and activation of P2X receptors also plays key roles during immune responses [17, 137, 142, 143], including clearance of apoptotic cells [144–146].

13.4 Mechanisms whereby P2X1 Receptors can generate Cytosolic Ca^{2+} Signals

There are three main mechanisms whereby P2X1 receptor activation can generate or potentiate an increase in intracellular Ca^{2+}. The first is direct Ca^{2+} entry via opening of the P2X1 trimeric channel by extracellular ATP (or possibly other agonists such as diadenosine polyphosphates, see above). The second is via potentiation of phospholipase-C (PLC)-stimulated Ca^{2+} mobilization downstream of GPCRs or TKRs. Finally, since P2X1 receptors will generate substantial cellular entry of Na^+, this may lead to secondary Ca^{2+} entry via reverse Na^+/Ca^{2+} exchange. This final route is a focus of Chap. 4 of this volume and is therefore only briefly considered.

13.4.1 Direct Entry of Ca^{2+} via the Ligand-Gated Ion Channel

P2X1 receptors show a high permeability to Ca^{2+} ($P_{Ca}/P_{Na} \approx 3.9$) [5, 147]. With a resting membrane potential of −50 to −70 mV in most non-excitable cells and a 10,000-fold concentration gradient for Ca^{2+} (1 mM outside and 100 nM inside), a large inward driving force exists for this cation. Selective stimulation of P2X1 receptors by ATP or the non-hydrolysable analogue α,β-methyleneATP can generate substantial increases in intracellular Ca^{2+}, typically in the range 100–500 nM following supramaximal agonist application to fura-2-loaded washed platelet suspensions [51, 52, 148, 149] (see Fig. 13.2Ai). Since P2X1 receptor responses show a steady decline from the outset within in vitro preparations [52], this may be an underestimation of maximal Ca^{2+} responses achievable in vivo through this ionotropic receptor. Increases in extracellular Ca^{2+} cause marked increases in P2X1-evoked intracellular Ca^{2+} responses [148], presumably via a combination of greater competition for the ion channel pore and enhanced driving force. In native tissues where P2X1 is the predominant isoform, it has been estimated that ≈6–10 % of current flow through this channel under physiological conditions is mediated by Ca^{2+} [150–152] (a similar estimate, ≈8 % has been reported for human P2X4 receptors in an expression system [153]). P2X1 receptors are permeable to multiple physiological cations, including Na^+, K^+, Ca^{2+}, Mg^{2+}, and under normal ionic conditions the reversal potential for current flow through this conductance is close to 0 mV [21, 147]. Its opening will therefore promote depolarization from the standard resting potential of ≈−50 to −70 mV in non-excitable cells. Thus, it can be considered that

Ca^{2+} influx, Na^{+} entry and membrane depolarization are the three "primary" signals resulting from P2X receptor activation.

When platelet preparations are generated in a manner that limits desentization of P2X1 receptors (e.g. by addition of sufficient apyrase, see Sect. 13.2.2), activation of this channel contributes to the Ca^{2+} response stimulated by multiple platelet agonists (see Fig. 13.2B) as a consequence of ATP released from granules or Panx1 channels (see Sects. 13.3.3 and 13.3.4). The percentage contribution by P2X1 depends upon the type and level of stimulation of the GPCR or TKR. For collagen-dependent GPVI and thrombin-evoked PAR receptors, P2X1 contributes more at low, compared to high, levels of agonist. For thromboxane A_2, similar enhancements due to P2X1 activation are observed at low and mid-range agonist concentrations and for ADP, P2X1 only contributes at high levels of P2Y receptor activation. This presumably reflects a combination of the amplitude of the PLC-dependent Ca^{2+} response, together with the extent to which secretion (and/or Panx1) is stimulated. Strikingly, at low levels of collagen, P2X1 can contribute up to 92 % of the peak response, which decreases to 45 % at high collagen levels [154] (see Fig. 13.2B *left panel*). This is consistent with the greater contribution of P2X1 to aggregation responses at low compared to high collagen concentrations [14, 155]. Ca^{2+} responses mediated via another tyrosine kinase-linked platelet receptor, Toll-like receptor 1/2, are also substantially amplified by ATP release and P2X1 activation. Greater amplification by P2X1 for TKRs compared to GPCRs likely reflects the efficiency with which the signalling cascade stimulates secretion (and Panx1 activation) compared to IP_3-dependent Ca^{2+} mobilization at low levels of agonist.

Two factors could explain the high level of efficiency with which P2X1 contributes to TKR or GPCR-evoked Ca^{2+} responses. Firstly, the sites of release of ATP could be closely located to P2X1 receptors. Evidence has been presented for co-localization of P2X1 receptors and Panx1 channels in platelets [129] and both proteins translocate (along with P2X4 receptors) to the immune synapse during interaction of T cells with antigen-presenting cells [137]. There is no direct evidence for localization of P2X1 and exocytotic sites of release, however this could potentially be a function of lipid rafts. These microdomains contain both P2X1 receptors and SNARE proteins and raft disruption leads to inhibition of P2X1 responses and exocytosis [43, 156]. Secondly, the speed of activation of P2X receptors means that ATP does not need to be in the proximity of the channel for very long to gate opening, and thus trigger significant Ca^{2+} entry, prior to being swept away from the cell surface as will occur within the circulation.

13.4.2 P2X-Dependent amplification of Phospholipase-C-stimulated Ca^{2+} Signals

Studies in platelets and a non-excitable cell line have shown that when P2X1 receptors and P2Y1 receptors are co-stimulated, the Ca^{2+} increase is super-additive, i.e. greater than the sum of the responses to each receptor alone [12, 157] (see

Fig. 13.2C). This effect is due to enhancement of the P2Y1 receptors by the P2X1-evoked Ca^{2+} influx without any evidence for a role for the Na^{+} entry or membrane depolarization that also occur following activation of the ionotropic receptor (see above). This effect is not restricted to the P2Y1 receptor as it also occurs with co-activation of P2X1 and thrombin-stimulated PAR1 or acetylcholine-stimulated muscarinic M1 receptors [157]. The underlying cause of the synergy is Ca^{2+} influx and a general increase in free Ca^{2+} (rather than a P2X1-specific subplasma membrane cytosolic Ca^{2+} response) as ionomycin similarly enhanced P2Y1 receptors. It is likely that Ca^{2+} works to enhance the GPCR-evoked Ca^{2+} mobilization by an effect on PLC and IP_3 receptors, both of which are known to be stimulated by Ca^{2+} increases in the concentration range observed following P2X1 stimulation [158–161], although the relative importance of these two targets was not resolved. Interestingly, the potentiation of GPCR Ca^{2+} signals by P2X1 declines with time after addition of the ionotropic receptor agonist, lasting for over a minute following a maximal response. This could represent an important means whereby P2X1 receptors potentiate platelet function since it can be argued that ATP will often be in excess of ADP immediately following release from damaged endothelial cells and ADP will subsequently rise in concentration due to ectonucleotidase activity [47]. Although no evidence was obtained for a role of Na^{+} influx or membrane depolarization in the synergy between P2X1 and P2Y1 receptors [157], Gq-protein-coupled receptors in some cell types are potentiated by membrane depolarizations [162–165] of the magnitude expected following opening of P2X receptors [166].

13.4.3 P2X1-Dependent Secondary Entry through Reverse Na^{+}/Ca^{2+} Exchange Activity

If 6–10 % of the P2X1-evoked inward current is carried by Ca^{2+} under physiological concentrations (see above), the remainder will be due to Na^{+} entry. Indeed, P2X1 receptors are known to generate significant increases in intracellular Na^{+} in platelets [167]. This may allow reverse Na^{+}/Ca^{2+} exchange (NCX) activity to generate Ca^{2+} entry, depending upon the membrane potential (as discussed in [168]). A NCX blocker has been used to suggest that reverse NCX activity occurs in platelets following P2X1 activation [169]. Readers are directed to another Chapter in this volume (Chap. 4) for more information on the role of reverse NCX activity as a pathway for Ca^{2+} entry.

13.5 Signalling through P2X1 Receptors

All P2X1-dependent functional responses require Ca^{2+} influx [14, 52, 148, 170, 171] and there is a strong correlation between the intracellular Ca^{2+} increase and shape change response to the P2X1 agonist α,β-meATP [52, 148]. The Ca^{2+} increase activates calmodulin dependent myosin light chain kinase (MLCK) leading to myosin light chain phosphorylation (MLC-P) that is presumably responsible for the cytoskeletal events underlying P2X1-evoked shape change and granule localization [171]. P2X1 does not stimulate shape change through Rho-kinase in the platelet [172]. Selective P2X1 activation has also been shown to stimulate transient phosphorylation of ERK2, a member of the mitogen-activated protein kinase family [173]. Based upon experiments with extracellular EGTA, the broad spectrum PKC inhibitor GF109203-X and the calmodulin inhibitor W-7, P2X1-evoked ERK2 phosphorylation requires extracellular calcium and activation of both calmodulin and PKC [171, 173]. However, since P2X1-evoked ERK2 phosphorylation is slow compared to the shape change, the main role of this signalling event is envisaged to be enhancement of aggregation at low to intermediate levels of collagen [171, 173, 174]. This role for ERK2 in P2X1-induced potentiation of collagen responses is further supported by studies using transgenic mice over-expressing the human P2X1 receptor in the megakaryocytic cell lineage [174]. It is, however, unclear why concentrations of ADP that normally stimulate Ca^{2+} responses of larger amplitude and longer duration compared to those generated by P2X1 receptors [148, 171, 173] fail to activate ERK2 phosphorylation [173]. Recently, it has been shown that ERK2 does not contribute to the potentiation of P2Y1 receptor-evoked Ca^{2+} responses by P2X1 [157]. However, pharmacological inhibition of ERK1/2 prevented the potentiation of aggregation responses resulting from co-activation of these two receptors in human platelets [157]. This concurs with the fact that P2X1-evoked ERK2 phosphorylation is relatively slow compared to the shape change [173] and that inhibition of ERK1/2 with the MEK1/2 antagonist U0126 does not alter α,β-meATP-evoked shape change [171]. However, U0126 does not influence the amplification of thromboxane A_2 responses by P2X1, including aggregation and P-selectin surface expression; instead this requires activation of the p38 MAPK [175]. In T lymphocytes stimulated by hypertonic stress, P2X1, along with P2X4 and P2X7, also stimulate interleukin-2 production through activation of p38 MAPK [176].

13.6 The Relevance of Non-Excitable Cell P2X1 Receptors in Health and Disease

13.6.1 Thrombosis and Haemostasis

Selective activation of P2X1 receptors in human and murine platelets leads to shape change and granule centralization without significant aggregation [14, 15, 52, 148, 155, 170, 173]. However, release of ATP and thus secondary activation of P2X1 receptors contributes to aggregation induced by low levels of collagen [14, 155], PAR1 agonist [15] or thromboxane A_2 [175]. In addition, P2X1 receptors potentiate aggregation when coactivated alongside low level stimulation of receptors for thrombin [15], ADP [157], thrombopoietin and adrenaline [170]. P2X1 receptors also amplify thrombus formation on a collagen-coated surface under flow, with the greatest contribution observed at levels of shear experienced within small arteries or arterioles and with little role at venous shear rates [14]. Genetic deletion or selective pharmacological block of P2X1 receptors in mice reduces thrombus formation in small arteries following laser injury [14, 15, 177] and also limits the fatal consequences of thromboembolism induced by co-injection of collagen and adrenaline [14, 177]. Furthermore, overexpression of human P2X1 receptors in a murine model enhances P2X1 receptor-dependent responses in vitro and in vivo [172, 174]. Neutrophils also accumulate at sites of vascular injury where they contribute to fibrin formation [178, 179] and indeed are one of the first cells to adhere to the inured vessel wall [180]. Interestingly, a recent study has reported that P2X1 receptors on both neutrophils and platelets are responsible for the reduced arterial thrombosis observed following vascular injury in P2X1$^{-/-}$ mice [19]. An increase in intracellular Ca^{2+} was observed in neutrophils activated at the sites of vascular injury prior to thrombus formation although whether this is entirely due to Ca^{2+} influx through P2X1 receptors was not studied, presumably because neutrophil adherence was virtually absent in P2X1$^{-/-}$ mice. Together, these studies suggest that P2X1 receptors may represent a useful target to prevent arterial thrombosis, particularly since bleeding times are little affected by loss of function of this receptor [14, 177].

An important property of P2X1 receptors that distinguishes them from other Ca^{2+} permeable ion channels in the platelet is their resistance to inhibition by cyclic nucleotides [4, 51, 113, 181]. Nitric oxide and prostacyclin are crucial signals whereby the endothelium can inhibit platelet function in the intact circulation by elevating cyclic GMP and cyclic AMP, respectively [182]. These cyclic nucleotides inhibit Ca^{2+} signals via actions on PLC and also IP_3 receptors, which will result in reduced activation of virtually all Ca^{2+} mobilization pathways (TRPC6, IP_3R, store-operated Orai1 channels) with the exception of ATP-gated P2X1 receptors [51, 182–184]. In addition, the platelet dense granule secretory responses downstream of GPVI collagen-receptors and the innate immune receptor TLR1/2 are also partially resistant to elevations of cytosolic cyclic nucleotides [51]. This may allow P2X1 receptors to act as a "primer" of platelet responses since they will be more active in the initial stages of vascular injury when residual effects of NO and PGI_2 are present.

13.6.2 Inflammation and Sepsis

Extracellular ATP is an important signal during the early stages of infection and inflammation [185]. In contrast to platelets, most leukocytes express multiple P2X receptor subtypes, particularly P2X7, 4 and 1 [143, 186]. P2X7 receptors, formerly known as P2Z receptors, have well-established roles in immune cell responses [121, 187, 188]. More recently, P2X1 receptors have also been shown to be important in activation of T cells by antigen-presenting cells [137] and to stimulate neutrophil chemotaxis [18, 189]. It is unclear why Rho kinase is responsible for P2X1-evoked neutrophil migration [189] but not involved in the shape change responses mediated by this receptor in the platelet, which instead depend upon calmodulin-dependent myosin light chain kinase (MLCK) [171]. Neutrophils are key mediators of the innate immune response and known to be recruited to infected tissues to fight bacterial infection, however when excessively activated they can lead to tissue damage, as occurs in sepsis [190, 191]. At sites of inflammation, neutrophils adhere to the activated endothelium prior to migration across the vascular wall [192]. In an animal model of sepsis, loss of P2X1 receptors reduced neutrophil migration from the circulation and protected against tissue damage following injection of LPS [18]. However, opposite results were obtained using similar LPS concentrations by another group [193], who observed enhanced neutrophil migration into the lungs and reduced animal survival in P2X1$^{-/-}$ compared to wild type mice. The reason for these contrasting results remains unclear. It is worth noting that whilst P2X1 receptors contribute to neutrophil signalling, they also promote thrombus formation (see above) which can either accentuate tissue damage or improve host protection by trapping and destroying invading organisms [194].

13.7 Conclusions

Extracellular nucleotides are important autocrine and intercellular signalling molecules, mediating their effects through surface receptors that fall into two categories, ligand gated P2X and G-protein-coupled P2Y receptors. Ca^{2+}-permeable P2X receptors play key roles in a variety of cell types, including non-excitable cells such as platelets and leukocytes, and thus represent therapeutic targets for treatment of thrombosis and modulation of immune responses. In the platelet it is well established that Ca^{2+} influx is the main signalling event that stimulates P2X1 receptor-dependent functional responses, alone or in synergy with other receptor signalling pathways, although Na^{+} influx will also occur. It is likely that the substantial contribution of ATP-gated P2X receptors to cellular responses results from their significant Ca^{2+} permeability, efficient stimulation by ATP release and rapid activation kinetics compared to other Ca^{2+} mobilization pathways.

References

1. Ralevic V, Burnstock G (1998) Receptors for purines and pyrimidines. Pharmacol Rev 50(3):413–492
2. North RA (2002) Molecular physiology of P2X receptors. Physiol Rev 82(4):1013–1067
3. Mahaut-Smith MP, Sage SO, Rink TJ (1992) Rapid ADP-evoked currents in human platelets recorded with the nystatin permeabilized patch technique. J Biol Chem 267(5):3060–3065
4. Sage SO, Rink TJ (1987) The kinetics of changes in intracellular calcium concentration in fura-2-loaded human platelets. J Biol Chem 262(34):16364–16369
5. Egan TM, Khakh BS (2004) Contribution of calcium ions to P2X channel responses. J Neurosci: 24(13):3413–3420

6. Sage SO, Reast R, Rink TJ (1990) ADP evokes biphasic Ca^{2+} influx in fura-2-loaded human platelets. Evidence for Ca^{2+} entry regulated by the intracellular Ca^{2+} store. Biochem J 265(3):675–680
7. Lohse MJ, Hein P, Hoffmann C, Nikolaev VO, Vilardaga JP, Bunemann M (2008) Kinetics o f
G-protein-coupled receptor signals in intact cells. Br J Pharmacol 153(Suppl 1):S125–S132
8. Morrell CN, Sun H, Ikeda M, Beique JC, Swaim AM, Mason E, Martin TV, Thompson LE, Gozen O, Ampagoomian D, Sprengel R, Rothstein J, Faraday N, Huganir R, Lowenstein CJ (2008) Glutamate mediates platelet activation through the AMPA receptor. J Exp Med 205(3):575–584
9. Sun H, Swaim A, Herrera JE, Becker D, Becker L, Srivastava K, Thompson LE, Shero MR, Perez-Tamayo A, Suktitipat B, Mathias R, Contractor A, Faraday N, Morrell CN (2009) Platelet kainate receptor signaling promotes thrombosis by stimulating cyclooxygenase activation. Circ Res 105(6):595–603
10. Sun B, Li J, Okahara K, Kambayashi J (1998) $P2X_1$ purinoceptor in human platelets. Molecular cloning and functional characterization after heterologous expression. J Biol Chem 273(19):11544–11547
11. Wang L, Ostberg O, Wihlborg AK, Brogren H, Jern S, Erlinge D (2003) Quantification of ADP and ATP receptor expression in human platelets. J Thromb Haemost 1(2):330–336
12. Vial C, Rolf MG, Mahaut-Smith MP, Evans RJ (2002) A study of $P2X_1$ receptor function in murine megakaryocytes and human platelets reveals synergy with P2Y receptors. Br J Pharmacol 135(2):363–372
13. Fung CY, Cendana C, Farndale RW, Mahaut-Smith MP (2007) Primary and secondary agonists can use $P2X_1$ receptors as a major pathway to increase intracellular Ca^{2+} in the human platelet. J Thromb Haemost 5(5):910–917
14. Hechler B, Lenain N, Marchese P, Vial C, Heim V, Freund M, Cazenave JP, Cattaneo M, Ruggeri ZM, Evans R, Gachet C (2003) A role of the fast ATP-gated $P2X_1$ cation channel in thrombosis of small arteries in vivo. J Exp Med 198(4):661–667
15. Erhardt JA, Toomey JR, Douglas SA, Johns DG (2006) $P2X_1$ stimulation promotes thrombin receptor-mediated platelet aggregation. J Thromb Haemost 4(4):882–890
16. Feske S, Skolnik EY, Prakriya M (2012) Ion channels and transporters in lymphocyte function and immunity. Nat Rev Immunol 12(7):532–547
17. Idzko M, Ferrari D, Eltzschig HK (2014) Nucleotide signalling during inflammation. Nature 509(7500):310–317
18. Maitre B, Magnenat S, Heim V, Ravanat C, Evans RJ, de la Salle H, Gachet C, Hechler B (2015) The $P2X_1$ receptor is required for neutrophil extravasation during lipopolysaccharide-induced lethal endotoxemia in mice. J Immunol 194(2):739–749
19. Darbousset R, Delierneux C, Mezouar S, Hego A, Lecut C, Guillaumat I, Riederer MA, Evans RJ, Dignat-George F, Panicot-Dubois L, Oury C, Dubois C (2014) P2X1 expressed on polymorphonuclear neutrophils and platelets is required for thrombosis in mice. Blood 124(16):2575–2585
20. Brake AJ, Wagenbach MJ, Julius D (1994) New structural motif for ligand-gated ion channels defined by an ionotropic ATP receptor. Nature 371(6497):519–523
21. Valera S, Hussy N, Evans RJ, Adami N, North RA, Surprenant A, Buell G (1994) A new class of ligand-gated ion channel defined by P2x receptor for extracellular ATP. Nature 371(6497):516–519
22. Kaczmarek-Hajek K, Lorinczi E, Hausmann R, Nicke A (2012) Molecular and functional properties of P2X receptors – recent progress and persisting challenges. Purinergic Signal 8(3):375–417
23. North RA, Surprenant A (2000) Pharmacology of cloned P2X receptors. Annu Rev Pharmacol Toxicol 40:563–580
24. Hattori M, Gouaux E (2012) Molecular mechanism of ATP binding and ion channel activation in P2X receptors. Nature 485(7397):207–212

25. Kawate T, Michel JC, Birdsong WT, Gouaux E (2009) Crystal structure of the ATP-gated $P2X_4$ ion channel in the closed state. Nature 460(7255):592–598
26. Evans RJ (2010) Structural interpretation of P2X receptor mutagenesis studies on drug action. Br J Pharmacol 161(5):961–971
27. Young MT (2010) P2X receptors: dawn of the post-structure era. Trends Biochem Sci 35(2):83–90
28. Rassendren F, Buell G, Newbolt A, North RA, Surprenant A (1997) Identification of amino acid residues contributing to the pore of a P2X receptor. EMBO J 16(12):3446–3454
29. Egan TM, Haines WR, Voigt MM (1998) A domain contributing to the ion channel of ATP-gated P2X2 receptors identified by the substituted cysteine accessibility method. J Neurosci 18(7):2350–2359
30. Li M, Chang TH, Silberberg SD, Swartz KJ (2008) Gating the pore of P2X receptor channels. Nat Neurosci 11(8):883–887
31. Kawate T, Robertson JL, Li M, Silberberg SD, Swartz KJ (2011) Ion access pathway to the transmembrane pore in P2X receptor channels. J Gen Physiol 137(6):579–590
32. Kracun S, Chaptal V, Abramson J, Khakh BS (2010) Gated access to the pore of a P2X receptor: structural implications for closed-open transitions. J Biol Chem 285(13):10110–10121
33. Migita K, Haines WR, Voigt MM, Egan TM (2001) Polar residues of the second transmembrane domain influence cation permeability of the ATP-gated $P2X_2$ receptor. J Biol Chem 276(33):30934–30941
34. Samways DS, Egan TM (2007) Acidic amino acids impart enhanced Ca^{2+} permeability and flux in two members of the ATP-gated P2X receptor family. J Gen Physiol 129(3):245–256
35. Lewis CJ, Evans RJ (2000) Lack of run-down of smooth muscle P2X receptor currents recorded with the amphotericin permeabilized patch technique, physiological and pharmacological characterization of the properties of mesenteric artery P2X receptor ion channels. Br J Pharmacol 131(8):1659–1666
36. Allsopp RC, Evans RJ (2011) The intracellular amino terminus plays a dominant role in desensitization of ATP-gated P2X receptor ion channels. J Biol Chem 286(52):44691–44701
37. Allsopp RC, Farmer LK, Fryatt AG, Evans RJ (2013) P2X receptor chimeras highlight roles of the amino terminus to partial agonist efficacy, the carboxyl terminus to recovery from desensitization, and independent regulation of channel transitions. J Biol Chem 288(29):21412–21421
38. Fryatt AG, Evans RJ (2014) Kinetics of conformational changes revealed by voltage-clamp fluorometry give insight to desensitization at ATP-gated human P2X1 receptors. Mol Pharmacol 86(6):707–715
39. Vial C, Tobin AB, Evans RJ (2004) G-protein-coupled receptor regulation of $P2X_1$ receptors does not involve direct channel phosphorylation. Biochem J 382(Pt 1):101–110
40. Ase AR, Raouf R, Belanger D, Hamel E, Seguela P (2005) Potentiation of P2X1 ATP-gated currents by 5-hydroxytryptamine 2A receptors involves diacylglycerol-dependent kinases and intracellular calcium. J Pharmacol Exp Ther 315(1):144–154
41. Wen H, Evans RJ (2009) Regions of the amino terminus of the P2X receptor required for modification by phorbol ester and mGluR1alpha receptors. J Neurochem 108(2):331–340
42. Vial C, Evans RJ (2005) Disruption of lipid rafts inhibits P2X1 receptor-mediated currents and arterial vasoconstriction. J Biol Chem 280(35):30705–30711
43. Vial C, Fung CY, Goodall AH, Mahaut-Smith MP, Evans RJ (2006) Differential sensitivity of human platelet $P2X_1$ and $P2Y_1$ receptors to disruption of lipid rafts. Biochem Biophys Res Commun 343(2):415–419
44. Allsopp RC, Lalo U, Evans RJ (2010) Lipid raft association and cholesterol sensitivity of P2X1-4 receptors for ATP: chimeras and point mutants identify intracellular amino-terminal residues involved in lipid regulation of P2X1 receptors. J Biol Chem 285(43):32770–32777
45. Lalo U, Roberts JA, Evans RJ (2011) Identification of human P2X1 receptor-interacting proteins reveals a role of the cytoskeleton in receptor regulation. J Biol Chem 286(35):30591–30599

46. Lalo U, Jones S, Roberts JA, Mahaut-Smith MP, Evans RJ (2012) Heat shock protein 90 inhibitors reduce trafficking of ATP-gated P2X1 receptors and human platelet responsiveness. J Biol Chem 287(39):32747–32754
47. Robson SC, Sevigny J, Zimmermann H (2006) The E-NTPDase family of ectonucleotidases: structure function relationships and pathophysiological significance. Purinergic Signal 2(2):409–430
48. Kukulski F, Levesque SA, Sevigny J (2011) Impact of ectoenzymes on p2 and p1 receptor signaling. Adv Pharmacol 61:263–299
49. Yegutkin GG (2008) Nucleotide- and nucleoside-converting ectoenzymes: important modulators of purinergic signalling cascade. Biochim Biophys Acta 1783(5):673–694
50. Handa M, Guidotti G (1996) Purification and cloning of a soluble ATP-diphosphohydrolase (apyrase) from potato tubers (*Solanum tuberosum*). Biochem Biophys Res Commun 218(3):916–923
51. Fung CY, Jones S, Ntrakwah A, Naseem KM, Farndale RW, Mahaut-Smith MP (2012) Platelet Ca^{2+} responses coupled to glycoprotein VI and Toll-like receptors persist in the presence of endothelial-derived inhibitors: roles for secondary activation of P2X1 receptors and release from intracellular Ca^{2+} stores. Blood 119(15):3613–3621
52. Rolf MG, Brearley CA, Mahaut-Smith MP (2001) Platelet shape change evoked by selective activation of P2X1 purinoceptors with α, β-methylene ATP. Thromb Haemost 85(2):303–308
53. Kauffenstein G, Pelletier J, Lavoie EG, Kukulski F, Martin-Satue M, Dufresne SS, Frenette J, Ribas Furstenau C, Sereda MJ, Toutain B, Henrion D, Sullivan R, Vial C, Sevigny J (2014) Nucleoside triphosphate diphosphohydrolase-1 ectonucleotidase is required for normal vas deferens contraction and male fertility through maintaining P2X1 receptor function. J Biol Chem 289(41):28629–28639
54. Kansas GS, Wood GS, Tedder TF (1991) Expression, distribution, and biochemistry of human CD39. Role in activation-associated homotypic adhesion of lymphocytes. J Immunol 146(7):2235–2244
55. Marcus AJ, Broekman MJ, Drosopoulos JH, Islam N, Alyonycheva TN, Safier LB, Hajjar KA, Posnett DN, Schoenborn MA, Schooley KA, Gayle RB, Maliszewski CR (1997) The endothelial cell ecto-ADPase responsible for inhibition of platelet function is CD39. J Clin Invest 99(6):1351–1360
56. Koziak K, Sevigny J, Robson SC, Siegel JB, Kaczmarek E (1999) Analysis of CD39/ATP diphosphohydrolase (ATPDase) expression in endothelial cells, platelets and leukocytes. Thromb Haemost 82(5):1538–1544
57. Atkinson B, Dwyer K, Enjyoji K, Robson SC (2006) Ecto-nucleotidases of the CD39/NTPDase family modulate platelet activation and thrombus formation: potential as therapeutic targets. Blood Cells Mol Dis 36(2):217–222
58. Banz Y, Beldi G, Wu Y, Atkinson B, Usheva A, Robson SC (2008) CD39 is incorporated into plasma microparticles where it maintains functional properties and impacts endothelial activation. Br J Haematol 142(4):627–637
59. Birk AV, Bubman D, Broekman MJ, Robertson HD, Drosopoulos JH, Marcus AJ, Szeto HH (2002) Role of a novel soluble nucleotide phospho-hydrolase from sheep plasma in inhibition of platelet reactivity: hemostasis, thrombosis, and vascular biology. J Lab Clin Med 139(2):116–124
60. Jones S, Evans RJ, Mahaut-Smith MP (2011) Extracellular Ca^{2+} modulates ADP-evoked aggregation through altered agonist degradation: implications for conditions used to study P2Y receptor activation. Br J Haematol 153(1):83–91
61. Cauwenberghs S, Feijge MA, Hageman G, Hoylaerts M, Akkerman JW, Curvers J, Heemskerk JW (2006) Plasma ectonucleotidases prevent desensitization of purinergic receptors in stored platelets: importance for platelet activity during thrombus formation. Transfusion 46(6):1018–1028
62. Enjyoji K, Sevigny J, Lin Y, Frenette PS, Christie PD, Esch JS 2nd, Imai M, Edelberg JM, Rayburn H, Lech M, Beeler DL, Csizmadia E, Wagner DD, Robson SC, Rosenberg RD

(1999) Targeted disruption of cd39/ATP diphosphohydrolase results in disordered hemostasis and thromboregulation. Nat Med 5(9):1010–1017
63. Buell G, Michel AD, Lewis C, Collo G, Humphrey PP, Surprenant A (1996) P2X1 receptor activation in HL60 cells. Blood 87(7):2659–2664
64. Mahaut-Smith MP, Sage SO, Rink TJ (1990) Receptor-activated single channels in intact human platelets. J Biol Chem 265(18):10479–10483
65. Sage SO, Pintado E, Mahaut-Smith MP, Merritt JE (1990) Rapid kinetics of agonist-evoked changes in cytosolic free Ca^{2+} concentration in fura-2-loaded human neutrophils. Biochem J 265(3):915–918
66. Mahaut-Smith MP, Ennion SJ, Rolf MG, Evans RJ (2000) ADP is not an agonist at $P2X_1$ receptors: evidence for separate receptors stimulated by ATP and ADP on human platelets. Br J Pharmacol 131(1):108–114
67. Vial C, Pitt SJ, Roberts J, Rolf MG, Mahaut-Smith MP, Evans RJ (2003) Lack of evidence for functional ADP-activated human $P2X_1$ receptors supports a role for ATP during hemostasis and thrombosis. Blood 102(10):3646–3651
68. Hechler B, Gachet C (2011) P2 receptors and platelet function. Purinergic Signal 7(3):293–303
69. Kahner BN, Shankar H, Murugappan S, Prasad GL, Kunapuli SP (2006) Nucleotide receptor signaling in platelets. J Thromb Haemost 4(11):2317–2326
70. Palmer RK, Boyer JL, Schachter JB, Nicholas RA, Harden TK (1998) Agonist action of adenosine triphosphates at the human P2Y1 receptor. Mol Pharmacol 54(6):1118–1123
71. Takasaki J, Kamohara M, Saito T, Matsumoto M, Matsumoto S, Ohishi T, Soga T, Matsushime H, Furuichi K (2001) Molecular cloning of the platelet $P2T_{AC}$ ADP receptor: pharmacological comparison with another ADP receptor, the $P2Y_1$ receptor. Mol Pharmacol 60(3):432–439
72. Waldo GL, Harden TK (2004) Agonist binding and Gq-stimulating activities of the purified human P2Y1 receptor. Mol Pharmacol 65(2):426–436
73. Hechler B, Vigne P, Leon C, Breittmayer JP, Gachet C, Frelin C (1998) ATP derivatives are antagonists of the $P2Y_1$ receptor: similarities to the platelet ADP receptor. Mol Pharmacol 53(4):727–733
74. Leon C, Hechler B, Vial C, Leray C, Cazenave JP, Gachet C (1997) The $P2Y_1$ receptor is an ADP receptor antagonized by ATP and expressed in platelets and megakaryoblastic cells. FEBS Lett 403(1):26–30
75. Mills DC (1996) ADP receptors on platelets. Thromb Haemost 76(6):835–856
76. Bianchi BR, Lynch KJ, Touma E, Niforatos W, Burgard EC, Alexander KM, Park HS, Yu H, Metzger R, Kowaluk E, Jarvis MF, van Biesen T (1999) Pharmacological characterization of recombinant human and rat P2X receptor subtypes. Eur J Pharmacol 376(1–2):127–138
77. Cinkilic O, King BF, van der Giet M, Schluter H, Zidek W, Burnstock G (2001) Selective agonism of group I P2X receptors by dinucleotides dependent on a single adenine moiety. J Pharmacol Exp Ther 299(1):131–136
78. Evans RJ, Lewis C, Buell G, Valera S, North RA, Surprenant A (1995) Pharmacological characterization of heterologously expressed ATP-gated cation channels (P2x purinoceptors). Mol Pharmacol 48(2):178–183
79. Lewis CJ, Gitterman DP, Schluter H, Evans RJ (2000) Effects of diadenosine polyphosphates (Ap_nAs) and adenosine polyphospho guanosines (Ap_nGs) on rat mesenteric artery P2X receptor ion channels. Br J Pharmacol 129(1):124–130
80. Sage SO, MacKenzie AB, Jenner S, Mahaut-Smith MP (1997) Purinoceptor-evoked calcium signalling in human platelets. Prostaglandins Leukot Essent Fatty Acids 57(4–5):435–438
81. Wildman SS, Brown SG, King BF, Burnstock G (1999) Selectivity of diadenosine polyphosphates for rat P2X receptor subunits. Eur J Pharmacol 367(1):119–123
82. Jankowski J, Hagemann J, Tepel M, van Der Giet M, Stephan N, Henning L, Gouni-Berthold I, Sachinidis A, Zidek W, Schluter H (2001) Dinucleotides as growth-promoting extracellular mediators. Presence of dinucleoside diphosphates Ap_2A, Ap_2G, and Gp_2G in releasable granules of platelets. J Biol Chem 276(12):8904–8909

83. Jankowski J, Potthoff W, van der Giet M, Tepel M, Zidek W, Schluter H (1999) High-performance liquid chromatographic assay of the diadenosine polyphosphates in human platelets. Anal Biochem 269(1):72–78
84. Jankowski J, Tepel M, van der Giet M, Tente IM, Henning L, Junker R, Zidek W, Schluter H (1999) Identification and characterization of P[1], P[7]-Di(adenosine-5′)-heptaphosphate from human platelets. J Biol Chem 274(34):23926–23931
85. Luthje J, Ogilvie A (1988) Catabolism of Ap_4A and Ap_3A in whole blood. The dinucleotides are long-lived signal molecules in the blood ending up as intracellular ATP in the erythrocytes. Eur J Biochem/FEBS 173(1):241–245
86. Schluter H, Tepel M, Zidek W (1996) Vascular actions of diadenosine phosphates. J Auton Pharmacol 16(6):357–362
87. Zimmermann H (2000) Extracellular metabolism of ATP and other nucleotides. Naunyn Schmiedebergs Arch Pharmacol 362(4–5):299–309
88. Tolle M, Jankowski V, Schuchardt M, Wiedon A, Huang T, Hub F, Kowalska J, Jemielity J, Guranowski A, Loddenkemper C, Zidek W, Jankowski J, van der Giet M (2008) Adenosine 5′-tetraphosphate is a highly potent purinergic endothelium-derived vasoconstrictor. Circ Res 103(10):1100–1108
89. Jankowski V, Tolle M, Vanholder R, Schonfelder G, van der Giet M, Henning L, Schluter H, Paul M, Zidek W, Jankowski J (2005) Uridine adenosine tetraphosphate: a novel endothelium- derived vasoconstrictive factor. Nat Med 11(2):223–227
90. Corriden R, Insel PA (2010) Basal release of ATP: an autocrine-paracrine mechanism for cell regulation. Sci Signal 3(104):re1
91. Hayoz S, Jia C, Hegg C (2012) Mechanisms of constitutive and ATP-evoked ATP release in neonatal mouse olfactory epithelium. BMC Neurosci 13:53
92. Sivaramakrishnan V, Bidula S, Campwala H, Katikaneni D, Fountain SJ (2012) Constitutive lysosome exocytosis releases ATP and engages P2Y receptors in human monocytes. J Cell Sci 125(Pt 19):4567–4575
93. Born GV, Kratzer MA (1984) Source and concentration of extracellular adenosine triphosphate during haemostasis in rats, rabbits and man. J Physiol 354:419–429
94. Praetorius HA, Leipziger J (2009) ATP release from non-excitable cells. Purinergic Signal 5(4):433–446
95. Golebiewska EM, Poole AW (2013) Secrets of platelet exocytosis – what do we really know about platelet secretion mechanisms? Br J Haematol 165(2):204–216
96. Koseoglu S, Flaumenhaft R (2013) Advances in platelet granule biology. Curr Opin Hematol 20(5):464–471
97. Bowser DN, Khakh BS (2007) Vesicular ATP is the predominant cause of intercellular calcium waves in astrocytes. J Gen Physiol 129(6):485–491
98. Holmsen H, Weiss HJ (1979) Secretable storage pools in platelets. Annu Rev Med 30:119–134
99. Locovei S, Bao L, Dahl G (2006) Pannexin 1 in erythrocytes: function without a gap. Proc Natl Acad Sci U S A 103(20):7655–7659
100. Dahl G (2015) ATP release through pannexon channels. Philos Trans R Soc Lond B Biol Sci 370(1672):20140191
101. Tomes CN (2015) The proteins of exocytosis: lessons from the sperm model. Biochem J 465(3):359–370
102. Wernersson S, Pejler G (2014) Mast cell secretory granules: armed for battle. Nat Rev Immunol 14(7):478–494
103. Montana V, Malarkey EB, Verderio C, Matteoli M, Parpura V (2006) Vesicular transmitter release from astrocytes. Glia 54(7):700–715
104. Messenger SW, Falkowski MA, Groblewski GE (2014) Ca^{2+}-regulated secretory granule exocytosis in pancreatic and parotid acinar cells. Cell Calcium 55(6):369–375
105. Higashi T, Isomoto A, Tyuma I, Kakishita E, Uomoto M, Nagai K (1985) Quantitative and continuous analysis of ATP release from blood platelets with firefly luciferase luminescence. Thromb Haemost 53(1):65–69

106. Mills DC, Robb IA, Roberts GC (1968) The release of nucleotides, 5-hydroxytryptamine and enzymes from human blood platelets during aggregation. J Physiol 195(3):715–729
107. Holmsen H, Day HJ (1971) Adenine nucleotides and platelet function. Ser Haematol 4(1):28–58
108. Detwiler TC, Feinman RD (1973) Kinetics of the thrombin-induced release of adenosine triphosphate by platelets. Comparison with release of calcium. Biochemistry 12(13):2462–2468
109. Ingerman CM, Smith JB, Silver MJ (1979) Direct measurement of platelet secretion in whole blood. Thromb Res 16(3–4):335–344
110. Beigi R, Kobatake E, Aizawa M, Dubyak GR (1999) Detection of local ATP release from activated platelets using cell surface-attached firefly luciferase. Am J Physiol 276(1):C267–C278
111. Hollins B, Ikeda SR (1997) Heterologous expression of a P2x-purinoceptor in rat chromaffin cells detects vesicular ATP release. J Neurophysiol 78(6):3069–3076
112. Kawa K (2004) Discrete but simultaneous release of adenine nucleotides and serotonin from mouse megakaryocytes as detected with patch- and carbon-fiber electrodes. Am J Physiol Cell Physiol 286(1):C119–C128
113. Mahaut-Smith MP, Tolhurst G, Evans RJ (2004) Emerging roles for P2X1 receptors in platelet activation. Platelets 15(3):131–144
114. Ge S, White JG, Haynes CL (2009) Quantal release of serotonin from platelets. Anal Chem 81(8):2935–2943
115. Ge S, Woo E, White JG, Haynes CL (2011) Electrochemical measurement of endogenous serotonin release from human blood platelets. Anal Chem 83(7):2598–2604
116. Costa JL, Detwiler TC, Feinman RD, Murphy DL, Patlak CS, Pettigrew KD (1977) Quantitative evaluation of the loss of human platelet dense bodies following stimulation by thrombin or A23187. J Physiol 264(1):297–306
117. Costa JL, Reese TS, Murphy DL (1974) Serotonin storage in platelets: estimation of storage-packet size. Science 183(4124):537–538
118. White JG (1968) The dense bodies of human platelets. Origin of serotonin storage particles from platelet granules. Am J Pathol 53(5):791–808
119. Cockcroft S, Gomperts BD (1979) Activation and inhibition of calcium-dependent histamine secretion by ATP ions applied to rat mast cells. J Physiol 296:229–243
120. Cockcroft S, Gomperts BD (1979) ATP induces nucleotide permeability in rat mast cells. Nature 279(5713):541–542
121. Surprenant A, Rassendren F, Kawashima E, North RA, Buell G (1996) The cytolytic P_{2Z} receptor for extracellular ATP identified as a P_{2X} receptor ($P2X_7$). Science 272(5262):735–738
122. Miklavc P, Thompson KE, Frick M (2013) A new role for $P2X_4$ receptors as modulators of lung surfactant secretion. Front Cell Neurosci 7:171
123. Bao Y, Chen Y, Ledderose C, Li L, Junger WG (2013) Pannexin 1 channels link chemoattractant receptor signaling to local excitation and global inhibition responses at the front and back of polarized neutrophils. J Biol Chem 288(31):22650–22657
124. Ma W, Compan V, Zheng W, Martin E, North RA, Verkhratsky A, Surprenant A (2012) Pannexin 1 forms an anion-selective channel. Pflugers Archiv: Eur J Physiol 463(4):585–592
125. Boassa D, Ambrosi C, Qiu F, Dahl G, Gaietta G, Sosinsky G (2007) Pannexin1 channels contain a glycosylation site that targets the hexamer to the plasma membrane. J Biol Chem 282(43):31733–31743
126. Boassa D, Qiu F, Dahl G, Sosinsky G (2008) Trafficking dynamics of glycosylated pannexin 1 proteins. Cell Commun Adhes 15(1):119–132
127. Scemes E, Spray DC, Meda P (2009) Connexins, pannexins, innexins: novel roles of "hemi-channels". Pflugers Archiv: Eur J Physiol 457(6):1207–1226
128. Bruzzone R, Hormuzdi SG, Barbe MT, Herb A, Monyer H (2003) Pannexins, a family of gap junction proteins expressed in brain. Proc Natl Acad Sci U S A 100(23):13644–13649

129. Taylor KA, Wright JR, Vial C, Evans RJ, Mahaut-Smith MP (2014) Amplification of human platelet activation by surface pannexin-1 channels. J Thromb Haemost 12(6):987–998
130. Locovei S, Wang J, Dahl G (2006) Activation of pannexin 1 channels by ATP through P2Y receptors and by cytoplasmic calcium. FEBS Lett 580(1):239–244
131. Thompson RJ, Zhou N, MacVicar BA (2006) Ischemia opens neuronal gap junction hemichannels. Science 312(5775):924–927
132. Sandilos JK, Chiu YH, Chekeni FB, Armstrong AJ, Walk SF, Ravichandran KS, Bayliss DA (2012) Pannexin 1, an ATP release channel, is activated by caspase cleavage of its pore-associated C-terminal autoinhibitory region. J Biol Chem 287(14):11303–11311
133. Taylor KA, Wright JR, Mahaut-Smith MP (2015) Regulation of Pannexin-1 channel activity. Biochem Soc Trans 43(3):502–507
134. Garre JM, Retamal MA, Cassina P, Barbeito L, Bukauskas FF, Saez JC, Bennett MV, Abudara V (2010) FGF-1 induces ATP release from spinal astrocytes in culture and opens pannexin and connexin hemichannels. Proc Natl Acad Sci U S A 107(52):22659–22664
135. Pelegrin P, Surprenant A (2006) Pannexin-1 mediates large pore formation and interleukin-1b release by the ATP-gated $P2X_7$ receptor. EMBO J 25(21):5071–5082
136. Romanov RA, Rogachevskaja OA, Bystrova MF, Jiang P, Margolskee RF, Kolesnikov SS (2007) Afferent neurotransmission mediated by hemichannels in mammalian taste cells. EMBO J 26(3):657–667
137. Woehrle T, Yip L, Elkhal A, Sumi Y, Chen Y, Yao Y, Insel PA, Junger WG (2010) Pannexin-1 hemichannel-mediated ATP release together with P2X1 and P2X4 receptors regulate T-cell activation at the immune synapse. Blood 116(18):3475–3484
138. Poornima V, Madhupriya M, Kootar S, Sujatha G, Kumar A, Bera AK (2012) $P2X_7$ receptor-pannexin 1 hemichannel association: effect of extracellular calcium on membrane permeabilization. J Mol Neurosci 46(3):585–594
139. Vessey DA, Li L, Kelley M (2011) $P2X_7$ receptor agonists pre- and postcondition the heart against ischemia-reperfusion injury by opening pannexin-1/$P2X_7$ channels. Am J Physiol Heart Circ Physiol 301(3):H881–H887
140. Xia J, Lim JC, Lu W, Beckel JM, Macarak EJ, Laties AM, Mitchell CH (2012) Neurons respond directly to mechanical deformation with pannexin-mediated ATP release and autostimulation of $P2X_7$ receptors. J Physiol 590(Pt 10):2285–2304
141. Molica F, Morel S, Meens MJ, Denis JF, Bradfield PF, Penuela S, Zufferey A, Monyer H, Imhof BA, Chanson M, Laird DW, Fontana P, Kwak BR (2015) Functional role of a polymorphism in the Pannexin1 gene in collagen-induced platelet aggregation. Thromb Haemost 114(2):325–336
142. Schenk U, Westendorf AM, Radaelli E, Casati A, Ferro M, Fumagalli M, Verderio C, Buer J, Scanziani E, Grassi F (2008) Purinergic control of T cell activation by ATP released through pannexin-1 hemichannels. Sci Signal 1(39):ra6
143. Di Virgilio F, Vuerich M (2015) Purinergic signaling in the immune system. Auton Neurosci: Basic Clin 191:117–123
144. Chekeni FB, Elliott MR, Sandilos JK, Walk SF, Kinchen JM, Lazarowski ER, Armstrong AJ, Penuela S, Laird DW, Salvesen GS, Isakson BE, Bayliss DA, Ravichandran KS (2010) Pannexin 1 channels mediate ‘find-me’ signal release and membrane permeability during apoptosis. Nature 467(7317):863–867
145. Elliott MR, Chekeni FB, Trampont PC, Lazarowski ER, Kadl A, Walk SF, Park D, Woodson RI, Ostankovich M, Sharma P, Lysiak JJ, Harden TK, Leitinger N, Ravichandran KS (2009) Nucleotides released by apoptotic cells act as a find-me signal to promote phagocytic clearance. Nature 461(7261):282–286
146. Poon IK, Chiu YH, Armstrong AJ, Kinchen JM, Juncadella IJ, Bayliss DA, Ravichandran KS (2014) Unexpected link between an antibiotic, pannexin channels and apoptosis. Nature 507(7492):329–334
147. Evans RJ, Lewis C, Virginio C, Lundstrom K, Buell G, Surprenant A, North RA (1996) Ionic permeability of, and divalent cation effects on, two ATP-gated cation channels (P2X receptors) expressed in mammalian cells. J Physiol 497(Pt 2):413–422

148. Rolf MG, Mahaut-Smith MP (2002) Effects of enhanced $P2X_1$ receptor Ca^{2+} influx on functional responses in human platelets. Thromb Haemost 88(3):495–502
149. MacKenzie AB, Mahaut-Smith MP, Sage SO (1996) Activation of receptor-operated cation channels via P_{2X1} not P_{2T} purinoceptors in human platelets. J Biol Chem 271(6):2879–2881
150. Benham CD (1989) ATP-activated channels gate calcium entry in single smooth muscle cells dissociated from rabbit ear artery. J Physiol 419:689–701
151. Benham CD, Tsien RW (1987) A novel receptor-operated Ca^{2+}-permeable channel activated by ATP in smooth muscle. Nature 328(6127):275–278
152. Schneider P, Hopp HH, Isenberg G (1991) Ca^{2+} influx through ATP-gated channels increments $[Ca^{2+}]_i$ and inactivates ICa in myocytes from guinea-pig urinary bladder. J Physiol 440:479–496
153. Garcia-Guzman M, Soto F, Gomez-Hernandez JM, Lund PE, Stuhmer W (1997) Characterization of recombinant human $P2X_4$ receptor reveals pharmacological differences to the rat homologue. Mol Pharmacol 51(1):109–118
154. Fung CY, Brearley CA, Farndale RW, Mahaut-Smith MP (2005) A major role for $P2X_1$ receptors in the early collagen-evoked intracellular Ca^{2+} responses of human platelets. Thromb Haemost 94(1):37–40
155. Oury C, Toth-Zsamboki E, Thys C, Tytgat J, Vermylen J, Hoylaerts MF (2001) The ATP-gated $P2X_1$ ion channel acts as a positive regulator of platelet responses to collagen. Thromb Haemost 86(5):1264–1271
156. Salaun C, James DJ, Chamberlain LH (2004) Lipid rafts and the regulation of exocytosis. Traffic 5(4):255–264
157. Jones S, Evans RJ, Mahaut-Smith MP (2014) Ca^{2+} influx through P2X1 receptors amplifies P2Y1 receptor-evoked Ca^{2+} signaling and ADP-evoked platelet aggregation. Mol Pharmacol 86(3):243–251
158. Eberhard DA, Holz RW (1988) Intracellular Ca^{2+} activates phospholipase C. Trends Neurosci 11(12):517–520
159. Foskett JK, White C, Cheung KH, Mak DO (2007) Inositol trisphosphate receptor Ca^{2+} release channels. Physiol Rev 87(2):593–658
160. Watson SP, Poole A, Asselin J (1995) Ethylene glycol bis(beta-aminoethyl ether)-N, N, N′, N′-tetraacetic acid (EGTA) and the tyrphostin ST271 inhibit phospholipase C in human platelets by preventing Ca^{2+} entry. Mol Pharmacol 47(4):823–830
161. Bezprozvanny I, Watras J, Ehrlich BE (1991) Bell-shaped calcium-response curves of Ins(1,4,5)P_3- and calcium-gated channels from endoplasmic reticulum of cerebellum. Nature 351(6329):751–754
162. Ohana L, Barchad O, Parnas I, Parnas H (2006) The metabotropic glutamate G-protein-coupled receptors mGluR3 and mGluR1a are voltage-sensitive. J Biol Chem 281(34):24204–24215
163. Parnas H, Parnas I (2007) The chemical synapse goes electric: Ca^{2+}- and voltage-sensitive GPCRs control neurotransmitter release. Trends Neurosci 30(2):54–61
164. Martinez-Pinna J, Gurung IS, Vial C, Leon C, Gachet C, Evans RJ, Mahaut-Smith MP (2005) Direct voltage control of signaling via $P2Y_1$ and other $G\alpha_q$-coupled receptors. J Biol Chem 280(2):1490–1498
165. Mahaut-Smith MP, Martinez-Pinna J, Gurung IS (2008) A role for membrane potential in regulating GPCRs? Trends Pharmacol Sci 29(8):421–429
166. Martinez-Pinna J, Tolhurst G, Gurung IS, Vandenberg JI, Mahaut-Smith MP (2004) Sensitivity limits for voltage control of P2Y receptor-evoked Ca^{2+} mobilization in the rat megakaryocyte. J Physiol 555(Pt 1):61–70
167. Sage SO, Rink TJ, Mahaut-Smith MP (1991) Resting and ADP-evoked changes in cytosolic free sodium concentration in human platelets loaded with the indicator SBFI. J Physiol 441:559–573
168. Mahaut-Smith MP (2013) A role for platelet TRPC channels in the Ca^{2+} response that induces procoagulant activity. Sci Signal 6(281):pe23

169. Harper MT, Mason MJ, Sage SO, Harper AG (2010) Phorbol ester-evoked Ca^{2+} signaling in human platelets is via autocrine activation of P_{2X1} receptors, not a novel non-capacitative Ca^{2+} entry. J Thromb Haemost 8(7):1604–1613
170. Erhardt JA, Pillarisetti K, Toomey JR (2003) Potentiation of platelet activation through the stimulation of $P2X_1$ receptors. J Thromb Haemost 1(12):2626–2635
171. Toth-Zsamboki E, Oury C, Cornelissen H, De Vos R, Vermylen J, Hoylaerts MF (2003) $P2X_1$-mediated ERK2 activation amplifies the collagen-induced platelet secretion by enhancing myosin light chain kinase activation. J Biol Chem 278(47):46661–46667
172. Oury C, Daenens K, Hu H, Toth-Zsamboki E, Bryckaert M, Hoylaerts MF (2006) ERK2 activation in arteriolar and venular murine thrombosis: platelet receptor GPIb vs. $P2X_1$. J Thromb Haemost 4(2):443–452
173. Oury C, Toth-Zsamboki E, Vermylen J, Hoylaerts MF (2002) $P2X_1$-mediated activation of extracellular signal-regulated kinase 2 contributes to platelet secretion and aggregation induced by collagen. Blood 100(7):2499–2505
174. Oury C, Kuijpers MJ, Toth-Zsamboki E, Bonnefoy A, Danloy S, Vreys I, Feijge MA, De Vos R, Vermylen J, Heemskerk JW, Hoylaerts MF (2003) Overexpression of the platelet $P2X_1$ ion channel in transgenic mice generates a novel prothrombotic phenotype. Blood 101(10):3969–3976
175. Huang Z, Liu P, Zhu L, Li N, Hu H (2014) $P2X_1$-initiated p38 signalling enhances thromboxane A2-induced platelet secretion and aggregation. Thromb Haemost 112(1):142–150
176. Woehrle T, Yip L, Manohar M, Sumi Y, Yao Y, Chen Y, Junger WG (2010) Hypertonic stress regulates T cell function via pannexin-1 hemichannels and P2X receptors. J Leukoc Biol 88(6):1181–1189
177. Hechler B, Magnenat S, Zighetti ML, Kassack MU, Ullmann H, Cazenave JP, Evans R, Cattaneo M, Gachet C (2005) Inhibition of platelet functions and thrombosis through selective or nonselective inhibition of the platelet P2 receptors with increasing doses of NF449 [4,4′,4″,4‴-(carbonylbis(imino-5,1,3-benzenetriylbis-(carbonylimino)))tetrakis -benzene-1,3-disulfonic acid octasodium salt]. J Pharmacol Exp Ther 314(1):232–243
178. Massberg S, Grahl L, von Bruehl ML, Manukyan D, Pfeiler S, Goosmann C, Brinkmann V, Lorenz M, Bidzhekov K, Khandagale AB, Konrad I, Kennerknecht E, Reges K, Holdenrieder S, Braun S, Reinhardt C, Spannagl M, Preissner KT, Engelmann B (2010) Reciprocal coupling of coagulation and innate immunity via neutrophil serine proteases. Nat Med 16(8):887–896
179. von Bruhl ML, Stark K, Steinhart A, Chandraratne S, Konrad I, Lorenz M, Khandoga A, Tirniceriu A, Coletti R, Kollnberger M, Byrne RA, Laitinen I, Walch A, Brill A, Pfeiler S, Manukyan D, Braun S, Lange P, Riegger J, Ware J, Eckart A, Haidari S, Rudelius M, Schulz C, Echtler K, Brinkmann V, Schwaiger M, Preissner KT, Wagner DD, Mackman N, Engelmann B, Massberg S (2012) Monocytes, neutrophils, and platelets cooperate to initiate and propagate venous thrombosis in mice in vivo. J Exp Med 209(4):819–835
180. Darbousset R, Thomas GM, Mezouar S, Frere C, Bonier R, Mackman N, Renne T, Dignat-George F, Dubois C, Panicot-Dubois L (2012) Tissue factor-positive neutrophils bind to injured endothelial wall and initiate thrombus formation. Blood 120(10):2133–2143
181. Geiger J, Nolte C, Butt E, Sage SO, Walter U (1992) Role of cGMP and cGMP-dependent protein kinase in nitrovasodilator inhibition of agonist-evoked calcium elevation in human platelets. Proc Natl Acad Sci U S A 89(3):1031–1035
182. Schwarz UR, Walter U, Eigenthaler M (2001) Taming platelets with cyclic nucleotides. Biochem Pharmacol 62(9):1153–1161
183. Varga-Szabo D, Braun A, Nieswandt B (2009) Calcium signaling in platelets. J Thromb Haemost 7(7):1057–1066
184. Mahaut-Smith MP (2012) The unique contribution of ion channels to platelet and megakaryocyte function. J Thromb Haemost 10(9):1722–1732
185. Junger WG (2011) Immune cell regulation by autocrine purinergic signalling. Nat Rev Immunol 11(3):201–212

186. Jacob F, Perez Novo C, Bachert C, Van Crombruggen K (2013) Purinergic signaling in inflammatory cells: P2 receptor expression, functional effects, and modulation of inflammatory responses. Purinergic Signal 9(3):285–306
187. Yip L, Woehrle T, Corriden R, Hirsh M, Chen Y, Inoue Y, Ferrari V, Insel PA, Junger WG (2009) Autocrine regulation of T-cell activation by ATP release and $P2X_7$ receptors. FASEB J 23(6):1685–1693
188. Miller CM, Boulter NR, Fuller SJ, Zakrzewski AM, Lees MP, Saunders BM, Wiley JS, Smith NC (2011) The role of the $P2X_7$ receptor in infectious diseases. PLoS Pathog 7(11):e1002212
189. Lecut C, Frederix K, Johnson DM, Deroanne C, Thiry M, Faccinetto C, Maree R, Evans RJ, Volders PG, Bours V, Oury C (2009) $P2X_1$ ion channels promote neutrophil chemotaxis through Rho kinase activation. J Immunol 183(4):2801–2809
190. Phillipson M, Kubes P (2011) The neutrophil in vascular inflammation. Nat Med 17(11):1381–1390
191. Grommes J, Soehnlein O (2011) Contribution of neutrophils to acute lung injury. Mol Med 17(3–4):293–307
192. Kolaczkowska E, Kubes P (2013) Neutrophil recruitment and function in health and inflammation. Nat Rev Immunol 13(3):159–175
193. Lecut C, Faccinetto C, Delierneux C, van Oerle R, Spronk HM, Evans RJ, El Benna J, Bours V, Oury C (2012) ATP-gated $P2X_1$ ion channels protect against endotoxemia by dampening neutrophil activation. J Thromb Haemost 10(3):453–465
194. Engelmann B, Massberg S (2013) Thrombosis as an intravascular effector of innate immunity. Nat Rev Immunol 13(1):34–45

Part III
Role of Cellular Microdomains and Organelles in Calcium Entry

Chapter 14
The Calcium Entry-Calcium Refilling Coupling

Ziane Elaib, Francois Saller, and Regis Bobe

Abstract Calcium ions (Ca^{2+}) are versatile messengers that need to be tidily regulated in time and space in order to create a large number of signals. The coupling between Ca^{2+} entry and Ca^{2+} refilling is playing a central role in this Ca^{2+} homeostasis. Since the capacitative Ca^{2+} entry has been described, different mechanisms have been proposed in order to explain how the Ca^{2+} entry could be under control of intracellular store Ca^{2+} depletion. Today, in addition of STIM1 and Orai1, the two major elements of SOCe, increasing attention is put on the role of the transient receptor potential canonical (TRPC), that can form protein clusters with Orai1, and Sarco/endoplasmic reticulum Ca^{2+}ATPases (SERCAs), that refill the stores and are also located in the same environment than SOC clusters. Altogether, these proteins elaborate either Ca^{2+} microdomains in the vicinity of the membrane or larger Ca^{2+} increases overtaking the whole cell. The coupling between Ca^{2+} entry and Ca^{2+} refilling can possibly act much further away from the plasma membrane. Ca^{2+}, uptaken by SERCAs, have been described to move faster and further in the ER than in the cytosol and to create specific signal that depends on Ca^{2+} entry but at longer distance from it. The complexity of such created Ca^{2+} currents resides in the heteromeric nature of channels as well as the presence of different intracellular stores controlled by SERCA2b and SERCA3, respectively. A role for mitochondria has also been explored. To date, mitochondria are other crucial compartments that play an important role in Ca^{2+} homeostasis. Although mitochondria mostly interact with intracellular stores, coupling of Ca^{2+} entry and mitochondria cannot be completely rule out.

Keywords SERCA • STIM1 • Orai1 • TRPC channels

Z. Elaib • F. Saller • R. Bobe (✉)
Inserm U1176, Le Kremlin Bicetre, France

UMR_S 1176, Univ Paris Sud, Le Kremlin Bicetre, France
e-mail: regis.bobe@inserm.fr

J.A. Rosado (ed.), *Calcium Entry Pathways in Non-excitable Cells*, Advances in Experimental Medicine and Biology 898, DOI 10.1007/978-3-319-26974-0_14

14.1 Introduction

Calcium ions (Ca^{2+}) are universal second messengers that create a ubiquitous signal transduction pathway that is functional in every cellular type and species. Ca^{2+} signal regulates various processes including cell proliferation, response to the environment, death and apoptosis…

In a review published in 2006, R.J.P. Williams presented a paradigm in which the role of calcium as second messenger could be the result of evolution to adapt cells to life in an oxidized environment [1]. When cyanobacteria changed the face of the world and killed nearly all living species (mostly prokaryotic cells), surviving cells evolved to internally compartmentalized eukaryotic cells. These cells needed protection, ability to recognize their environment as well as to coordinate the internal activity of their new compartments; a new messenger was needed. As life had a long experience in dealing with calcium ions, widely common molecules that were rejected out cells for billions of years because of their ability to form insoluble salts (precipitates) with inorganic and organic anions carbon and phosphate, calcium ions became the messengers that not only coordinated the action of intracellular compartments but also triggered cell responses to the environment.

As Ca^{2+} signaling is very versatile, the question is to figure out how a single cation can code for a multitude of cellular responses, (i.e.) how distinct signals can be generated and how cell compartments can decipher specific messages.

The Ca^{2+} signal has, in order to be efficient and specific, to be tightly modulated in time and space. Multiple partners exist, forming cell specific calcium toolkits that are organized to control calcium flux and to translate Ca^{2+} signals into cellular activity. At rest, cytosolic Ca^{2+} is maintained around 50–200 nM. Upon activation its concentration will increase to create a global Ca^{2+} signal that can propagate over large distances (10–100 μm) in the range of μM and/or formation of microdomains with very high level of Ca^{2+} concentration (50–100 μM) in the vicinity of the Ca^{2+} channels. These latter events only spread over 20 nm and, Ca^{2+} concentration drops rapidly as ions are buffered and diluted in the cytosol [2, 3].

The increase in cytosolic calcium is due to its influx from the extracellular medium through the plasma membrane (PM) (extracellular Ca^{2+} concentration ranges between 1 and 2 mM in aqueous conditions) and its depletion from intracellular stores, mainly (but not only) the endoplasmic reticulum (ER). The storage capacity of the ER is limited and Ca^{2+} reuptake is performed by Ca^{2+} pumps termed Sarco/Endoplasmic Reticulum Ca^{2+}ATPases (SERCA) that control the Ca^{2+} concentration in the ER lumen. As SERCAs compete with other calcium transporters, such as Na/Ca exchangers or plasma membrane Ca^{2+}ATPases (PMCA) that extrude Ca^{2+} to the extracellular medium, a continuously loss of intracellular calcium storage would be observed if a subsequent calcium entry from extracellular medium was not organized to help cells to refill their intracellular stores in calcium. This Ca^{2+} entry is controlled by the ER lumen concentration and termed SOCe (Store operated calcium entry). If the SOC process was initially designed to maintain the concentration of stored Ca^{2+}, it now appears as a crucial actor that shapes the calcium signal within the cells.

14.2 Store Operated Calcium Entry (SOCe)

The classical way of non-excitable cell stimulation associates G protein coupled receptor and phospholipase activation. The newly formed inositol 1,4,5-trisphosphate (IP_3) will open its receptor channel (IP_3R) inserted in the endoplasmic reticulum (ER) membrane and will allow the mobilization of Ca^{2+} [4–6]. SOCe is a ubiquitous pathway in non-excitable cells as well as in some excitable cell types [7]. This event is associated to Ca^{2+} influx across the plasma membrane (PM). This paradigm of the existence of a store-dependent Ca^{2+} influx was proposed 30 years ago by J. W. Putney as Capacitive Calcium Entry (CCE) [8]. The use of Thapsigargin (Tg), a selective noncompetitive inhibitor of the SERCAs that is isolated from the plant *Thapsia garganica* and triggers a "passive" depletion of the ER stored Ca^{2+} without receptor activation, have permitted to establish that the store depletion by itself was responsible for a Ca^{2+} influx similar to those recorded as SOCe and, more important, independently of IP_3 [9, 10].

While the idea of having a Ca^{2+} entry from the extracellular medium under the control of intracellular organelles catches on, it was very difficult to have a clear insight into the underlying mechanisms until recently. Since, a number of studies have focused on the current specificities of this SOCe [11–14].

The first question was to understand how Ca^{2+} mobilization was able to activate the Ca^{2+} entry. At least three major hypothetic mechanisms have been advanced.

First, the existence of a soluble mediator from intracellular stores was proposed and termed "calcium influx factor" (CIF) [15]. This messenger was described to be released into the cytoplasm upon Ca^{2+} mobilization in Jurkat cells and is still unknown. Probably due to the lack of its molecular identity, very few studies are since focusing on CIF. Additionally, as CIF is "purified" or isolated after a long process, it is difficult to ascertain its specificity [16]. However, a recent publication showed that only CIF produced by agonists or Thapsigargin activation is able to induce a 2-APB (SOCe inhibitor) sensitive Ca^{2+} influx when injected in non-activated cells. Similar treatments, with components isolated in the same way than CIF but from non-activated cells, have no effect. Such data suggest in contrast, that the existence of a putative CIF is not ruled out [17].

The second proposed hypothesis was a conformational coupling between IP_3 receptor (IP_3R) and SOC channels in an ER-PM junction [18]. In this model, upon activation, IP_3R is able to activate Ca^{2+} channels and even to create a long channel that could transport Ca^{2+} directly from the extracellular medium into the ER lumen. While the idea of having the IP_3R working in a reverse mode was not established, it was later proposed that signaling clusters containing Transient receptor potential canonical channel type 1 (TRPC1) (see below for the roles of TRPC in SOCe) and IP_3R exist [19], both being able to bind directly through a CIRB domain positioned in the C-Terminal part of TRPC1 [20]. TRPC1 was described to associate with IP_3R in platelets upon store depletion, and to participate to extracellular Ca^{2+} influx into the cell [21, 22].

The last and most successful model was the existence of a direct interaction between ER resident proteins and Ca^{2+} channels.

14.2.1 Molecular Identity and Mechanism of Store Operated Channel

It took nearly 20 years to validate the whole concept of the SOC channel formation and the two key proteins involved in SOCe were almost concomitantly identified. First, the role of the Ca^{2+} sensor protein STIM1 (stromal interacting molecule 1) was highlighted in 2005 using a RNA interference-based screening in Drosophila and HeLa cells that showed that its knockdown strongly reduced SOCe [23, 24]. The second player, Orai1, was identified a year later also in Drosophila and human cells. Both RNAi screens in *Drosophila* and gene mapping on lymphocytes from a family with a severe combined immunodeficiency (SCID) revealed that Orai1 deficiency was associated to a defect in SOCe [25–27].

14.2.1.1 STIM1

STIM1 is a ubiquitously expressed protein which is widely expressed from Drosophila to mammalian cells. STIM1 is a single transmembrane protein of 77 kDa inserted predominantly in the ER [28] and distributed evenly along its membrane [29]. This protein is the crucial link between ER and Ca^{2+} channel in the PM. The luminal part of the protein contains two EF-hand (canonical and hidden EF-hand domains (cEF) and (hEF), respectively) domains and an adjacent sterile α motif (SAM) (Fig. 14.1a). Only the cEF domain can bind Ca^{2+} and confers to STIM1 its sensor function [23, 30, 31].

When Ca^{2+} stores are at rest, the estimated luminal Ca^{2+} concentration is about 500 μM [32]. At this level, Ca^{2+} binding to cEF allows the formation of a stable complex between EF and SAM domains. Upon ER depletion, Ca^{2+} dissociates from the STIM1 EF-hand domain, thus destabilizing the EF-hand-SAM complex (Fig. 14.1c). Both EF-hand and SAM domain will then expose their hydrophobic regions leading to the oligomerization of STIM1 [31, 33, 34]. It seems that the clustering of

Fig. 14.1 (continued) are composed of four transmembrane domains (TM1 (*blue*), TM2 (*purple*), TM3 *dark red*) and TM4 (*green*)). Two regions (CAD binding domains) are exposed into the cytosol and localized at both extremities of the proteins. On the right part of the figure using the same color code for the transmembrane domains of Orai is a top down schematic view of the hexameric Orai structure inserted into the plasma membrane with a Ca^{2+} (*red dot*) in the channel. (**c**) STIM1/Orai1 activation model. (*a*) Under resting conditions, the STIM1 Ca^{2+} binding to EF-hand domain masks SAM domains and keeps cytosolic domains folded and inactive. (*b*) Depletion of Ca^{2+} within the ER induces structural reorganization that uncovers SAM domains allowing oligomerization of STIM1. This also extends binding/activating domains towards the plasma membrane and (*c*) leads to clustering of Orai through interaction between SOAR and CAD domains

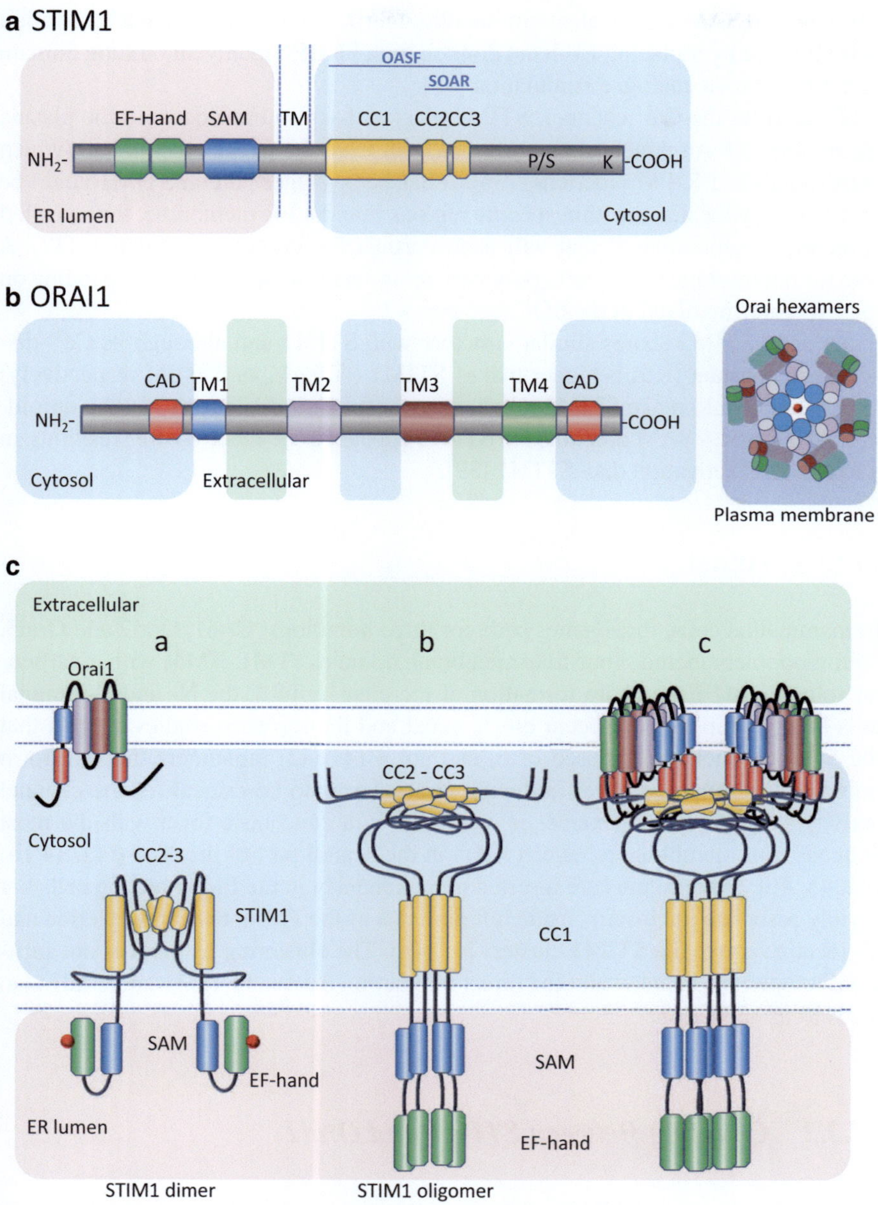

Fig. 14.1 Functional domains of STIM1 and Orai1 and model of coupling between STIM1 and Orai1. (**a**) Schematic representation of the main functional domains of STIM1. The N-terminal part of STIM1 (amino acids 1–214) is in the ER lumen and includes both EF-Hand domains (canonical and hidden EF-hand) and a sterile α motif (*SAM*) followed by transmembrane domain. In the cytosolic part of the protein are the coiled coil domains (CC1, CC2 and CC3) that form the region involved in the binding and activation of Orai (STIM-Orai activation region (SOAR, amino acids 345–444) or Orai activating small fragment (OASF, amino acids 233–450) and. The C-terminal part of STIM1 includes a serine/proline rich domain (P/S) and a lysine-rich domain (K). (**b**) Schematic representation of the main functional domains of Orai1. Orai monomers

EF-hand and SAM is critical, as similar oligomerization resulting in SOC activation was obtained by replacing EF-hand domain with FKBP-rapamycin binding domain under rapamicyn analogue stimulation [35].

To activate the Ca^{2+} channel, STIM1 clusters have to translocate to the plasma membrane to form punctate structures <25 nm from the plasma membrane (which corresponds to ER-PM junctions) [36]. It has been suggested that STIM1 could be preferentially localized within specific regions into the ER membrane, the so called "precortical subdomains" that will form cortical ER upon Ca^{2+} depletion [37]. A role for microtubule and F-actin polymerization seems to be involved, depending on the ER stores involved in the SOC (see below).

Of note, STIM2 shares similar structure with STIM1 and although its Ca^{2+} dissociation constant (Kd) is higher that of STIM1 (400 μM vs 200 μM respectively) [38], which could make STIM2 a better sensor for Ca^{2+} depletion [38], the unfolding of EF-hand – SAM domains for STIM2 appears to be much slower resulting in a lower SOC activation than STIM1 [39].

14.2.1.2 ORAI1

In mammalian cells, three genes code for three homologs: Orai1, Orai2 and Orai3. Orai monomers include four transmembrane domains (TM1–TM4) with a particular role of TM1 in the pore formation of the channel [40], the N- and C-terminal tails being cytoplasmic. Recent biochemical and fluorescence studies suggest that the SOC channel is composed of 6, and not 4 [41, 42] monomers that can form homo or heteromers of Orai (although Orai1 seems to be critical for the channel activity [43, 44]). The hexamer is organized in a concentric form with the most aqueous transmembrane α helice (TM1) in the central part of the pore (Fig. 14.1b) [42, 45, 46]. Orai proteins are inserted homogenously in the PM in resting cells but rapidly associate in clusters inside raft domains of the PM upon Ca^{2+} depletion and co-localize above the STIM1 clusters [47, 48]. The clustering by itself is not sufficient to open the channel and the direct interaction between STIM1 and Orai1 also leads to its activation.

14.2.2 Coupling Between STIM1 and Orai1

The coupling between STIM1 and Orai1 involves several domains of these proteins, all localized in the cytosol. Various regions named CAD, SOAR, OASF or Ccb9, all similarly located in the cytosolic part of STIM, have been identified in 2009 as being critical for the Orai activation (Fig. 14.1a) [48–51]. All of them cover the CC2 (aa 363–389) and CC3 (aa 399–423) regions of STIM1 that form a hairpin motif.

Two regions termed "CRAC activation domain" (CAD) have been proposed to bind STIM1 based on mutations in Orai and in silico studies; one is N-terminal (aa 73–85) and one is C-Terminal (aa 272–292) [52–54]. Whether this two CAD form distinct binding regions or can form a unique binding site for STIM1 is still unknown and has to be determined. Interestingly, these regions seem to share structural homology with the STIM1 CC2 and CC3 domains, suggesting the possible formation of complex structure between the domains of both proteins [54].

The proposed mechanism suggests that, at rest, activation sites of STIM1 are masked by interaction with other flanking domains (possibly CC1 domain) of STIM1, thus preventing any interaction with Orai1. Upon Ca^{2+} depletion, lack of Ca^{2+} binding to EF-hand leads to homo association of CC1 (similarly to the SAM domains). This structural reorganization unmasks the CC2–CC3 domains and allows a physical extension of these domains in the cytoplasm [55] letting them to become available for Orai1 binding [52]. When binding to STIM1, Orai proteins will also change their structural conformation and either open their gates to Ca^{2+} or become highly specific for Ca^{2+} [56, 57] (Fig. 14.1c). Indeed, Orai channels display a selectivity for Ca^{2+} that is 1000 times higher than that for Na^+ [58] when they are activated by STIM1. However, the mutant V102C/A, which appears to be constitutively activated, displays a poorly selective activity in absence of STIM1 and can regain its selectivity when co-expressed with STIM1 [59].

This first part of the chapter explained the relationship between ER and SOCe activation via STIM1 and ORAI1. Although STIM1 and Orai1 are crucial in coupling between ER and SOCe activation, other players participate in SOCe. First, a number of regulators of SOCe interact with either STIM1 or Orai1 (for review see [60, 61]). In addition other proteins can also form Ca^{2+} channels. A special attention was given to TRPC, that role as SOCe was strongly considered, until Orai1 was identified, a matter of debate.

14.2.3 TRPC

Transient receptor potential canonical (TRPC) are members of the superfamily of Transient receptor potential proteins. Since the discovery of its role as Ca^{2+} channel downstream of PLCγ2 activation, the TRPC proteins were proposed to be part of the SOC channel complex [62]. Probably because TRPC family is large and presents members with distinct functional capacities, whether TRPC function as SOCe has been highly debated [63, 64]. Among the TRPC family (TRPC1-7), TRPC1 is the most often associated to SOCe and was described to act as a STIM1-dependent Ca^{2+} channel in many studies [65–69]. Its expression in cell lines potentiates SOCe induced either by PLC activation or pharmacological (Tg) Ca^{2+} depletion [70].

It is however possible that TRPC form a macro-complex with Orai1 which therefore would explain their impact on SOCe without forming the channel by itself. Accordingly, direct association of Orai1 with TRPC3 and TRPC6 has been observed in HEK293 and COS1 cells [71, 72].

Most likely, as it has been described in blood platelets for TRPC1, TRPC play multiple role in the Ca^{2+} entry and their participation implies a higher level of complexity in the regulation of the Ca^{2+} entry controlled by store depletion [73].

Activation of TRPC1 downstream of Ca^{2+} depletion and their direct association with STIM1 has very recently been confirmed with a new single ion channel detection technique which further ascertains their possible role as SOCe [74].

Additionally to their action as SOCe, some members of the family, TRPC3, 4 6 and 7 have been described to be activated via diacylglycerol (DAG) upon PLC activation by G-protein coupled receptor stimulation. These activations are parallel to the Ca^{2+} mobilization but independent of Ca^{2+} depletion [67, 75, 76] and behave as ROC. Moreover, some authors suggest a direct link between IP_3R and TRPC, which lets suppose another way of store dependent Ca^{2+} entry regulation [19, 65, 77].

Thus, these three distinct pathways all activate TRPC channels, two being dependent on Ca^{2+} store depletion and one being parallel to the activation pathway of the Ca^{2+} depletion, might co-exist in the same cells (Fig. 14.2). Therefore, some TRPC members appear to be more inclined to participate to the SOC, like TRPC1, while some others seem to be more sensitive to DAG or IP_3R depending on their level of expression, their stoichiometry with STIM1 expression and the cellular models.

Finally, the whole system complexity increases again as TRPC can form heterodimer, thus sharing the properties of the different subunits that constitute the channels.

The paradigm of the crosstalk between Ca^{2+} entry and Ca^{2+} refilling is based on the observation that the fundamental reason for store-operated Ca^{2+} entry is that cells loose Ca^{2+} during signaling, Ca^{2+} mobilized from the ER to the cytoplasm is later transported outside the cell by plasma membrane Ca^{2+}ATPases (PMCA) or by the Na^+/Ca^{2+} exchanger (NCX). To ensure sustained signaling, the Ca^{2+} ions that have been extruded must return back to the store. Direct measurements of Ca^{2+} concentrations inside different organelles using nucleus, mitochondria or ER targeted aequorins have established that the impact of Ca^{2+} influx (by increasing extracellular Ca^{2+}) on Ca^{2+} uptake was 30 times larger in the ER than in other organelles [78]. Therefore, the third (or fourth taking into account TRPC) player in the coupling between Ca^{2+} entry – Ca^{2+} refilling has to be the Ca^{2+} pumps inserted in the ER membrane, the sarco/endoplasmic reticulum Ca^{2+}ATPase (SERCA).

14.3 SERCA

The SERCA family is composed of three genes giving rise to multiple isoforms though alternative splicing. SERCA2b isoform is ubiquitously expressed and plays a housekeeping function in the Ca^{2+} homeostasis. SERCA3 isoforms were firstly observed in non-muscular cells but some SERCA3 isoforms have now been described in smooth muscle as well [79–81].

The function of SERCA is to transport cytosolic Ca^{2+} into the ER, an action that will affect SOCe in two different ways depending on the point of view. From the ER point view, SERCA increase luminal ER Ca^{2+} concentration and allow STIM1 to return back to its inactivate status [82]. At the same time, Ca^{2+} uptake from the

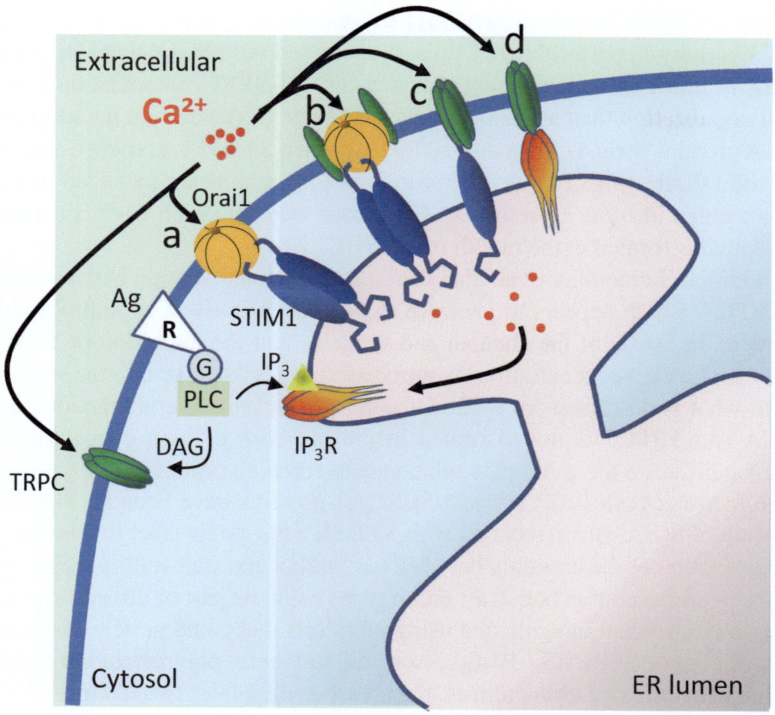

Fig. 14.2 Possible mechanism of store operated Ca^{2+} influx involving Orai1 and TRPc. Upon agonist stimulation of its receptor, PLC activation generates IP_3 and DAG. The latter is able to directly activate Ca^{2+} influx via TRPC channels. Produced IP_3 results in the depletion of Ca^{2+} from the ER via IP3R. (*a*) STIM1, the Ca^{2+} sensor inserted in the ER membrane oligomerizes and translocated to the ER-MP junction where it binds to and activates Orai1. (*b*) TRPC have been proposed to associate with Orai1 in a multiprotein complex or (*c*) to act as SOC channel when directly activated by STIM1. (*d*) Finally it has also been suggested that TRPC could be directly activated by IP_3R to directly refill the ER store. Abbr. 1,2-diacylglycerol (*DAG*); inositol trisphosphate (*IP_3*); inositol trisphophaste receptor (*IP_3R*); G protein-couple receptors (*R*); Ca^{2+} release-activate Ca^{2+} channel protein 1 (*Orai1*); phophoslipase (*PLC*); stromal interaction molecule 1 (*STIM1*); transient receptor potential cation channel (*TRPC*)

cytosol will avoid formation of high Ca^{2+} concentration microdomains at the mouth of channel and therefore will prevent channel inactivation [83, 84]. The combination of both actions explains why SERCA are so important in the shaping of Ca^{2+} signals. When over-expressed in cells, SERCA proteins are very efficient in reducing the need for Ca^{2+} entry. In agreement with this paradigm, it has been observed in smooth muscle cells that increased in SERCA2a expression leads in higher Ca^{2+} uptake and decrease in Ca^{2+} entry, showing a direct link between cell Ca^{2+} uptake capacity and SOCe activation [85].

Interestingly, recent reports show evidence that SERCA2 or SERCA3 associated with STIM1 upon Ca^{2+} depletion. This was inferred by confocal colocalization and co-immunoprecipitation [63, 86–88], although no FRET could have been obtained when SERCA-GFP and STIM1-CFP were co-expressed in HEK293T [89].

Additionally, while reduction in STIM1 expression using siRNA does not regulate SERCA activity in permeabilized cells, it decreased both SOCe and Ca^{2+} transport into ER in intact cells, which suggests that STIM1-SERCA association is more a spatial organization than a functional cluster [78, 90]. The current paradigm is that SERCA proteins form a crown around STIM1 when STIM1 is activated and associated with ORAI1 (Fig. 14.3b). Therefore, the Ca^{2+} pumps are localized just under the Ca^{2+} entry in order to refile the Ca^{2+} store from the high Ca^{2+} concentration microdomains formed at the mouth of the SOC.

In addition, Lemonnier et al. also demonstrated an association between SERCA and TRPC7, which appears to preserve the non capacitative Ca^{2+} influx (which is dependent on DAG) of the channel and suggests that the inhibition of TRPC7 by high cytosolic Ca^{2+} concentrations is suppressed by Ca^{2+} uptake into the stores similarly to what is described for SOC. Of note, the fact that the interaction between SERCA and TRPC7 is not disrupted in the presence of high concentrations of BAPTA indicates a more intimate relation between both proteins [83].

The fact that both SERCA2 and SERCA3 proteins have been involved in the regulation of SOCe and associated with STIM1, adds a new level of complexity in the organization of the crosstalk between Ca^{2+} influx and Ca^{2+} refilling. Since 1997, it has been proposed that both Ca^{2+} pump types could be part of distinct stores [91]. Based on electronical imaging and using inhibitors that could preferentially inhibit SERCA2 (Tg) or SERCA3 (tBHQ) associated to F-actin polymerization disruption by Cytochalisin-D or Latrunculin-A, the presence of at least two distinct SOCe has been observed in platelets (Fig. 14.4) [92, 93].

Acidic granules have been associated to SERCA3 [87, 94]. These granules are thought to be localized near the plasma membrane even in basal conditions. Therefore, formation of the ER-MP junction, necessary for activation of SOCe, does not need cytoskeleton reorganization. This was enlightened by the absence of effect (although some potentialization was actually observed) of F-actin polymerization inhibitors, on the SOCe induced by Ca^{2+} depletion using tBHQ that specifically inhibits SERCA3 [93]. These acidic granules have also been proposed to be sensitive to nicotinic acid adenine dinucleotide phosphate (NAADP), whose associated channel seems to be the two pores channels (TPC) [95, 96]. These granules are supposed to play a role of trigger of Ca^{2+} signaling starting before intracellular Ca^{2+} events that are sensitive IP_3R [97].

It is possible to imagine that these granules also act as SOCe trigger. Their submembrane localization can lead to a rapid association between STIM1 and Orai1 and the fast SERCA3 Ca^{2+} pumps can also participate preventing to reach high Ca^{2+} concentration in microdomains that could block the channel (Fig. 14.4a).

In contrast, SERCA2b-associated stores are strongly dependent on the actin polymerization and the similar experiments (pretreatments of cells with F-actin polymerization inhibitors) result in a decreased SOCe in response to Ca^{2+} depletion by Tg which preferentially inhibits SERCA2b [93]. This suggests a deeper insertion in the cytosol that could be associated to a slower process of activation in which STIM1 proteins has to relocalize to the ER-PM junctions, as observed in RBL cells (a tumor mast cell line) wherein STIM1 oligomers diffusion rate was evaluated to

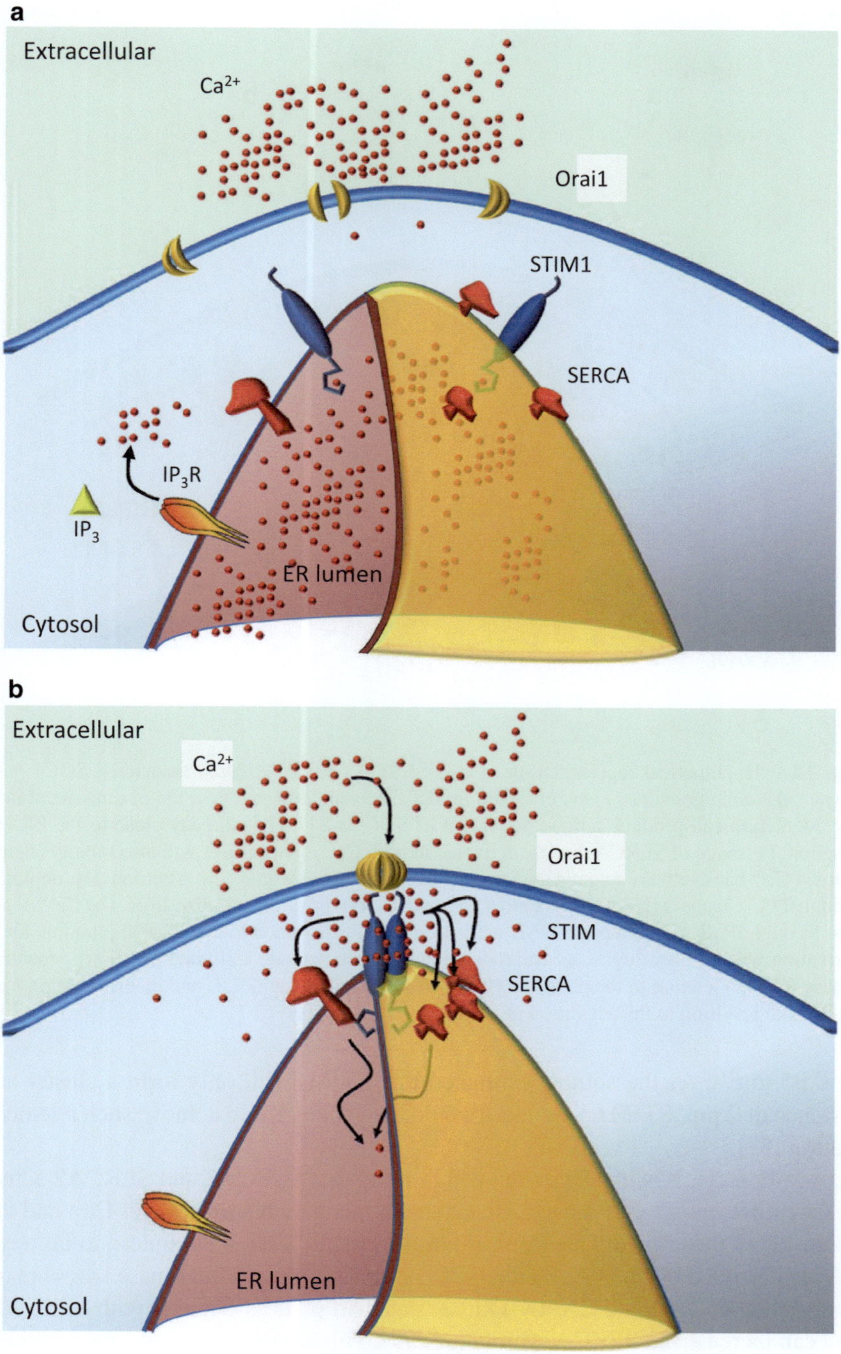

Fig. 14.3 Spatial organization of the SOC partners Orai1, STIM1 and SERCA. (**a**) Under resting conditions, high Ca^{2+} concentrations in the ER lumen keep SERCA and STIM1 homogenously at the surface of the ER membrane. Similarly, Orai is diffusely organized on the plasma membrane (*PM*). (**b**) Upon Ca^{2+} depletion though IP3R, STIM1 proteins oligomerise and translocate to form cluster under PM raft domains wherein Orai1 also form hexamers. SERCA pumps are also concentrated around STIM1 oligomers. Such organisation will avoid massive diffusion of Ca^{2+} ions in the cytosol and Ca^{2+} are re-uptaken by the pumps as soon as they reach the cytosol

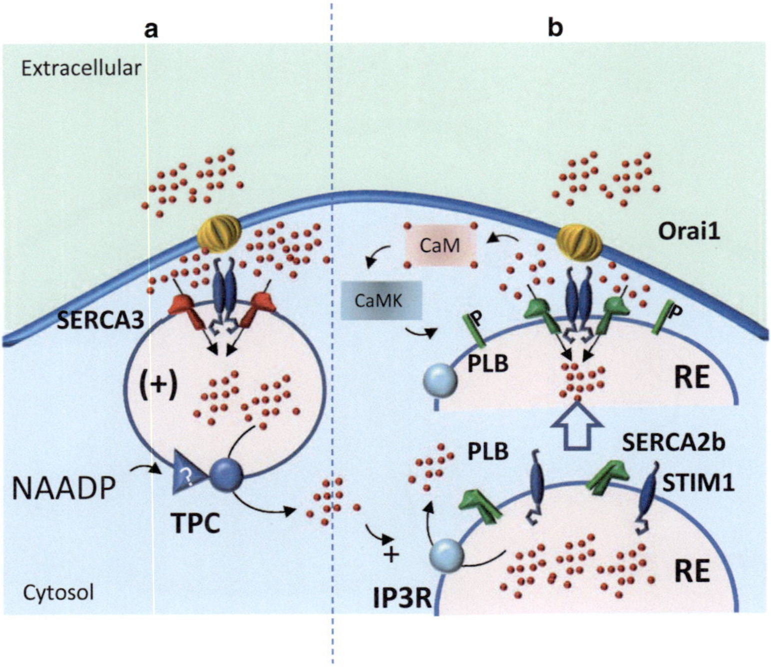

Fig. 14.4 Hypothetical representation of two SERCA3 and SERCA2b associated SOCe pathways. (**a**) Acidic granules (+) are enriched in SERCA3 and localized near the plasma membrane. Ca^{2+} depletion can produce a rapid activation of SOC as STIM1 is already close to the ER-PM junction. Presence of SERCA that are activated in high Ca^{2+} environment will also rapidly capture cytosol Ca^{2+} back into the granule avoiding its diffusion to the whole cell. Alternatively, depletion though TPC channels (NAADP receptor?) can act as trigger to potentialize IP_3R. (**b**) Ca^{2+} depletion through IP3R will induced STIM1 activation and translocation to the ER-PM junction. Orai1 activation will produce high Ca^{2+} microdomains that can activate Calmodulin (*Cam*)-dependent kinase (*CamK*) leading to the phosphorylation of phospholamban (*PLB*) and its dissociation from SERCA2b resulting in the activation of SERCA

be 0.05 μm²/s. As the authors estimated that STIM1 will only form a cluster at a distance of 2 μm, STIM1 oligomers would take about 40s to achieve such a relocalization [82].

Additionally, it is important to keep in mind that SERCA3 and SERCA2 activities are differently regulated. SERCA2b has a much higher affinity for Ca^{2+} and this could allow those pumps to work at basal cytosolic Ca^{2+} level unless to be negatively controlled by phospholamban (PLB), a cAMP-dependent kinase whose phosphorylation by PKA and CAMKII kinase can disrupt its association with SERCA2b and can increase the SERCA affinity [98, 99]. This signaling cascade can provide a useful mechanism that specifically controls the SERCA2b Ca^{2+} uptake (Fig. 14.4b). Here, the Ca^{2+} uptake involves different signaling cascades including protein movements on membrane and in the cytosol, which could take a longer time to be functional. It is therefore possible that the resulting SOCe is larger than the one depending on SERCA3-associated granule and overtakes the whole cytosol.

14.4 Coupling Response Between Mitochondria and SOCe

Mitochondria are multifunctional organelles that control a large number of cellular processes. Mitochondria are the cell energy factories that burn oxygen during oxidative phosphorylation to produce the cellular ATP used for biochemical reactions. They contain two membranes. In contact with the cytosol, is an outer membrane that is permeable, due to the abundant expression of voltage dependent anion selective channel (VDAC), acting as a general diffusion pore for small hydrophilic molecules such as ADP and ATP [100]. The larger Ca^{2+} uptake into the mitochondrial matrix occurs predominantly through the impermeable inner membrane via the ruthenium red-sensitive mitochondrial Ca^{2+} uniporter (MCU) complex.

Therefore, mitochondria are another crucial compartment that plays an important role in Ca^{2+} homeostasis.

Early models even proposed mitochondria as intermediate stores for Ca^{2+} originating from the extracellular medium before being transferred into the ER [101–103]. To do so, mitochondria should be located between the PM and ER. With the latter demonstration of the ER-PM junction and the direct interaction between STIM1 and Orai1 it is now hard to imagine such a role [104, 105].

Nevertheless, Mitochondria adjacent to ER and in the proximity of the SOCe channel, could take large amount of Ca^{2+} which dissipates high Ca^{2+} concentration in the ER-PM environment, facilitating STIM1 oligomerization and preventing Orai1 inactivation by Ca^{2+} [106, 107].

Ca^{2+} pumping by both SERCA and PMCA is an ATP-dependent process that is favored by the presence of activated mitochondria. By supplying ATP to the PMCA, mitochondria favor the transport of Ca^{2+} ions from the lumen of the ER to the extracellular medium when signaling is triggered, and thus favor SOCE activation. In contrast, by energizing SERCA, mitochondria improve store refilling and prevent SOCE activation (for review see [108]. Whether mitochondria enhance or reduce the level of ER depletion thus depends on the spatial organization of the organelles and on the relative contribution of mitochondria in buffering Ca^{2+} action (in the vicinity of Ca^{2+} depletion and Ca^{2+} entry channels) and in supplying ATP for SERCA and PMCA.

However, In more recent reports, it seems that Ca^{2+} uptake by mitochondria is essentially independent of SOC [78] and is much more sensitive to Ca^{2+} release by the ER than Ca^{2+} originating from the extracellular medium [109, 110]. Both organelles are finely tuned and interact one with each other. Although no direct link has ever been established between STIM1 and mitochondria, mitochondria and ER stores have been reported to be in close contact and to be able to exchange Ca^{2+} [111–113].

Additionally, recent findings also suggest that mitochondria can control the SOC independently of the Ca^{2+} buffering action through its action on STIM1 trafficking. Some authors have proposed that under depolarization, mitochondria might have a positive effect of STIM1 migration and subsequent SOC activation, a mechanism that appears to be dependent on mitofusin 2 (Mfn2) [114].

14.5 Ca^{2+} Tunnelling: Link Between SOCe and Store Depletion

Lastly, an important part of the relationship between ER refilling and SOC Ca^{2+} entry could rely on the Ca^{2+} tunnelling which has been described preferentially in highly polarized cells; i.e. pancreatic acinar cells. Such a mechanism allows the intracellular propagation of Ca^{2+} signals from a local entry at the basolateral membrane to distant targets several μm away. In these cells, Ca^{2+} across the cytosol would have been blocked by the greater buffering capacity of the cytosol, including mitochondria [115]. This supports the concept that entering Ca^{2+} is taken up into the ER by the SERCA, where it rapidly diffuses to the apical region and it is released through IP_3R. In the ER, Ca^{2+} ions can travel much easier (lower Ca^{2+} buffer activities and lower capacities of Ca^{2+} binding proteins of the ER lumen compared to the cytosol) [116, 117]. A similar mechanism is also proposed for induction of SOCe dependent Ca^{2+} signalling activation, such as NFAT, without induction of a global Ca^{2+} signal and faraway of the SOCe Ca^{2+} microdomains [118].

14.6 Conclusions

In a great number of cell types, Ca^{2+} signal is a message coded by change in Ca^{2+} concentration from steady increase to high oscillation frequencies with a period ranging from a few seconds to a few minutes. These Ca^{2+} oscillations are thought to control a wide variety of cellular processes, and are often organized into intracellular and intercellular Ca^{2+} waves.

SOCe plays a crucial part in those processes. Although the first goal of this Ca^{2+} influx into the cytosol via SOCe is to refill intracellular stores after their depletion, the impact of SOCe is major under physiological conditions. The existence of the SOC process is based on the crosstalk between distinct compartments. SERCA, the ER Ca^{2+} pump, STIM1 and Orai1 and TRPC1 can be localized at the ER-PM junction. This results in a very efficient refilling of intracellular stores.

The crosstalk between SOCe and intracellular stores allows a complex and finely tuned regulation of Ca^{2+} events either at the mouth of the channel, between distinct Ca^{2+} stores including mitochondria, or even far more distant from the SOC clusters into the cell via the diffusion of the message inside the ER. This is why this crosstalk appears much more than a simple process that refills intracellular stores but is recognized as a crucial member of the Ca^{2+} signalling.

References

1. Williams RJ (2006) The evolution of calcium biochemistry. Biochim Biophys Acta 1763(11):1139–1146
2. Parekh AB (2008) Ca2+ microdomains near plasma membrane Ca2+ channels: impact on cell function. J Physiol 586(13):3043–3054
3. McCarron JG, Chalmers S, Bradley KN, MacMillan D, Muir TC (2006) Ca2+ microdomains in smooth muscle. Cell Calcium 40(5–6):461–493
4. Taylor CW, Berridge MJ, Cooke AM, Potter BV (1989) Inositol 1,4,5-trisphosphorothioate, a stable analogue of inositol trisphosphate which mobilizes intracellular calcium. Biochem J 259(3):645–650
5. Berridge MJ (2004) Calcium signal transduction and cellular control mechanisms. Biochim Biophys Acta 1742(1–3):3–7
6. Berridge MJ (2006) Calcium microdomains: organization and function. Cell Calcium 40(5–6):405–412
7. Tojyo Y, Morita T, Nezu A, Tanimura A (2014) Key components of store-operated Ca2+ entry in non-excitable cells. J Pharmacol Sci 125(4):340–346
8. Putney JW (1986) A model for receptor-regulated calcium entry. Cell Calcium 7:1–12
9. Takemura H, Hughes AR, Thastrup O, Putney JWJ (1989) Activation of calcium entry by the tumor promoter thapsigargin in parotid acinar cells. J Biol Chem 264:12266–12271
10. Thastrup O, Cullen PJ, Drobak BK, Hanley MR, Dawson AP (1990) Thapsigargin, a tumor promotor, discharges intracellular Ca^{2+} stores by specific inhibition of the endoplasmic reticulum Ca^{2+} -ATPase. Proc Natl Acad Sci U S A 87:2466–2470
11. Lewis RS, Cahalan MD (1989) Mitogen-induced oscillations of cytosolic Ca2+ and transmembrane Ca2+ current in human leukemic T cells. Cell Regul 1(1):99–112
12. von Tscharner V, Prod'hom B, Baggiolini M, Reuter H (1986) Ion channels in human neutrophils activated by a rise in free cytosolic calcium concentration. Nature 324(6095):369–372
13. Kuno M, Gardner P (1987) Ion channels activated by inositol 1,4,5-trisphosphate in plasma membrane of human T-lymphocytes. Nature 326(6110):301–304
14. Penner R, Matthews G, Neher E (1988) Regulation of calcium influx by second messengers in rat mast cells. Nature 334(6182):499–504
15. Randriamampita C, Tsien RY (1993) Emptying of intracellular Ca2+ stores releases a novel small messenger that stimulates Ca2+ influx. Nature 364(6440):809–814
16. Smani T, Zakharov SI, Csutora P, Leno E, Trepakova ES, Bolotina VM (2004) A novel mechanism for the store-operated calcium influx pathway. Nat Cell Biol 6(2):113–120
17. Dramane G, Akpona S, Besnard P, Khan NA, Pt A (2014) Cell mechanisms of gustatory lipids perception and modulation of the dietary fat preference. Biochimie 107 Pt A:11–14
18. Irvine R (1990) 'Quantal' Ca^{2+} release and the control of Ca^{2+} entry by inositol phosphates. A possible mechanism. FEBS Lett 263:5–9
19. Redondo PC, Jardin I, Lopez JJ, Salido GM, Rosado JA (2008) Intracellular Ca(2+) store depletion induces the formation of macromolecular complexes involving hTRPC1, hTRPC6, the type II IP(3) receptor and SERCA3 in human platelets. Biochim Biophys Acta 1783(6):1163–1176
20. Tang J, Lin Y, Zhang Z, Tikunova S, Birnbaumer L, Zhu MX (2001) Identification of common binding sites for calmodulin and inositol 1,4,5-trisphosphate receptors on the carboxyl termini of trp channels. J Biol Chem 276(24):21303–21310
21. Rosado JA, Sage SO (2001) Activation of store-mediated calcium entry by secretion-like coupling between the inositol 1,4,5-trisphosphate receptor type II and human transient receptor potential (hTrp1) channels in human platelets. Biochem J 356(Pt 1):191–198
22. Rosado JA, Brownlow SL, Sage SO (2002) Endogenously expressed Trp1 is involved in store-mediated Ca2+ entry by conformational coupling in human platelets. J Biol Chem 277(44):42157–42163

23. Liou J, Kim ML, Heo WD, Jones JT, Myers JW, Ferrell JE Jr, Meyer T (2005) STIM is a Ca2+ sensor essential for Ca2+-store-depletion-triggered Ca2+ influx. Curr Biol 15(13):1235–1241
24. Roos J, DiGregorio PJ, Yeromin AV, Ohlsen K, Lioudyno M, Zhang S, Safrina O, Kozak JA, Wagner SL, Cahalan MD, Velicelebi G, Stauderman KA (2005) STIM1, an essential and conserved component of store-operated Ca2+ channel function. J Cell Biol 169(3):435–445
25. Feske S, Gwack Y, Prakriya M, Srikanth S, Puppel SH, Tanasa B, Hogan PG, Lewis RS, Daly M, Rao A (2006) A mutation in Orai1 causes immune deficiency by abrogating CRAC channel function. Nature 441(7090):179–185
26. Vig M, Beck A, Billingsley JM, Lis A, Parvez S, Peinelt C, Koomoa DL, Soboloff J, Gill DL, Fleig A, Kinet JP, Penner R (2006) CRACM1 multimers form the ion-selective pore of the CRAC channel. Curr Biol 16(20):2073–2079
27. Zhang SL, Yeromin AV, Zhang XH, Yu Y, Safrina O, Penna A, Roos J, Stauderman KA, Cahalan MD (2006) Genome-wide RNAi screen of Ca(2+) influx identifies genes that regulate Ca(2+) release-activated Ca(2+) channel activity. Proc Natl Acad Sci U S A 103(24):9357–9362
28. Spassova MA, Soboloff J, He LP, Xu W, Dziadek MA, Gill DL (2006) STIM1 has a plasma membrane role in the activation of store-operated Ca(2+) channels. Proc Natl Acad Sci U S A 103(11):4040–4045
29. Shen WW, Frieden M, Demaurex N (2011) Local cytosolic Ca2+ elevations are required for stromal interaction molecule 1 (STIM1) de-oligomerization and termination of store-operated Ca2+ entry. J Biol Chem 286(42):36448–36459
30. Zhang SL, Yu Y, Roos J, Kozak JA, Deerinck TJ, Ellisman MH, Stauderman KA, Cahalan MD (2005) STIM1 is a Ca2+ sensor that activates CRAC channels and migrates from the Ca2+ store to the plasma membrane. Nature 437(7060):902–905
31. Stathopulos PB, Zheng L, Li GY, Plevin MJ, Ikura M (2008) Structural and mechanistic insights into STIM1-mediated initiation of store-operated calcium entry. Cell 135(1):110–122
32. Demaurex N, Frieden M (2003) Measurements of the free luminal ER Ca(2+) concentration with targeted "cameleon" fluorescent proteins. Cell Calcium 34(2):109–119
33. Stathopulos PB, Li GY, Plevin MJ, Ames JB, Ikura M (2006) Stored Ca2+ depletion-induced oligomerization of stromal interaction molecule 1 (STIM1) via the EF-SAM region: an initiation mechanism for capacitive Ca2+ entry. J Biol Chem 281(47):35855–35862
34. Zheng L, Stathopulos PB, Schindl R, Li GY, Romanin C, Ikura M (2011) Auto-inhibitory role of the EF-SAM domain of STIM proteins in store-operated calcium entry. Proc Natl Acad Sci U S A 108(4):1337–1342
35. Luik RM, Wang B, Prakriya M, Wu MM, Lewis RS (2008) Oligomerization of STIM1 couples ER calcium depletion to CRAC channel activation. Nature 454(7203):538–542
36. Wu MM, Buchanan J, Luik RM, Lewis RS (2006) Ca2+ store depletion causes STIM1 to accumulate in ER regions closely associated with the plasma membrane. J Cell Biol 174(6):803–813
37. Orci L, Ravazzola M, Le Coadic M, Shen WW, Demaurex N, Cosson P (2009) STIM1-induced precortical and cortical subdomains of the endoplasmic reticulum. Proc Natl Acad Sci U S A 106(46):19358–19362
38. Brandman O, Liou J, Park WS, Meyer T (2007) STIM2 is a feedback regulator that stabilizes basal cytosolic and endoplasmic reticulum Ca2+ levels. Cell 131(7):1327–1339
39. Stathopulos PB, Zheng L, Ikura M (2009) Stromal interaction molecule (STIM) 1 and STIM2 calcium sensing regions exhibit distinct unfolding and oligomerization kinetics. J Biol Chem 284(2):728–732
40. McNally BA, Yamashita M, Engh A, Prakriya M (2009) Structural determinants of ion permeation in CRAC channels. Proc Natl Acad Sci U S A 106(52):22516–22521
41. Penna A, Demuro A, Yeromin AV, Zhang SL, Safrina O, Parker I, Cahalan MD (2008) The CRAC channel consists of a tetramer formed by Stim-induced dimerization of Orai dimers. Nature 456(7218):116–120

42. Madl J, Weghuber J, Fritsch R, Derler I, Fahrner M, Frischauf I, Lackner B, Romanin C, Schutz GJ (2010) Resting state Orai1 diffuses as homotetramer in the plasma membrane of live mammalian cells. J Biol Chem 285(52):41135–41142
43. Vig M, Peinelt C, Beck A, Koomoa DL, Rabah D, Koblan-Huberson M, Kraft S, Turner H, Fleig A, Penner R, Kinet JP (2006) CRACM1 is a plasma membrane protein essential for store-operated Ca2+ entry. Science 312(5777):1220–1223
44. Prakriya M, Feske S, Gwack Y, Srikanth S, Rao A, Hogan PG (2006) Orai1 is an essential pore subunit of the CRAC channel. Nature 443(7108):230–233
45. Mercer JC, Dehaven WI, Smyth JT, Wedel B, Boyles RR, Bird GS, Putney JW Jr (2006) Large store-operated calcium selective currents due to co-expression of Orai1 or Orai2 with the intracellular calcium sensor, Stim1. J Biol Chem 281(34):24979–24990
46. Hou X, Pedi L, Diver MM, Long SB (2012) Crystal structure of the calcium release-activated calcium channel Orai. Science 338(6112):1308–1313
47. Smyth JT, Dehaven WI, Bird GS, Putney JW Jr (2008) Ca2+-store-dependent and -independent reversal of Stim1 localization and function. J Cell Sci 121(Pt 6):762–772
48. Park CY, Hoover PJ, Mullins FM, Bachhawat P, Covington ED, Raunser S, Walz T, Garcia KC, Dolmetsch RE, Lewis RS (2009) STIM1 clusters and activates CRAC channels via direct binding of a cytosolic domain to Orai1. Cell 136(5):876–890
49. Yuan JP, Zeng W, Dorwart MR, Choi YJ, Worley PF, Muallem S (2009) SOAR and the polybasic STIM1 domains gate and regulate Orai channels. Nat Cell Biol 11(3):337–343
50. Muik M, Fahrner M, Derler I, Schindl R, Bergsmann J, Frischauf I, Groschner K, Romanin C (2009) A cytosolic homomerization and a modulatory domain within STIM1 C terminus determine coupling to ORAI1 channels. J Biol Chem 284(13):8421–8426
51. Kawasaki T, Lange I, Feske S (2009) A minimal regulatory domain in the C terminus of STIM1 binds to and activates ORAI1 CRAC channels. Biochem Biophys Res Commun 385(1):49–54
52. Korzeniowski MK, Manjarres IM, Varnai P, Balla T (2010) Activation of STIM1-Orai1 involves an intramolecular switching mechanism. Sci Signal 3(148):ra82
53. Zhou Y, Meraner P, Kwon HT, Machnes D, Oh-hora M, Zimmer J, Huang Y, Stura A, Rao A, Hogan PG (2010) STIM1 gates the store-operated calcium channel ORAI1 in vitro. Nat Struct Mol Biol 17(1):112–116
54. Stathopulos PB, Schindl R, Fahrner M, Zheng L, Gasmi-Seabrook GM, Muik M, Romanin C, Ikura M (2013) STIM1/Orai1 coiled-coil interplay in the regulation of store-operated calcium entry. Nat Commun 4:2963
55. Zhou Y, Srinivasan P, Razavi S, Seymour S, Meraner P, Gudlur A, Stathopulos PB, Ikura M, Rao A, Hogan PG (2013) Initial activation of STIM1, the regulator of store-operated calcium entry. Nat Struct Mol Biol 20(8):973–981
56. Derler I, Plenk P, Fahrner M, Muik M, Jardin I, Schindl R, Gruber HJ, Groschner K, Romanin C (2013) The extended transmembrane Orai1 N-terminal (ETON) region combines binding interface and gate for Orai1 activation by STIM1. J Biol Chem 288(40):29025–29034
57. McNally BA, Somasundaram A, Yamashita M, Prakriya M (2012) Gated regulation of CRAC channel ion selectivity by STIM1. Nature 482(7384):241–245
58. Hoth M, Penner R (1992) Depletion of intracellular calcium stores activates a calcium current in mast cells. Nature 355(6358):353–356
59. Demuro A, Penna A, Safrina O, Yeromin AV, Amcheslavsky A, Cahalan MD, Parker I (2011) Subunit stoichiometry of human Orai1 and Orai3 channels in closed and open states. Proc Natl Acad Sci U S A 108(43):17832–17837
60. Shim AH, Tirado-Lee L, Prakriya M (2015) Structural and functional mechanisms of CRAC channel regulation. J Mol Biol 427(1):77–93
61. Majewski L, Kuznicki J (2015) SOCE in neurons: signaling or just refilling? Biochim Biophys Acta 1853(9):1940–1952
62. Montell C (2005) The TRP superfamily of cation channels. Sci STKE 2005(272):re3
63. DeHaven WI, Jones BF, Petranka JG, Smyth JT, Tomita T, Bird GS, Putney JW Jr (2009) TRPC channels function independently of STIM1 and Orai1. J Physiol 587(Pt 10):2275–2298

64. Putney JW (2009) Capacitative calcium entry: from concept to molecules. Immunol Rev 231(1):10–22
65. Beech DJ (2005) TRPC1: store-operated channel and more. Pflugers Arch 451(1):53–60
66. Vaca L, Sampieri A (2002) Calmodulin modulates the delay period between release of calcium from internal stores and activation of calcium influx via endogenous TRP1 channels. J Biol Chem 277(44):42178–42187
67. Salido GM, Jardin I, Rosado JA (2011) The TRPC ion channels: association with Orai1 and STIM1 proteins and participation in capacitative and non-capacitative calcium entry. Adv Exp Med Biol 704:413–433
68. Zeng W, Yuan JP, Kim MS, Choi YJ, Huang GN, Worley PF, Muallem S (2008) STIM1 gates TRPC channels, but not Orai1, by electrostatic interaction. Mol Cell 32(3):439–448
69. Lopez JJ, Salido GM, Pariente JA, Rosado JA (2006) Interaction of STIM1 with endogenously expressed human canonical TRP1 upon depletion of intracellular Ca2+ stores. J Biol Chem 281(38):28254–28264
70. Liu X, Wang W, Singh BB, Lockwich T, Jadlowiec J, O'Connell B, Wellner R, Zhu MX, Ambudkar IS (2000) Trp1, a candidate protein for the store-operated Ca(2+) influx mechanism in salivary gland cells. J Biol Chem 275(5):3403–3411
71. Liao Y, Erxleben C, Yildirim E, Abramowitz J, Armstrong DL, Birnbaumer L (2007) Orai proteins interact with TRPC channels and confer responsiveness to store depletion. Proc Natl Acad Sci U S A 104(11):4682–4687
72. Liao Y, Erxleben C, Abramowitz J, Flockerzi V, Zhu MX, Armstrong DL, Birnbaumer L (2008) Functional interactions among Orai1, TRPCs, and STIM1 suggest a STIM-regulated heteromeric Orai/TRPC model for SOCE/Icrac channels. Proc Natl Acad Sci U S A 105(8):2895–2900
73. Redondo PC, Harper AG, Harper MT, Brownlow SL, Rosado JA, Sage SO (2007) hTRPC1-associated alpha-actinin, and not hTRPC1 itself, is tyrosine phosphorylated during human platelet activation. J Thromb Haemost 5(12):2476–2483
74. Asanov A, Sampieri A, Moreno C, Pacheco J, Salgado A, Sherry R, Vaca L (2015) Combined single channel and single molecule detection identifies subunit composition of STIM1-activated transient receptor potential canonical (TRPC) channels. Cell Calcium 57(1):1–13
75. Hofmann T, Obukhov AG, Schaefer M, Harteneck C, Gudermann T, Schultz G (1999) Direct activation of human TRPC6 and TRPC3 channels by diacylglycerol. Nature 397(6716):259–263
76. Gudermann T, Mederos y Schnitzler M, Dietrich A (2004) Receptor-operated cation entry – more than esoteric terminology? Sci STKE 2004(243):pe35
77. Jardin I, Lopez JJ, Salido GM, Rosado JA (2008) Functional relevance of the de novo coupling between hTRPC1 and type II IP3 receptor in store-operated Ca2+ entry in human platelets. Cell Signal 20(4):737–747
78. Alonso MT, Manjarres IM, Garcia-Sancho J (2012) Privileged coupling between Ca(2+) entry through plasma membrane store-operated Ca(2+) channels and the endoplasmic reticulum Ca(2+) pump. Mol Cell Endocrinol 353(1–2):37–44
79. Burk SE, Lytton J, MacLennan DH, Shull GE (1989) cDNA cloning, functional expression, and mRNA tissue distribution of a third organellar Ca^{2+} pump. J Biol Chem 264(31):18561–18568
80. Bobe R, Bredoux R, Wuytack F, Quarck R, Kovacs T, Papp B, Corvazier E, Magnier C, Enouf J (1994) The rat platelet 97-kDa Ca^{2+}ATPase isoform is the sarcoendoplasmic reticulum Ca^{2+}ATPase 3 protein. J Biol Chem 269(2):1417–1424
81. Dally S, Monceau V, Corvazier E, Bredoux R, Raies A, Bobe R, del Monte F, Enouf J (2009) Compartmentalized expression of three novel sarco/endoplasmic reticulum Ca^{2+}ATPase 3 isoforms including the switch to ER stress, SERCA3f, in non-failing and failing human heart. Cell Calcium 45(2):144–154
82. Liou J, Fivaz M, Inoue T, Meyer T (2007) Live-cell imaging reveals sequential oligomerization and local plasma membrane targeting of stromal interaction molecule 1 after Ca2+ store depletion. Proc Natl Acad Sci U S A 104(22):9301–9306

83. Lemonnier L, Trebak M, Lievremont JP, Bird GS, Putney JW Jr (2006) Protection of TRPC7 cation channels from calcium inhibition by closely associated SERCA pumps. FASEB J 20(3):503–505
84. Redondo PC, Salido GM, Pariente JA, Sage SO, Rosado JA (2008) SERCA2b and 3 play a regulatory role in store-operated calcium entry in human platelets. Cell Signal 20(2):337–346
85. Bobe R, Hadri L, Lopez JJ, Sassi Y, Atassi F, Karakikes I, Liang L, Limon I, Lompre AM, Hatem SN, Hajjar RJ, Lipskaia L (2011) SERCA2a controls the mode of agonist-induced intracellular Ca^{2+} signal, transcription factor NFAT and proliferation in human vascular smooth muscle cells. J Mol Cell Cardiol 50(4):621–633
86. Sampieri A, Zepeda A, Asanov A, Vaca L (2009) Visualizing the store-operated channel complex assembly in real time: identification of SERCA2 as a new member. Cell Calcium 45(5):439–446
87. Lopez JJ, Jardin I, Bobe R, Pariente JA, Enouf J, Salido GM, Rosado JA (2008) STIM1 regulates acidic Ca2+ store refilling by interaction with SERCA3 in human platelets. Biochem Pharmacol 75(11):2157–2164
88. Jousset H, Frieden M, Demaurex N (2007) STIM1 knockdown reveals that store-operated Ca2+ channels located close to sarco/endoplasmic Ca2+ ATPases (SERCA) pumps silently refill the endoplasmic reticulum. J Biol Chem 282(15):11456–11464
89. Vaca L (2010) SOCIC: the store-operated calcium influx complex. Cell Calcium 47(3):199–209
90. Manjarres IM, Alonso MT, Garcia-Sancho J (2011) Calcium entry-calcium refilling (CECR) coupling between store-operated Ca(2+) entry and sarco/endoplasmic reticulum Ca(2+)-ATPase. Cell Calcium 49(3):153–161
91. Kovacs T, Berger G, Corvazier E, Paszty K, Brown A, Bobe R, Papp B, Wuytack F, Cramer EM, Enouf J (1997) Immunolocalization of the multi-sarco/endoplasmic reticulum Ca^{2+}ATPase system in human platelets. Br J Haematol 97(1):192–203
92. Lopez JJ, Redondo PC, Salido GM, Pariente JA, Rosado JA (2006) Two distinct Ca(2+) compartments show differential sensitivity to thrombin, ADP and vasopressin in human platelets. Cell Signal 18(3):373–381
93. Rosado JA, Lopez JJ, Harper AG, Harper MT, Redondo PC, Pariente JA, Sage SO, Salido GM (2004) Two pathways for store-mediated calcium entry differentially dependent on the actin cytoskeleton in human platelets. J Biol Chem 279(28):29231–29235
94. Lopez JJ, Camello-Almaraz C, Pariente JA, Salido GM, Rosado JA (2005) Ca2+ accumulation into acidic organelles mediated by Ca2+- and vacuolar H+-ATPases in human platelets. Biochem J 390(Pt 1):243–252
95. Churchill GC, Okada Y, Thomas JM, Genazzani AA, Patel S, Galione A (2002) NAADP mobilizes Ca(2+) from reserve granules, lysosome-related organelles, in sea urchin eggs. Cell 111(5):703–708
96. Ruas M, Davis LC, Chen CC, Morgan AJ, Chuang KT, Walseth TF, Grimm C, Garnham C, Powell T, Platt N, Platt FM, Biel M, Wahl-Schott C, Parrington J, Galione A (2015) Expression of Ca2+-permeable two-pore channels rescues NAADP signalling in TPC-deficient cells. EMBO J 34(13):1743–1758
97. Patel S (2015) Function and dysfunction of two-pore channels. Sci Signal 8(384):re7
98. Simmerman HK, Jones LR (1998) Phospholamban: protein structure, mechanism of action, and role in cardiac function. Physiol Rev 78(4):921–947
99. Bhupathy P, Babu GJ, Periasamy M (2007) Sarcolipin and phospholamban as regulators of cardiac sarcoplasmic reticulum Ca2+ ATPase. J Mol Cell Cardiol 42(5):903–911
100. Betz C, Stracka D, Prescianotto-Baschong C, Frieden M, Demaurex N, Hall MN (2013) Feature Article: mTOR complex 2-Akt signaling at mitochondria-associated endoplasmic reticulum membranes (MAM) regulates mitochondrial physiology. Proc Natl Acad Sci U S A 110(31):12526–12534
101. Bernardi P (1999) Mitochondrial transport of cations: channels, exchangers, and permeability transition. Physiol Rev 79(4):1127–1155

102. Malli R, Frieden M, Osibow K, Zoratti C, Mayer M, Demaurex N, Graier WF (2003) Sustained Ca2+ transfer across mitochondria is Essential for mitochondrial Ca2+ buffering, sore-operated Ca2+ entry, and Ca2+ store refilling. J Biol Chem 278(45):44769–44779
103. Hoth M, Button DC, Lewis RS (2000) Mitochondrial control of calcium-channel gating: a mechanism for sustained signaling and transcriptional activation in T lymphocytes. Proc Natl Acad Sci U S A 97(19):10607–10612
104. Park MK, Ashby MC, Erdemli G, Petersen OH, Tepikin AV (2001) Perinuclear, perigranular and sub-plasmalemmal mitochondria have distinct functions in the regulation of cellular calcium transport. EMBO J 20(8):1863–1874
105. Lewis RS (2007) The molecular choreography of a store-operated calcium channel. Nature 446(7133):284–287
106. Gilabert JA, Parekh AB (2000) Respiring mitochondria determine the pattern of activation and inactivation of the store-operated Ca(2+) current I(CRAC). EMBO J 19(23):6401–6407
107. Parekh AB, Putney JW Jr (2005) Store-operated calcium channels. Physiol Rev 85(2):757–810
108. Demaurex N, Poburko D, Frieden M (2009) Regulation of plasma membrane calcium fluxes by mitochondria. Biochim Biophys Acta 1787(11):1383–1394
109. Csordas G, Varnai P, Golenar T, Roy S, Purkins G, Schneider TG, Balla T, Hajnoczky G (2010) Imaging interorganelle contacts and local calcium dynamics at the ER-mitochondrial interface. Mol Cell 39(1):121–132
110. Giacomello M, Drago I, Bortolozzi M, Scorzeto M, Gianelle A, Pizzo P, Pozzan T (2010) Ca2+ hot spots on the mitochondrial surface are generated by Ca2+ mobilization from stores, but not by activation of store-operated Ca2+ channels. Mol Cell 38(2):280–290
111. Rizzuto R, Pinton P, Carrington W, Fay FS, Fogarty KE, Lifshitz LM, Tuft RA, Pozzan T (1998) Close contacts with the endoplasmic reticulum as determinants of mitochondrial Ca^{2+} responses. Science 280:1763–1766
112. Arnaudeau S, Kelley WL, Walsh JV Jr, Demaurex N (2001) Mitochondria recycle Ca^{2+} to the endoplasmic reticulum and prevent the depletion of neighboring endoplasmic reticulum regions. J Biol Chem 276(31):29430–29439
113. Li B, Xiao L, Wang ZY, Zheng PS (2014) Knockdown of STIM1 inhibits 6-hydroxydopamine-induced oxidative stress through attenuating calcium-dependent ER stress and mitochondrial dysfunction in undifferentiated PC12 cells. Free Radic Res 48(7):758–768
114. Singaravelu K, Nelson C, Bakowski D, de Brito OM, Ng SW, Di Capite J, Powell T, Scorrano L, Parekh AB (2011) Mitofusin 2 regulates STIM1 migration from the Ca2+ store to the plasma membrane in cells with depolarized mitochondria. J Biol Chem 286(14):12189–12201
115. Tinel H, Cancela JM, Mogami H, Gerasimenko JV, Gerasimenko OV, Tepikin AV, Petersen OH (1999) Active mitochondria surrounding the pancreatic acinar granule region prevent spreading of inositol trisphosphate-evoked local cytosolic Ca(2+) signals. EMBO J 18(18):4999–5008
116. Mogami H, Nakano K, Tepikin AV, Petersen OH (1997) Ca2+ flow via tunnels in polarized cells: recharging of apicl Ca2+ stores by focal Ca2+ entry through basal membrane patch. Cell 88:49–55
117. Mogami H, Tepikin AV, Petersen OH (1998) Termination of cytosolic Ca^{2+} signals: Ca^{2+} reuptake into intracellular stores is regulated by the free Ca^{2+} concentration in the store lumen. EMBO J 17(2):435–442
118. Courjaret R, Machaca K (2014) Mid-range Ca2+ signalling mediated by functional coupling between store-operated Ca2+ entry and IP3-dependent Ca2+ release. Nat Commun 5:3916

Chapter 15
Microdomains Associated to Lipid Rafts

Jonathan Pacheco, Josué O. Ramírez-Jarquín, and Luis Vaca

Abstract Store Operated Ca^{2+} Entry (SOCE), the main Ca^{2+} influx mechanism in non-excitable cells, is implicated in the immune response and has been reported to be affected in several pathologies including cancer. The basic molecular constituents of SOCE are Orai, the pore forming unit, and STIM, a multidomain protein with at least two principal functions: one is to sense the Ca^{2+} content inside the lumen of the endoplasmic reticulum(ER) and the second is to activate Orai channels upon depletion of the ER. The link between Ca^{2+} depletion inside the ER and Ca^{2+} influx from extracellular media is through a direct association of STIM and Orai, but for this to occur, both molecules have to interact and form clusters where ER and plasma membrane (PM) are intimately apposed. In recent years a great number of components have been identified as participants in SOCE regulation, including regions of plasma membrane enriched in cholesterol and sphingolipids, the so called lipid rafts, which recruit a complex platform of specialized microdomains, which cells use to regulate spatiotemporal Ca^{2+} signals.

Keywords Lipid rafts • SOCE • TRPC channels • STIM1 • Orai1 • ER/PM junction

Abbreviations

AC8	Adenylyl cyclase 8
APC	Adenomatous polyposis coli
ARC channels	Arachidonic acid gated channels
ASM	Airway smooth muscle
CaM	Calmodulin
cAMP	Cyclic adenosine monophosphate
$Ca_v1.2$	Voltage gated Ca^{2+} channel 1.2
Cav-1	Caveolin 1
CNS	Central nervous system

J. Pacheco • J.O. Ramírez-Jarquín • L. Vaca (✉)
Instituto de Fisiología Celular, Universidad Nacional Autónoma de México,
Ciudad Universitaria, México, DF 04510, Mexico
e-mail: lvaca@ifc.unam.mx

J.A. Rosado (ed.), *Calcium Entry Pathways in Non-excitable Cells*, Advances in Experimental Medicine and Biology 898, DOI 10.1007/978-3-319-26974-0_15

CRAC	Cholesterol recognition/amino acid consensus
CRACR2A	Ca^{2+} release-activated Ca^{2+} channel regulator 2A
DRMs	Detergent-resistance membranes
EAE	Experimental autoimmune encephalomyelitis
ER	Endoplasmic reticulum
ERM	Ezrin/radixin/moesin domain
ER-PM junctions	Endoplasmic reticulum-plasma membrane junctions
FCS	Fluorescence correlation spectroscopy
FRET	Fluorescence resonance energy transfer
GPI	Glycosylphosphatidyl inositol
HEK293	Human embryonic kidney 293
IBD	Inflammatory bowel disease
Icrac	Ca^{2+} release-activated Ca^{2+} current
IP_3R	Inositol trisphosphate receptor
IS	Immunological synapse
MS	Multiple sclerosis
MβCD	Methyl-β-cyclodextrin
NFAT	Nuclear factor of activated T-cells
PDGF	Platelet-derived growth factor
PI_4P	Phosphatidylinositol 4-phosphate
PIP_2	Phosphatidylinositol 4,5-biphosphate
PIP_3	Phosphatidylinositol (3,4,5)-trisphosphate
$PIP_5KI\beta$	Phosphatidylinositol 4-phosphate-5-kinase I isoform β
$PIP_5KI\gamma$	Phosphatidylinositol 4-phosphate-5-kinase I isoform γ
PKB	Protein kinase B
PM	Plasma membrane
PMCA	Plasma membrane Ca^{2+}-ATPase
POST	Partner of STIM1
RYR	Ryanodine receptor
SCID	Severe combined immunodeficiency
SERCA	Sarco/endoplasmic reticulum Ca^{2+}-ATPase
SG	Salivary gland
SOCE	Store operated Ca^{2+} entry
SOCIC	Store operated Ca^{2+} influx complex
SPCA2	Secretory pathway Ca^{2+}-ATPase
TCR	T-cells receptors
TIRFM	Total internal reflexion fluorescence microscopy
TRPC	Transient receptor potential canonical
VGCC	Voltage gated Ca^{2+} channels

15.1 Introduction

The vision that we had not long ago about the plasma membrane (PM) was that proposed by the Singer and Nicolson model, a fluid mosaic where proteins and lipids convey their dynamics following Brownian motion [1]. This intrinsic fluidity excludes ordered phases. However, coincident with the development of new technologies, novel concepts have emerged and new attributes have appeared decorating a different view of the PM [2, 3]. Perhaps one of the greatest changes in our original idea about the PM is that of lateral heterogeneity, which presents the lipid bilayer as formed by heterogeneous patches or domains enriched in selective types of lipids and proteins [4].

Organization of the PM domains demands a great diversity of lipids that change structural and functional features of the membrane. Eukaryote cells possess three main classes of membrane lipids: glycerophospholipids, sphingolipids and sterols. Glycerophospholipids and sphingolipids can vary to produce a combinatorial diversity of more than 1,000 different structures, meanwhile sterols in mammalian cell membranes are formed only by cholesterol [4] (Fig. 15.1a).

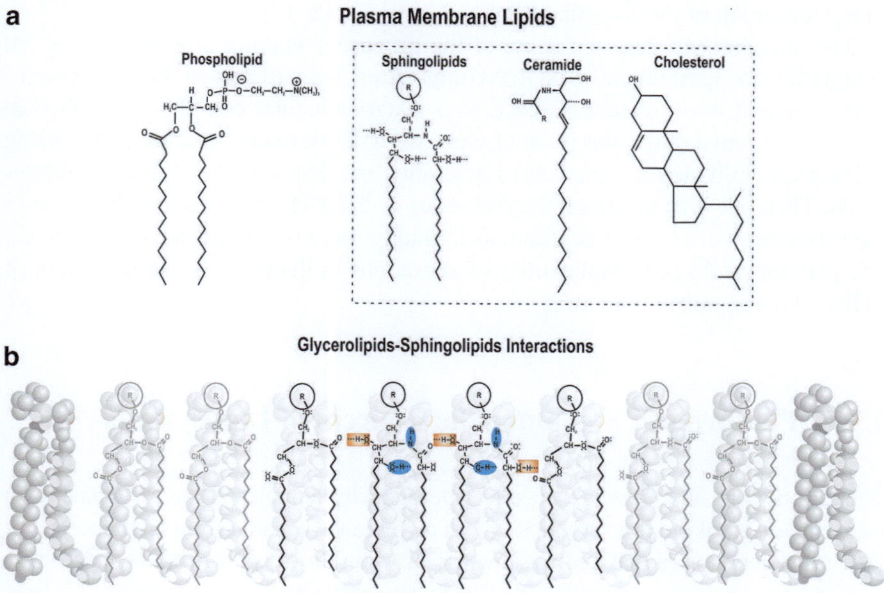

Fig. 15.1 Lipid components of rafts. (**a**) The drawing illustrates the four main lipids from the plasma membrane (PM). Outside the box are the phospholipids, the general components of membranes. Inside the box, the three major constituents of lipid rafts. Head polar groups are on *top*, the *R* group of sphingolipid and ceramide backbone results in a great diversity of lipids. (**b**) Inter and intra molecular interactions forming dynamic lipid rafts. In the center of the membrane two molecules of sphingolipids interact between them and with contiguous glycerolipids through hydrogen bonds (*rectangles*). In *ovals* intramolecular hydrogen bonds of sphingolipids constrain the lipid rafts

PM domains enriched in sphingolipids and cholesterol are components of the so called lipid rafts. Highly ordered and dynamic regions, restricted at the nanoscale size and with life spanning of only a few nanoseconds [5]. Lipid rafts plasticity results from thermodynamically favorable lipid-lipid, protein-protein and protein-lipid interactions [6, 7], and represent a recruitment center and a signaling platform for a large number of proteins. Lipid rafts have been shown to play important roles in diverse cellular functions, like trafficking, adhesion and Ca^{2+} signaling [8, 9]. This last, of great importance in the immune system, since T-cell activation requires a sustained rise of Ca^{2+} in order to activate the transcription factor NFAT, involved in antigen recognition [10].

The molecular mechanism in charge of increasing intracellular Ca^{2+} in T-cells is ubiquitous in non-excitable cells, the so-called Store Operated Ca^{2+} Entry (SOCE), a macromolecular protein complex for Ca^{2+} signaling associated to lipid rafts [9, 11–13]. The molecular assembled of SOCE begins immediately after the depletion of Ca^{2+} from the ER, followed by Ca^{2+} influx from the extracellular space [9]. The minimum proteins in charge of this mechanism are Orai and STIM. Orai by forming the pore channel and STIM by sensing Ca^{2+} which produce oligomerization and activation of Orai, when Ca^{2+} levels drop inside the ER [14]. Activation of Orai by STIM is an intricate and complex mechanism, which involves the physical movement of segments of the ER to bring them in close proximity with the PM, resulting in the formation of the so-called ER-PM junctions [15–17].

This chapter will describe some of the properties making lipid rafts excellent centers for the recruitment for SOCE components. We will review the recent methodologies that have identified SOCE as a macromolecular complex assembled in lipid rafts. From a molecular point of view, we will discuss the evidence supporting a complex multi-step mechanism of assembly and disassembly of SOCE components. Then, we will tackle on the dynamics of ER-PM junctions and their role in the establishment of Ca^{2+} microdomains. Finally, we will address the physiological and pathophysiological implications of alterations in the restricted organization of SOCE components in lipid rafts.

15.2 Platforms of Signaling, the Concept of Lipids Rafts

The first notions about the presence of heterogeneity domains at the PM came from the comparison of the lipid composition in membranes isolated from epithelial polarized cells, showing lateral and asymmetric distribution of sphingolipids and cholesterol from glycerophospholipids [18, 19]. Simons and van Meer hypothesized a lipid sorting process initiated at the Golgi complex, as responsible for the observed asymmetric distribution of sphingolipids in the PM. They proposed also a fundamental role of intermolecular interactions between hydrogen bonds of sphingolipids to maintain restricted domains [20] (Fig. 15.1b). Later, the advent of physicochemical experiments on model membranes came to recognize cholesterol as a key component promoting liquid ordered phases in artificial membranes [21, 22]. Finally, the concept of raft emerged for first time in 1992, with the studies of Brown and Rose, who developed a new assay consisting in the treatment of cells with nonionic Triton X-100 detergent at 4 °C. Through this technique these authors showed that the association of sphingolipids and glycosylphosphatidylinositol (GPI) anchored proteins was found on the Triton X-100-insoluble membrane fractions [23]. Since that initial study, isolation of Triton X-100-insoluble membrane fractions would become the major biochemical procedure to examine proteins embedded in detergent-resistance membranes (DRMs), which are considered in many studies as synonymous of lipid rafts [24].

DRMs extraction is a widespread technique to identify proteins residing in lipid rafts. However, the biological significance of DRMs must be taken with caution. The evidence shows robust correlation regarding chemical composition of membrane domains and DRMs [25]. Liposomes and cholesterol enriched model membranes have shown to be detergent insoluble [26, 27]. Conversely, cells that have been depleted of cholesterol present an increment in membrane detergent solubility [28]. However, evidence supporting DRMs in the biology context is limited because is unknown if the lipid content resembles exactly those of lipid rafts in living cells. Other criticism to DRMs isolation are the need to use detergents and the fact that such domains cannot be isolated at 37 °C (physiological temperature at which such domains would reside in a biological environment) [29]. In addition to the unknown process that takes place during the detergent extraction, which causes a drastic perturbation of the PM and may alter lipid organization, limiting the conclusions obtained by using this technique [30].

In the last decade our compression of lipid rafts has been improved by the use of new powerful techniques, in particular by new microscopy technology, which takes lipid rafts from being an in vitro elusive domain to well-defined regions in the PM of living cells. Liposomes and artificial membranes were left aside and exchanged for complete cells to study biological membranes. By using a wide range of methodologies, covering from elegant colocalization assays [31] to atomic force microscopy [32] and a great diversity of spectroscopy techniques, covering FRET [33, 34], FCS [35–37], TIRFM [38], molecular tracking and super resolution microscopy [39, 40], all together condense general attributes of lipid rafts, describing them as discrete regions with short time lifespans (<1 s), covering small areas (10–200 nm) and being highly dynamic [40]. In the rest of the chapter we will refer to DRMs as lipid rafts, taking in consideration the limitations of using a biochemical procedure for its isolation.

15.3 Store Operated Ca^{2+} Entry Components, Assembling the Complex

A plethora of molecules participate and regulate SOCE. We have previously referred to these multiprotein interactions as the Store Operated Ca^{2+} Influx Complex (SOCIC) [9]. Several proteins from the cytoskeleton [41, 42], endoplasmic reticulum [9, 43, 44], plasma membrane [45] and cytoplasm [46] in conjunction with lipid rafts at the PM play a role orchestrating the correct assembly and regulation of SOCE (Figs. 15.2 and 15.3).

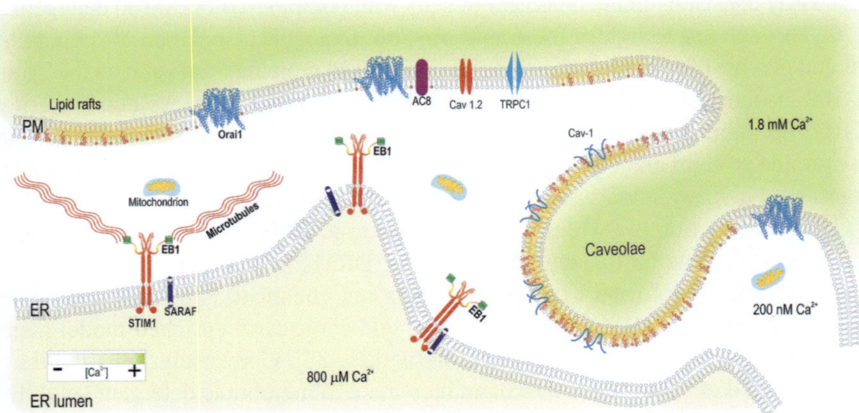

Fig. 15.2 SOCE at resting state. Model of SOCE components when the endoplasmic reticulum (*ER*) is replenish of Ca^{2+}. *Shadow patches* in PM depicts lipid rafts, Orai1, TRPC1 and adenylyl cyclase 8 (*AC8*) are inactive. $Ca_v1.2$, a voltage gated Ca^{2+} channel may be active, depending on the membrane potential. STIM1 moves continuously through the ER membrane by tracking microtubules via direct binding to EB1, who in turn tracks the tip of microtubules. Caveolae shows abundance expression of caveolin-1 and cholesterol and is presented as an invagination in the PM. Cholesterol molecules are shown in major abundance in the inner face of the PM. Solid polar heads of lipids in the PM represent PIP_2

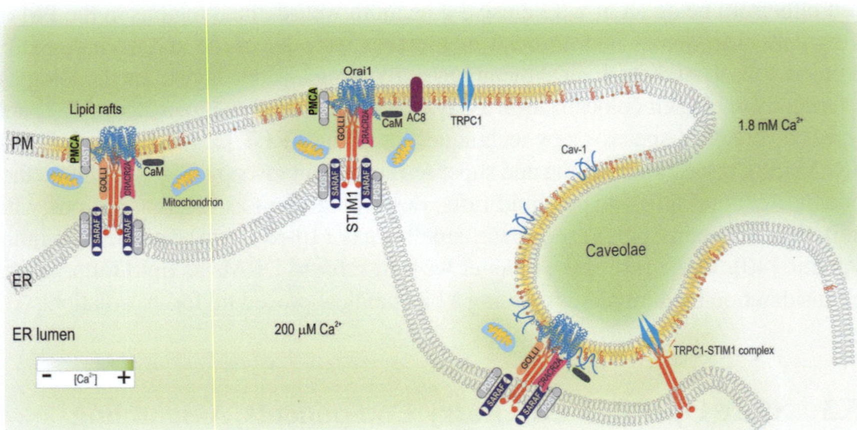

Fig. 15.3 SOCE in ER Ca^{2+} depletion conditions. SOCIC complex is illustrated associated to lipid rafts. STIM1 interacts with Orai1 to recruit a large number of regulators as Golli, CRACR2A, calmodulin (*CaM*), SARAF, POST and mitochondria. High density of proteins is supported in ER-PM junctions. In addition the SOCIC complex can be found in caveolae. Here, STIM1 may interact with and activate TRPC1. Cholesterol molecules are shown in major abundance in the inner face of the PM. Solid polar heads of lipids in the PM represent PIP_2. Focalized Ca^{2+} microdomains are developed beneath lipid rafts as signaling centers

STIM1 and Orai1 are the basic components of SOCE [47]. Together they generate a highly Ca^{2+} selective current with strong inward rectification, which resembles several of the biophysical properties of the so-called I_{crac} (Ca^{2+} release-activated Ca^{2+} current) [48]. However, depending of the cell type, the ionic current activated upon depletion of the ER does not recapitulate completely the I_{crac} initially described in mast cells, suggesting a heterogeneous molecular identity of the protein involved in ion conduction [49]. Indeed, several studies point to other possible players, like members from the transient receptor potential canonical (TRPC) channels, as putative ER-depletion activated channels [50]. Furthermore, there is experimental evidence showing direct interactions and gating of TRPC1 by STIM1 [51, 52]. Even more, a study shows that Orai1 may be required for the interaction between TRPC1 and STIM1 [53]. More recently, using a novel method combining single molecule imaging with single channel electrophysiological recordings we have shown the stoichiometry of STIM1-activated TRPC channels [54].

STIM1 appears to interact with a wide variety of proteins [55]. In resting state, when ER is filled of Ca^{2+}, STIM1 interacts with the microtubule tracking protein EB1, traveling continuously throughout the ER [41, 56]. Indeed, cytoskeleton plays an important role in SOCE regulation [57]. Short exposition to colchicine, to disrupt microtubules, enhances SOCE. Conversely, microtubules stabilization attenuates it [58]. At resting state, STIM1 interacts with a luminal ER oxidoreductase (ERp57), which associates to the amino terminal region of STIM1. Abolishing such interaction by using cells deficient in ERp57, results in an increment of Ca^{2+} entry when SOCE is activated [59]. The precise molecular role of ERp57 in SOCE modulation has not been elucidated to this date.

On the other hand, when ER depletion occurs, STIM1 undergoes a series of molecular rearrangements in order to activate Orai1. The active conformation of STIM1 is visualized as STIM1 clusters known as puncta [60]. These structures drive the recruitment of a great variety of molecules, including channels different from Orai and TRPC. STIM1 has been shown to associate and activate ARC channels [61, 62], leukotriene C_4-regulated Ca^{2+} channels [63] and mediates suppression of voltage-gated Ca^{2+} $Ca_v1.2$ activity [64].

The C-terminal of STIM1 is the target of intense modulation, aside from the autoinhibitory domain in its resting state [65, 66]. Golli-BG21, a myelin basic protein present in T-cells and oligodendrocytes, interacts with this STIM1 region, down regulating Ca^{2+} entry. Conversely, golli-deficient cells show increase Ca^{2+} entry mediated by SOCE [67]. Furthermore, other studies have shown a multitude of molecules regulating this cytosolic region of STIM1, including SARAF, an ER resident protein, which is found bound to STIM1 in resting conditions and having a role during depletion, where SARAF relocates to puncta to down regulate SOCE [44]. Conversely, the protein adenomatous polyposis coli (APC) facilitates translocation of STIM1 to ER-PM junctions, by binding to STIM1 upon depletion of the ER and inducing the dissociation of STIM1 from EB1 [42]. Later on, the translocation of STIM1 to ER-PM junctions is stabilized by septin4 and E-Syt1 [68].

Orai1 associates to a broad spectrum of proteins that modulate its function. CRACR2A associates to Orai1 and stabilizes STIM1-Orai1 puncta. On the other hand, calmodulin modulates Ca^{2+}-dependent inactivation of Orai1, by binding in the same region of CRACR2A [46, 69] (Fig. 15.3).

Other novel proteins have been identified recently, for instance Junctate, an ER resident protein, plays a critical role in clustering Orai1 and STIM1 at ER-PM junctions [70]. Partner of STIM1 (POST), a ten transmembranal protein localized in the PM and ER, constitutively binds to Orai1 and interacts with STIM1 only after ER depletion, while promoting the formation of a complex integrated by SERCA, PMCA, Na^+/K^+-ATPase, importins and exportins [71]. Interestingly, while POST reduction does not alter SOCE, is reported that the activity of PMCA is notably reduced [71, 72]. The biological significance of decreasing Ca^{2+} clearance by PMCA results in a sustained NFAT activity in T-cells, a preliminary step in the immunological response [73].

In human airway smooth muscle (ASM), Caveolin-1 (Cav-1), a principal component of caveolae, has an important role organizing SOCE elements. Overexpressing Cav-1 results in an increment of Orai1 expression [74]. Cav-1 also targets TRPC1 to the PM and promotes its association to STIM1 [11, 75, 76]. In addition, during *Xenopus laevis* oocyte meiosis, Orai1 internalizes by a mechanism dependent on caveolin and dynamin [77].

Caveolae and the immunological synapse (IS) convey a remarkable molecular organization in the cell. Caveolae serve as a center for trafficking and signal transduction [8] and assembly of the IS represents a key step in T-cells activation and antigen-presenting cells recognition. Both structures trigger a well-modulated signaling cascade [78] which brings together several proteins in charge of processing all the signals. Another interesting example is adenylyl cyclase 8 (AC8) which produces cyclic adenosine monophosphate (cAMP) and whose activation works synergistically with specific Ca^{2+} signals arising from SOCE [79, 80]. AC8 interacts with N-terminal of Orai1, which facilitates its own dynamic microdomain formation involving the cortical cytoskeleton; such microdomain assembly occurs in cholesterol rich domains of the PM [81].

15.4 SOCE in Lipid Rafts

The first evidence pointing out to SOCE assembly in lipid rafts domains come from the studies of Murata et al. and Prakash et al. in endothelial cells and human ASM cells, respectively. Endothelial cells lacking Cav-1 showed a drastic misallocation of TRPC1 and TRPC4 and a diminished Ca^{2+} entry. Conversely, Cav-1 reconstituted cells showed a functional recovery of Ca^{2+} influx and adequate localization of TRPC1 and TRPC4. In addition, functional Cav-1 microdomains incorporate IP3R [82]. On the other hand, silencing Cav-1 in ASM resulted in the attenuation of SOCE [83]. The group of Rosado reported a reduction of thapsigargin-evoked Ca^{2+} entry and the dissociation of STIM1, Orai1 and TRPC1 in cells exposed to MβCD, an agent that reduces cholesterol from the plasma membrane [84, 85].

Noteworthy, caveolae and lipid rafts differ at different levels. First, caveolae are well defined structures that conform invaginations of the cell surface varying from 25 to 150 nm [86], whereas lipid rafts do not produce membrane invaginations. Second, the presence of caveolae is cell-specific, for example in endothelia and muscle cells caveolae are highly abundant, while in lymphocytes and neurons are absent [87]. Caveolae are also rich in caveolin-1, whose function is to stabilize these structures, but in addition are enriched in a variety of molecular components present in lipid rafts, and hence the biochemical similarity of both. Lipid rafts and caveolae are detergent-insoluble and enriched in cholesterol. Because of these similarities caveolae is considered a subtype of lipid raft [8].

There is abundant experimental evidence showing that TRPC1 interacts with Cav-1 [82, 83, 88–90]. However, the main regulator of TRPC1 inside rafts seems to be STIM1 [11, 13], whose interaction carry on important changes on TRPC1 channel properties [11]. Our group has shown in HEK293 cells a dual activity of TRPC1, which is determined by its interaction with STIM1 at PM microdomains. TRPC1 function as a store-operated channel when is associated to STIM1 and this interaction promotes TRPC1 insertion into lipid rafts domains. However, when TRPC1 is not associated to STIM1, the channel functions as an agonist-activated channel, outside of lipid rafts [11]. Pani et al. confirmed these results showing that interaction between TRPC1 and STIM1 is performed inside lipid rafts, where they function as platforms for gathering TRPC1 and STIM1 only after ER depletion [13]. In addition, activation of SOCE promotes TRPC1 dissociation from Cav-1 in a STIM1-dependent form [76]. Such interaction supposes a scaffold function of Cav-1 to partition proteins into lipid rafts domains.

On the other hand, partitioning of SOCE components in discrete domains at the PM, enriched in cholesterol, have been shown as important in the modulation of SOCE [11, 13, 84, 91]. The precise mechanism responsible for this observation remains largely unexplored. Nevertheless, one repetitive observation is that cells treated with MβCD (an agent that reduces cholesterol levels at the PM) results in the attenuation of SOCE in a wide variety of cells [12, 13, 91]. Most surprisingly, the cholesterol depletion effect on SOCE may be recovered by overexpression of Orai1 and STIM1 [91]. Noteworthy, STIM1 puncta formation, which is a prerequisite for SOCE activation, is inhibited in cells treated with MβCD [13].

Even though activation of SOCE occurs in lipid rafts, seems that lipid rafts are required only during SOCIC assembly, once the STIM1-Orai1-TRPC1 complex is formed; MβCD has no effect on SOCE or the complex itself [12]. STIM1 appears to have a determinant role orchestrating SOCE components into lipid rafts, the actual model of how STIM1 interacts with Orai1 suggests a diffusion trap, which postulate that a STIM1-PM association precedes the interaction with Orai1, probably by a mechanism facilitated by PIP_2 and PIP_3 [92], which are particular abundant in lipid rafts [93] (Fig. 15.3).

15.5 Targeting SOCE to Lipid Rafts

Assuming that lipid rafts are a highly ordered arrangement of membrane domains, independently of the protein content and that occur spontaneously, as with liquid-ordered phases in model membranes, this suggest the existence of discrete regions in charge of directing proteins to the rafts. Even though, all the evidence of raft-targeting motifs was obtained in liquid-ordered phases and DRMs [94, 95], well-defined raft-targeting motifs may recognize three types of lipidations, which could increase the probability of finding a protein associated to lipid rafts, especially on proteins without transmembrane domains [94, 96]. Acylation with myristic or palmitic acid, GPI-linkage and direct interaction with cholesterol are the three main mechanisms associated to lipid raft targeting [97, 98]. There are several reports showing a clear effect of myristate/palmitate modifications in directing integrins [99] and G proteins subunits [94] to lipid rafts. Regarding SOCE, an important myristoylation on the N terminus of golli-BG21 allows its PM association, resulting in the reduction of SOCE when golli-BG21 is overexpressed. Eliminating the myristoylation domain produces a recovery of Ca^{2+} entry in the same way than golli-deficient cells have an enhanced Ca^{2+} influx [100]. This important finding suggests that golli must be present in rafts in order to operate properly.

Meanwhile, cholesterol has been shown to have a carrier effect gathering proteins in cholesterol-rich domains. Many cholesterol-binding motifs have been identified in different proteins [101]. Perhaps one of the most prominent of such binding domains is the Cholesterol Recognition/Amino acid consensus (CRAC) motif, a consensus sequence first identified in the benzodiazepine receptor [102]. For example, Cav-1, on top of its palmitoylation residues [103], presents three well characterized CRAC domains, which may function reorganizing cholesterol from plasma membrane in a first step of raft assembly [104, 105]. However, the main function of caveolin is to recruit proteins to caveolae [106], and Orai1 possesses a caveolin consensus-binding site in its N terminus, which is important to internalize the channel in *X. laevis* oocytes [77]. Cav-1 and Orai1 interact forming a complex [74]. Similarly, salivary gland (SG) cells of a knockout mouse lacking Cav-1 showed a drastic reduction in SOCE. Interestingly, the mechanism for SOCE reduction appears to be the translocation of the TRPC1-STIM1 complex outside of the lipid rafts [90]. Furthermore, TRPC channels have a conserved Cav-1 binding domain within N and C termini, which could serves to direct this protein to lipid rafts by Cav-1 [88].

STIM1 possesses an ERM (ezrin/radixin/moesin) homology domain [107]. This region is located in the cytosolic part of STIM1 and covers the fragment in charge of activating Orai1 and TRPC1 [51]. ERM domains are coiled-coil structures that mediate interactions between cytoskeleton and PM proteins. A dynamic interplay between active and inactive forms is regulated by phosphorylation, which results in the exposition of binding sites for proteins of the cytoskeleton and PM [108]. However, the best-characterized motif to bind lipids in STIM1 is the polybasic region, a lysine rich motif proximal to its C-terminal region, which interacts with the PM via anionic phospholipids [109, 110] (Fig. 15.4). The polybasic region of

STIM1 interacts with PIP_2 and PIP_3 [111]. Deletion mutants of STIM1 lacking this polybasic region oligomerize normally forming puncta but are not translocated near the PM after ER depletion [112]. Using phosphatidyl kinases inhibitors, Korzeniowski et al. showed that depletion of PI4P, a precursor of PIP_2, causes a decrease in Orai1-mediated Ca^{2+} entry, but STIM1 puncta formation was not affected, which suggest that SOCE is also sensitive to PI4P in addition to PIP_2 depletion [113]. Something interesting is that overexpression of Orai1 results in recovery of STIM1 translocation to the PM, supporting the idea that STIM1 puncta involves interaction with the PM, but a stronger binding with Orai1 rescues STIM1 translocation when phosphoinositides are depleted [114].

In contrast, very little is known about how transmembrane proteins are targeted to lipid rafts [94]. Calloway et al. studied an arginine rich sequence on the N-terminal region of Orai1, suggesting that this polybasic region may be involved in directing the channel to distinct PIP2 pools [115] (Fig. 15.4).

Other example of a SOCE member with raft-targeting motifs is adenylyl cyclase 8, which presents a well-identified raft-targeting motif. Cooper's group, using an elegant approach with combinatorial chimeras from adenylyl cyclases containing raft and non-raft-targeted isoforms, showed that the membrane targeting sequence in adenylyl cyclases lies on cytoplasmic domains, illustrating an example of the major role of cytoplasmic and not transmembrane domains in directing a protein to lipid rafts [116] (Fig. 15.4).

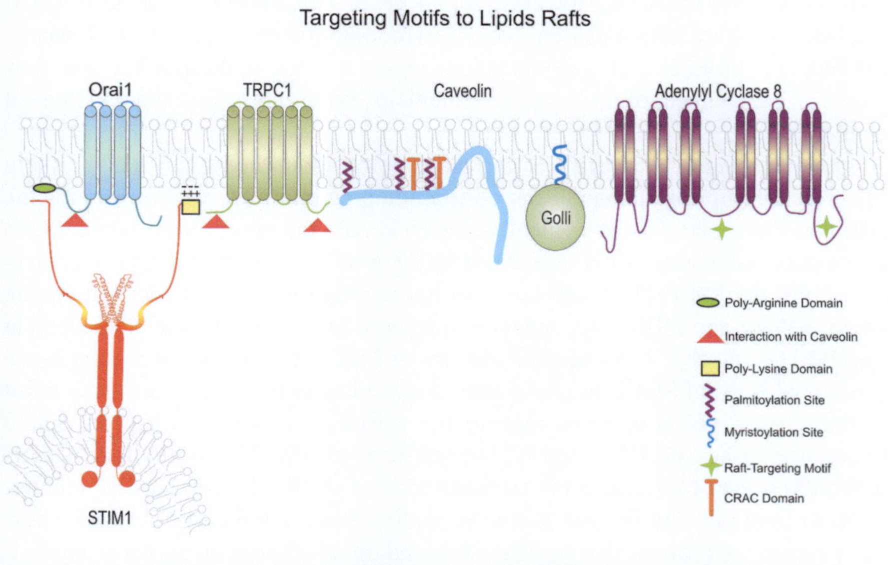

Fig. 15.4 Targeting motifs to lipid rafts. Domains reported to guide SOCIC components to lipid rafts. STIM1 possesses a poly-lysine domain that interacts with PIP_2 (*square*). Orai1 and TRPC1 present caveolin-1 binding domains (*triangle*) and in addition Orai1 has a poly-arginine domain suggested to interact with phospholipids (*oval*). Acylation of Cav-1 and Golli, palmitoylation and myristoylation are illustrated, respectively. Cav-1 also presents two CRAC domains and AC8 raft-targeting motifs

Nonetheless raft-targeting motifs assume the pre-existence of lipid rafts, a body of evidence indicates that rafts formation may be induced by proteins that will be part of the raft later on. Indeed, caveolin, flotillins and cytoskeleton regulate the assembling of caveolae and rafts [94, 117].

15.6 Methodologies Supporting SOCE in Rafts

The challenges of study lipid rafts are related to the very short life span and very small size of these PM domains. The definition of raft is intimately related to the methodology utilized to study them. For this reason we have highlighted the problems of using DRMs as synonymous of lipid rafts. While the lipid rafts is a general concept, DRMs are linked to the methodology employed to isolate such structures. For these reasons, it is important to discuss the diverse methodologies used to identify SOCE components in lipid rafts.

Using DRMs is a powerful and fast approach to identify proteins in detergent-resistant PM domains. However it is important to include a variety of controls in such studies. One important control frequently overlooked is the use of proteins that are not found in lipid rafts domains, as negative controls. That is the case of the transferrin receptor or actin, for example [13, 82].

DRMs isolation have permitted identify TRPC1, TRPC4 [82] and STIM1 [11, 13] on insoluble fractions. However, not all TRPC and STIM1 is present in DRM fractions, there are subpopulations recovered in soluble portions [13]. Unfortunately, the functional role of each subpopulation is unknown and understanding how they work in the context of its own microdomain represents a big methodological challenge.

Other biochemical assays yielding information about the presence of a protein of interest in lipid rafts have been co-immunoprecipitations, which involves the pull down of a bait protein to identify a second one (the prey) in conjunction with treatments that disrupts lipid rafts, like MβCD or filipin, agents that deplete cholesterol from the PM [91]. Rosado's group has reported that the interactions among Orai1, STIM1 and TRPC1 depend on the presence of normal levels of cholesterol at the PM [12, 84, 85]. At same time, the use of MβCD has permitted to investigate functional roles of SOCE in cholesterol depleted conditions, an interesting point is that cholesterol is required just for the activation and recruitment of SOCE components (the SOCIC assembly) but not for sustaining Ca^{2+} influx [12]. Others approaches used to examine the intricate relation of SOCE with specific components in lipid rafts are the use of commercially available lipid arrays, which allow the screening of proteins that may bind selected lipids, Zhou et al. used this approximation to confirm the interaction of STIM1 with PIP2 [111].

Today, major advances in understand lipid rafts are coming by spectroscopic approaches, in particular by microscopy techniques [39, 40]. These techniques provide a view of the unperturbed lipid environment on the cell, with high spatial and temporal resolution. Using a combined approach with electrophysiology and

FRET imaging, we have observed the dynamic interaction between STIM1-TRPC1. The association of both proteins precedes the ionic current trough TRPC1 channels. Removing cholesterol from the cell via incubation with MβCD abolished FRET signal and prevented the ionic current activation [11]. Calloway et al. showed with FRET an enhancement of STIM1-Orai1 interaction in mast cells overexpressing $PIP_5KI\beta$, which previously was reported to increase lipid rafts enriched in PIP_2 [118]. In contrast, overexpression of $PIP_5KI\gamma$, which also increases the production of PIP_2, did not enhance STIM1-Orai1 interaction. These results suggest that STIM1-Orai1 complex is established in membrane domains enriched in a subpopulation of PIP_2 and cholesterol, while a different subpopulation of PIP_2 contains less of the STIM1-Orai1 complex [115]. Recently, Maléth et al. confirmed this finding, showing that PIP_2 enriched domains support interactions between STIM1 and SARAF in membrane regions conformed by E-Syt1 and septin4 [68]. Lewis's group obtained similar results through single molecule tracking of STIM1-Orai1 complex in ER-PM junctions. Their results showed restriction of the STIM1-Orai1 complex to "corrals" that confined their diffusion. In addition, reduced events of free diffusible molecules were observed, suggesting a reversible binding of STIM1-Orai1 complex that escape to extrajunctional domains [92]. Such elegant techniques, including single molecule tracking, have provided a wealth of information about restricted movement of macromolecular complexes including those of STIM1-Orai1, STIM1-SARAF and STIM-TRPC1.

Recently, we have developed a combined single molecule-single channel determination technique, which allowed us to identify the stoichiometry of the STIM1-TRPC complex for several TRPC channels. Using this method we have determined also the stoichiometry for the TRPC-calmodulin complex. The power of this method is that allow us to study the stoichiometry of functional channels [54].

Other excellent tool to explore dynamic PM events of SOCE components is by TIRF microscopy, due to its high surface selectivity, which permits exciting fluorophores at distances not greater than 100 nm from the coverslip [38]. TIRF microscopy led to the identification of how clusters of STIM1 are inhibited when cholesterol was depleted from the PM [13], providing a possible explanation for SOCE reduction when lipid raft are disrupted. In a similar way, Sampieri et al. used TIRF to measure individual association events of STIM1 with PM microdomains in vivo [11]. By the use of the cholera toxin β subunit conjugated to a fluorophore, patches decorated with this conjugate represent areas with high concentration of the ganglioside GM1, an abundant constituent of lipid rafts [119]. The results show a clear preference of STIM1 to attach in areas decorated with the cholera toxin, suggesting an important role of lipid rafts in first steps of SOCIC assembly and SOCE activation.

Through TIRF microscopy Quintana et al. analyzed subplasmalemmal long-term Ca^{2+} signals in T-cells at IS or outside this domain. Taking advantage of T-cell function, IS was induced on coverslips coated with anti-CD3 antibodies. Interestingly, local Ca^{2+} at IS vicinity remained lower when compared to not-induced cells, suggesting a mechanism that prevents Ca^{2+}-dependent inactivation of Orai1 channels to sustain global Ca^{2+} signals and to activate the transcription factor NFAT [73].

With the advancement of methods like fluorescence correlation spectroscopy (FCS) and super-resolution microscopy more information regarding the elusive lipid rafts will be available in the near future, overcoming limitations of more traditional methods for the identification of lipid rafts elements, such as isolation of DRMs.

15.7 Plasma Membrane-Reticulum Endoplasmic Junctions

The endoplasmic reticulum (ER) is the main reservoir of Ca^{2+} in the cell [120]. In addition, proteins and lipid synthesis occurs here, positioning this organelle as a critical cellular structure in charge of Ca^{2+} signaling, protein traffic and lipid transfer [121]. To perform these functions, ER maintains highly modulated physical contact with cellular structures like mitochondria and the PM. In the case of the plasma membrane, these structures are known as ER-PM junctions [122].

ER-PM junctions are ubiquitous in eukaryote cells [121] and involves close apposition between two membranes by means of protein interactions, which provide the function of molecular bridges. These structures are highly dynamic, assembling and disassembling continuously. Some of the proteins functioning as molecular bridges are calsequestrin [123], RYR [124], junctions [123], mitsugumins [124], synaptotagmins [125], E-Syt1, Nir2 [126] and junctophilins. These last proteins are located in the ER, interacting with the PM most likely by binding to PIP_2 [121, 127].

The best characterized junctions are those for cardiac and skeletal striated muscles, where RYR and VGCC are key players in the junction formation [128]. Most interestingly, several SOCE components have been shown to play a role in these junctions. For instance, knockout mice lacking STIM1 present muscle fatigue and premature death due to myopathies, suggesting an important role for STIM1 in muscle ER-PM junction [129].

Translocation of STIM1 near to PM is a critical step prior to the activation of SOCE, thus Orai1 localization in ER-PM junctions is determined by interactions with STIM1 [130, 131]. Noteworthy, overexpression of STIM1 increases the number of ER-PM contacts, highlight the role of this protein as a molecular bridge [132]. PIP_2 and PIP_3 at the PM may have an important role in anchoring STIM1 molecules by its polybasic tail. However, the precise mechanism is poorly understood and appears to be very complex. Recently, the role of septins [133], ERp57, P100, golli, calmodulin and CRACR2A [134] in the translocation of STIM1 to ER-PM junctions has been evidenced. All these proteins perform a cooperative recruitment role to stabilize and regulate diverse SOCE components at ER-PM junctions.

SOCE in junctions is highly regulated by Ca^{2+}. The complex E-Syt1-septins have been shown to help in recruiting SARAF to ER-PM junctions and facilitate its interaction with STIM1, promoting slow Ca^{2+} dependent inactivation of Orai1 [68].

Calmodulin, CRACR2A and junctate possess Ca^{2+} binding domains important to inactivate or induce the disassembly of SOCE [46, 69, 70]. Junctate, for example, is a partner of Orai1-STIM1 complex and possess a Ca^{2+}-binding EF-hand relevant to initiate the clustering of STIM1 independently of ER depletion [70]. Overexpression of junctate increases the number of ER-PM junctions, while IP3R and TRPC3 channels help to stabilize them [135]. Another element that facilitates STIM1 translocation is the protein APC, which interacts with EB1 at the tip of microtubules [136]. Reducing APC in the cell does not affect STIM1 oligomerization, but favors puncta assembly outside ER-PM junctions [42].

Electron microscopy studies on lymphocytes have shown the distance between PM and ER membranes during the ER-PM junction assembly, the distance determined was of ~17 nm [16]. It is at this junction where the SOCE macromolecular complex is recruited, but in addition, distribution of STIM1 puncta at ER-PM junctions can be highly polarized. Pancreatic acinar cells and lymphocytes are examples of this polarization in STIM protein. STIM1 is present in the basolateral membrane, while Orai1 distributes in all the plasma membrane [137], similarly to what takes place at the IS in lymphocytes, where STIM1 distributes in the vicinity of the IS while Orai1 is observed throughout the PM [138].

The proper ER structure is a critical for functional SOCE and morphological changes that undergoes after ER depletion are required for STIM1 redistribution [139]. Holowka et al. showed recently that acute addition of linoleic acid prevents STIM1 puncta formation and Orai1 interactions; these authors hypothesized that this effect is the result of perturbations in the ER membrane structure [140].

STIM1 distribution in ER-PM junctions brings important physiological consequences; the most relevant may be the control of Ca^{2+} increments through localized signals between ER and PM, which detonates particular physiological responses. This subject will be further explored in the next section.

15.8 Physiological and Pathophysiological Significance of SOCE Association to Microdomains

SOCE is a ubiquitous mechanism in non-excitable cells, providing Ca^{2+} to trigger a plethora of functions including proliferation, migration, differentiation, secretion and many others. In the immune system, the main role of SOCE is to trigger antigen recognition in T-cells, which is a central event in the immune response [10]. Microdomains play a role redirecting and restricting SOCE components to sites were T-cells receptors (TCR) contact with antigen-presenting B cells, in the so-called immunological synapse (IS) [138].

Through different methodologies, including biochemical approaches like isolation of DRMs [141, 142], or spectroscopy techniques like FRET [143], TIRF [144], fluorescence lifetime imaging [144, 145], evidence accumulates to provide a fantastic picture about the role of microdomains in forming the IS, where lipid rafts have a central role as platforms to recruit proteins that conform the machinery of the immune response [78].

SOCE maintenance in these microdomains promotes a sustained signal of Ca^{2+} to induce an effective NFAT activation [146], a key event to induce transcription of cytokine genes and other genes involved in the immune response [147]. NFAT activation results by a highly regulated Ca^{2+} signaling pattern, in which appropriate frequency and amplitude promotes activation of different transcription factors [148]. Dolmetsch et al. showed that activation of pro-inflammatory transcription factors, NF-κB and c-Jun N-terminal kinase depend on large and transient increments of Ca^{2+}, whereas NFAT activation occurs by low and sustained Ca^{2+} increments [149]. In this way, a unique messenger (Ca^{2+}) can promote different and selective responses [148]. Indeed, NFAT activation is highly selective to local Ca^{2+} increments resulting from the Orai1 microdomain distribution, meanwhile TRPC1 or TRPC3-mediated Ca^{2+} entry cannot activate NFAT, but instead induce the activation of NF-κB [150, 151].

Lioudyno et al. showed for first time the polarization of STIM1 and Orai1 during IS formation of human T-cells [138]. Polarization of SOCE components is accompanied by focalized increments of Ca^{2+}. In an elegant study, Quintana et al. showed in great detail the intricate mechanism that modulates Ca^{2+} microdomains in T-cells following IS formation. The authors observed accumulation of Orai1, STIM1, PMCA as well mitochondria in the vicinity of the IS. STIM1-Orai1 are assembled at IS and mitochondria plays a fundamental role on the clearance of Ca^{2+} to prevent Ca^{2+} dependent inactivation of Orai1 channels in the microdomain. Mitochondria inhibit PMCA, avoiding extrusion of Ca^{2+}, providing a continuous Ca^{2+} signal at the domain [73].

Disruption of Ca^{2+} entry mediated by Orai1 in T-Cells results in severe combined immunodeficiency (SCID). The SCID syndrome is characterized by propensity for fungal and viral infections and a deficiency in NFAT activation [45], as a consequence of mutations in Orai1 and STIM1 genes [45, 152].

Also interesting, knockout mice lacking STIM1 and Orai1 genes display T-cell-mediated autoimmunity problems, resulting in autoimmune inflammatory bowel disease (IBD) [153]. In the same way, SOCE is required to induce autoimmune CNS inflammation, mice lacking STIM1 and STIM2 present resistance to acquire experimental autoimmune encephalomyelitis (EAE), a model for multiple sclerosis (MS) [154, 155].

Other SOCE players have been shown to play a role in the modulation of immune cells. Golli-deficient T-cells increase Ca^{2+} entry mediated by SOCE followed by TCR stimulation, which produce in vitro hyperproliferation. On other hand, in vivo studies have shown tolerance to produce EAE. Authors hypothesize that Golli-deficient T-cells present excessive Ca^{2+} entry and subsequent alterations in NFAT signaling [100].

SOCE recently has been implicated in carcinogenesis processes [156], Fedida-Metula et al. showed an increment in proliferation promoted by SOCE in B16-BL6 cells, a melanoma model. SOCE activates calmodulin (CaM) and SRC kinase, which in turn inactivates PP2A phosphatase, a negative regulator of protein kinase B (PKB) [157]. An interesting observation is that SOCE promotes proliferation via CaM and SRC kinase in a cholesterol dependent manner, resulting in a high sensitivity of PKB in cells depleted from cholesterol. The same study showed in vivo and in vitro retardation of tumor growth in cells exposed to MβCD [157]. SOCE also plays a role in the growth of mammary tumor cells, were SPCA2 gates Orai1 independently from STIM1 and ER Ca^{2+} content [158].

Lipid rafts and SOCE have been implicated in the proliferation of smooth muscle in the respiratory system where STIM1 and Orai1 modulate proliferation but also contraction [74, 159, 160]. Interesting data show that migration induced by PDGF is upregulated by the STIM1-Orai1 complex, having important consequences in remodeling airway epithelium in asthma [161]. SOCE activation in air smooth muscle cells is regulated by STIM1 but not STIM2 [160]. In addition, Cav-1 has been shown as an important component in SOCE. During inflammation processes Ca^{2+} entry-mediated by Orai1 contributes in a Cav-1 dependent-manner [74] and disruption of caveola domains with propofol, a lipid agent used for bronchodilatation, downregulates Ca^{2+} entry [159].

Cav-1 plays a role in intestinal epithelial cells as well, where TRPC1 functions as a store-operated channel, stimulating cell migration and producing rapid mucosal restitution after injury [89].

According to all the evidence obtained with cell lines, SOCIC components present an exquisite regulation by partitioning in lipid rafts, suggesting important consequences in the context of their native cell types. Notwithstanding SOCE partners and its regulation by rafts remains poorly explored, which could contribute in the future to better understand T-cell immunoreactivity, asthma, cancer and other pathophysiological conditions.

15.9 Conclusion

Since the original model proposed by Putney in 1986 [162], our understanding about the store-operated Ca^{2+} influx has evolved significantly to a sophisticated mechanism of intercommunication between the main Ca^{2+} reservoir (ER) and the PM. The use of novel methodologies helped us in the identification of many players in this complex signaling process. Among these players are a handful of proteins and lipids, including lipid-delimited domains such as lipid rafts.

An orchestrated symphony of intracellular signaling events transforms the initial trigger (depletion of the ER) into a profound rearrangement of the intracellular milieu, including the generation of de novo structures (ER-PM junctions) and the recruitment of a macromolecular complex into specialized microdomains (what we have named SOCIC). This sophisticated mechanism ensures the spatio-temporal modulation of intracellular Ca^{2+} concentrations, which in turn determines the type of cellular response.

Here we have reviewed the role of lipid rafts as centers involved in the modulation of Ca^{2+} signaling, we have analyzed the current evidence indicating that many of the proteins responsible for SOCE are found in lipid rafts. Finally, we discussed some of the physiological repercussions of SOCE association to rafts domains and how alterations in these microdomains may lead to pathological disorders.

References

1. Singer SJ, Nicolson GL (1972) The fluid mosaic model of the structure of cell membranes. Science 175(4023):720–731
2. Goni FM (2014) The basic structure and dynamics of cell membranes: an update of the Singer-Nicolson model. Biochim Biophys Acta 1838(6):1467–1476
3. Vereb G, Szollosi J, Matko J, Nagy P, Farkas T, Vigh L, Matyus L, Waldmann TA, Damjanovich S (2003) Dynamic, yet structured: the cell membrane three decades after the Singer-Nicolson model. Proc Natl Acad Sci U S A 100(14):8053–8058
4. Blom T, Somerharju P, Ikonen E (2011) Synthesis and biosynthetic trafficking of membrane lipids. Cold Spring Harb Perspect Biol 3(8):a004713
5. Pike LJ (2006) Rafts defined: a report on the keystone symposium on lipid rafts and cell function. J Lipid Res 47(7):1597–1598
6. Epand RM (2006) Cholesterol and the interaction of proteins with membrane domains. Prog Lipid Res 45(4):279–294
7. Quinn PJ (2010) A lipid matrix model of membrane raft structure. Prog Lipid Res 49(4):390–406
8. Laude AJ, Prior IA (2004) Plasma membrane microdomains: organization, function and trafficking. Mol Membr Biol 21(3):193–205
9. Vaca L (2010) SOCIC: the store-operated calcium influx complex. Cell Calcium 47(3):199–209
10. Hogan PG, Lewis RS, Rao A (2010) Molecular basis of calcium signaling in lymphocytes: STIM and ORAI. Annu Rev Immunol 28:491–533
11. Alicia S, Angelica Z, Carlos S, Alfonso S, Vaca L (2008) STIM1 converts TRPC1 from a receptor-operated to a store-operated channel: moving TRPC1 in and out of lipid rafts. Cell Calcium 44(5):479–491
12. Galan C, Woodard GE, Dionisio N, Salido GM, Rosado JA (2010) Lipid rafts modulate the activation but not the maintenance of store-operated Ca(2+) entry. Biochim Biophys Acta 1803(9):1083–1093
13. Pani B, Ong HL, Liu X, Rauser K, Ambudkar IS, Singh BB (2008) Lipid rafts determine clustering of STIM1 in endoplasmic reticulum-plasma membrane junctions and regulation of store-operated Ca2+ entry (SOCE). J Biol Chem 283(25):17333–17340
14. Shim AH, Tirado-Lee L, Prakriya M (2015) Structural and functional mechanisms of CRAC channel regulation. J Mol Biol 427(1):77–93
15. Baba Y, Hayashi K, Fujii Y, Mizushima A, Watarai H, Wakamori M, Numaga T, Mori Y, Iino M, Hikida M, Kurosaki T (2006) Coupling of STIM1 to store-operated Ca2+ entry through its constitutive and inducible movement in the endoplasmic reticulum. Proc Natl Acad Sci U S A 103(45):16704–16709
16. Wu MM, Buchanan J, Luik RM, Lewis RS (2006) Ca2+ store depletion causes STIM1 to accumulate in ER regions closely associated with the plasma membrane. J Cell Biol 174(6):803–813
17. Luik RM, Wang B, Prakriya M, Wu MM, Lewis RS (2008) Oligomerization of STIM1 couples ER calcium depletion to CRAC channel activation. Nature 454(7203):538–542
18. Forstner GG, Tanaka K, Isselbacher KJ (1968) Lipid composition of the isolated rat intestinal microvillus membrane. Biochem J 109(1):51–59

19. Kawai K, Fujita M, Nakao M (1974) Lipid components of two different regions of an intestinal epithelial cell membrane of mouse. Biochim Biophys Acta 369(2):222–233
20. Simons K, van Meer G (1988) Lipid sorting in epithelial cells. Biochemistry 27(17):6197–6202
21. Sankaram MB, Thompson TE (1991) Cholesterol-induced fluid-phase immiscibility in membranes. Proc Natl Acad Sci U S A 88(19):8686–8690
22. Brown DA, London E (1998) Structure and origin of ordered lipid domains in biological membranes. J Membr Biol 164(2):103–114
23. Brown DA, Rose JK (1992) Sorting of GPI-anchored proteins to glycolipid-enriched membrane subdomains during transport to the apical cell surface. Cell 68(3):533–544
24. Sonnino S, Prinetti A (2013) Membrane domains and the "lipid raft" concept. Curr Med Chem 20(1):4–21
25. Brown DA, London E (1997) Structure of detergent-resistant membrane domains: does phase separation occur in biological membranes? Biochem Biophys Res Commun 240(1):1–7
26. Schroeder R, London E, Brown D (1994) Interactions between saturated acyl chains confer detergent resistance on lipids and glycosylphosphatidylinositol (GPI)-anchored proteins: GPI-anchored proteins in liposomes and cells show similar behavior. Proc Natl Acad Sci U S A 91(25):12130–12134
27. Ahmed SN, Brown DA, London E (1997) On the origin of sphingolipid/cholesterol-rich detergent-insoluble cell membranes: physiological concentrations of cholesterol and sphingolipid induce formation of a detergent-insoluble, liquid-ordered lipid phase in model membranes. Biochemistry 36(36):10944–10953
28. Cerneus DP, Ueffing E, Posthuma G, Strous GJ, van der Ende A (1993) Detergent insolubility of alkaline phosphatase during biosynthetic transport and endocytosis. Role of cholesterol. J Biol Chem 268(5):3150–3155
29. Melkonian KA, Chu T, Tortorella LB, Brown DA (1995) Characterization of proteins in detergent-resistant membrane complexes from Madin-Darby canine kidney epithelial cells. Biochemistry 34(49):16161–16170
30. Munro S (2003) Lipid rafts: elusive or illusive? Cell 115(4):377–388
31. Harder T, Scheiffele P, Verkade P, Simons K (1998) Lipid domain structure of the plasma membrane revealed by patching of membrane components. J Cell Biol 141(4):929–942
32. Pralle A, Keller P, Florin EL, Simons K, Horber JK (2000) Sphingolipid-cholesterol rafts diffuse as small entities in the plasma membrane of mammalian cells. J Cell Biol 148(5):997–1008
33. Sohn HW, Tolar P, Jin T, Pierce SK (2006) Fluorescence resonance energy transfer in living cells reveals dynamic membrane changes in the initiation of B cell signaling. Proc Natl Acad Sci U S A 103(21):8143–8148
34. Varma R, Mayor S (1998) GPI-anchored proteins are organized in submicron domains at the cell surface. Nature 394(6695):798–801
35. Lasserre R, Guo XJ, Conchonaud F, Hamon Y, Hawchar O, Bernard AM, Soudja SM, Lenne PF, Rigneault H, Olive D, Bismuth G, Nunes JA, Payrastre B, Marguet D, He HT (2008) Raft nanodomains contribute to Akt/PKB plasma membrane recruitment and activation. Nat Chem Biol 4(9):538–547
36. Lenne PF, Wawrezinieck L, Conchonaud F, Wurtz O, Boned A, Guo XJ, Rigneault H, He HT, Marguet D (2006) Dynamic molecular confinement in the plasma membrane by microdomains and the cytoskeleton meshwork. EMBO J 25(14):3245–3256
37. Wenger J, Conchonaud F, Dintinger J, Wawrezinieck L, Ebbesen TW, Rigneault H, Marguet D, Lenne PF (2007) Diffusion analysis within single nanometric apertures reveals the ultrafine cell membrane organization. Biophys J 92(3):913–919
38. Asanov A, Zepeda A, Vaca L (2010) A novel form of Total Internal Reflection Fluorescence Microscopy (LG-TIRFM) reveals different and independent lipid raft domains in living cells. Biochim Biophys Acta 1801(2):147–155

39. Lingwood D, Kaiser HJ, Levental I, Simons K (2009) Lipid rafts as functional heterogeneity in cell membranes. Biochem Soc Trans 37(Pt 5):955–960
40. Owen DM, Magenau A, Williamson D, Gaus K (2012) The lipid raft hypothesis revisited--new insights on raft composition and function from super-resolution fluorescence microscopy. Bioessays 34(9):739–747
41. Grigoriev I, Gouveia SM, van der Vaart B, Demmers J, Smyth JT, Honnappa S, Splinter D, Steinmetz MO, Putney JW Jr, Hoogenraad CC, Akhmanova A (2008) STIM1 is a MT-plus-end-tracking protein involved in remodeling of the ER. Curr Biol 18(3):177–182
42. Asanov A, Sherry R, Sampieri A, Vaca L (2013) A relay mechanism between EB1 and APC facilitate STIM1 puncta assembly at endoplasmic reticulum-plasma membrane junctions. Cell Calcium 54(3):246–256
43. Roos J, DiGregorio PJ, Yeromin AV, Ohlsen K, Lioudyno M, Zhang S, Safrina O, Kozak JA, Wagner SL, Cahalan MD, Velicelebi G, Stauderman KA (2005) STIM1, an essential and conserved component of store-operated Ca2+ channel function. J Cell Biol 169(3):435–445
44. Palty R, Raveh A, Kaminsky I, Meller R, Reuveny E (2012) SARAF inactivates the store operated calcium entry machinery to prevent excess calcium refilling. Cell 149(2):425–438
45. Feske S, Gwack Y, Prakriya M, Srikanth S, Puppel SH, Tanasa B, Hogan PG, Lewis RS, Daly M, Rao A (2006) A mutation in Orai1 causes immune deficiency by abrogating CRAC channel function. Nature 441(7090):179–185
46. Srikanth S, Jung HJ, Kim KD, Souda P, Whitelegge J, Gwack Y (2010) A novel EF-hand protein, CRACR2A, is a cytosolic Ca2+ sensor that stabilizes CRAC channels in T cells. Nat Cell Biol 12(5):436–446
47. Zhou Y, Meraner P, Kwon HT, Machnes D, Oh-hora M, Zimmer J, Huang Y, Stura A, Rao A, Hogan PG (2010) STIM1 gates the store-operated calcium channel ORAI1 in vitro. Nat Struct Mol Biol 17(1):112–116
48. Hoth M, Penner R (1993) Calcium release-activated calcium current in rat mast cells. J Physiol 465:359–386
49. Liu X, Groschner K, Ambudkar IS (2004) Distinct Ca(2+)-permeable cation currents are activated by internal Ca(2+)-store depletion in RBL-2H3 cells and human salivary gland cells, HSG and HSY. J Membr Biol 200(2):93–104
50. Birnbaumer L (2009) The TRPC class of ion channels: a critical review of their roles in slow, sustained increases in intracellular Ca(2+) concentrations. Annu Rev Pharmacol Toxicol 49:395–426
51. Huang GN, Zeng W, Kim JY, Yuan JP, Han L, Muallem S, Worley PF (2006) STIM1 carboxyl-terminus activates native SOC, I(crac) and TRPC1 channels. Nat Cell Biol 8(9):1003–1010
52. Yuan JP, Zeng W, Huang GN, Worley PF, Muallem S (2007) STIM1 heteromultimerizes TRPC channels to determine their function as store-operated channels. Nat Cell Biol 9(6):636–645
53. Cheng KT, Liu X, Ong HL, Ambudkar IS (2008) Functional requirement for Orai1 in store-operated TRPC1-STIM1 channels. J Biol Chem 283(19):12935–12940
54. Asanov A, Sampieri A, Moreno C, Pacheco J, Salgado A, Sherry R, Vaca L (2015) Combined single channel and single molecule detection identifies subunit composition of STIM1-activated transient receptor potential canonical (TRPC) channels. Cell Calcium 57(1):1–13
55. Soboloff J, Rothberg BS, Madesh M, Gill DL (2012) STIM proteins: dynamic calcium signal transducers. Nat Rev Mol Cell Biol 13(9):549–565
56. Sampieri A, Zepeda A, Asanov A, Vaca L (2009) Visualizing the store-operated channel complex assembly in real time: identification of SERCA2 as a new member. Cell Calcium 45(5):439–446
57. Smyth JT, DeHaven WI, Bird GS, Putney JW Jr (2007) Role of the microtubule cytoskeleton in the function of the store-operated Ca2+ channel activator STIM1. J Cell Sci 120(Pt 21):3762–3771

58. Galan C, Dionisio N, Smani T, Salido GM, Rosado JA (2011) The cytoskeleton plays a modulatory role in the association between STIM1 and the Ca2+ channel subunits Orai1 and TRPC1. Biochem Pharmacol 82(4):400–410
59. Prins D, Groenendyk J, Touret N, Michalak M (2011) Modulation of STIM1 and capacitative Ca2+ entry by the endoplasmic reticulum luminal oxidoreductase ERp57. EMBO Rep 12(11):1182–1188
60. Liou J, Kim ML, Heo WD, Jones JT, Myers JW, Ferrell JE, Meyer T (2005) STIM is a Ca2+ sensor essential for Ca2+ -store-depletion-triggered Ca2+ influx. Curr Biol 15(13):1235–1241
61. Thompson JL, Mignen O, Shuttleworth TJ (2013) The ARC channel – an endogenous store-independent Orai channel. Curr Top Membr 71:125–148
62. Thompson JL, Shuttleworth TJ (2012) A plasma membrane-targeted cytosolic domain of STIM1 selectively activates ARC channels, an arachidonate-regulated store-independent Orai channel. Channels 6(5):370–378
63. Zhang X, Gonzalez-Cobos JC, Schindl R, Muik M, Ruhle B, Motiani RK, Bisaillon JM, Zhang W, Fahrner M, Barroso M, Matrougui K, Romanin C, Trebak M (2013) Mechanisms of STIM1 activation of store-independent leukotriene C4-regulated Ca2+ channels. Mol Cell Biol 33(18):3715–3723
64. Wang Y, Deng X, Mancarella S, Hendron E, Eguchi S, Soboloff J, Tang XD, Gill DL (2010) The calcium store sensor, STIM1, reciprocally controls Orai and CaV1.2 channels. Science 330(6000):105–109
65. Yu F, Sun L, Hubrack S, Selvaraj S, Machaca K (2013) Intramolecular shielding maintains the ER Ca(2)(+) sensor STIM1 in an inactive conformation. J Cell Sci 126(Pt 11):2401–2410
66. Stathopulos PB, Schindl R, Fahrner M, Zheng L, Gasmi-Seabrook GM, Muik M, Romanin C, Ikura M (2013) STIM1/Orai1 coiled-coil interplay in the regulation of store-operated calcium entry. Nat Commun 4:2963
67. Walsh CM, Doherty MK, Tepikin AV, Burgoyne RD (2010) Evidence for an interaction between Golli and STIM1 in store-operated calcium entry. Biochem J 430(3):453–460
68. Maleth J, Choi S, Muallem S, Ahuja M (2014) Translocation between PI(4,5)P2-poor and PI(4,5)P2-rich microdomains during store depletion determines STIM1 conformation and Orai1 gating. Nat Commun 5:5843
69. Mullins FM, Park CY, Dolmetsch RE, Lewis RS (2009) STIM1 and calmodulin interact with Orai1 to induce Ca2+ -dependent inactivation of CRAC channels. Proc Natl Acad Sci U S A 106(36):15495–15500
70. Srikanth S, Jew M, Kim KD, Yee MK, Abramson J, Gwack Y (2012) Junctate is a Ca2+-sensing structural component of Orai1 and stromal interaction molecule 1 (STIM1). Proc Natl Acad Sci U S A 109(22):8682–8687
71. Krapivinsky G, Krapivinsky L, Stotz SC, Manasian Y, Clapham DE (2011) POST, partner of stromal interaction molecule 1 (STIM1), targets STIM1 to multiple transporters. Proc Natl Acad Sci U S A 108(48):19234–19239
72. Cully TR, Edwards JN, Friedrich O, Stephenson DG, Murphy RM, Launikonis BS (2012) Changes in plasma membrane Ca-ATPase and stromal interacting molecule 1 expression levels for Ca(2+) signaling in dystrophic mdx mouse muscle. Am J Physiol Cell Physiol 303(5):C567–C576
73. Quintana A, Pasche M, Junker C, Al-Ansary D, Rieger H, Kummerow C, Nunez L, Villalobos C, Meraner P, Becherer U, Rettig J, Niemeyer BA, Hoth M (2011) Calcium microdomains at the immunological synapse: how ORAI channels, mitochondria and calcium pumps generate local calcium signals for efficient T-cell activation. EMBO J 30(19):3895–3912
74. Sathish V, Abcejo AJ, Thompson MA, Sieck GC, Prakash YS, Pabelick CM (2012) Caveolin-1 regulation of store-operated Ca(2+) influx in human airway smooth muscle. Eur Respir J 40(2):470–478

75. Brazer SC, Singh BB, Liu X, Swaim W, Ambudkar IS (2003) Caveolin-1 contributes to assembly of store-operated Ca2+ influx channels by regulating plasma membrane localization of TRPC1. J Biol Chem 278(29):27208–27215
76. Pani B, Ong HL, Brazer SC, Liu X, Rauser K, Singh BB, Ambudkar IS (2009) Activation of TRPC1 by STIM1 in ER-PM microdomains involves release of the channel from its scaffold caveolin-1. Proc Natl Acad Sci U S A 106(47):20087–20092
77. Yu F, Sun L, Machaca K (2010) Constitutive recycling of the store-operated Ca2+ channel Orai1 and its internalization during meiosis. J Cell Biol 191(3):523–535
78. Dustin ML (2014) The immunological synapse. Cancer Immunol Res 2(11):1023–1033
79. Martin AC, Willoughby D, Ciruela A, Ayling LJ, Pagano M, Wachten S, Tengholm A, Cooper DM (2009) Capacitative Ca2+ entry via Orai1 and stromal interacting molecule 1 (STIM1) regulates adenylyl cyclase type 8. Mol Pharmacol 75(4):830–842
80. Willoughby D, Everett KL, Halls ML, Pacheco J, Skroblin P, Vaca L, Klussmann E, Cooper DM (2012) Direct binding between Orai1 and AC8 mediates dynamic interplay between Ca2+ and cAMP signaling. Sci Signal 5(219), ra29
81. Ayling LJ, Briddon SJ, Halls ML, Hammond GR, Vaca L, Pacheco J, Hill SJ, Cooper DM (2012) Adenylyl cyclase AC8 directly controls its micro-environment by recruiting the actin cytoskeleton in a cholesterol-rich milieu. J Cell Sci 125(Pt 4):869–886
82. Murata T, Lin MI, Stan RV, Bauer PM, Yu J, Sessa WC (2007) Genetic evidence supporting caveolae microdomain regulation of calcium entry in endothelial cells. J Biol Chem 282(22):16631–16643
83. Prakash YS, Thompson MA, Vaa B, Matabdin I, Peterson TE, He T, Pabelick CM (2007) Caveolins and intracellular calcium regulation in human airway smooth muscle. Am J Physiol Lung Cell Mol Physiol 293(5):L1118–L1126
84. Jardin I, Salido GM, Rosado JA (2008) Role of lipid rafts in the interaction between hTRPC1, Orai1 and STIM1. Channels 2(6):401–403
85. Dionisio N, Galan C, Jardin I, Salido GM, Rosado JA (2011) Lipid rafts are essential for the regulation of SOCE by plasma membrane resident STIM1 in human platelets. Biochim Biophys Acta 1813(3):431–437
86. Pike LJ (2004) Lipid rafts: heterogeneity on the high seas. Biochem J 378(Pt 2):281–292
87. Parton RG, Richards AA (2003) Lipid rafts and caveolae as portals for endocytosis: new insights and common mechanisms. Traffic 4(11):724–738
88. Lockwich TP, Liu X, Singh BB, Jadlowiec J, Weiland S, Ambudkar IS (2000) Assembly of Trp1 in a signaling complex associated with caveolin-scaffolding lipid raft domains. J Biol Chem 275(16):11934–11942
89. Rathor N, Chung HK, Wang SR, Wang JY, Turner DJ, Rao JN (2014) Caveolin-1 enhances rapid mucosal restitution by activating TRPC1-mediated Ca2+ signaling. Physiol Rep 2(11):1–11
90. Pani B, Liu X, Bollimuntha S, Cheng KT, Niesman IR, Zheng C, Achen VR, Patel HH, Ambudkar IS, Singh BB (2013) Impairment of TRPC1-STIM1 channel assembly and AQP5 translocation compromise agonist-stimulated fluid secretion in mice lacking caveolin1. J Cell Sci 126(Pt 2):667–675
91. Gwozdz T, Dutko-Gwozdz J, Schafer C, Bolotina VM (2012) Overexpression of Orai1 and STIM1 proteins alters regulation of store-operated Ca2+ entry by endogenous mediators. J Biol Chem 287(27):22865–22872
92. Wu MM, Covington ED, Lewis RS (2014) Single-molecule analysis of diffusion and trapping of STIM1 and Orai1 at ER-plasma membrane junctions. Mol Biol Cell 25:3672–3685
93. Liu Y, Casey L, Pike LJ (1998) Compartmentalization of phosphatidylinositol 4,5-bisphosphate in low-density membrane domains in the absence of caveolin. Biochem Biophys Res Commun 245(3):684–690
94. Brown DA (2006) Lipid rafts, detergent-resistant membranes, and raft targeting signals. Physiology 21:430–439
95. Brown DA, London E (1998) Functions of lipid rafts in biological membranes. Annu Rev Cell Dev Biol 14:111–136

96. Zacharias DA, Violin JD, Newton AC, Tsien RY (2002) Partitioning of lipid-modified monomeric GFPs into membrane microdomains of live cells. Science 296(5569):913–916
97. Epand RM (2008) Proteins and cholesterol-rich domains. Biochim Biophys Acta 1778(7–8):1576–1582
98. Levental I, Grzybek M, Simons K (2010) Greasing their way: lipid modifications determine protein association with membrane rafts. Biochemistry 49(30):6305–6316
99. Gagnoux-Palacios L, Dans M, van't Hof W, Mariotti A, Pepe A, Meneguzzi G, Resh MD, Giancotti FG (2003) Compartmentalization of integrin alpha6beta4 signaling in lipid rafts. J Cell Biol 162(7):1189–1196
100. Feng JM, Hu YK, Xie LH, Colwell CS, Shao XM, Sun XP, Chen B, Tang H, Campagnoni AT (2006) Golli protein negatively regulates store depletion-induced calcium influx in T cells. Immunity 24(6):717–727
101. Fantini J, Barrantes FJ (2013) How cholesterol interacts with membrane proteins: an exploration of cholesterol-binding sites including CRAC, CARC, and tilted domains. Front Physiol 4:31
102. Li H, Papadopoulos V (1998) Peripheral-type benzodiazepine receptor function in cholesterol transport. Identification of a putative cholesterol recognition/interaction amino acid sequence and consensus pattern. Endocrinology 139(12):4991–4997
103. Dietzen DJ, Hastings WR, Lublin DM (1995) Caveolin is palmitoylated on multiple cysteine residues. Palmitoylation is not necessary for localization of caveolin to caveolae. J Biol Chem 270(12):6838–6842
104. Yang G, Xu H, Li Z, Li F (2014) Interactions of caveolin-1 scaffolding and intramembrane regions containing a CRAC motif with cholesterol in lipid bilayers. Biochim Biophys Acta 1838(10):2588–2599
105. Epand RM, Sayer BG, Epand RF (2005) Caveolin scaffolding region and cholesterol-rich domains in membranes. J Mol Biol 345(2):339–350
106. Fridolfsson HN, Roth DM, Insel PA, Patel HH (2014) Regulation of intracellular signaling and function by caveolin. FASEB J 28:3823–3831
107. Dziadek MA, Johnstone LS (2007) Biochemical properties and cellular localisation of STIM proteins. Cell Calcium 42(2):123–132
108. Parameswaran N, Gupta N (2013) Re-defining ERM function in lymphocyte activation and migration. Immunol Rev 256(1):63–79
109. Heo WD, Inoue T, Park WS, Kim ML, Park BO, Wandless TJ, Meyer T (2006) PI(3,4,5)P3 and PI(4,5)P2 lipids target proteins with polybasic clusters to the plasma membrane. Science 314(5804):1458–1461
110. Collins SR, Meyer T (2011) Evolutionary origins of STIM1 and STIM2 within ancient Ca2+ signaling systems. Trends Cell Biol 21(4):202–211
111. Zhou Y, Srinivasan P, Razavi S, Seymour S, Meraner P, Gudlur A, Stathopulos PB, Ikura M, Rao A, Hogan PG (2013) Initial activation of STIM1, the regulator of store-operated calcium entry. Nat Struct Mol Biol 20(8):973–981
112. Liou J, Fivaz M, Inoue T, Meyer T (2007) Live-cell imaging reveals sequential oligomerization and local plasma membrane targeting of stromal interaction molecule 1 after Ca2+ store depletion. Proc Natl Acad Sci U S A 104(22):9301–9306
113. Korzeniowski MK, Popovic MA, Szentpetery Z, Varnai P, Stojilkovic SS, Balla T (2009) Dependence of STIM1/Orai1-mediated calcium entry on plasma membrane phosphoinositides. J Biol Chem 284(31):21027–21035
114. Walsh CM, Chvanov M, Haynes LP, Petersen OH, Tepikin AV, Burgoyne RD (2010) Role of phosphoinositides in STIM1 dynamics and store-operated calcium entry. Biochem J 425(1):159–168
115. Calloway N, Owens T, Corwith K, Rodgers W, Holowka D, Baird B (2011) Stimulated association of STIM1 and Orai1 is regulated by the balance of PtdIns(4,5)P(2) between distinct membrane pools. J Cell Sci 124(Pt 15):2602–2610

116. Crossthwaite AJ, Seebacher T, Masada N, Ciruela A, Dufraux K, Schultz JE, Cooper DM (2005) The cytosolic domains of Ca2+-sensitive adenylyl cyclases dictate their targeting to plasma membrane lipid rafts. J Biol Chem 280(8):6380–6391
117. Head BP, Patel HH, Insel PA (2014) Interaction of membrane/lipid rafts with the cytoskeleton: impact on signaling and function: membrane/lipid rafts, mediators of cytoskeletal arrangement and cell signaling. Biochim Biophys Acta 1838(2):532–545
118. Vasudevan L, Jeromin A, Volpicelli-Daley L, De Camilli P, Holowka D, Baird B (2009) The beta- and gamma-isoforms of type I PIP5K regulate distinct stages of Ca2+ signaling in mast cells. J Cell Sci 122(Pt 14):2567–2574
119. Schon A, Freire E (1989) Thermodynamics of intersubunit interactions in cholera toxin upon binding to the oligosaccharide portion of its cell surface receptor, ganglioside GM1. Biochemistry 28(12):5019–5024
120. Somlyo AP, Bond M, Somlyo AV (1985) Calcium content of mitochondria and endoplasmic reticulum in liver frozen rapidly in vivo. Nature 314(6012):622–625
121. Carrasco S, Meyer T (2011) STIM proteins and the endoplasmic reticulum-plasma membrane junctions. Annu Rev Biochem 80:973–1000
122. Lam AK, Galione A (2013) The endoplasmic reticulum and junctional membrane communication during calcium signaling. Biochim Biophys Acta 1833(11):2542–2559
123. Lee KW, Maeng JS, Choi JY, Lee YR, Hwang CY, Park SS, Park HK, Chung BH, Lee SG, Kim YS, Jeon H, Eom SH, Kang C, Kim do H, Kwon KS (2012) Role of Junctin protein interactions in cellular dynamics of calsequestrin polymer upon calcium perturbation. J Biol Chem 287(3):1679–1687
124. Pan Z, Yang D, Nagaraj RY, Nosek TA, Nishi M, Takeshima H, Cheng H, Ma J (2002) Dysfunction of store-operated calcium channel in muscle cells lacking mg29. Nat Cell Biol 4(5):379–383
125. Giordano F, Saheki Y, Idevall-Hagren O, Colombo SF, Pirruccello M, Milosevic I, Gracheva EO, Bagriantsev SN, Borgese N, De Camilli P (2013) PI(4,5)P(2)-dependent and Ca(2+)-regulated ER-PM interactions mediated by the extended synaptotagmins. Cell 153(7):1494–1509
126. Chang CL, Hsieh TS, Yang TT, Rothberg KG, Azizoglu DB, Volk E, Liao JC, Liou J (2013) Feedback regulation of receptor-induced Ca2+ signaling mediated by E-Syt1 and Nir2 at endoplasmic reticulum-plasma membrane junctions. Cell Rep 5(3):813–825
127. Takeshima H, Komazaki S, Nishi M, Iino M, Kangawa K (2000) Junctophilins: a novel family of junctional membrane complex proteins. Mol Cell 6(1):11–22
128. Stefan CJ, Manford AG, Emr SD (2013) ER-PM connections: sites of information transfer and inter-organelle communication. Curr Opin Cell Biol 25(4):434–442
129. Stiber J, Hawkins A, Zhang ZS, Wang S, Burch J, Graham V, Ward CC, Seth M, Finch E, Malouf N, Williams RS, Eu JP, Rosenberg P (2008) STIM1 signalling controls store-operated calcium entry required for development and contractile function in skeletal muscle. Nat Cell Biol 10(6):688–697
130. Xu P, Lu J, Li Z, Yu X, Chen L, Xu T (2006) Aggregation of STIM1 underneath the plasma membrane induces clustering of Orai1. Biochem Biophys Res Commun 350(4):969–976
131. Luik RM, Wu MM, Buchanan J, Lewis RS (2006) The elementary unit of store-operated Ca2+ entry: local activation of CRAC channels by STIM1 at ER-plasma membrane junctions. J Cell Biol 174(6):815–825
132. Orci L, Ravazzola M, Le Coadic M, Shen WW, Demaurex N, Cosson P (2009) From the cover: STIM1-induced precortical and cortical subdomains of the endoplasmic reticulum. Proc Natl Acad Sci U S A 106(46):19358–19362
133. Sharma S, Quintana A, Findlay GM, Mettlen M, Baust B, Jain M, Nilsson R, Rao A, Hogan PG (2013) An siRNA screen for NFAT activation identifies septins as coordinators of store-operated Ca2+ entry. Nature 499(7457):238–242
134. Srikanth S, Gwack Y (2012) Orai1, STIM1, and their associating partners. J Physiol 590(Pt 17):4169–4177

135. Treves S, Vukcevic M, Griesser J, Armstrong CF, Zhu MX, Zorzato F (2010) Agonist-activated Ca2+ influx occurs at stable plasma membrane and endoplasmic reticulum junctions. J Cell Sci 123(Pt 23):4170–4181
136. Nathke IS (2004) The adenomatous polyposis coli protein: the Achilles heel of the gut epithelium. Annu Rev Cell Dev Biol 20:337–366
137. Lur G, Haynes LP, Prior IA, Gerasimenko OV, Feske S, Petersen OH, Burgoyne RD, Tepikin AV (2009) Ribosome-free terminals of rough ER allow formation of STIM1 puncta and segregation of STIM1 from IP(3) receptors. Curr Biol 19(19):1648–1653
138. Lioudyno MI, Kozak JA, Penna A, Safrina O, Zhang SL, Sen D, Roos J, Stauderman KA, Cahalan MD (2008) Orai1 and STIM1 move to the immunological synapse and are up-regulated during T cell activation. Proc Natl Acad Sci U S A 105(6):2011–2016
139. Jozsef L, Tashiro K, Kuo A, Park EJ, Skoura A, Albinsson S, Rivera-Molina F, Harrison KD, Iwakiri Y, Toomre D, Sessa WC (2014) Reticulon 4 is necessary for endoplasmic reticulum tubulation, STIM1-Orai1 coupling, and store-operated calcium entry. J Biol Chem 289(13):9380–9395
140. Holowka D, Korzeniowski MK, Bryant KL, Baird B (2014) Polyunsaturated fatty acids inhibit stimulated coupling between the ER Ca(2+) sensor STIM1 and the Ca(2+) channel protein Orai1 in a process that correlates with inhibition of stimulated STIM1 oligomerization. Biochim Biophys Acta 1841(8):1210–1216
141. Montixi C, Langlet C, Bernard AM, Thimonier J, Dubois C, Wurbel MA, Chauvin JP, Pierres M, He HT (1998) Engagement of T cell receptor triggers its recruitment to low-density detergent-insoluble membrane domains. EMBO J 17(18):5334–5348
142. Xavier R, Brennan T, Li Q, McCormack C, Seed B (1998) Membrane compartmentation is required for efficient T cell activation. Immunity 8(6):723–732
143. Hashimoto-Tane A, Yokosuka T, Ishihara C, Sakuma M, Kobayashi W, Saito T (2010) T-cell receptor microclusters critical for T-cell activation are formed independently of lipid raft clustering. Mol Cell Biol 30(14):3421–3429
144. Owen DM, Oddos S, Kumar S, Davis DM, Neil MA, French PM, Dustin ML, Magee AI, Cebecauer M (2010) High plasma membrane lipid order imaged at the immunological synapse periphery in live T cells. Mol Membr Biol 27(4–6):178–189
145. Wheeler G, Tyler KM (2011) Widefield microscopy for live imaging of lipid domains and membrane dynamics. Biochim Biophys Acta 1808(3):634–641
146. Feske S, Giltnane J, Dolmetsch R, Staudt LM, Rao A (2001) Gene regulation mediated by calcium signals in T lymphocytes. Nat Immunol 2(4):316–324
147. Rao A, Luo C, Hogan PG (1997) Transcription factors of the NFAT family: regulation and function. Annu Rev Immunol 15:707–747
148. Srikanth S, Gwack Y (2013) Orai1-NFAT signalling pathway triggered by T cell receptor stimulation. Mol Cells 35(3):182–194
149. Dolmetsch RE, Lewis RS, Goodnow CC, Healy JI (1997) Differential activation of transcription factors induced by Ca2+ response amplitude and duration. Nature 386(6627):855–858
150. Ong HL, Jang SI, Ambudkar IS (2012) Distinct contributions of Orai1 and TRPC1 to agonist-induced [Ca(2+)](i) signals determine specificity of Ca(2+)-dependent gene expression. PLoS One 7(10), e47146
151. Kar P, Nelson C, Parekh AB (2011) Selective activation of the transcription factor NFAT1 by calcium microdomains near Ca2+ release-activated Ca2+ (CRAC) channels. J Biol Chem 286(17):14795–14803
152. Picard C, McCarl CA, Papolos A, Khalil S, Luthy K, Hivroz C, LeDeist F, Rieux-Laucat F, Rechavi G, Rao A, Fischer A, Feske S (2009) STIM1 mutation associated with a syndrome of immunodeficiency and autoimmunity. N Engl J Med 360(19):1971–1980
153. McCarl CA, Khalil S, Ma J, Oh-hora M, Yamashita M, Roether J, Kawasaki T, Jairaman A, Sasaki Y, Prakriya M, Feske S (2010) Store-operated Ca2+ entry through ORAI1 is critical for T cell-mediated autoimmunity and allograft rejection. J Immunol 185(10):5845–5858

154. Schuhmann MK, Stegner D, Berna-Erro A, Bittner S, Braun A, Kleinschnitz C, Stoll G, Wiendl H, Meuth SG, Nieswandt B (2010) Stromal interaction molecules 1 and 2 are key regulators of autoreactive T cell activation in murine autoimmune central nervous system inflammation. J Immunol 184(3):1536–1542
155. Ma J, McCarl CA, Khalil S, Luthy K, Feske S (2010) T-cell-specific deletion of STIM1 and STIM2 protects mice from EAE by impairing the effector functions of Th1 and Th17 cells. Eur J Immunol 40(11):3028–3042
156. Bergmeier W, Weidinger C, Zee I, Feske S (2013) Emerging roles of store-operated Ca(2)(+) entry through STIM and ORAI proteins in immunity, hemostasis and cancer. Channels 7(5):379–391
157. Fedida-Metula S, Feldman B, Koshelev V, Levin-Gromiko U, Voronov E, Fishman D (2012) Lipid rafts couple store-operated Ca2+ entry to constitutive activation of PKB/Akt in a Ca2+/calmodulin-, Src- and PP2A-mediated pathway and promote melanoma tumor growth. Carcinogenesis 33(4):740–750
158. Feng M, Grice DM, Faddy HM, Nguyen N, Leitch S, Wang Y, Muend S, Kenny PA, Sukumar S, Roberts-Thomson SJ, Monteith GR, Rao R (2010) Store-independent activation of Orai1 by SPCA2 in mammary tumors. Cell 143(1):84–98
159. Grim KJ, Abcejo AJ, Barnes A, Sathish V, Smelter DF, Ford GC, Thompson MA, Prakash YS, Pabelick CM (2012) Caveolae and propofol effects on airway smooth muscle. Br J Anaesth 109(3):444–453
160. Peel SE, Liu B, Hall IP (2006) A key role for STIM1 in store operated calcium channel activation in airway smooth muscle. Respir Res 7:119
161. Suganuma N, Ito S, Aso H, Kondo M, Sato M, Sokabe M, Hasegawa Y (2012) STIM1 regulates platelet-derived growth factor-induced migration and Ca^{2+} influx in human airway smooth muscle cells. PLoS One 7(9), e45056
162. Putney JW Jr (1986) A model for receptor-regulated calcium entry. Cell Calcium 7(1):1–12

Chapter 16
Role of Scaffolding Proteins in the Regulation of TRPC-Dependent Calcium Entry

Bruno Constantin

Abstract Plasma membrane ion channels, and in particular TRPC channels need a specific membrane environment and association with scaffolding, signaling, and cytoskeleton proteins in order to play their important functional role. The molecular composition of TRPC channels is an important factor in determining channel activation mechanisms. TRPC proteins are incorporated in macromolecular complexes including several key Ca^{2+} signaling proteins as well as proteins involved in vesicle trafficking, cytoskeletal interactions, and scaffolding. Evidence has been provided for association of TRPC with calmodulin (CaM), IP_3R, PMCA, $G_{q/11}$, RhoA, and a variety of scaffolding proteins. The interaction between TRPC channels with adaptor proteins, determines their mode of regulation as well as their cellular localization and function. Adaptor proteins do not display any enzymatic activity but act as scaffold for the building of signaling complexes. The scaffolding proteins are involved in the assembling of these Ca^{2+} signaling complexes, the correct subcellular localization of protein partners, and the regulation of the TRPC channelosome. In particular, these proteins, via their multiple protein–protein interaction motifs, can interact with various ion channels involved in the transmembrane potential, and membrane excitability. Scaffolding proteins are key components for the functional organization of TRPC channelosomes that serves as a platform regulating slow Ca^{2+} entry, spatially and temporally controlled $[Ca^{2+}]_i$ signals and Ca^{2+} -dependent cellular functions.

Keywords TRPC channels • Caveolin • Homer • NHERF • Syntrophin

B. Constantin (✉)
Laboratory STIM, ERL-7368 CNRS-Université de Poitiers,
1, rue Georges Bonnet, Bat. B36, Pôle Biologie-Santé, 86000 Poitiers, France
e-mail: bruno.constantin@univ-poitiers.fr

J.A. Rosado (ed.), *Calcium Entry Pathways in Non-excitable Cells*, Advances in Experimental Medicine and Biology 898, DOI 10.1007/978-3-319-26974-0_16

16.1 Introduction

The ability of a cell to perceive and correctly respond to its microenvironment is depending on very complex signaling systems. This includes various membrane and transmembrane proteins, which are at the interface between the extracellular and intracellular medium. An efficient cellular response is asking for a precise localization and coupling of various signaling components, such as transmembrane receptors, ion channels and transporters, signal transduction proteins, and the secondary effectors. This coordinated and highly controlled signaling system is depending on the assembling at the plasma membrane of macromolecular complexes including adaptor proteins or scaffolding proteins [1–3]. This class of proteins is responsible for the correct clustering and anchoring of signaling proteins at specific sub-cellular areas and membrane domains. Scaffolding proteins contains several binding sites, which confer the ability of interacting with multiple signaling proteins, membrane lipids and cytoskeleton proteins. Adaptor proteins do not display any enzymatic activity but act as scaffold for the building of signaling complexes. The scaffolding proteins are thus involved in the assembling of these complexes, the correct subcellular localization of protein partners, and the regulation of signaling proteins. In particular, these proteins, via their multiple protein–protein interaction motifs, can interact with various ion channels involved in the transmembrane potential, membrane excitability.

Adaptor proteins, through their interaction with calcium channels can be implicated in regulation of intracellular calcium signaling. Plasma membrane calcium channels support the entry of Ca^{2+} along its concentration gradient across the plasma membrane into the cell. Entry of Ca^{2+} is driven by the presence of a large electrochemical gradient across the plasma membrane. Various calcium channels, such as voltage-gated Ca^{2+} channels (Ca_v family), can be involved in this Ca^{2+} influx [4]. The expression of this family of proteins is a characteristic of "excitable cells," and these channels do require depolarization of the plasma membrane (PM) for activation. However, some of these, such as members of the low voltage-activated (low threshold of activation) Ca_v3 subfamily, are expressed in "non-excitable cell". Despite Ca_v expression, Ca^{2+} entry in non-excitable cells mostly occurs through non-voltage-gated channels [5]. These include ligand-gated channels (P2X purinergic ionotropic receptor families, for instance); Secondary messenger-operated channels (SMOC) linked to GPCR (G-protein-Coupled Receptor) activation and the production of secondary messengers; Store-Operated Channels (SOC); and Stretch-Activated Channels (SAC). The opening of these calcium channels is contributing to calcium entry in response to cell stimulation, and cells use a transient increase in cytosolic free calcium concentration ($[Ca^{2+}]_i$) for intracellular signaling. A tight control of intracellular $[Ca^{2+}]$ is essential for the survival and normal function of cells, which maintain resting calcium activity at a low level (around 100 nM) in order to keep a large dynamic range for the calcium signal. Membrane ionic channels and transporters, cytosolic calcium buffers and calcium buffering organelles regulate calcium influx, storage and extrusion to maintain $[Ca^{2+}]_i$ below the activation thresholds and extraphysiological values. This precise control is essential for differential modulation, in an individual cell, of various signaling pathways and intracellular Ca^{2+}-regulated proteins involved in specific cellular processes. These include regulation of metabolism, proliferation, death, gene transcription, cell migration, exocytosis, and contraction [6].

16.2 Scaffolding Proteins and TRPC Channels

16.2.1 TRP Channels

The founding member of the TRP superfamily of cation channels is the *Drosophila* TRP channel. This superfamily consists of a large number of cation channels [7–9]. All TRPs contain six putative transmembrane domains, which are thought to assemble as homo- or hetero-tetramers to form cation selective channels. It is admitted that TRP channels display intracellular NH2 and COOH termini, as well as six transmembrane domains and an ion-selectivity P loop involved in the building of the pore. TRP channels form tetrameric assemblies around a central selectivity filter and gate, consistent with the structure of voltage-dependent K^+ channels [7–9] (Fig. 16.1).

On the basis of sequence homology, the TRP family can be divided in seven main subfamilies: the

TRPC ('Canonical') family; the TRPV ('Vanilloid') family; the TRPM ('Melastatin') family; the TRPP ('Polycystin') family; the TRPML ('Mucolipin') family; the TRPA ('Ankyrin') family; and the TRPN ('NOMPC') family. All functionally characterized TRP channels are cation channels and are permeable to Ca^{2+} with the exceptions of TRPM4 and TRPM5, which are only permeable to monovalent cations.

TRP channels participate in changes in $[Ca^{2+}]_i$ either by supporting Ca^{2+} entry through the plasma membrane, or via changes in membrane polarization, modulating the driving force for Ca^{2+} entry.

TRP channels are activated by a wide range of stimuli that include the binding of intra- and extracellular messengers, changes in temperature, chemical agents, mechanical stimuli, and osmotic stress [10, 11].

16.2.2 Scaffolding Proteins and TRP in Drosophila Photoreceptor

The first evidence for a TRP-containing macromolecular complex was demonstrated in *drosophila* photoreceptor, where the Inactive No After Potential D (INAD) provides the molecular scaffold [12, 13]. Interestingly, the *InaD* mutant

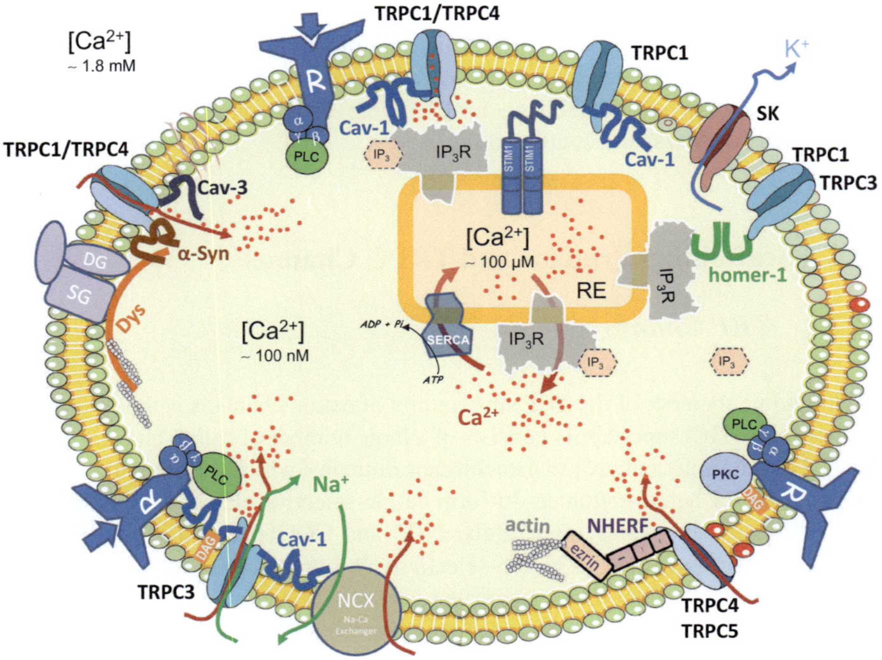

Fig. 16.1 Overview of the interplay between TRPC channels and scaffolding proteins. TRPC channels require scaffolding proteins for plasma membrane targeting, and incorporation in macromolecular signalplex controlling the opening of the channels and/or regulating their activity. It is proposed that Caveolin-1 regulates TRPC1 membrane targeting/retention as well as its association with STIM1. This is proposed to control SOCE activation through TRPC1 channels. Caveolin-1 is also involved in building a complex between TRPC1/TRPC4 heterotetramers and IP_3R, which is necessary for the proper activation of ROCE. The association of TRPC1 or TRPC3 channels with IP3R was also shown to be dependent on homer-1 and to maintain the channels in a close state. It was also suggested that caveolin-1 may associate TRPC3 channels and the Na^+/Ca^{2+} exchanger NCX in a same complex. TRPC4 and TRPC5 channels were also shown to associate with the PDZ domain of NHERF, the scaffolding protein regulating the membrane expression of the channels and the anchorage to ERM and actin cytoskeleton. Similarly, TRPC1/TRPC4 channels associates to the PDZ domain of α1-syntrophin in muscle, which anchors the channels to the dystrophin cytoskeleton and regulates the calcium entry

(*i*nactivation *no* *a*fterpotential) is a drosophila phototransduction mutant that exhibits altered responses to light, suggesting a crucial role of the scaffolding protein in the signal transduction. INAD contains five PDZ domains and has the capacity of binding multiple proteins for constituting a signaling complex. Named "signalplex" by Craig Montell, this macromolecular complex is consisting in INAD binding with TRP [14], Phospholipase C (PLC) [15, 16], rhodopsin [16], protein kinase C (PKC) [16], calmodulin [17], and the NINAC (neither inactivation nor-afterpotential C) myosin III [18], and an eye-specific homologue of TRP, TRPL (TRP-like). Drosophila photoreceptors respond to light with an increase in membrane conductance, leading to a light-induced current (LIC) and membrane depolarization. Cells from double mutants are almost totally unresponsive to light, indicating that TRP and TRPL account for most, of the LIC. Other proteins interact with this signalplex such as rhodopsin, calmodulin, FKBP59 and the nonconventional myosin III (NINAC), which is involved in efficiently terminating the photoresponse. The interaction with the scaffolding protein INAD is necessary for the proper localization of signaling proteins in the light-sensitive structures, the rhabdomeres. In *inaD* mutant flies, the localization of three INAD partners, TRP, PKC, and PLC, is severely altered [17, 19]. INAD plays an essential role in retention of these proteins to the rhabdomeres [20]. In addition, elimination of INAD or the INAD-binding sites in TRP, PKC, or PLC results in instability of these INAD-binding proteins [19, 20]. The interaction of the TRP/TRPL signalplex with the cortical actin cytoskeleton is mediated by the non-conventional myosin NINAC. This interaction also controls the gating of the channels [18] and regulates the translocation of TRPL [21]. The signalplex is thought to maintain the proper stoichiometry of signaling proteins in the rhabdomeres and to provide efficient transmission and termination of the signal.

The interaction between PDZ-containing protein, INAD and TRPC channels (see below) was also examined by heterologous expression of TRPC proteins in Sf9 insect cells [22]. The TRPC1, TRPC4, and TRPC5 channels co-immunoprecipitated with INAD, whereas TRPC3, TRPC6, and TRPC7 did not. This interaction could be explained by the homology of the PDZ domains (27.5 % similarity) of INAD and the Na+/H+-exchanger regulatory factor 1 (NHERF1), which is known to interact with TRL motif of TRPC4 and TRPC5 (see below). Thus INAD could form a binding domain comparable to NHERF1, allowing the binding of the TRL motif of TRPC4 and TRPC5.

16.2.3 TRPC Channels

Of the TRP families, the *Drosophila* TRP channel exhibits the greatest homology to the TRPCs, the first member of which, TRPC1, was cloned in 1995 [23]. A total of seven TRPC proteins have been described in mammals (TRPC1-TRPC7). In contrast to mice and rats, humans express only six TRPCs, because TRPC2 is a pseudogene [23]. TRPC channels are constituted of six predicted transmembrane domains (TM1-TM6), including a putative region between TM5 and TM6 involved in the constitution of the pore and assemble both as homo-tetramers or hetero-tetramers complexes. The cytosolic N-terminus is composed of three to four ankyrin repeats, a coiled-coil region involved in homo- or hetero-tetramerization of channels, and a putative caveolin binding domain. The cytoplasmic C-terminus region contains the TRP signature (EWKFAR), a highly conserved proline-riched motif (LPXPFXXXPSPK), the Calmodulin/IP_3 Receptor Binding (CIRB) region, and a coiled-coil region. All mammalian TRPCs build non-selective Ca^{2+}-permeable cation channels that demonstrate variable Ca^{2+}/Na^{+} permeability ratio, and support relatively non-selective cation currents [11, 24]. Several TRPC channels were earlier described as SOC and SMOC [25, 26]. Upon activation of transmembrane receptors by extracellular ligand, stimulation of PLC and production of inositol 1,4,5-trisphosphate (IP3) and Diacylglycerol (DAG) activates or potentiates TRPC-dependent current and calcium entry, referred as receptor-operated Ca^{2+} entry (ROCE). Following IP3-stimulated calcium release through the opening of IP3R, some TRPC can be activated upon depletion of intracellular Ca^{2+} stores from the endoplasmic or sarcoplasmic reticulum (ER and SR) and contribute to Store-Operated Calcium Entry (SOCE). The use of antibodies against TRPC1 [27, 30], TRPC3 [27, 28], TRPC5 [29, 30], TRPC6 [30] and TRPC7 [30] inhibited SOCE and ROCE after stimulation of cells with an extracellular agonist. The Knockdown by siRNA of several native TRPC channels also reduced SOCE [26] including knockdown of TRPC4 in adrenal cells and corneal epithelial cells [31, 32], or of TRPC1 and TRPC6 in granulocytes [33]. In particular, the knockdown of TRPC1 was consistently reported to significantly reduce store depletion-dependent Ca^{2+} entry in a lot of different cell types, such as salivary gland [34], B-lymphocytes [35], skeletal muscle fibers [36], endothelial cells [37] embryonic neural stem cells [38], keratinocytes [39] and intestinal epithelial cells [40]. Endogenous heteromeric TRPC channels have also been suggested to contribute to SOCE or ROCE by using siRNA knock-down or antibodies, such as endogenous TRPC1/TRPC3 in human parotid gland ductal cells [41] and rat H19-7 hippocampal cell lines [42]; TRPC1/TRPC5 in vascular smooth muscles [43, 44]; TRPC1/TRPC6 in intestinal smooth muscles [45]; TRPC1/TRPC4 in endothelial cells [46], and in skeletal muscle cells [47]; and TRPC1/TRPC3/TRPC7 in HEK293 cells [48].

Gene deletion in mice of TRPC1 [49, 50], TRPC3 [51] and TRPC4 [52, 53] also highlighted the contribution of TRPC proteins in non-voltage activated Ca^{2+} entries such as SOCE and ROCE. Cells isolated from knock-out mice displayed alteration of cell function that could be related to the reduction of SOCE and of Ca^{2+}-dependent regulation. For instance, studies using salivary gland and pancreatic acinar cells from TRPC1$^{-/-}$ mice reported a strong reduction in SOCE and an alteration of fluid secretion and Ca^{2+}-dependent K^+ channel activation [49]. Acinar cells isolated from the submandibular glands of TRPC1 knockout mice showed significant inhibition of the outward Cl^- currents, suggesting TRPC1 channels are essential for the activation of calcium-activated chloride channel (CaCC) [50]. Trpc1$^{-/-}$ mast cells derived from the bone marrow (BMMCs) responded to antigen with enhanced calcium signaling but with little defect in degranulation or associated signaling. In contrast, antigen-mediated production of TNF-α, and other cytokines, was enhanced in the Trpc1$^{-/-}$ BMMCs, as were calcium-dependent events required for these responses [54]. These data suggest that TRPC1 promotes recovery from the anaphylactic response by repressing antigen-mediated TNF-α release from MCs. Smooth muscle cells from TRPC4$^{-/-}$ mice display loss of agonist-induced Ca^{2+} entry, resulting in impaired vasorelaxation, suggesting that TRPC4 is contributing to the regulation of blood vessel tone [52]. The defect in Ca^{2+} influx in TRPC4(–/–) endothelial cells was associated with a lack of thrombin-induced actin-stress fiber formation and a reduced endothelial cell retraction response [53]. TRPC4-dependent Ca2+

entry in mouse lung vascular endothelial cells was shown to be crucial for increased microvascular permeability [53]. In the gastric smooth muscle cells of TRPC4 knockout mice, the observations suggest that TRPC4 is an essential component of the NSCC activated by muscarinic stimulation in the murine stomach [55]. A constitutive background cation entry was also shown to be dependent on TRPC1/TRPC4 heteromers and TRPC1/C4-gene inactivation protects against development of maladaptive cardiac remodelling without altering cardiac or extracardiac functions [56].

Although TRPC channels has been shown to contribute to ROCE and SOCE, the core component of calcium selective Store-operated channels have been demonstrated to be the ER Ca^{2+} sensor protein STIM1 (STromal Interaction Molecule 1), and the calcium channel Orai1 [57, 58]. STIM1 is a type 1, single transmembrane-span protein that transmits the information of the ER Ca^{2+} load to the Ca^{2+} channels at the plasma membrane [59, 60]. A minimal sequence of STIM1(344–442) termed STIM1 Orai-activating region (SOAR) [61] is sufficient to fully activate the Orai channels and has a coiled-coil domain that interacts with the C-terminal coiled-coil domain of the Orais [61–64]. Although it is now clear that Orai1 is required for all forms of SOCE [57, 58], it was shown that TRPC1, TRPC4 and TRPC5, but not TRPC3, TRPC6 and TRPC7, interact with STIM1 and that STIM1 is required for the activity of TRPC1 [65, 66]. The ability of TRPC heteromers to respond to ER-Ca^{2+} store depletion has been suggested to depend on the presence of STIM1-activated TRPC, such as TRPC1 [67]. Other studies have shown that TRPC1 activation by STIM1, following store depletion, also requires functional Orai1 [68–71]. Moreover it was suggested that local Ca^{2+} entry mediated by Orai1 controls the recruitment of TRPC1 into the plasma membrane where TRPC1 channel is activated by STIM1 [72]. In different cell types, native Orai and TRPC1 channels were shown to contribute to SOCE after activation of STIM1 by store depletion [71, 73]. For instance, in developing myotubes SOCE were demonstrated to be dependent on STIM1, Orai1 and TRPC1 [74]. It was also shown that Ca^{2+} entry mediated by TRPC1 and TRPC4 participate in SOCE, and specifically allow the formation of normal-sized myotubes [75], but also is involved in alteration of calcium entry in dystrophic cells deficient in the dystrophin/syntrophin scaffold [47].

In addition to interacting with the STIM1 protein and Orai channels, TRPC proteins are incorporated in macromolecular complexes including several key Ca^{2+} signaling proteins as well as proteins involved in vesicle trafficking, cytoskeletal interactions, and scaffolding. Evidence has been provided for association of TRPC with calmodulin (CaM), IP_3R, PMCA, $G_{q/11}$, RhoA, and a variety of scaffolding proteins [76, 77]. The interaction between TRPC channels with accessory proteins, determines their mode of regulation as well as their cellular localization and function. Scaffolding proteins are key components for the functional organization of TRPC channelosomes that serves as a platform regulating Ca^{2+} entry, and spatially and temporally controlled intracellular Ca^{2+} signals and Ca^{2+} -dependent cellular functions.

16.2.4 Caveolin and TRPC Channels

TRPC1 and TRPC3 have been found in glycosphingolipid- and cholesterol-enriched membrane microdomains referred to as caveolae. Caveolae play a key role in organizing receptor signaling complexes containing receptors and their accessory regulatory protein. Caveolae are also enriched with various proteins that participate in Ca^{2+} regulation and with Caveolin, a transmembrane scaffolding protein that nucleates signaling complexes [78]. Caveolin-1 has been described to interact with TRPC1, TRPC4 and TRPC3 channels (see below).

16.2.4.1 Caveolin-1 and TRPC1

TRPC1 and caveolin-1 were colocalized in the plasma membrane region of human submandibular gland and Madin-Darby canine kidney cells [79]. This work has shown Full-length caveolin-1 bound to both the N and C termini of TRPC1. Amino acids 271–349, which includes a caveolin-1 binding motif (amino acids 322–349), was identified as the caveolin-1 binding domain in the TRPC1 N terminus. They suggested a role for caveolin-1 interaction in membrane translocation of TRPC1. All TRPC members conserve a similar motif at the N-terminal part close to the first transmembrane domain TM1. Deletion of this region (amino acids 271–349 or 322–349) prevents the targeting of TRPC1 to the plasma membrane [79]. Importantly, TRPC1Δ271-349 exerted a dominant negative effect on endogenous SOCE and was associated with wild-type TRPC1. In addition, expression of a truncated caveolin-1 (Cav1Δ51-169), lacking its protein scaffolding and membrane anchoring domains (but not expression of full-length caveolin-1), disrupted plasma membrane targeting of TRPC1 [79]. Cav1Δ51-169 also suppressed thapsigargin- and carbachol-stimulated Ca^{2+} entry. These data demonstrate that plasma membrane localization of TRPC1 depends on an interaction between its N terminus and Cav1. Conversely, another study proposed a role of the TRPC1 C terminus domain in the regulation of calcium entry by the scaffolding protein [80]. Caveolin-1 associates with TRPC1 cation channels in human pulmonary artery endothelial cells (HPAECs). The role of the caveolin-1 scaffolding domain (CSD) in regulating thrombin-induced Ca^{2+} entry was demonstrated by using the cell-permeant antennapedia (AP)-conjugated CSD peptide, which competes for protein binding partners with caveolin-1. Immunoprecipitation studies demonstrated an interaction between endogenous TRPC1 and ectopically expressed hemagglutinin-tagged CSD. Moreover, Streptavidin-bead pull-down assay indicated strong binding of biotin-labeled AP-CSD peptide to TRPC1. A CSD binding consensus sequence was identified in the TRPC1 C terminus, and biotin-labeled AP-TRPC1 C terminus peptide interacts with caveolin-1. Finally, AP-TRPC1 peptide containing the CSD binding sequence markedly reduced the thrombin-induced Ca^{2+} entry. These results demonstrated a crucial role of caveolin-1 scaffolding domain interaction with TRPC1 in regulating thrombin-induced Ca^{2+} entry.

Caveolin-1 interaction is critical for TRPC1 targeting/retention and the scaffolding protein has been shown to play a dual role in regulation of Ca^{2+} entry. Overexpression of Cav-1ΔCSD construct, which decreases TRPC1– caveolin-1 interaction, actually showed an increase in SOCE [81]. In contrast, the TRPC1 COOH-terminal residues 781–789 truncated (TRPC1-CΔ781-789) mutant expression abolished SOCE in human dermal microvascular endothelial cell line (HMEC). TRPC1-CΔ781-789 binding to Cav-1 was markedly suppressed, but effectively interacted with IP_3R3. This work describes that CSD interacts with TRPC1 but also with IP_3R3. The negative regulation of SOCE could be explained on the basis of CSD interaction with IP_3R3.

Using mice deficient in caveolin-1, Murata et al. [82] demonstrated that the scaffolding protein governs in endothelial cells the localization and the interactions of TRPC1 and TRPC4. Cav-1 is associated with a dynamic protein complex consisting of TRPC4, TRPC1, and IP_3Rs, and the loss of Cav-1 impairs the localization of TRPC4 and agonist-simulated complex formation. Moreover, results with mice deficient in caveolin-1, showed that caveolin-1 is essential for ACh-mediated calcium entry and the co-precipitation of TRPCs with IP_3Rs. Both were rescued in caveolin-1 knock-out reconstituted with endothelium-specific caveolin-1. These data suggest that caveolin-1 is responsible for the assembly of an activated calcium-influx protein complex consisting of TRPCs and IP_3Rs in endothelial cells. More generally, caveolae are thought to organize a calcium signaling complex containing TRPC1 anchored to caveolin-1 where it associates with signaling proteins including IP_3R, calmodulin (CaM), plasma membrane calcium pump (PMCA), and $G_{\alpha q/11}$ [77, 83].

Caveolin-1 association with TRPC1 is thus important for membrane localization of the channel at microdomain but also for its activation by signaling pathways. It appears also to play a role in the interaction of TRPC1 with STIM1 and in participation of TRPC1 in SOCE. It was reported that caveolin-1 retains TRPC1 within the plasma membrane regions where STIM1 puncta are localized following store depletion [84]. This enables the interaction of TRPC1 with STIM1 that is required for the activation of TRPC1-SOCE. Silencing caveolin-1 in human submandibular gland (HSG) cells decreased plasma membrane retention of TRPC1 and TRPC1-STIM1 clustering. This also reduced TRPC1-dependent SOCE, without altering the formation of STIM1 puncta. Importantly, the work is describing a balance where activation of TRPC1-dependent SOCE was associated with an increase in TRPC1-STIM1 association and a decrease in TRPC1-caveolin-1 clustering. Overexpression of caveolin-1 decreased TRPC1-STIM1 clustering and SOCE, while silencing STIM1 or expression of STIM1 mutant prevented dissociation of TRPC1-caveolin-1 and activation of TRPC1-dependent SOCE [84]. It is thus proposed that the scaffolding protein associates with inactive TRPC1 at the plasma membrane and dissociates from the channel upon activation of TRPC1 by STIM1. Caveolin-1 seems to be essential for the selective organization of TRPC1-STIM1 channel assembly, but not for Orai1-STIM1 channels: Plasma membrane localization of TRPC1, its association with lipid raft microdomains and interaction with STIM1, are also disrupted in SG acinar cells from caveolin-1 −/− mice [85]. Forced expression of caveolin-1 in

SG acinar cells of caveolin-1 –/– mice restored agonist-stimulated TRPC1-STIM1 association, and SOCE. In contrast, Orai1–STIM1 interaction is not affected in SG acinar cells from caveolin-1 –/– mice, whereas agonist-induced and store-dependent calcium entries are significantly reduced [85].

In vascular smooth muscle cells (VSMC), Oxidized low-density lipoprotein (oxLDL) enhanced the cell surface expression of TRPC1, as shown by biotinylation of cell surface proteins, and induced TRPC1 translocation into caveolar compartment [86], which was dependent on cytoskeleton. Caveolin-1 silencing induced concomitant decrease of TRPC1 expression and reduced oxLDL-induced apoptosis of VSMC. The reported results show that caveolin-1-dependent translocation of TRPC1 channels are involved in apoptotic processes that play a major role in atherosclerosis. TRPC1 and caveolin-1 association has been also shown to participate in the formation of cell polarity [87]. Human bone osteosarcoma U2OS cells exhibit distinct morphological polarity during directional migration. The scaffolding protein caveolin-1, and the calcium channels TRPC1, and Orai1 were found to be concentrated at the rear end of polarized cells. The inhibition of store-operated Ca^{2+} entry, or Knockdown of TRPC1, but not knockdown of Orai1, reduced cell polarization. Furthermore, disruption of lipid rafts or overexpression of caveolin-1 contributed to the down-regulation of cell polarity.

16.2.4.2 Caveolin-1 and TRPC3

Similarly to TRPC1, TRPC3 is assembled in macromolecular complex containing $G_{\alpha q/11}$, PLCβ, IP_3R, SERCA, Ezrin and caveolin-1 [88]. TRPC3 is proposed to be included in a caveolar Ca^{2+} signaling complex linked to the cortical actin cytoskeleton. Interestingly, the TRPC3- IP_3R signaling complex has been shown to be internalized by conditions which stabilize the cortical actin cytoskeleton.

Coupling of TRPC3 channels with IP_3R-1 has been suggested to activate a cation current in arterial smooth muscle cells that induces vasoconstriction. Interestingly, Caveolin-1 knockdown using shRNA attenuates IP_3-induced vasoconstriction [89]. Moreover, a synthetic peptide corresponding to CSD sequence (amino acids 82–101) also attenuates vasoconstriction as well as IP_3-induced activation of cation current. Caveolin-1 has been shown to co-immunoprecipitate with TRPC3, and IP_3R-1 from cerebral artery lysate. TRPC3 channels co-localize with IP_3R-1 and Caveolin-1 in arterial smooth muscle cells and this localization is disrupted by the CSD peptide. These data indicate that Caveolin-1 co-localizes SR IP_3R1 and plasma membrane TRPC3 channels in close spatial proximity thereby enabling IP_3-induced coupling of these proteins, leading to calcium and sodium entry and in turn vasoconstriction of arterial smooth muscle cells. The same kind of macromolecular complex associating Caveolin-1 IP_3R-1 and TRPC3 channels has been found in mesenteric artery lysate [89].

In plasma membrane lysates from human prostate cancer cells, caveolin co-immunoprecipitated with p-PLCγ1 and TRPC3 [90]. Interestingly this macromolecular complex also contained the cell surface GRP78 that regulates apoptosis, and

TFII-I, a transcription factor, which regulates cellular proliferation and apoptosis. In addition to its transcriptional functions, cytosolic TFII-I regulates agonist-induced Ca^{2+} entry and cell surface localization of TFII-I is dependent on cell surface-associated GRP78, binding α_2 -macroglobulin. Down regulation of TFII-I expression activates agonist-induced Ca^{2+} entry and is associated with decreased cell surface expression of GRP78 and the induction of pro-apoptotic signaling.

Upon Angiotensine II-stimulation, a TRPC3-dependent NCX1-mediated Ca^{2+} influx was described into cardiomyocytes and was associated with an increased cell surface presentation of the two ion transport proteins [91]. Receptor/PLC-dependent membrane recruitment of cardiac TRPC3/NCX1 complexes was suggested to activate TRPC3 cation conductances, TRPC3-mediated local rises in $[Na^+]_i$ and reverse mode Na^+/Ca^{2+} exchangers-mediated Ca^{2+} entry. Interestingly, co-immunoprecipitation experiments revealed an interaction of NCX1 with TRPC3. GST pulldown-experiments revealed only a weak association of NCX with an N-terminal fragment of TRPC3, while a C-terminal domain efficiently retained NCX1. It is tempting to speculate that TRPC3 and NCX1 are co-transported as a preformed protein complex via vesicular trafficking and the interaction with caveolin. The TRPC3-NCX1 signaling complexes described here, are likely to play a significant role in cardiac physiology or/and pathophysiology.

16.2.4.3 Caveolin-3 and TRPC1

Gervasio and collaborators [92] reported the association of the scaffolding protein caveolin-3 with TRPC1, by showing co-localization at the sarcolemma and co-immunoprecipitation of endogenous proteins. Moreover, this work suggested by FRET assay and expression of TRPC1-CFP and caveolin-3-YFP, that the latter is necessary for localization of TRPC1 at the plasma membrane of myoblasts. This suggested the scaffolding protein caveolin-3 is necessary for localization of TRPC1 at muscle sarcolemma. These observations are in accordance with the idea that TRPC1 is part of a Dystrophin-Associated protein complex at the sarcolemma of skeletal muscle cells [93], which includes caveolin-3 but also syntrophin (see below).

16.2.5 Syntrophin and TRPC Channels

TRPC1 was found at the sarcolemma of adult mouse fibres [36, 47, 92, 94] and of developing mouse myotubes [47, 95]. This suggested that the cationic channels could be anchored to the subsarcolemmal cytoskeleton including spectrin and/or dystrophin. Indeed, TRPC1 isolated from adult mouse skeletal muscle and from developing myotubes was found to co-immunoprecipitate with dystrophin [95]. Dystrophin is providing in striated muscle the scaffold for multiple dystrophin-associated proteins (DAPs) and cytosolic adaptors, such as syntrophin. Through syntrophins, the DAP-complex is thought to anchor various signaling molecules such as enzymes and channels near their functional site at the membrane. Syntrophins are multigene family of intracellular membrane-associated adaptor proteins. The syntrophin family consists of five homologous isoforms, α1-syntrophin, β1-syntrophin, β2-syntrophin, γ1-syntrophin and γ2-syntrophin [96, 97]. The α1-syntrophin is the predominant syntrophin isoform in skeletal and cardiac muscles [98]. These scaffold proteins are characterized by the presence of a N-terminal PH-1 domain (Plekstrin Homology) split in two halves (PH_N and PH_C) by insertion of a PDZ domain (Post Synaptic Density protein-95, Drosophila discs large protein, and the Zona occludens protein 1), a second Plekstrin Homology domain (PH-2) and a C-terminal domain unique to syntrophins (SU). The SU domain and PH-2 domain interact with the carboxy terminus of dystrophin [96, 97]. This adaptor protein binds transmembrane channels through a PDZ domain such as voltage-gated sodium channels [99–101] or the potassium channel Kir4.1 [102]. We demonstrated that native TRPC1 and TRPC4 could be co-immunoprecipitated with endogenous and recombinant Alpha1-syntrophin [47, 95]. Moreover, GST-pull down assays showed that the PDZ domain of the Alpha1-syntrophin could be involved in this association, and native TRPC1 and TRPC4 proteins could be precipitated with the PDZ-GST domain as bait. This suggested that TRPC1 and TRPC4 could bind to Alpha1-syntrophin and be a constituent of a costameric macromolecular complex anchored to the dystrophin-based cytoskeleton. As shown by Si RNA knock-down of Alpha1-syntrophin, TRPC-mediated Ca^{2+} entry evoked by depletion of SR Ca^{2+} stores were greatly increased in wild type skeletal myotubes [47, 95]. The TRPC-dependent Ca^{2+} entry was similar to the one observed in dystrophin-deficient myotubes. This suggests that the scaffolding protein is a necessary component of the dystrophin-associated protein complex for moderating the calcium entry through TRPC1 and TRPC4 channels. Moreover, forced expression of recombinant Alpha1-syntrophin, but not of Alpha1-syntrophin lacking the N-terminal part containing the PDZ domain, restored normal TRPC-dependent calcium entry in dystrophic myotubes. TRPC-mediated Ca^{2+} entry were increased by a PLC/PKC-dependent pathway in dystrophic muscle cells and was restored at normal levels by Alpha1-syntrophin expression both in mouse and human skeletal muscle cells [74, 103]. In addition, silencing α1-syntrophin also increased cation influx in a PLC/PKC-dependent pathway. We also showed that α1-syntrophin and PLCβ are part of a same protein complex [74] reinforcing the idea of their interaction in a calcium signaling complex. This alteration of TRPC-mediated Ca^{2+} entry may be achieved by the loss of a α1-syntrophin-mediated inhibitory mechanism. As α1-syntrophin is present in the same complex than PLCβ, it could act directly on PLC pathway as a regulatory molecule or indirectly through the modulation of trimeric G proteins, which can also interact with α1-syntrophin [96, 97]. These observations were providing the idea that TRPC association with the scaffolding protein and DAPs in skeletal muscle could constitute a signaling complex at costameres, which regulates the TRPC-mediated Ca^{2+} entry, and spatially and temporally controlled intracellular Ca^{2+} signals [93]. The lack of Alpha1-syntrophin at the plasma membrane was shown to dysregulate TRPC-mediated Ca^{2+} entry, which may contribute to Ca^{2+} -dependent cellular necrosis.

16.2.6 Homer and TRPC Channels

TRPC1 also interacts with the highly conserved class II EVH1 [Ena (Enabled)/VASP (vasodilator-stimulated phosphoprotein) homology 1] domain [104] of the scaffolding protein Homer. A proline-rich motif (LPXPFXXXPSPK), downstream of the TRP domain (EWKFAR) is conserved in all members of the TRPC subfamily and TRPC channels exhibit two Homer-binding sites named type 1 (PPXXF or PXXF) [105] interacting with EVH1. Homer is able to dimerize through a Coiled-Coil motif, and the EVH domain of each monomer binds to a consensus PPXXF motif [106]. The coiled-coils support the assembling of Homer monomers into elongated tetramers [107]. The tetrameric Homer complex is thus thought to form a lattice with other scaffolds that bind Ca^{2+} signaling proteins in cellular microdomains [108–110]. The interaction of Homer-1 with TRPC channels and regulation of channel activity by Homer1 have been explored extensively.

TRPC1 is associated with Homer in brain protein extracts [105], and localizes in a complex containing IP_3 receptors and group 1 metabotropic glutamate receptors (mGluR1). Yuan and collaborators found a second binding site of Homer in the N-terminus, LPSSP, which does not include a phenylalanine, unlike the previous Homer binding sites (PPXXF). Homer expression was found to be crucial for mediating a TRPC1-IP_3R complex necessary for responses to G-protein-coupled receptor activation [105]. On the contrary expression of TRPC1 with mutation of the proline-rich motif disrupted Homer binding and resulted interestingly in a constitutive activity of channels with reduced agonist regulation. Homer was thus proposed to permit the assembly of an agonist responsive TRPC1-IP_3R complex. To further demonstrate the role of Homer in native cells, Ca^{2+} influx were measured in acini from Homer-1^{-}/$^{-}$ mice [105]. Deletion of Homer1 resulted in the increase of spontaneous Ca^{2+} in pancreatic acinar cells. This study also demonstrated that the filling state of the ER dynamically controls the association of TRPC1 with Homer and of TRPC1 with IP_3R. The activation of TRPC1-dependent SOCE by depletion of the intracellular Ca^{2+} stores was associated with the dissociation of TRPC1–Homer complex, while store refilling led to reassembly of the TRPC1–Homer–IP_3R complex.

TRPC3 was also found in complex with IP3R and Homer1b, which is suggested to mediate both gating by IP3R and TRPC3 trafficking [111]. The authors proposed a model in which the assembly of the TRPC3-Homer-1b/c-IP_3Rs complexes by Homer1b/c mediates both the translocation of TRPC3-containing vesicles to the plasma membrane and gating of TRPC3 by IP_3Rs. In resting cells, complexes containing TRPC3, Homer-1b/c and IP_3R were found in part at the PM and in part in intracellular vesicles. The binding of IP_3 to the IP_3Rs dissociated the interaction between IP_3Rs and Homer1 but not between Homer-1 and TRPC3 to form

IP_3R-TRPC3-Homer-1b/c. The store-dependent translocation of the TRPC3 to the PM was dependent on IP_3R [111].

These studies suggest that Homer1 is essential to maintain the TRPC channels in the closed state. At the basal state, TRPC channels are present in a complex with IP_3Rs dependent on the long Homer-1b/c, which could keep the channel inactive. The N-terminal domain of the IP_3R is in the conformation that exposes its Homer-1-binding ligand. This allows the binding of IP_3R to Homer-1, which also binds to the Homer-1-binding site of the TRPC channels. It is proposed that activation of TRPC channels following cell stimulation is dependent on the dissociations of the complexes within Homer.

Interestingly, the homer-binding motif of the TRPC channels' C-terminal domains is separated by only four residues from the conserved DD/E residues interacting with the two terminal lysines K684 and K685 of the STIM1. A function of this K-domain in the ER protein STIM1 is gating of TRPC channels [65, 67, 112], although the K-domain is not required for activation of Orai1 by STIM1 [113]. These findings lead to propose a model for store-dependent activation of TRPC channels by STIM1 [114]. When the ER is filled with Ca^{2+}, which binds to the EF hand of STIM1, sequestration of STIM1 in the ER and formation of the TRPC-Homer-IP_3R channel complex keep the TRPC channels in the closed state. Upon cell stimulation, binding of IP_3 to the IP_3Rs dissociates the IP_3R-Homer-1-TRPC channel complexes, triggers the calcium release through IP_3Rs, which in turn dissociates Ca^{2+} from STIM1 EF hand and results in the clustering of STIM1 with the TRPC channels. Finally, the interaction of STIM1(K684,K685) with the DD/E residues of the TRPC channels is proposed to stabilize the TRPC channels open state.

The involvement of Homer 1 in maintaining TRPC channels in the closed state, could explain the consequences observed in skeletal muscle cells from mice lacking Homer-1 [95]. Homer-1 knockout myotubes displayed increased basal current density and spontaneous cation influx that was blocked by silencing of TRPC1. This spontaneous TRPC1-dependent cation entry was blocked by reexpression of Homer-1b, but not Homer-1a, suggesting Homer-1b is specifically necessary for preventing a constitutive activation of TRPC1 channels. Interestingly, In HEK293, dissociating the Homer-1b/c-IP_3R complex resulted in TRPC3 translocation to the plasma membrane, where it was spontaneously active [106].

In human platelets, a study suggested a role of Homer proteins in association of TRPC1 with the type II IP3 receptor (IP3RII), as well as in the binding of with Orai1 [115]. However, Treatment of human platelets with thapsigargin or thrombin enhanced the association of Homer-1 with STIM1 and Orai1 in a Ca^{2+}-dependent manner, while it resulted in a Ca^{2+}-independent association of Homer-1 with TRPC1 and IP3RII. The interference with Homer PPXXF motif by using a synthetic PPKKFR peptide reduced both STIM1-Orai1 and TRPC1- IP3RII associations and inhibited thrombin-evoked Ca^{2+} entry and thapsigargin-induced store-operated Ca^{2+} entry.

16.2.7 NHERF and TRPC4, C5 Channels

Several studies have identified different components of the signaling machinery involved in TRPC4 and TRPC5 mediated Ca^{2+} entry. These include the Na+/H+-exchanger regulatory factor 1 (NHERF) a scaffolding protein with PDZ domain. NHERF was first isolated as a cofactor required for protein kinase A-mediated inhibition of type 3 Na1/H1 exchanger localized in the renal brush-border membrane [116]. It was later determined to bind members of the ezrin/radixin/moesin (ERM) proteins [117–119]. The ERM proteins are known to link cell surface receptors to actin cytoskeleton and in many cases such link is mediated through NHERF. NHERF1, or ezrin/radixin/moesin(ERM)-Binding Protein 50 (EBP50), is a relatively small PDZ domain-containing protein characterized by two PDZ domains and a carboxyl terminal ezrin-binding domain. NHERF2 is quite similar to NHERF1 as it shares 44 % sequence homology with NHERF1 and contains two PDZ domains and a carboxyl terminal ezrin-binding domain [120, 121]. The two PDZ domains of NHERF bind to target proteins while its C terminus binds to the N terminus of ERMs [117, 122]. It has been shown that the first PDZ domain of NHERF interacts with a number of G-protein-coupled receptors and ion transporting proteins [121].

Murine TRPC4 and TRPC5 were shown to bind to the first PDZ domain of NHERF [123]. The same PDZ domain binds to the C termini of PLC-β1 and PLC-β2. The Association of PLC-β1, TRPC4, and NHERF was demonstrated in an HEK293 cell line stably expressing TRPC4 and in adult mouse brain by co-immunoprecipitation experiments. This was an indication that NHERF is bringing together the signaling molecules involved in the TRPC-mediated calcium entry pathway in mammalian cells.

TRPC4 and TRPC5, have a TRL COOH-terminal amino acid sequence, which represents a class 1 PDZ-binding domain. The binding of two partners to the same PDZ domain suggests that NHERF can form a homodimer via PDZ2 and the PDZ1 domains, bringing TRPC4 and TRPC5 in vicinity of the PLCβ1 and PLCβ2 [123, 124]. The role of the TRL motif was examined on the subcellular distribution of TRPC4 in HEK293 cell line [125], as well as the consequences of the interaction between EBP50 and the membrane-cytoskeletal adaptors of ERM family for the cell surface expression of TRPC4. Confocal immunofluorescence microscopy analysis showed that the mutant lacking the TRL motif accumulated into cell outgrowths with a punctate distribution pattern whereas the wild-type channel was distributed on the cell surface [125]. In Cells co-expressing TRPC4 and an EBP50 mutant lacking the ERM-binding site, TRPC4 was not present in the plasma membrane but co-localized with the truncated EBP50 in a perinuclear compartment. Deletion of the PDZ-interacting domain also decreased the expression of TRPC4 in the plasma membrane, as assessed by cell surface biotinylation experiments. These data demonstrated that the PDZ-interacting domain of TRPC4 controls its localization and surface expression in transfected HEK293 cells.

Immunohistochemical studies demonstrated expression of TRPC4 and NHERF-2 proteins in both the endothelial cells and pericytes [126] of microdissected descending vasa recta (DVR). A complex containing both TRPC4 and NHERF-2 was also found in microdissected in renal medullary descending vasa recta. TRPC4 coimmunoprecipitated with NHERF-2 from renal medullary lysates, and NHERF-2 coimmunoprecipitated with TRPC4. The function of TRPC4 and the significance of its binding to NHERF-2 are unknown at present. Studies in TRPC4$^{-/-}$ mice showed a marked attenuation of agonist-induced increases in intracellular Ca^{2+} in endothelial cells and impaired agonist-induced vasorelaxation in preconstricted aortic rings from TRPC4-null mice [52]. These findings suggest that TRPC4 and NHERF-2 may play a role in the control of Ca^{2+} entry and in the regulation of medullary blood flow.

The functional effects of EBP50 on TRPC5 activity have been investigated with rat TRPC5 (rTRPC5), expressed in HEK293 cell [127]. Both rTRPC5 and the VTTRL deletion mutant were localized to the plasma membrane, and deletion of the VTTRL motif had no detectable effect on the biophysical properties of the channel when studied with patch-clamp technique. Co-expression of EBP50 with rTRPC5 led to a significant delay in the time-to-peak of the histamine-evoked, transient large inward current. Conversely, EBP50 did not modify the activation kinetics of the VTTRL-deletion mutant. It suggests that the VTTRL motif is not necessary for activation of TRPC5, but may mediate the modulatory effect of EBP50 on TRPC5 activation kinetics.

16.3 Concluding Remarks

Plasma membrane ion channels, and in particular TRPC channels need a specific membrane environment and association with scaffolding, signaling, and cytoskeleton proteins in order to play their important functional role. The molecular composition of TRPC channels is an important factor in determining channel activation mechanisms. Homomeric and heteromeric TRPC channels composed of TRPC3, TRPC6, and TRPC7 subunits are often termed classical ROCs, as they are activated by DAG through a PKC-independent mechanism that does not involve intracellular Ca^{2+} stores. TRPC channels containing TRPC1 subunits are gated by DAG via a PKC-dependent action, and also by STIM1 following stimulations that deplete intracellular Ca^{2+} stores. A recent work [44] showed that agents that deplete intracellular Ca^{2+} stores activated native heteromeric TRPC1/C5 channels in freshly isolated, contractile mesenteric artery vascular smooth muscle cells (VSMCs). It was shown that TRPC1 subunits are important for conferring activation by PKC, PIP_2 and PIP_3 on TRPC1/C5 channels. This work suggested that PKC is necessary for activation of native TRPC1/C5 channels by agents that deplete Ca^{2+} stores in VSMCs. In addition, PIP_2 and PIP_3 also activate native TRPC1/C5 channels and PKC activity is also important for this activation. The excitatory effects of PKC, PIP_2, and PIP_3 on TRPC1/C5 channel activities were absent in TRPC1$^{-/-}$ VSMCs. Moreover, TRPC1$^{-/-}$ VSMCs express homomeric TRPC5 channels, which are inhibited by PKC, PIP_2, and PIP_3. This illustrates that many cells and tissues coexpress several members of the seven TRPCs, which can result in various channels with significantly different current properties, stimulation mechanisms and functional interactions. Moreover, Previous reports showed that the movement of TRPC1 to the plasma membrane is dependent on co-expression with TRPC4 or TRPC5 [128]. It suggests that TRPC1 is not able to reach the plasma membrane alone and at least without association with TRPC4 or TRPC5, or without association of a scaffolding protein such as caveolin-3 as observed by FRET assay and coexpression of TRPC1-CFP and caveolin-3-YFP in myoblasts [93]. Caveolin-1 interaction is also critical for TRPC1 [79] and for TRPC4 targeting/retention [82].

TRPC homotetramers and heterotetramers associate with cytoskeleton constituents, such as non-muscle actin and microtubules [76] and with various scaffolding proteins in order to be addressed to specific membrane domains, and to be incorporated in macromolecular complexes playing a role in cell signaling. In these "signalplex", the interaction of scaffolding and adaptor proteins plays a relevant role in the association of TRPC with other signaling molecules, which regulate the channel properties and the cation entry. Moreover, scaffolding proteins may be part of the mechanism triggering TRPC channel opening. A lack of the adaptor protein due to genetic defect may lead to TRPC channels over-activation or constitutive activation as observed with α-syntrophin and homer-1. Since the cation and calcium entry through TRPC channels is involved in the modulation of membrane excitability and of submembrane microdomains, this lack of regulation can have major consequences in calcium homeostasis and cell survival. Given the variety of cellular responses linked to the activation of these channels, many physiological or physiopathological functions have been shown to depend on TRPCs [129, 130]. A recent example is a pivotal role of TRPC channels, in particular TRPC1, in the striated muscle tissue and in the development of calcium mishandling observed in dystrophin-deficient skeletal and cardiac muscle cells [93]. Interestingly, enhanced TRPC1-mediated Ca^{2+} influx were dependent of PKC, supported by TRPC1/TRPC4 heteromers and regulated by the scaffolding protein α1-syntrophin. When incorporated in various signalplex, TRPC tetramers, can play different roles depending of the specific associations with different scaffolding proteins, signaling molecules, and with other channels, such as Orai or calcium-dependent K^+ channels. The presence of multiple TRPC signalplexs in the same cells can serves to mediate selective cellular functions. For example, it was shown in human salivary gland cell line HSG cells that TRPC1-mediated Ca^{2+} influx stimulates Ca^{2+}-activated K^+ channels and NFκB, while Orai1-mediated calcium entry activates NFAT but not TRPC-mediated Ca^{2+} influx [72]. The differential association of TRPC channels was also shown to lead to Antagonistic regulation of actin dynamics and cell motility [131]. It was shown that TRPC5 was in a molecular complex with Rac1, whereas TRPC6 was in a molecular complex with RhoA. TRPC5-mediated Ca^{2+} influx induced Rac1 activation, thereby promoting cell migration, whereas TRPC6-mediated Ca^{2+} influx increased RhoA activity, thereby inhibiting cell migration. Depending on the signaling repertory of a given cell and on the complement TRPC channels co-expressed, the various stimuli activating TRPC may lead to complex changes in membrane-potential, various mode of calcium entry, slow sustained increase in intracellular calcium concentration and numerous Ca^{2+}-induced responses. Scaffolding proteins also play a pivotal role by incorporating TRPC channels in signalplex and specific membrane domain, leading to calcium microdomains and localized calcium signals with different cell function. For example, Ca^{2+} influx through TRPC1 channels has been shown to promote in *Xenopus* spinal neuron growth cones filopodial Ca^{2+} transients, and to activate calpain, which cleaves talin at the tips of filopodia [132]. Mislocalization of the TRPC channels can thus alter their function and consequently downstream localized Ca^{2+} signaling events that are dependent on these channels.

In conclusion, TRPC channels require scaffolding proteins for plasma membrane targeting, and incorporation in macromolecular signalplex controlling the opening of the channels and/or regulating their activity. An important role of Caveolins has been reported by several studies for TRPC membrane targeting/retention. This family of scaffolding protein was also involved in the association of TRPC channels with the calcium release channel IP3R, which seems to control the opening of the channels and the resulting calcium signals evoked by cell stimulation. The coupling of an agonist responsive TRPC-IP3R complex was also shown to be dependent on Homer scaffolding proteins. These studies suggest that Homer1 is essential to maintain the TRPC channels in the closed state. Clearly, Scaffolding proteins also help to build signaling complex comparable to the TRP/INAD/ and bring TRPC channels in the vicinity of signaling proteins that may regulate their activity. Scaffolding proteins such as NHERF were suggested to associate TRPC4 and TRPC5 with PLC, and we described also a putative TRPC/syntrophin/PLC complex in muscle.

References

1. Harris BZ, Lim WA (2001) Mechanism and role of PDZ domains in signaling complex assembly. J Cell Sci 114:3219–3231
2. Garbett D, Bretscher A (2014) The surprising dynamics of scaffolding proteins. Mol Biol Cell 25(16):2315–2319
3. Walther C, Ferguson SS (2015) Minireview: role of intracellular scaffolding proteins in the regulation of endocrine g protein-coupled receptor signaling. Mol Endocrinol 29(6):814–830
4. Catterall WA (2011) Voltage-gated calcium channels. Cold Spring Harb Perspect Biol 3(8):a003947
5. Roderick HL, Cook SJ (2008) Ca2+ signalling checkpoints in cancer: remodelling Ca2+ for cancer cell proliferation and survival. Nat Rev Cancer 8(5):361–375
6. Berridge MJ, Bootman MD, Roderick HL (2003) Calcium signalling: dynamics, homeostasis and remodelling. Nat Rev Mol Cell Biol 4(7):517–529
7. Benemei S, Patacchini R, Trevisani M, Geppetti P (2015) TRP channels. Curr Opin Pharmacol 22:18–23
8. Hellmich UA, Gaudet R (2014) Structural biology of TRP channels. Handb Exp Pharmacol 223:963–990
9. Montell C (2011) The history of TRP channels, a commentary and reflection. Pflugers Arch 461(5):499–506
10. Zheng J (2013) Molecular mechanism of TRP channels. Compr Physiol 3(1):221–242
11. Venkatachalam K, Montell C (2007) TRP channels. Annu Rev Biochem 76:387–417

12. Montell C (2012) Drosophila visual transduction. Trends Neurosci 35(6):356–363
13. Hardie RC, Juusola M (2015) Phototransduction in Drosophila. Curr Opin Neurobiol 34C:37–45
14. Shieh BH, Zhu MY (1996) Regulation of the TRP Ca2+ channel by INAD in Drosophila photoreceptors. Neuron 16(5):991–998
15. Huber A, Sander P, Bähner M, Paulsen R (1998) The TRP Ca2+ channel assembled in a signaling complex by the PDZ domain protein INAD is phosphorylated through the interaction with protein kinase C (ePKC). FEBS Lett 425(2):317–322
16. Xu XZ, Choudhury A, Li X, Montell C (1998) Coordination of an array of signaling proteins through homo- and heteromeric interactions between PDZ domains and target proteins. J Cell Biol 142(2):545–555
17. Chevesich J, Kreuz AJ, Montell C (1997) Requirement for the PDZ domain protein, INAD, for localization of the TRP store-operated channel to a signaling complex. Neuron 18(1):95–105
18. Wes PD, Xu XZ, Li HS, Chien F, Doberstein SK, Montell C (1999) Termination of phototransduction requires binding of the NINAC myosin III and the PDZ protein INAD. Nat Neurosci 2(5):447–453
19. Tsunoda S, Sierralta J, Sun Y, Bodner R, Suzuki E, Becker A, Socolich M, Zuker CS (1997) A multivalent PDZ-domain protein assembles signalling complexes in a G-protein-coupled cascade. Nature 388(6639):243–249
20. Li HS, Montell C (2000) TRP and the PDZ protein, INAD, form the core complex required for retention of the signalplex in Drosophila photoreceptor cells. J Cell Biol 150(6):1411–1422
21. Bähner M, Frechter S, Da Silva N, Minke B, Paulsen R, Huber A (2002) Light-regulated subcellular translocation of Drosophila TRPL channels induces long-term adaptation and modifies the light-induced current. Neuron 34(1):83–93
22. Goel M, Sinkins WG, Schilling WP (2002) Selective association of TRPC channel subunits in rat brain synaptosomes. J Biol Chem 277(50):48303–48310
23. Wes PD, Chevesich J, Jeromin A, Rosenberg C, Stetten G, Montell C (1995) TRPC1, a human homolog of a Drosophila store-operated channel. Proc Natl Acad Sci U S A 92(21):9652–9656
24. Owsianik G, Talavera K, Voets T, Nilius B (2006) Permeation and selectivity of TRP channels. Annu Rev Physiol 68:685–717
25. Parekh AB, Putney JW Jr (2005) Store-operated calcium channels. Physiol Rev 85(2):757–810
26. Cheng KT, Ong HL, Liu X, Ambudkar IS (2013) Contribution and regulation of TRPC channels in store-operated Ca^{2+} entry. Curr Top Membr 71:149–179
27. Xu SZ, Beech DJ (2001) TrpC1 is a membrane-spanning subunit of store-operated Ca^{2+} channels in native vascular smooth muscle cells. Circ Res 88(1):84–87
28. Chen J, Crossland RF, Noorani MM, Marrelli SP (2009) Inhibition of TRPC1/TRPC3 by PKG contributes to NO-mediated vasorelaxation. Am J Physiol Heart Circ Physiol 297(1):H417–H424
29. Xu SZ, Boulay G, Flemming R, Beech DJ (2006) E3-targeted anti-TRPC5 antibody inhibits store-operated calcium entry in freshly isolated pial arterioles. Am J Physiol Heart Circ Physiol 291(6):H2653–H2659
30. Saleh SN, Albert AP, Peppiatt-Wildman CM, Large WA (2008) Diverse properties of store-operated TRPC channels activated by protein kinase C in vascular myocytes. J Physiol 586(10):2463–2476
31. Philipp S, Trost C, Warnat J, Rautmann J, Himmerkus N, Schroth G, Kretz O, Nastainczyk W, Cavalie A, Hoth M, Flockerzi V (2000) TRP4 (CCE1) protein is part of native calcium release-activated Ca^{2+}-like channels in adrenal cells. J Biol Chem 275(31):23965–23972
32. Yang H, Mergler S, Sun X, Wang Z, Lu L, Bonanno JA, Pleyer U, Reinach PS (2005) TRPC4 knockdown suppresses epidermal growth factor-induced store-operated channel activation and growth in human corneal epithelial cells. J Biol Chem 280(37):32230–32237

33. Bréchard S, Melchior C, Plançon S, Schenten V, Tschirhart EJ (2008) Store-operated Ca2+ channels formed by TRPC1, TRPC6 and Orai1 and non-store-operated channels formed by TRPC3 are involved in the regulation of NADPH oxidase in HL-60 granulocytes. Cell Calcium 44(5):492–506
34. Liu X, Wang W, Singh BB, Lockwich T, Jadlowiec J, O'Connell B, Wellner R, Zhu MX, Ambudkar IS (2000) Trp1, a candidate protein for the store-operated Ca^{2+} influx mechanism in salivary gland cells. J Biol Chem 275(5):3403–3411
35. Mori Y, Wakamori M, Miyakawa T, Hermosura M, Hara Y, Nishida M, Hirose K, Mizushima A, Kurosaki M, Mori E, Gotoh K, Okada T, Fleig A, Penner R, Iino M, Kurosaki T (2002) Transient receptor potential 1 regulates capacitative Ca2+ entry and Ca2+ release from endoplasmic reticulum in B lymphocytes. J Exp Med 195(6):673–681
36. Vandebrouck C, Martin D, Colson-Van Schoor M, Debaix H, Gailly P (2002) Involvement of TRPC in the abnormal calcium influx observed in dystrophic (mdx) mouse skeletal muscle fibers. J Cell Biol 158(6):1089–1096
37. Mehta D, Ahmmed GU, Paria BC, Holinstat M, Voyno-Yasenetskaya T, Tiruppathi C, Minshall RD, Malik AB (2003) RhoA interaction with inositol 1,4,5-trisphosphate receptor and transient receptor potential channel-1 regulates Ca2+ entry. Role in signaling increased endothelial permeability. J Biol Chem 278(35):33492–33500
38. Fiorio Pla A, Maric D, Brazer SC, Giacobini P, Liu X, Chang YH, Ambudkar IS, Barker JL (2005) Canonical transient receptor potential 1 plays a role in basic fibroblast growth factor (bFGF)/FGF receptor-1-induced Ca2+ entry and embryonic rat neural stem cell proliferation. J Neurosci 25(10):2687–2701
39. Cai S, Fatherazi S, Presland RB, Belton CM, Roberts FA, Goodwin PC, Schubert MM, Izutsu KT (2006) Evidence that TRPC1 contributes to calcium-induced differentiation of human keratinocytes. Pflugers Arch 452(1):43–52
40. Rao JN, Platoshyn O, Golovina VA, Liu L, Zou T, Marasa BS, Turner DJ, Yuan JX, Wang JY (2006) TRPC1 functions as a store-operated Ca2+ channel in intestinal epithelial cells and regulates early mucosal restitution after wounding. Am J Physiol Gastrointest Liver Physiol 290(4):G782–G792
41. Liu X, Bandyopadhyay BC, Singh BB, Groschner K, Ambudkar IS (2005) Molecular analysis of a store-operated and 2-acetyl-sn-glycerol-sensitive non-selective cation channel. Heteromeric assembly of TRPC1-TRPC3. J Biol Chem 280(22):21600–21606
42. Wu X, Zagranichnaya TK, Gurda GT, Eves EM, Villereal ML (2004) A TRPC1/TRPC3-mediated increase in store-operated calcium entry is required for differentiation of H19-7 hippocampal neuronal cells. J Biol Chem 279(42):43392–43402
43. Shi J, Ju M, Saleh SN, Albert AP, Large WA (2010) TRPC6 channels stimulated by angiotensin II are inhibited by TRPC1/C5 channel activity through a Ca2+- and PKC-dependent mechanism in native vascular myocytes. J Physiol 588(Pt 19):3671–3682
44. Shi J, Ju M, Abramowitz J, Large WA, Birnbaumer L, Albert AP (2012) TRPC1 proteins confer PKC and phosphoinositol activation on native heteromeric TRPC1/C5 channels in vascular smooth muscle: comparative study of wild-type and $TRPC1^{-/-}$ mice. FASEB J 26(1):409–419
45. Tsvilovskyy VV, Zholos AV, Aberle T, Philipp SE, Dietrich A, Zhu MX, Birnbaumer L, Freichel M, Flockerzi V (2009) Deletion of TRPC4 and TRPC6 in mice impairs smooth muscle contraction and intestinal motility in vivo. Gastroenterology 137(4):1415–1424
46. Sundivakkam PC, Freichel M, Singh V, Yuan JP, Vogel SM, Flockerzi V, Malik AB, Tiruppathi C (2012) The Ca2+ sensor stromal interaction molecule 1 (STIM1) is necessary and sufficient for the store-operated Ca(2+) entry function of transient receptor potential canonical (TRPC) 1 and 4 channels in endothelial cells. Mol Pharmacol 81(4):510–526
47. Sabourin J, Lamiche C, Vandebrouck A, Magaud C, Rivet J, Cognard C, Bourmeyster N, Constantin B (2009) Regulation of TRPC1 and TRPC4 cation channels requires an alpha1-syntrophin-dependent complex in skeletal mouse myotubes. J Biol Chem 284(52):36248–36261

48. Zagranichnaya TK, Wu X, Villereal ML (2005) Endogenous TRPC1, TRPC3, and TRPC7 proteins combine to form native store-operated channels in HEK-293 cells. J Biol Chem 280(33):29559–29569
49. Liu X, Cheng KT, Bandyopadhyay BC, Pani B, Dietrich A, Paria BC, Swaim WD, Beech D, Yildrim E, Singh BB, Birnbaumer L, Ambudkar IS (2007) Attenuation of store-operated Ca2+ current impairs salivary gland fluid secretion in TRPC1(-/-) mice. Proc Natl Acad Sci U S A 104(44):17542–17547
50. Sun Y, Birnbaumer L, Singh BB (2015) TRPC1 regulates calcium-activated chloride channels in salivary gland cells. J Cell Physiol 230(11):2848–2856
51. Kim MS, Hong JH, Li Q, Shin DM, Abramowitz J, Birnbaumer L, Muallem S (2009) Deletion of TRPC3 in mice reduces store-operated Ca2+ influx and the severity of acute pancreatitis. Gastroenterology 137(4):1509–1517
52. Freichel M, Suh SH, Pfeifer A, Schweig U, Trost C, Weissgerber P, Biel M, Philipp S, Freise D, Droogmans G, Hofmann F, Flockerzi V, Nilius B (2001) Lack of an endothelial store-operated Ca2+ current impairs agonist-dependent vasorelaxation in TRP4-/- mice. Nat Cell Biol 3(2):121–127
53. Tiruppathi C, Freichel M, Vogel SM, Paria BC, Mehta D, Flockerzi V, Malik AB (2002) Impairment of store-operated Ca2+ entry in TRPC4(-/-) mice interferes with increase in lung microvascular permeability. Circ Res 91(1):70–76
54. Medic N, Desai A, Olivera A, Abramowitz J, Birnbaumer L, Beaven MA, Gilfillan AM, Metcalfe DD (2013) Knockout of the Trpc1 gene reveals that TRPC1 can promote recovery from anaphylaxis by negatively regulating mast cell TNF-α production. Cell Calcium 53(5–6):315–326
55. Lee KP, Jun JY, Chang IY, Suh SH, So I, Kim KW (2005) TRPC4 is an essential component of the nonselective cation channel activated by muscarinic stimulation in mouse visceral smooth muscle cells. Mol Cells 20(3):435–441
56. Camacho Londoño JE, Tian Q, Hammer K, Schröder L, Camacho Londoño J, Reil JC, He T, Oberhofer M, Mannebach S, Mathar I, Philipp SE, Tabellion W, Schweda F, Dietrich A, Kaestner L, Laufs U, Birnbaumer L, Flockerzi V, Freichel M, Lipp P (2015) A background Ca^{2+} entry pathway mediated by TRPC1/TRPC4 is critical for development of pathological cardiac remodelling. Eur Heart J. doi:10.1093/eurheartj/ehv250
57. Hogan PG, Rao A (2015) Store-operated calcium entry: mechanisms and modulation. Biochem Biophys Res Commun 460(1):40–49
58. Hogan PG (2015) The STIM1-ORAI1 microdomain. Cell Calcium 58(6):357–367, pii: S0143-4160(15)00117-7
59. Liou J, Kim ML, Heo WD, Jones JT, Myers JW, Ferrell JE Jr, Meyer T (2005) STIM is a Ca2+ sensor essential for Ca2+-store-depletion-triggered Ca2+ influx. Curr Biol 15(13):1235–1241
60. Roos J, DiGregorio PJ, Yeromin AV, Ohlsen K, Lioudyno M, Zhang S, Safrina O, Kozak JA, Wagner SL, Cahalan MD, Veliçelebi G, Stauderman KA (2005) STIM1, an essential and conserved component of store-operated Ca2+ channel function. J Cell Biol 169(3):435–445
61. Yuan JP, Zeng W, Dorwart MR, Choi YJ, Worley PF, Muallem S (2009) SOAR and the polybasic STIM1 domains gate and regulate Orai channels. Nat Cell Biol 11(3):337–343
62. Muik M, Fahrner M, Derler I, Schindl R, Bergsmann J, Frischauf I, Groschner K, Romanin C (2009) A cytosolic homomerization and a modulatory domain within STIM1 C terminus determine coupling to ORAI1 channels. J Biol Chem 284(13):8421–8426
63. Schindl R, Muik M, Fahrner M, Derler I, Fritsch R, Bergsmann J, Romanin C (2009) Recent progress on STIM1 domains controlling Orai activation. Cell Calcium 46(4):227–232
64. Korzeniowski MK, Manjarrés IM, Varnai P, Balla T (2010) Activation of STIM1-Orai1 involves an intramolecular switching mechanism. Sci Signal 3(148):ra82
65. Huang GN, Zeng W, Kim JY, Yuan JP, Han L, Muallem S, Worley PF (2006) STIM1 carboxyl-terminus activates native SOC, I(crac) and TRPC1 channels. Nat Cell Biol 8(9):1003–1010

66. Yuan JP, Kim MS, Zeng W, Shin DM, Huang G, Worley PF, Muallem S (2009) TRPC channels as STIM1-regulated SOCs. Channels (Austin) 3(4):221–225
67. Yuan JP, Zeng W, Huang GN, Worley PF, Muallem S (2007) STIM1 heteromultimerizes TRPC channels to determine their function as store-operated channels. Nat Cell Biol 9(6):636–645
68. Ong HL, Cheng KT, Liu X, Bandyopadhyay BC, Paria BC, Soboloff J, Pani B, Gwack Y, Srikanth S, Singh BB, Gill DL, Ambudkar IS (2007) Dynamic assembly of TRPC1-STIM1-Orai1 ternary complex is involved in store-operated calcium influx. Evidence for similarities in store-operated and calcium release-activated calcium channel components. J Biol Chem 282(12):9105–9116
69. Liao Y, Erxleben C, Abramowitz J, Flockerzi V, Zhu MX, Armstrong DL, Birnbaumer L (2008) Functional interactions among Orai1, TRPCs, and STIM1 suggest a STIM-regulated heteromeric Orai/TRPC model for SOCE/Icrac channels. Proc Natl Acad Sci U S A 105(8):2895–2900
70. Liao Y, Plummer NW, George MD, Abramowitz J, Zhu MX, Birnbaumer L (2009) A role for Orai in TRPC-mediated Ca2+ entry suggests that a TRPC:Orai complex may mediate store and receptor operated Ca2+ entry. Proc Natl Acad Sci U S A 106(9):3202–3206
71. Kim MS, Zeng W, Yuan JP, Shin DM, Worley PF, Muallem S (2009) Native store-operated Ca2+ influx requires the channel function of Orai1 and TRPC1. J Biol Chem 284(15):9733–9741
72. Cheng KT, Liu X, Ong HL, Swaim W, Ambudkar IS (2011) Local Ca2+ entry via Orai1 regulates plasma membrane recruitment of TRPC1 and controls cytosolic Ca2+ signals required for specific cell functions. PLoS Biol 9(3):e1001025
73. Ong HL, Ambudkar IS (2015) Molecular determinants of TRPC1 regulation within ER-PM junctions. Cell Calcium. doi:10.1016/j.ceca.2015.03.008
74. Sabourin J, Harisseh R, Harnois T, Magaud C, Bourmeyster N, Déliot N, Constantin B (2012) Dystrophin/α1-syntrophin scaffold regulated PLC/PKC-dependent store-operated calcium entry in myotubes. Cell Calcium 2012(6):445–456
75. Antigny F, Koenig S, Bernheim L, Frieden M (2013) During post-natal human myogenesis, normal myotube size requires TRPC1- and TRPC4-mediated Ca2+ entry. J Cell Sci 126(Pt 11):2525–2533
76. Smani T, Dionisio N, López JJ, Berna-Erro A, Rosado JA (2014) Cytoskeletal and scaffolding proteins as structural and functional determinants of TRP channels. Biochim Biophys Acta 1838(2):658–664
77. Ambudkar IS, Ong HL (2007) Organization and function of TRPC channelosomes. Pflugers Arch 455(2):187–200
78. Fridolfsson HN, Roth DM, Insel PA, Patel HH (2014) Regulation of intracellular signaling and function by caveolin. FASEB J 28(9):3823–3831
79. Brazer SC, Singh BB, Liu X, Swaim W, Ambudkar IS (2003) Caveolin-1 contributes to assembly of store-operated Ca2+ influx channels by regulating plasma membrane localization of TRPC1. J Biol Chem 278(29):27208–27215
80. Kwiatek AM, Minshall RD, Cool DR, Skidgel RA, Malik AB, Tiruppathi C (2006) Caveolin-1 regulates store-operated Ca2+ influx by binding of its scaffolding domain to transient receptor potential channel-1 in endothelial cells. Mol Pharmacol 70(4):1174–1183
81. Sundivakkam PC, Kwiatek AM, Sharma TT, Minshall RD, Malik AB, Tiruppathi C (2009) Caveolin-1 scaffold domain interacts with TRPC1 and IP3R3 to regulate Ca2+ store release-induced Ca2+ entry in endothelial cells. Am J Physiol Cell Physiol 296(3):C403–C413
82. Murata T, Lin MI, Stan RV, Bauer PM, Yu J, Sessa WC (2007) Genetic evidence supporting caveolae microdomain regulation of calcium entry in endothelial cells. J Biol Chem 282(22):16631–16643
83. Lockwich TP, Liu X, Singh BB, Jadlowiec J, Weiland S, Ambudkar IS (2000) Assembly of Trp1 in a signaling complex associated with caveolin-scaffolding lipid raft domains. J Biol Chem 275(16):11934–11942

84. Pani B, Ong HL, Brazer SC, Liu X, Rauser K, Singh BB, Ambudkar IS (2009) Activation of TRPC1 by STIM1 in ER-PM microdomains involves release of the channel from its scaffold caveolin-1. Proc Natl Acad Sci U S A 106(47):20087–20092
85. Pani B, Liu X, Bollimuntha S, Cheng KT, Niesman IR, Zheng C, Achen VR, Patel HH, Ambudkar IS, Singh BB (2012) Impairment of TRPC1-STIM1 channel assembly and AQP5 translocation compromise agonist-stimulated fluid secretion in mice lacking caveolin1. J Cell Sci 126(Pt 2):667–675
86. Ingueneau C, Huynh UD, Marcheix B, Athias A, Gambert P, Nègre-Salvayre A, Salvayre R, Vindis C (2009) TRPC1 is regulated by caveolin-1 and is involved in oxidized LDL-induced apoptosis of vascular smooth muscle cells. J Cell Mol Med 13(8B):1620–1631
87. Huang YW, Chang SJ, I-Chen Harn H, Huang HT, Lin HH, Shen MR, Tang MJ, Chiu WT (2015) Mechanosensitive store-operated calcium entry regulates the formation of cell polarity. J Cell Physiol 230(9):2086–2097
88. Lockwich T, Singh BB, Liu X, Ambudkar IS (2001) Stabilization of cortical actin induces internalization of transient receptor potential 3 (Trp3)-associated caveolar Ca2+ signaling complex and loss of Ca2+ influx without disruption of Trp3-inositol trisphosphate receptor association. J Biol Chem 276(45):42401–42408
89. Adebiyi A, Narayanan D, Jaggar JH (2011) Caveolin-1 assembles type 1 inositol 1,4,5-trisphosphate receptors and canonical transient receptor potential 3 channels into a functional signaling complex in arterial smooth muscle cells. J Biol Chem 286(6):4341–4348
90. Misra UK, Mowery YM, Gawdi G, Pizzo SV (2011) Loss of cell surface TFII-I promotes apoptosis in prostate cancer cells stimulated with activated α_2-macroglobulin. J Cell Biochem 112(6):1685–1695
91. Eder P, Probst D, Rosker C, Poteser M, Wolinski H, Kohlwein SD, Romanin C, Groschner K (2007) Phospholipase C-dependent control of cardiac calcium homeostasis involves a TRPC3-NCX1 signaling complex. Cardiovasc Res 73(1):111–119
92. Gervásio OL, Whitehead NP, Yeung EW, Phillips WD, Allen DG (2008) TRPC1 binds to caveolin-3 and is regulated by Src kinase – role in Duchenne muscular dystrophy. J Cell Sci 121(Pt 13):2246–2255
93. Sabourin J, Cognard C, Constantin B (2009) Regulation by scaffolding proteins of canonical transient receptor potential channels in striated muscle. J Muscle Res Cell Motil 30(7–8):289–297
94. Stiber JA, Zhang ZS, Burch J, Eu JP, Zhang S, Truskey GA, Seth M, Yamaguchi N, Meissner G, Shah R, Worley PF, Williams RS, Rosenberg PB (2008) Mice lacking Homer 1 exhibit a skeletal myopathy characterized by abnormal transient receptor potential channel activity. Mol Cell Biol 28(8):2637–2647
95. Vandebrouck A, Sabourin J, Rivet J, Balghi H, Sebille S, Kitzis A, Raymond G, Cognard C, Bourmeyster N, Constantin B (2007) Regulation of capacitative calcium entries by alpha1-syntrophin: association of TRPC1 with dystrophin complex and the PDZ domain of alpha1-syntrophin. FASEB J 21(2):608–617
96. Bhat HF, Adams ME, Khanday FA (2013) Syntrophin proteins as Santa Claus: role(s) in cell signal transduction. Cell Mol Life Sci 70(14):2533–2554
97. Constantin B (2014) Dystrophin complex functions as a scaffold for signalling proteins. Biochim Biophys Acta 1838(2):635–642
98. Adams ME, Butler MH, Dwyer TM, Peters MF, Murnane AA, Froehner SC (1993) Two forms of mouse syntrophin, a 58 kd dystrophin-associated protein, differ in primary structure and tissue distribution. Neuron 11:531–540
99. Gee SH, Madhavan R, Levinson SR, Caldwell JH, Sealock R, Froehner SC (1998) Interaction of muscle and brain sodium channels with multiple members of the syntrophin family of dystrophin-associated proteins. J Neurosci 18:128–137
100. Gavillet B, Rougier JS, Domenighetti AA, Behar R, Boixel C, Ruchat P, Lehr HA, Pedrazzini T, Abriel H (2006) Cardiac sodium channel Nav1.5 is regulated by a multiprotein complex composed of syntrophins and dystrophin. Circ Res 99(4):407–414

101. Shy D, Gillet L, Ogrodnik J, Albesa M, Verkerk AO, Wolswinkel R, Rougier JS, Barc J, Essers MC, Syam N, Marsman RF, van Mil AM, Rotman S, Redon R, Bezzina CR, Remme CA, Abriel H (2014) PDZ domain-binding motif regulates cardiomyocyte compartment-specific NaV1.5 channel expression and function. Circulation 130(2):147–160
102. Connors NC, Adams ME, Froehner SC, Kofuji P (2004) The potassium channel Kir4.1 associates with the dystrophin-glycoprotein complex via alpha-syntrophin in glia. J Biol Chem 279:28387–28392
103. Harisseh R, Chatelier A, Magaud C, Déliot N, Constantin B (2013) Involvement of TRPV2 and SOCE in calcium influx disorder in DMD primary human myotubes with a specific contribution of α1-syntrophin and PLC/PKC in SOCE regulation. Am J Physiol Cell Physiol 304(9):C881–C894
104. Xiao B, Tu JC, Petralia RS, Yuan JP, Doan A, Breder CD, Ruggiero A, Lanahan AA, Wenthold RJ, Worley PF (1998) Homer regulates the association of group 1 metabotropic glutamate receptors with multivalent complexes of homer-related, synaptic proteins. Neuron 21(4):707–716
105. Yuan JP, Kiselyov K, Shin DM, Chen J, Shcheynikov N, Kang SH, Dehoff MH, Schwarz MK, Seeburg PH, Muallem S, Worley PF (2003) Homer binds TRPC family channels and is required for gating of TRPC1 by IP3 receptors. Cell 114(6):777–789
106. Beneken J, Tu JC, Xiao B, Nuriya M, Yuan JP, Worley PF, Leahy DJ (2000) Structure of the Homer EVH1 domain-peptide complex reveals a new twist in polyproline recognition. Neuron 26(1):143–154
107. Hayashi MK, Ames HM, Hayashi Y (2006) Tetrameric hub structure of postsynaptic scaffolding protein homer. J Neurosci 26(33):8492–8501
108. Worley PF, Zeng W, Huang G, Kim JY, Shin DM, Kim MS, Yuan JP, Kiselyov K, Muallem S (2007) Homer proteins in Ca2+ signaling by excitable and non-excitable cells. Cell Calcium 42(4–5):363–371
109. Salanova M, Volpe P, Blottner D (2013) Homer protein family regulation in skeletal muscle and neuromuscular adaptation. IUBMB Life 65(9):769–776
110. Jardin I, López JJ, Berna-Erro A, Salido GM, Rosado JA (2013) Homer proteins in Ca^{2+} entry. IUBMB Life 65(6):497–504
111. Kim JY, Zeng W, Kiselyov K, Yuan JP, Dehoff MH, Mikoshiba K, Worley PF, Muallem S (2006) Homer 1 mediates store- and inositol 1,4,5-trisphosphate receptor-dependent translocation and retrieval of TRPC3 to the plasma membrane. J Biol Chem 281(43):32540–32549
112. Lee KP, Yuan JP, So I, Worley PF, Muallem S (2010) STIM1-dependent and STIM1-independent function of transient receptor potential canonical (TRPC) channels tunes their store-operated mode. J Biol Chem 285(49):38666–38673
113. Zeng W, Yuan JP, Kim MS, Choi YJ, Huang GN, Worley PF, Muallem S (2008) STIM1 gates TRPC channels, but not Orai1, by electrostatic interaction. Mol Cell 32(3):439–448
114. Yuan JP, Lee KP, Hong JH, Muallem S (2012) The closing and opening of TRPC channels by Homer1 and STIM1. Acta Physiol (Oxf) 204(2):238–247
115. Jardin I, Albarrán L, Bermejo N, Salido GM, Rosado JA (2012) Homers regulate calcium entry and aggregation in human platelets: a role for Homers in the association between STIM1 and Orai1. Biochem J 445(1):29–38
116. Yun CH, Oh S, Zizak M, Steplock D, Tsao S, Tse CM, Weinman EJ, Donowitz M (1997) cAMP-mediated inhibition of the epithelial brush border Na+/H+ exchanger, NHE3, requires an associated regulatory protein. Proc Natl Acad Sci U S A 94(7):3010–3015
117. Murthy A, Gonzalez-Agosti C, Cordero E, Pinney D, Candia C, Solomon F, Gusella J, Ramesh V (1998) NHE-RF, a regulatory cofactor for Na^+-H^+ exchange, is a common interactor for merlin and ERM (MERM) proteins. J Biol Chem 273(3):1273–1276
118. Yun CH, Lamprecht G, Forster DV, Sidor A (1998) NHE3 kinase A regulatory protein E3KARP binds the epithelial brush border Na+/H+ exchanger NHE3 and the cytoskeletal protein ezrin. J Biol Chem 273(40):25856–25863
119. Terawaki S, Maesaki R, Hakoshima T (2006) Structural basis for NHERF recognition by ERM. Structure 14(4):777–789

120. Ardura JA, Friedman PA (2011) Regulation of G protein-coupled receptor function by Na+/H+ exchange regulatory factors. Pharmacol Rev 63(4):882–900
121. Dunn HA, Ferguson SS (2015) PDZ protein regulation of GPCR trafficking and signaling pathways. Mol Pharmacol 88:624–639, pii: mol.115.098509
122. Reczek D, Berryman M, Bretscher A (1997) Identification of EBP50: a PDZ-containing phosphoprotein that associates with members of the ezrin-radixin-moesin family. J Cell Biol 139(1):169–179
123. Tang Y, Tang J, Chen Z, Trost C, Flockerzi V, Li M, Ramesh V, Zhu MX (2000) Association of mammalian trp4 and phospholipase C isozymes with a PDZ domain-containing protein. NHERF J Biol Chem 275(48):37559–37564
124. Suh PG, Hwang JI, Ryu SH, Donowitz M, Kim JH (2001) The roles of PDZ-containing proteins in PLC-beta-mediated signaling. Biochem Biophys Res Commun 288(1):1–7
125. Mery L, Strauss B, Dufour JF, Krause KH, Hoth M (2002) The PDZ-interacting domain of TRPC4 controls its localization and surface expression in HEK293 cells. J Cell Sci 115(Pt 17):3497–3508
126. Lee-Kwon W, Wade JB, Zhang Z, Pallone TL, Weinman EJ (2005) Expression of TRPC4 channel protein that interacts with NHERF. Am J Physiol Cell Physiol 288(4):C942–C949
127. Obukhov AG, Nowycky MC (2004) TRPC5 activation kinetics are modulated by the scaffolding protein ezrin/radixin/moesin-binding phosphoprotein-50 (EBP50). J Cell Physiol 201(2):227–235
128. Hoffman T, Schaefer M, Schultz G, Gudermann T (2002) Subunit composition of mammalian ransient receptor potential channels in living cells. Proc Natl Acad Sci U S A 99:7461–7466
129. Abramowitz J, Birnbaumer L (2009) Physiology and pathophysiology of canonical transient receptor potential channels. FASEB J 23(2):297–328
130. Ong HL, de Souza LB, Cheng KT, Ambudkar IS (2014) Physiological functions and regulation of TRPC channels. Handb Exp Pharmacol 223:1005–1034
131. Tian D, Jacobo SM, Billing D, Rozkalne A, Gage SD, Anagnostou T, Pavenstädt H, Hsu HH, Schlondorff J, Ramos A, Greka A (2010) Antagonistic regulation of actin dynamics and cell motility by TRPC5 and TRPC6 channels. Sci Signal 3(145):ra77
132. Kerstein PC, Jacques-Fricke BT, Rengifo J, Mogen BJ, Williams JC, Gottlieb PA, Sachs F, Gomez TM (2013) Mechanosensitive TRPC1 channels promote calpain proteolysis of talin to regulate spinal axon outgrowth. J Neurosci 33(1):273–285

Chapter 17
Modulation of Calcium Entry by Mitochondria

Rosalba Fonteriz, Jessica Matesanz-Isabel, Jessica Arias-del-Val, Pilar Alvarez-Illera, Mayte Montero, and Javier Alvarez

Abstract The role of mitochondria in intracellular Ca^{2+} signaling relies mainly in its capacity to take up Ca^{2+} from the cytosol and thus modulate the cytosolic [Ca^{2+}]. Because of the low Ca^{2+}-affinity of the mitochondrial Ca^{2+}-uptake system, this organelle appears specially adapted to take up Ca^{2+} from local high-Ca^{2+} microdomains and not from the bulk cytosol. Mitochondria would then act as local Ca^{2+} buffers in cellular regions where high-Ca^{2+} microdomains form, that is, mainly close to the cytosolic mouth of Ca^{2+} channels, both in the plasma membrane and in the endoplasmic reticulum (ER). One of the first targets proposed already in the 1990s to be regulated in this way by mitochondria were the store-operated Ca^{2+} channels (SOCE). Mitochondria, by taking up Ca^{2+} from the region around the cytosolic mouth of the SOCE channels, would prevent its slow Ca^{2+}-dependent inactivation, thus keeping them active for longer. Since then, evidence for this mechanism has accumulated mainly in immunitary cells, where mitochondria actually move towards the immune synapse during T cell activation. However, in many other cell types the available data indicate that the close apposition between plasma and ER membranes occurring during SOCE activation precludes mitochondria from getting close to the Ca^{2+}-entry sites. Alternative pathways for mitochondrial modulation of SOCE, both Ca^{2+}-dependent and Ca^{2+}-independent, have also been proposed, but further work will be required to elucidate the actual mechanisms at work. Hopefully, the recent knowledge of the molecular nature of the mitochondrial Ca^{2+} uniporter will allow soon more precise studies on this matter.

Keywords SOCE • Store-operated Ca^{2+} entry • Mitochondria • Endoplasmic reticulum • Orai • STIM • MCU • Ca^{2+} uniporter • Ca^{2+} microdomain • CRAC channels

R. Fonteriz • J. Matesanz-Isabel • J. Arias-del-Val • P. Alvarez-Illera • M. Montero
J. Alvarez (✉)
Departamento de Bioquímica y Biología Molecular y Fisiología, Instituto de Biología y Genética Molecular (IBGM), Facultad de Medicina, Universidad de Valladolid y Consejo Superior de Investigaciones Científicas, Ramon y Cajal 7, 47005 Valladolid, Spain
e-mail: rfonteri@ibgm.uva.es; realidadjessi@gmail.com; jessica.arias@ibgm.uva.es; pilar_alvill@hotmail.com; mmontero@ibgm.uva.es; jalvarez@ibgm.uva.es

J.A. Rosado (ed.), *Calcium Entry Pathways in Non-excitable Cells*, Advances in Experimental Medicine and Biology 898, DOI 10.1007/978-3-319-26974-0_17

Abbreviations

$[Ca^{2+}]_M$	Mitochondrial matrix free $[Ca^{2+}]$
CICR	Ca^{2+}-induced Ca^{2+} release
CRAC channels	Ca^{2+}-release activated Ca^{2+} channels
ER	Endoplasmic reticulum
GDAP1	Ganglioside-induced differentiation-associated protein 1
$InsP_3$	Inositol 1,4,5-trisphosphate
IS	Immunological synapse or Immune synapse
JPH1	Junctophilin 1
MCU	Mitochondrial calcium uniporter
NFAT	Nuclear factor of activated T-cells
SERCA	Sarcoplasmic/endoplasmic-reticulum Ca^{2+}-ATPase
SOCE	Store-operated Ca^{2+} entry
STIM	Stromal interaction molecule
TRP	Transient receptor potential

17.1 Mitochondrial Ca^{2+} Transport

Ca^{2+} is an intracellular second messenger able to fulfill a variety of different functions in all cell types with no exception, both excitable and non-excitable cells. Let's just mention muscle contraction, neurotransmitter secretion, fertilization, proliferation, development, learning, memory or cell death [1]. Most of these functions are triggered by changes in cytosolic Ca^{2+}, but several intracellular organelles also play a very important role in cellular Ca^{2+} homeostasis, either because Ca^{2+} plays specific roles inside the organelle or because Ca^{2+}-fluxes in and out the organelle are able to regulate cytosolic $[Ca^{2+}]$. The first organelle that was known to have an important role in Ca^{2+} homeostasis was the endoplasmic reticulum (ER), that was shown to release Ca^{2+} in the presence of inositol 1,4,5-trisphosphate (IP_3) [2]. This organelle behaves as a Ca^{2+} store and in steady-state conditions has a very high $[Ca^{2+}]$ in its lumen, around 1–2 mM, as we have recently reported using a new double-mutated aequorin with very low Ca^{2+} affinity [3]. Different kinds of cell stimulation can induce a rapid release of the stored Ca^{2+} to the cytosol through several Ca^{2+} channels present in its membrane (IP_3 and ryanodine receptors, mainly).

Another organelle with an important role in cell Ca^{2+} homeostasis is mitochondria, the organelle responsible for aerobic energy production in the cell. The ability of mitochondria to take up large amounts of Ca^{2+} was known since the 1960s [4]. In fact, until the end of the 1970s, mitochondria were thought to be the main cellular Ca^{2+} store. Several findings changed then the view on this point. First, it was shown that Ca^{2+} uptake by mitochondria was carried out by a system (the uniporter) which had a very low Ca^{2+}-affinity (K_M above 10 μM) [5, 6]. In the same dates, the first measurements of cytosolic $[Ca^{2+}]$ in intact cells obtained with loaded dyes showed that cytosolic [Ca^{2+}] under resting conditions was very low, around 100 nM, and that even after cell stimulation, it did not usually rise above 1–2 μM, that is, still tenfold below the K_M of the uniporter. Moreover, the finding that the endoplasmic reticulum had Ca^{2+} channels activated by IP_3 [2] suggested that the main responsible for intracellular Ca^{2+} release was the endoplasmic reticulum. As a consequence, mitochondria were thought not to have a significant role in Ca^{2+} homeostasis, at least under physiological conditions (although perhaps under pathological conditions). Then, in the beginning of the 1990s, the general view on this point changed again when experiments performed in the laboratory of T. Pozzan and R. Rizzuto using recombinant aequorin targeted to the mitochondria showed that mitochondrial [Ca^{2+}] ([Ca^{2+}]$_M$) was able to undergo rapid changes during cell activation [7]. That was the origin of a series of works by several research groups, demonstrating that mitochondria are very active players indeed in the control of the global cell Ca^{2+} homeostasis.

The rationale developed in those years to explain the apparent contradiction among the low Ca^{2+}-affinity of the uniporter and the fast mitochondrial Ca^{2+} uptake observed in intact cells was based in the concept of high-Ca^{2+} microdomains. So, it is right that mitochondria would be unable to take up much Ca^{2+} when exposed to the mean [Ca^{2+}] of the cytosol, around 1 μM. However, during cell stimulation, some mitochondria may be placed very close to the Ca^{2+} channels responsible for either Ca^{2+} release from the ER or Ca^{2+} entry from the extracellular medium. These particular mitochondria would be exposed to a much higher [Ca^{2+}] (a hot-spot or high-Ca^{2+} microdomain), enough to activate the uniporter and trigger fast mitochondrial Ca^{2+} uptake. A corollary of this hypothesis is that Ca^{2+} uptake by the mitochondrial network is heterogeneous, because only mitochondria close enough to those channels will take up significant amounts of Ca^{2+}, while the rest will remain with low [Ca^{2+}]. And, of course, the precise intracellular localization of mitochondria with respect to the Ca^{2+} channels becomes critical to determine the rate and amount of Ca^{2+} uptake. This point will be particularly important with regards to the interaction between mitochondria and capacitative Ca^{2+} entry.

Evidence for the multiple roles played by mitochondria in cell Ca^{2+} homeostasis has been continuously growing in the last years. First, it was already known that Ca^{2+} activates several Krebs cycle enzymes: pyruvate dehydrogenase, α-ketoglutarate dehydrogenase and isocitrate dehydrogenase [8], and it has been conclusively shown that an increase in [Ca^{2+}]$_M$ immediately leads to an increase in ATP production [9, 10]. This mechanism constitutes a basic homeostatic response that associates cell activation to energy production, in order to cover the energy requirements of the response to the stimulus. In addition, it has been recently shown that mitochondrial metabolism is also modulated by extramitochondrial [Ca^{2+}], thanks to the activity of several Ca^{2+}-dependent mitochondrial carriers, including both the aspartate/glutamate carriers from the malate/aspartate shuttle and the ATP-Mg/Pi carriers [11–13]. These mitochondrial carriers are in the inner mitochondrial membrane but sense cytosolic Ca^{2+} through EF-hand motifs placed in the intermembrane space. Another Ca^{2+}-dependent mitochondrial dehydrogenase is the FAD-glycerol phosphate dehydrogenase, a key component of the glycerol-3-phosphate shuttle, which is also placed in the inner mitochondrial membrane and senses cytosolic Ca^{2+} in the intermembrane space [14]. This dehydrogenase has been recently shown to be the target of metformin, one of the most effective therapeutics for treating type 2 diabetes [15].

On the other hand, the fast mitochondrial Ca^{2+}-accumulation that occurs during cell stimulation constitutes in fact a mechanism of transient Ca^{2+} buffering (see [16] for a review). Because of the low-Ca^{2+}-affinity of their Ca^{2+} uptake mechanism, mitochondria are very well adapted to take up Ca^{2+} from local high-Ca^{2+} microdomains, such as those formed after the activation of plasma membrane or ER Ca^{2+} channels. These high-Ca^{2+} cytosolic microdomains are responsible for a variety of Ca^{2+}-dependent phenomena, such as neurotransmitter secretion or cardiac cell contraction. Mitochondria can therefore modulate these important cellular functions by acting as local Ca^{2+} sinks. For example, we have shown that mitochondria can modulate local cytosolic Ca^{2+} concentrations and Ca^{2+}-dependent catecholamine secretion in chromaffin cells [17, 18]. In these cells, a large fraction of the Ca^{2+} entering the cell during stimulation appears to be transiently buffered by mitochondria [18–20]. In fact, during chromaffin cell stimulation, mitochondria can undergo reversible and repetitive near millimolar $[Ca^{2+}]_M$ transients [17], as a consequence of the generation of local high Ca^{2+} microdomains of up to 20–40 μM, compared with a mean cytosolic $[Ca^{2+}]$ of 2 μM. The ability of mitochondria to buffer local high-Ca^{2+} microdomains has also been used to explain the modulation by mitochondria of the activity of several Ca^{2+} channels showing Ca^{2+}-dependent inactivation. The role of mitochondria in these cases would be to control the size of the local $[Ca^{2+}]$ microdomain close to the cytosolic mouth of the channel. Capacitative Ca^{2+} entry channels were the first ones shown to be modulated in this way by mitochondria [21]. It was proposed that mitochondrial Ca^{2+} uptake in the vicinity of the channels in the plasma membrane reduced the local $[Ca^{2+}]$ and avoided their slow Ca^{2+}-dependent inactivation, thus keeping the channels active for longer. The same idea was then used to explain the modulation by mitochondria of plasma membrane voltage-dependent Ca^{2+} channels [22, 23] and ER IP_3 receptor channels [24, 25] (see [26] for a review).

Largely in parallel with the history of the capacitative calcium entry, most of the mitochondrial Ca^{2+} transport systems have resisted for many years all the attempts to characterize them molecularly, in spite of the efforts of many research groups. However, the last 5 years have been very fruitful at this respect and we have now a much clearer picture of the molecules involved. First the mitochondrial Na^+/Ca^{2+} exchanger [27] and then the mitochondrial Ca^{2+} uniporter [28, 29] and all its different regulatory components have been described in these years (for comprehensive reviews, see [30, 31]). Among the unexpected findings obtained, let us mention two main points. First, the large complexity of the mitochondrial Ca^{2+} uniporter, which has not only a pore subunit (the MCU protein, that most probably forms tetramers in the membrane), but also a paralog with dominant negative function (MCUb), and at least five additional regulatory subunits: MICU1, MICU2, MICU3, EMRE and MICUR1, all of them with different and in some cases opposite functions, which are still not completely clarified. The second surprise came about 1 year ago, when the group of T. Finkel developed a transgenic mouse with a knockout of the pore subunit (MCU), and it was not only viable, but in fact showed a very mild phenotype [32]. It is true that these mice are indeed only viable if they are maintained on a mixed genetic background, and that even in this case there appears to be still a significant amount of embryonic lethality [33]. But the survivors, which had no detectable protein expression of MCU and showed a complete abrogation of any fast mitochondrial calcium uptake in isolated mitochondria, were nevertheless nearly normal.

17.2 Capacitative Ca^{2+} Entry

Capacitative Ca^{2+} entry was first described nearly 30 years ago [34, 35] as a mechanism able to activate Ca^{2+} entry from the extracellular medium after the depletion of Ca^{2+} of the intracellular stores, namely, the endoplasmic reticulum. The pathway was later renamed store-operated Ca^{2+} entry (SOCE, [36]) and also Ca^{2+}-release activated Ca^{2+} channels (CRAC channels or I_{CRAC}, [37]) when referred to the Ca^{2+} current measured in the plasma membrane of immunitary cells after depletion of the intracellular Ca^{2+} stores. A large series of works has showed that SOCE is not only the predominant Ca^{2+}-entry pathway in non-excitable cells, but it is also present and very important in excitable cells, where it coexists with many different types of voltage-dependent and receptor-operated Ca^{2+} channels in the plasma membrane.

In spite of the long history of the SOCE channels, the molecular mechanism responsible for Ca^{2+} entry and the mechanism of signal transduction from the emptying of the Ca^{2+} stores to the activation of Ca^{2+} entry has been only elucidated in the last 10 years. The discovery in 2005 and 2006 of the main constituents of the CRAC channels, the family of STIM proteins as sensors of the ER [Ca^{2+}] and the family of Orai proteins as Ca^{2+} channels in the plasma membrane, finally provided the molecular tools required to explore the importance of the SOCE mechanism (for reviews, see [38–43]).

Co-expression of STIM1 and Orai1 generates Ca^{2+} currents identical to those of native I_{CRAC}, store-operated, with very high Ca^{2+} selectivity, inward rectifying and with extremely small single-channel conductance. However, although I_{CRAC} is certainly the best characterized Ca^{2+} current responsible for SOCE, it may not be the only one. Store-operated Ca^{2+} currents with different properties (particularly with lower Ca^{2+} selectivity or non-selective) have also been reported and named I_{SOC} to differentiate them from I_{CRAC} [44]. TRP channels, and particularly the TRPC subfamily, could play an important role in the I_{SOC} currents. Most of the members of the TRPC subfamily have been reported to generate store-operated Ca^{2+} currents and some authors sustain that these currents might be carried through homo or heteromeric combinations of different channel subunits involving TRPCs and/or Orai. Interaction of these plasma membrane homo/heteromeric channels with STIM1 in the ER would confer them store-operated properties. STIM1 has in fact been reported to interact physically both with Orai1 and TRPC channels [45–47].

17.3 Effect of Mitochondrial Ca^{2+} Buffering on SOCE Activation

The classical mechanism proposed for regulation by mitochondria of the activity of SOCE channels relies in the ability of mitochondria to reduce the size of the local Ca^{2+} microdomain around the inner mouth of the channel and thus avoid their Ca^{2+}-dependent inactivation. CRAC channels have two types of Ca^{2+}-dependent inactivation. Fast inactivation occurs with time constants of 10–100 ms and is due to Ca^{2+} binding at a site within less that 10 nm of the pore. The molecular determinants of this fast Ca^{2+}-dependent inactivation have been found both in Orai1 and STIM1, and a role of calmodulin has also been suggested [39, 48]. Because of the extremely small spatio-temporal scale of this phenomenon, mitochondria are not considered to be able to modulate the Ca^{2+} microdomain responsible for this fast inactivation. However, there is also a slow Ca^{2+}-dependent inactivation of CRAC channels that develops with a time course of tens of seconds and involves a larger high-Ca^{2+} spatial domain. It is this slow inactivation mechanism the one that could be modulated by mitochondria. The mechanism and molecular determinant of this slow Ca^{2+}-dependent inactivation is unknown, but it seems to be placed at around 100 nm distance from the CRAC channel pore [44]. In the case of TRPC1 channels, which also have Ca^{2+}-dependent inactivation, there is evidence that it may be mediated by calmodulin [49].

The first evidences for the modulation of SOCE by mitochondria through this mechanism were obtained in 1997 [21]. Both mitochondrial depolarization and inhibition of the electron transport chain enhanced the slow inactivation of the CRAC channel current in Jurkat T lymphocytes. Therefore, energized mitochondria were necessary to prevent the slow inactivation of the channels and facilitate a sustained Ca^{2+} entry through CRAC channels. T cell activation after exposure to antigen or other stimuli requires this prolonged Ca^{2+} entry in order to activate the Ca^{2+}-dependent protein phosphatase calcineurin. Then, dephosphorylation of the transcription factor NFAT by calcineurin unmasks the nuclear localization sequence of NFAT, which then translocates to the nucleus to activate gene transcription of T-cell activation genes [50]. In addition, mitochondria could also have a direct role in the activation of the CRAC current. Removal of cytosolic Ca^{2+} by mitochondria was proposed to compete with store refilling by SERCA pumps, increasing IP_3-induced ER depletion and leading to faster CRAC activation [51]. Thus, mitochondrial Ca^{2+} uptake would be essential both to activate and to sustain CRAC activity [52]. On the other hand, inhibition of mitochondrial Ca^{2+} efflux also diminished SOCE, suggesting that both uptake and release of Ca^{2+} from mitochondria were important to control SOCE [53].

In any case, some uncertainties remain on the role of mitochondria to control SOCE and the possible mechanism. The main mechanism proposed for mitochondrial modulation of SOCE involves local Ca^{2+} buffering to prevent the slow Ca^{2+}-inactivation of the channels. However, local Ca^{2+} microdomains in the SOCE Ca^{2+}-entry sites in T lymphocytes have been reported to be also essential to activate ryanodine receptors via Ca^{2+}-induced Ca^{2+} release (CICR), thus inducing further Ca^{2+}-release to avoid store refilling and keep SOCE active [54]. Coupling of ryanodine receptors and CRAC channels through this Ca^{2+} microdomain would therefore be critical to keep SOCE active in T cells. A similar mechanism has been described for the selective activation of adenylate cyclase by SOCE in defined microdomains. Ca^{2+}-sensitive adenylate cyclase would interact with Orai1 in order to be activated by the local Ca^{2+} microdomain created by channel activation [55]. In addition, local Ca^{2+} microdomains around Orai1 channels are also essential for translocation of NFAT to the nucleus, probably because NFAT dephosphorylation depends on the local recruitment of calcineurin in the microdomain region [56–58]. Thus, the local $[Ca^{2+}]$ near CRAC channels has to rise above a certain threshold in order to trigger NFAT migration. It is therefore difficult to envision the role of mitochondria controlling the architecture of this local Ca^{2+} microdomain, which would be at the same time responsible for SOCE inactivation and essential to trigger CICR and calcineurin activation.

Other mechanisms for Ca^{2+}-dependent mitochondrial regulation of SOCE have also been proposed, among them local ATP production, which could chelate Ca^{2+} close to mitochondria, competition for Ca^{2+} uptake with the plasma membrane Ca^{2+} ATPase, which could contribute to sustain intracellular $[Ca^{2+}]$ increase and direct modulation of ER Ca^{2+}-release through the Ca^{2+}-dependence of IP_3 receptors [59–61]. In addition, Ca^{2+}-independent mechanisms for mitochondrial regulation of SOCE may also be present. In RBL-1 cells and mouse embryonic fibroblasts, store depletion did not induce any change in the pattern of mitochondrial distribution, and most mitochondria were far from the plasma membrane, between 500 nm and 1 μm. However, mitochondrial depolarization impaired the trafficking of STIM1 oligomers to ER-plasma membrane junctions and correspondingly inhibited SOCE [62]. This effect depended on the mitochondrion-shaping protein mitofusin-2. In the absence of mitofusin-2, STIM1 traffic and SOCE were no longer affected by mitochondrial depolarization. Therefore, mitochondria may act on SOCE by mechanisms unrelated to Ca^{2+} fluxes.

17.4 Mitochondrial Subcellular Localization and SOCE

In order for mitochondria to be able to control the local $[Ca^{2+}]$ around the SOCE channels in the plasma membrane, they should be placed very close to them. As mentioned above, the low Ca^{2+}-affinity of the mitochondrial Ca^{2+} uniporter puts a lower limit to the cytosolic $[Ca^{2+}]$ around mitochondria required to trigger fast Ca^{2+} uptake by the organelle. Thus, although mitochondria have been shown to take up Ca^{2+} at submicromolar $[Ca^{2+}]$ [63, 64], high rates of Ca^{2+} uptake are only obtained when extramitochondrial $[Ca^{2+}]$ rises above 2–3 μM [17, 64]. Given that most cellular stimuli induce under physiological conditions mean cytosolic $[Ca^{2+}]$ increases below 1 μM, mitochondria should be quite close to the Ca^{2+}-entry sites in order to activate fast Ca^{2+} uptake through the MCU. In chromaffin cells, for example, Ca^{2+}

entry through voltage-dependent Ca^{2+} channels generates local Ca^{2+} microdomains of 20–40 μM that trigger fast mitochondrial Ca^{2+} uptake in a mitochondrial population. However, more than 50 % of the mitochondria in each cell, those placed far away from the channels, hardly take up any Ca^{2+} [17, 18].

In immune cells, considerable evidence suggests that mitochondria actually move towards the SOCE sites during cell activation. In the Jurkat T cell line, stimulation of Ca^{2+} entry facilitated translocation along microtubules of the complete mitochondrial network towards the plasma membrane, in a Ca^{2+} and kinesin-dependent manner [65]. In T cells, the interface between an antigen-presenting cell or target cell and the lymphocyte is known as the immunological synapse or immune synapse (IS). The junction between T cells and antigen-presenting cells is characterized by close membrane appositions, separated from one another by 'synaptic' regions. Inside the T cell, membrane microdomains containing signaling proteins reorganize and several organelles redistribute toward the IS, including mitochondria and endoplasmic reticulum [66]. Thus, T cell stimulation causes a redistribution of the mitochondrial network towards the plasma membrane and the immunological synapse (IS), and some mitochondria approach to distances of 200 nm of the SOCE channels. In addition, mitochondria close to the IS took up more Ca^{2+} than those farther away, and redistribution of mitochondria to the IS was necessary to maintain Ca^{2+} influx across the plasma membrane and Ca^{2+}-dependent T-cell activation [67–69]. In fact, once formed the IS, mitochondria move towards them in a Ca^{2+}-dependent manner, and maintain SOCE active specifically at the IS, even in the absence of Orai1 accumulation there. Mitochondrial localization would therefore determine where in the plasma membrane should Ca^{2+} entry be active [70] (for reviews see [60, 71]).

However, there are still some findings awaiting explanation. Redistribution of mitochondria to the IS was impaired by knockdown of the mitochondrial fission factor dynamin-related protein 1 (Drp1). Moreover, Drp1 knockdown induced mitochondrial depolarization. However, Drp1 knockdown increased T-cell receptor signal strength and produced a larger and more durable intracellular [Ca^{2+}] increase, in spite of the absence of mitochondria close to the IS [72]. On the other hand, silencing of the outer mitochondrial membrane protein GDAP1, mutations of which produce Charcot-Marie-Tooth neuropathy, induced abnormal distribution of the mitochondrial network in SH-SY5Y human neuroblastoma cells and decreased Ca^{2+} influx through SOCE [73]. Interestingly, junctophilin 1 (JPH1) overexpression was able to restore the SOCE activity in GDAP1-silenced cells [74]. JPH1 is a protein placed in the ER within the mitochondria-associated membrane (MAM) complexes (as well as GDAP1 from the mitochondrial side), but relocates during SOCE activation to co-localize with STIM1 in ER-plasma membrane junctions. Surprisingly, although mitochondrial blockers produced severe inactivation of SOCE in control cells, SOCE activity in JPH1-overexpressing cells was no longer inhibited by mitochondrial blockers, indicating that the restoration of SOCE activity by JPH1 in GDAP1 deficient cells was not related to the recovery of mitochondrial distribution or function [74].

In other non-excitable cell types, the evidence for the regulation of SOCE by mitochondria is weaker. In HeLa cells, expression of the mitochondrial fission promoter protein hFis1 induced a rapid and complete fragmentation of mitochondria, which redistributed away from the plasma membrane and clustered around the nucleus, even though they maintained a normal transmembrane potential and pH, and took up normally the Ca^{2+} released from intracellular stores upon agonist stimulation. Despite the dramatic morphological alteration, SOCE was only marginally reduced. However, disruption of mitochondrial potential with uncouplers or oligomycin/rotenone still reduced SOCE by 35 % [75]. Therefore, either close contact of mitochondria and Ca^{2+} influx channels is not required for SOCE modulation in these cells or mitochondrial inhibitors have additional effects directly on the channels. Also in HeLa cells, activation of SOCE has been shown not to produce significant local $[Ca^{2+}]$ microdomains. While release of Ca^{2+} from the ER generated $[Ca^{2+}]_M$ levels five to tenfold higher in neighboring mitochondria than in the bulk cytosol, Ca^{2+} entry through SOCE generated similar $[Ca^{2+}]$ levels in plasma membrane mitochondria and in deep mitochondria [76, 77]. These results suggest that mitochondria are excluded in these cells from the regions where SOCE occurs. In fact, in COS-7 cells, mitochondria were located either in the gaps between STIM1 puncta or in remote, STIM1-free regions, and the increase in $[Ca^{2+}]_M$ after Ca^{2+} addition was independent of the mitochondrion–STIM1 distance [78]. Similarly, in the cell lines RBL-2H3 and H9c2 cells, all mitochondria were shown to have contacts with the ER, but contacts between mitochondria and the plasma membrane were less frequent due to the presence of interleaving ER stacks [79, 80]. In fact, these findings provide structural basis for the observation that in some conditions Ca^{2+} reaches mitochondria only after crossing the ER [81].

These observations indicate that in these cells mitochondria are only exposed to Ca^{2+} diffusing laterally from the SOCE Ca^{2+}-entry sites, suggesting that the close apposition among ER and plasma membrane in the sites of STIM-Orai interaction represents indeed a physical barrier for access of mitochondria. These STIM-Orai interacting sites have also been shown to co-localize with SERCA pumps in non-excitable cells [82, 83]. For example, in HEK293 cells, SOCE activation generates subplasmalemmal high Ca^{2+} microdomains, but most of the entering Ca^{2+} is taken by the ER and not by mitochondria. This may also be due to the small size of the microdomains generated by SOCE, as SERCA pumps have a much higher Ca^{2+} affinity than mitochondrial uniporters [83].

In summary, mitochondrial localization with respect to SOCE appears to change considerably among different cell types [48]. In cells such as HeLa, RBL or HEK293, mitochondria are far away from the SOCE Ca^{2+} entry sites, while in T cells or Jurkat T lymphocytes cell activation triggers a fast movement of mitochondria toward the IS. Regarding excitable cells, there is also a significant variability. In vascular smooth muscle A10 cells, mitochondrial Ca^{2+} uptake was triggered by SOCE activation, but not by activation of voltage-dependent Ca^{2+} channels [84]. In other excitable cells, instead, mitochondrial Ca^{2+} uptake was much larger after voltage-dependent Ca^{2+} channels activation [17, 76].

17.5 Effect of MCU Knockout on SOCE Activation

The recent discovery of the main constituents of the mitochondrial Ca^{2+} uniporter provides a much better tool to assay the effects of mitochondrial Ca^{2+} uptake on Ca^{2+} entry through SOCE. Knockout of the pore component of the uniporter, the MCU, has been recently used to study the effect on SOCE. In RBL-1 cells, knockout of MCU considerably reduced Ca^{2+} entry after store depletion with thapsigargin [85]. MCU silencing by siRNA in MDA-MB-231 breast cancer cells also reduced serum- or thapsigargin-induced SOCE [86]. Similarly, in HeLa cells, MCU knockout reduced the rate of Ca^{2+} entry through SOCE and favored the development of slow inactivation [87].

We have also studied this point by using specific shRNA to silence the MCU protein in HeLa cell clones. However, our data do not show any inhibition of SOCE in the MCU-silenced clones with respect to scrambled clones. Figure 17.1 shows studies carried out using aequorin targeted to mitochondria or endoplasmic reticulum. The histamine-induced $[Ca^{2+}]_M$ peak was nearly abolished in the MCU-silenced cells (Fig. 17.1a), and addition of Ca^{2+} to permeabilized cells produced no increase in $[Ca^{2+}]_M$ in the MCU-silenced cells (Fig. 17.1b), showing that MCU activity was almost fully abolished in the silenced clone. In Ca^{2+}-depleted cells, refilling with Ca^{2+} of the ER was in fact faster in the silenced cells (Fig. 17.1c, d), suggesting that Ca^{2+}-entry through SOCE channels was not reduced in the MCU-silenced cells. Regarding Ca^{2+}-release through IP_3 receptors induced by the agonist histamine, it was somewhat slower in the MCU –silenced cells (Fig. 17.1e). This observation may be due to the increased feed-back inhibition by Ca^{2+} of IP_3 receptors in the absence of mitochondrial Ca^{2+} uptake.

Imaging studies of cytosolic $[Ca^{2+}]$ (Fig. 17.2) showed that the histamine induced cytosolic $[Ca^{2+}]_c$ increase was significantly larger in the MCU-silenced cells (Fig. 17.2a), probably because of the lack of mitochondrial Ca^{2+} uptake. When SOCE was explored directly by looking at the cytosolic $[Ca^{2+}]$ increase induced by Ca^{2+} addition to Ca^{2+}-depleted cells, Ca^{2+} entry through SOCE was found to be significantly increased in the MCU-silenced cells (Fig. 17.2b). Consistently, the rate of Mn^{2+} entry in Ca^{2+}-depleted cells (Fig. 17.2c) was also significantly increased in the MCU-silenced cells. Therefore, our data indicate that the absence of mitochondrial Ca^{2+} uptake did not inhibit SOCE in HeLa cells. In fact, a significant stimulation was observed. The reasons for the discrepancy with previously reported data [87] are unknown.

17.6 Conclusions

Intracellular Ca^{2+} signaling is mainly controlled by ER and mitochondria, two organelles with opposite roles in terms of Ca^{2+} homeostasis. The ER is the main intracellular Ca^{2+}-store, releases Ca^{2+} to the cytosol during cell activation and then

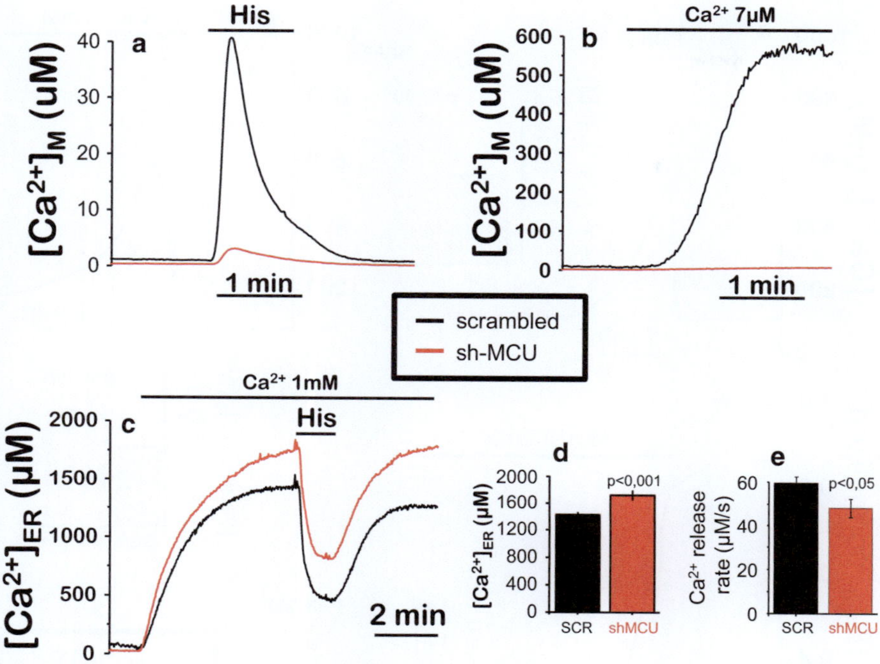

Fig. 17.1 Effect of MCU knockout on mitochondrial and ER [Ca^{2+}] dynamics. HeLa cell clones expressing either shRNA for the MCU protein or scrambled shRNA were transfected with either mitochondrially targeted mutated (**a**) or double mutated (**b**) aequorin or ER-targeted double-mutated aequorin (**c**). (**a**) Effect of stimulation with 100 μM histamine on $[Ca^{2+}]_M$. (**b**) Effect of perfusion of a 7 μM [Ca^{2+}] buffer in permeabilized cells (see [88] for other details). (**c**) Refilling of the ER with Ca^{2+} and Ca^{2+} release induced by 100 μM histamine (see Ref. [3] for other details). (**d**) Increase in the steady-state ER [Ca^{2+}] in MCU-silenced cells. (**e**) Decrease in the Ca^{2+} release rate in MCU-silenced cells

helps to end the Ca^{2+} transient by taking up Ca^{2+} back after the stimulation. In contrast, mitochondria have little [Ca^{2+}] under resting conditions, but have a large capacity to take up Ca^{2+} during cell stimulation and then release it again to the cytosol at the end of the Ca^{2+} transient. In addition, because of the low-Ca^{2+} affinity of the mitochondrial Ca^{2+} uptake mechanism, this organelle is specially adapted to buffer Ca^{2+} from high-Ca^{2+} microdomains such as those formed around the cytosolic mouth of open Ca^{2+} channels. The classical mechanism proposed for mitochondrial regulation of SOCE activity relies in the effect of mitochondrial Ca^{2+} uptake controlling the local Ca^{2+} microdomain around the SOCE channels, in order to prevent its Ca^{2+}-dependent inactivation. To be effective, this mechanism requires very close contacts between mitochondria and SOCE channels. However, mitochondrial subcellular localization with respect to SOCE channels appears to be highly variable among different cell types. While in immunitary cells mitochondria move to the IS during cell activation, in other cells there is little evidence for that. Figure 17.3 shows diagrams summarizing these mechanisms. In addition, activation of SOCE

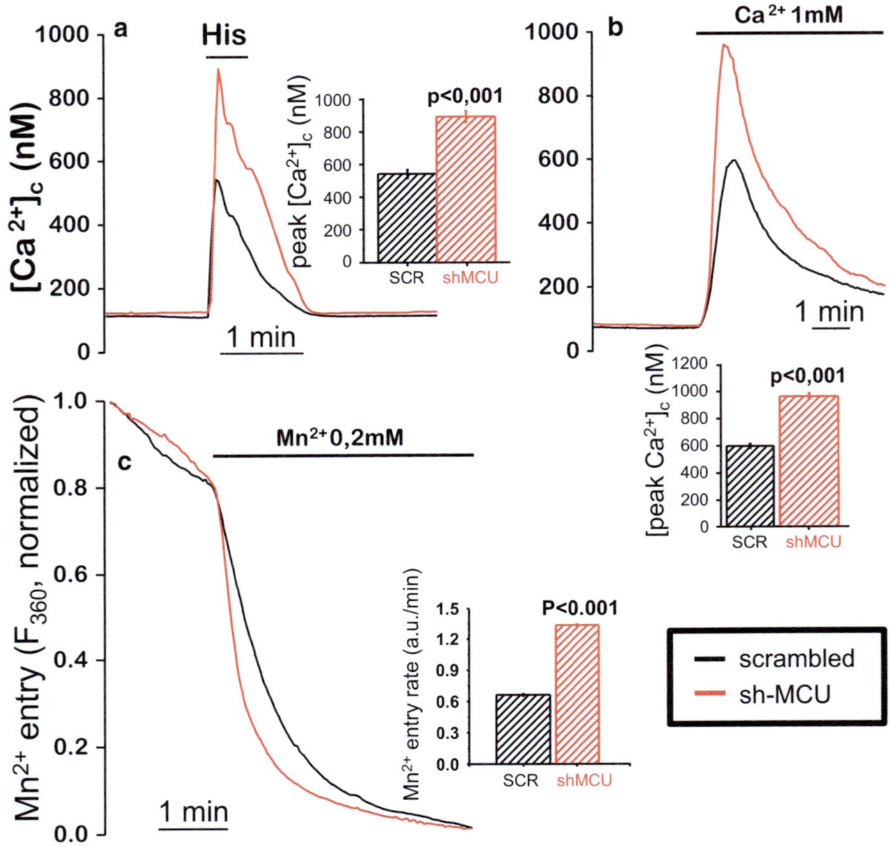

Fig. 17.2 Effect of MCU knockout on cytosolic [Ca^{2+}] dynamics and Mn^{2+} entry. The shMCU and scrambled HeLa cell clones were loaded with fura-2 and prepared for single-cell imaging experiments as described previously [25]. (**a**) Cells were stimulated with 100 μM histamine. Data are mean±s.e. from five experiments in each condition (221 cells scrambled, 191 cells shMCU). (**b**) Cells were depleted of Ca^{2+} with 10 μM benzohydroquinone and then SOCE was induced with 1 mM extracellular [Ca^{2+}]. Data are mean±s.e. from nine experiments in each condition (700 cells scrambled, 602 cells shMCU). **c** Cells were depleted of Ca^{2+} in Ca^{2+}-free medium and then Mn^{2+} entry was monitored at 360 nm fluorescence excitation. Data are mean±s.e. from five experiments in each condition (611 cells scrambled, 768 cells shMCU)

requires a close interaction among Orai (or TRPs) channels in the plasma membrane and STIM sensors in the ER. The steric hindrance due to the close apposition of plasma and ER membranes in the SOCE sites constitutes a barrier that limits the approach of mitochondria. Even in immune cells, this makes the distance of mitochondria to the SOCE sites to be at least 200 nm. In other cells the distance is even larger. On the other hand, the local Ca^{2+} microdomain created in the SOCE sites in immune cells seems to be also essential for T cell activation and NFAT dephosphorylation. Therefore, many uncertainties still remain on the role of mitochondria to control SOCE and the possible mechanism. Alternative pathways for mitochondrial modulation of SOCE, both Ca^{2+}-dependent and Ca^{2+}-independent, have also been proposed, but clear evidence for a specific mechanism is still lacking. The

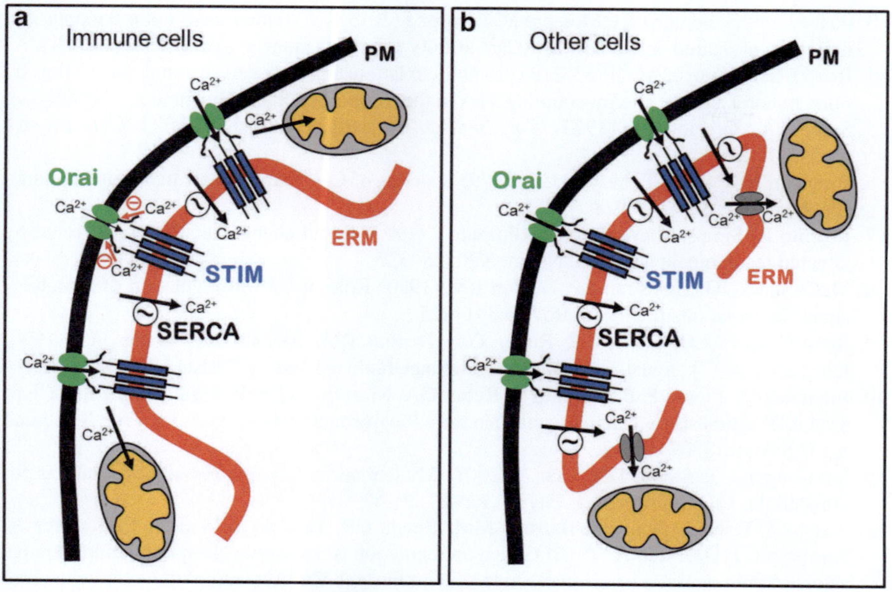

Fig. 17.3 A model of the role of mitochondria in the regulation of SOCE. (**a**) In immune cells, mitochondria move to the plasma membrane during cell activation and considerable evidence indicates that, after SOCE activation, mitochondrial Ca^{2+} uptake reduces the size of the local [Ca^{2+}] microdomain around nearby Orai-STIM complexes and protects Orai1 channels from Ca^{2+}-dependent inactivation. (**b**) In many other cell types, the available evidence suggests that mitochondria remain too far from the Orai-STIM complexes to be able to modulate directly the local Ca^{2+} microdomain. Mitochondria could anyway take up some of the incoming Ca^{2+} after passing through the ER, but the possible modulation of Ca^{2+} entry by mitochondria remains unclear. Alternative mechanisms of modulation based on mitochondrial ATP production, ER depletion induced by mitochondrial Ca^{2+} uptake, or others, have also been proposed. *PM* plasma membrane, *ERM* ER membrane, *SERCA* ER Ca^{2+} ATPase

recent advances in the knowledge of the molecular nature of the mitochondrial Ca^{2+} uptake mechanism open new ways to approach this problem directly, avoiding the use of mitochondrial inhibitors or protonophors and all their large side-effects.

Acknowledgements This work was supported by a grant from the Spanish Ministerio de Ciencia e Innovación (BFU2011-25763). Jessica Matesanz-Isabel holds a FPI (Formación de Personal Investigador) fellowship from the Spanish Government. Jessica Arias-del-Val holds a fellowship from Junta de Castilla y León.

References

1. Berridge MJ, Lipp P, Bootman MD (2000) The versatility and universality of calcium signalling. Nat Rev Mol Cell Biol 1:11–21
2. Streb H, Irvine RF, Berridge MJ, Schulz I (1983) Release of Ca^{2+} from a nonmitochondrial intracellular store by $InsP_3$. Nature 306:67–69

3. de la Fuente S, Fonteriz RI, Montero M, Alvarez J (2013) Ca^{2+} homeostasis in the endoplasmic reticulum measured with a new low-Ca^{2+}-affinity targeted aequorin. Cell Calcium 54:37–45
4. Rossi CS, Lehninger AL (1963) Stoichiometric relationships between accumulation of ions by mitochondria and the energy-coupling sites in the respiratory chain. Biochem Z 338:698–713
5. Scarpa A, Graziotti P (1973) Ca^{2+} uptake by cardiac mitochondria. J Gen Physiol 62:756–772
6. Bragadin M, Pozzan T, Azzone GF (1979) Kinetics of Ca^{2+} carrier in rat liver mitochondria. Biochemistry 18:5972–5978
7. Rizzuto R, Simpson AW, Brini M, Pozzan T (1992) Rapid changes of $[Ca^{2+}]_M$ revealed by targeted recombinant aequorin. Nature 358:325–327
8. McCormack JG, Halestrap AP, Denton RM (1990) Role of Ca^{2+} in regulation of intramitochondrial metabolism. Physiol Rev 70:391–425
9. Robb-Gaspers LD, Burnett P, Rutter GA, Denton RM, Rizzuto R, Thomas AP (1998) Integrating $[Ca^{2+}]_c$ signals into mitochondrial metabolic responses. EMBO J 17:4987–4989
10. Jouaville LS, Pinton P, Bastianutto C, Rutter GA, Rizzuto R (1999) Regulation of mitochondrial ATP synthesis by calcium: evidence for a long-term metabolic priming. Proc Natl Acad Sci U S A 96:13807–13812
11. Satrústegui J, Pardo B, Del Arco A (2007) Mitochondrial transporters as novel targets for intracellular calcium signaling. Physiol Rev 87:29–67
12. Amigo I, Traba J, González-Barroso MM, Rueda CB, Fernández M, Rial E, Sánchez A, Satrústegui J, Del Arco A (2013) Glucagon regulation of oxidative phosphorylation requires an increase in matrix adenine nucleotide content through Ca^{2+} activation of the mitochondrial ATP-Mg/Pi carrier SCaMC-3. J Biol Chem 288:7791–7802
13. Llorente-Folch I, Rueda CB, Amigo I, del Arco A, Saheki T, Pardo B, Satrústegui J (2013) Calcium-regulation of mitochondrial respiration maintains ATP homeostasis and requires ARALAR/AGC1-malate aspartate shuttle in intact cortical neurons. J Neurosci 33:13957–13971
14. Denton RM (2009) Regulation of mitochondrial dehydrogenases by calcium ions. Biochim Biophys Acta 1787:1309–1316
15. Madiraju AK, Erion DM, Rahimi Y, Zhang XM, Braddock DT, Albright RA, Prigaro BJ, Wood JL, Bhanot S, MacDonald MJ, Jurczak MJ, Camporez JP, Lee HY, Cline GW, Samuel VT, Kibbey RG, Shulman GI (2014) Metformin suppresses gluconeogenesis by inhibiting mitochondrial glycerophosphate dehydrogenase. Nature 510:542–546
16. Santo-Domingo J, Demaurex N (2010) Biochim Biophys Acta 1797:907–912
17. Montero M, Alonso MT, Carnicero E, Cuchillo-Ibáñez I, Albillos A, García AG, García-Sancho J, Alvarez J (2000) Chromaffin-cell stimulation triggers fast millimolar mitochondrial $[Ca^{2+}]$ transients that modulate secretion. Nat Cell Biol 2:57–61
18. Villalobos C, Nuñez L, Montero M, García AG, Alonso MT, Chamero P, Alvarez J, García-Sancho J (2002) Redistribution of Ca^{2+} among cytosol and organella during stimulation of bovine chromaffin cells. FASEB J 16:343–353
19. Park YB, Herrington J, Babcock DF, Hille B (1996) Ca^{2+} clearance mechanisms in isolated rat adrenal chromaffin cells. J Physiol 492:329–346
20. White RJ, Reynolds IJ (1997) Mitochondria accumulate Ca^{2+} following intense glutamate stimulation of cultured rat forebrain neurones. J Physiol 498:31–47
21. Hoth M, Fanger CM, Lewis RS (1997) Mitochondrial regulation of store-operated calcium signaling in T lymphocytes. J Cell Biol 137:633–648
22. Hernandez-Guijo JM, Maneu-Flores VE, Ruiz-Nuno A, Villarroya M, Garcia AG, Gandia L (2001) Calcium-dependent inhibition of L, N, and P/Q Ca^{2+} channels in chromaffin cells: role of mitochondria. J Neurosci 21:2553–2560
23. Tekmen M, Gleason E (2010) Multiple Ca^{2+}-dependent mechanisms regulate L-type Ca^{2+} current in retinal amacrine cells. J Neurophysiol 104:1849–1866
24. Montero M, Barrero MJ, Alvarez J (1997) $[Ca^{2+}]$ microdomains control agonist-induced Ca^{2+} release. FASEB J 11:881–885

25. Vay L, Hernández-Sanmiguel E, Santo-Domingo J, Lobatón CD, Moreno A, Montero M, Alvarez J (2007) Modulation of Ca^{2+}-release and Ca^{2+} oscillations by Ca^{2+} uniporter stimulation. J Physiol 580:39–49
26. Olson ML, Chalmers S, McCarron JG (2012) Mitochondrial organization and Ca^{2+} uptake. Biochem Soc Trans 40:158–167
27. Palty R, Silverman WF, Hershfinkel M, Caporale T, Sensi SL, Parnis J, Nolte C, Fishman D, Shoshan-Barmatz V, Herrmann S, Khananshvili D, Sekler I (2010) NCLX is an essential component of mitochondrial Na^+/Ca^{2+} exchange. Proc Natl Acad Sci U S A 107:436–441
28. Baughman JM, Perocchi F, Girgis HS, Plovanich M, Belcher-Timme CA, Sancak Y, Bao XR, Strittmatter L, Goldberger O, Bogorad RL, Koteliansky V, Mootha VK (2011) Integrative genomics identifies MCU as an essential component of the mitochondrial calcium uniporter. Nature 476:341–345
29. De Stefani D, Raffaello A, Teardo E, Szabo I, Rizzuto R (2011) A forty-kilodalton protein of the inner membrane is the mitochondrial calcium uniporter. Nature 476:336–340
30. Pendin D, Greotti E, Pozzan T (2014) The elusive importance of being a mitochondrial Ca^{2+} uniporter. Cell Calcium 55:139–145
31. Marchi S, Pinton P (2014) The mitochondrial calcium uniporter complex: molecular components, structure and physiopathological implications. J Physiol 592:829–839
32. Pan X, Liu J, Nguyen T, Liu C, Sun J, Teng Y, Fergusson MM, Rovira II, Allen M, Springer DA, Aponte AM, Gucek M, Balaban RS, Murphy E, Finkel T (2013) The physiological role of mitochondrial calcium revealed by mice lacking the mitochondrial calcium uniporter. Nat Cell Biol 15:1464–1472
33. Murphy E, Pan X, Nguyen T, Liu J, Holmström KM, Finkel T (2014) Unresolved questions from the analysis of mice lacking MCU expression. Biochem Biophys Res Commun 449:384–385
34. Putney JW (1986) A model for receptor-regulated calcium entry. Cell Calcium 7:1–12
35. Takemura H, Putney JW (1989) Capacitative calcium entry in parotid acinar cells. Biochem J 258:409–412
36. Montero M, Garcia-Sancho J, Alvarez J (1993) Transient inhibition by chemotactic peptide of a store-operated Ca^{2+} entry pathway in human neutrophils. J Biol Chem 268:13055–13061
37. Hoth M, Penner R (1992) Depletion of intracellular calcium stores activates a calcium current in mast cells. Nature 355:353–356
38. Putney JW (2009) Capacitative calcium entry: from concept to molecules. Immunol Rev 231:10–22
39. Smyth JT, Hwang SY, Tomita T, DeHaven WI, Mercer JC, Putney JW (2010) Activation and regulation of store-operated calcium entry. J Cell Mol Med 14:2337–2349
40. Lewis RS (2011) Store-operated calcium channels: new perspectives on mechanism and function. Cold Spring Harb Perspect Biol 3:a003970
41. Srikanth S, Gwack Y (2013) Orai1-NFAT signalling pathway triggered by T cell receptor stimulation. Mol Cells 35:182–194
42. Cheng KT, Ong HL, Liu X, Ambudkar IS (2013) Contribution and regulation of TRPC channels in store-operated Ca^{2+} entry. Curr Top Membr 71:149–179
43. Shaw PJ, Qu B, Hoth M, Feske S (2013) Molecular regulation of CRAC channels and their role in lymphocyte function. Cell Mol Life Sci 70:2637–2656
44. Parekh AB, Putney JW Jr (2005) Store-operated calcium channels. Physiol Rev 85:757–810
45. Dionisio N, Redondo PC, Jardin I, Rosado JA (2012) Transient receptor potential channels in human platelets: expression and functional role. Curr Mol Med 12:1319–1328
46. Berna-Erro A, Redondo PC, Rosado JA (2012) Store-operated Ca^{2+} entry. In: Islam MS (ed) Calcium signaling series: advances in experimental medicine and biology, vol 740. Springer, Heidelberg, pp 349–382
47. Albarrán L, Dionisio N, Lopez E, Salido GM, Redondo PC, Rosado JA (2014) STIM1 regulates TRPC6 heteromultimerization and subcellular location. Biochem J 463:373–381
48. Watson R, Parekh AB (2012) Mitochondrial regulation of CRAC channel-driven cellular responses. Cell Calcium 52:52–56

49. Singh BB, Liu X, Tang J, Zhu MX, Ambudkar IS (2002) Calmodulin regulates Ca^{2+}-dependent feedback inhibition of store-operated Ca^{2+} influx by interaction with a site in the C terminus of TrpC1. Mol Cell 9:739–750
50. Hoth M, Button DC, Lewis RS (2000) Mitochondrial control of calcium-channel gating: a mechanism for sustained signaling and transcriptional activation in T lymphocytes. Proc Natl Acad Sci U S A 97:10607–10612
51. Gilabert JA, Parekh AB (2000) Respiring mitochondria determine the pattern of activation and inactivation of the store-operated Ca^{2+} current I_{CRAC}. EMBO J 19:6401–6407
52. Parekh AB (2003) Store-operated Ca^{2+} entry: dynamic interplay between endoplasmic reticulum, mitochondria and plasma membrane. J Physiol 547:333–348
53. Malli R, Frieden M, Osibow K, Zoratti C, Mayer M, Demaurex N, Graier WF (2003) Sustained Ca^{2+} transfer across mitochondria is Essential for mitochondrial Ca^{2+} buffering, sore-operated Ca^{2+} entry, and Ca^{2+} store refilling. J Biol Chem 278:44769–44779
54. Thakur P, Dadsetan S, Fomina AF (2012) Bidirectional coupling between ryanodine receptors and Ca^{2+} release-activated Ca^{2+} (CRAC) channel machinery sustains store-operated Ca^{2+} entry in human T lymphocytes. J Biol Chem 287:37233–37244
55. Willoughby D (2012) Organization of cAMP signalling microdomains for optimal regulation by Ca^{2+} entry. Biochem Soc Trans 40:246–250
56. Kar P, Nelson C, Parekh AB (2011) Selective activation of the transcription factor NFAT1 by calcium microdomains near Ca^{2+} release-activated Ca^{2+} (CRAC) channels. J Biol Chem 286:14795–14803
57. Kar P, Nelson C, Parekh AB (2012) CRAC channels drive digital activation and provide analog control and synergy to Ca^{2+}-dependent gene regulation. Curr Biol 22:242–247
58. Kar P, Samanta K, Kramer H, Morris O, Bakowski D, Parekh AB (2014) Dynamic assembly of a membrane signaling complex enables selective activation of NFAT by Orai1. Curr Biol 24:1361–1368
59. Duszyński J, Kozieł R, Brutkowski W, Szczepanowska J, Zabłocki K (2006) The regulatory role of mitochondria in capacitative calcium entry. Biochim Biophys Acta 1757:380–387
60. Joseph N, Reicher B, Barda-Saad M (2014) The calcium feedback loop and T cell activation: how cytoskeleton networks control intracellular calcium flux. Biochim Biophys Acta 1838:557–568
61. Huang HM, Chen HL, Gibson GE (2014) Interactions of endoplasmic reticulum and mitochondria Ca^{2+} stores with capacitative calcium entry. Metab Brain Dis 29:1083–1093
62. Singaravelu K, Nelson C, Bakowski D, de Brito OM, Ng SW, Di Capite J, Powell T, Scorrano L, Parekh AB (2011) Mitofusin 2 regulates STIM1 migration from the Ca^{2+} store to the plasma membrane in cells with depolarized mitochondria. J Biol Chem 286:12189–12201
63. Spät A, Szanda G, Csordás G, Hajnóczky G (2008) High- and low-calcium-dependent mechanisms of mitochondrial calcium signalling. Cell Calcium 44:51–63
64. de la Fuente S, Matesanz-Isabel J, Fonteriz RI, Montero M, Alvarez J (2014) Dynamics of mitochondrial Ca^{2+} uptake in MICU1-knockdown cells. Biochem J 458:33–40
65. Quintana A, Schwarz EC, Schwindling C, Lipp P, Kaestner L, Hoth M (2006) Sustained activity of calcium release-activated calcium channels requires translocation of mitochondria to the plasma membrane. J Biol Chem 281:40302–40309
66. Martín-Cófreces NB, Baixauli F, Sánchez-Madrid F (2014) Immune synapse: conductor of orchestrated organelle movement. Trends Cell Biol 24:61–72
67. Quintana A, Schwindling C, Wenning AS, Becherer U, Rettig J, Schwarz EC, Hoth M (2007) T cell activation requires mitochondrial translocation to the immunological synapse. Proc Natl Acad Sci U S A 104:14418–14423
68. Contento RL, Campello S, Trovato AE, Magrini E, Anselmi F, Viola A (2010) Adhesion shapes T cells for prompt and sustained T-cell receptor signalling. EMBO J 29:4035–4047
69. Quintana A, Pasche M, Junker C, Al-Ansary D, Rieger H, Kummerow C, Nuñez L, Villalobos C, Meraner P, Becherer U, Rettig J, Niemeyer BA, Hoth M (2011) Calcium microdomains at the immunological synapse: how ORAI channels, mitochondria and calcium pumps generate local calcium signals for efficient T-cell activation. EMBO J 30:3895–3912

70. Schwindling C, Quintana A, Krause E, Hoth M (2010) Mitochondria positioning controls local calcium influx in T cells. J Immunol 184:184–190
71. Quintana A, Hoth M (2012) Mitochondrial dynamics and their impact on T cell function. Cell Calcium 52:57–63
72. Baixauli F, Martín-Cófreces NB, Morlino G, Carrasco YR, Calabia-Linares C, Veiga E, Serrador JM, Sánchez-Madrid F (2011) The mitochondrial fission factor dynamin-related protein 1 modulates T-cell receptor signalling at the immune synapse. EMBO J 30:1238–1250
73. Pla-Martín D, Rueda CB, Estela A, Sánchez-Piris M, González-Sánchez P, Traba J, de la Fuente S, Scorrano L, Renau-Piqueras J, Alvarez J, Satrústegui J, Palau F (2013) Silencing of the Charcot-Marie-Tooth disease-associated gene GDAP1 induces abnormal mitochondrial distribution and affects Ca^{2+} homeostasis by reducing store-operated Ca^{2+} entry. Neurobiol Dis 55:140–151
74. Pla-Martín D, Calpena E, Lupo V, Márquez C, Rivas E, Sivera R, Sevilla T, Palau F, Espinós C (2015) Junctophilin-1 is a modifier gene of GDAP1-related Charcot-Marie-Tooth disease. Hum Mol Genet 24:213–229
75. Frieden M, James D, Castelbou C, Danckaert A, Martinou JC, Demaurex N (2004) Ca^{2+} homeostasis during mitochondrial fragmentation and perinuclear clustering induced by hFis1. J Biol Chem 279:22704–22714
76. Giacomello M, Drago I, Bortolozzi M, Scorzeto M, Gianelle A, Pizzo P, Pozzan T (2010) Ca^{2+} hot spots on the mitochondrial surface are generated by Ca^{2+} mobilization from stores, but not by activation of store-operated Ca^{2+} channels. Mol Cell 38:280–290
77. Contreras L, Drago I, Zampese E, Pozzan T (2010) Mitochondria: the calcium connection. Biochim Biophys Acta 1797:607–618
78. Korzeniowski MK, Szanda G, Balla T, Spät A (2009) Store-operated Ca^{2+} influx and subplasmalemmal mitochondria. Cell Calcium 46:49–55
79. Csordás G, Várnai P, Golenár T, Roy S, Purkins G, Schneider TG, Balla T, Hajnóczky G (2010) Imaging interorganelle contacts and local calcium dynamics at the ER-mitochondrial interface. Mol Cell 39:121–132
80. Shin DM, Muallem S (2010) What the mitochondria see. Mol Cell 39:6–7
81. Demaurex N, Poburko D, Frieden M (2009) Regulation of plasma membrane calcium fluxes by mitochondria. Biochim Biophys Acta 1787:1383–1394
82. Alonso MT, Manjarrés IM, García-Sancho J (2012) Privileged coupling between Ca^{2+} entry through plasma membrane store-operated Ca^{2+} channels and the endoplasmic reticulum Ca^{2+} pump. Mol Cell Endocrinol 353:37–44
83. García-Sancho J (2014) The coupling of plasma membrane calcium entry to calcium uptake by endoplasmic reticulum and mitochondria. J Physiol 592:261–268
84. Muñoz E, Valero RA, Quintana A, Hoth M, Núñez L, Villalobos C (2011) Nonsteroidal anti-inflammatory drugs inhibit vascular smooth muscle cell proliferation by enabling the Ca^{2+}-dependent inactivation of calcium release-activated calcium/Orai channels normally prevented by mitochondria. J Biol Chem 286:16186–16196
85. Samanta K, Douglas S, Parekh AB (2014) Mitochondrial calcium uniporter MCU supports cytoplasmic Ca^{2+} oscillations, store-operated Ca^{2+} entry and Ca^{2+}-dependent gene expression in response to receptor stimulation. PLoS One 9:e101188
86. Tang S, Wang X, Shen Q, Yang X, Yu C, Cai C, Cai G, Meng X, Zou F (2015) Mitochondrial Ca^{2+} uniporter is critical for store-operated Ca^{2+} entry-dependent breast cancer cell migration. Biochem Biophys Res Commun. doi:10.1016/j.bbrc.2015.01.092
87. Deak AT, Blass S, Khan MJ, Groschner LN, Waldeck-Weiermair M, Hallström S, Graier WF, Malli R (2014) IP_3-mediated STIM1 oligomerization requires intact mitochondrial Ca^{2+} uptake. J Cell Sci 127:2944–2955
88. de la Fuente S, Fonteriz RI, de la Cruz PJ, Montero M, Alvarez J (2012) Mitochondrial free $[Ca^{2+}]$ dynamics measured with a novel low-Ca^{2+} affinity aequorin probe. Biochem J 445:371–376

Chapter 18
Modulation of Calcium Entry by the Endo-lysosomal System

G. Cristina Brailoiu and Eugen Brailoiu

Abstract Endo-lysosomes are acidic organelles that besides the role in macromolecules degradation, act as intracellular Ca^{2+} stores. Nicotinic acid adenine dinucleotide phosphate (NAADP), the most potent Ca^{2+}-mobilizing second messenger, produced in response to agonist stimulation, activates Ca^{2+}-releasing channels on endo-lysosomes and modulates a variety of cellular functions. NAADP-evoked signals are amplified by Ca^{2+} release from endoplasmic reticulum, via the recruitment of inositol 1,4,5-trisphosphate and/or ryanodine receptors through a Ca^{2+}-induced Ca^{2+}- release (CICR) mechanism. The endo-lysosomal Ca^{2+} channels activated by NAADP were recently identified as the two-pore channels (TPCs). In addition to TPCs, endo-lysosomes express another distinct family of Ca^{2+}- permeable channels, namely the transient receptor potential mucolipin (TRPML) channels, functionally distinct from TPCs. TPCs belong to the voltage-gated channels, resembling voltage-gated Na^{+} and Ca^{2+} channels. TPCs have important roles in vesicular fusion and trafficking, in triggering a global Ca^{2+} signal and in modulation of the membrane excitability. Depletion of acidic Ca^{2+} stores has been shown to activate store-operated Ca^{2+} entry in human platelets and mouse pancreatic β-cells. In human platelets, Ca^{2+} influx in response to acidic stores depletion is facilitated by the tubulin-cytoskeleton and occurs through non-selective cation channels and transient receptor potential canonical (TRPC) channels. Emerging evidence indicates that activation of intracellular receptors, situated on endo-lysosomes, elicits canonical and non-canonical signaling mechanisms that involve CICR and activation of non-selective cation channels in plasma membrane. The ability of endo-lysosomal Ca^{2+} stores to modulate the Ca^{2+} release from other organelles and the Ca^{2+} entry increases the diversity and complexity of cellular signaling mechanisms.

G.C. Brailoiu (✉)
Department of Pharmaceutical Sciences, Jefferson School of Pharmacy, Thomas Jefferson University, 901 Walnut St, Rm 916, Philadelphia, PA 19107, USA
e-mail: gabriela.brailoiu@jefferson.edu

E. Brailoiu
Center for Substance Abuse Research, Temple University School of Medicine, 3500 N. Broad Street, Room 848, Philadelphia, PA 19140, USA
e-mail: eugen.brailoiu@temple.edu

J.A. Rosado (ed.), *Calcium Entry Pathways in Non-excitable Cells*, Advances in Experimental Medicine and Biology 898, DOI 10.1007/978-3-319-26974-0_18

Keywords Acidic calcium stores • NAADP • Two-pore channels (TPCs) • Intracellular GPCR • Store-operated calcium entry

Abbreviations

$[Ca^{2+}]_i$	Cytosolic Ca^{2+} concentration, Ca_V, voltage-gated Ca^{2+} channels
cADPR	Cyclic ADP ribose
CCE	Capacitative Ca^{2+} entry
CICR	Ca^{2+}-induced Ca^{2+} release
CRAC	Ca^{2+}-release-activated Ca^{2+}
DTS	Dense tubular system
ER	Endoplasmic reticulum
IP_3	Inositol 1,4,5-trisphosphate
IP_3R	Inositol 1,4,5-trisphosphate receptor
LPI	L-α-lysophosphatidylinositol
MDCK	Madin-Darby canine kidney
NAADP	Nicotinic acid adenine dinucleotide
Na_V	Voltage-gated sodium channels
PARs	Protease-activated receptors
PI(3,5)P2	Phosphatidylinositol 3,5-bisphosphate
PLC	Phospholipase C
RyR	Ryanodine receptor
SOCE	Store-operated calcium entry
STIM1	Stromal interaction molecule-1
TBHQ	2,5-di-(tert-butyl)-1,4-hydroquinone
TPCs	Two-pore channels
TRPC	Transient receptor potential canonical
TRPML	Transient receptor potential mucolipin.

18.1 Endo-lysosomal System as Ca^{2+} Store

Lysosomes are intracellular organelles discovered in the mid-1950s by Christian de Duve [1, 2], commonly associated with the intracellular digestion and macromolecules degradation [3]. Lysosomes were identified as vacuolar structures with an acidic pH, containing hydrolytic enzymes, surrounded by a membrane [1, 2]. Four digestive processes, namely receptor-mediated endocytosis, pinocytosis, phagocytosis and autophagy are mediated by the lysosomes [4]. The lysosomal system is a heterogeneous system, comprising of: primary/nascent lysosomes, early autophagic vacuoles, intermediate/late endosomes, pinocytic/phagocytic vacuoles and

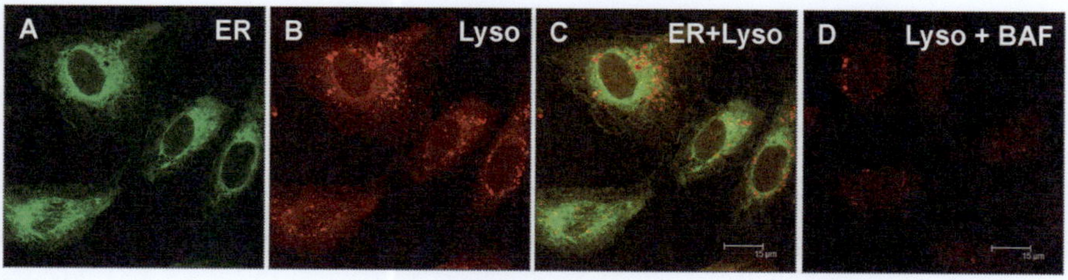

Fig. 18.1 Localization of the acidic compartments and the endoplasmic reticulum in human aortic endothelial cells (*HAEC*). Confocal images of endoplasmic reticulum labeled with ER-Tracker Green (**A**) and of endo-lysosomes labeled with LysoTracker Red (**B**) in HAEC. (**C**), a merged image of **a** and **b**. (**D**), pretreatment of HAEC cells with bafilomycin A1 (*BAF*), a V-type H^+ ATPase inhibitor, abolished the LysoTracker Red fluorescence

multivesicular bodies. Lysosomes play critical roles in proteolysis, an important regulatory mechanism, through an ubiquitin-proteasome system [5].

In the last few years, increasing evidence has established endo-lysosomes as a Ca^{2+} store. The endoplasmic reticulum (ER) is the best characterized intracellular Ca^{2+} store, endowed with a well-defined system of Ca^{2+} channels and pumps [6, 7]. In addition, mitochondria, Golgi apparatus, nucleus, caveolae, and acidic stores contain variable amount of Ca^{2+} and contribute to Ca^{2+} homeostasis [8, 9]. Acidic Ca^{2+} stores, in addition to the endo-lysosomal system, include acidocalcisomes, vacuoles, lysosome-related organelles, secretory vesicles and the Golgi complex [10].

The use of LysoTracker Red, a weak base that accumulates in acidic organelles [11] and of lysosomotropic agents that interfere with acidic compartments, such as bafilomycin A1 (Fig. 18.1), glycyl-L-phenylalanine-β-napthylamide (GPN), and monensin, facilitated the understanding of lysosomes as Ca^{2+} stores. Bafilomycin A1, a V-type ATPase inhibitor, prevents the acidification of lysosomes [12, 13]; the endo-lysosomal H^+ gradient is the driving force for most transporters [14]. GPN, a basic amine, is hydrolyzed by the lysosomal enzyme cathepsin C; accumulation of GPN in acidic organelles leads to osmotic swelling and, loss of lysosomal membrane barrier [15–17]. Monensin acts by collapsing the pH gradients across acidic organelles by forming a Na^+/H^+ exchanger [18]. The Ca^{2+} uptake by the endo- lysosomes occurs via a mechanism incompletely characterized; it requires the H^+ gradient generated by the V-type H^+ ATPase pump together with Na^+/H^+, Ca^{2+}/H^+ and perhaps Na^+/Ca^{2+} exchangers whose identity is still controversial [14, 19–21].

Lysosomal Ca^{2+} stores were identified in several cell models from various species from invertebrates to mammalians: snail neurons [22], human neutrophils [23], Drosophila melanogaster S2 cell lines [24], trypanosomatids parasites [25], mouse macrophages [26], Madin-Darby canine kidney (MDCK) cell lines [16, 27], bovine corneal endothelial cells [28], pancreatic acinar and beta cells [29], rat neurons [30, 31], human aortic endothelial cells [32], and human platelets [33].

On the other hand, defective Ca^{2+} handing by lysosomes disrupts endo-lysosomal membrane trafficking and has been involved in the pathogenesis of lysosomal storage disorders such as mucolipidosis type IV, Niemann-Pick disease, Chediak–Higashi syndrome [21, 34–37], in acute pancreatitis [38, 39] and Alzheimer's disease [40].

18.2 NAADP Mobilizes Ca^{2+} from Endo-lysosomes

Nicotinic acid adenine dinucleotide phosphate (NAADP) is the most potent Ca^{2+} mobilizing second messenger [41–44]. NAADP is synthesized by the same enzyme, ADP-ribosyl cyclase and its homolog CD38, as cyclic ADP-ribose (cADPR) [45]. NAADP was first reported to produce Ca^{2+} mobilization from Ca^{2+} stores insensitive to inositol trisphosphate (IP_3) and cADPR in sea urchin eggs [46–49]. Early studies provide pharmacological and biochemical evidence that this Ca^{2+} store is the reserve granule, the functional equivalent of a lysosome in the sea urchin eggs [50].

Other studies reported that NAADP also released Ca^{2+} from acidic stores in pancreatic acinar cells [51], brain microsomes [52], pancreatic beta cells [29, 53], arterial smooth muscle cells [54], platelets [55] or astrocytes [56]. Our previous work indicates that NAADP releases Ca^{2+} from acidic lysosomal-like Ca^{2+} stores, distinct from endoplasmic reticulum, in the frog neuromuscular junction [57], intact mammalian neurons [31] or human aortic endothelial cells (HAEC) [32]. An example of NAADP-induced increased in cytosolic Ca^{2+} concentration $[Ca^{2+}]_i$, produced by intracellular microinjection of NAADP in preganglionic vagal neurons of nucleus ambiguus, is shown in Fig. 18.2.

NAADP-dependent Ca^{2+} release is characterized pharmacologically by its insensitivity to specific inhibitors of IP_3 receptors, ryanodine receptors or cADPR-mediated Ca^{2+} release [47, 49, 58, 59]. The study of NAADP-mediated Ca^{2+} release was facilitated by the synthesis of the cell-permeant form of NAADP, NAADP-acetoxymethylester (NAADP-AM) [60] and by the identification of Ned-19, a chemical probe that inhibits NAADP signaling (Fig. 18.2) while fluorescently labeling the NAADP receptors [61].

The second messenger role of NAADP was first proven in sea urchin eggs [62]; additional evidence supports NAADP as a second messenger in several other cells/tissues. In the nervous system, NAADP is a second messenger [63, 64] that modulates neurosecretion [65], neurite outgrowth [30], neuronal differentiation [66] and excitability [31]. NAADP has been involved in several other cellular functions such as T-cell activation [67], insulin secretion [68, 69], smooth muscle contraction [70, 71], fertilization [62, 72], cardiac contractility [73], platelets aggregation [55, 74], skeletal muscle differentiation [75], and angiogenesis [76].

In pancreatic secretory cells, vascular smooth muscle cells, endothelial cells, neurons, or myometrium, an increase in NAADP levels or NAADP-induced calcium response occurred during stimulation with agonists such as: cholecystokinin [29], glucagon-like peptide-1 (GLP-1) [77], insulin [78, 79], endothelin-1 [54, 80, 81], histamine [82, 83], glutamate [84], acetylcholine [32] and oxytocin [85]. More

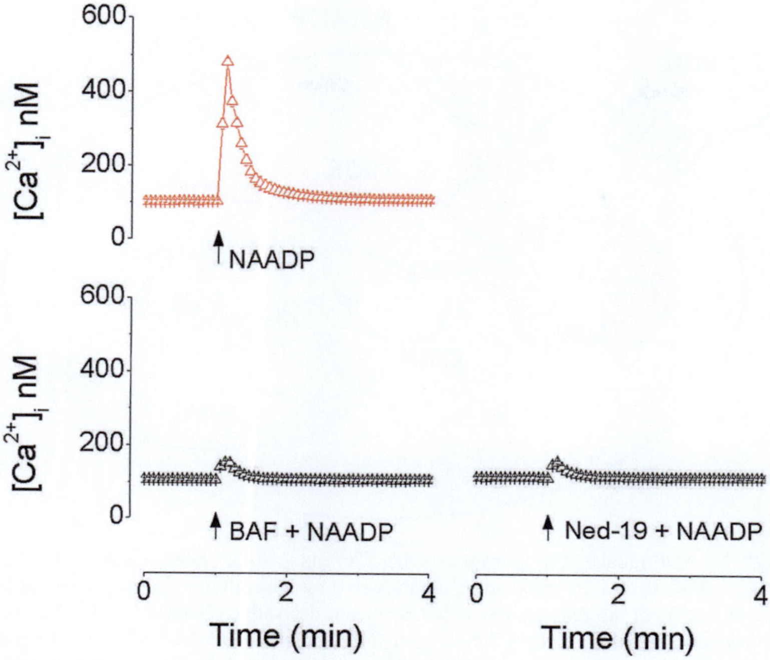

Fig. 18.2 NAADP mobilizes Ca^{2+} from acidic Ca^{2+} stores in vagal preganglionic neurons of nucleus ambiguus. Intracellular microinjection of NAADP (300 nM estimated intracellular concentration) (*red trace*) elicits a fast and transient increase in cytosolic Ca^{2+} concentration, $[Ca^{2+}]_i$. The NAADP-induced increase in $[Ca^{2+}]_i$ was abolished by bafilomycin A1 (BAF, 1 μM, 1 h), a V-type H^+ ATPase inhibitor, and by Ned-19 (1 μM, 15 min), an antagonist of NAADP signaling

recently, NAADP has been reported to be a second messenger in tracheal smooth muscle [86].

Intriguingly, in some cellular models, NAADP appears to have different roles depending on the agonist generating this second messenger. For example, in pancreatic β-cells, NAADP produced by activation of the cells with insulin modulates insulin synthesis and β-cells proliferation, while GLP-1-induced NAADP production potentiates insulin secretion and glucose-mediated Ca^{2+} signaling [77–79].

18.3 Functional Coupling of Acidic Ca^{2+} Stores with Other Intracellular Ca^{2+} Stores

One remarkable feature of the intracellular Ca^{2+} stores is that despite their structural segregation, they are functionally interacting [9, 87, 88]. Interactions between different Ca^{2+} stores and channels shape the Ca^{2+} signal and contribute to the diversity of cellular responses [89].

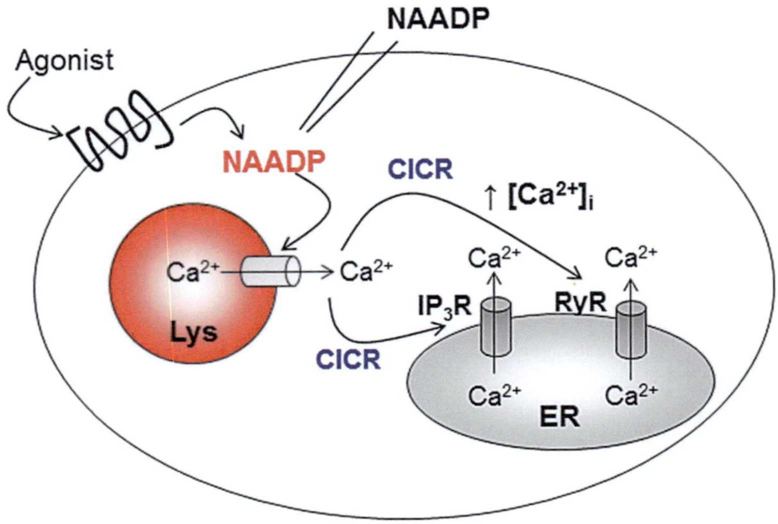

Fig. 18.3 NAADP-induced Ca^{2+} release activates Ca^{2+} induced Ca^{2+} release (*CICR*). NAADP produced in response to agonist stimulation, administered by intracellular microinjection, photolysis of caged NAADP, or administered as a cell permeant derivative (NAADP-AM), activates Ca^{2+} release from endo-lysosomes (*Lys*). This trigger Ca^{2+} is amplified by Ca^{2+} release from endoplasmic reticulum (*ER*) via inositol 1,4,5-trisphosphate receptors (*IP_3R*) or via ryanodine receptors (*RyR*) through a CICR mechanism

An early identified characteristic of NAADP-induced lysosomal Ca^{2+} release was its association with the recruitment of inositol trisphosphate and/or ryanodine receptors via Ca^{2+}-induced Ca^{2+} release (CICR) [30, 51, 54, 70, 87, 90–92]. Through this mechanism, a small, local release of Ca^{2+} from lysosomes is amplified via release of Ca^{2+} from ER leading to a global increase in Ca^{2+}. NAADP is thus thought to act as trigger during agonist-evoked Ca^{2+} signaling; the Ca^{2+} signal is then amplified by Ca^{2+}-sensitive Ca^{2+} release channels on the ER [90, 93, 94].

Early studies demonstrate that the lysosomal Ca^{2+} pool is functionally coupled with the IP_3-sensitive Ca^{2+} pool in MDCK cells [27]. At the frog neuromuscular junction, NAADP increases neurosecretion by releasing Ca^{2+} from an internal store via a two-pool model [65], supporting the functional coupling between the NAADP-sensitive store and ER stores [57]. In some cellular models that express both IP_3R and RyR, such as vascular smooth muscle [54] or SKBR3 breast cancer cells [95], Ca^{2+} release from acidic stores couples selectively to RyR. Conversely, in pancreatic acinar cells [51], starfish oocytes [92] and sea urchin eggs [90], NAADP appears to recruit both IP_3R and RyR; a diagram summarizing these mechanisms is illustrated in Fig. 18.3.

Recently, total internal reflection fluorescence (TIRF) microscopy revealed that lysosomes were closely associated with the ER, allowing them also to accumulate Ca^{2+} released by IP_3 receptors [17]. Furthermore, Ca^{2+} released from the ER may activate the NAADP pathway by stimulating Ca^{2+}-dependent NAADP synthesis and by activating NAADP-regulated channels, supporting a bidirectional Ca^{2+} signaling between the ER and acidic organelles [96].

Intriguingly, the three main Ca^{2+}-mobilizing second messengers, IP_3, cADPR and NAADP were ascribed different functions in various systems, such as invertebrate Ascidian oocytes [97], mammalian pancreatic cells [51] or mouse smooth muscle [71]. A messenger-specific role of NAADP was identified in neuronal differentiation [66] neurite outgrowth [30] and endothelial function [32].

The differential recruitment of intracellular Ca^{2+}-mobilizing messengers and their target Ca^{2+} stores contribute to the specificity and versatility of Ca^{2+} signaling [98]. The functional coupling ("chatter") [99] between NAADP, IP_3, and cADPR has been involved in several Ca^{2+}-dependent events, including fertilization [62, 97], glucose sensing [29, 69, 77] or smooth muscle contraction [71].

18.4 Ca^{2+} Release Channels in the Endo-lysosomal System

Despite the early studies indicating that NAADP releases Ca^{2+} from endo-lysosomes, the molecular identity of endo-lysosomal Ca^{2+} release channels targeted by NAADP remained elusive for several years [100, 101]. Candidates NAADP receptors included RyRs [102, 103] and endo-lysosomal transient receptor potential mucolipin 1(TRPML1) channels [104]. In 2009, members of the two-pore channel (TPC) family, namely TPC1 [95] and TPC2 [105] or TPCN2 [106] were identified to mediate NAADP-evoked Ca^{2+} release from acidic organelles and thus, considered the long-sought after target channels for NAADP [14, 101, 107].

The role of TPCs as a target for NAADP is supported by several findings: overexpression of TPC1 or TPC2 increased the Ca^{2+} response to NAADP [56, 95, 107], and increased [^{32}P]NAADP binding [105]; while knocking down the TPCs expression abolished the response to NAADP [95, 105, 108]. Moreover, the pharmacology of TPCs [95, 105] was in agreement to that reported in earlier studies for the "NAADP receptors" [96, 101, 104, 109–111].

In addition, activation of NAADP-induced currents was identified in plasma membrane patches obtained from cells expressing plasma membrane-targeted TPC2 [112] and in bilayers reconstituted with TPC1 [113] or TPC2 [114–116]. Furthermore, the response to NAADP was abolished in cells from TPC2 knockout mice [71, 105].

Genomic analysis indicated that the animal kingdom express up to three family members (TPC1–3) [95, 105]. The three TPCN genes (*TPCN1*, *TPCN2* and *TPCN3*) are expressed in most vertebrate species [107, 117]. All three TPC isoforms were cloned from the sea urchin, while TPC3 is a pseudogene in humans and other primates [107].

TPC2 is expressed primarily in lysosomes, while TPC1 and TPC3 appear also to be expressed in endosomes [95, 105, 118, 119]. The restricted subcellular distribution of TPC1 compared to TPC2 leads to NAADP-induced localized Ca^{2+} transients in TPC1-expressing cells and global Ca^{2+} increases in cells overexpressing TPC2 [105].

In addition to TPCs, another distinct family of Ca^{2+} permeable channels, namely the transient receptor potential mucolipin (TRPML) channels are expressed in endo-lysosomes [120]. A segregated distribution has been reported: TRPML3, similarly to TPC1, are present mainly in endosomes, while TRPML1, similarly to TPC2 are expressed in lysosomes [14, 101, 120, 121]. Despite the fact that TPCs and TRPMLs are expressed in the same complex, they are functionally independent [120]. TRPMLs act as phosphatidylinositol 3,5-bisphosphate [PI(3,5)P2]-activated non-selective cation channels [34, 122], while TPCs were proposed as NAADP-gated channels [95, 105, 106]. Several aspects such as the functional relationships between TPCs and TRPMLs remain to be explored [14].

The role of TPCs as NAADP-regulated Ca^{2+} channels has been recently challenged [123, 124] leading to some controversies [101]. One study [124] suggested that TPCs are Na^{+}-selective channels regulated by PI(3,5)P2 and by ATP and not by NAADP. Another study [123] found that TPCs couple the cell's metabolic state to endo-lysosomal function.

However, an elegant study [19] has recently contributed to the reconciliation of the findings, by demonstrating that TPC2 is activated by NAADP and essential for NAADP-mediated Ca^{2+} release. The study confirmed the activation of TPC2 by PI(3,5)P2 and its permeability to Na^{+}, while identifying Mg^{2+}, MAPKs, JNK and P38 as novel regulators of TPC2 [19]. Changes in cytosolic Mg^{2+} associated with receptor activation and with changes in cytosolic ATP are sensed by TPC2 and affect the lysosomal membrane potential [19]. On the other hand, mTORC1, MAPKs, JNK and P38 kinases inhibit TPC2 [19].

While, several studies demonstrate indubitably that NAADP activates TPCs [19, 95, 105–107, 113–116], the actual site of NAADP binding remained unclear. Photoaffinity labeling studies indicate that binding of NAADP occurs at an accessory component within a larger TPC complex [125].

18.5 TPCs Regulation and Roles

TPCs are a family of intracellular ion channels, residing in the membranes of acidic organelles [10], initially cloned from rat kidney [126]. The TPC family is a member of voltage-gated superfamily of cation channels, resembling ancient voltage-gated Na^{+} (Na_V) and Ca^{2+} channels (Ca_V) [127, 128]. Unlike Ca_Vs and Na_Vs channels that are composed of four homologous domains, TPCs consist of two-linked six transmembrane domains [128]. TPCs assemble as homo- and heterodimers through differential interactions between transmembrane regions [128–130].

Biophysical properties such as single-channel gating and conductance properties of purified human TPC2 have been recently reported [114]. The luminal Ca^{2+} concentration and the luminal pH control the sensitivity of TPCs to NAADP: low NAADP levels and low pH activate TPC2, while high NAADP levels and neutral pH inhibit the channel [114] Interestingly, Ned-19, an inhibitor of NAADP signaling [61], in the absence of NAADP and at high concentration (1 μM), closes TPC2 while at lower concentration (100 nM) potentiates the effect of NAADP (10 nM) by binding at a site distinct from NAADP [114].

To explore the evolution of TPCs, a TPC1 model was recently constructed using the crystal structure of a bacterial Na_V that only has one, instead of four, of the repeated domains [128]. Molecular docking analysis indicated that Ca_Vs or Na_Vs antagonists bind the pore region of TPC1 through common sites [128]. In addition, in functional, calcium imaging studies, the Ca_Vs or Na_Vs antagonists blocked the Ca^{2+} signals elicited by NAADP in TPC1-expressing cells [128].

TPCs play an important role in vesicular fusion and trafficking in endo-lysosomes, by promoting the local Ca^{2+} release [91, 119, 131, 132]. Emerging evidence support the role of TPC2 in endo-lysosomal trafficking [133, 134]. Fibroblasts and hepatocytes from TPC2 KO mice present a profound impairment of LDL- cholesterol and of EGF-receptor trafficking [133]. Additional, morphological and functional changes were identified in lysosomal fibroblasts from patients with Parkinson disease with a common mutation in LRRK2; the changes could be reversed by inhibition of TPC2 indicating that TPC2 acts downstream of pathogenic LRRK2 to regulate endo-lysosomal trafficking [134].

TPCs also trigger a global increase in cytosolic Ca^{2+} concentration and contribute to the amplification of a local Ca^{2+} signal by recruiting CICR at the lysosomal-ER junction [91, 131]. To dissociate the role of trigger of NAADP-TPCs from that of amplifier played by the ER, we examined the ability of NAADP to evoke Ca^{2+} signals in cells expressing plasma membrane-targeted TPC2 [112]. Redirecting TPCs to the plasma membrane produced a dissociation of NAADP-induced Ca^{2+} release from the ER-dependent Ca^{2+} release [112], indicating the requirement for TPCs targeting to the endo-lysosomal system. Another role described for TPCs is to regulate the cellular excitability via Ca^{2+} release from endo-lysosomal stores from the close proximity of the plasma membrane and the consequent activation of Ca^{2+}-activated channels at the plasma membrane [10, 78, 131, 132].

TPCs interact with many proteins with critical roles in Ca^{2+} homeostasis, trafficking, and membrane organization leading to important isoform-specific roles in organelle proliferation and cellular pigmentation [135]. Both TPC1 and TPC2 interact with several Rab isoforms, however, pigmentation phenotype required TPC2, but not TPC1 activity [135]. This is in agreement with the results of a genetic screen that identified coding variants of human TPC2 that are associated with hair color [136]. Similar to some other channels, TPCs are regulated by N-glycosylation [91]. Glycosylation serves to inhibit TPC function, as mutation of the N-glycosylation sites in TPC1 enhances NAADP-evoked Ca^{2+} release [137].

NAADP via TPC2-mediated Ca^{2+} release has been involved in angiogenesis in response to vascular endothelial growth factor (VEGF) [76]. In skeletal muscle C2C12 cells, that express both TPC1 and TPC2, NAADP promotes skeletal muscle differentiation; the effect was inhibited by NAADP antagonist, Ned-19 [61] or by the down-regulation of TPC1 and TPC2 [75]. TPCs have been proposed to contribute to autophagy signaling and protein homeostasis in skeletal muscle via mTOR pathway [123, 138].

In the megakaryoblastic cell line MEG01, expressing TPC1 and TPC2, NAADP releases Ca^{2+} from acidic lysosomal-like stores that promotes subsequent activation of ryanodine-sensitive Ca^{2+} release and amplification of Ca^{2+} signals via a CICR mechanism [108]. NAADP-induced Ca^{2+} release was impaired by TPC1 and TPC2 expression silencing, while TPC1 and TPC2 overexpression produced opposite results [108].

TPC2 has a critical role in coupling of the muscarinic receptors with Ca^{2+} release from acidic stores in mouse detrusor muscle [71]. NAADP-evoked contractions are abolished in detrusor muscle from TPC2 KO mice indicating that TPC2 is required for NAADP-mediated contractility. TPC2 contributes, in addition to IP_3R and RyR, to the response to the agonist carbachol [71]. Moreover, tissues from TPC2 KO mice have a reduced sensitivity in the IP_3-induced contraction, indicating a cross-talk between acidic stores and ER Ca^{2+} stores [71].

The endo-lysosomes are dynamic organelles and the TPCs are highly mobile proteins [139], facilitating the interactions with ER Ca^{2+} channels [140]. Emerging evidence indicates intimate functional and physical coupling between acidic stores and ER Ca^{2+} stores [140, 141]; membrane contact sites were identified between the ER and late endosomes [142], multivesicular bodies [143] or lysosomes [141]. The membrane contact sites between acidic stores and the ER are similar to those between the ER and other organelles, supporting local Ca^{2+} microdomain signaling [140, 141]. These findings support previous reports that NAADP signaling occurred at discrete “trigger” zones [54].

A recent computational model of the microdomains between lysosomes and ER suggests that the Ca^{2+} oscillations in the microdomains may produce global Ca^{2+} signals even in the absence of high microdomains Ca^{2+} concentrations [144, 145]. Moreover, the IP_3R-dependent global Ca^{2+} signals can be regulated by the distribution and density of TPCs [144, 145].

18.6 Ca^{2+} Release from Endo-lysosomes and Ca^{2+} Entry

In non-excitable cells, depletion of intracellular Ca^{2+} stores by IP_3 leads to store-operated calcium entry (SOCE) or capacitative calcium entry (CCE) through Ca^{2+} release-activated Ca^{2+} (CRAC) channels [146, 147]. There is limited information whether or not release of Ca^{2+} from endo-lysosomes leads to Ca^{2+} influx. Earlier studies indicate that depletion of lysosomal Ca^{2+} stores with GPN induced Ca^{2+} influx via CCE in bovine corneal endothelium [28]. On the other hand, depletion of acidic stores by fatty acids in amoeba [148] induced CCE.

A number of non-selective Ca^{2+}-permeable channels with different biophysical properties has been considered to carry the SOCE current, I_{CRAC} [147, 149–151]. The protein Orai1 (also named CRACM1 for CRAC modulator 1) has been proposed to form the pore of the channel mediating I_{CRAC} [152–154]. The activation of SOCE involves the participation of the stromal interaction molecule-1 (STIM1), an endoplasmic reticulum (ER) Ca^{2+} sensor [155, 156]. STIM-1 is present in the membranes of the ER [157] and acidic Ca^{2+} stores [158]. When the intraluminal Ca^{2+} concentration decreases, STIM1 redistributes to ER-plasma membrane sites where the Orai1 subunit of the CRAC channel accumulates in the plasma membrane and CRAC channels open, activating SOCE [159–161].

The transient receptor potential (TRP) channels, particularly members of the TRP canonical (TRPC) subfamily have been involved in Ca^{2+} entry during SOCE [162]. In addition to interacting with Orai1, emerging evidence indicates that STIM1 also interact with TRPC channels [163]. Unlike STIM1, STIM2, a closely related protein, is considered a feedback regulator that inhibits SOCE [164].

Human platelets are endowed with two distinct Ca^{2+} stores: the dense tubular system (DTS) and acidic organelles (Fig. 18.4). Two mechanisms of SOCE, regulated by each Ca^{2+} compartment, were identified [33, 55, 165–167]. Human platelets express SERCA2b in the DTS; it has high sensitivity to thapsigargin, while it is insensitive to 2,5-di-(tert-butyl)-1,4-hydroquinone (TBHQ) [168]. Conversely, the platelet acidic Ca^{2+} stores express different isoforms of SERCA3 Ca^{2+} pump, sensitive to TBHQ and with low affinity for thapsigargin [74, 165, 169, 170]. In addition, the acidic stores express the vacuolar H^+-ATPase [165, 170].

In platelets, both mechanisms of SOCE are modulated by the actin cytoskeleton and microtubules [171]. Tubulin-cytoskeleton plays a dual role in SOCE: it facili-

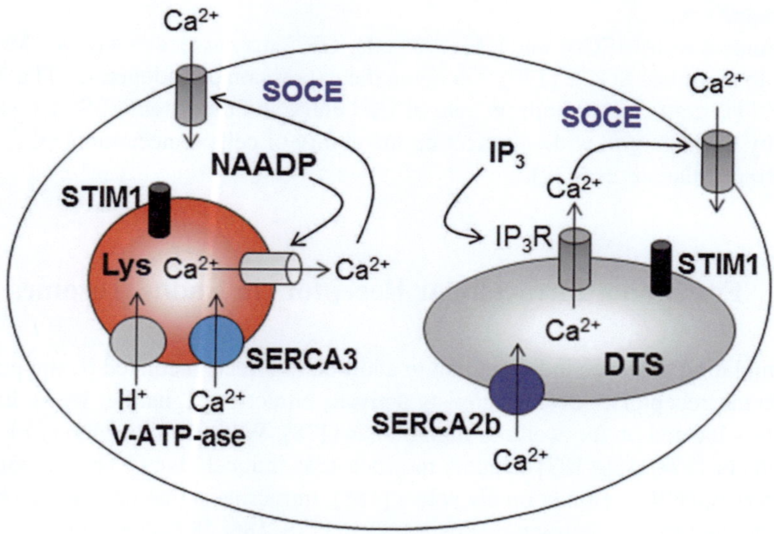

Fig. 18.4 Acidic stores-induced SOCE pathways in human platelets. In human platelets, SOCE may be induced by two mechanisms: by depletion of the acidic Ca^{2+} stores (*Lys*) or by depletion of the dense tubular system (*DTS*) [33]. NAADP releases Ca^{2+} from acidic stores; depletion of acidic organelles stimulates Ca^{2+} entry via plasma membrane Ca^{2+}-permeable channels via a SOCE mechanism. Depletion of DTS via IP_3 acting on IP_3R activates SOCE. STIM1 located on the membrane of acidic organelles or DTS, interacts with Orai 1 channels or with TRPC channels on plasma membrane, promoting the Ca^{2+} influx. SOCE is modulated by the actin cytoskeleton and microtubules. Acidic stores express SERCA3 and vacuolar H^+-ATPase, and DTS express SERCA2b pump [33, 167]

tates SOCE mediated by depletion of the acidic stores, while acting as a barrier that prevents constitutive SOCE regulated by DTS [172].

SOCE may be modulated by the interaction between the IP_3R type II situated in the membrane of DTS and the TRPC channels in the plasma membrane [173, 174]. Moreover, selective depletion of the intracellular Ca^{2+} stores in human platelets, induces the formation of heteromeric complexes between hTRPC1 and hTRPC6 and the interaction of both hTRPC1 and hTRPC6 with the IP_3R type II, SERCA2b and SERCA3 [171].

NAADP has been shown to release Ca^{2+} in human platelets from TBHQ-sensitive stores [55, 74], and NAADP-binding sites was demonstrated using a radioreceptor-binding assay [175]. Proposed mechanisms involved in SOCE activated by the depletion of acidic Ca^{2+} stores include the translocation of STIM1 and STIM2 and their association with Orai1 as well association of STIM1 with TRPC1 and TRPC6 [158, 176]. SOCE controlled by the acidic stores has been shown to be regulated by phosphatidylinositol 4,5-bisphosphate (PIP_2) [33].

In mouse pancreatic β-cells, NAADP, in addition to releasing Ca^{2+} via TPCs, evoked oscillatory non-selective cation currents; the effect was abolished by Ned-19 or in cells from TPC2 knockout mice [61, 105]. The underlying mechanism coupling the NAADP-induced Ca^{2+} release from acidic stores with the membrane depolarization was considered the activation of TRPM4/5 channels at the plasma membrane [78].

In human cells MEG01 and HEK293 cells, TPC2 may associate with STIM1 and Orai1 to modulate SOCE [170]. Silencing the expression of endogenous TPC2, but not of TPC1, attenuated both the rate of Ca^{2+} entry and the extent of SOCE stimulated by thapsigargin, without affecting the ability of cells to accumulate Ca^{2+} into the intracellular stores [177].

18.7 Functional Intracellular Receptors in Endo-lysosomes

Accumulating evidence indicates that in addition to effects mediated by the plasma membrane receptors, several agonists activate intracrine signaling, by acting on receptors located on intracellular membranes [178]. Work from our lab [179–184] and others [178, 185–187] support the idea that intracellular G protein-coupled receptors (GPCR) are functionally active [188]. Intracrine signaling has been better characterized for the renin-angiotensin system [178, 186, 187, 189, 190].

Intracrine signaling plays important roles in the physiology and pharmacology of GPCR ligands and expands the diversity and complexity of signaling mechanisms [178, 186]. Several lines of evidence, such as the presence of heterotrimeric G proteins in endoplasmic reticulum, Golgi, endo-lysosomes and nucleus, indicate that classical receptor signaling occurs at intracellular membranes [191–195]. The enzymes involved in NAADP and cADPR synthesis are also present at the membrane of acidic organelles [196, 197] and the release of second messengers from organelles contributes to Ca^{2+} signaling [198].

While receptor endocytosis is generally seen as distinct from cell signaling [199], relatively recent, endosomes-localized GPCRs were demonstrated to be present in the active form, supporting the view that canonical and non-canonical GPCR signaling occurs from endosomes, in addition to the plasma membrane [200–203]. For example, internalized beta-adrenoreceptors, thyroid stimulating hormone (TSH) or parathyroid hormone (PTH) receptors can initiate Gs-mediated signal transduction and consequent increase in cAMP from endosomes, in addition to the signal initiated from the plasma membrane [200, 202, 203]. Moreover, internalized protease-activated receptors (PARs), GPCRs that are activated by proteolysis, signal from endosomes through the recruitment of β-arrestins [201, 204]. Receptor signaling from endosomes is not surprising, since the endo-lysosomal membranes are organized into specialized domains where molecules can assemble into specific signaling complexes [199, 205].

Furthermore, peptides such angiotensin II [182, 184] or endothelin-1 [183], can be readily transferred from the cytosol inside the endo-lysosomal system via microautophagy [206]. In excitable cells, like neurons [184] or myometrium [182], as well as in U2OS cells transfected with AT_1 receptor [184], we identified that angiotensin II activates AT_1 receptors located on endo-lysosomes and activates phospholipase C (PLC) at the lysosomal membrane. The subsequent production of IP_3 activates IP_3R followed by the Ca^{2+} release from the ER. In neurons, the local Ca^{2+} release activates TRPC and produces membrane depolarization [184], while in myometrial cells, the Ca^{2+} response is further augmented by a CICR mechanism via ryanodine receptors activation [182].

Similarly, in endothelial cells that express endothelin-1 receptors type B (ET_B), or in cells transfected with ET_B, endothelin-1 (ET-1) activates ET_B receptors on endo-lysosomes, followed by activation of PLC, release of IP_3 and activation of IP_3R from the ER. The subsequent IP_3-induced increase in cytosolic Ca^{2+} activates endothelial nitric oxide synthase to produce nitric oxide [183].

The emerging paradigm of functional intracellular GPCRs is particularly significant in the case of lipid messengers that are generated intracellularly "on-demand" [207–209]. We examined the distribution and function of CB_1 and CB_2 cannabinoid receptors, and of the atypical cannabinoid receptors GPR55, GPCRs activated by lipids [180, 181, 210]. We monitored the cytosolic Ca^{2+} concentration during intracellular microinjection and extracellular application of the receptor agonists and identified distinct signaling mechanisms initiated by the intracellular versus plasma membrane receptors [180, 181, 210].

In endothelial cells, we provided functional and morphological evidence that CB_2 receptors are localized, in addition to the plasma membrane, at the endo-lysosomes, and their activation mobilizes NAADP-sensitive acidic-like Ca^{2+} stores and IP_3-sensitive Ca^{2+} stores [181]. Our results support the functionality of intracellular CB_2 receptors and their ability to couple to Gq proteins and to elicit Ca^{2+} signaling [181]. In CB_1-transfected cells, anandamide, acting on CB_1 receptors situated on acid-filled Ca^{2+} stores, activates IP_3-dependent and NAADP-dependent Ca^{2+} pathways [180]. Similarly, activation of intracellular GPR55 by L-α-lysophosphatidylinositol (LPI) elicited NAADP-dependent Ca^{2+} mobilization from

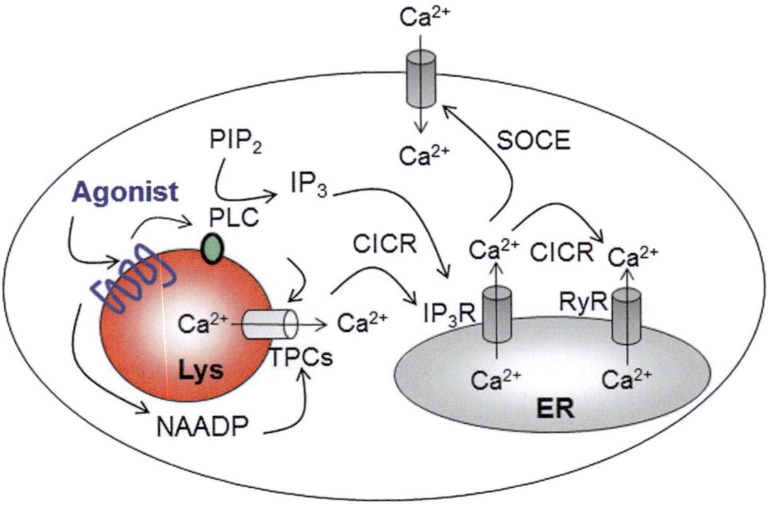

Fig. 18.5 Proposed mode of action of agonists via activation of intracellular receptors. Agonists such as angiotensin II [182, 184], endothelin-1 [183], anandamide [180], 2-arachidonoyl glycerol (2-AG) [181], and L-α-lysophosphatidylinositol (*LPI*) [210], act on receptors (AT_1, ET_B, CB_1, CB_2, and GPR55, respectively) situated on acid-filled Ca^{2+} stores (endo-lysosomes, *Lys*). Receptor activation may lead to the activation of lysosomal phospholipase C (*PLC*), and generation of IP_3 from phosphatidylinositol 4,5-bisphosphate (*PIP_2*), or to NAADP production. IP_3 releases Ca^{2+} via IP_3 receptors (*IP_3R*) situated on endoplasmic reticulum (*ER*). NAADP releases Ca^{2+} from endo-lysosomes via TPCs. IP_3- or NAADP-induced increase in $[Ca^{2+}]_i$ may activate ryanodine receptors (*RyR*) and further potentiates Ca^{2+} release from the ER via a Ca^{2+}-induced Ca^{2+} release (*CICR*) mechanism. Depletion of Ca^{2+} from organelles may also activate the store-operated Ca^{2+} entry (*SOCE*) mechanism and promote the Ca^{2+} influx via plasma membrane Ca^{2+} channels

acidic-like Ca^{2+} stores, mechanisms distinct from those elicited by the activation of plasma membrane GPR55 [210]. A diagram summarizing mechanisms elicited by activation of endo-lysosomal GPCRs is presented in Fig. 18.5.

18.8 Conclusions

Emerging evidence supports endo-lysosomes as intracellular Ca^{2+} stores targeted by NAADP, the most potent Ca^{2+}-mobilizing second messenger. NAADP, produced in response to several agonists, elicits Ca^{2+} release via TPCs, and modulates a variety of cellular functions such as fertilization, neurosecretion, contraction and differentiation. Ca^{2+} release from endo-lysosomes is amplified by Ca^{2+} release from the ER; depletion of acidic Ca^{2+} stores, similar to the ER, elicits store-operated Ca^{2+} entry. The physiological significance of the Ca^{2+} signaling originating from endo-lysosomes and the pathophysiological consequences of defective Ca^{2+} handling at this level are just beginning to be understood.

Acknowledgements We apologize to the many researchers whose work we were not able to credit due to space limitations.

References

1. De Duve C, Gianetto R, Appelmans F, Wattiaux R (1953) Enzymic content of the mitochondria fraction. Nature 172(4390):1143–1144
2. Gianetto R, De Duve C (1955) Tissue fractionation studies. 4. Comparative study of the binding of acid phosphatase, beta-glucuronidase and cathepsin by rat-liver particles. Biochem J 59(3):433–438
3. Luzio JP, Pryor PR, Bright NA (2007) Lysosomes: fusion and function. Nat Rev Mol Cell Biol 8(8):622–632
4. Ciechanover A (2005) Proteolysis: from the lysosome to ubiquitin and the proteasome. Nat Rev Mol Cell Biol 6(1):79–87
5. Hershko A, Heller H, Elias S, Ciechanover A (1983) Components of ubiquitin-protein ligase system. Resolution, affinity purification, and role in protein breakdown. J Biol Chem 258(13):8206–8214
6. Blaustein MP, Golovina VA (2001) Structural complexity and functional diversity of endoplasmic reticulum Ca(2+) stores. Trends Neurosci 24(10):602–608
7. Berridge MJ (2002) The endoplasmic reticulum: a multifunctional signaling organelle. Cell Calcium 32(5–6):235–249
8. Clapham DE (2007) Calcium signaling. Cell 131(6):1047–1058
9. Lee HC (2000) Multiple calcium stores: separate but interacting. Sci STKE 2000(40):pe1
10. Patel S, Muallem S (2011) Acidic Ca(2+) stores come to the fore. Cell Calcium 50(2):109–112
11. Bucci C, Thomsen P, Nicoziani P, McCarthy J, van Deurs B (2000) Rab7: a key to lysosome biogenesis. Mol Biol Cell 11(2):467–480
12. Drose S, Altendorf K (1997) Bafilomycins and concanamycins as inhibitors of V-ATPases and P-ATPases. J Exp Biol 200(Pt1):1–8
13. Yoshimori T, Yamamoto A, Moriyama Y, Futai M, Tashiro Y (1991) Bafilomycin A1, a specific inhibitor of vacuolar-type H(+)-ATPase, inhibits acidification and protein degradation in lysosomes of cultured cells. J Biol Chem 266(26):17707–17712
14. Jha A, Brailoiu E, Muallem S (2014) How does NAADP release lysosomal Ca(2+)? Channels 8(3):174–175
15. Berg TO, Stromhaug E, Lovdal T, Seglen O, Berg T (1994) Use of glycyl-L-phenylalanine 2-naphthylamide, a lysosome-disrupting cathepsin C substrate, to distinguish between lysosomes and prelysosomal endocytic vacuoles. Biochem J 300(Pt 1):229–236
16. Haller T, Dietl P, Deetjen P, Volkl H (1996) The lysosomal compartment as intracellular calcium store in MDCK cells: a possible involvement in InsP3-mediated Ca2+ release. Cell Calcium 19(2):157–165
17. Lopez-Sanjurjo CI, Tovey SC, Prole DL, Taylor CW (2013) Lysosomes shape Ins(1,4,5)P3-evoked Ca2+ signals by selectively sequestering Ca2+ released from the endoplasmic reticulum. J Cell Sci 126(Pt1):289–300
18. Vercesi AE, Docampo R (1996) Sodium-proton exchange stimulates Ca2+ release from acidocalcisomes of Trypanosoma brucei. Biochem J 315(Pt 1):265–270
19. Jha A, Ahuja M, Patel S, Brailoiu E, Muallem S (2014) Convergent regulation of the lysosomal two-pore channel-2 by Mg(2)(+), NAADP, PI(3,5)P(2) and multiple protein kinases. EMBO J 33(5):501–511
20. Lemons RM, Thoene JG (1991) Mediated calcium transport by isolated human fibroblast lysosomes. J Biol Chem 266(22):14378–14382

21. Morgan AJ, Platt FM, Lloyd-Evans E, Galione A (2011) Molecular mechanisms of endolysosomal Ca2+ signalling in health and disease. Biochem J 439(3):349–374
22. Sugaya E, Onozuka SE (1978) Intracellular calcium: its release from granules during bursting activity in snail neurons. Science 202(4373):1195–1197
23. Styrt B, Schwartz MA, Klempner MS (1988) Calcium release associated with discharge of specific granule contents from human neutrophils. Biochem Biophys Res Commun 154(3):1308–1313
24. Yagodin S, Pivovarova NB, Andrews SB, Sattelle DB (1999) Functional characterization of thapsigargin and agonist-insensitive acidic Ca2+ stores in Drosophila melanogaster S2 cell lines. Cell Calcium 25(6):429–438
25. Docampo R, Moreno SN (1999) Acidocalcisome: a novel Ca2+ storage compartment in trypanosomatids and apicomplexan parasites. Parasitol Today 15(11):443–448
26. Christensen KA, Myers JT, Swanson JA (2002) pH-dependent regulation of lysosomal calcium in macrophages. J Cell Sci 115(Pt3):599–607
27. Haller T, Volkl H, Deetjen P, Dietl P (1996) The lysosomal Ca2+ pool in MDCK cells can be released by ins(1,4,5)P3-dependent hormones or thapsigargin but does not activate store-operated Ca2+ entry. Biochem J 319(Pt 3):909–912
28. Srinivas SP, Ong A, Goon L, Goon L, Bonanno JA (2002) Lysosomal Ca(2+) stores in bovine corneal endothelium. Invest Ophthalmol Vis Sci 43(7):2341–2350
29. Yamasaki M, Masgrau R, Morgan AJ, Churchill GC, Patel S, Ashcroft SJ, Galione A (2004) Organelle selection determines agonist-specific Ca2+ signals in pancreatic acinar and beta cells. J Biol Chem 279(8):7234–7240
30. Brailoiu E, Hoard JL, Filipeanu CM, Brailoiu GC, Dun SL, Patel S, Dun NJ (2005) Nicotinic acid adenine dinucleotide phosphate potentiates neurite outgrowth. J Biol Chem 280(7):5646–5650
31. Brailoiu GC, Brailoiu E, Parkesh R, Galione A, Churchill GC, Patel S, Dun NJ (2009) NAADP-mediated channel 'chatter' in neurons of the rat medulla oblongata. Biochem J 419(1):91–97
32. Brailoiu GC, Gurzu B, Gao X, Parkesh R, Aley PK, Trifa DI, Galione A, Dun NJ, Madesh M, Patel S, Churchill GC, Brailoiu E (2010) Acidic NAADP-sensitive calcium stores in the endothelium: agonist-specific recruitment and role in regulating blood pressure. J Biol Chem 285(48):37133–37137
33. Rosado JA (2011) Acidic Ca(2+) stores in platelets. Cell Calcium 50(2):168–174
34. Dong XP, Shen D, Wang X, Dawson T, Li X, Zhang Q, Cheng X, Zhang Y, Weisman LS, Delling M, Xu H (2010) PI(3,5)P(2) controls membrane trafficking by direct activation of mucolipin Ca(2+) release channels in the endolysosome. Nat Commun 1:38
35. Kiselyov K, Yamaguchi S, Lyons CW, Muallem S (2010) Aberrant Ca2+ handling in lysosomal storage disorders. Cell Calcium 47(2):103–111
36. Lloyd-Evans E, Morgan AJ, He X, Smith DA, Elliot-Smith E, Sillence DJ, Churchill GC, Schuchman EH, Galione A, Platt FM (2008) Niemann-pick disease type C1 is a sphingosine storage disease that causes deregulation of lysosomal calcium. Nat Med 14(11):1247–1255
37. Lloyd-Evans E, Platt FM (2011) Lysosomal Ca(2+) homeostasis: role in pathogenesis of lysosomal storage diseases. Cell Calcium 50(2):200–205
38. Petersen OH, Gerasimenko OV, Tepikin AV, Gerasimenko JV (2011) Aberrant Ca(2+) signalling through acidic calcium stores in pancreatic acinar cells. Cell Calcium 50(2):193–199
39. Gerasimenko J, Peng S, Gerasimenko O (2014) Role of acidic stores in secretory epithelia. Cell Calcium 55(6):346–354
40. Coen K, Flannagan RS, Baron S, Carraro-Lacroix LR, Wang D, Vermeire W, Michiels C, Munck S, Baert V, Sugita S, Wuytack F, Hiesinger PR, Grinstein S, Annaert W (2012) Lysosomal calcium homeostasis defects, not proton pump defects, cause endo-lysosomal dysfunction in PSEN-deficient cells. J Cell Biol 198(1):23–35

41. Cancela JM, Charpentier G, Petersen OH (2003) Co-ordination of Ca(2+) signalling in mammalian cells by the new Ca(2+)-releasing messenger NAADP. Pflugers Arch 446(3):322–327
42. Lee HC (2005) Nicotinic acid adenine dinucleotide phosphate (NAADP)-mediated calcium signaling. J Biol Chem 280(40):33693–33696
43. Galione A (2006) NAADP, a new intracellular messenger that mobilizes Ca2+ from acidic stores. Biochem Soc Trans 34(Pt 5):922–926
44. Guse AH, Ernst IM, Fliegert R (2013) NAADP signaling revisited. Curr Top Med Chem 13(23):2978–2990
45. Lee HC, Aarhus R (1991) ADP-ribosyl cyclase: an enzyme that cyclizes NAD+ into a calcium-mobilizing metabolite. Cell Regul 2(3):203–209
46. Berridge G, Dickinson G, Parrington J, Galione A, Patel S (2002) Solubilization of receptors for the novel Ca2+ -mobilizing messenger, nicotinic acid adenine dinucleotide phosphate. J Biol Chem 277(46):43717–43723
47. Chini EN, Beers KW, Dousa TP (1995) Nicotinate adenine dinucleotide phosphate (NAADP) triggers a specific calcium release system in sea urchin eggs. J Biol Chem 270(7):3216–3223
48. Clapper DL, Walseth TF, Dargie PJ, Lee HC (1987) Pyridine nucleotide metabolites stimulate calcium release from sea urchin egg microsomes desensitized to inositol trisphosphate. J Biol Chem 262(20):9561–9568
49. Lee HC, Aarhus R (1995) A derivative of NADP mobilizes calcium stores insensitive to inositol trisphosphate and cyclic ADP-ribose. J Biol Chem 270:2152–2157
50. Churchill GC, Okada Y, Thomas JM, Genazzani AA, Patel S, Galione A (2002) NAADP mobilizes Ca(2+) from reserve granules, lysosome-related organelles, in sea urchin eggs. Cell 111(5):703–708
51. Cancela JM, Churchill GC, Galione A (1999) Coordination of agonist-induced Ca2+ -signalling patterns by NAADP in pancreatic acinar cells. Nature 398(6722):74–76
52. Bak J, White P, Timar G, Missiaen L, Genazzani AA, Galione A (1999) Nicotinic acid adenine dinucleotide phosphate triggers Ca2+ release from brain microsomes. Curr Biol 9(14):751–754
53. Mitchell KJ, Lai FA, Rutter GA (2003) Ryanodine receptor type I and nicotinic acid adenine dinucleotide phosphate receptors mediate Ca2+ release from insulin-containing vesicles in living pancreatic beta-cells (MIN6). J Biol Chem 278(13):11057–11064
54. Kinnear NP, Boittin FX, Thomas JM, Galione A, Evans AM (2004) Lysosome-sarcoplasmic reticulum junctions. A trigger zone for calcium signaling by nicotinic acid adenine dinucleotide phosphate and endothelin-1. J Biol Chem 279(52):54319–54326
55. Lopez JJ, Redondo PC, Salido GM, Pariente JA, Rosado JA (2006) Two distinct Ca2+ compartments show differential sensitivity to thrombin, ADP and vasopressin in human platelets. Cell Signal 18(3):373–381
56. Pereira GJ, Hirata H, Fimia GM, do Carmo LG, Bincoletto C, Han SW, Stilhano RS, Ureshino RP, Bloor-Young D, Churchill G, Piacentini M, Patel S, Smaili SS (2011) Nicotinic acid adenine dinucleotide phosphate (NAADP) regulates autophagy in cultured astrocytes. J Biol Chem 286(32):27875–27881
57. Brailoiu E, Patel S, Dun NJ (2003) Modulation of spontaneous transmitter release from the frog neuromuscular junction by interacting intracellular Ca(2+) stores: critical role for nicotinic acid-adenine dinucleotide phosphate (NAADP). Biochem J 373(Pt 2):313–318
58. Chini EN, Dousa TP (1996) Nicotinate-adenine dinucleotide phosphate-induced Ca(2+)-release does not behave as a Ca(2+)-induced Ca(2+)-release system. Biochem J 316(Pt 3):709–711
59. Genazzani AA, Galione A (1996) Nicotinic acid-adenine dinucleotide phosphate mobilizes Ca2+ from a thapsigargin-insensitive pool. Biochem J 315(Pt 3):721–725

60. Parkesh R, Lewis AM, Aley PK, Arredouani A, Rossi S, Tavares R, Vasudevan SR, Rosen D, Galione A, Dowden J, Churchill GC (2008) Cell-permeant NAADP: a novel chemical tool enabling the study of Ca2+ signalling in intact cells. Cell Calcium 43(6):531–538
61. Naylor E, Arredouani A, Vasudevan SR, Lewis AM, Parkesh R, Mizote A, Rosen D, Thomas JM, Izumi M, Ganesan A, Galione A, Churchill GC (2009) Identification of a chemical probe for NAADP by virtual screening. Nat Chem Biol 5(4):220–226
62. Churchill GC, O'neill JS, Masgrau R, Patel S, Thomas JM, Genazzani AA, Galione A (2003) Sperm deliver a new second messenger: NAADP. Curr Biol 13(2):125–128
63. Bezin S, Charpentier G, Fossier P, Cancela JM (2006) The Ca2+ -releasing messenger NAADP, a new player in the nervous system. J Physiol 99(2–3):111–118
64. Petersen OH, Cancela JM (1999) New Ca2+ -releasing messengers: are they important in the nervous system? Trends Neurosci 22(11):488–495
65. Brailoiu E, Miyamoto MD, Dun NJ (2001) Nicotinic acid adenine dinucleotide phosphate enhances quantal neurosecretion at the frog neuromuscular junction: possible action on synaptic vesicles in the releasable pool. Mol Pharmacol 60(4):718–724
66. Brailoiu E, Churamani D, Pandey V, Brailoiu GC, Tuluc F, Patel S, Dun NJ (2006) Messenger-specific role for nicotinic acid adenine dinucleotide phosphate in neuronal differentiation. J Biol Chem 281(23):15923–15928
67. Berg I, Potter BV, Mayr GW, Guse AH (2000) Nicotinic acid adenine dinucleotide phosphate (NAADP(+)) is an essential regulator of T-lymphocyte Ca(2+)-signaling. J Cell Biol 150(3):581–588
68. Johnson JD, Misler S (2002) Nicotinic acid-adenine dinucleotide phosphate-sensitive calcium stores initiate insulin signaling in human beta cells. Proc Natl Acad Sci U S A 99(22):14566–14571
69. Masgrau R, Churchill GC, Morgan AJ et al (2003) NAADP: a new second messenger for glucose-induced Ca2+ responses in clonal pancreatic beta cells. Curr Biol 13(3):247–251
70. Boittin FX, Galione A, Evans AM (2002) Nicotinic acid adenine dinucleotide phosphate mediates Ca2+ signals and contraction in arterial smooth muscle via a two-pool mechanism. Circ Res 91(12):1168–1175
71. Tugba Durlu-Kandilci N, Ruas M, Chuang KT, Brading A, Parrington J, Galione A (2010) TPC2 proteins mediate nicotinic acid adenine dinucleotide phosphate (NAADP)- and agonist-evoked contractions of smooth muscle. J Biol Chem 285(32):24925–24932
72. Galione A, Evans AM, Ma J, Parrington J, Arredouani A, Cheng X, Zhu MX (2009) The acid test: the discovery of two-pore channels (TPCs) as NAADP-gated endolysosomal Ca(2+) release channels. Pflugers Arch 458(5):869–876
73. Macgregor A, Yamasaki M, Rakovic S, Sanders L, Parkesh R, Churchill GC, Galione A, Terrar DA (2007) NAADP controls cross-talk between distinct Ca2+ stores in the heart. J Biol Chem 282(20):15302–15311
74. Jardin I, Lopez JJ, Pariente JA, Salido GM, Rosado JA (2008) Intracellular calcium release from human platelets: different messengers for multiple stores. Trends Cardiovasc Med 18(2):57–61
75. Aley PK, Mikolajczyk AM, Munz B, Churchill GC, Galione A, Berger F (2010) Nicotinic acid adenine dinucleotide phosphate regulates skeletal muscle differentiation via action at two-pore channels. Proc Natl Acad Sci U S A 107(46):19927–19932
76. Favia A, Desideri M, Gambara G, D'Alessio A, Ruas M, Esposito B, Del Bufalo D, Parrington J, Ziparo E, Palombi F, Galione A, Filippini A (2014) VEGF-induced neoangiogenesis is mediated by NAADP and two-pore channel-2-dependent Ca2+ signaling. Proc Natl Acad Sci U S A 111(44):E4706–E4715
77. Kim BJ, Park KH, Yim CY, Takasawa S, Okamoto H, Im MJ, Kim UH (2008) Generation of nicotinic acid adenine dinucleotide phosphate and cyclic ADP-ribose by glucagon-like peptide-1 evokes Ca2+ signal that is essential for insulin secretion in mouse pancreatic islets. Diabetes 57(4):868–878

78. Arredouani A, Evans AM, Ma J, Parrington J, Zhu MX, Galione A (2010) An emerging role for NAADP-mediated Ca2+ signaling in the pancreatic beta-cell. Islets 2(5):323–330
79. Shawl AI, Park KH, Kim UH (2009) Insulin receptor signaling for the proliferation of pancreatic beta-cells: involvement of Ca2+ second messengers, IP3, NAADP and cADPR. Islets 1(3):216–223
80. Gambara G, Billington RA, Debidda M, D'Alessio A, Palombi F, Ziparo E, Genazzani AA, Filippini A (2008) NAADP-induced Ca2+ signaling in response to endothelin is via the receptor subtype B and requires the integrity of lipid rafts/caveolae. J Cell Physiol 216(2):396–404
81. Zhang F, Zhang G, Zhang AY, Koeberl MJ, Wallander E, Li PL (2006) Production of NAADP and its role in Ca2+ mobilization associated with lysosomes in coronary arterial myocytes. Am J Physiol Heart Circ Physiol 291(1):H274–H282
82. Soares S, Thompson M, White T, Isbell A, Yamasaki M, Prakash Y, Lund FE, Galione A, Chini EN (2007) NAADP as a second messenger: neither CD38 nor base-exchange reaction are necessary for in vivo generation of NAADP in myometrial cells. Am J Physiol Cell Physiol 292(1):C227–C239
83. Esposito B, Gambara G, Lewis AM, Palombi F, D'Alessio A, Taylor LX, Genazzani AA, Ziparo E, Galione A, Churchill GC, Filippini A (2011) NAADP links histamine H1 receptors to secretion of von Willebrand factor in human endothelial cells. Blood 117(18):4968–4977
84. Pandey V, Chuang CC, Lewis AM, Aley PK, Brailoiu E, Dun NJ, Churchill GC, Patel S (2009) Recruitment of NAADP-sensitive acidic Ca2+ stores by glutamate. Biochem J 422(3):503–512
85. Aley PK, Noh HJ, Gao X, Tica AA, Brailoiu E, Churchill GC (2010) A functional role for nicotinic acid adenine dinucleotide phosphate in oxytocin-mediated contraction of uterine smooth muscle from rat. J Pharmacol Exp Ther 333(3):726–735
86. Aley PK, Singh N, Brailoiu GC, Brailoiu E, Churchill GC (2013) Nicotinic acid adenine dinucleotide phosphate (NAADP) is a second messenger in muscarinic receptor-induced contraction of guinea pig trachea. J Biol Chem 288(16):10986–10993
87. Galione A, Churchill GC (2002) Interactions between calcium release pathways: multiple messengers and multiple stores. Cell Calcium 32(5–6):343–354
88. Lee HC, Aarhus R (2000) Functional visualization of the separate but interacting calcium stores sensitive to NAADP and cyclic ADP-ribose. J Cell Sci 113(Pt 24):4413–4420
89. Taylor CW, Dale P (2012) Intracellular Ca(2+) channels – a growing community. Mol Cell Endocrinol 353:21–28
90. Churchill GC, Galione A (2001) NAADP induces Ca2+ oscillations via a two-pool mechanism by priming IP3- and cADPR-sensitive Ca2+ stores. EMBO J 20(11):2666–2671
91. Galione A, Parrington J, Funnell T (2011) Physiological roles of NAADP-mediated Ca2+ signaling. Sci China Life Sci 54(8):725–732
92. Santella L, Kyozuka K, Genazzani AA, De Riso L, Carafoli E (2000) Nicotinic acid adenine dinucleotide phosphate-induced Ca(2+) release. Interactions among distinct Ca(2+) mobilizing mechanisms in starfish oocytes. J Biol Chem 275(12):8301–8306
93. Guse AH, Lee HC (2008) NAADP: a universal Ca2+ trigger. Sci Signal 1(44):re10
94. Guse AH (2012) Linking NAADP to ion channel activity: a unifying hypothesis. Sci Signal 5(221):pe1
95. Brailoiu E, Churamani D, Cai X, Schrlau MG, Brailoiu GC, Gao X, Hooper R, Boulware MJ, Dun NJ, Marchant JS, Patel S (2009) Essential requirement for two-pore channel 1 in NAADP-mediated calcium signaling. J Cell Biol 186(2):201–209
96. Morgan AJ, Davis LC, Wagner SK, Lewis AM, Parrington J, Churchill GC, Galione A (2013) Bidirectional Ca(2)(+) signaling occurs between the endoplasmic reticulum and acidic organelles. J Cell Biol 200(6):789–805
97. Albrieux M, Lee HC, Villaz M (1998) Calcium signaling by cyclic ADP-ribose, NAADP, and inositol trisphosphate are involved in distinct functions in ascidian oocytes. J Biol Chem 273(23):14566–14574

98. Berridge MJ, Bootman MD, Roderick HL (2003) Calcium signalling: dynamics, homeostasis and remodelling. Nat Rev Mol Cell Biol 4(7):517–529
99. Patel S, Churchill GC, Galione A (2001) Coordination of Ca2+ signalling by NAADP. Trends Biochem Sci 26(8):482–489
100. Galione A, Ruas M (2005) NAADP receptors. Cell Calcium 38(3–4):273–280
101. Morgan AJ, Galione A (2014) Two-pore channels (TPCs): current controversies. BioEssays 36(2):173–183
102. Dammermann W, Guse AH (2005) Functional ryanodine receptor expression is required for NAADP-mediated local Ca2+ signaling in T-lymphocytes. J Biol Chem 280(22):21394–21399
103. Langhorst MF, Schwarzmann N, Guse AH (2004) Ca2+ release via ryanodine receptors and Ca2+ entry: major mechanisms in NAADP-mediated Ca2+ signaling in T-lymphocytes. Cell Signal 16(11):1283–1289
104. Zhang F, Li PL (2007) Reconstitution and characterization of a nicotinic acid adenine dinucleotide phosphate (NAADP)-sensitive Ca2+ release channel from liver lysosomes of rats. J Biol Chem 282(35):25259–25269
105. Calcraft PJ, Ruas M, Pan Z, Cheng X, Arredouani A, Hao X, Tang J, Rietdorf K, Teboul L, Chuang KT, Lin P, Xiao R, Wang C, Zhu Y, Lin Y, Wyatt CN, Parrington J, Ma J, Evans AM, Galione A, Zhu MX (2009) NAADP mobilizes calcium from acidic organelles through two-pore channels. Nature 459(7246):596–600
106. Zong X, Schieder M, Cuny H, Fenske S, Gruner C, Rotzer K, Griesbeck O, Harz H, Biel M, Wahl-Schott C (2009) The two-pore channel TPCN2 mediates NAADP-dependent Ca(2+)-release from lysosomal stores. Pflugers Arch 458(5):891–899
107. Brailoiu E, Hooper R, Cai X, Brailoiu GC, Keebler MV, Dun NJ, Marchant JS, Patel S (2010) An ancestral deuterostome family of two-pore channels mediates nicotinic acid adenine dinucleotide phosphate-dependent calcium release from acidic organelles. J Biol Chem 285(5):2897–2901
108. Dionisio N, Albarran L, Lopez JJ, Berna-Erro A, Salido GM, Bobe R, Rosado JA (2011) Acidic NAADP-releasable Ca(2+) compartments in the megakaryoblastic cell line MEG01. Biochim Biophys Acta 1813(8):1483–1494
109. Genazzani AA, Mezna M, Dickey DM, Michelangeli F, Walseth TF, Galione A (1997) Pharmacological properties of the Ca2+ -release mechanism sensitive to NAADP in the sea urchin egg. Br J Pharmacol 121(7):1489–1495
110. Mandi M, Toth B, Timar G, Bak J (2006) Ca2+ release triggered by NAADP in hepatocyte microsomes. Biochem J 395(2):233–238
111. Zhang F, Jin S, Yi F, Li PL (2009) TRP-ML1 functions as a lysosomal NAADP-sensitive Ca2+ release channel in coronary arterial myocytes. J Cell Mol Med 13(9B):3174–3185
112. Brailoiu E, Rahman T, Churamani D, Prole DL, Brailoiu GC, Hooper R, Taylor CW, Patel S (2010) An NAADP-gated two-pore channel targeted to the plasma membrane uncouples triggering from amplifying Ca2+ signals. J Biol Chem 285(49):38511–38516
113. Rybalchenko V, Ahuja M, Coblentz J, Churamani D, Patel S, Kiselyov K, Muallem S (2012) Membrane potential regulates nicotinic acid adenine dinucleotide phosphate (NAADP) dependence of the pH- and Ca2+ -sensitive organellar two-pore channel TPC1. J Biol Chem 287(24):20407–20416
114. Pitt SJ, Funnell TM, Sitsapesan M, Venturi E, Rietdorf K, Ruas M, Ganesan A, Gosain R, Churchill GC, Zhu MX, Parrington J, Galione A, Sitsapesan R (2010) TPC2 is a novel NAADP-sensitive Ca2+ release channel, operating as a dual sensor of luminal pH and Ca2+. J Biol Chem 285(45):35039–35046
115. Schieder M, Rotzer K, Bruggemann A, Biel M, Wahl-Schott C (2010) Planar patch clamp approach to characterize ionic currents from intact lysosomes. Sci Signal 3(151):pl3
116. Schieder M, Rotzer K, Bruggemann A, Biel M, Wahl-Schott CA (2010) Characterization of two-pore channel 2 (TPCN2)-mediated Ca2+ currents in isolated lysosomes. J Biol Chem 285(28):21219–21222

117. Zhu MX, Evans AM, Ma J, Parrington J, Galione A (2010) Two-pore channels for integrative Ca signaling. Commun Integr Biol 3(1):12–17
118. Ogunbayo OA, Zhu Y, Shen B, Agbani E, Li J, Ma J, Zhu MX, Evans AM (2014) Organelle-specific subunit interactions of the vertebrate two pore channel family. J Biol Chem 290(2):1086–1095
119. Ruas M, Rietdorf K, Arredouani A, Davis LC, Lloyd-Evans E, Koegel H, Funnell TM, Morgan AJ, Ward JA, Watanabe K, Cheng X, Churchill GC, Zhu MX, Platt FM, Wessel GM, Parrington J, Galione A (2010) Purified TPC isoforms form NAADP receptors with distinct roles for Ca(2+) signaling and endolysosomal trafficking. Curr Biol 20(8):703–709
120. Yamaguchi S, Jha A, Li Q, Soyombo AA, Dickinson GD, Churamani D, Brailoiu E, Patel S, Muallem S (2011) Transient receptor potential mucolipin 1 (TRPML1) and two-pore channels are functionally independent organellar ion channels. J Biol Chem 286(26):22934–22942
121. Wang W, Zhang X, Gao Q, Xu H (2014) TRPML1: an ion channel in the lysosome. Handb Exp Pharmacol 222:631–645
122. Kim HJ, Yamaguchi S, Li Q, So I, Muallem S (2010) Properties of the TRPML3 channel pore and its stable expansion by the Varitint-Waddler-causing mutation. J Biol Chem 285(22):16513–16520
123. Cang C, Zhou Y, Navarro B, Seo YJ, Aranda K, Shi L, Battaglia-Hsu S, Nissim I, Clapham DE, Ren D (2013) mTOR regulates lysosomal ATP-sensitive two-pore Na(+) channels to adapt to metabolic state. Cell 152(4):778–790
124. Wang X, Zhang X, Dong XP, Samie M, Li X, Cheng X, Goschka A, Shen D, Zhou Y, Harlow J, Zhu MX, Clapham DE, Ren D, Xu H (2012) TPC proteins are phosphoinositide- activated sodium-selective ion channels in endosomes and lysosomes. Cell 151(2):372–383
125. Lin-Moshier Y, Walseth TF, Churamani D, Davidson SM, Slama JT, Hooper R, Brailoiu E, Patel S, Marchant JS (2012) Photoaffinity labeling of nicotinic acid adenine dinucleotide phosphate (NAADP) targets in mammalian cells. J Biol Chem 287(4):2296–2307
126. Ishibashi K, Suzuki M, Imai M (2000) Molecular cloning of a novel form (two-repeat) protein related to voltage-gated sodium and calcium channels. Biochem Biophys Res Commun 270(2):370–376
127. Zakon HH (2012) Adaptive evolution of voltage-gated sodium channels: the first 800 million years. Proc Natl Acad Sci U S A 109(Suppl 1):10619–10625
128. Rahman T, Cai X, Brailoiu GC, Abood ME, Brailoiu E, Patel S (2014) Two-pore channels provide insight into the evolution of voltage-gated Ca2+ and Na+ channels. Sci Signal 7(352):ra109
129. Churamani D, Hooper R, Brailoiu E, Patel S (2012) Domain assembly of NAADP-gated two-pore channels. Biochem J 441(1):317–323
130. Rietdorf K, Funnell TM, Ruas M, Heinemann J, Parrington J, Galione A (2011) Two-pore channels form homo- and heterodimers. J Biol Chem 286(43):37058–37062
131. Galione A (2011) NAADP receptors. Cold Spring Harb Perspect Biol 3(1):a004036
132. Patel S, Ramakrishnan L, Rahman T, Hamdoun A, Marchant JS, Taylor CW, Brailoiu E (2011) The endo-lysosomal system as an NAADP-sensitive acidic Ca(2+) store: role for the two-pore channels. Cell Calcium 50(2):157–167
133. Grimm C, Holdt LM, Chen CC, Hassan S, Muller C, Jors S, Cuny H, Kissing S, Schroder B, Butz E, Northoff B, Castonguay J, Luber CA, Moser M, Spahn S, Lullmann-Rauch R, Fendel C, Klugbauer N, Griesbeck O, Haas A, Mann M, Bracher F, Teupser D, Saftig P, Biel M, Wahl-Schott C (2014) High susceptibility to fatty liver disease in two-pore channel 2-deficient mice. Nat Commun 5:4699
134. Hockey LN, Kilpatrick BS, Eden ER, Lin-Moshier Y, Brailoiu GC, Brailoiu E, Futter CE, Schapira AH, Marchant JS, Patel S (2014) Dysregulation of lysosomal morphology by pathogenic LRRK2 is corrected by two-pore channel 2 inhibition. J Cell Sci 128(2):232–238
135. Lin-Moshier Y, Keebler MV, Hooper R, Boulware MJ, Liu X, Churamani D, Abood ME, Walseth TF, Brailoiu E, Patel S, Marchant JS (2014) The Two-pore channel (TPC) interac-

tome unmasks isoform-specific roles for TPCs in endolysosomal morphology and cell pigmentation. Proc Natl Acad Sci U S A 111(36):13087–13092
136. Sulem P, Gudbjartsson DF, Stacey SN, Helgason A, Rafnar T, Jakobsdottir M, Steinberg S, Gudjonsson SA, Palsson A, Thorleifsson G, Palsson S, Sigurgeirsson B, Thorisdottir K, Ragnarsson R, Benediktsdottir KR, Aben KK, Vermeulen SH, Goldstein AM, Tucker MA, Kiemeney LA, Olafsson JH, Gulcher J, Kong A, Thorsteinsdottir U, Stefansson K (2008) Two newly identified genetic determinants of pigmentation in Europeans. Nat Genet 40(7):835–837
137. Hooper R, Churamani D, Brailoiu E, Taylor CW, Patel S (2011) Membrane topology of NAADP-sensitive two-pore channels and their regulation by N-linked glycosylation. J Biol Chem 286(11):9141–9149
138. Lin PH, Duann P, Komazaki S, Park KH, Li H, Sun M, Sermersheim M, Gumpper K, Parrington J, Galione A, Evans AM, Zhu MX, Ma J (2014) Lysosomal two-pore channel subtype 2 (TPC2) regulates skeletal muscle autophagic signaling. J Biol Chem 290(6):3377–3389
139. Patel S, Marchant JS, Brailoiu E (2010) Two-pore channels: regulation by NAADP and customized roles in triggering calcium signals. Cell Calcium 47(6):480–490
140. Patel S, Brailoiu E (2012) Triggering of Ca2+ signals by NAADP-gated two-pore channels: a role for membrane contact sites? Biochem Soc Trans 40(1):153–157
141. Kilpatrick BS, Eden ER, Schapira AH, Futter CE, Patel S (2013) Direct mobilisation of lysosomal Ca2+ triggers complex Ca2+ signals. J Cell Sci 126(Pt1):60–66
142. Rocha N, Kuijl C, Van Der Kant R, Janssen L, Houben D, Janssen H, Zwart W, Neefjes J (2009) Cholesterol sensor ORP1L contacts the ER protein VAP to control Rab7-RILP-p150 Glued and late endosome positioning. J Cell Biol 185(7):1209–1225
143. Eden ER, White IJ, Tsapara A, Futter CE (2010) Membrane contacts between endosomes and ER provide sites for PTP1B-epidermal growth factor receptor interaction. Nat Cell Biol 12(3):267–272
144. Penny CJ, Kilpatrick BS, Han JM et al (2014) A computational model of lysosome-ER Ca2+ microdomains. J Cell Sci 127(Pt13):2934–2943
145. Penny CJ, Kilpatrick BS, Min Han J, Sneyd J, Patel S (2014) A "mix-and-match" approach to designing Ca(2+) microdomains at membrane-contact sites. Commun Integr Biol 7:e29586
146. Rosado JA (2006) Discovering the mechanism of capacitative calcium entry. Am J Physiol Cell Physiol 291(6):C1104–C1106
147. Putney JW Jr (2007) Recent breakthroughs in the molecular mechanism of capacitative calcium entry (with thoughts on how we got here). Cell Calcium 42(2):103–110
148. Schaloske R, Sonnemann J, Malchow D, Schlatterer C (1998) Fatty acids induce release of Ca2+ from acidosomal stores and activate capacitative Ca2+ entry in Dictyostelium discoideum. Biochem J 332(Pt 2):541–548
149. Hogan PG, Rao A (2007) Dissecting ICRAC, a store-operated calcium current. Trends Biochem Sci 32(5):235–245
150. Parekh AB, Putney JW Jr (2005) Store-operated calcium channels. Physiol Rev 85(2):757–810
151. Salido GM, Sage SO, Rosado JA (2009) Biochemical and functional properties of the store-operated Ca2+ channels. Cell Signal 21(4):457–461
152. Feske S, Gwack Y, Prakriya M, Srikanth S, Puppel SH, Tanasa B, Hogan PG, Lewis RS, Daly M, Rao A (2006) A mutation in Orai1 causes immune deficiency by abrogating CRAC channel function. Nature 441(7090):179–185
153. Peinelt C, Vig M, Koomoa DL, Beck A, Nadler MJ, Koblan-Huberson M, Lis A, Fleig A, Penner R, Kinet JP (2006) Amplification of CRAC current by STIM1 and CRACM1 (Orai1). Nat Cell Biol 8(7):771–773
154. Vig M, Peinelt C, Beck A, Koomoa DL, Rabah D, Koblan-Huberson M, Kraft S, Turner H, Fleig A, Penner R, Kinet JP (2006) CRACM1 is a plasma membrane protein essential for store-operated Ca2+ entry. Science 312(5777):1220–1223

155. Liou J, Kim ML, Heo WD, Jones JT, Myers JW, Ferrell JE Jr, Meyer T (2005) STIM is a Ca2+ sensor essential for Ca2+ -store-depletion-triggered Ca2+ influx. Curr Biol 15(13):1235–1241
156. Zhang SL, Yu Y, Roos J, Kozak JA, Deerinck TJ, Ellisman MH, Stauderman KA, Cahalan MD (2005) STIM1 is a Ca2+ sensor that activates CRAC channels and migrates from the Ca2+ store to the plasma membrane. Nature 437(7060):902–905
157. Luik RM, Wang B, Prakriya M, Wu MM, Lewis RS (2008) Oligomerization of STIM1 couples ER calcium depletion to CRAC channel activation. Nature 454(7203):538–542
158. Zbidi H, Jardin I, Woodard GE, Lopez JJ, Berna-Erro A, Salido GM, Rosado JA (2011) STIM1 and STIM2 are located in the acidic Ca2+ stores and associates with Orai1 upon depletion of the acidic stores in human platelets. J Biol Chem 286(14):12257–12270
159. Hewavitharana T, Deng X, Soboloff J, Gill DL (2007) Role of STIM and Orai proteins in the store-operated calcium signaling pathway. Cell Calcium 42(2):173–182
160. Soboloff J, Spassova MA, Tang XD, Hewavitharana T, Xu W, Gill DL (2006) Orai1 and STIM reconstitute store-operated calcium channel function. J Biol Chem 281(30):20661–20665
161. Spassova MA, Soboloff J, He LP, Xu W, Dziadek MA, Gill DL (2006) STIM1 has a plasma membrane role in the activation of store-operated Ca(2+) channels. Proc Natl Acad Sci U S A 103(11):4040–4045
162. Pedersen SF, Owsianik G, Nilius B (2005) TRP channels: an overview. Cell Calcium 38(3–4):233–252
163. Worley PF, Zeng W, Huang GN, Yuan JP, Kim JY, Lee MG, Muallem S (2007) TRPC channels as STIM1-regulated store-operated channels. Cell Calcium 42(2):205–211
164. Soboloff J, Spassova MA, Hewavitharana T, He LP, Xu W, Johnstone LS, Dziadek MA, Gill DL (2006) STIM2 is an inhibitor of STIM1-mediated store-operated Ca2+ entry. Curr Biol 16(14):1465–1470
165. Juska A, Jardin I, Rosado JA (2008) Physical properties of two types of calcium stores and SERCAs in human platelets. Mol Cell Biochem 311(1–2):9–18
166. Lopez JJ, Jardin I, Bobe R, Pariente JA, Enouf J, Salido GM, Rosado JA (2008) STIM1 regulates acidic Ca2+ store refilling by interaction with SERCA3 in human platelets. Biochem Pharmacol 75(11):2157–2164
167. Rosado JA, Lopez JJ, Harper AG, Harper MT, Redondo PC, Pariente JA, Sage SO, Salido GM (2004) Two pathways for store-mediated calcium entry differentially dependent on the actin cytoskeleton in human platelets. J Biol Chem 279(28):29231–29235
168. Papp B, Enyedi A, Paszty K, Kovacs T, Sarkadi B, Gardos G, Magnier C, Wuytack F, Enouf J (1992) Simultaneous presence of two distinct endoplasmic-reticulum-type calcium-pump isoforms in human cells. Characterization by radio-immunoblotting and inhibition by 2,5-di-(t-butyl)-1,4-benzohydroquinone. Biochem J 288(Pt 1):297–302
169. Bobe R, Bredoux R, Wuytack F, Quarck R, Kovacs T, Papp B, Corvazier E, Magnier C, Enouf J (1994) The rat platelet 97-kDa Ca2+ ATPase isoform is the sarcoendoplasmic reticulum Ca2+ ATPase 3 protein. J Biol Chem 269(2):1417–1424
170. Lopez JJ, Camello-Almaraz C, Pariente JA, Salido GM, Rosado JA (2005) Ca2+ accumulation into acidic organelles mediated by Ca2+ – and vacuolar H+ -ATPases in human platelets. Biochem J 390(Pt1):243–252
171. Redondo PC, Jardin I, Lopez JJ, Salido GM, Rosado JA (2008) Intracellular Ca2+ store depletion induces the formation of macromolecular complexes involving hTRPC1, hTRPC6, the type II IP3 receptor and SERCA3 in human platelets. Biochim Biophys Acta 1783(6):1163–1176
172. Redondo PC, Harper AG, Sage SO, Rosado JA (2007) Dual role of tubulin-cytoskeleton in store-operated calcium entry in human platelets. Cell Signal 19(10):2147–2154
173. Jardin I, Lopez JJ, Salido GM, Rosado JA (2008) Functional relevance of the de novo coupling between hTRPC1 and type II IP3 receptor in store-operated Ca2+ entry in human platelets. Cell Signal 20(4):737–747

174. Rosado JA, Sage SO (2000) Coupling between inositol 1,4,5-trisphosphate receptors and human transient receptor potential channel 1 when intracellular Ca2+ stores are depleted. Biochem J 350(Pt 3):631–635
175. Coxon CH, Lewis AM, Sadler AJ, Vasudevan SR, Thomas A, Dundas KA, Taylor L, Campbell RD, Gibbins JM, Churchill GC, Tucker KL (2012) NAADP regulates human platelet function. Biochem J 441(1):435–442
176. Jardin I, Redondo PC, Salido GM, Rosado JA (2008) Phosphatidylinositol 4,5-bisphosphate enhances store-operated calcium entry through hTRPC6 channel in human platelets. Biochim Biophys Acta 1783(1):84–97
177. Lopez J, Dionisio N, Berna-Erro A, Galan C, Salido GM, Rosado JA (2012) Two-pore channel 2 (TPC2) modulates store-operated Ca(2+) entry. Biochim Biophys Acta 1823(10):1976–1983
178. Re RN, Cook JL (2008) The basis of an intracrine pharmacology. J Clin Pharmacol 48:344–350
179. Brailoiu E, Filipeanu CM, Tica A, Toma CP, de Zeeuw D, Nelemans SA (1999) Contractile effects by intracellular angiotensin II via receptors with a distinct pharmacological profile in rat aorta. Br J Pharmacol 126(5):1133–1138
180. Brailoiu GC, Oprea TI, Zhao P, Abood ME, Brailoiu E (2011) Intracellular cannabinoid type 1 (CB1) receptors are activated by anandamide. J Biol Chem 286(33):29166–29174
181. Brailoiu GC, Deliu E, Marcu J, Hoffman NE, Console-Bram L, Zhao P, Madesh M, Abood ME, Brailoiu E (2014) Differential activation of intracellular versus plasmalemmal CB2 cannabinoid receptors. Biochemistry 53(30):4990–4999
182. Deliu E, Tica AA, Motoc D, Brailoiu GC, Brailoiu E (2011) Intracellular angiotensin II activates rat myometrium. Am J Physiol Cell Physiol 301(3):C559–C565
183. Deliu E, Brailoiu GC, Mallilankaraman K, Wang H, Madesh M, Undieh AS, Koch WJ, Brailoiu E (2012) Intracellular endothelin type B receptor-driven Ca2+ signal elicits nitric oxide production in endothelial cells. J Biol Chem 287(49):41023–41031
184. Deliu E, Brailoiu GC, Eguchi S, Rabinowitz JE, Tilley DG, Madesh M, Koch WJ, Brailoiu E (2014) Direct evidence of intracrine angiotensin II signaling in neurons. Am J Physiol Cell Physiol 306(8):C736–C744
185. Haller H, Lindschau C, Erdmann B, Quass P, Luft FC (1996) Effects of intracellular angiotensin II in vascular smooth muscle cells. Circ Res 79(4):765–772
186. Re RN, Cook JL (2007) Mechanisms of disease: intracrine physiology in the cardiovascular system. Nat Clin Pract Cardiovasc Med 4(10):549–557
187. Re RN, Cook JL (2011) Noncanonical intracrine action. J Am Soc Hypertens 5(6):435–448
188. Fiaschi-Taesch NM, Stewart AF (2003) Minireview: parathyroid hormone-related protein as an intracrine factor – trafficking mechanisms and functional consequences. Endocrinology 144(2):407–411
189. Carey RM (2012) Functional intracellular renin-angiotensin systems: potential for pathophysiology of disease. Am J Physiol Regul Integr Comp Physiol 302(5):R479–R481
190. Zhuo JL, Li XC (2011) New insights and perspectives on intrarenal renin-angiotensin system: focus on intracrine/intracellular angiotensin II. Peptides 32(7):1551–1565
191. Boivin B, Vaniotis G, Allen BG, Hebert TE (2008) G protein-coupled receptors in and on the cell nucleus: a new signaling paradigm? J Recept Signal Transduct Res 28(1–2):15–28
192. Hewavitharana T, Wedegaertner PB (2012) Non-canonical signaling and localizations of heterotrimeric G proteins. Cell Signal 24(1):25–34
193. Jalink K, Moolenaar WH (2010) G protein-coupled receptors: the inside story. BioEssays 32(1):13–16
194. Marrari Y, Crouthamel M, Irannejad R, Wedegaertner P (2007) Assembly and trafficking of heterotrimeric G proteins. Biochemistry 46(26):7665–7677
195. Wedegaertner PB (2012) G protein trafficking. Subcell Biochem 63:193–223
196. Cosker F, Cheviron N, Yamasaki M, Menteyne A, Lund FE, Moutin MJ, Galione A, Cancela JM (2010) The ecto-enzyme CD38 is a nicotinic acid adenine dinucleotide phosphate

(NAADP) synthase that couples receptor activation to Ca2+ mobilization from lysosomes in pancreatic acinar cells. J Biol Chem 285(49):38251–38259
197. Kim SY, Cho BH, Kim UH (2010) CD38-mediated Ca2+ signaling contributes to angiotensin II-induced activation of hepatic stellate cells: attenuation of hepatic fibrosis by CD38 ablation. J Biol Chem 285(1):576–582
198. Davis LC, Morgan AJ, Ruas M, Wong JL, Graeff RM, Poustka AJ, Lee HC, Wessel GM, Parrington J, Galione A (2008) Ca(2+) signaling occurs via second messenger release from intraorganelle synthesis sites. Curr Biol 18(20):1612–1618
199. Sorkin A, Von Zastrow M (2009) Endocytosis and signalling: intertwining molecular networks. Nat Rev Mol Cell Biol 10(9):609–622
200. Calebiro D, Nikolaev VO, Lohse MJ (2010) Imaging of persistent cAMP signaling by internalized G protein-coupled receptors. J Mol Endocrinol 45(1):1–8
201. Grimsey N, Lin H, Trejo J (2014) Endosomal signaling by protease-activated receptors. Methods Enzymol 535:389–401
202. Irannejad R, Tomshine JC, Tomshine JR, Chevalier M, Mahoney JP, Steyaert J, Rasmussen SG, Sunahara RK, El-Samad H, Huang B, von Zastrow M (2013) Conformational biosensors reveal GPCR signalling from endosomes. Nature 495(7442):534–538
203. Irannejad R, Von Zastrow M (2014) GPCR signaling along the endocytic pathway. Curr Opin Cell Biol 27:109–116
204. Lin H, Trejo J (2013) Transactivation of the PAR1-PAR2 heterodimer by thrombin elicits beta-arrestin-mediated endosomal signaling. J Biol Chem 288(16):11203–11215
205. Gruenberg J (2001) The endocytic pathway: a mosaic of domains. Nat Rev Mol Cell Biol 2(10):721–730
206. Li WW, Li J, Bao JK (2012) Microautophagy: lesser-known self-eating. Cell Mol Life Sci 69(7):1125–1136
207. Di Marzo V (2008) Targeting the endocannabinoid system: to enhance or reduce? Nat Rev Drug Discov 7(5):438–455
208. Hashimotodani Y, Ohno-Shosaku T, Tanimura A, Kita Y, Sano Y, Shimizu T, Di Marzo V, Kano M (2013) Acute inhibition of diacylglycerol lipase blocks endocannabinoid-mediated retrograde signalling: evidence for on-demand biosynthesis of 2-arachidonoylglycerol. J Physiol 591(Pt19):4765–4776
209. Zhu T, Gobeil F, Vazquez-Tello A, Leduc M, Rihakova L, Bossolasco M, Bkaily G, Peri K, Varma DR, Orvoine R, Chemtob S (2006) Intracrine signaling through lipid mediators and their cognate nuclear G-protein-coupled receptors: a paradigm based on PGE2, PAF, and LPA1 receptors. Can J Physiol Pharmacol 84(3–4):377–391
210. Yu J, Deliu E, Zhang XQ, Hoffman NE, Carter RL, Grisanti LA, Brailoiu GC, Madesh M, Cheung JY, Force T, Abood ME, Koch WJ, Tilley DG, Brailoiu E (2013) Differential activation of cultured neonatal cardiomyocytes by plasmalemmal versus intracellular G protein-coupled receptor 55. J Biol Chem 288(31):22481–22492

Chapter 19
Remodeling of Calcium Entry Pathways in Cancer

Carlos Villalobos, Diego Sobradillo, Miriam Hernández-Morales, and Lucía Núñez

Abstract Ca^{2+} entry pathways play important roles in control of many cellular functions, including long-term proliferation, migration and cell death. In recent years, it is becoming increasingly clear that, in some types of tumors, remodeling of Ca^{2+} entry pathways could contribute to cancer hallmarks such as excessive proliferation, cell migration and invasion as well as resistance to cell death or survival. In this chapter we briefly review findings related to remodeling of Ca^{2+} entry pathways in cancer with emphasis on the mechanisms that contribute to increased store-operated Ca^{2+} entry (SOCE) and store-operated currents (SOCs) in colorectal cancer cells. Finally, since SOCE appears critically involved in colon tumorogenesis, the inhibition of SOCE by aspirin and other NSAIDs and its possible contribution to colon cancer chemoprevention is reviewed.

Keywords Store-operated Ca^{2+} entry • Cancer • Colorectal cancer • Aspirin • NSAID

19.1 Intracellular Ca^{2+} Homeostasis and Ca^{2+} Entry Pathways

Intracellular Ca^{2+} is a very versatile second messenger involved in the control of many different physiological and cellular processes in the short and the long-term. Unlike other messengers, Ca^{2+} is not created or destroyed by cells, but transported down electrochemical gradients through specific channels, or transported back against gradients at the expense of the energy stored as ATP (Ca^{2+} pumps) or coupled with transport of other ions (Ca^{2+} exchangers). Ca^{2+} channels, pumps and exchangers are located in the plasma membrane and endomembranes of the endoplasmic reticulum (ER), mitochondria and other cell organelles. Unlike other ions,

C. Villalobos (✉) • D. Sobradillo • M. Hernández-Morales • L. Núñez
Instituto de Biología y Genética Molecular (IBGM), Universidad de Valladolid y Consejo Superior de Investigaciones Científicas (CSIC), C/Sanz y Forés 3, 47003 Valladolid, Spain
e-mail: carlosv@ibgm.uva.es

J.A. Rosado (ed.), *Calcium Entry Pathways in Non-excitable Cells*, Advances in Experimental Medicine and Biology 898, DOI 10.1007/978-3-319-26974-0_19

the cytosolic Ca^{2+} concentration ($[Ca^{2+}]_{cyt}$) is so little (100 nM) that even minimal changes in channel activity may induce large increases in $[Ca^{2+}]_{cyt}$. Thus, this cation is unique in the sense that has been selected by nature to carry signals inside cells and organelles. Changes in $[Ca^{2+}]_{cyt}$ may be restricted in time and space resulting in elementary events or Ca^{2+} microdomains that can regulate specifically located cellular functions such as exocytosis in plasma membrane, cell respiration and ATP synthesis in mitochondria or gene transcription in the nucleus. Alternatively, elementary events may give rise to regenerative waves leading to sustained, global changes associated to long-term events like cell growth, differentiation or death. Thus, the study of the intracellular Ca^{2+} homeostasis is a matter of ion transport across boundaries that requires the use of sophisticated methodologies for recording Ca^{2+}-driven currents (patch-clamp) or measuring the tiny concentrations of Ca^{2+} in the cytosol or the very variable Ca^{2+} concentrations in organelles and/or subcellular environments using live cell imaging and targeted calcium probes with different affinities for Ca^{2+}.

Ca^{2+} pumps and transporters contribute significantly to the maintenance of resting $[Ca^{2+}]_{cyt}$ and to the recovery of basal $[Ca^{2+}]_{cyt}$ after stimulation. However, most increases in $[Ca^{2+}]_{cyt}$, are rather due to activation of Ca^{2+} entry pathways at the plasma membrane and Ca^{2+} release channels at the ER. In the RE, IP_3 receptors and ryanodine receptors are ligand-gated Ca^{2+} channels that mediate Ca^{2+} release from stores. Ca^{2+} release channels induce transient increases in $[Ca^{2+}]_{cyt}$ but their activity may secondarily activate Ca^{2+} channels in plasma membrane that are gated by the filling state of Ca^{2+} stores. In mitochondria, the main Ca^{2+} channel is the so-called mitochondrial Ca^{2+} uniporter (MCU), a Ca^{2+}-activated Ca^{2+} channel recently characterized at the molecular level [1, 2]. Activation of this channel in physiological conditions leads to mitochondrial Ca^{2+} uptake and removal of Ca^{2+} from cytosol. This is due to the fact that mitochondria inner membrane shows a strong mitochondrial potential, negative inside the mitochondrial matrix, thus favoring Ca^{2+} influx into mitochondria provided that cytosolic Ca^{2+} is large enough to activate the MCU [3, 4]. At the plasma membrane there are many different types of Ca^{2+} channels, including receptor-operated and voltage-operated Ca^{2+} channels that are widely expressed in excitable cells (ROCCs and VOCCs) together with voltage-independent channels that are particularly relevant in non-excitable cells. The most important Ca^{2+} entry pathway in non-excitable cells is the store-operated Ca^{2+} entry (SOCE), a Ca^{2+} entry pathway ubiquitous and responsible for the entry of Ca^{2+} after agonist-induced activation of phospholipase C and emptying of intracellular Ca^{2+} stores.

SOCE is activated physiologically after the emptying of the intracellular Ca^{2+} stores induced by physiological agonists producing IP_3 or, pharmacologically, after inhibition of the sarcoplasmic and endoplasmic reticulum Ca^{2+} ATPase (SERCA) pump with thapsigargin or similar SERCA antagonists. SOCE usually remains active until Ca^{2+} stores become filled again. Interestingly, SOCE is also regulated by mitochondria. Mitochondrial control of SOCE is due to the ability of these organelles to take up Ca^{2+}, thus preventing refilling of Ca^{2+} stores and preventing also the Ca^{2+}-dependent inactivation of IP_3 receptors and the own Ca^{2+} release activated channels responsible for SOCE. This mechanism ensures the efficient emptying of

Ca^{2+} stores and SOCE activation and is believed to be critical to maintain Ca^{2+} entry in those signaling pathways in which a sustained activation is required. This is the case, for instance, of the activation of the nuclear factor of activated T (NFAT) cells during the immunological synapsis. In this case, a sustained entry of Ca^{2+} through SOCE is needed to promote IL2 gene expression and the clonal expansion of the activated T cell [5].

For years, the molecular basis of SOCE had remained elusive. However, the molecular players involved in SOCE began to be crack after the discovery of the TRP superfamily of ion channels. Some TRP channels were held during quite some time responsible for SOCE. However, more recently, the protein families STIM and ORAI were described to be the cornerstone of SOCE becoming now fully accepted as responsible for Icrac and SOCE in multiple cell types. At the molecular level, SOCE begins with the emptying of intracellular Ca^{2+} stores. Emptying, in this case, means that Ca^{2+} concentration inside the ER decreases from around 700 μM before stimulation to about 200 μM after agonist-induced Ca^{2+} release, as revealed by ER targeted probes with very low affinity for Ca^{2+} [6]. This "emptying" is detected by a sensor named Stromal Interaction Molecule 1 (STIM1) which, upon dissociation of Ca^{2+} ions from Ca^{2+} binding sites, undergoes oligomerization and its interaction with ORAI1, a Ca^{2+} channel located in specific places of the plasma membrane. This interaction opens Ca^{2+} specific CRAC channels, thus enabling the entry of Ca^{2+} into the cytosol. The whole mechanism is reversed when Ca^{2+} stores become filled again [7–9]. There is another Ca^{2+} sensor at the ER called STIM2, which has a lower affinity for Ca^{2+} and is believed to be activated only after moderate depletion of Ca^{2+} stores [10]. However, its role in SOCE remains controversial. In addition, two other ORAI family members (ORAI2 and ORAI3) may also be involved in SOCE but their role is also poorly known [11]. Finally, other types of channels, including some of the TRP superfamily, particularly TRPC channels may contribute to SOCE as well by forming channel complexes with STIM1 and ORAI1 or simply forming alternative store-operated channels less selective for Ca^{2+} [12, 13].

Other Ca^{2+} entry pathways also widely expressed are now collectively termed as store-independent Ca^{2+} entry (SICE) pathways. They are less known and most of them are characterized only at the functional level but not at the molecular level. They include channels gated by araquidonic acid, diacylglycerol (DAG) and stretch activated channels. Most of these channels may well be mediated by TRP channels and/or channel complexes made thereof. Other chapters in this issue describe in more detail characteristics and molecular basis related to store-independent Ca^{2+} entry pathways.

19.2 Ca^{2+} Entry Remodeling in Cancer

A series of recent reports suggest that changes in intracellular Ca^{2+} homeostasis (remodeling) may be critically involved in various forms of cancer. This is not surprising if we consider that the so many cellular processes exacerbated in the

Table 19.1 Changes in expression levels of voltage-operated Ca^{2+} channels [VOCCs] in tumor cells relative to normal cells or tissue samples

Channel isoform	Type of tumor	Research model	Change	Reference
$Ca_v1.1$	Colorectal	Cell lines and tissue samples from patients	↑ mRNA	[20]
$Ca_v1.2$	Colon	Cell lines and tissue samples from patients	↑ mRNA	[21]
$Ca_v3.1$	Colorectal	Cell lines and tissue samples from patients	↓ mRNA	[22]
	Gastric	Cell lines and tissue samples from patients	↓ mRNA	[22]
$Ca_v3.2$	Prostate	Tissue samples from patients	↑ Protein	[23]

transformed cell such as excessive cell proliferation, migration and invasion capabilities as well as apoptosis resistance and cell survival are regulated by intracellular Ca^{2+} [14–16]. A Ca^{2+} entry pathway that has been implicated in cancer is SOCE [17, 18] which could offer new therapeutic possibilities against cancer as suggested earlier [19]. However, some other Ca^{2+} entry pathways and channels have been also related to tumorogenesis. For example, it has been reported that voltage-gated Ca^{2+} channels (VOCCs) are overexpressed in various types of tumors (Table 19.1).

Thus, epithelial tumors express L-type [21] and T type VOCCs and their inhibition may prevent cell proliferation [24]. It is not clear the role of VOCCs in epithelial cells or even if they may work as channels. For example, plasma depolarization with medium containing a high concentration of K^+, a typical experimental maneuver intended to activate voltage-gated Ca^{2+} entry in Ca^{2+} imaging experiments in excitable cells [25], has no effect in epithelial cells, either normal or tumoral. Accordingly, increased expression of VOCCs in cancer cells may not be related to intracellular Ca^{2+} homeostasis. Expression of other Ca^{2+} channels of the TRP superfamily, particularly canonical TRP channels not so selective for Ca^{2+}, has been reported as well to be altered in different tumors (Table 19.2). Other channels of the same superfamily have been also reported to be either overexpressed or downregulated in different tumors, particularly TRPV4, TRPV6 and TRPM8 in carcinomas and other tumors [34, 37, 38]. The role of these channels in cancer remains unclear but recent evidence suggests a role for some of these channels in cancer hallmarks. For instance, it has been shown that the calcium selective channel TRPV6 is able to translocate to the plasma membrane via Orai1-mediated mechanism controlling cancer cell survival [39]. This expanding issue is being covered by excellent recent reviews [40]. See Table 19.3 for further details on changes in the expression of these channels in different tumors.

More recently, it has proposed an essential role for SOCE and/or its molecular players in tumorigenesis. For example, STIM1 and Orai1 may be critical for migra-

Table 19.2 Changes in expression levels of TRPC channels in tumor cells relative to normal cells or tissue samples

Channel isoform	Type of tumor	Research model	Change	Reference
TRPC1	Breast	Tissue samples from patients	↑ mRNA and protein	[26]
TRPC3	Ovary	Tissue samples from patients	↑ Proteína	[27]
	Breast	Tissue samples from patients	↑ mRNA	[28]
TRPC4	Kidney	Cell lines	↓ mRNA	[29]
TRPC6	Esophagus	Tissue samples from patients	↑ mRNA and protein	[30, 31]
	Gastric	Tissue samples from patients	↑ mRNA and protein	[32]
	Glioma	Tissue samples from patients	↑ mRNA and protein	[33]
	Liver	Tissue samples from patients	↑ mRNA and protein	[34]
	Breast	Tissue samples from patients	↑ mRNA and protein	[26, 28]
		Cell lines and Tissue samples from patients	↑ mRNA and protein	[35]
	Prostate	Cell lines	↑ mRNA	[36]

tion of breast cancer cells and metastasis [27]. STIM1 plays an important role in cell growth and migration in cervical cancer [54]. Indeed STIM1 is overexpressed in 70 % of cervical cancers, which has been associated with an increased risk of metastasis. As a matter of fact, the suppression of STIM1 inhibits human glioblastoma cell proliferation and induces G0/G1 phase arrest [55]. Moreoever, STIM1 and Orai1 mediate CRAC channel activity and are essential for human glioblastoma invasion [56]. Orai1 could be also overexpressed in breast cancer [57]. The entry of Ca^{2+} mediated by Orai1 regulates also proliferation and survival in glioblastoma cells and hepatoma [34].

The role of other molecular players involved in SOCE in tumorogenesis is unknown except for the case of Orai3 that has been proposed to be an estrogen receptor α-regulated Ca^{2+} channel that promotes tumorigenesis [58]. Likewise, Orai3 has been also reported to constitute a native SOCE regulating non-small cell lung adenocarcinoma cell proliferation [59]. Therefore, multiple recent evidence suggests an unexpected role for intracellular Ca^{2+} remodeling, particularly SOCE in cancer [15, 16]. See Table 19.4 for further details on changes in expression of molecular players involved in SOCE in different forms of cancer.

Table 19.3 Changes in expression of TRPV4, TRPV6 and TRPM8 in tumor cells relative to normal cells or tissue samples

Channel isoform	Type of tumor	Research model	Change	Reference
TRPV4	Skin	Tissue samples from patients	↓ mRNA and protein	[41]
	Bladder	Tissue simple and mouse cell line	↓ mRNA and protein	[42]
TRPV6	Breast	Tissue samples from patients	↑ mRNA and protein	[26, 37, 43]
	Prostate	Tissue samples from patients	↑ mRNA	[44, 45]
		Tissue samples from patients	↑ Proteína	[37]
	Thyroid	Muestras de tejido de pacientes	↑ Proteína	[37]
	Colon	Tissue samples from patients	↑ Proteína	[37]
	Ovary	Tissue samples from patients	↑ Proteína	[37]
	Lung	Tissue samples from patients	↓ mRNA and protein	[46]
TRPM8	Pancreas	Cell lines and tissue samples from patients	↑ mRNA and protein	[47]
	Prostate	Tissue samples from patients	↑ mRNA and protein	[48–51]
	Androgen-independent prostate tumor	Tissue samples from patients	↓ mRNA	[51]
	Breast	Tissue samples from patients	↑ mRNA and protein	[26, 48]
		Cell lines and tissue samples from patients	↑ mRNA and protein	[52]
		Tissue samples from patients	↑ mRNA and protein	[48]
		Tissue samples from patients	↑ mRNA	[48]
		Cell lines and tissue samples from patients	↑ Protein	[53]
	Colorectal	Tissue samples from patients	↑ mRNA	[48]
	Lung	Tissue samples from patients	↑ mRNA	[48]

Table 19.4 Changes in expression of ORAI and STIM family members in tumor cells relative to normal cells or tissues

Channel isoform	Type of tumor	Research model	Change	Reference
ORAI1	Breast	Cell lines	↑ mRNA	[57]
	Glioblastoma	Cell lines and tissue samples from patients	↑ mRNA	[56]
	Glioma	Tissue samples from patients	↑ Protein	[60]
	Melanoma	Cell lines	↑ Protein	[61]
	Kidney	Tissue samples from patients	↑ Protein	[62]
ORAI3	Breast	Cell lines and tissue samples from patients	↑ mRNA	[63]
	Prostate	Tissue samples from patients	↓ mRNA	[64]
		Tissue samples from patients	↑ mRNA	[65]
STIM1	Glioblastoma	Tissue samples from patients	↑ mRNA	[66]
	Cervix	Tissue samples from patients	↑ protein	[54]
	Breast	Cell lines	↑ mRNA	[57]
	Lung	Tissue samples from patients	↑ protein	[55]
	Liver	Cell lines and Tissue samples from patients	↑ mRNA	[27]
	Melanoma	Cell lines	↑ protein	[61]
	Colon	Tissue samples from patients	↑ mRNA and protein	[21]
STIM2	Glioblastoma	Tissue samples from patients	↑ mRNA	[67]
	Colon	Tissue samples from patients	↑ mRNA	[68]

19.3 Ca^{2+} Entry Pathways in Normal and Colon Cancer Cells

We have recently reported a deep remodeling of SOCE in colorectal cancer [69]. For investigating Ca^{2+} remodeling in colon cancer we have compared Ca^{2+} entry pathways in a series of normal human colonic mucosa cell lines and human colon adenocarcinoma cells. We found that SOCE is significantly larger in colorectal cancer cells than in normal cells [69]. Both normal and tumor cells differed also in their rate of cell proliferation with tumor cells showing always larger rates of cell proliferation. Interestingly, there was a clear correlation between SOCE and the rate of cell proliferation suggesting that the larger rate of cell proliferation of tumor cells follows changes in SOCE [69]. Consistently, SOCE inhibition with antagonists prevents not only tumor cell proliferation but also cell invasion as tested by Matrigel invasion assays. Therefore, enhancement of SOCE in colon cancer cells may contribute not only to increased cell proliferation characteristic of tumor cells but also to cell invasion, both being critical hallmarks of cancer [69].

Differences between normal and tumor cells regarding Ca^{2+} entry and Ca^{2+} release induced by physiological agonists were also studied in detail. ATP and

Carbachol, two physiological agonists that activate G protein-coupled receptors and phospholipase C, increase $[Ca^{2+}]_{cyt}$ in both normal and tumor cells. However, the rises in $[Ca^{2+}]_{cyt}$ are much larger in tumor cells compared with normal colonic epithelium cells. Analysis in Ca^{2+} free medium revealed that the two agonists released more Ca^{2+} in tumor cells than in normal cells. Results on agonist-induced Ca^{2+} entry were surprising: In fact, both agonists induced Ca^{2+} entry only in tumor cells but not in normal cells, despite that both released Ca^{2+} from intracellular stores. To avoid contribution of possible differences in expression of GPCRs in normal and tumor cells, experiments were carried out using caged IP_3. Again, the IP_3-induced increase of $[Ca^{2+}]_{cyt}$ was larger in tumor cells than normal cells [69]. However, these results could be due to either differences in the level of expression and/or activity of IP_3 receptors and/or to differences in the extent of Ca^{2+} store content. Experiments using ionomycin or cyclopiazonic acid in Ca^{2+} free medium revealed that Ca^{2+} stores were paradoxically larger in normal cells than in tumor cells. In other words, Ca^{2+} stores in tumor cells are partially empty. These data together with the data derived form Ca^{2+} release experiments, always much larger in tumor cells, suggest that in tumor cells Ca^{2+} stores are nearly depleted and any minimal stimulation is sufficient to reach the threshold for SOCE activation. However, in normal cells, where Ca^{2+} stores are full and physiological stimulation releases only a limited amount of Ca^{2+}, SOCE threshold is beyond reach leading to no Ca^{2+} entry. These data may explain why agonists do activate Ca^{2+} entry only in tumor cells but not in normal mucosa cells [69]. Of course, we have to take into account that these data derive from the analysis of a few, non isogenic cell lines serving as models of normal and tumor colon cells. Results must be confirmed in additional normal and tumor samples derived from the same specimen.

19.4 Store-Operated Currents in Normal and Colon Cancer Cells

To identify functional and pharmacologically SOCs in normal and tumor cells we have used planar patch-clamp in the voltage clamp configuration [69]. The depletion of Ca^{2+} stores with thapsigargin in normal colonocytes induces a small, voltage-independent, inward rectifying current that is highly selective for Ca^{2+} and sensitive to La^{3+} and to low concentrations of 2-APB [69]. Therefore, this current is very similar, if not identical, to the Icrac current originally described as responsible for SOCE in mast and T cells. In tumor cells, however, SOCs are very different. Emptying of Ca^{2+} stores induces two types of currents: First, a current similar to Icrac recorded in normal cells with the only exception that current density is larger in tumor cells. Second, we observe a different Isoc, absent in normal cells, with a large outward component, not selective for Ca^{2+} and sensitive to high concentrations of 2-APB [69]. The data suggest that SOCE in normal cells is mediated by Orai1 and STIM1, the molecular players previously reported to be involved in Icrac in mast and T cells. However, tumor cells would have two streams, one should be Icrac but with larger current density probably mediated by changes in expression of Orai1 and Stim1. The other current could be mediated by a TRP channel. Accordingly, data suggest that store-operated currents may be mediated by different channels in normal and tumor cells.

19.5 Molecular Basis of Remodeling of Ca^{2+} Entry Pathways in Colorectal Cancer

Molecular candidates involved in SOCE and SOCs in normal and colon cancer cells have been investigated using conventional and quantitative RT-PCR. All members of the STIM and Orai families (Orai1, 2, 3 and STIM1 and 2) are expressed in normal and tumor cells. Other TRPC channels including TRPC1 and TRPC4 channels are expressed as well in both normal and tumor cells. However, other genes that have been related to SOCE are expressed only in normal cells but not in tumor cells including TRPV6 and TRPM8 suggesting a loss of function during tumorogenesis. Finally, some other related channels are missing in both cell types. Quantitative RT-PCR and Western blotting show significant increases in the expression of many of the above genes. Interestingly, the expression pattern of the different molecular players involved in SOCE and SOCs is roughly similar in normal and tumor cells except that most genes are rather increased in tumor cells relative to normal cells. At the protein level, Orai1, Orai2, and Orai3 proteins are increased significantly in tumor cells. Similar results are obtained with TRPC1 and STIM1. Surprisingly, although STIM2 gene expression is increased in tumor cells, STIM2 protein is nearly lost in tumor cells [69].

Using siRNA against each of these specific molecular players, we have established that SOCE and SOCs in normal cells are mediated by the interactions between ORAI1, STIM1 and STIM2. However, in tumor cells, SOCs and SOCE are more complex and include not only STIM1 and Orai1 but also TRPC1. The increased expression of both may contribute to explain the increase in SOCE and agonist-induced Ca^{2+} observed in tumor cells relative to normal cells (Fig. 19.1).

Therefore, SOCs in colon carcinoma cells are made of different molecular players than in colon normal mucosa cells. Differences involve likely changes in channel complexes including the appearance of TRPC1 and the disappearance of STIM2 from those complexes. This view is supported by silencing experiments. TRPC1 silencing has no influence on Icrac in normal cells but it decreases not only the outward component associated to the non-selective current present only in tumor cells but also the inward component. Moreover, ORAI1 silencing decreases Icrac in normal cells and both components inward and outward in tumor cells. Thus, these data suggest strongly that ORAI1 and TRPC1 form a channel complex in tumor cells but not in normal cells. The data invite speculation as to what is the role played by TRPC1 in SOCE. Silencing experiments reveal that TRPC1 knockdown does not

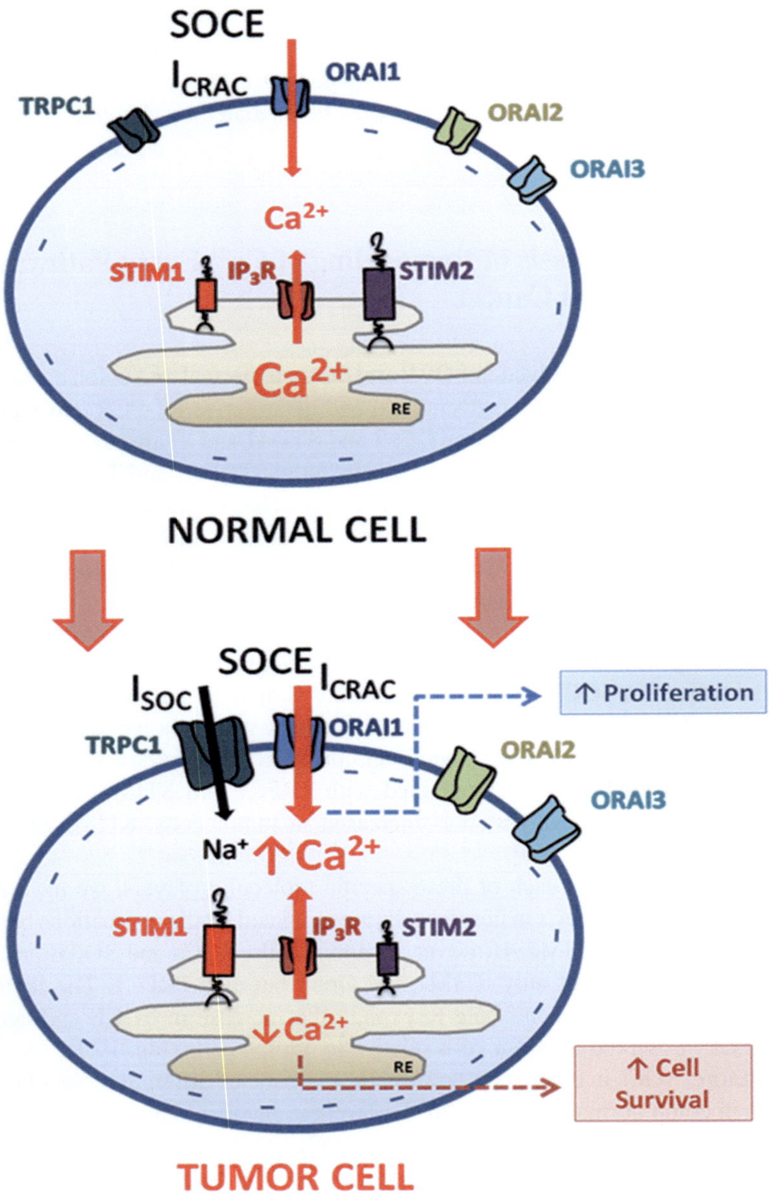

Fig. 19.1 Ca^{2+} entry remodeling in colorectal cancer. Normal human colon cells (*above*) show a small SOCE mediated Orai1 and STIM1 and a large Ca^{2+} store content associated with the large expression of STIM2. The cells also express TRPC1, TRPC4, TRPC7, TRPV6 and TRPM8. Cells of human colon adenocarcinoma (*below*) present large and modified SOCE mediated by increases in expression of Orai1, TRPC1 and STIM1, and partial depletion tank contents mediated decreased expression of STIM2. These cells lack TRPC7, TRPV6 and TRPM8. Remodeling may contribute to increased proliferative capacity, invasion and survival of tumor cells [69]

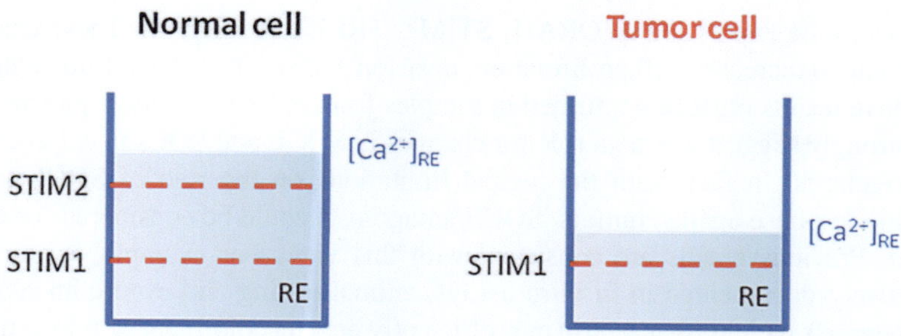

Fig. 19.2 Loss of Stim2 in tumor cells decreases Ca^{2+} store content in tumor cells. Normal cells express both Stim1 and Stim2, ER Ca^{2+} sensors with different affinities for Ca^{2+}. Loss of Stim2 in colon cancer cells leaves Stim1 as the only sensor available that refills Ca^{2+} stores poorly leading to partially depleted stores in tumor cells

decrease SOCE induced by thapsigargin in tumor cells. Therefore, TRPC1 might play roles different from supporting Ca^{2+} entry in tumor cells. Since this channel permeates mostly Na^+, a possible role could be to depolarize plasma membrane and limit Ca^{2+} entry. Alternatively, TRPC1 may be involved in changes in cell volume related to cell cycle.

What is the biological significance of STIM2 loss in tumor cells? STIM2 is a Ca^{2+} sensor inside the ER with low affinity for Ca^{2+}. This means that STIM2 should sense Ca^{2+} concentrations inside the ER when Ca^{2+} stores are filled at around >500 μM. A little decrease below this value should activate STIM2 and likely SOCE to keep the store filled. Consistently with this view, STIM2 knockout in normal cells decreases SOCE. In tumor cells, the loss of STIM2, leaves STIM1 in charge of sensing and refilling the stores. However, STIM1 is a different sensor with lower affinity for Ca^{2+} than STIM2 (Kd values around 300 μM). In this scenario, in which STIM1 would became activated only if stores are substantially depleted, the loss of STIM2 should lead to a partial depletion of Ca^{2+} stores (Fig. 19.2).

The above possibility was tested experimentally using siRNA against STIM2 in normal cells [69]. The decrease in STIM2 expression leads to the partial depletion of Ca^{2+} stores suggesting that STIM2 loss in tumor cells may contribute to Ca^{2+} remodeling by modifying Ca^{2+} store content. This partial depletion may have two relevant functional consequences. First, it may move Ca^{2+} store content close to the threshold for SOCE activation. Second, as Ca^{2+} store content has been related to the intrinsic pathway for cell death, loss of STIM2 in tumor cells may favor resistance to cell death and cell survival. Consistently, STIM2 silencing in normal cells decreases Ca^{2+} store content and increases resistance to apoptosis induced by H_2O_2 [69]. Therefore, the reciprocal shift in STIM1/STIM2 observed in colon cancer cells may be critical in colorectal tumorogenesis. Consistently with this view, it has been recently reported that STIM1 overexpression promotes colorectal cancer progression, cell motility and COX-2 expression [36].

In summary, colon cancer cells display enhanced and modified store and agonist-induced Ca^{2+} entry together with enhanced agonist-induced Ca^{2+} release and decreased Ca^{2+} store content. These differences are likely mediated by reciprocal changes in the expression of ORAI1, STIM1, TRPC1 and STIM2. These changes contribute to increased cell proliferation, invasion and survival. Therefore, although the above results must be confirmed in samples from colorectal cancer patients, the data strongly suggest a critical role for changes in SOCE and SOCs in colon cancer tumorigenesis. In fact, with the logical limitations on the role of SOCE in the immune response against tumors, SOCE antagonists could be considered for colon cancer. Previous results are consistent with this view. For example, it has been extensively documented in in vitro assays, animal testing and even clinical trials with high risk patients that aspirin may efficiently prevent colon cancer. Interestingly, our previous results clearly showed that the main aspirin metabolite salicylate inhibits SOCE off site and colon cancer cell growth in a mitochondria-dependent manner suggesting that aspirin may prevent colon cancer acting on SOCE [70].

19.6 Aspirin Prevents Cancer

A large series of epidemiological evidences suggest that aspirin and other non-steroidal anti-inflammatory drugs (NSAIDs) prevent colorectal cancer and other forms of cancer including breast cancer [71]. A recent meta-analysis showed that NSAIDs decreased the frequency of adenomas, colorectal cancer and deaths related to colorectal cancer in 57 out of the 59 studies carried out between 1988 and 2006 [72]. Yet, the lack of clinical trials and the risk of secondary effects associated to chronic NSAID use prevented recommendation of aspirin. Recently, a few clinical trials have been completed in high risk patients including Lynch syndrome patients, a type familial colorectal cancer with 100 % chances of developing colorectal cancer. Aspirin protected largely (63 %) against cancer in these high risk patients which has resulted in recommendation of aspirin for high risk patients of colorectal cancer [73]. Ongoing clinical trials suggest that combinations of aspirin or other NSAIDs with additional chemopreventive compounds may be highly efficient in preventing polyp formation and cancer death in patients that had undergone surgery for tumor removal. These are high risk patients with a 50 % chance of recurrence and death. Similar trials are underway in other forms of familial cancer including breast cancer associated to driving mutations in BRCA1 and 2.

Basic studies carried out in cell lines and animal models of cancer indicate that aspirin and other NSAIDs inhibit tumor cell proliferation and growth, cell migration and invasion and tumor growth in animal models of cancer. Interest on the action mechanism is growing since the realization that a large part of the effects are largely independent of the anti-inflammatory activity of these drugs. This view is based in that antitumor activity remains in tumor cells lacking expression of COX, the classic target of anti-inflammatory compounds. Moreover, structural analogues like R-flurbiprofen that lack ant-inflammatory activity are also efficient in preventing tumor cell growth. Therefore, even though COX-mediated synthesis or prostanoids may contribute to inflammation and colon tumorogenesis and NSAIDs may act partially by preventing inflammation, other targets of aspirin and other NSAIDs are likely involved in the antitumor actions of these drugs. We and others have shown previously that the aspirin metabolite and other NSAIDs may inhibit SOCs and SOCE in colon cancer cells likely providing a candidate mechanism for cancer chemoprevention by these compounds and a novel target for cancer chemoprevention.

19.7 Aspirin and Other NSAIDs Inhibit SOCE

NSAIDs inhibit SOCE and colon cancer cell proliferation [74]. However, no mechanism of action was provided. A few years ago, our group proposed that the main metabolite of aspirin, salicylate, could prevent tumor cell growth by inhibiting SOCE off site in a mitochondria-dependent manner [70, 75]. Salicylate is a mild mitochondrial uncoupler. This is due to the fact that the negative charge of the carboxylic residue is delocalized throughout the aromatic ring of salicylate. Thus, salicylic acid is neutral and capable of entering the cell and the mitochondrial matrix down a chemical gradient without restriction. However, once inside mitochondria, the matrix pH favors salicylic acid dissociation leading to salicylate formation that can exit the matrix favored by the negative mitochondrial potential. The net result is the release of protons inside mitochondria and mitochondrial uncoupling, partially depolarizing mitochondria and limiting the electromotive force for mitochondrial Ca^{2+} uptake [70]. Inasmuch as SOCE is strongly regulated by mitochondria, salicylate effect promotes the Ca^{2+} -dependent inactivation of Icrac and inhibition of SOCE. Since a sustained SOCE pathway is required for cell proliferation, the result is that salicylate inhibits cell proliferation via inactivation of a cancer-relevant Ca^{2+} channel. This mechanism may contribute to explain the antiproliferative effects of aspirin on T lymphocytes and vascular smooth muscle cells [70, 76, 77]. Interestingly, it has been recently reported that STIM1 overexpression promotes colorectal cancer progression, cell motility and COX-2 expression [36]. Accordingly, the effects of NSAIDs on SOCE may inhibit also colorectal cancer progression by preventing SOCE-mediated expression of COX-2.

Acknowledgements This work has been funded by grants from Ministerio de Economía y competitividad, Spain (BFU2012-37146) and Junta de Castilla y León, Spain [BIO/VA46/14]. DS was supported by a predoctoral fellowship from the JAE program, National Research Council (CSIC), Spain.

References

1. Baughman JM, Perocchi F, Girgis HS, Plovanich M, Belcher-Timme CA, Sancak Y, Bao XR, Strittmatter L, Goldberger O, Bogorad RL, Koteliansky V, Mootha VK (2011) Integrative genomics identifies MCU as an essential component of the mitochondrial calcium uniporter. Nature 476:341–345

2. De Stefani D, Raffaello A, Teardo E, Szabò I, Rizzuto R (2011) A forty-kilodalton protein of the inner membrane is the mitochondrial calcium uniporter. Nature 476:336–340
3. Montero M, Alonso MT, Carnicero E, Cuchillo-Ibáñez I, Albillos A, García AG, García-Sancho J, Alvarez J (2000) Chromaffin-cell stimulation triggers fast millimolar mitochondrial Ca^{2+} transients that modulate secretion. Nat Cell Biol 2(2):57–61
4. Villalobos C, Núñez L, Montero M, García AG, Alonso MT, Chamero P, Alvarez J, García-Sancho J (2002) Redistribution of Ca^{2+} among cytosol and organella during stimulation of bovine chromaffin cells. FASEB J 16(3):343–353
5. Quintana A, Pasche M, Junker C, Al-Ansary D, Rieger H, Kummerow C, Nuñez L, Villalobos C, Meraner P, Becherer U, Rettig J, Niemeyer BA, Hoth M (2011) Calcium microdomains at the immunological synapse: how ORAI channels, mitochondria and calcium pumps generate local calcium signals for efficient T-cell activation. EMBO J 30:3895–3912
6. Alvarez J, Montero M (2002) Measuring $[Ca^{2+}]$ in the endoplasmic reticulum with aequorin. Cell Calcium 32(5–6):251–260
7. Hewavitharana T, Deng X, Soboloff J, Gill DL (2007) Role of STIM and Orai proteins in the store-operated calcium signaling pathway. Cell Calcium 42:173–182
8. Putney JW (2009) Capacitative calcium entry: from concept to molecules. Immunol Rev 231:10–22
9. Muik M, Schindl R, Fahrner M, Romanin C (2012) Ca^{2+} release-activated Ca^{2+} (CRAC) current, structure, and function. Cell Mol Life Sci 69:4163–4176
10. Brandman O, Liou J, Park WS, Meyer T (2007) STIM2 is a feedback regulator that stabilizes basal cytosolic and endoplasmic reticulum Ca^{2+} levels. Cell 131:1327–1339
11. Hoth M, Niemeyer BA (2013) The neglected CRAC proteins: Orai2, Orai3, and STIM2. Curr Top Membr 1:237–271
12. Ma X, Nilius B, Wong JW, Huang Y, Yao X (2011) Electrophysiological properties of heteromeric TRPV4-C1 channels. Biochim Biophys Acta 1808:2789–2797
13. Cheng KT, Ong HL, Liu X, Ambudkar IS (2013) Contribution and regulation of TRPC channels in store-operated Ca^{2+} entry. Curr Top Membr 71:149–179
14. Monteith GR, Davis FM, Roberts-Thomson SJ (2012) Calcium channels and pumps in cancer: changes and consequences. J Biol Chem 287:31666–31673
15. Prevarskaya N, Ouadid-Ahidouch H, Skryma R, Shuba Y (2014) Remodelling of Ca^{2+} transport in cancer: how it contributes to cancer hallmarks? Philos Trans R Soc Lond B Biol Sci 369:20130097
16. Prevarskaya N, Skryma R, Shuba Y (2011) Calcium in tumor metastasis: new roles for known actors. Nat Rev Cancer 11:609–618
17. Bergmeier W, Weidinger C, Zee I, Feske S (2013) Emerging roles of store-operated Ca^{2+} entry through STIM and ORAI proteins in immunity, hemostasis and cancer. Channels 7:379–391
18. Stewart TA, Yapa KT, Monteith GR (2015) Altered calcium signaling in cancer cells. Biochim Biophys Acta 1848 (10PtB):2502–11
19. Roderick HL, Cook SJ (2008) Ca^{2+} signalling checkpoints in cancer: remodelling Ca^{2+} for cell proliferation and survival. Nat Cancer Rev 8:361–375
20. Zhang L, Zhou W, Velculescu VE, Kern SE, Hruban RH, Hamilton SR, Vogelstein B, Kinzler KW (1997) Gene expression profiles in normal and cancer cells. Science 276:1268–1272
21. Wang JY, Sun J, Huang MY, Wang YS, Hou MF, Sun Y, He H, Krishna N, Chiu SJ, Lin S, Yang S, Chang WC. STIM1 overexpression promotes colorectal cancer progression, cell motility and COX-2 expression. Oncogene Nov 10. Wang XT, Nagaba Y, Cross HS, Wrba F, Zhang L, Guggino SE (2000) The mRNA of L-type calcium channel elevated in colon cancer: protein distribution in normal and cancerous colon. Am J Pathol 157:1549–1562
22. Toyota M, Ho C, Ohe-Toyota M, Baylin SB, Issa JP (1999) Inactivation of CACNA1G, a T-type calcium channel gene, by aberrant methylation of its 5′ CpG island in human tumors. Cancer Res 59:4535–4541
23. Gackiere F, Bidaux G, Delcourt P, Van Coppenolle F, Katsogiannou M, Dewailly E, Bavencoffe A, Van Chuoi-Mariot MT, Mauroy B, Prevarskaya N, Mariot P (2008) $Cav_{3.2}$ T-type calcium

channels are involved in calcium-dependent secretion of neuroendocrine prostate cancer cells. J Biol Chem 283:10162–10173
24. Taylor JT, Zeng XB, Pottle JE, Lee K, Wang AR, Yi SG, Scruggs JA, Sikka SS, Li M (2008) Calcium signaling and T-type calcium channels in cancer cell cycling. World J Gastroenterol 14:4984–4991
25. Villalobos C, Fonteriz R, López MG, García AG, García-Sancho J (1992) Inhibition of voltage-gated Ca^{2+} entry into GH_3 and chromaffin cells by imidazole antimycotics and other cytochrome P_{450} blockers. FASEB J 6(9):2742–2747
26. Dhennin-Duthille I, Gautier M, Faouzi M, Guilbert A, Brevet M, Vaudry D, Ahidouch A, Sevestre H, Ouadid-Ahidouch H (2011) High expression of transient receptor potential channels in human breast cancer epithelial cells and tissues: correlation with pathological parameters. Cell Physiol Biochem 28:813–822
27. Yang S, Zhang JJ, Huang XY (2009) Orai1 and STIM1 are critical for breast tumor cell migration and metastasis. Cancer Cell 5:124–134
28. Aydar E, Yeo S, Djamgoz M, Palmer C (2009) Abnormal expression, localization and interaction of canonical transient receptor potential ion channels in human breast cancer cell lines and tissues: a potential target for breast cancer diagnosis and therapy. Cancer Cell Int 9:23
29. Veliceasa D, Ivanovic M, Hoepfner FT, Thumbikat P, Volpert OV, Smith ND (2007) Transient potential receptor channel 4 controls thrombospondin-1 secretion and angiogenesis in renal cell carcinoma. FEBS J 274:6365–6377
30. Shi Y, Ding X, He ZH, Zhou KC, Wang Q, Wang YZ (2009) Critical role of TRPC6 channels in G2 phase transition and the development of human oesophageal cancer. Gut 58:1443–1450
31. Zhang SS, Wen J, Yang F, Cai XL, Yang H, Luo KJ, Liu QW, Hu RG, Xie X, Huang QY, Chen JY, Fu JH, Hu Y (2013) High expression of transient potential receptor C6 correlated with poor prognosis in patients with esophageal squamous cell carcinoma. Med Oncol 30:607
32. Cai R, Ding X, Zhou K, Shi Y, Ge R, Ren G, Jin Y, Wang Y (2009) Blockade of TRPC6 channels induced G2/M phase arrest and suppressed growth in human gastric cancer cells. Int J Cancer 125:2281–2287
33. Ding X, He Z, Zhou K, Cheng J, Yao H, Lu D, Cai R, Jin Y, Dong B, Xu Y, Wang Y (2010) Essential role of TRPC6 channels in G2/M phase transition and development of human glioma. J Natl Cancer Inst 102:1052–1068
34. El Boustany C, Bidaux G, Enfissi A, Delcourt P, Prevarskaya N, Capiod T (2008) Capacitative calcium entry and transient receptor potential canonical 6 expression control human hepatoma cell proliferation. Hepatology 47:2068–2077
35. Guilbert A, Dhennin-Duthille I, Hiani YE, Haren N, Khorsi H, Sevestre H, Ahidouch A, Ouadid-Ahidouch H (2008) Expression of TRPC6 channels in human epithelial breast cancer cells. BMC Cancer 8:125
36. Wang JY, Sun J, Huang MY, Wang YS, Hou MF, Sun Y, He H, Krishna N, Chiu SJ, Lin S, Yang S, Chang WC (2015) STIM1 overexpression promotes colorectal cancer progression, cell motility and COX-2 expression. Oncogene 34(33):4358–4367
37. Zhuang L, Peng JB, Tou L, Takanaga H, Adam RM, Hediger MA, Freeman MR (2002) Calcium-selective ion channel, CaT1, is apically localized in gastrointestinal tract epithelia and is aberrantly expressed in human malignancies. Lab Invest 82:1755–1764
38. Capiod T, Shuba Y, Skryma R, Prevarskaya N (2007) Calcium signaling and cancer cell growth. Subcell Biochem 45:405–427
39. Raphaël M, Lehen'kyi V, Vandenberghe M, Beck B, Khalimonchyk S, Vanden Abeele F, Farsetti L, Germain E, Bokhobza A, Mihalache A, Gosset P, Romanin C, Clézardin P, Skryma R, Prevarskaya N (2014) TRPV6 calcium channel translocates to the plasma membrane via Orai1-mediated mechanism and controls cancer cell survival. Proc Natl Acad Sci U S A 111(37):E3870–E3879
40. Déliot N, Constantin B (2015) Plasma membrane calcium channels in cancer: alterations and consequences for cell proliferation and migration. Biochim Biophys Acta 1848:2512–2522

41. Fusi C, Materazzi S, Minocci D, Maio V, Oranges T, Massi D, Nassini R (2014) Transient receptor potential vanilloid 4 (TRPV4) is downregulated in keratinocytes in human non-melanoma skin cancer. J Invest Dermatol 134:2408–2417
42. Mizuno H, Suzuki Y, Watanabe M, Sokabe T, Yamamoto T, Hattori R, Gotoh M, Tominaga M (2014) Potential role of transient receptor potential (TRP) channels in bladder cancer cells. J Physiol Sci 64(4):305–314
43. Bolanz KA, Hediger MA, Landowski CP (2008) The role of TRPV6 in breast carcinogenesis. Mol Cancer Ther 7:271–279
44. Peng JB, Zhuang L, Berger UV, Adam RM, Williams BJ, Brown EM, Hediger MA, Freeman MR (2001) CaT1 expression correlates with tumor grade in prostate cancer. Biochem Biophys Res Commun 282:729–734
45. Wissenbach U, Niemeyer BA, Fixemer T, Schneidewind A, Trost C, Cavalie A, Reus K, Meese E, Bonkhoff H, Flockerzi V (2001) Expression of CaT-like, a novel calcium-selective channel, correlates with the malignancy of prostate cancer. J Biol Chem 276(22):19461–19468
46. Fan H, Shen YX, Yuan YF (2014) Expression and prognostic roles of TRPV5 and TRPV6 in non-small cell lung cancer after curative resection. Asian Pac J Cancer Prev 15:2559–2563
47. Yee NS, Zhou W, Lee M (2010) Transient receptor potential channel TRPM8 is over-expressed and required for cellular proliferation in pancreatic adenocarcinoma. Cancer Lett 297:49–55
48. Tsavaler L, Shapero MH, Morkowski S, Laus R (2001) Trp-p8, a novel prostate-specific gene, is up-regulated in prostate cancer and other malignancies and shares high homology with transient receptor potential calcium channel proteins. Cancer Res 61:3760–3769
49. Fuessel S, Sickert D, Meye A, Klenk U, Schmidt U, Schmitz M, Rost AK, Weigle B, Kiessling A, Wirth MP (2003) Multiple tumor marker analyses (PSA, hK2, PSCA, trp-p8) in primary prostate cancers using quantitative RT-PCR. Int J Oncol 23:221–228
50. Schmidt U, Fuessel S, Koch R, Baretton GB, Lohse A, Tomasetti S, Unversucht S, Froehner M, Wirth MP, Meye A (2006) Quantitative multi-gene expression profiling of primary prostate cancer. Prostate 66(14):1521–1534
51. Prevarskaya N, Flourakis M, Bidaux G, Thebault S, Skryma R (2007) Differential role of TRP channels in prostate cancer. Biochem Soc Trans 35(Pt 1):133–135
52. Chodon D, Guilbert A, Dhennin-Duthille I, Gautier M, Telliez MS, Sevestre H, Ouadid-Ahidouch H (2010) Estrogen regulation of TRPM8 expression in breast cancer cells. BMC Cancer 10:212
53. Liu J, Chen Y, Shuai S, Ding D, Li R, Luo R (2014) TRPM8 promotes aggressiveness of breast cancer cells by regulating EMT via activating AKT/GSK-3β pathway. Tumour Biol 35(9):8969–8977
54. Chen YF, Chiu WT, Chen YT, Lin PY, Huang HJ, Chou CY, Chang HC, Tang MJ, Shen MR (2011) Calcium store sensor stromal-interaction molecule1-dependent signaling plays an important role in cervical cancer growth, migration and angiogenesis. Proc Natl Acad Sci U S A 108:15225–15230
55. Li G, Zhang Z, Wang R, Ma W, Yang Y, Wei J, Wei Y (2013) Suppression of STIM1 inhibits human glioblastoma cell proliferation and induces G0/G1 phase arrest. J Exp Clin Cancer Res 32:20
56. Motiani RK, Hyzinski-García MC, Zhang X, Henkel MM, Abdullaev IF, Kuo YH, Matrougui K, Mongin AA, Trebak M (2013) STIM1 and Orai1 mediate CRAC channel activity and are essential for human glioblastoma invasion. Pflugers Arch 465(9):1249–1260
57. McAndrew D, Grice DM, Peters AA, Davis FM, Stewart T, Rice M, Smart CE, Brown MA, Kenny PA, Roberts-Thomson SJ, Monteith GR (2011) ORAI1-mediated calcium influx in lactation and in breast cancer. Mol Cancer Ther 10:448–460
58. Motiani RK, Zhang X, Harmon KE, Keller RS, Matrougui K, Bennett JA, Trebak M (2013) Orai3 is an estrogen receptor α-regulated Ca^{2+} channel that promotes tumorigenesis. FASEB J 27(1):63–75

59. Ay AS, Benzerdjeb N, Sevestre H, Ahidouch A, Ouadid-Ahidouch H (2013) Orai3 constitutes a native store-operated calcium entry that regulates non small cell lung adenocarcinoma cell proliferation. PLoS One 8(9):e72889
60. Zhu M, Chen L, Zhao P, Zhou H, Zhang C, Yu S, Lin Y, Yang X (2014) Store-operated Ca^{2+} entry regulates glioma cell migration and invasion via modulation of Pyk2 phosphorylation. J Exp Clin Cancer Res 33:98
61. Umemura M, Baljinnyam E, Feske S, De Lorenzo MS, Xie LH, Feng X, Oda K, Makino A, Fujita T, Yokoyama U, Iwatsubo M, Chen S, Goydos JS, Ishikawa Y, Iwatsubo K (2014) Store-operated Ca^{2+} entry (SOCE) regulates melanoma proliferation and cell migration. PLoS One 9:e89292
62. Kim JH, Lkhagvadorj S, Lee MR, Hwang KH, Chung HC, Jung JH, Cha SK, Eom M (2014) Orai1 and STIM1 are critical for cell migration and proliferation of clear cell renal cell carcinoma. Biochem Biophys Res Commun 448:76–82
63. Faouzi M, Hague F, Potier M, Ahidouch A, Sevestre H, Ouadid-Ahidouch H (2011) Down-regulation of Orai3 arrests cell-cycle progression and induces apoptosis in breast cancer cells but not in normal breast epithelial cells. J Cell Physiol 226:542–551
64. Holzmann C, Kilch T, Kappel S, Armbruster A, Jung V, Stockle M, Bogeski I, Schwarz EC, Peinelt C (2013) I_{CRAC} controls the rapid androgen response in human primary prostate epithelial cells and is altered in prostate cancer. Oncotarget 4:2096–2107
65. Dubois C, Vanden Abeele F, Lehen'kyi V, Gkika D, Guarmit B, Lepage G, Slomianny C, Borowiec AS, Bidaux G, Benahmed M, Shuba Y, Prevarskaya N (2014) Remodeling of channel-forming ORAI proteins determines an oncogenic switch in prostate cancer. Cancer Cell 26:19–32
66. Scrideli CA, Carlotti CG Jr, Okamoto OK, Andrade VS, Cortez MA, Motta FJ, Lucio-Eterovic AK, Neder L, Rosemberg S, Oba-Shinjo SM, Marie SK, Tone LG (2008) Gene expression profile analysis of primary glioblastomas and non-neoplastic brain tissue: identification of potential target genes by oligonucleotide microarray and real-time quantitative PCR. J Neurooncol 88:281–291
67. Ruano Y, Mollejo M, Ribalta T, Fiano C, Camacho FI, Gomez E, de Lope AR, Hernandez-Moneo JL, Martinez P, Melendez B (2006) Identification of novel candidate target genes in amplicons of glioblastoma multiforme tumors detected by expression and CGH microarray profiling. Mol Cancer 5:39
68. Aytes A, Mollevi DG, Martinez-Iniesta M, Nadal M, Vidal A, Morales A, Salazar R, Capella G, Villanueva A (2012) Stromal interaction molecule 2 (STIM2) is frequently overexpressed in colorectal tumors and confers a tumor cell growth suppressor phenotype. Mol Carcinog 51:746–753
69. Sobradillo D, Hernández-Morales M, Ubierna D, Moyer MP, Núñez L, Villalobos C (2014) A reciprocal shift in transient receptor potential channel 1 (TRPC1) and stromal interaction molecule 2 (STIM2) contributes to Ca^{2+} remodeling and cancer hallmarks in colorectal carcinoma cells. J Biol Chem 289:28765–28782
70. Núñez L, Valero RA, Senovilla L, Sanz-Blasco S, García-Sancho J, Villalobos C (2006) Cell proliferation depends on mitochondrial Ca^{2+} uptake: inhibition by salicylate. J Physiol (Lond) 571:57–73
71. Rothwell PM, Fowkes FG, Belch JF, Ogawa H, Warlow CP, Meade TW (2011) Effect of daily aspirin on long-term risk of death due to cancer: analysis of individual patient data from randomized trials. Lancet 377:31–41
72. Arber N, Levin B (2008) Chemoprevention of colorectal neoplasia: the potential for personalized medicine. Gastroenterology 134(4):1224–1237
73. Burn J, Mathers JC, Bishop DT (2013) Chemoprevention in Lynch syndrome. Fam Cancer 12(4):707–718

74. Weiss H, Amberger A, Widschwendter M, Margreiter R, Ofner D, Dietl P (2001) Inhibition of store-operated calcium entry contributes to the anti-proliferative effect of non-steroidal anti-inflammatory drugs in human colon cancer cells. Int J Cancer 92(6):877–882
75. Valero RA, Senovilla L, Núñez L, Villalobos C (2008) The role of mitochondrial potential in control of calcium signals involved in cell proliferation. Cell Calcium 44:259–269
76. Muñoz E, Valero RA, Quintana A, Hoth M, Núñez L, Villalobos C (2011) Nonsteroidal anti-inflammatory drugs inhibit vascular smooth muscle cell proliferation by enabling the Ca^{2+}-dependent inactivation of calcium release-activated calcium/Orai channels normally prevented by mitochondria. J Biol Chem 286:16186–16196
77. Muñoz E, Hernández-Morales M, Sobradillo D, Rocher A, Núñez L, Villalobos C (2013) Intracellular Ca^{2+} remodeling during the phenotypic journey of human coronary smooth muscle cells. Cell Calcium 54:375–385

Index

J.A. Rosado (ed.), *Calcium Entry Pathways in Non-excitable Cells*, Advances in Experimental Medicine and Biology 898, DOI 10.1007/978-3-319-26974-0

MIX
Papier aus verantwortungsvollen Quellen
Paper from responsible sources
FSC® C105338

If you have any concerns about our products,
you can contact us on
ProductSafety@springernature.com

In case Publisher is established outside the EU,
the EU authorized representative is:
Springer Nature Customer Service Center GmbH
Europaplatz 3, 69115 Heidelberg, Germany

Printed by Libri Plureos GmbH
in Hamburg, Germany